Advances in Intelligent Systems and Computing

Volume 1084

The series "Advances in Intelligent Systems and Computing" contains publications on theory, applications, and design methods of Intelligent Systems and Intelligent Computing. Virtually all disciplines such as engineering, natural sciences, computer and information science, ICT, economics, business, e-commerce, environment, healthcare, life science are covered. The list of topics spans all the areas of modern intelligent systems and computing such as: computational intelligence, soft computing including neural networks, fuzzy systems, evolutionary computing and the fusion of these paradigms, social intelligence, ambient intelligence, computational neuroscience, artificial life, virtual worlds and society, cognitive science and systems, Perception and Vision, DNA and immune based systems, self-organizing and adaptive systems, e-Learning and teaching, human-centered and human-centric computing, recommender systems, intelligent control, robotics and mechatronics including human-machine teaming, knowledge-based paradigms, learning paradigms, machine ethics, intelligent data analysis, knowledge management, intelligent agents, intelligent decision making and support, intelligent network security, trust management, interactive entertainment, Web intelligence and multimedia.

The publications within "Advances in Intelligent Systems and Computing" are primarily proceedings of important conferences, symposia and congresses. They cover significant recent developments in the field, both of a foundational and applicable character. An important characteristic feature of the series is the short publication time and world-wide distribution. This permits a rapid and broad dissemination of research results.

** Indexing: The books of this series are submitted to ISI Proceedings, EI-Compendex, DBLP, SCOPUS, Google Scholar and Springerlink **

More information about this series at http://www.springer.com/series/11156

Fatos Xhafa · Srikanta Patnaik ·
Madjid Tavana
Editors

Advances in Intelligent Systems and Interactive Applications

Proceedings of the 4th International
Conference on Intelligent, Interactive
Systems and Applications (IISA2019)

Editors
Fatos Xhafa
Dept De Ciències De La Computació
Universitat Politècnica De Catalunya
Barcelona, Spain

Srikanta Patnaik
Department of Computer Science
and Engineering
SOA University
Bhubaneswar, Odisha, India

Madjid Tavana
Department of Business Systems
and Analytics
La Salle University
Philadelphia, PA, USA

ISSN 2194-5357   ISSN 2194-5365   (electronic)
Advances in Intelligent Systems and Computing
ISBN 978-3-030-34386-6   ISBN 978-3-030-34387-3   (eBook)
https://doi.org/10.1007/978-3-030-34387-3

This Springer imprint is published by the registered company Springer Nature Switzerland AG
The registered company address is: Gewerbestrasse 11, 6330 Cham, Switzerland

4th International Conference on Intelligent and Interactive Systems and Applications (IISA2019)

Editorial

The advancement in technologies like AI, cognitive science, robotics, IoT, smart-phones and big data analytics has led to the fusion of these techniques and convergence into intelligent and interactive systems that are designed to interact with human beings as well as other entities and the environment. These systems are embodied with several humanlike capabilities such as the ability to perceive, interpret, reason, learn, decide, plan and act. Intelligent interactive systems exhibit these capabilities by merging the physical world with the cyber world. Further, the widespread adoption of IoT has brought heterogeneous systems and systems of systems to interact with each other according to requirements. Moreover, these intelligent systems can sense and detect contextual changes in respective environments and automate IoT-based systems to accelerate development by facilitating interaction of these systems with corresponding domain experts for making better decisions wherever needed. The entire ecosystem of intelligent interactive systems is based on intelligent connectivity. Again the interaction between heterogeneous systems in several applications like object recognition, speech recognition, search engine optimization and recommender systems makes the system and the corresponding environment more complex and gives rise to new challenges and issues which involves human intelligence as well as machine intelligence or artificial intelligence. To address these complexities and challenges, it is required to understand to learn the task itself to be done, the environmental setup in which it has to be done and how it should be done optimally from both human and machine perspectives.

However, the five major areas where intelligent interactive systems play an important role include: transportations and logistics; residential sector; industrial and manufacturing operation sector; healthcare sector; and public safety and security. Some of the most active research domains of intelligent interactive systems include: recommender systems, information retrieval, intelligent learning environment, natural language processing, decision support systems, computer graphics, conversational agents, cognitive science, Web intelligence, test automation, smart wearable devices, mobile computing, ubiquitous computing, human–robot interaction, interactive communication technology, game designed using AI, context awareness, knowledge discovery-based systems, development of disruptive digital services, smart navigation techniques and finally semantics-based applications and techniques. Also, some of the major issues being faced during carrying out research work include: how to integrate human intelligence with artificial intelligence; how intelligent processing can perform better than traditional processes; why careful design is important for these cognitive as well as autonomic systems; what could be the consequences of negligence; privacy and security issues; and many more.

This volume of IISA attempts to present a diversified collection of research papers focusing on the novel design and analysis section as well as the application-oriented deployment section of the interactive intelligent systems. It further attempts to explore the state of the art for the most recent developments of intelligent interactive products incorporating new ideas and features as these systems are hard to design. Also, intelligent interactive systems require proactive technological novelty to match human abilities to learn from past experiences and predict future actions while overcoming human limitations. Most of the selected papers address both interactivity and intelligence along with design, analysis and development of agents and robotics-based applications that are capable of interacting with human beings as well as other agents to perceive, plan and communicate for making critical decisions and relevant suggestions. Some recent trends of such interactive intelligent system-based applications include: (i) Recommender systems use personalized information, and attempts explore a large information space consisting user's interests and preferences to identify the best-suited option for the user from other available alternatives; (ii) smart navigation solutions include another set of sub-application which are being widely used as a part for a variety of smart applications. In smart navigation systems, users can interact with the system to choose the best-suited route for a particular destination out of a set of offered options; (iii) again, lots of context-based applications have also been developed that takes user's requirement-based context into account such as user's location, mood, current climatic or weather context for accomplishing the task at hand; (iv) further, interactive Web-based lecture session-based applications have been designed to bridge the gap between the educators and learners across the globe, thus improving the performance of the students. These interactive educational application tools are usually supported by interactive contents and visualizations to improve the learning rate and comprehension of students; (v) also, applications with bio-monitoring capabilities have been developed for studying the environmental health by researchers in complex environments where it is difficult for human beings to conduct physical research; (vi) another set of intelligent interactive system-based applications include healthcare-based wearable products that monitor heart rates, blood pressure and sugar levels; these products monitor the patients at regular intervals and interact with users to generate alerts and suggestions to regulate things using medical informatics; (vii) other applications include interactive intelligent applications for crowd-sourcing, predicting consumer behavior and customer retention, understanding Web site aesthetics, performance evaluation of new products overcoming global constraints and so on.

The 4th International Conference on IISA2019 aims to bring together the researchers, academia and practitioners to exchange their ideas about recent trends and shed light on these issues and limitations and how they can be resolved. Several novel algorithms and methodologies as well as various combinations of existing intelligent algorithms have been considered to explore interoperability, flexibility and platform independence among intelligent interactive systems with the goal to make this volume a base reference for the scholars interested to work in this area. This volume contains 102 contributions from diverse areas of IISs which have been categorized into seven sections, namely:

  (i) Intelligent systems;
 (ii) Autonomous systems;
(iii) Pattern recognition and vision systems;
 (iv) E-enabled systems;
  (v) Internet and cloud computing;
 (vi) Mobile computing and intelligent networking; and
(vii) Various applications.

Fatos Xhafa
Srikanta Patnaik
Madjid Tavana

Acknowledgements

The contributions covered in this proceeding are the outcome of the contributions from more than hundred researchers. We are thankful to the authors and paper contributors of this volume.

We are thankful to the Editor-in-Chief of the Springer Book Series on **"Advances in Intelligent Systems and Computing"** Prof Janusz Kacprzyk for his support to bring out the fourth volume of the conference, i.e., IISA2019. It is noteworthy to mention here that constant support from the editor-in-chief and the members of the publishing house makes the conference fruitful for the second edition. We would like to appreciate the encouragement and support of Dr. Thomas Ditzinger, Executive Editor, and his Springer publishing team.

We are also thankful to the experts and reviewers who have worked for this volume despite the veil of their anonymity.

We look forward to your valued contribution and support to next editions of the International Conference on Intelligent and Interactive Systems and Applications.

We are sure that the readers shall get immense benefit and knowledge from the second volume of the area of intelligent and interactive systems and applications.

Fatos Xhafa's work is partially supported by Spanish Ministry of Science, Innovation and Universities, Programme *"Estancias de profesores e investigadores senior en centros extranjeros, incluido el Programa Salvador de Madariaga 2019"*, PRX19/00155.

Contents

Contents

Applications

Intelligent Systems

Recognition of Voiceprint Using Deep Neural Network Combined with Support Vector Machine

Lv Han, Tianxing Li, Weijie Zheng, Tao Ma, Wenlian Ma,
Sanzhi Shi, Xiaoning Jia, and Linhua Zhou[(✉)]

School of Science, Department of Applied Mathematics,
Changchun University of Science and Technology, Changchun 130022, China
chowlhl718@163.com

Abstract. With the rapid development of artificial intelligence technology of deep learning method, it has been applied to many fields, especially to life science. In this paper, a novel approach for the task of voiceprint recognition was proposed. The combination of deep belief network (DBN) and support vector machine (SVM) was used to identify the voiceprint of 10 different individuals. Based on a 24-dimension Mel Frequency Cepstrum Coefficient (MFCC), the authors extracted 256-dimension deep voiceprint features via DBN model that is developed by stacking three layers of Restricted Boltzmann Machine (RBM), and conducted voiceprint recognition by approach of SVM. According to the recognition results, the new approach can significantly improve both of accuracy and efficiency when it was compared with traditional voiceprint features and recognition models. The low dimensional features of voiceprint was extracted into higher dimensions by DBN model, while SVM can avoid the elevation of computation complexity caused by the increases of feature dimension. The combined strengths have been fully expressed by the experimental results.

Keywords: Voiceprint recognition · Deep Neural Network · Restricted Boltzmann Machine · Support vector machine

1 Introduction

The research on voiceprint recognition that started in the 1960s [1] mainly refers to the technology used to identify the identity of a speaker by extracting the voiceprint features from the speech signals of the speaker. It has attracted extensive attentions from many scholars. As regards the problems in voiceprint recognition, the decisive factor of recognition effect is to select appropriate voiceprint characteristic signal and suitable acoustic recognition model. Mel Frequency Cepstrum Coefficient (MFCC) that imitates human auditory perception developed by Davis et al. [2] can embody the personality characteristics of a speaker by verification and is regarded as the most common voiceprint feature among the existing voiceprint recognitions. In terms of acoustic modeling, the initial artificial neural network models have obtained significant recognition effects [3–6]. In the following, the voiceprint recognition technology based

© Springer Nature Switzerland AG 2020
F. Xhafa et al. (Eds.): IISA 2019, AISC 1084, pp. 3–11, 2020.
https://doi.org/10.1007/978-3-030-34387-3_1

on Gaussian mixture model-universal background model (GMM-UBM) has achieved further growth, such as Gaussian mixture model and super vector-support vector machine (GSV-SVM) [7], Joint factor analysis (JFA) [8, 9] and Identity vector-cosine distance scoring (IVECCDS) [10] etc. With the development of such voiceprint recognition technology, the accuracy of voiceprint recognition has been enhanced greatly, yet there are some considerable problems, such as poor distinguishing ability of models and inadequate representational ability.

Hinton et al. put forward a deep belief network model [11] in 2006. In view of its powerful abilities to extract deep information, the deep learning theory has been widely used in many machine learning fields. For example, the deep learning model that is applied in speech recognition has dramatically increased the accuracy of recognition [12, 13] and broken through the requirements for speech recognition performance in some practical application scenes, which makes the speech recognition technology become more practical.

In recent years, many scholars have introduced the deep learning model into the modeling framework of voiceprint recognition, established the voiceprint recognition systems based on GMM-DNN [14] and performed voiceprint recognition via the mixed model of noise reduction automatic coding machine and restricted boltzmann machine (HDAE-RBMM) [15]. Considering that the features obtained from deep learning generally have higher dimensions, the support vector machine enjoys prominent advantages with respect to the problem. The terminal decision function of support vector machine is confirmed with the minority of support vector, and the calculation complexity lies in the quantity of support vector rather than the dimensions of sample space, which in a sense, avoids "the curse of dimensionality" and possesses simple algorithm and better robustness. The authors have extracted 256-dimension deep voiceprint features from 24-dimension MFCC voiceprint information based on DBN and performed the study of voiceprint recognition via SVM. According to the experimental results, such voiceprint recognition system has experienced an improvement in terms of the recognition accuracy rate, especially the efficiency of recognition.

2   Methods

As a Deep Neural Network (DNN), Deep Belief Network (DBN) is comprised by a stack of Restricted Boltzmann Machines (RBM) and has achieved great success in the fields of speech recognition and image recognition. DBN has fully reflected the effectiveness of unsupervised learning at all layers of training, in which each layer can engage in the unsupervised training again based on the previous layer of training results. The hierarchical nonlinear mapped deep structure can accomplish complex functional approximation [16].

The Restricted Boltzmann Machine (RBM) is a neural network model with two-layer structure (the visible layer and hidden layer), symmetrical connection and no sub-feedback, in which the different layers are fully connected and there is no any connection within the layers. Suppose one RBM consists of n visible units and m hidden units. The states of visible unit and hidden unit are expressed by vector-v and h, respectively, in which v_i refers to the state of ith visible unit, while h_j stands for the state of jth hidden unit. For a group of given states (v, h), the energy of RBM as a system shall be defined as [17]:

$$E(v, h|\theta) = -\sum_{i=1}^{n} \alpha_i v_i - \sum_{j=1}^{m} b_j h_j - \sum_{i=1}^{n} \sum_{j=1}^{m} v_i W_{ij} h_j. \tag{1}$$

$\theta = \{W_{ij}, \alpha_i, b_j\}$ in the above formula is a parameter of RBM, in which W_{ij} refers to the link weight between visible unit i and hidden unit j, α_i refers to the offset of visible unit i, while b_j refers to the offset of hidden unit j. When the parameter is confirmed, the joint probability distribution of (v, h) is shown as follows based on the energy function:

$$P(v, h|\theta) = \frac{e^{-E(v,h|\theta)}}{Z(\theta)}, \quad Z(\theta) = \sum_{v,h} e^{-E(v,h|\theta)}, \tag{2}$$

In which $Z(\theta)$ is a normalization factor. The distribution of RBM about observation data $v(P(v|\theta))$ that namely is the boundary distribution of joint probability distribution about v (also known as the likelihood function) shall be expressed as:

$$P(v|\theta) = \frac{1}{Z(\theta)} \sum_{h} e^{-E(v,h|\theta)}. \tag{3}$$

During the learning process of RBM, suppose there are N samples, and the value of parameter θ can be confirmed by maximizing the log-likelihood function:

$$\theta^* = \arg \max_{\theta} L(\theta) = \arg \max_{\theta} \sum_{n=1}^{N} \log P(v^{(n)}|\theta). \tag{4}$$

DBN can be available by increasing the quantity of hidden layer and stacking several RBMs. After completing the unsupervised learning, DBN usually uses the supervised learning set up on the top layer to adjust and optimize the parameters of whole neural network based on the back propagation algorithm, which can achieve a complete DBN structure [18].

3 Voiceprint Recognition Based on DBN-SVM

3.1 Experimental Procedures and Processes

The experiment includes three procedures: At first, Mel Frequency Cepstrum Coefficient (MFCC) is extracted from the original speech signal; secondly, obtain the deep features of extracted MFCC via DBN; finally, make classification by using SVM. The complete training and recognition process is shown as Fig. 1. The authors have compared and analyzed the recognition performance of SVM and BP neural network.

3.2   Corpus Description and MFCC Extraction

The experimental corpus is Chinese speech which is the speech data (THCHS-30) issued by CSLT in Tsinghua University. In order to analyze the performance of voiceprint recognition system under different sizes of corpus in the "training set", three groups of corpus set will be used in the study (see Table 1).

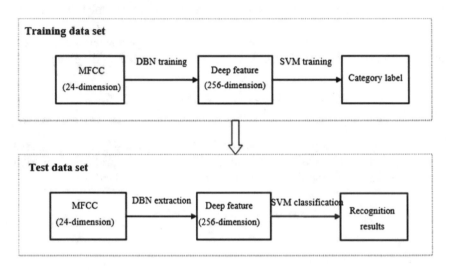

Fig. 1. Experimental procedures of voiceprint recognition

Table 1. Three groups of corpus set

Corpus set	Number of people to be selected	Number of sentences selected by everyone	Number of training sentences used by everyone	Number of test sentences used by everyone
A	10 people	12 sentences	3 sentences	9 sentences
B	10 people	12 sentences	4 sentences	8 sentences
C	10 people	12 sentences	6 sentences	6 sentences

MFCC is developed based on human auditory characteristics. The extraction of MFCC from original speech is composed of the following procedures: Pre-filtering; pre-emphasis; framing; windowing; fast Fourier transform; triangle window filtering; logarithm settlement; discrete cosine transform; cepstral mean subtraction and difference, etc. During the extraction of MFCC in this paper, the frame size is 15 ms, while the frame shift is 7.5 ms. 24 filter banks are used to filter, and each frame of MFCC is 24-dimision data.

Suppose one sentence spoken by someone has n frames in total, and MFCC extracted from each frame is expressed as $X_k(k = 1, 2, \ldots, n)$, then MFCC of this sentence shall be recalculated as:

$$\sum_{k=1}^{n} X_k \Big/ n, \tag{9}$$

Each frame of MFCC vectors is added together and averaged to achieve MFCC mean vector.

3.3 DBN Training and Deep Feature Extraction

DBN is developed by a stack of RBMs. The weight of each RBM can be estimated by using Gibbs sampling [18]. DBN used in this experiment is formed by stacking three layers of RBMs (network node: 24-256-256-256). The training process of DBN is divided into two steps: (1) Independent training of RBM conducted from the bottom to the top without supervision; the pre-training of each layer is performed by using unlabeled data, and the pre-training result of each layer shall be used as the input of the higher layer. Thus, the unsupervised learning of features can be completed so as to obtain an independent intra-layer optimization parameter for each RBM; (2) The supervised training conducted from the top to the bottom; the labeled data is used to adjust the weights and thresholds at all layers and transmitted from the top to the bottom based on the error back propagation algorithm. The whole network parameters are adjusted slightly so as to obtain the overall optimization parameter of the stacked RBM.

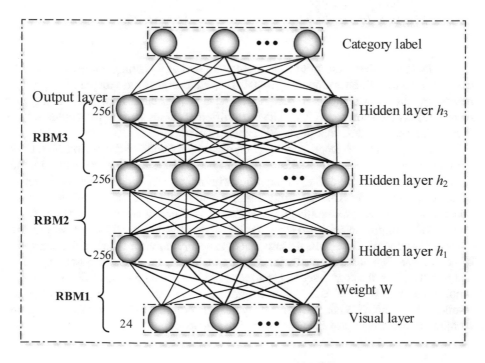

Fig. 2. The structure of DBN model.

DBN training structure applied in this experiment is shown in Fig. 2. The output value of 256 nodes at the last layer shall be used as the deep feature obtained by 24-dimension MFCC via DBN..

3.4   Experimental Result and Analysis

When engaging in classification, SVM shall not only select a kernel function, but adjust the penalty parameter c and the parameter of kernel function g so as to achieve better classification results. Taking the corpus set C as an example, some experimental results of hyper-parameter adjustment are shown in Table 2. The best classification results can be obtained when $s = 1$ and $t = 3$ (i.e. sigmoid function). Besides, c and g can be confirmed via cross validation.

According to the experimental results, the voiceprint recognition system based on DBN and SVM performs well. After the cross validation in the corpus set A and B, the average recognition accuracy rates are 96.12% and 98.17%, respectively. The recognition accuracy rate in the corpus set C is up to 100% (100% of recognition accuracy rate after several repeated experiments).

Table 2. Hyper-parameter adjustment experiment of SVM

s	t	c	g	Accuracy rate of training set	Accuracy rate of test rate
1	1	0.25	0.0625	100%	75%
1	2	0.25	0.625	100%	92.5%
1	3	0.25	0.0625	100%	100%

In view of the deep voiceprint feature obtained by DBN learning, three layers of BP network structure (network node: 256-20-10) are constructed to compare the performance of voiceprint recognition between SVM and BP neural network. The experimental results are shown in Fig. 3. The recognition accuracy rates in the corpus set A, B and C are higher than that of the BP neural network.

In addition, in the voiceprint recognition experiment with 24-dimension MFCC as the identity feature vector of speakers (retraining of SVM and BP neural network), the recognition accuracy rate of SVM is significantly higher (see Fig. 3). With the increase of feature dimensions (Change from 24-dimension MFCC to 256-dimension deep voiceprint feature), the recognition accuracy rate of BP neural network has experienced a remarkable improvement (Increase from less than 80% to over 95%). However, the computation complexity suffers a sharp increase accordingly, and the running duration of calculation is extended greatly. According to Fig. 4, the running duration of BP neural network in the corpus set A, B and C is not lower than 545 s, while the average running duration of SVM is 2 s at most under the same data and experimental environment. SVM holds better performance with respect to the reduction of computation complexity and the improvement of operation efficiency.

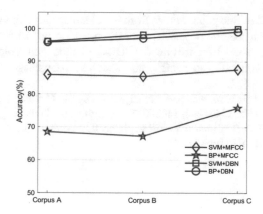

Fig. 3. The recognition accuracy of SVM and BP neural network with respect to the deep voiceprint features and MFCC features.

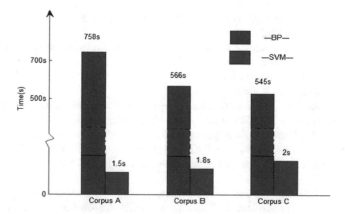

Fig. 4. Comparison of time spent between different approaches of SVM and BP neural network to classify deep voiceprint features.

4   Conclusion

The authors have conducted relevant researches by using "DBN + SVM" as the voiceprint recognition model framework. Based on a 24-dimension Mel Frequency Cepstrum Coefficient (MFCC), the authors have extracted 256-dimension deep voiceprint features via DBN that is developed by stacking three layers of RBMs, and conducted voiceprint recognition by means of SVM. According to the experimental results, the voiceprint recognition model framework designed by the authors in the paper has experienced an improvement in terms of the recognition accuracy and efficiency of recognition time, especially when compared with the traditional voiceprint features and recognition models. DNN allows the voiceprint features to be extracted and presented into the higher dimensions, while SVM can avoid the elevation of computation complexity caused by the increases of feature dimension. The combined strengths have been fully expressed by the experimental results.

The corpus used in this paper comes from the clean corpus from the laboratory. The future directions of research are to deal with the noisy corpus in natural environments and to develop the voiceprint feature extraction method and voiceprint recognition model with strong generalization ability and excellent robustness.

Acknowledgments. This work was partially supported by the National Natural Science Foundation of P. R. China (No. 11401092, 11426045), Scientific and Technological Planning Project of Ji Lin Province of P. R. China (No. 20180101229JC), Foundation of Ji Lin Educational Committee (No. JJKH20181100KJ).

References

1. Campbell, J.P.J.: Speaker recognition: a tutorial. Proc. IEEE **85**(9), 1437–1462 (2008)
2. Davis, S.B., Mermelstein, P.: Comparison of parametric representations for monosyllabic word recognition in continuously spoken sentences. IEEE Trans. Acoust. Speech Signal Process. **28**(4), 357–366 (1980)
3. Doddington, G.R.: Speaker recognition-identifying people by their voices. Proc. IEEE **73**(11), 1651–1664 (1985)
4. Shikano, K.: Text-independent speaker recognition experiments using codebooks in vector quantization. J. Acoust. Soc. Am. **77**(S11), 57–68 (1985)
5. Waibel, A.: Modular Construction of Time-Delay Neural Networks for Speech Recognition. MIT Press, Cambridge (1989)
6. Reynolds, D.A., Rose, R.C.: Robust text-independent speaker identification using Gaussian mixture speaker models. IEEE Trans. Speech Audio Process. **3**(1), 72–83 (1995)
7. Campbell, W.M., Sturim, D.E., Reynolds, D.A., et al.: SVM based speaker verification using a GMM super vector kernel and NAP variability compensation. In: IEEE International Conference on Acoustics, Speech and Signal Processing. ICASSP 2006 Proceedings, 2006:I-I. IEEE (2006)
8. Kenny, P., Boulianne, G., Ouelletm, P., Dumouchel, P.: Speaker and session variability in GMM-based speaker verification. IEEE Trans. Audio Speech Lang. Process. **15**(4), 1448 (2007)
9. Kenny, P., Boulianne, G., Ouellet, P., Dumouchel, P.: Speaker and session variability in GMM-based speaker verification. IEEE Trans. Audio Speech Lang. Process. **15**(4), 1435–1447 (2007)
10. Dehak, N., Kenny, P.J., Dehak, R., Dumouchel, P., Ouellet, P.: Front-end factor analysis for speaker verification. IEEE Trans. Audio Speech Lang. Process. **19**(4), 788–798 (2011)
11. Hinton, G.E., Osindero, S., Teh, Y.-W.: A fast learning algorithm for deep belief nets. Neural Comput. **18**(7), 1527–1554 (2006)
12. Mohamed, A., Dahl, G., Hinton, G.: Deep belief networks for phone recognition. In: Nips Workshop on Deep Learning for Speech Recognition and Related Applications, vol. 1(9), pp. 39 (2009)
13. Hinton, G., Deng, L., Yu, D., et al.: Deep neural networks for acoustic modeling in speech recognition: the shared views of four research groups. IEEE Signal Process. Mag. **29**(6), 82–97 (2012)
14. Wu, M.: A Study of Text-Independent Speaker Recognition Based on Deep Learning. University of Science and Technology of China (2016)
15. Lv, L.: A Study of Voiceprint Recognition Methods based on Deep Learning. Southeast University (2016)

16. Zhijun, S., Lei, X., Yangming, X., et al.: The research overview of deep learning. Appl. Res. Comput. **29**(8), 2806–2810 (2012)
17. Chunxia, Z., Nannan, J., Guanwei, W.: Restricted boltzmann machine. Chin. J. Eng. Math. **32**(2), 161–175 (2015)
18. Jinhui, L., Junan, Y., Yi, W.: New feature extraction method based on bottleneck deep belief networks and its application in language recognition. Comput. Sci. **41**(3), 263–266 (2014)

Research on Intelligent Patent Classification Scheme Based on Title Analysis

Yao Zhang[1] and Gening Zhang[2(✉)]

[1] School of Mechanical, Electrical and Information Engineering,
Shandong University, Weihai, China
[2] School of Computer Science, University of Manchester, Manchester, UK
gening96@163.com

Abstract. With the rapid increasing number of patents, it is becoming more significant but difficult to mine underlying information from huge patent data in database. By integrating Latent Dirichlet Allocation (LDA) topic mode with text mining algorithms, we propose two patent classification schemes: topic-based patent classification and title word-frequency-based patent classification, which can be applied in the areas of patent retrieval, patent evaluation and patent recommendation. The process and implementation methods of proposed schemes are discussed, and the examples to intelligently classify patent records in the area of railway transportation in international patent database are given, the results can adequately verify effectiveness of our proposed schemes.

Keywords: Patent classification · Text mining · Topic analysis · LDA model · Clustering algorithm · Railway transportation

1 Introduction

Patent data almost contains all scientific and technical achievements in every field, and has proved to be the most important information resources to promote economic development, technological innovation and enterprise strategy, so it is most significant to deeply explore underlying information from patent data in international patent database. However, with the dramatic increasing number of patents year by year, it is becoming a more challenging work to mine and analyze valuable information from huge patent data. The technologies of big data processing are an efficient way to solve this problem [1].

Patent classification has wide application prospect in the areas of patent retrieval, patent value evaluation and patent recommendation. The traditional classification algorithms such as logistic regression, support vector machine, K-means clustering or random forests are all based on quantified data analysis, which are not suitable for requirements of our content-based patent classification. In this paper, we propose two intelligent patent classification schemes: topic-based patent classification and title word-frequency-based patent classification, which integrate Latent Dirichlet Allocation (LDA) topic analysis model with text mining algorithms. The process and implementation methods of proposed patent classification schemes are discussed, and the examples to dynamically classify patents in the field of railway transportation in international patent database produced by the European Patent Office are given to verify effectiveness of our proposed patent classification schemes.

© Springer Nature Switzerland AG 2020
F. Xhafa et al. (Eds.): IISA 2019, AISC 1084, pp. 12–20, 2020.
https://doi.org/10.1007/978-3-030-34387-3_2

2   Topic-Based Patent Classification Scheme

2.1   The Process of Topic-Based Patent Classification Scheme

The process of topic-based patent classification scheme is shown in algorithm 1, which employs LDA topic model [2, 3].

Algorithm 1: topic-based patent classification scheme

#define following variables:

$D=\{d_1, d_2, \ldots, d_m\}$: dataset of patent titles in database;

$W_{ds}=\{w_1, w_2, \ldots, w_n\}$: word set of d_s (s=1-m);

$W_d=\{w_1, w_2, \ldots, w_q\}$: word set of D;

$\theta_{ds}=\{t_1, t_2, \ldots, t_p\}$: topic probability distribution of d_s (s=1-m);

$\Phi_{tj}=<p_{w1}, p_{w2}, \ldots, p_{wq}>$: word weight coefficients of extracted topic t_j (j=1-p);

N: The maximum iterations of LDA model;

stop_list: word set including all meaningless words such as prepositions, conjunctions, modal particles, auxiliary word and adverbs;

public_list: word set of public words, which includes all characterless words in patent title dataset D.

Step1: Build patent title dataset D by statistics to sample patent records.

Step 2: Clean patent title dataset D.

   1. Turn all capital letters into lowercase letters.
   2. Delete punctuations.
   3. Remove stop words in stop_list .
   4. Remove public words in public_list.
   5. Stemming by removing prefix and postfix of words.

Step 3: Extract topics from patent title dataset D by LDA model.

   1. randomly set initial values of θ_{ds} and Φ_{tj};
   2. read d_s from D (s=1~m) {
   3.   read w_i from W_{ds} (i=1~n) {
   4.   let $w_i \in t_j$ (j=1~p) {

$$p_{tj} = p(t_j / d_s) = \frac{the\ number\ of\ d_s\ words\ in\ t_j}{the\ number\ of\ d_s\ words}$$

$$p_{wi} = p(w_i / t_j) = \frac{the\ number\ of\ t_j\ word\ in\ w_i}{the\ number\ of\ t_j\ word} \qquad (1)\}$$

$$p(w_i / d_s) = p(t_j / d_s) \cdot p(w_i / t_j)$$

   5.   if ($p(w_i/d_s)$ is bigger than previous value)
            update values of θ_{ds} and Φ_{tj} according to formula (1); }}
   6. if (less than maximum iterations N)   go to 2;
   7. return final values of θ_{ds} and Φ_{tj}

Step 4: classify patent records by extracted topic words.

   1. Select higher weight coefficient's topic words as classification key words.
   2. Classify patent records according to classification key words.

In LDA model of step 3, random sampling should abide by following rulers [3]:

- Generate topic distribution by Dirichlet sampling;
- Obtain relevant words by polynomial sampling to topic distribution;
- Generate word distribution of topics by Dirichlet sampling;
- Obtain words by polynomial sampling to word distribution.

2.2 Results and Discussion

We analyze sample data in international patent database which contains more than 60 thousand patent records in the area of railway transportation from Jan 2013 to Dec 2018. In international patent database, railway transportation patents are divided into following different types [4]:

- B61B: railway systems and equipment;
- B61C: locomotives, motor, rail-cars;
- B61D: body details or kinds of railway vehicles;
- B61F: rail vehicles suspensions;
- B61G: coupling, draught and buffing appliances;
- B61H: brakes or other retarding apparatus peculiar to rail vehicles, arrangements or dispositions of brakes or other retarding apparatus in rail vehicles;
- B61 J: shifting or shunting of rail vehicles;
- B61 K: other auxiliary equipment for railways;
- B61L: guiding railway traffic, ensuring the safety of railway traffic.

However, these classification types are rough and factitious, and should be classified into more accurate sub-types. As an example, we further classify B61D patents into sub-types by our topic-based patent classification scheme, the results are shown in Tables 1 and 2.

Table 1 shows extracted key words and weight coefficients of different topics, Table 2 shows classification results of B61D patents. Considering that public words can't reflect characteristics of relevant topics, so they should be removed from patent title dataset. From Tables 1 and 2, we can obtain that:

(1) Our topic-based patent classification scheme has better performance on dynamically classifying patents according to patent title contents.
(2) Topic key words with high weight coefficient have more influence on topics, the bigger the word weight coefficients are, the more relevant the patent records belong to this topic class.
(3) The weight coefficient of topic key words decreases with the increase of the number of topics, and topics with smaller key word weight coefficients should introduce more key words to sufficiently describe their characteristics.

Table 1. The extracted topics from title dataset of B61D patents (The maximum iterations of LDA model is 50)

Topic	Patent title words	
	public word	Topic key word/ weight coefficient
Topic 1	Vehicle, rail, device, car, railway,	Door, air/0.018, seat/0.013, high/0.009, passenger, control, speed/0.008/
Topic 2	train, system, body, type, track, carriage,	Mine/0.008, Use, floor, assemble, wall/0.007, Load/0.006, Electric, roof, locomotive, contain, connect, cover, apparatus, hopper, support, plate/0.005
Topic 3	unit, wagon, side, railcar, road,	Tank, window, detect, platform, operate/0.005, trolley, manufacture, freight, mount, equip, frame, steel, material, light, berth, table, transit, automat, bogie/0.004
Topic 4	structure, method, transport, mechanism, railroad	Energy/0.005, rack, parallel, protect, water, luggage, lock, guide, module, motor/0.004
Topic 5		Head, subway, cargo, maintenance, display, box, compart, power, heat, roll/0.004
Topic 6		Mitigate, drive, flat, vacuum collect/0.004, truck, novel, wind, emerge, pedal, wheel, open, end, generate, safety, cab, move, front, comprise, double, particular, handle, special, cable, prevent, work, transfer, monitor, self, construct function, cart, inspect, toilet/0.003
Topic 7		Improve, composite, panel, tip, interior, deform/0.004, provide, board, dump, tramcar, traction, traffic, articulation, arrange, zone, ventilator, part, bucket, hood, formula, ladder, cool, element/0.003
Topic 8		Illuminate, chassis, lamp, travel, piggyback, mean, trailer/0.004, suction, valve, aerodynamic, brake, conveyor, carrier, bathroom, evacuate, reduce, alloy, automobile, gondola, access, holder, include, bottom, integrate, duct, fire, bridge, dollies, refrigerate, sleep, glass, section, turn, cell, suspension, car-body, convey, couple, two, shape/0.003
Topic 9		coach/0.004, Inform, install, adjust, stop, ballast, pipeline, ramp, slide, wire, secure, fasten, good, facility, extern, exhaust, top/0.003
Others		Antique, acoustics, roadway, cushion, anti, catch, combine, overhaul, armrest, coal, multi, crane, hand, bed, lavatory, tool, man, drainage, slide, barrier, mobile, monorail, incline, tram/0.002

Table 2. The classification results of B61D patents by topic-based scheme

Topic	Amount	The examples of patent title
Topic 1	4244	A high-efficient air diffuser for railway vehicle
		Automatic temperature-regulating seat cushion
		Self-service booking catering equipment used in high-speed train and working method of equipment
Topic 2	3837	Mining method for two-side tipping and discharging and mining U-shaped car tipper
		Wheeled vehicle loading strengthening device for railway flatcar
		Urban railway vehicle side roof structure assembly
Topic 3	1925	Bracket for auto-passing neutral section ground-sensor vehicle-mounted detection device
		Chain transmission mechanism in train examination trolley
		Device and method for automatically locking and fastening cars and applied to train
Topic 4	875	Means of transport with a waste water system
		Device for pivoting one or more front flaps of a track-guided vehicle
		Backrest, vehicle, and guide track transportation system
Topic 5	499	Display screen suspending device for sleeper compartment of train
		Heating system of breakdown rail vehicle
		Multi-deck rolling stock including at least two ramps for accessing an upper deck
Topic 6	1172	Winding body treating system, winding body supply method, and winding body transport device
		Vacuum keeping type excrement-collecting machine system
		Toilet cabin for public transport vehicle, intended to receive person with reduced mobility
Topic 7	682	Upper apron board system on rail vehicle and rail vehicle
		Suspension type tramcar road and suspension type tramcar
		Cooling aeraulics device for a rail vehicle element and corresponding rail vehicle
Topic 8	347	Transporting means having manual hydraulic brake for railway
		Rail vehicle, illuminating lamp mechanism and illuminating lamp angle adjustment device
		Chassis of piggyback transport vehicle
Topic 9	86	Engineering vehicle with ballast discharging device
		Mechanism for raising ramps of wagon for car transport
		Be applied to security alarm device of hillside orchard single -rail transportation machine
Others	599	Acoustic insulation device for mobile vehicle
		Drainage device for a vehicle body and vehicle body substructure with a drainage device and a corresponding vehicle body
		Rail sliding vehicle device

3 Title Word-Frequency-Based Patent Classification Scheme

3.1 The Process of Title Word-Frequency-based Patent Classification Scheme

Title word frequency is closely related to research hotspots of patents. The process of title word-frequency-based patent classification scheme is shown in algorithm 2, which integrates word frequency statistics with K-means clustering algorithm [5].

Algorithm 2: title word-frequency-based patent classification scheme

#define following variables:
N: maximum iterations of K-means clustering algorithm;
d: Euclidean Distance between cluster node and cluster center;
SSE: Sum of Squared Errors;
D: dataset of patent titles in database;
k: the cluster number in K-means algorithm.

Step 1: clean patent title dataset D.
Step 2: word frequency statistics to patent title dataset D.
Step 3: build patent title word frequency sheet.
Step 4: ascertain optimal value of k to classify patent title word frequency sheet.
Step 5: patent classification by K-means cluster algorithm.
    1. randomly select k records as initial cluster centers μ_i (i=0~k-1);
    2. calculate Euclidean Distance between sample data x to clustering center by

$$d = \sum_{i=1}^{k} \sqrt{\sum_{x_j \in C_i} (x_j - \mu_i)^2} \qquad (2)$$

    3. cluster all records according to the principle of minimum Euclidean Distance;
    4. update cluster center μ_i by calculating mean values of clusters i (i=0~k-1);
    5. if iterations are less than N go to 2;
    6. return cluster results
Step 6: analysis of cluster results

In step 4 of algorithm 2, the optimal value of cluster number k can be selected by

$$SSE = \sum_{i=1}^{k} \sum_{x \in C_i} |x - \mu_i|^2 \qquad (3)$$

In formula (3), C_i denotes sample dataset of cluster i ($i = 0 - k - 1$), μ_i denotes center of cluster i.

The optimal k can be ascertained according to following principle [6]:

- SSE nonlinearly decreases with the increase of k;
- When k is smaller than optimal value, the decline of SSE is rapidly;
- When k is bigger than optimal k, the decline of SSE is smoothly.

3.2   Results and Discussion

We analyze all sample data of B61 patents by algorithm 2. Table 3 shows some high frequency words in title field of B61 patents in database. Figure 1 shows relationship between k and SSE in K-means algorithm. Because the curve of Fig. 1 has an obvious inflection point at $k = 5$, the selected optimal value of k is 5 in which suitable SSE and the number of clusters can be both obtained.

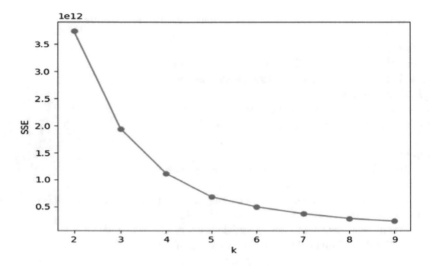

Fig. 1. Relationship between k and SSE

Table 3. Some high frequency words in title field of B61 patent records

Word	Frequency	Word	Frequency	Word	Frequency
Vehicle	14258	Bogie	2916	High	1634
System	13592	Structure	2650	Electric	1553
Device	13082	Locomotive	2556	Monitoring	1504
Rail	9932	Type	2514	Wheel	1495
Method	9618	Door	2117	Unit	1491
Railway	8182	Brake	2082	Detection	1412
Train	7632	Air	1916	Railroad	1408
Car	4910	Apparatus	1776	Platform	1393
Control	4298	Transport	1710	Automatic	1384
Track	3087	Speed	1700	Power	1353

Tables 4 and 5 are clustering results by algorithm 2, the patent hotspot levels are classified into five classes according to different clusters. We can observe that among these clusters, cluster 2 has maximum average of patent title word frequency, it means that patents belong to cluster 2 have the highest hotspot level.

Table 4. Cluster results by K-means algorithm (The maximum iterations of K-means is 1000)

Cluster	Mean value of clusters	Total number	Hotspot level
0	3178.999065	16022	5
1	25351.06122	14322	3
2	57425.88131	1744	1
3	14618.39227	17445	4
4	37784.79008	6342	2

Table 5. The part of classification results of B61 patents by title word-frequency-based scheme

Cluster	Patent title	Title word frequency/hotspot level
0	An aggregate erection seat structure for installing tread cleaner and braking clamp	5058/5
1	Rail flaw detection vehicle capable of detecting flaws on top surfaces of rails	27343/3
2	Rail vehicle and clutch assembly for the rail vehicle, and method for coupling the railway vehicle with a second rail vehicle	92196/1
3	Train footboard used for boarding and alighting and train	16065/4
4	Wheel type probe following mechanism of double rail type rail ultrasonic inspection car	33404/2

4 Conclusions

In this paper, we propose two intelligent patent classification schemes: topic-based scheme and title word-frequency-based scheme. The results to classify transportation patents show that our schemes not only can dynamically classify patent records with better performance on distinguishing patents, but also have advantages on mining research spots, technology trajectories and development tendency of patents.

As our further work, we will focus on improving our patent classification scheme in aspects of patent topic extracting algorithm, correlations between key words and topics, and indicators to evaluate hotspot levels of patents. Besides, we will extend our research to patent value evaluation model, text mining algorithm of patent documents and patent intelligent recommendation scheme by deep learning frameworks.

References

1. Liu, F., Ma, R.: Growth patterns of national innovation capacity: international comparison based on technology development path. Sci. Sci. Manag. S. & T. **34**(4), 70–79 (2013)
2. Zhu, H.: Research on Several Core Techniques of Text Mining. Beijing Institute of Technology Press, Beijng (2017)
3. Li, B.: Machine Learning Practice & Application. Poster and Telecom Press, Beijing (2017)
4. Patent database of European Patent Office. https://worldwide.espacenet.com/classification
5. Yuan, M.: Foundation of Machine Learning Principle, Algorithm & Practice. Tsinghua University Press, Beijing (2018)
6. Kubat, M.: Introduction to Machine Learning. China Machine Press, Beijing (2018)

Study on Intelligent Recognition Matching and Fusion Interaction Technology of Electrical Equipment Model

Zhimin He[1,2]($\boxtimes$), Xiaodong Zhang[3], Chengzhi Zhu[4], Lin Peng[1], and Min Xu[1]

[1] Global Energy Interconnection Research Institute,
Nanri Road 8, Nanjing 210003, China
825097034@qq.com
[2] State Grid Key Laboratory of Information and Network Security, Beiqijia,
Beijing 100000, China
[3] State Grid Corporation of China, Beijing 100031, China
[4] State Grid Zhejiang Electric Power Co., Ltd., Hangzhou 310007, China

Abstract. After investigation and analysis, this project studies the intelligent identification matching technology of the physical device model, the overlay technology of the lightweight model and the physical equipment, and realized the panoramic layered display of the internal and external three-dimensional models of the physical device. According to the specific needs of the operation and maintenance of the physical device, the component-level monitoring data (operating conditions, thyristors, water level, temperature and humidity, etc.) of the physical device are overlaid onto the corresponding equipment model.

Keywords: Intelligent recognition · Overlay display · Fusion interaction

1 Introduction

1.1 Background

In recent years, the state has intensified its efforts in power grid construction. In particular, it has built a series of long-distance transmission networks. While continuously expanding the size of transmission lines, it has also advanced the management and inspection of current operation and maintenance operations. Claim. At present, most of the domestic power line inspections use manual paper media recording, strip identification, RF acquisition, and semi-automated inspections. The inspection results mainly depend on the personal qualities of the inspection personnel. The line inspection results are difficult. Ensure that there are no accurate data records for the location, video, sound, and gas concentrations of the inspectors on site. This often results in missed inspections and wrong inspections, and it does not guarantee the in-place rate of inspectors. At the same time, there is also irregular information entry. The management is not scientific enough, and these methods can no longer meet the requirements of current power system development.

© Springer Nature Switzerland AG 2020
F. Xhafa et al. (Eds.): IISA 2019, AISC 1084, pp. 21–28, 2020.
https://doi.org/10.1007/978-3-030-34387-3_3

1.2   Purpose and Significance

The research content of this project is an important direction for the intelligent operation and maintenance of power equipment in the future. Through the subsequent research and development of intelligent equipment manufacturing, assembly, operation and maintenance systems based on the panoramic three-dimensional model, it can serve the large-scale equipment manufacturing, assembly, operation and maintenance of the power grid. In patrol assist operations, the efficiency of on-site operations and the accuracy of fault diagnosis of field devices are improved.

1.3   Overseas Research Levels

Augmented reality came into being during the development of virtual reality. On the one hand, it was born with the development of the times. On the other hand, it was to make up for the insufficiency of virtual reality that was completely separated from the reality. Foreign countries are at the leading level in algorithm research. Since Boeing's researcher Thomas Caudell created augmented reality in 1990, more and more research institutes, universities, and companies have begun to carry out extensive research on augmented reality and achieved certain scientific research. Achievements. One of the more famous is the Computer Vision Laboratory of Lausanne Institute of Technology in Switzerland. Its three-dimensional registration algorithm based on natural plane image and three-dimensional object tracking is considered to represent the leading level in the industry; the Interactive Mul-timedia Lab of National University of Singapore focuses on the Research on human-computer interaction technology of Augmented Reality technology; German BMW Lab is researching and developing an augmented reality auxiliary vehicle mechanical maintenance project. The goal is to realize a first-view augmented reality program based on a wearable computer. Currently, the Steve Feiner team of Columbia University has used AR to manufacture a laser printer maintenance system. Users can use the system to maintain the printer through the operation instructions in the system, such as to remove the paper tray, and earlier to propose enhancements. Boeing, a real-world concept, is using this technology to help technicians install wiring harnesses that make up the aircraft's electrical system, develop a digital set of operational procedures through augmented reality, and replace the huge physical layout boards to make wiring harnesses, thereby saving time and money. If this technology can be successfully applied to the factory, it will bring huge benefits. On April 4, 2012, Google formally announced the product plan of Google Glasses. This augmented reality product has all kinds of services that smart phones can provide. The lenses have the function of a micro-display, which can transmit information to the lenses. And allows wearable users to control the sending and receiving of information by voice. In addition, with the help of the famous printer manufacturer Epson, the startup company Meta has also developed a reality enhanced glasses META PRO. It is mainly based on the unique two-dimensional space gesture recognition function, equipped with Epson production Wearover BT-100 wearable dual-eye display, and equipped with a low-delay 3D camera on the top of the device. Although this device is low in configuration, it has revolutionized the field of augmented reality devices. Subsequently, Epson also officially launched its own augmented reality glasses

EPSON BT-200/BT-2000 glasses, of which BT-2000 glasses specifically tailored for industrial customers, with higher endurance and computing speed, wear it More comfortable.

1.4 Domestic Research Levels

Relative to developed countries, China's augmented reality technology started comparatively later than other countries. However, in recent years, both the Chinese government and the scientific community have attached great importance to it, and have formulated corresponding plans for scientific research and continuously catch up with advanced technologies. Major universities have also conducted research on augmented reality technology and made some progress. Huazhong University of Science and Technology has studied the key technologies of remote operation based on augmented reality, proposed a vision-based augmented reality tracking registration method and a real-time scaling strategy based virtual reality registration method, and designed a marker-based global A three-dimensional registration method combining homography matrix; Shanghai Jiaotong University proposed the spatial augmented reality pipeline concept; Zhejiang University based on the positioning mark-based video detection, starting from the shadow generation method and the light detection algorithm in the augmented reality environment. ARSGF, an augmented reality software framework based on scene management. In addition, China has adopted augmented reality technology in many aspects. Since the beginning of the 2012 Spring Festival Gala, augmented reality has begun to be applied more to the media industry, which has made media programs even more admirable. In addition, with the increasing popularity of mobile Internet application development, application developers have started to develop applications that meet user needs around large platforms, and have increasingly enriched the App Store of major platforms. The rise of small hand-held mobile devices (handheld PDAs, smart phones) has provided a new development platform for mobile augmented reality technology. Five-dimensional space-time launched an entertainment and interactive mobile community based on augmented reality - the OSEE game. After users enter the OSEE world, they can understand the new things that are happening around the world and share their feelings. The heart's emotional expressions are shared with other friends on the spot, and all this gives the user a completely new visual experience in real life situations.

2 Solutions and Key Technologies

From the power equipment model identification technology, panoramic model and background real-time monitoring data fusion interaction technology, physical equipment model superposition and virtual reality integration technology, model control interaction technology, to carry out the development of panoramic model identification and physical equipment overlay interaction program development, as shown in Fig. 1.

Taking the physical device as the object, using the identifiable model of the scanning model, identifying and registering the on-site power equipment, designing a convenient and natural model control interaction method, and overlaying the three-

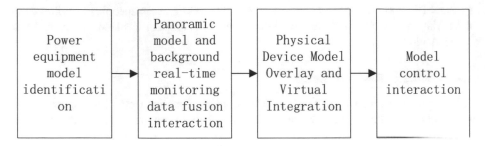

Fig. 1. Solutions

dimensional virtual model and the real-time monitoring data in a superimposed manner around the physical power equipment, so that The on-site personnel understand the structure and operating status of the power equipment more clearly and intuitively.

2.1   Power Equipment Model Identification

First, the point cloud of the scene depth camera capturing the scene is turned on and sent to the scene recognition server in the background. The server that has completed the sample training in advance quickly recognizes the received point cloud information and returns the hot spot number and the scene feature information to the terminal. The terminal judges whether the recognition module continues or stops sending the point cloud of the hotspot according to the hotspot number, and then the server tracks the sensor data of the terminal device continuously after the server calculates the pose data using the scene feature information and the sequence frame subsequently delivered by the camera. Passed to the rendering module to achieve the virtual tag rendering and real-time tracking display, so as to realize the identification and tracking of on-site physical power equipment.

The specific processing process is mainly divided into depth-of-field sensor point cloud image acquisition, point cloud image processing, construction of a model database, and real-time identification algorithms. The cloud image of the current scene captured by the depth sensor contains noise effects, passes through filtering and other preprocessing stages, passes into the point cloud image processing stage, and uses the European clustering segmentation method to extract the point cloud containing the recognized object. Clustering, followed by real-time identification.

In the real-time recognition stage, the CVFH feature of the point cloud clustering is first calculated, KNN algorithm is used to find the k nearest neighbor model features from the model database, and after matching and recognition, it is determined whether it is an object to be identified. If yes, the current point cloud clustering is returned. The three-dimensional center point coordinate value; otherwise, continue to calculate the CVFH feature descriptor and find neighbors.

2.2 Panoramic Model and Background Real-Time Monitoring Data Fusion Interaction

The visual modeling system based on depth of field sensor is composed of hardware and software. Through the calibration of sensors, the conversion formulas of depth data and point cloud data are obtained. After analyzing the sensor accuracy, it is determined that the ICP point cloud splicing algorithm and the MC surface modeling algorithm are used as the main algorithm. The scanned three-dimensional model can be used for subsequent model identification, and the function of registration superposition and tracking can be achieved.

Power equipment model and background real-time monitoring data fusion technology are mainly divided into data layer fusion and feature layer fusion technology. The data layer fusion is mainly used for the filtering technology of multi-source sensing data to remove noise data. Feature layer fusion technology can classify field job data to facilitate the standardized storage and use of data. In this project, according to the specific needs of the operation and maintenance of the physical device, the component level monitoring data (operating conditions, thyristors, water level, temperature and humidity, etc.) of the physical device are fused and superimposed on the corresponding equipment model.

The data from the multi-source sensor is processed by the corresponding algorithm in the data fusion engine, and the externally processed auxiliary information and external knowledge are combined to improve the accuracy of the fusion processing. The fusion processing result can be used as a decision, and can also be fed back as a kind of auxiliary information to the fusion processing process, so that the fusion system can adaptively optimize the fusion processing. The data fusion processing unit structure of this project is shown in Fig. 2.

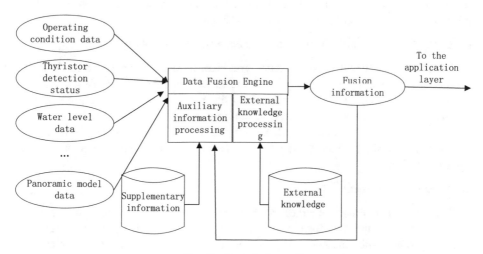

Fig. 2. Data fusion unit

The core part of data fusion is the data fusion engine and its various internal algorithms.

Firstly, a special system interface is developed to extract the information collected by the back-end monitoring system of the power equipment, the type of information collected, and the frequency of acquisition. Based on the individual characteristics of the extracted monitoring information, the on-site monitoring data is divided into categories including operating data and equipment status data.

Secondly, the acquired data is accurately registered, and the on-site power equipment monitoring data (including the DC valve operating condition data, thyristor monitoring status, physical device water level data, temperature and humidity data, etc.) and the three-dimensional model of the equipment are established. The logical connection, combined with historical data, completes the denoising of the monitoring data of the underlying equipment and avoids random disturbances to interfere with the perceived data.

Finally, after the background fusion processed monitoring data is classified and integrated with the panoramic model, the overlay is displayed on the relevant parts of the field power equipment, and the combination of the panoramic equipment component model and its corresponding monitoring data is realized, making the on-site operator more intuitive. Understand the internal and external structure of the equipment and the overall equipment and component operating status.

2.3   Physical Device Model Overlay and Virtual Integration

When the device model is superimposed on a real scene, using model superposition and virtual reality fusion techniques, virtual reality integration can establish a close relationship between the user, the virtual world and the real world, and realize an integrated perceptual awareness system through natural human-computer interaction. In order to enhance display efficiency, the following methods are used:

(1)   Using the AR hybrid tracking strategy for virtual integration

Virtual reality fusion uses computer graphics, human-computer interaction technology, image display technology, and multimedia technology to fuse computer-generated virtual objects or information into the real environment that users want to experience. Head pose tracking is one of the core technologies, and the only effective way for head tracking in outdoor AR systems is to use hybrid tracking techniques to obtain high-precision, robust head poses by fusing measurement data from different types of trackers.

(2)   Virtual and synthetic display

The real scene image obtained by the camera is set as the background in the environment, and then the position and posture of the virtual object are determined according to the calibrated internal and external camera parameters and the placement point of the virtual object specified by the user interaction, and the virtual object is converted for perspective projection. In the matrix, the virtual object is projected into a planar image, and then the background of the projection plane image is made transparent. Finally, the projected image is superimposed on the real scene image captured by the camera, and the real image and the virtual model can be seen in a window. At the same time display, that is, virtual synthesis display.

2.4   Model Control Interaction

So far, the real world is the space and world of human activities, and the virtual world is a computer-generated world. These two spaces are separated. As mentioned earlier, augmented reality technology is the natural interface and interface of these two spaces, enabling humans in the real world to understand and drive the dynamics of the virtual world. In order to break the limitations of traditional AR mobile applications in human-computer interaction, it is necessary to expand the interactive experience space. Mainly from two aspects to conduct interactive technology research:

(1) Interactive operation: Take different strategies for the interaction of different objects.

This project divides interactive operations into three categories: interaction with the real world, interaction with virtual objects, and interaction between virtual objects and the real world. Use the depth camera to obtain the depth information of the object, obtain the power worker's gesture information from the field work, use the middleware technology to analyze and process the data, convert the data into corresponding gesture commands, study the dynamic characteristics of the gesture, and track the movement of the gesture, And then to identify the complex movements that combine gestures and hand movements, to realize the interactive technology of the electric power worker's gestures to control the virtual objects.

(2) seamless switch: based on the device model to interact with the user's needs as the center of the design, using a seamless space conversion mode to achieve the natural switching of two types of interactive space.

In the first type of active interaction space, the interaction part actively pushes related information, such as equipment ledger, equipment drawings, maintenance records, equipment models, and current conditions, to the terminal when viewing the equipment in the form of enhanced information. When a user selects a piece of information, he or she generally desires to be able to in-depth view the internal detailed information. At this time, only the corresponding information option is selected, and the interaction space naturally transitions to the second type of passive interaction space. In the enlarged three-dimensional environment, the user can not only randomly roam within the device model from the first person's perspective, but can also locally adjust or zoom in and view the interior of the device model according to the operation requirements.

3   Conclusion

Firstly, the theoretical investigation, analysis and comparison of intelligent identification and virtual reality fusion involved in the process of intelligent identification match and overlay interaction of the panoramic scanning model are introduced. After image input, image preprocessing, image feature extraction, and identification classification, the recognition result is finally output. In virtual reality integration, virtual scenery needs to share a unified space with the real world. The interaction between the scene as a virtual object and the scene in the real scene needs to conform to the laws of nature, avoid space conflicts and visual conflicts, and build models of real scenes. Land assists in the coordination and integration of virtual and real space.

Through the research of the real equipment panoramic scanning model intelligent identification matching, panoramic model and background real-time monitoring data fusion interaction, full-size panoramic model and physical equipment stacking interaction and model control interaction technology, form a complete program, and achieve panoramic panoramic layered display of the physical device The internal and external three-dimensional model. According to the specific needs of the operation and maintenance of the physical device, the component-level monitoring data (operating conditions, thyristors, water level, temperature and humidity, etc.) of the physical device are overlaid onto the corresponding equipment model.

Acknowledgments. This work was financially supported by the science and technology project to State Grid Corporation 'Research on 3D Lightweight Engine Technology for Power Grid Service Scenarios'.

I would like to express my heartfelt gratitude to my colleagues and friends, they gave me a lot of useful advices during the process of studying this topic research and also provided enthusiastic help in the process of typesetting and writing thesis! At the same time, I want to thank all the scholars who are quoted in this paper. Due to my limited academic level, there are some disadvantages to writing a paper, and I will solicit the criticism and corrections from experts and scholars.

References

1. Chen, B., Qin, X.: Mixture of Virtual Reality and Human-Computer Intelligence in Mixed Reality, SCIENTIA SINICA Information (2016)
2. Li, Y.: Research on Key Techniques of 3D Surface Reconstruction Based on Depth Camera, Doctoral Dissertation, Zhejiang University (2015)
3. He, Q., Cheng, H., Yang, X.: A review of 3D model process technology oriented to 3D printing. Manuf. Technol. Mach. Tools **6**, 54–61 (2016)
4. Fan, L., Tang, J.: Expectation and review on overhead transmission lines 3D modeling methods. South. Energy Constr. **4**(2), 120–125 (2017)
5. Zhao, C.: Research and Application of Substation Visualization Based on 3D Panorama and Rapid Modeling Technology, master's degree Dissertation, Shandong University (2013)
6. Qiang, Z.: Research of Visual Recognition Algorithm for Service Robot based on PCL, master's degree Dissertation, University of Science and Technology of China (2014)

Deep Knowledge Tracing Based on Bayesian Neural Network

Li Donghua[1], Jia Yanming[2], Zhou Jian[3], Wang Wufeng[4],
and Xu Ning[3(✉)]

[1] School of Computer Science and Technology,
Wuhan University of Technology, Wuhan, China
[2] Artificial Intelligence and Big Data Department, Beijing Global Wisdom Inc,
Beijing, China
[3] School of Information Engineering,
Wuhan University of Technology, Wuhan, China
xuning@whut.edu.cn
[4] Zhongshan Institute of Advanced Engineering Technology of WUT
Zhongshan, China

Abstract. Knowledge Tracing (KT) is the de-facto standard for inferring student knowledge from performance data, modeling students' changing knowledge state during skill acquisition. The existing knowledge tracing models have problems such as incomplete features consideration, poor effect and easy over-fitting in the process of simulating students' knowledge acquisition. In order to better model students' learning and track the process of students' knowledge acquisition, a Bayesian neural network Deep Knowledge Tracing (BDKT) is proposed. The model combines Bayesian neural network with Deep Knowledge Tracing (DKT), it can not only model feature-rich students' behavior data, but also effectively prevent over-fitting, enhance the generalization ability of the model, and accelerate the convergence speed of the model. The above performances of the model are tested by the experiments on the data sets of KDD Cup and Assistments.

Keywords: Knowledge tracing · Bayesian neural network · Deep Knowledge Tracing · Over-fitting · Generalization ability · Convergence speed

1 Introduction

The personalized adaptive learning is an important research topic of Intelligent Tutoring Systems (ITS). The premise of implementing personalized adaptive learning is to have a precise assessment of the students' knowledge acquisition state. The existing knowledge tracing method based on the deep neural network does not require the explicit encoding of human domain knowledge and can capture more complex representations of student knowledge, but the model didn't outperform the traditional method [1].

In view of the shortcomings of the existing research methods, this paper proposes a Bayesian neural network Deep Knowledge Tracing(BDKT) based on the Bayesian RNN

© Springer Nature Switzerland AG 2020
F. Xhafa et al. (Eds.): IISA 2019, AISC 1084, pp. 29–37, 2020.
https://doi.org/10.1007/978-3-030-34387-3_4

[2]. The contributions of this paper are as follows: (1) we analyze and compare the performance of BDKT and DKT on different datasets. It is found that BDKT performs better on small datasets in students' behavior modeling. (2) we apply Bayes by Backprop (BBB) to DKT. The BBB affords two advantages: explicit representations of uncertainty and regularization [3]. (3) we get low-dimensional and efficient students' behavior vectors by train and apply t-SNE (t-distributed Stochastic Neighbor Embedding) [4] for dimensionality reduction and visualization of student behavior vectors which can be used to analyze the relationship and similarity between students' behaviors and knowledge components.

2   Related Works

In this section, we briefly introduce and analyze some knowledge tracing methods.

Item Response Theory (IRT) is a standard framework for modeling student responses dating back to the 1950s [5] , IRT is essentially a structured logistic regression that assesses the potential quantitative relationship between the students' abilities and the problems' difficulties. It is assumed that this proficiency isn't changing during the examination. It is not suitable for modeling the dynamics of the student learning process. The BKT model is an important model for tracking the changes in the knowledge acquisition status in the learning process, introduced in the field of intelligent education [6]. BKT assumes that the knowledge components in the knowledge system are independent of each other, and the difficulty is the same under the same knowledge component. For each knowledge component, a set of exclusive parameters needs to be trained. Instead of mining attributes that are relevant to assessing student abilities, BKT uses a dynamic representation of student abilities. However, in fact, the division of knowledge components can't be so precise, and there is a certain correlation between the knowledge components. Therefore, BKT is not suitable for modeling feature-rich student behavior data.

According to the characteristics of IRT and BKT, the combinations of the two models are derived, Feature Aware Student Knowledge Tracing (FAST) [7] and Latent Factor and Knowledge Tracing (LFKT) [8]. The improvement is that the students' ability and difficulty of the problems are introduced into the BKT. FAST and LFKT make up for the lack of IRT and BKT models in tracking students' knowledges. However, FAST and LFKT assume that knowledge components are independent of each other, ignoring the impact of the relationship between knowledge components and student behaviors on the students' performance. And as the feature behavior increases, it is extremely difficult to reconstruct the multi-feature logic function.

Chris Piech et al. have proposed a deep knowledge tracing model - DKT model [9]. DKT doesn't require the explicit encoding of human domain knowledge and can capture more complex representations of students' knowledge components. However, the current deep knowledge tracing with Simple RNN applies Back Propagation Through Time (BPTT) algorithm and Batch Gradient Descent (BGD) to adjust parameters, which is easy to over-fitting and is prone to gradient disappearance and gradient explosion for long-dependent data. And Long Short-Term Memory(LSTM) is better than RNN, but it adjusts the parameters according to the training loss and a small amount of noise data seriously affects parameter learning.

3 BDKT

The Bayesian neural network subtly applies the Bayesian method in probabilistic programming to deep learning. It realizes the combination of probabilistic programming and deep learning, which brings a huge innovation to deep learning [10]. In this paper, a Bayesian neural network is applied to students' behavior analysis and knowledge tracing. The uncertainty of predicted students' behavior and model representation are given well, and better prediction results are obtained.

3.1 BDKT Input Data

Processing the BDKT model input data is to digitize students' behavior records. The commonly used method is one-hot coding. Students' behavior record data contains many attributes, the two main characteristics of the students' answer results and the questions corresponding to the knowledge components are selected to code students' behavior data. Before the coding, the students' knowledge components are sorted first, the serial numbers represent knowledge components, the students' behavior.

r coding is calculated according to (1). As is shown in (1), i, knowledge component number; N, the total number of knowledge components; u_i, the answer result ($u_i = 0$, wrong; $u_i = 1$, right).

$$r = i + N * u_i \tag{1}$$

3.2 BDKT Model Structure

The model structure is shown in Fig. 1. The BDKT model has made the following two improvements in the structure of the DKT model: (1) An embedded layer is introduced. In order to get a better students' behavior vectors, an embedded layer is added between the model input and the hidden layer, and a more vivid students' behavior representation vectors are obtained by training the embedded layer. (2) Apply Bayesian LSTM in the DKT. Simple RNN encounters great difficulties in dealing with long-term dependencies. The calculation of the connection between distant nodes involves multiple multiplications of the Jacobian matrix, which is prone to gradient disappearance and gradient explosion. The input threshold, forget the threshold and output threshold are added to the LSTM to solve the gradient disappearance and gradient explosion problems. At present, LSTM has been proven to be an effective technique for dealing with long dependency problems. The students' behavior data sequences are long and there are many long-term dependencies. LSTM is more suitable than Simple RNN. But LSTM is a certain model, it can't express the uncertainty relationship between the inputs and the outputs.

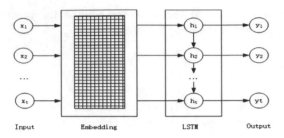

Fig. 1. Structure of BDKT model

3.3   BDKT Parameter Learning

In the BDKT model, the BDKT model's parameters are set to the form of distribution. According to the Bayesian formula, the parameter learning of the model can be described as (2):

$$p(W|X,Y) = \frac{p(W)p(Y|X,W)}{p(Y|X)} \quad (2)$$

X is the input data, Y is the output data, $p(W)$ is the priors of the parameters W, $p(Y|X,W)$ is the probability of the output Y for a given weight parameter W and input X. As the training set (X, Y) is known, $p(Y|X)$ is a constant. The goal $p(W|X,Y) \propto p(W)p(W|X,Y)$, Since the probability distribution of p(W|X,Y) on the training data set (X, Y) is complicated, it is difficult to process with the normalized constant, and the posterior distribution of the model parameters cannot be directly calculated. To solve this problem, Variational Inference is used. the variational distribution $q(W) \sim N(W|\mu,\sigma^2)$ is used to approximate the true posterior distribution p(W|X,Y). The distance between p and q is Calculated as follows:

$$KL(q\|p) = \sum_{i=1}^{n} q(W) \log \frac{q(W)}{p(W|X,Y)}$$
$$= E_q[\log q(W)] - E_q[\log p(W)] - E_q[\log p(Y|X,W)] + E_q[\log p(Y|X)] \quad (3)$$

In (3), $E_q[\log p(Y|X)]$ is a constant, $E_q[\log p(Y|X,W)]$ can get from the train set. In BDKT, The loss of the model is determined by the training loss and the KL term.: $loss = \hat{y} - y + E_q[\log q(W)] - E_q[\log p(W)]$. The Bayes by Backprop(BBB) algorithm gives the parameter learning process of general neural networks. LSTM uses BPTT algorithm for parameter learning. IN BDKT, Applying the BBB algorithm to the RNN requires consideration of the timing of parameter sampling and the contribution of the KL term. KL penalty in LSTM is only once. BDKT backpropagation process as follows (Let B be the number of mini-batches and C the number of truncated sequences):

1. sample $\xi \sim N(0, 1)$, $\xi \in R^d$;
2. Set network parameters to $W = \mu + \sigma\xi$, so $N(W|u, \sigma)$;
3. Sample a minibatch of truncated sequences;
4. Do forward propagation and backpropagation as normal;
5. Let g be the gradient with respect to W from backpropagation;
6. Let g_W^{KL}, g_μ^{KL}, g_σ^{KL} be the gradients of $\log N(W|\mu, \sigma) - \log p(W)$ w.r.t. W, μ and σ respectively;
7. Update μ according to the gradient $\frac{1}{B}(g + \frac{1}{C}g_W^{KL}) + \frac{1}{BC}g_\mu^{KL}$;
8. Update σ according to the gradient $\frac{1}{B}(g + \frac{1}{C}g_W^{KL})\xi + \frac{1}{BC}g_\sigma^{KL}$.

4   Results and Discussion

In order to fully prove that BDKT is better at knowledge tracing. This paper uses two public datasets KDD Cup (Bridge_to_Algebra_2006_2007_train.txt) [11] and Assistments (skill_builder_data.csv) [12] to train and test the performance of the model. The data sets are widely used in knowledge tracing. In this paper, two data sets KDD Cup and Assistments are used to train and test the BDKT model. Two different metrics are applied for comparing the predicted correctness probabilities with the observed correctness values. Accuracy (Acc) is computed as the percent of responses in which the correctness coincides with the probability being greater than 50%. AUC is the Area Under the ROC Curve of the probability of correctness for each response.

4.1   BDKT Vs DKT

In order to compare the performance of BDKT model and DKT model, and reduce the influence of variable factors on experimental results, this paper uses the control variable method to conduct comparative experiments. (1) Model structure, BDKT model and DKT model both adopt the model structure as shown in Fig. 1. (2) Hyperparameters, the parameter dimension in the model structure and the learning rate and the learning rate attenuation coefficient are same. The experiments compare the model ACC and AUC of the BDKT and DKT models on different size datasets. The data set size is controlled by the length of the intercepted student interaction records. The length of the students' interaction records is from 50 to 500 to represent different size datasets. The longer the step size is, the larger the data set is. The students' interaction records are shorter in the Assistments dataset. Intercept the first 50 entries of the students' interaction records' length from 50 to 500 data to represent different size datasets. The shorter the step size is, the larger the data set is. Figures 2 and 3 show ACC and AUC histograms of the BDKT and DKT models in the different size datasets of KDD Cup and Assistments.

It can be known in Figs. 2 and 3 that the performance of the BDKT model is better than that of the DKT. As the data set increases, the BDKT model performs more stable, and the DKT model is getting better. It can be known from the experimental results in Fig. 2 that, in the BDKT model, the generalization ability is stronger on the small data set. As the model data increases, the influence of prior knowledge in the model decreases, and the performance of BDKT and DKT models show a close trend.

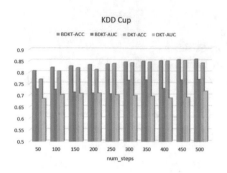

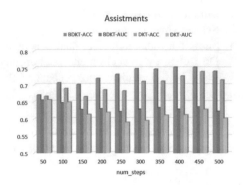

Fig. 2. BDKT VS DKT in KDD Cup   **Fig. 3.** BDKT VS DKT in Assistments

The ACC value of the model on the KDD Cup dataset has been similar, which is due to the uneven distribution of data. In the Assistments dataset, the difference between the number of correct answers and the number of correct answers is small, and the data set is small. The difference between BDKT and DKT performance is more obvious. Comparing BDKT and DKT, it can be found that the performance of BDKT model is obviously better than DKT, and the smaller the data set is, the more obvious it is.

4.2   BDKT Parameter Randomness and Model Uncertainty

In the BDKT model, the probability distribution of the parameters makes the model have better anti-noise ability. The experiment takes the first 500 interaction records of 996 students on the KDD Cup dataset as a sample. The sample contained 289 knowledge components and 578 student behaviors. Figure 4(a) shows the accuracy of the data training during the 1st and 2nd iterations and Fig. 4(b) shows the accuracy of the 69th and 70th iterations.

It can be known in Fig. 4(a) that the accuracy rate of the model training set is large, and the accuracy of the same data in the adjacent iterations of the same data is also large. The model parameters are sharply adjusted during initial training. In addition to the random selection of the model parameters on the training set, the corresponding loss of the model in the training changes greatly, and the parameters are randomly optimized and adjusted, which causes the accuracy of the model to fluctuate greatly in the two rounds of iteration. Stochastic optimization in model parameter training can avoid model training falling into a local optimum. It can be known in the comparison of Fig. 4(a) and (b) that as the number of iterations increases, the accuracy of the model on the training set gradually becomes stable. After multiple rounds of training, the mean and variance of the weight parameters tend to be stable, and the deviation of the weight parameters is small. The accuracy of the same batch of data in the two iterations is basically the same, but the sampling parameters will have large deviations with small probability, and the results of the two iterations will be biased, which reflects the uncertainty of the output of the model.

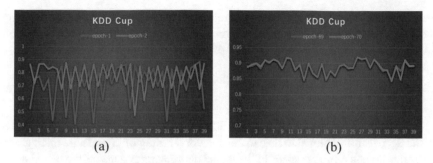

Fig. 4. ACC in KDD Cup's training set

4.3 Relationship Between the Students' Behaviors

The student behavior vector is trained while training the model. The nonlinear dimension reduction visualization of the trained students' behavior vectors is carried out by t-SNE. Figure 5 shows the students' behavior vector visualization graph is obtained by training the first 50 records of the students' interactions in the KDD Cup dataset. The dataset contains 183 items, and 366 different behavioral records. According to the initial students' behavior coding (1), each number of the knowledge components and the answer results of the knowledge components in the figure can be known, the potential relationship of the student's answer behaviors can be reflected. Table 1 gives a description of the student behavior corresponding to the circled points. According to the distance between the scatter points in Fig. 5 and the students' behaviors description in Table 1, it can be known that the points of the knowledge components are related or similar, and the closer the distance on the scatter plot is, the more similar the corresponding students' behavior vectors are. The trained student

Table 1. Students' behavior description table

Behavior number	Skill number	Right or wrong	Description
133	133	N	Identifying units
189	6	Y	Calculate difference – non contextual
250	67	Y	Enter thousandths multiplier as fraction
296	113	Y	Identify number of equal divisions (vertical bar)

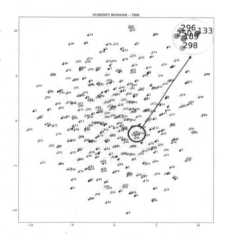

Fig. 5. Students' behavior vector in KDD Cup

behavior vector is the mining of potential relationships in the data, the relationships that are difficult to find in the students' behaviors can be dug out by training the students' behavior vectors through a large amount of data.

5 Conclusion

In this paper, we propose a BDKT model. The Bayesian prior knowledge is introduced into the deep learning model to realize the transfer learning of knowledge tracing, which accelerates the convergence speed of the model parameters and makes the model perform better on small datasets. The combination of deep learning and probabilistic programming solves the over-fitting and local-optimal problems that often occur in the traditional DKT. Besides, we propose the students' behavior vector learning, which better encodes the students' behaviors and improves the performance of the model. The student's answer records have rich features, and the paper only considers some basic features' information. Taking the performance of the model and the information contained in the students' behavior vectors into account, further work can be done on the improvement of the model's performance and the analysis of the students' behavior vectors.

Acknowledgment. Supported by the Foundation of Zhongshan Institute of Advanced Engineering Technology of WUT (Grant No. WUT-2017-01).

References

1. Wilson, K.H., Karklin, Y., Han, B., et al.: Back to the basics: Bayesian extensions of IRT outperform neural networks for proficiency estimation. arXiv preprint arXiv:1604.02336 (2016)
2. Fortunato, M., Blundell, C., Vinyals, O.: Bayesian recurrent neural networks. arXiv preprint arXiv:1704.02798 (2017)
3. McDermott, P.L., Wikle, C.K.: Bayesian recurrent neural network models for forecasting and quantifying uncertainty in spatial-temporal data. Entropy **21**(2), 184 (2019)
4. Maaten, L., Hinton, G.: Visualizing data using t-SNE. J. Mach. Learn. Res. **9**(Nov), 2579–2605 (2008)
5. Lord, F.: A theory of test scores. Psychom. Monogr. (1952)
6. Corbett, A.T., Anderson, J.R.: Knowledge tracing: modeling the acquisition of procedural knowledge. User Model. User-Adapt. Interact. **4**(4), 253–278 (1994)
7. González-Brenes, J., Huang, Y., Brusilovsky, P.: General features in knowledge tracing to model multiple subskills, temporal item response theory, and expert knowledge. In: The 7th International Conference on Educational Data Mining, pp. 84–91. University of Pittsburgh (2014)
8. Khajah, M., Wing, R., Lindsey, R., et al.: Integrating latent-factor and knowledge-tracing models to predict individual differences in learning. In: Educational Data Mining 2014 (2014)
9. Piech, C., Bassen, J., Huang, J., et al.: Deep knowledge tracing. In: Advances in Neural Information Processing Systems, pp. 505–513 (2015)

10. Hernández-Lobato, J.M., Adams, R.: Probabilistic backpropagation for scalable learning of bayesian neural networks. In: International Conference on Machine Learning, pp. 1861–1869 (2015)
11. Stamper, J., Niculescu-Mizil, A., Ritter, S., Gordon, G.J., Koedinger, K.R., [Data set name].: [Challenge/Development] data set from KDD Cup 2010 Educational Data Mining Challenge (2009). http://pslcdatashop.web.cmu.edu/KDDCup/downloads.jsp
12. Feng, M., Heffernan, N.T., Koedinger, K.R.: Addressing the assessment challenge in an intelligent tutoring system that tutors as it assesses. J. User Model. User-Adapt. Interact. **19**, 243–266 (2009)

Co-simulation of Sub-frame of the Intelligent Fire Truck Based on SolidWorks and ANSYS Workbench

Bailin Fan[✉], Wei Zhang, and Fan Chen

School of Mechanical Engineering, University of Science and Technology
Beijing, 30 Xueyuan Road, Haidian District, Beijing, China
Fanbailin868@sina.cn

Abstract. The structure of this simulation is the assembly structure established by SolidWorks, so it is more convenient to use ANSYS workbench for simulation. It mainly involves the combination of ANSYS workbench and Solid-Works, the processing and mesh division of the assembly contact part, and the division of the entire structure mesh. Mainly performed static simulation and modal analysis. Provide a theoretical basis for the subsequent manufacture of the sub-frame.

Keywords: Assembly · ANSYS · SolidWorks · Meshing · Simulation

1 Introduction

The development of society brings about the special development of social activities. The leakage and burning of some chemicals is extremely harmful. However, if it is treated manually, it will cause injury or even danger to firefighters. Therefore, fire-fighting robots are urgently required to enter dangerous fires to extinguish fires [1, 2]. The successful development of intelligent fire trucks can reduce the burden on fire-fighters to some extent and reduce their safety hazards. Some fire scenes containing toxic gases do not have to be firefighters to enter the scene to extinguish the fire. They can manually control the smart fire trucks to enter the scene. Fire fighting has greatly improved the safety and security of firefighters, which has great practical significance [3]. Co-simulation based on SolidWorks and ANSYS workbench assemblies is actually an extension of the product design process. The basic structure of the device needs to be designed in SolidWorks. After the components are designed, assemble the assembly in SolidWorks. After the co-simulation environment is installed, it can be directly analyzed in ANSYS workbench. Co-simulation avoids format conversion and a large number of contact definitions [4]. This design uses SolidWorks 14 and ANSYS 16.0 for co-simulation. Specific joint operations can refer to ANSYS 16.0 Help.

© Springer Nature Switzerland AG 2020
F. Xhafa et al. (Eds.): IISA 2019, AISC 1084, pp. 38–45, 2020.
https://doi.org/10.1007/978-3-030-34387-3_5

2 Main Technical Difficulties

Complex assemblies have the following problems when performing ANSYS simulation: connection problems of various components, mesh refinement problems, and the degree to which the simulation results fit the actual situation.

2.1 The Connection Problem

The connection between the various parts of the assembly is a very complicated problem. Generally, it depends on specific working conditions. For the general connection ANSYS workbench provides two options which are connect and joint. Among them, connect is a relatively common processing method. ANSYS workbench provides the function of automatic contact analysis. Automatically perform contact analysis on imported models to establish contact pairs. The designer only needs to modify the contact parameters according to the actual situation. This joint simulation mainly uses the contact pair to analyze. Joint is used to deal with the joints of pins, shafts, universal joints and other structures commonly used in engineering. It is generally suitable for dealing with multi-flexible dynamics [5].

2.2 Contact Settings

In fact, when parts are assembled into parts or parts assembled into machines or equipment, the deformation of the contact body will change the contact condition of the contact surface under the load, and the force between the contact surfaces is closely related to the contact state, if conventional The finite element analysis method will inevitably bring obvious errors to the mechanical calculation, and even affect the judgment of the strength safety of the part. This is especially obvious in the stress analysis near the contact surface [6, 7]. When ANSYS workbench and SolidWorks are combined, the default contact of the open geometry model is bonded, and the contact settings can be changed in the option named connections in the workbench after importing the model. Firstly, the tolerance needs to be modified. The Tolerance Type generally uses the slider and the vale. If the value is used, the tolerance value can be directly input. If the value is smaller than the value, the contact is determined as contact. With the Tolerance slider, determine the contact tolerance by adjusting the coefficient. The coefficient range is between −100 and 100. The larger the value, the larger the tolerance and the more the contact will be. How to set it up according to the specific needs of own modifications. Another important option is Group By. The workbench default is bodies option. If you are an assembly, you can use parts by referring to Help. This will reduce some unnecessary contacts. For the next mesh, you will have more calculations. After the contacts are generated, in order to determine the quality of contacts, the Contact Tool can be used to calculate the various indicators of each contact, and the calculation results will be displayed in the worksheet. Once the calculation results in orange and red, special attention must be paid to the local contact verification process, which can be modified by changing the contact type and specific contact algorithm. In the static analysis, the individual contact type can be changed to the friction contact to get close to the actual situation, but for modal analysis, the bonded contact is generally used directly to reduce the calculation time and memory.

2.3   Meshing

Complex assembly topologies are complex and have large size spans. When meshing, you should minimize the number of meshes while ensuring accuracy. Two solutions can be used for complex cavities and thin shells: The extracted midplane is simulated using 2D shell elements and directly simulated using 3D solid elements. It is a better solution to adopt hybrid meshing as a whole. Hybrid meshing means that on the geometric model, according to the characteristics of each part, various meshing methods such as free, mapping and sweeping are adopted respectively to form a finite model with the best comprehensive effect [8–10]. For parts that can be swept, the sweeping method is preferred, so that the mesh is better. Different grid sizes are generally used for different structures. In the part where the data gradient is large (such as stress concentration), in order to better reflect the data variation law, a dense mesh is needed. In some areas where the data gradient is small, a larger grid can be used, which can speed up the calculation and also has little effect on the accuracy [11]. Adjust the relevance under the default option when meshing, and adjust the specific options under the sizing option to get a better quality and a smaller number of meshes [12]. The specific adjustment is shown in Fig. 2 below (Fig. 1).

Defaults	
Physics Preference	Mechanical
□ Relevance	79
Sizing	
Use Advanced Size Function	Off
Relevance Center	Coarse
□ Element Size	Default
Initial Size Seed	Active Assembly
Smoothing	Medium
Transition	Fast
Span Angle Center	Fine
Minimum Edge Length	5.5956e-004 m

Fig. 1. Grid option adjustment

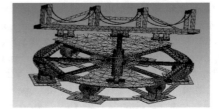

Fig. 2. Meshing results

The quality of the grid thus drawn is about 0.7, barely reaching expectations, the number of nodes is about 700,000, and the calculation process is also faster. The results of the method of hybrid meshing are shown in Fig. 2. Part of the shape rule uses the sweep method to divide the mesh, and the other parts are mesh refined at the relative concentration of stress.

3   Load Loading

For the frame structure, it is equipped with three fire-fighting tanks on the upper side. Since the mounted part is designed to conform to the arc shape of the fire-fighting tank, the force acting on the upper side can be simplified to a uniform load. After calculation, the pressure is 22000 pa or 8000 pa; Note that the connecting part of the pin generally needs to add a load. After the reference data is combined with the actual situation, cylindrical support is added, It should be noted that the cylindrical support constraint can be further set in the definition option to determine whether the three directions are constrained; after the fixed support constraint and gravity are added, the first case can be simulated: the impact force at the time of launch is not added. Considering the transient dynamics simulation when simulating the structural performance with impact force, the calculation process and complexity will be simplified, first simplified into a static structure, and the static simulation results will be used as a reference to further provide transient dynamics simulation. The modal analysis only needs to load the constraints, and the external load does not need to be added.

4 Simulation Results

4.1 Static Simulation Without Impact Force

The overall calculation results of the static simulation without impact force are shown in Figs. 3 and 4.

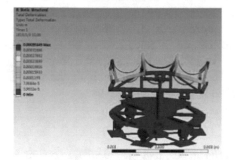

Fig. 3. Total deformation   **Fig. 4.** Equivalent stress

For partial stress analysis of part of the stressed component, you can further check the simulation results, you can select the components you want to view. Right-click on the option you want to view, generally check the stress and strain. This simulation shows further results for the pin and polyurethane wheel,As shown in Figs. 5, 6, 7 and 8.

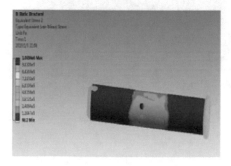
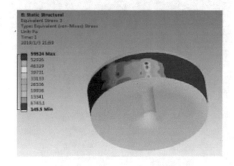

Fig. 5. Pin equivalent stress   **Fig. 6.** The polyurethane wheel

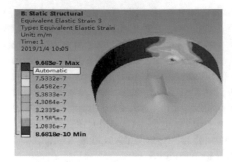   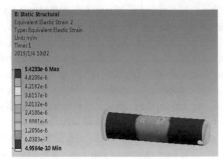

Fig. 7. The polyurethane wheel equivalent elastic strain

Fig. 8. The Pin equivalent elastic strain

It can be seen from the calculation results that the maximum deformation amount appears at the corner of the frame on the upper side of the structure, and the maximum deformation amount is 0.00035849 m. The deformation amount is small and the overall result stiffness is sufficient; The maximum stress is the part of the hydraulic cylinder support frame. There is generally welding treatment here. Therefore, the safety factor should be large enough. The maximum stress is about 24 MPa. The ordinary carbon steel is used as the material and the safety factor is close to 10. It can be preliminarily determined. The strength of the structure is also sufficient. For pin and polyurethane wheels: the maximum force at the pin is calculated to be only about 1 MPa and the strain is about 5.4233e-6; The calculation results for the polyurethane wheel are similar to those of the pin. It can be concluded that the strength and stiffness of the overall structure and components are sufficient for practical use.

4.2   Static Simulation with Impact Force

The calculation process is similar to the previous statics, and the calculation results are shown in Figs. 9 and 10.

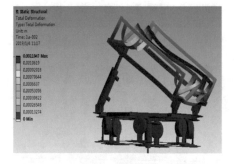   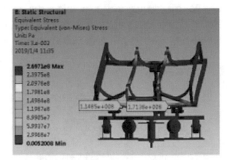

Fig. 9. Total deformation

Fig. 10. Equivalent stress

After adding the impact force, the overall deformation amount and stress will have a large increment, wherein the maximum deformation is 0.0011947 m, the strain is about 0.0015, and the overall stiffness meets the actual use requirements; the maximum stress is 260 MPa, but the maximum stress appears in a very small position, the maximum stress in a large area is about 170 MPa, and the safety factor is about 2 when using ordinary carbon steel. Combined with the previous static simulation, the solution can be found: replace the part of the frame with high impact strength with high-strength steel to improve the overall safety factor and increase the strength of the overall structure. Insufficient situation. The static simulation results are shown in the Table 1.

Table 1. The simulation data summary

Working condition	Maximum stress position	Maximum stress	Maximum strain position	Maximum strain
No impact	Hydraulic cylinder connection	23.75 MPa	Frame beam	0.0012
Impact	Corner of the frame	260 MPa	Frame vertical beam	0.0015

4.3   Modal Analysis of the Overall Structure

The modal analysis of an assembly gives the dynamic behavior of multiple components as a whole. When the assembly is subjected to a dynamic load, the combined portion of the assembly exhibits both elastic and damped dynamic characteristics, and the dynamic characteristics exhibited by the joint are related to the form, function, and medium condition of the joint portion. The elasticity in the dynamic characteristics of the joint can be replaced by an equivalent spring. The damping can be replaced by equivalent damping. Any joint can be simplified into a dynamic model composed of a series of equivalent springs and equivalent dampers [13, 14]. For the determination of the dynamic parameters, the experimental results of Japanese scholar Yoshimura Yunomi can be used to calculate:

$$K_n = \iint k_n(P)dS, C_n = \iint c_n(P)dS \tag{1}$$

$$K_t = \iint k_t(P)dS, C_t = \iint c_t(P)dS \tag{2}$$

Where $k_n(P)$ and $c_n(P)$ are the normal stiffness and damping coefficient of the unit area value of the joint; $k_t(P)$ and $c_t(P)$ are the tangential stiffness and damping coefficient of the unit area value of the joint; K_n and C_n are the normal equivalent stiffness and damping coefficient of the joint; K_t and C_t are the tangential equivalent stiffness and damping coefficient of the joint [15].

In the calculation of modal analysis, the basic settings such as meshing and static mechanics are basically the same, except that the equivalent spring and damping need to be added according to the actual situation and calculation results at some joints. The modal analysis results after setting the conditions are shown in Fig. 11 and Table 2 (Calculated 1 to 6 modes):

Fig. 11. Modal analysis histogram

Table 2. Modal analysis specific numerical table

Mode	Frequency [Hz]
1.	69.94
2.	95.803
3.	109.11
4.	109.23
5.	109.45
6.	163.76

The main excitation force of the car can be considered as two source: 1. Vibration caused by ground irregularities. 2. Vibration caused by engine imbalance. Since the sub-frame is used on an electric vehicle, the second type of vibration can generally be ignored. The frequency of the first excitation is generally below 6–8 Hz. Comparing the calculated frequency of the first 6-stage sub-frame, it can be seen that during normal driving, the natural frequency is much larger than the external excitation frequency, and there is no resonance, and the structure is reasonable.

5 Conclusion

1. The maximum stress of the sub-frame without impact is 23.76 MPa to meet the strength requirements. The maximum 0.00012 meets the stiffness requirements. The maximum impact stress is 260 MPa, and the higher stress part uses high-strength steel to improve the strength of the sub-frame. The maximum strain is 0.0015, which meets the stiffness requirements.
2. The frequency of the 6th-order modal analysis of the modal analysis avoids the resonance frequency of the fire truck and meets the design requirements, and the structure is reasonable.
3. Through the design of the sub-frame by SolidWorks, combined with ANSYS workbench simulation, the stress cloud and strain cloud diagram of the whole structure under two working conditions are obtained. According to the simulation results, the structural design is evaluated and the suitable materials are selected to provide a powerful material for the physical production. reference.

Acknowledgments. Cooperation Project between Beijing University of Science and Technology and Technology Industry Co., Ltd. USTB, Contract No. 2018-345.

References

1. Jia, Y.-Z.: Design and research of small crawler fire. In: IEEE Chinese Automation Congress, pp. 4120–4122 (2018)
2. Almonacid, M.: Motion planning of a climbing parallel robot. IEEE Trans. Robot. Autom. **19**(3), 485–489 (2003)
3. Yintang, N.: Summary and prospect of research status of fire fighting robots. Autom. Appl. **2**, 28–29 (2017)
4. Weiping, Z.: Bond-slip numerical simulation based on ANSYS contact analysis. J. Architect. Civil Eng. **128**(2), 45–51 (2011)
5. Han, J.: Co-simulation and optimization of complicated assembled part based on Pro/E and ANSYS WORKBENCH. Mach. Des. Manuf. **1**, 190–192 (2010)
6. Vale, T.D.O.: Methodology for structural integrity analysis of gas turbine blades. J. Aerosp. Technol. Manag. **4**(1), 51–59 (2012)
7. Hwang, S.-C., Lee, J.-H., Lee, D.-H., Han, S.-H., Lee, K.-H.: Contact stress analysis for a pair of mating gears. Math. Comput. Model. **57**(1), 40–49 (2013)
8. Waseem, M.G., Elahi, N.: Structural integrity assessment and stress measurement of CHASNUPP-1 fuel assembly. Nuclear Engineering and Design **280**, 130–136 (2014)
9. Belhocine, A., Abu Bakar, A.R., Abdullah, O.I.: Structural and contact analysis of disc brake assembly during single stop braking event. Trans. Indian Inst. Met. **68**(3), 403–410 (2014)
10. Shankar, M., Kumar, K., Ajit Prasad, S.L.: T-root blades in a steam turbine rotor: a case study. Eng. Fail. Anal. **17**(5), 1205–1212 (2010)
11. Duan, W., Joshi, S.: Structural behavior of large-scale triangular and trapezoidal threaded steel tie rods in assembly using finite element analysis. Eng. Fail. Anal. **34**, 150–165 (2013)
12. Liu, Y., Saputra, A.A., Wang, J., Tin-Loi, F., Song, C.: Automatic polyhedral mesh generation and scaled boundary finite element analysis of STL models. Comput. Methods Appl. Mech. Eng. **313**, 106–132 (2017)
13. Ooi, J., Wang, X., Tan, C., Ho, J.-H., Lim, Y.P.: Modal and stress analysis of gear train design in portal axle using finite element modeling and simulation. J. Mech. Sci. Technol **26**(2), 575–589 (2012)
14. Armentani, E., Citarella, R.: DBEM and FEM analysis on non-linear multiple crack propagation in an aeronautic doubler-skin assembly. Int. J. Fatigue **28**(5), 598–608 (2006)
15. Han, J.: ANSYS modal analysis and optimization of working table components of FM6324 vertical machining center. J. Hefei Univ. Technol. **35**(8), 1039–1042 (2012)

Design of Framework on Intelligent Simulation and Test Platform for High Power Microwave Sources

Shuangshi Zhang[1,2(✉)] and Chaojun Lei[2]

[1] University of Electronics Science and Technology of China, Chengdu, China
shuangshizhang@qq.com
[2] China People's Police University, Beijing, China

Abstract. In order to attract more scientific and technological talents to participate in the research of high power microwave sources and accelerate the research iteration process in this field, the idea of establishing an open intelligent platform for high power microwave source simulation and testing is put forward, and the framework design is given. It is hoped that a sustainable, evolving and iterating technological ecology of study, simulation, design, and fault diagnosis and test operation of high power microwave sources will be established on the basis of this platform.

Keywords: High power microwave source · Numerical simulation · Artificial intelligence · Big data · Cloud computing

1 Introduction

HPM (High power microwave) has been applied to many fields, such as nuclear fusion plasma heating, current drive, particle acceleration, material structure exploration, medical imaging, security detection, radar, communication and so on. It has very good prospects for development. However, the development of HPMS (high power microwave source), especially gyro oscillator, is difficult in theory, involving many fields, and is expensive, which has become the bottleneck of technology development. Among them, numerical simulation plays an increasingly important role. It not only promotes and enriches the development of theory, but also greatly shortens the development cycle and saves the development cost. At present, there are still many problems in numerical simulation, and there is still much room for development. Many physical phenomena such as the interaction mechanism between particles and electromagnetic waves, the nonlinear relativistic effect of high-speed particles and the starting process of working mode are still unclear. There are many software and programming languages used to simulate, design and monitor HPMS. But no software or language can solve a series of theoretical and engineering problems of HPMS perfectly. The famous general software includes Magic, CST, HFSS, Egun, etc. The common programming languages include FORTRAN, C, C++, Java and matlab. But they are not satisfactory [1–4].

© Springer Nature Switzerland AG 2020
F. Xhafa et al. (Eds.): IISA 2019, AISC 1084, pp. 46–54, 2020.
https://doi.org/10.1007/978-3-030-34387-3_6

For the HPMS numerical simulation requires a large number of efficient numerical calculations, the support of algorithm library and the requirement of computational power are very high. It is necessary to establish an open, efficient, iterative and evolving platform for learning, design simulation and test operation for HPMS. The framework design of this intelligent platform is proposed, in line with the following principles, efficient scientific computing and numerical computing algorithm library support; distributed parallel and cloud computing support; artificial intelligence and large data algorithm library support; maintainability, scalability; strong three-dimensional visualization support. It is different from the traditional electromagnetic field simulation software, which only uses numerical analysis method to establish simulation process, and is different from other professional HPMS simulation systems, which can only do analysis and calculation for part of the system, but regards HPMS as a whole system from particle emission to microwave power output. It can not only analyze and calculate part of the system, but also improve the performance of the whole tube. It will be a learning research assistant a simulation design assistant and a simulation control expert. It is discussed from three aspects: (1) requirement analysis; (2) framework design; (3) technical support.

2 Requirement Analyses

Powerful intelligent platform which provides researchers, designers and engineers with an open virtual simulation computing environment for theoretical study, engineering simulation, and design and test operation of HPMS. ISTP-HPMS should fully learn from the excellent characteristics of the existing HPMS numerical simulation software, and fully adopt the core ideas, concepts and technical methods of the current development of computer software and hardware. Here is a relatively rough overview of requirements analysis from the perspective of framework design. In actual development, it is necessary to write software specifications, and give clear definitions and standards for all parts of software.

2.1 User Requirements

ISTP-HPMS users are divided into three categories according to the design, production and use process of HPMS, namely researchers, designers and users, as is shown in Fig. 1.

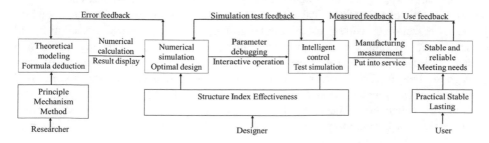

Fig. 1. Cycle of HPMS design, manufacture and use

Beam-wave interaction model, self-consistent transient beam-wave interaction model, depressurized collector model and particle simulation PIC model, etc. By adjusting various parameters, HPMS simulation with high frequency, high power and high efficiency is realized. It can provide reference for the production and use of HPMS.

Engineering practitioners are more concerned about the practicability, stability, persistence and portability of HPMS, which is also the goal HPMS has been pursuing. Figure 1 shows the iterative evolution of HPMS. This process has been going on, never ending. If HPMS can be equipped with an intelligent "heart", it can always sense the working state of the tube and make corresponding adjustments and controls. This is the ideal state and the core idea and goal of intelligent HPMS. This is not only the need of actual testing, but also the need of simulation testing. Therefore, with the continuous development of intelligent HPMS, three types of users of ISTP-HPMS may need to be integrated, that is, researchers also need to design, simulate and test to verify the correctness and practicability of their theory. The designer is not only a skilled worker in software use, but also knows the parameter configuration and performance debugging method of the specific tube design. He also needs to be familiar with the theory, understand the principle, mechanism and calculation method.

If the users of HPMS can understand the relevant principles, mechanisms and simulation design methods, it will be helpful to better maintain and use the normal operation of HPMS, to better discover the problems, defects and deficiencies in the use of tubes, to better feedback and communicate with designers and theoretical researchers. As shown in Fig. 1, every participant, especially the core participant, should experience the whole life cycle of HPMS on a platform.

2.2 Functional Requirements

Figure 2 shows the flow chart of HPWS simulation and test. It is necessary for every user to have a clear understanding of the whole system including electronic optics system, beam-wave interaction resonator, microwave transmission system, step-down collection of waste electrons, output window. Such skills as measurement of microwave output field strength, working parameters, etc. require mastery by all.

ISTP-HPMS should provide the function of theoretical deduction and algorithm analysis, such as matrix operation, calculus operation, symbolic operation, special function operation, solution of linear equations, interpolation and fitting and other numerical calculation; also need to provide the drawing of two-dimensional function curves, isoclines, vector lines and the description function of three-dimensional curves, surfaces and volumes of physical quantities; provide symbolic representation, operation and visualization of mathematical formulas and equations. Representation, solution and visualization of boundary-like problems and initial problems; providing the motion state, trajectory and control of electrons in electric field, magnetic field and electromagnetic field in HPMS research. Some analytical formulas and equations written by some theories cannot be solved analytically, but can only be carried out by means of numerical calculation. Therefore, ISTP-HPMS system needs to provide some basic formulas, error evaluation process and methods of numerical solutions of equations.

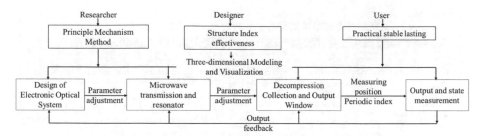

Fig. 2. Flow chart of HPW tube simulation & test

2.3   Interface Requirements

There are many contents of interface. For software, there are data interface, man machine interface, and software interface and so on. It is related to data input efficiency, algorithm efficiency, operation speed and calculation speed of the whole system, stability, usability and robustness of the system, operability, expansibility and maintainability of the system.

3   Frame Design

Framework design is very important in the development of ISTP-HPMS, which is to establish some rules and put forward some concepts and ideas pertinently, such as programming language selection, data representation, data interface, man-machine interface, modular and component design mode, combination mode and communication mode, so as to make the system compatible, stable, usable, scalable and maintainable. For this reason, the framework design of ISTP-HPMS is established, as is shown in Fig. 3.

Firstly, we need to design a distributed cloud computing hardware platform, on which we can deploy Hadoop-based open source distributed storage and computing environment for large data, which can aggregate and interoperate various data, including relational databases, text files, non-relational databases, graph databases and other non-relational unstructured databases. Secondly, all kinds of resources, documents and expert experience related to HPMS are sorted out and built to form a knowledge base for search engine, intelligent design and fault diagnosis, and an algorithm library for various interactive formula deduction, algorithm testing, numerical analysis, microwave numerical simulation template library, tube library, design case library, software module component library and log data are formed. At the same time, a controller providing intelligent computing service is designed, which can flexibly connect various databases and users' various business needs, including data warehouse and data mart for large data online analysis and intelligent decision-making, various autonomous programming interfaces, and processing various computing requests, etc. Thirdly, a one-stop graphical user interface based on Web is designed to provide four main functional service groups: HPMS related resource management, learning research, design simulation and test monitoring.

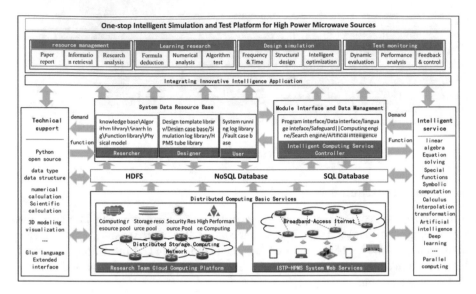

Fig. 3. ISTP-HPMS design framework

Based on the design principles of loose coupling, high cohesion, easy expansion, easy migration, easy integration, cross-platform and high adaptability, Python as the main development language makes full use of open source community and open source software resources to provide technical support for hardware and software design and form a benign interaction with user services. Based on distributed storage and cloud computing architecture, library and library operations are relatively separated, parallel development and dynamic operation, providing powerful and rich intelligent services for the design and manufacture of intelligent HPMS, and constituting the evolution iteration of requirements and services.

3.1 Development Language Selection

Python programming languages have rich object libraries for process-oriented, object-oriented and data-oriented programming supported by a large number of open source third-party libraries. When it comes to fast numerical computation, existing C language libraries and Fortran libraries are available. There are ready-made open source machine learning libraries and artificial intelligence frameworks available, as well as special open source VTK three-dimensional description libraries. In scientific computing, compared with matlab, except for a few professional toolboxes, most of the commonly used functions have corresponding extended library support.

The whole framework of Python is elegant in design, strong in algorithmic expression and open source, so it is quickly accepted by researchers from all walks of life and developed interfaces and class libraries for different applications. Therefore, the core programming technology of Python language is the flexible application of data types and the use of various class library interfaces, as is shown in Fig. 4 for Python programming technology system and ecology. In a word, programming with Python has strong customization and autonomy. It requires a deep understanding of its data structure and flexible use of its libraries and modules. According to the requirement analysis of ISTP-HPMS, Python is chosen as the main language of system development [5–7].

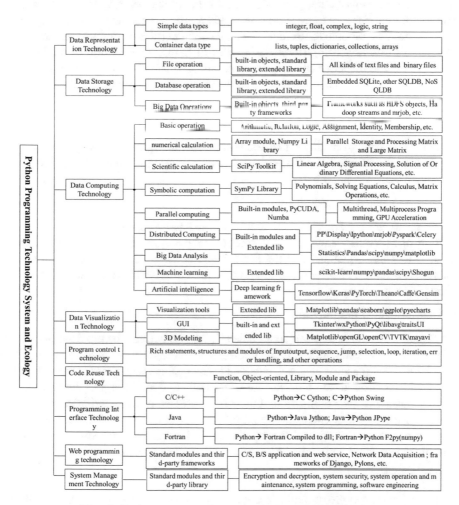

Fig. 4. Python programming technology system and ecology

3.2 Design Model

The basic workflow of the computer is input, calculation and output. The design model of MVC is abbreviated as Model-view-controller. The user interface of input and output is defined as View, which is made into an independent module. The update of the interface is completely separated from the background business processing, and only a clear data interface can be done. Business addition and deletion will not affect the normal operation of other parts; Controller is the bridge between business logic and input-output interface logic, receiving interface requests, decoding and sending to the response business module for execution, and receiving business module feedback to the interface display. According to the design idea of MVC, the cohesion of modules can be improved, and the loose coupling between modules can be increased to improve the scalability of the system and facilitate the maintenance and upgrading in the later stage.

3.3 Design Method

Up to now, the design method of software has gone through several stages of process-oriented, object-oriented and data-oriented, ordinarily several methods are indispensable in software development.

1. Process-oriented programming

As far as the research, design, simulation and test of HPWS are concerned, the solution of Maxwell's equations, the solution of boundary initial value problems of partial differential equations and the fast calculation of various functions are needed. The fastest and most effective programming method to solve these problems is process-oriented programming.

2. Object-oriented programming

According to the idea of object-oriented, it will be helpful for the development of ISTP-HPWS to classify all kinds of problem solving algorithms, and to analyze and describe in detail the attributes, behaviors and relationships with other objects. This not only facilitates the full and efficient use of traditional algorithms, but also provides a new understanding and sublimation of the whole research process and theoretical rules. Therefore, the object-oriented programming system is an efficient system with high cohesion and low coupling, which is conducive to the spark transmission of theory and technology and the iterative evolution of simulation test system.

3. Data-oriented

The data-oriented programming method is an important direction of future software development, discovering rules and relations from data through machine learning. All kinds of well-used design schemes are valuable experiences, which are worth learning, inheriting and developing, and avoid duplication of work. For this reason, the data-oriented programming method needs to be introduced into the design of ISTP-HPWS, and the machine learning method can greatly improve the design efficiency.

3.4 Computing Model

So far, as a tool for human to understand and transform nature, computer has two main ways. One is numerical simulation based on mathematical equation, the other is machine learning algorithm based on neural network and big data. Both technologies need to be fully utilized in development of ISTP-HPMS.

4 Technical Support

Good ideas and methods need feasible technology to support the function of ISTP-HPW system. ISTP-HPW system involves a variety of technologies, such as numerical analysis technology, input and output technology, database technology, storage technology, three-dimensional modeling technology and three-dimensional visualization technology. Typically, there are three items, programming technology, big data technology and artificial intelligence technology.

There are not less than 1000 existing programming languages. Each programming language has its design purpose, and has its good operation, so the programming technology involved is different. While, Python can meet all the requirements of the ISTP-HPW system

Distributed storage and calculation are indispensable to the analysis of large log data and the precise control of process in each stage of intelligent design, simulation, test and operation of high power microwave tubes. At present, Hadoop has grown into a huge system. As long as there are problems related to massive data, no field can lack the image of Hadoop. The design of high power microwave tube needs to build a practical and powerful Hadoop large data ecosphere. The asynchronous parallel computation of transient field simulation of high power microwave tube is expected to be implemented on Hadoop platform.

Artificial intelligence will play an important role in HPWS design, fault diagnosis and output pattern recognition. The non-linear probability model of input and output is attempted to be established through machine learning, which is another way of thinking and method of design without simulation.

Acknowledgments. Thanks for the support of the National Natural Science Foundation Project (61571078); Ministry of Education's Industry-University Cooperation and Education Project (201702043061).

References

1. Thumm, M.: Recent advances in the worldwide fusion gyrotron development. IEEE Trans. Plasma Sci. **42**(3), 590–599 (2014)
2. Thumm, M., Alberti, S., Arnold, A., et al.: EU megawatt-class 140-GHz CW gyrotron. IEEE Trans. Plasm. Sci. **32**(5), 143–153 (2007)
3. Luce, T.C.: Applications of high-power millimeter wave in fusion energy research. IEEE Trans. Plasma Sci. **30**(3), 734–754 (2002)

4. Barker, R.J., Schamiloglu, E.D.L.: High-Power Microwave Sources and Technologies. The Institute of Electrical and Electronics Engineers Inc. (2001)
5. Haiqun, H.: Quick Start of Zero-Start Python Machine Learning. Electronic Industry Press, Beijing (2017)
6. Mehta, H.K.: Python Basic Course of Scientific Computing. People's Posts and Telecommunications Press, Beijing, January 2017 (reprinted in June 2018). Translated by Junjie, T., Xiaoli, C
7. Ruoyu, Z.: Python Scientific Computing, 2nd edn. Tsinghua University Press, Beijing (2016). (reprinted on November 2017)

Design of Intelligent Meter Life Cycle Management System Based on RFID

Xiao Xiangqi[1(✉)], Chen Xiangqun[1], Huang Yanjiao[2],
Xiao Wencheng[3], Yang Maotao[1], and Huang Rui[1]

[1] State Grid Hunan Electric Power Company Limited Power Supply Service
Center(Metrology Center), Hunan Province Key Laboratory of Intelligent
Electrical Measurement and Application Technology, Changsha 410004, China
84578610@qq.com
[2] State Grid Shao Yang Power Supply Company, Shaoyang 422000,
Hunan Province, China
[3] China Resources Power Holdings Company Limited, Shenzhen 518001,
Guangdong, China

Abstract. Based on RFID technology, the key technologies of smart meter life cycle management system are researched. Through RFID and cloud platform technology, the purpose of closed-loop management covering the entire life cycle of RFID smart meter design, production, parameter setting, factory inspection, warehouse management, on-site meter reading and maintenance scrapped was realized. RFID features include anti-collision mechanism and contactless, etc. are used to achieve multi-radio chip, multi-channel smart meter factory settings, factory testing, on-site meter reading, data backup and other functions. The cloud platform big data analysis mechanism is utilized to guide the watch factory in material selection and optimization design to improve the quality of the meter and reduce the failure rate of the meter so as to ensure the safety of the key data in the table and the purpose of fast and accurate data exchange with RFID intelligent meter is realized. The RFID tag antenna is optimized and simulated. The simulation results show that the maximum antenna gain is 1.6304 dBi, and the coverage characteristic reaches 360° along the θ direction, which can effectively improve the data reading rate and distance.

Keywords: RFID · UHVDC · Cloud platform · Smart meters · Lifecycle management

1 Introduction

Intelligent electric meters life-cycle management industry in China is in a stage of rapid development, but started late compared with developed countries, the degree of automation and intelligence of the management system, as well as information level still have a large gap comparing with developed country [1–3], most of them still stagnate in the simple business function application stage, the intelligent electric meter design, production and site operation have not achieve the aim of life-cycle link closed-loop management and feedback guidance, and could not meet the needs of manufacturers and asset owners for real-time monitoring of the status of smart meters [4, 5].

© Springer Nature Switzerland AG 2020
F. Xhafa et al. (Eds.): IISA 2019, AISC 1084, pp. 55–63, 2020.
https://doi.org/10.1007/978-3-030-34387-3_7

RFID is a non-contact automatic identification technology, It realizes the non-contact information transmission through the spatial coupling of the radio frequency signal, and automatically identifies the target object and obtains the relevant data.

This design applies RFID and cloud platform management technology to the life-cycle management system of smart electricity meters. By making use of the anti-collision mechanism and contactless features of RFID technology, it can not only ensure the information security, but also quickly and accurately read the data information of RFID smart electricity meters, use big data cloud platform analysis and feedback mechanism to filtrate useful data information for guiding plant to conduct material selection and optimization design, gradually improve the quality of electricity meters and reduce the failure rate of electricity meters, to realize the purpose of accurate and efficient management of the life-cycle of mass intelligent meters.

2   Anti-collision Algorithm Processing of RFID Signal

As an important part of the life-cycle management system, the storage and on-site meter reading processes of smart meters have high requirements for fast and accurate identification of smart meter information. In order to achieve this goal, it is necessary to study the anti-collision algorithm of RFID signals.

In this chapter, a special efficient tag collision processing algorithm based on adaptive binary tree search (ABS) algorithm is used. This kind of algorithm is complex and takes a long time to recognize, but the algorithm itself is deterministic, there is no label in a long time no response, so this kind of algorithm is also called deterministic algorithm. In the binary search algorithm (ABS), the encoding method of the algorithm is Manchester encoding, which can identify collisions by bits [6, 7] ,the concrete diagram is shown in Fig. 1.

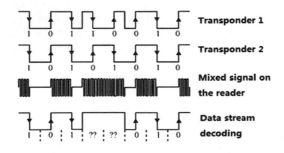

Fig. 1. Collision of Manchester encoding

Suppose the number of tags in the system is N, the number of branches the system allocates is M, when the search depth of the system is 1, the recognition probability is:

$$p(1) = \left[1 - \frac{1}{M}\right]^{N-1} \tag{1}$$

When the search depth is d, the probability of system identification is:

$$p(d) = p(1)[1 - p(1)]^{d-1} \tag{2}$$

Then in a complete identification process, the average depth required for searching N tags is:

$$E(d) = \sum_{d=1}^{\infty} p(d) = \sum_{d=1}^{\infty} p(1)[1 - p(1)]^{d-1} \tag{3}$$

Simplify and summarize above formula, and the sum formula can be obtained according to the geometric series:

$$E(d) = \frac{1}{\left[1 - \frac{1}{M}\right]^{N-1}} \tag{4}$$

In the above formula, is the common ratio of a geometric progression, i.e. the number of branches that the system has allocated. In this way, we can obtain the average number of time slots required in the system search process as:

$$T = E(d)M = \frac{M}{\left[1 - \frac{1}{M}\right]^{N-1}} \tag{5}$$

When the search algorithm has two branches, the average number of time slots required is $T_2 = 2/(1 - 1/2)^{N-1}$. When the search algorithm has four branches, the average number of time slots required is $T_4 = 4/(1 - 1/4)^{N-1}$. By comparing the above two formulas, we can know that, when N is greater than 3, the four-branch search algorithm uses less average time slot and the system performance is better than the two-branch search algorithm; When N is less than or equal to 3, the two-branch search algorithm uses less average time slot and the system performance is better than the four-branch search algorithm.

Based on this algorithm, Design RFID electronic tags with appropriate inventory cycle, by setting the parameters of the dwell time, power and Q factor, it is possible to maintain a high identification rate and at the same time, to read and write electronic tags quickly and improve work efficiency.

3   Composition of the Life Cycle Management System

The life-cycle management system of intelligent electricity meters is studied based on RFID and cloud platform technology, core data of all process for life-cycle management, from the design of RFID smart electricity meters to intelligent production, parameter setting, factory testing, warehouse management, on-site meter reading, interact with cloud platform data through RFID, take advantage of RFID conflict prevention mechanism and contactless to achieve multi-radio chip, multi-channel intelligent electricity meter parameter Settings, factory rapid detection, supply chain management, on-site meter reading, fault detection, data security backup, etc., combined with cloud platform, integrate the producing stage and manufacture factory into the cloud platform management, establish big data analytics mechanism, make use of lifecycle management data to guide the meter factory in their material selection, optimization design and other intelligent production, to realize the whole process closed-loop life-cycle management of intelligent electricity meters in industry 4.0 mode.

3.1 Cloud Platform Management System

The cloud service platform, designed based on the business requirements of the industry 4.0 intelligent electricity meter life-cycle management system, can connect with the three existing marketing systems of the grid company, and bring the production process of the manufacturer under the control of the cloud platform, at the same time, calibration detect intelligent electric meters systematically, monitor and analyze the data in the process of experiment and verification, establish intelligent electricity meters life cycle management system, control intelligent electricity meters from the source, reduce the failure rate. Manufacturers can also use this system to guide the meter production process, do well out of management of production, testing and rotation of meter products, ensure that the meter products meet the standard requirements of material selection and optimization design of power grid company. Grid companies can also understand the production process and level of each manufacturer in real time through this system, providing data support for grid companies to select manufacturers and other work, and forming a closed loop feedback management mechanism of intelligent production in industry 4.0 mode.

3.2 RFID Intelligent Meter Storage System

As shown in Fig. 2, the storage information management system based on RFID technology is a new type of storage management system. Compared with RFID which is in other frequency band, semi-active RFID chips are used in smart meters, its electronic label system has the obvious characteristics of long recognition distance, multi-label simultaneous recognition, large information storage, corrosion resistance, strong resistance to harsh environment and so on, combined with the cloud platform mechanism, it can fully meet the need of reliable identification of large quantities and multiple labels in the storage management process of smart electricity meters, and realize the purpose of rapid logistics flow and management of smart meters.

As shown in Fig. 3, a semi-active RIFD read-write chip is designed for the read-write module of the semi-active RFID chip and the intelligent warehouse management system based on the cloud platform, it makes the warehouse management system has the advantages of farther identification distance, multi-label batch, accurate identification and so on.

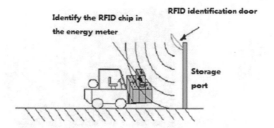

Fig. 2. RFID schematic diagram of storage process of intelligent electricity meters

Fig. 3. Semi-active RIFD put out/in storage read and write chip

3.3 RFID Smart Meter On-Site Data Copying System

At present, the on-site data copying of electricity meters is based on infrared equipment, which has many limitations and is inconvenient to operate. To make it easier to copy, research and development RFID intelligent meter reading system and handheld terminals (as shown in Figs. 4 and 5), this terminal is based on S3C6410 chip and CLRC632 radio frequency chip which has high performance and low power consumption to achieve non-contact RFID reading and writing operation. It uses RFID chip technology, embedded system design, remote communication technology and many other technologies, to carry out the software and hardware design of HMI module, power management module, read-write module and so on of semi-active RFID meter reading terminal, so that it can identify a large number of internal RFID chips in a wide range, establish the communication path, automatically identify all electricity meters file information, one-key copy electricity meters metering data, realize the electricity meters fast and accurate meter reading, make the field copy work more simple and fast.

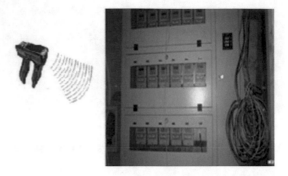

Fig. 4. On-site data copying system based on RFID

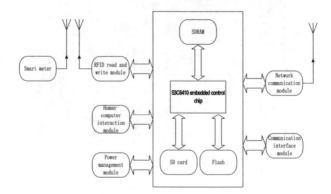

Fig. 5. Overall structure of the handheld terminal

4   Design of RFID Electronic Tag Antenna

RFID electronic tag is one of the key components of the information exchange between RFID electricity meter and the whole life management system. The data transmission rate and distance of electronic tag directly affect the work efficiency of the whole system. Therefore, this paper aims to optimize the design of electronic tag.

In this paper, dipole oscillator antenna is adopted, which is characterized by strong radiation capacity, simple manufacturing, low cost and mature application theory. It is an ideal choice for RFID electronic tag antenna. The optimal design of the bent dipole antenna with t-shaped structure is shown in Fig. 6.

(1)   antenna impedance

The diagram of the antenna impedance is shown in Fig. 7, the two curves are the resistance value and reactance value of the designed tag antenna changing with frequency. As can be seen from the Fig. 7, the input impedance at the antenna resonance frequency of 922.5 MHz is 11.87 + 164.43 O, the matching error of tag antenna resistance value is 7.90%, and the reactance matching error is 14.98% of the overall impedance matching rate. The power transmission coefficient formula is shown as

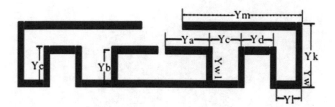

Fig. 6. The tag antenna prototype

Fig. (7).The relevant parameters of the antenna are substituted into the energy transmission coefficient formula (6), and formula (7) is obtained.

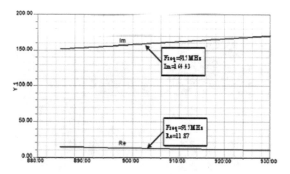

Fig. 7. Antenna impedance diagram

$$\tau = \frac{4R_aR_c}{(R_a + R_c)^2 + (Z_a + Z_c)^2} \tag{6}$$

$$\tau = \frac{4 \times 11 \times 11.87}{(11 + 11.87)^2 + (164.43 - 143)^2} = 0.5317 \tag{7}$$

That is, the final impedance matching rate is 53.17% and the matching state is good. And the change is gentle in a wide band range, it shows that the impedance matching band between the tag antenna and the chip is wide and has strong robustness.

(2) gain of antenna

Figure 8 is the h-plane ($\theta = -180° \sim 180°$, $\phi = 90°$) directional gain diagram of the antenna. As can be seen from the diagram, after optimization, dipole antenna with meander-line cover features 360° in H plane, and have the biggest gain in $\phi = 90°$ direction.

As Fig. 9 shows, the optimized antenna 3D gain pattern can directly observe that the designed antenna gain direction have 360-degree coverage in the direction of theta, and its maximum gain is 1.6304 dBi.

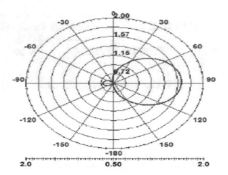

Fig. 8. Directional gain of the antenna in H plane diagram

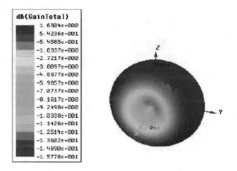

Fig. 9. 3D directional gain of the antenna diagram

5 Conclusion

(1) Design of the intelligent watt-hour meter based on RFID the life-cycle management system using RFID anti-collision mechanism and characteristics of contactless radio frequency chip, multichannel intelligent watt-hour meter factory Settings, the factory inspection, on-site meter reading, data backup, and other functions, using cloud data analysis mechanism, guide meter factory to do the material selection and optimization design, which can be helpful for profit the electric meter quality step by step, and reduce the failure rate meter.

(2) After optimize the design of RFID electronic tag antenna, the simulation results show superior performance in antenna impedance and gain, and can effectively improve the data reading rate and reading distance.

Acknowledgements. The project funds are derived from the science and technology projects of China State Grid Corporation (5216A01600VZ).

References

1. Feng, H.: Practical application of life cycle management system for electric safety tools. Commun. World **21**, 108–109 (2017)
2. Yongzhu, Z., Genzhou, Z., Xiaolong, R., et al.: RFID-based smart grid asset life cycle management system design. Smart Power **45**(11), 57–61 (2017)
3. Weng Liguo, X., Jinyu, M.K., et al.: Whole life cycle management system of transmission and distribution equipment based on guided wave technology. Rural Electrif. **06**, 45–46 (2017)
4. Min, L.: Research on the life Cycle Management of Fixed Assets of Power Grid Companies. Inner Mongolia University (2011)
5. Shilin, W., Yun, Y.: Analysis of the relationship between asset life management and "three episodes and five majors" system construction. Electric Power Inf. **11**(02), 15–20 (2013)
6. Yonggang, L.: Research on Mine Personnel Tracking and Positioning System Based on Binary Search Algorithm. Shandong University of Science and Technology (2009)
7. Xiaohua, C.: Research on Some Key Technologies of Container Electronic Label System. Wuhan University of Technology (2008)

Design of Single-Phase Intelligent Meter Based on RFID Semi Active Technology

Xiao Xiangqi[1]([⊠]), Chen Xiangqun[1], Yang Maotao[1], Huang Rui[1], Huang Yanjiao[2], and Xiao Wencheng[3]

[1] State Grid Hunan Electric Power Company Limited Power Supply Service Center(Metrology Center), Hunan Province Key Laboratory of Intelligent Electrical Measurement and Application Technology, Changsha 410004, China
84578610@qq.com
[2] State Grid Shao Yang Power Supply Company, Shaoyang 422000, Hunan Province, China
[3] China Resources Power Holdings Company Limited, Shenzhen 518001, Guangdong, China

Abstract. When the smart meter remote meter reading failure, the staff need one-on-one meter reading on-site, resulting in great workload, a semi-active RFID-based smart single-phase smart meters are designed. A single-phase smart meter based on RFID semi-active technology was designed to achieve the goal of non-contact fast, efficient and batch meter reading. The FM3307 MCU produced by Fudan Microelectronics Co., Ltd., the ATT7053BU high-precision measurement chip produced by Juquan Company and the RENSAS (R5F212BC) 8-bit machine produced by Fudan Microelectronics Co., Ltd. were adopted to design the power supply unit, the metering unit, the load control unit, the communication unit, Clock and memory cells and other core circuits. Three-tier architecture, modular design ideas are used to divide the software functions, and through a unified interface, data exchange between modules is achieved, this method is easy to mount and remove any module, which will help the expansion and deletion of features. Hardware and software debugging, machine performance test results show that the batch, fast meter reading function of the smart meter is realized, effectively reducing the workload on-site, which has a certain value of promotion and application.

Keywords: RFID · Single-phase smart meter · ATT7053BU

1 Introduction

With the development of economy and society and the progress of power science and technology, the traditional power grid is an inevitably trend of develop into smart grid. Common electronic watt-hour meters have been gradually eliminated due to many defects, such as single function, poor anti-stealing effect, backward meter reading method, and easy damage to IC card. which have been gradually eliminated [1–3]. As an important part of smart power grid, smart meters have been widely used. Smart meters to overcome some shortcomings of electronic watt-hour meter, has realized the remote read, fee control, and other functions, but there is still a certain proportion of

© Springer Nature Switzerland AG 2020
F. Xhafa et al. (Eds.): IISA 2019, AISC 1084, pp. 64–71, 2020.
https://doi.org/10.1007/978-3-030-34387-3_8

meter reading, the failure rate at this time still need staff to one-on-one meter reading on-site. According to statistics, the number of smart meters installed amount exceeds 500 million pieces, meter reading after the failure of the meter reading workload is great. How to quickly and efficiently read the data of electric energy meter needs to be solved urgently. As a kind of rapid development of advanced non-contact recognition technology, RFID technology has many advantages in terms of reading quantity, read distance and reading accuracy. It has been maturely applied in the fields of security, transportation, logistics, medical treatment and environmental protection. The RFID technology used in the design of smart meters will effectively improve the efficiency of the smart meter information. At present, the country is planning to promote the object-oriented protocol 698 and the IR46 standard. With the new standard to be implemented, the new generation of smart meters will be updated and the market prospect is broad [4–6].

A single-phase smart meter based on RFID semi-active technology is designed. The RFID communication module is added on the basis of the original smart meter to realize the purpose of fast transmission of various data of the smart energy meter through high-frequency wireless signals. The FID communication module includes an RFID tag chip and an RFID radio frequency antenna. It is connected to the smart electricity meter MCU through the I2C bus to achieve two-way data communication. All the devices are placed in the electric energy meter casing, without changing the external structure of the meter, not only the transformation cost is low, but also easy to promote the application [7–9].

2 Overall Design of Smart Meter

A single-phase smart meter based on RFID semi-active technology is designed. In terms of hardware, the key core circuits are designed according to the modular design ideas, including power supply unit, metering unit, load control unit, communication unit, clock and memory cells and other core circuits. The FM3307 MCU produced by Fudan Microelectronics Co., Ltd., the ATT7053BU high-precision measurement chip produced by Juquan Company and the RENSAS (R5F212BC) 8-bit machine produced by Fudan Microelectronics Co., Ltd. were adopted to design [10].

In terms of software system, three-tier architecture, modular design ideas are used to divide the software functions. The three-tier architecture includes the driver layer, the platform layer, and the service layer. The modules include driver modules, platform modules, and service modules. Through a unified interface, data exchange between modules is achieved, this method is easy to mount and remove any module, which will help the expansion and deletion of features.

Finally, the hardware and software debugging of the single-phase smart meter and the performance of the whole machine are tested and verified.

3 Hardware Design of Smart Meter

The block diagram is shown in Fig. 1. The FM3307 MCU produced by Fudan Microelectronics Co., Ltd., the ATT7053BU high-precision measurement chip produced by Juquan Company, and has peripheral circuit modules such as clock, storage, communication, ESAM certification and display.

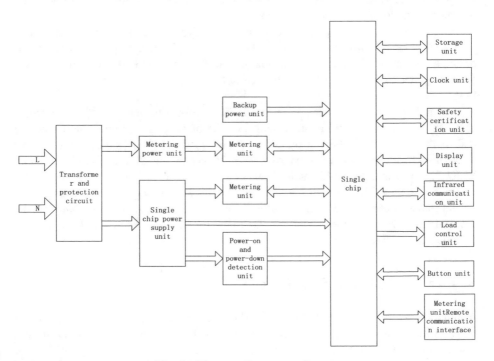

Fig. 1. The overall structure diagram

3.1 CPU Module

The low-power microprocessor RENSAS (R5F212BC) 8-bit machine produced by Fudan Microelectronics Co., Ltd. be adopted to the CPU, with a main frequency is 8 MHz, a flash is 128 K, a RAM is 3 K, and an external memory EEPROM of 512 K. The CPU is the core of the whole product. It processes the information transmitted by the metering module unit, completes the energy judgment, accumulates the energy of each rate, stores the data, and compares the voltage, current, power and other measurement signals, as well as the power, time, and communication address, through the display module. Its are equipped with a query button to facilitate users to understand the working information of the watt-hour meter. The watt-hour meter has the function of safety certification. Only the certification of the safety certification module can operate and set the watt-hour meter to ensure the safety of power consumption.

3.2 Metering Module

Due to the metering module is the key module of the smart energy meter, so that this design adopts the high-precision metering chip ATT7053BU produced by Juquan Company. The circuit is mainly composed of A/D conversion, power calculation, energy accumulation, anti-submarine movement, RFID communication and other parts, with the characteristics of high reliability, high precision and low power consumption.

Its analog input port supports differential signal input, the current input signal is provided by current transformer or shunt, the voltage input signal is provided by voltage divider network, and it has two current channels to measure the current of phase and neutral simultaneously, as well as with anti-stealing real-time monitoring.

The metering module collects voltage and current signals through the sampling circuit to complete energy metering, stores power-related data and real-time voltage, current, power factor and other data and outputs them in pulses. The data is exchanged with the CPU through the SPI communication interface to complete the power pulse collection, power distribution, infrared and 485 communication, LCD display and other functions. The metering sampling circuit is shown in Fig. 2.

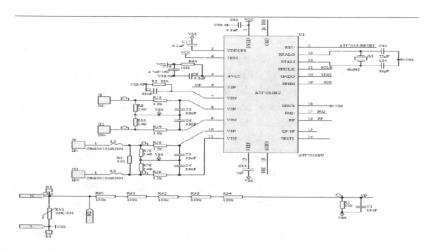

Fig. 2. Metering sampling circuit

3.3 Communication Module

Based on the reservation of infrared communication and 485 communication, the design scheme increases the RFID communication module to exchange various data information with the power supply department through high-frequency wireless signals. In order to realizes the functions of asset management and meter reading of the smart watt-hour meter. The reading distance is more than 4 meters, and the number of meter readings reaches more than 100, which effectively improves the efficiency of meter reading.

The structural design is shown in Fig. 3. The RFID communication module used in this design is mainly composed of two parts: an RFID tag chip and an RFID radio frequency antenna. It is connected to the smart watt-hour meter MCU through the I2C bus to realize two-way data communication. The RFID communication module is disposed in the electric energy meter housing and does not change the external structure of the electric energy meter, so the transformation cost is low.

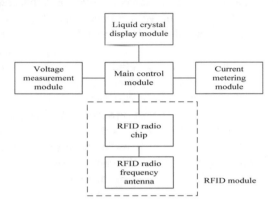

Fig. 3. The module design RFID

The RFID system is mainly composed of three parts: electronic tag (RF card), reader and application supporting software hardware. According to the power supply source of the electronic tag, it can be divided into active, passive and semi-active forms. Compared with passive tags, semi-active tags solve the energy supply problem required for chip operation because they carry batteries themselves. The requirements for the power intensity of input RF signals are greatly reduced, and the effect of reading distance and recognition rate is better. Compared with the active tags, semi-active tag because do not need to be active power, do not need to be active network communication between labels and tags, circuit complexity and the requirement of battery capacity and current driving ability are relatively active tags are lower, the manufacturing cost is low.

After analysis and comparison, the design uses the Monza® x chip produced by American Infineon to design semi-active electronic tags, and is connected to the device processor through the standard IIC bus to ensure UHF when the electronic device is powered off, the reader can also read and write the memory of Monza X chip.

3.4   Clock and Storage Module

This module is mainly used to provide accurate clock signal and data information, event record storage and other functions for the system. The clock module adopts RX8025T's new real-time clock chip produced by Epson. It has I2C interface and temperature compensation function. It can maintain high precision and high stability from −40 °C to above 85 °C. The clock timing error is less than 0.5 s/Days, the monthly error does not exceed 25 s, to meet the requirements of the relevant national departments. As shown in Fig. 4, the clock circuit and the external memory are bidirectionally communicated with the microcontroller through an I2C bus composed of a CL serial clock line and an SDA serial data line.

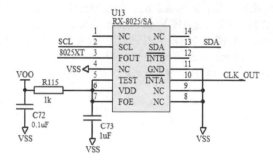

Fig. 1. The clock module of circuit schematic

4 Software Design

4.1 Overall Structure of the Software

The overall architecture of the software is hierarchical, and the functions are divided into individual modules with low coupling. The modification of the design business code will not have an impact on the lower layer, which is beneficial to the team division and cooperation development, and the function can be convenient increased or decreased; the code structure in the module is consistent, easy to expand and reuse, and the unit test can be facilitated. Static use of memory greatly improves system reliability; The platform module and the business module call the business processing functions through the message mechanism. The driver layer encapsulates the driver function as a unified interface through the intermediate file, which reduces the coupling between the layers. The data involved in the business module is visible only in this module. External access needs to pass through a unified interface, which protects data security, and data reading and storage are transparent. The separation of the driver layers significantly enhances software portability.

4.2 RFID Interface Design Based on Three-Tier Architecture

The traditional RFID interface model is closely integrated with the underlying hardware. Although the access speed is fast, it is not convenient for the upper layer developers. To compensate for this defect, the design of the RFID interface model based on the three-tier architecture is designed.

The RFID interface model based on the three-tier architecture consists of three layers: Low Level API, Web Platform Service and High Level API. The Low Level API and Web Platform service layer are deployed inside the RF reader, while the High Level API is deployed on the application side. The layers are relatively independent but closely linked. The Low Level API layer is the basic data layer, which implements the encapsulation of the underlying hardware of the RF reader. It is similar to the device driver layer in the Windows operating system, through which all data exchange with the reader is completed.

Web platform service layer is similar to the Windows operating system, it follows the EPC standard, define the standard operation interface and data communication format, extending the Low Level API Web applications, for High Level API provides a support platform, and the High Level API is similar to the Win32 SDK, it will be the standard operation interface and data communication format further encapsulation, provides a simple and easy to use for the upper developers, powerful API functions.

5 System Debugging

After the product design is completed, the hardware and software of the watt-hour meter are jointly adjusted, and the watt-hour meter after the processing is completed is shown in Fig. 5.

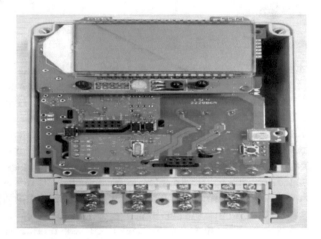

Fig. 5. Front view of the watt-hour meter

After the watt-hour meter is processed, hardware debugging should be carried out first to check whether all functional modules of the watt-hour meter work normally. Firstly, add the rated voltage to the watt-hour meter, test the power supply test points of each module reserved on the printed board with a multimeter, ensure that the power supply circuit of the watt-hour meter can work normally. After that, burn the test program, and then install the prototype on the automated test tool. The functional test of each hardware module is automatically completed by the test tooling. After each functional module has passed the test, the standard table is used to correct the error of the prototype. The meter used in the watt-hour meter is a software calibration table, and the automatic calibration software is used to complete the automatic calibration process of the watt-hour meter error through RS485 communication.

After the hardware debugging is completed, the software function compliance test is performed.

The results show that the performance index and related functions of the single-phase smart meter designed in this paper can completely satisfy the requirements of smart meter, and realize the function of non-contact fast, efficient and batch meter reading.

6 Conclusion

A new generation of single-phase smart meter based on RFID semi-active technology. The FM3307 MCU produced by Fudan Microelectronics Co., Ltd., the ATT7053BU high-precision measurement chip produced by Juquan Company and the RENSAS (R5F212BC) 8-bit machine produced by Fudan Microelectronics Co., Ltd. were adopted to achieve the purpose of rapid communication.

The experimental results show that the designed smart meter performance indicators and related functions can completely satisfy the requirements of the smart meter. In the case of no power, the various parameter information in the smart energy meter can still be read in batches by RFID and is not affected by distance, position, intensity, effectively reducing the workload on-site, which has a certain value of promotion and application.

Acknowledgements. The project funds are derived from the science and technology projects of China State Grid Corporation (5216A01600VZ).

References

1. Minghu, Z., Wei, Z., Xin, Y., Zhan, Q.: Design and implementation of the RFID traceability management system for hand - held devices. Process Autom. Instrum. **38**(01), 57–60 (2017)
2. Centscents, H., Junqiao, X.: Design of wireless handheld terminal based on SX1233. Electron. Des. Eng. **19**(24), 4–6 (2011)
3. Yuejiao, W.: RFID-Based Batch Identification of the Application of RF Gate. Zhengzhou University, Zhengzhou (2015)
4. Xu Renheng, Q., Jingzhi, L.D., Puzhi, Y., Longdi, G.: Research on the full life cycle management system of smart meters. Electr. Meas. Instrum. **54**(01), 67–70 (2017)
5. Bo, C., Kaihua, L.: STM32 RFID handheld terminal hardware design. SCM Embed. Syst. Appl. **12**(04), 45–48 (2012)
6. Guowei, W., Zongpu, J., Weiping, P.: Two-way authentication protocol for mobile RFID based on dynamic shared key. J. Electron. **45**(03), 612–618 (2017)
7. Bo, C.: Based on CLRC632 and STM32 RFID Reader Circuit Design. Tianjin University, Tianjin (2012)
8. Hang, Z., Wei, H., Chen, Yu., Bing, L., Junhong, Z., Huanghuang, Z.: Design of point to point communication performance testing platform for broadband power line carrier. Electr. Meas. Instrum. **53**(21), 100–105 (2016)
9. Xinchun, L., Yongxin, Y.: Design of mobile 13.5 MHz RFID reader. J. Comput. Appl. **20**(08), 229–232 (2011)
10. Yufeng, Z., Chengwei, C.: Design of handheld RFID reader based on CLRC632. J. Wuxi Vocat. Tech. Coll. **12**(01), 25 (2013)

Research on Anti-fall Algorithms of Substation Intelligent Inspection Robot

Xiangqian Wu[1], Dongsong Li[2], Tianli Liu[2], Xuesong Li[2],
Chuanyou Zhang[2], Jian Li[2], Guangting Shao[2], Yafei Wang[2],
Yan Deng[2], and Guoqing Yang[2(✉)]

[1] Production Technology Department, Guizhou Power Grid Co., Ltd.,
Guiyang, China
[2] Robotics Cause Department,
Shandong Luneng Intelligence Technology Co., Ltd., Jinan, China
sdygq2004@163.com

Abstract. A fall protection method of intelligent patrol robot based on binocular vision is presented in the paper, and the safe running of robot is realized in substation. A binocular camera is installed on the intelligent inspection robot of the substation, and the front road surface of the inspection robot is photographed by a binocular camera. The disparity map is computed by the current road image obtained by the left and right camera, and the actual distance between the two cameras is obtained. By comparing with the preset distance threshold to determine whether the front is a dangerous area. Experimental results show that the proposed method has high accuracy and strong robustness. It can effectively avoid the fall protection of the inspection robot, reduce the damage caused by the drop of the inspection robot and improve the security during the patrolling process of the inspection robot.

Keywords: Substation · Inspection robot · Fall protection · Binocular vision · Disparity map

1 Introduction

Substation is an important facility of power grid. It is important to ensure the safe operation of substation for the stability of the whole power grid. In order to monitor the operation status of substations, it is necessary to inspect substations regularly [1–3].

At present, many unattended substations use inspection robots to inspect outdoor high-voltage equipment [4]. However, there are still many imperfections in the current inspection robot. In the inspection process of inspection robot, there are occasional problems such as deviation from track and depression of track cover plate. When the trajectory deviates, if the patrol robot travels on the narrow cover, it may fall and cause damage to the patrol robot. The maintenance and removal of track cover plate without reset or damage will lead to pits, which will easily cause the falling damage of inspection robot. Therefore, a kind of anti-falling technology of inspection robot is urgently needed.

© Springer Nature Switzerland AG 2020
F. Xhafa et al. (Eds.): IISA 2019, AISC 1084, pp. 72–77, 2020.
https://doi.org/10.1007/978-3-030-34387-3_9

In view of this, a binocular vision anti-falling algorithm based on intelligent inspection robot platform of substation is proposed in this paper. It realizes the identification of dangerous area and avoids falling.

2   Design of Inspection Platform

Patrol inspection of substation equipment is an important work that perplexes the operation and maintenance personnel of power system. It is the primary task of station staff to accomplish patrol inspection task with high quality to ensure the normal operation of power grid. Due to the particularity of substation, live detection technology plays an important role in the maintenance of substation equipment. With the gradual development of electronic equipment and the improvement of its performance, the use of visible light camera, infrared camera and ultraviolet equipment for the detection of substation equipment has achieved great results. With the gradual maturity of artificial intelligence technology, it is possible for intelligent inspection robot detection equipment in substation.

Robot instead of manual inspection effectively solves the problems of heavy patrol task, data cannot be uploaded in real time and in high-risk environment, and robot instead of manual inspection ensures the personal safety of staff. The simplified diagram of substation inspection robot is shown in Fig. 1.

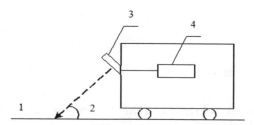

Fig. 1. The simplified diagram of substation inspection robot

Among them, in the figure, 1 is the ground, 2 is the angle of the binocular camera to take the ground image, 3 is the binocular vision camera, and 4 is the robot data processor. That is, a binocular camera set on the outside surface of the front of the robot and a processor connected with the binocular camera. The system includes image acquisition unit, image processing unit, motion control unit and early warning control unit.

3   Ranging Principle of Binocular Camera

Binocular vision is an important form of machine vision. Using parallax principle and imaging equipment, two images of the target object (one image from left and right cameras) are acquired, and then the position deviation of the target point in the image is calculated to obtain the geometric information of the target point.

Binocular vision ranging is an important part of binocular vision technology. In optical ranging, it is a passive ranging. Figure 2 shows the principle and structure of binocular camera ranging.

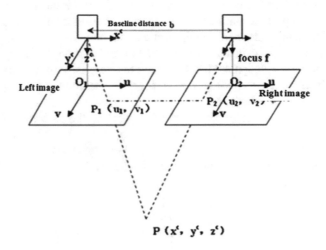

Fig. 2. Ranging principle of binocular camera

The distance between the lens center and optical center of left and right cameras is called the baseline B of binocular vision. Using binocular vision imaging system, the actual position coordinates of three-dimensional point P with image plane coordinates (u_1, v_1) and (u_2, v_2) can be determined [6, 7].

It is found that the parallax size corresponds to the depth, and the parallax contains the spatial information of the target point in the real environment [8, 9]. Therefore, the disparity image generated by binocular vision can obtain the distance between the point in the world coordinate system and the camera, and the corresponding coordinates can also be obtained.

4   Application Principle of Anti-falling

First, we define a line-of-interest (*LOI*). When installing and debugging binocular cameras, we need to determine a preset location in advance, that is, the distance from the camera. This distance is defined as *LOI*. Figure 3 is a comparative diagram of normal driving and deviating driving routes. The dashed line is Lane line, which is wider than the robot, and the solid line is *LOI* (the width of the robot). Figure 4 illustrates the corresponding relationship.

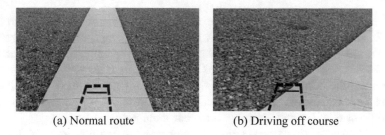

(a) Normal route   (b) Driving off course

Fig. 3. Off-line detection sketch of driving route

The principle of application is as follows. Binocular camera can obtain the disparity image in the field of vision, the length L of the camera (the front segment of the robot body) at any position in the field of vision, and calculate the Δl of each point on *LOI*. Among them, L is the actual distance.

$$\Delta l = |L - l| \tag{1}$$

The Δl of each point on *LOI* is counted,

$$f = \frac{n_{\Delta l > T}}{n_{LOI}} \tag{2}$$

Among them, $n_{\Delta l > T}$ is the point set of $\Delta l > T$ (the number of points exceeding the error limit), T is the fixed value, n_{LOI} is the total number of points on *LOI*. When detecting, the distance value corresponding to the points on the line segment is counted. If the proportion of the points exceeding the limit is greater than f, it is considered that the front is a dangerous area and alarm parking is needed.

Fig. 4. Diagram of robot, Lane line and LOI relations

5   Design and Analysis of Algorithms

As mentioned above, the shooting direction of the binocular camera is at an acute angle to the horizontal plane, which is used to shoot the road ahead of the inspection robot in the substation. The image of the front road surface in the left and right cameras is obtained, and the disparity map of the current left and right cameras road surface image is calculated. Binocular cameras need to be calibrated first to obtain the relevant parameters of the camera. Then they are installed on the platform of intelligent inspection robot in substation and connected with the relevant system of the robot. Figure 5 is a flow chart of road image processing, and the algorithm steps are as follows:

Step1: Image acquisition: Get the current road image captured by binocular camera.
Step2: Parallax image acquisition: According to the right and left images of the current road collected by binocular camera, the road parallax image is calculated.
Step3: ROI selection: Intercepting ROI region of road disparity image: not only guarantees the accuracy of the algorithm, but also improves the operation speed.
Step4: Data calculation statistics: According to formula (1), formula (2) to calculate the ratio of relevant data as a measurement parameter.
Step5: Determine whether the ratio is greater than the preset proportional threshold, and if so, identify the dangerous area ahead

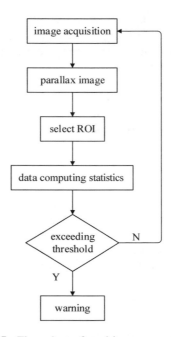

Fig. 5. Flow chart of road image processing

In the experiment, flat asphalt circuit and cable trench cover plate are selected as test data for the road of substation. The test results are shown in Table 1.

From the table, we can see that the accuracy of the algorithm is very high, especially on the flat road surface, the accuracy is as high as 95.6%. The main reason for the low accuracy of the cover road is that the cover road is uneven and the robot fluctuates up and down in the course of walking, which results in the deviation of the measurement results and leads to the decrease of the accuracy.

Table 1. Statistics of experimental results

Data type	Number of frames	Correct	Error	Accuracy rate
Asphalt road	270	258	12	95.6%
Cable trench cover plate	324	301	23	92.9%

6 Conclusion

On the intelligent inspection robot of substation, binocular camera and binocular vision technology are used to measure the distance of target points. By comparing with the preset distance threshold, we can determine whether the front is a dangerous area, thus avoiding the falling of the inspection robot, reducing the damage caused by the falling of the inspection robot, and improving the security of the inspection robot in the process of inspection. The experimental results show that the proposed algorithm can effectively solve the problem of robot falling in the course of walking, and can meet the security requirements of substation inspection robot. It enhances the automation and intellectualization ability of substation inspection robot and promotes the process of unattended substation.

References

1. Zhou, L.H., Zhang, Y.S., Sun, Y., et al.: Development and application of equipment inspection robot for smart substations. Autom. Electr. Power Syst. **35**(19), 85–88, 96 (2011)
2. Yang, X.D., Huang, Y.Z., Li, J.G., et al.: Research status review of robots applied in substations for equipment inspection. Shandong Electr. Power **42**(1), 30–34 (2015)
3. Peng, X.Y., Jin, L., Wang, K., et al.: Design and application of robot inspection system in substation. Electr. Power **51**(2), 82–89 (2018)
4. Ding, S.K., Li, J.: Analysis and countermeasures of substation inspection robot in the practical application. Distrib. Utilization **33**(1), 80–82 (2016)
5. Zhang, E.S.: Research on ranging algorithm based on binocular stereo vision algorithm. Inner Mongolia University, pp. 1–62 (2018)
6. Guo, P., Du, H.: Study on the robotic binocular distance measurement. Wirel. Internet Technol. **15**(5), 99–101 (2018)
7. Wang, Y.X., Zhang, J.M., Kan, J.M.: Hardware platform of target positioning and ranging system based on binocular vision. Comput. Eng. **7**, 214–218, 223 (2013)
8. Zhang, Y.J., Pan, Y., Wu, C.: Distance measurement of binocular CCD camera on vehicle-mounted system. Inf. Secur. Technol. **7**(1), 57–62 (2016)
9. Xu, S.S., Wang, Y.Q., Zhang, Z.Y.: Extracting disparity map from bifocal monocular stereo vision in a novel way. J. Comput. Appl. **32**(2), 341–343, 378 (2011)

Research and Application of Process Object Modeling Method Based on Deep Learning

Feng Yin, Su Ye, Quan Li[⊠], Jiandong Sun, and Junyu Cai

Institute of Electric Power Science, State Grid Zhejiang Electric Power Co., Ltd.,
Hangzhou, China
lq_lq_2001@163.com

Abstract. Predictive control is often used to improve control performance because most industrial sites are inertial lag objects, but it is difficult to establish predictive model. In this paper, artificial intelligence method is applied to control object modeling. A deep learning based modeling method is proposed, which includes open-loop modeling and closed-loop modeling. Closed-loop modeling is based on open-loop modeling. Firstly, the DNN deep learning algorithm used in this paper is introduced. Aiming at open-loop data of multi-order inertial system, a deep learning DNN network system based on Multiple Inertial filters is designed. After training step disturbance data, a deep learning neural network model with object characteristics is obtained. Secondly, a closed-loop modeling method based on two DNN models is proposed for multi-order inertial closed-loop systems. Firstly, forward and backward step disturbances are added to the control variables, then model 1 is obtained after training the DNN network system with deep learning based on Multiple Inertial filters, and then forward and backward step disturbances are added to the set values. The output data of model 1 and the output data of controlled objects are trained as inputs of deep learning DNN random inactivation network, and the closed-loop object model is finally obtained.

Keywords: Deep learning · Open-loop modeling · Closed loop modeling · Intelligent model

1 Introduction

In industrial process control systems, most of them are inertial lagging objects, which often affect the control quality of the system. Predictive control is often used to improve the quality of the industrial control system. However, due to the more noise in the field, the conventional identification algorithm needs more complex excitation signals. The disturbance to the system is large, and the prediction model is difficult to establish accurately, which often affects the control performance of the system. Intelligent modeling method based on in-depth learning only needs input and output data to train the network to obtain the object model, which can accurately predict the output of the actual model. In this paper, artificial intelligence method is applied to control object modeling. A deep learning based modeling method is proposed, which includes

F. Xhafa et al. (Eds.): IISA 2019, AISC 1084, pp. 78–88, 2020.
https://doi.org/10.1007/978-3-030-34387-3_10

open-loop modeling and closed-loop modeling. Closed-loop modeling is based on open-loop modeling.

Firstly, the DNN deep learning algorithm used in this paper is introduced. Aiming at open-loop data of multi-order inertial system, a deep learning DNN network system based on Multiple Inertial filters is designed. After training step disturbance data, a deep learning neural network model with object characteristics is obtained. Secondly, a closed-loop modeling method based on two DNN models is proposed for multi-order inertial closed-loop systems. Firstly, forward and backward step disturbances are added to the control variables, then model 1 is obtained after training the DNN network system with deep learning based on Multiple Inertial filters, and then forward and backward step disturbances are added to the set values, The output data of model 1 and the output data of object are trained as inputs of deep learning random inactivation DNN network, and finally the closed-loop object model is obtained. Finally, the validity of the deep learning model is verified by the simulation experiment data. This method has great practical significance for the design and application of intelligent models.

2 DNN Deep Learning Algorithms

Neural network is based on the extension of perceptron. DNN can be understood as a neural network with many hidden layers. Its multiple hidden layers can simulate extremely complex decision-making functions. DNN can be regarded as a combination of multiple regression models. Given a set of inputs, DNN can provide a more flexible output combination than a single model and can fit any function. There are three kinds of neural network layers in DNN: input layer, hidden layer and output layer. The first layer is input layer, the last layer is output layer, and the number of layers in the middle is hidden layer. Layer to layer is fully connected, that is, any neuron in layer I must be connected with any neuron in layer I + 1. It is a fully connected DNN network, as shown in Fig. 1. Sometimes, in order to improve the generalization ability of the network, some hidden layer neurons are randomly removed during training, and then the DNN network is randomly inactivated, as shown in Fig. 2.

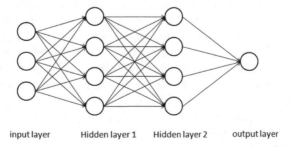

input layer   Hidden layer 1   Hidden layer 2   output layer

Fig. 1. Fully connected DNN deep learning network diagram

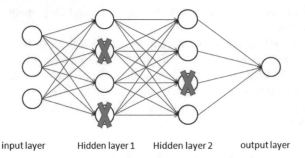

input layer   Hidden layer 1   Hidden layer 2   output layer

Fig. 2. Random deactivated DNN deep learning network graph

In this paper, DNN deep learning network is used to model the characteristics of multi-order inertial objects. In the process of modeling, data standardization is needed, variables are standardized, attributes are scaled between given maximum and minimum values, or absolute values of each attribute are scaled to unit size; ReLU function is chosen as activation function, which can improve the sparsity of the network and make the network have better generalization ability. In order to reduce the complexity of the deep neural network and improve the generalization ability, regularization is used to restrict the network, so as to avoid the network weight becoming very large, which is realized by adding attenuation parameters to punish the larger weight.

3 Modeling of DNN Deep Learning Network

3.1 Open-Loop Modeling Method Based on Multiple Inertial Filters

The modeling method proposed in this paper is based on a deep learning DNN network with multiple inertial filters. An open-loop object identifier based on deep learning is constructed after the first-order and second-order inertial filters with inertial time constants of 30 s, 60 s and 160 s are set up at the input end of the network; When identifying the open-loop object model, step input is added to the input of the object model, and the corresponding data is output. Then input and output data are simultaneously input into the identifier. After in-depth learning of the DNN network through off-line training, the characteristics of the model can be effectively identified; In order to verify the validity of the identifier network, sine wave signals can be added to the actual model and the input of the identifier model respectively, and the accuracy of identification can be verified by comparing the coincidence degree of the actual model output and the output data of the identifier. The structure of DN network based on Multiple Inertial filtering is shown in Fig. 3.

In Fig. 3, G is the transfer function of the model object, A is the input step forward and backward excitation signal data set, B is the output data set of the model, C is the output data set of the deep learning identifier, D is the data set generated by A through the first-order inertia link with the inertia time of 30 s; E is the data set generated by A passing through the first inertial link of 60 s, F is the data set generated by A passing through the second inertial link of 30 s, H is the data set generated by A passing

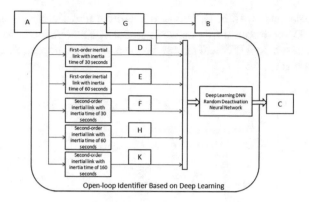

Fig. 3. Open-loop modeling diagram of DN network based on multiple inertial filters

through the second inertial link of 60 s, K is the data set generated by A passing through the second inertial link of 160 s.

The in-depth learning DNN neural network designed in this paper standardizes the input data to a given maximum and minimum. ReLU activation function is selected and the network is set in the form of multi-variable input. For SISO multi-order inertial model objects, the input signals are five sets of data set D, E, F, H and K in Fig. 3, and six sets of data set B in object output are input into the deep learning DNN network at the same time. After training, the identified model can better reflect the characteristics of the actual object model.

In this paper, DNN deep learning network is used to train and model multi-step inertial objects. Firstly, two or three step disturbances are made to the objects, and step input signals and object output signals are used as the modeling data of DNN deep learning network. Secondly, the specific parameters of DNN network, including hidden layer and learning rate, are set up. The input signal of the network is filtered, and the input data and output data of the object after processing are sent to the DNN network for training. Finally, the trained model is validated, the input signal is changed to sine wave signal, and a set of data is input into the object model, At the same time, a set of data is sent to the trained DNN network to verify the fitting degree of the two data.

Taking the second-order inertial system as an example, the object model is as follows:

$$G(s) = \frac{0.6}{(50s+1)*(40s+1)} \tag{1}$$

A series of step perturbations of 1, −1, 1 and −1 are simulated in 0 s, 800 s, 1300 s and 2200 s respectively. A set of input data A and output data of model G are obtained through 3000 steps simulation. Five groups of data, D, E, F, H and K, are obtained after filtering the input data A according to the method of Fig. 3.

Using the method of Fig. 3, the DNN network is set as two hidden layers, the first layer is 6 neurons, the second layer is 14 neurons, and ReLU is chosen as the transfer function. The learning rate is set to 0.02, and the number of iterations is 100 generations. Data sets D, E, F, H, K and B are input into DNN network for training. After training, DNN deep learning network model DW is obtained. Data sets D, E, F, H, K are input into model DW and output data sets C. The comparison curves of C and B are as follows (Fig. 4):

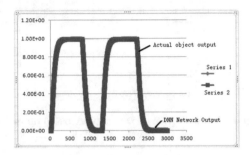

Fig. 4. DNN network model output and actual output contrast curve 1

To verify the validity of the trained deep learning network model DW, the input signal is transformed into sinusoidal wave, The output data B1 is obtained by the object model G (s) in the signal input formula (1), and the output data C1are obtained by the input to the network model DW. The comparison curves of C1 and B1 are shown in Fig. 5.

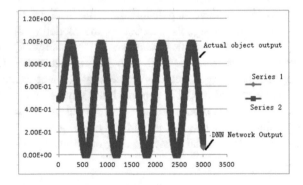

Fig. 5. DNN network model output and actual output contrast curve 2

3.2   Closed-Loop Modeling Method Based on Two DNN Models

Closed-loop modeling is based on open-loop modeling. For multi-order inertial closed-loop systems, a closed-loop modeling method based on two DNN models is proposed. Firstly, the forward and reverse step disturbances are added to the closed-loop control variables. After training, the model 1 is obtained by using the deep learning DNN network system based on Multiple Inertial filters (as shown in Fig. 3). Then, the forward and reverse step disturbances are added to the set values. The output data of model 1 and the output data of the controlled object are trained as the input of the deep learning stochastic inactivation DNN network, Finally, the closed-loop object model is obtained. The detailed schematic diagram is shown in Figs. 6 and 7.

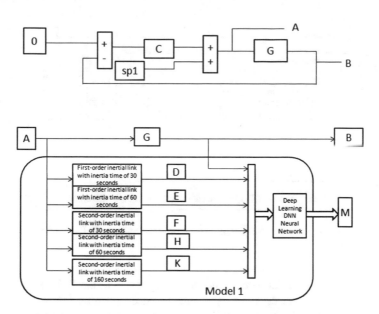

Fig. 6. Principle diagram of DNN model 1 based on internal disturbance of control variables

In Fig. 6, C is the controller, SP1 is the positive and negative step disturbance added to the control quantity, G is the controlled object, A is the input data of G, B is the output data of G, M is the output data of deep learning network. D is the data produced by A passing through the first inertial link of 30 s, E is the data produced by A passing through the first inertial link of 60 s, F is the data produced by A passing through the second inertial link of 30 s, H is the data produced by A passing through the second inertial link of 60 s, K is the data produced by A passing through the second inertial link of 160 s.

In Fig. 7, C is the controller, SP2 is the positive and negative step disturbance added to the set value, and G is the controlled object. A is the input data of G, B is the output data of G, D is the data of A passing through the first inertial link of 30 s, E is the data of A passing through the first inertial link of 60 s, F is the data of A passing through the second inertial link of 30 s, H is the data of A passing through the second inertial link of 60 s and K is the data of A passing through the second inertial link of 60 s. The second-order inertial link with 160 s over-inertia time produces data, J is the output data of model 1 and M is the output data of deep learning network.

The design scheme of DNN model 1 based on internal disturbance control standardizes the input data between the given maximum and minimum values. ReLU activation function is selected and the network is set up as a form of multi-variable input.

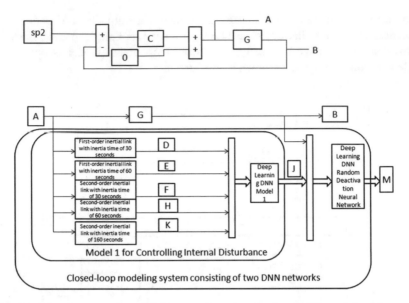

Fig. 7. Principle diagram of establishing random deactivation DNN model with set value external disturbance

For SISO multi-order inertial model objects, the input signals are B, D, E, F, H and K in Fig. 3. At the same time, they are input into the deep learning DNN network. After training, DNN model 1 is obtained. When establishing random inactivated DNN model based on set-point external disturbance, the neurons in each hidden layer are randomly ignored from the network with a 2% probability in the training process. The output data J of model 1 and output data B of controlled object are trained as inputs of random inactivated DNN network, and the closed-loop object model is finally obtained.

Taking the second-order inertial system as an example, the object model is as follows:

$$G(s) = \frac{0.6}{(90s+1) * (90s+1)} \tag{2}$$

A closed-loop control system is designed for the model. The structure is shown in Fig. 8.

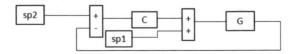

Fig. 8. Structure of closed-loop control system

In Fig. 8, C is the controller, G is the object model, Sp1 is the internal disturbance of the control quantity, SP2 is the set value disturbance, and for model G of formula (2), the controller C adopts PI algorithm, the ratio is 1, and the integral coefficient is 0.01. Firstly, the set value disturbance SP2 is set to 0, and the continuous step disturbance SP1 of the control quantity is added. Step perturbations of 1, −1, 1 and −1 are done in 0, 800, 1600 and 2400 s respectively. A 3000-step simulation is carried out according to the way shown in Fig. 6. A set of input data D, E, F, H, K and a set of output data B of model G (s) are obtained.

Using the method of 3.1, DNN network is set as two hidden layers, the first layer is 6 neurons, the second layer is 14 neurons, ReLU is chosen as the transfer function, the learning rate is set to 0.02, and the number of iterations is 100 generations. Data D, E, F, H, K and B are input into DNN network for training, and DNN network model 1 is obtained after training. Data D, E, F, H, K are input into DNN model 1 and output data M. The comparison curves of M and B are as follows (Fig. 9):

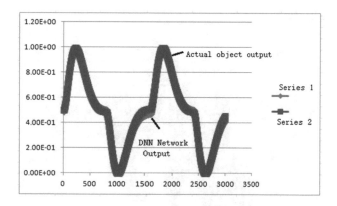

Fig. 9. DNN model 1 output and actual output contrast curve

Secondly, the set-point disturbance Sp1 is set to 0, and the set-point continuous step disturbance SP2 is added. The step disturbances of 1, −1, 1 and −1 are made in 0 s, 800 s, 1600 s and 2400 s respectively. In the way shown in Fig. 7, 3000 steps of simulation are carried out to obtain a set of input data D, E, F, H, K, a set of output data B of model G (s), and a set of output data J of DNN network model 1.

When establishing the random inactivation DNN model shown in Fig. 7, the input data are B and J. During the training process, the neurons in each hidden layer are randomly ignored from the network with a 2% probability. Others are the same as the design of DNN model 1. After the training, the closed-loop object model DW based on two DNN networks is finally obtained. Data D, E, F, H, K are input into model DW and output data set M. The comparison curves of M and B are as follows (Fig. 10):

To verify the validity of the trained in-depth learning network model DW, the set value SP2 signal is transformed into sinusoidal wave. The set value Sp1 is set to 0, the output data B1 is obtained by the object model G (s), and the output data M1 is

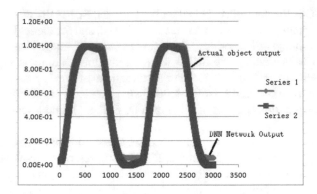

Fig. 10. Contrast curve between output and actual output of two DNN networks

obtained by the input to the network model DW. The comparison curves of M1 and B1 are shown in Fig. 11.

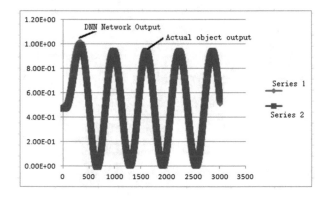

Fig. 11. DNN network model verification output and actual output contrast curve 3

To further verify the accuracy of the deep learning network model DW, open-loop perturbation with step 1 is applied to the model shown in formula (2). The perturbation signal is input to model G to obtain the output data B2, and the perturbation signal is input to DNN model DW to obtain the output data M2. The comparison curves of M2 and B2 are shown in Fig. 12.

As can be seen from Fig. 12, DNN network can well reflect the characteristics of object model and is suitable for closed-loop system modeling.

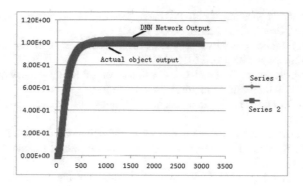

Fig. 12. DNN network model verification output and actual output contrast curve 4

4 Conclusion

In this paper, a deep learning based modeling method is studied. An open-loop modeling method based on Multiple Inertial filters and a closed-loop modeling method based on two DNN networks are proposed. In open-loop modeling, only forward and backward step perturbations are needed, and the data are sent to the deep learning network for training. In closed-loop modeling, we need to first add forward and reverse step disturbances to the control variables, train DNN model 1 through data, then add forward and reverse step disturbances to the set values, and train the second DNN model through the output of DNN model 1 and the output data of the object. The two DNN models constitute a closed-loop modeling model. The validity of the deep learning model is verified by the simulation experiment data. This method has great practical significance for the design and application of intelligent model.

References

1. Hornik, K., Stichcombe, M., White, H.: Multilayer feedforward networks are universal approximators. Neural Netw. **2**, 359–366 (1989)
2. Keshmiri, S., et al.: Application of deep neural network in estimation of the weld bead parameters. arXiv preprint arXiv:1502.04187 (2015)
3. Shaw, A.M., Doyle, F.J., Schwaber, J.S.: A dynamic neural network approach to nonlinear process modeling. Comput. Chem. Eng. **21**(4), 371–385 (1997)
4. Freitas, F.D., De Souza, A.F., de Almeida, A.R.: Prediction-based portfolio optimization model using neural networks. Neurocomputing **72**(10), 2155–2170 (2009)
5. Rogers, S.K., et al.: Neural networks for automatic target recognition. Neural Netw. **8**(7), 1153–1184 (1995)
6. Zhu, B.: Thermal Automation System Test of Thermal Power Plant. China Electric Power Press (2006)
7. Xi, Y.: Predictive Control. National Defense Industry Press (1993)
8. Wang, W.: A new direct method of generalized predictive adaptive control. J. Autom. **22**(3) (1996)

9. Wang, G., et al.: Application of PFC-PID cascade control strategy in main steam temperature control system. Proc. CSEE **22**(12), 50–55 (2002)
10. Li, Q., et al.: Model predictive control of denitrification system for supercritical units. Zhejiang Electr. Power **35**(11), 34–36 (2016)
11. Jin, X., Wang, S., Rong, G.: Predictive functional control (PFC) – a new predictive control strategy. Chem. Autom. Instrum. **26**(2), 74–80 (1999)
12. Schmidhuber, J.: Deep learning in neural in neural networks: an overview. Neural Netw. **61**, 85–117 (2015)
13. Beale, R., Jackson, T.: Neural Computing-An Introduction. CRC Press, Boca Raton (1990)
14. Berniker, M., Kording, K.P.: Deep networks for motor control functions. Front. Comput. Neurosci. **9**, 32 (2015)
15. Moed, M.C., Saridis, G.N.: A Boltzmann machine for the organization of intelligent machines. IEEE Trans. Syst. Man Cybern. **20**(5), 1094–1102 (1990)

Performance Assessment of Smart Electricity Meters Based on Max Margin Bayesian Classifiers

Haibo Yu[1], Helong Li[1], Yungang Zhu[2(✉)], and Yang Wang[3]

[1] China Electric Power Research Institute, Beijing, China
[2] College of Computer Science and Technology, Jilin University, Changchun, China
zhuyungang@jlu.edu.cn
[3] China National Accreditation Service for Conformity Assessment, Beijing, China

Abstract. There are large amount of quality and monitoring data of electricity meters, it is crucial and valuable to assess the operating performance of the smart electricity meters automatically with this data. In this paper, we propose an intelligent operating performance assessment method for smart electric meters based on selective ensemble of max margin Bayesian classifiers. The genetic algorithm is firstly used to select most relevant attributes for assessment, and max margin Bayesian classifiers is utilized to make the assessment. We use the bagging and clustering to ensemble multiple classifiers to obtain better results. The experimental results illustrate the efficiency and effectiveness of the proposed method.

Keywords: Ensemble learning · Smart electricity meter · Performance assessment

1 Introduction

In recent years, large amount of quality data and monitoring data collected from electricity meters has not been fully exploited and utilized [1, 2]. It is valuable to use aforementioned data to assess the operating performance of the electricity meter, so as to treat the meters with different quality in different way. The meters with poor performance can be replaced in advance, and the replacement of meters with high performance can be delayed.

To address above issue, in this paper, we propose an intelligent operating performance assessment method for smart electric meters based on selective ensemble of max margin Bayesian classifiers. We present the feature selection, max margin Bayesian classifier training, and clustering based classifiers ensemble.

The remainder of this paper is organized as follows: Sect. 2 presents the method we proposed. Section 3 presents the experiments to verity the proposed method. Finally, conclusions and recommendations for future work are summarized in Sect. 4.

© Springer Nature Switzerland AG 2020
F. Xhafa et al. (Eds.): IISA 2019, AISC 1084, pp. 89–95, 2020.
https://doi.org/10.1007/978-3-030-34387-3_11

2 Methodology

2.1 Feature Selection Based on Genetic Algorithm

There are lots of attribute indicators related to the status assessment of smart electric meter. For this reason, based on the pre-acquired history data of smart electric meters, the genetic algorithm is used to select the set of n attributes $X = \{X_1, \ldots X_i, \ldots X_n\}$ which is most relevant to the performance assessment of the smart meter. The optimal attribute subset is represented as a binary string composed of 0 and 1, 0 means that the corresponding attribute is not selected, and 1 means that the corresponding attribute is selected. For example, 10010100110 indicates that X_1, X_4, X_6, X_9, and X_{10} are attributes most relevant to the performance assessment of the smart meter.

Firstly, the pre-acquired history data set of smart meters is divided into two parts: training data and test data. λ binary individuals are randomly generated as the initial population Q_k, and the initial value k is 0.

For each binary string individual S_i in the initial population Q_k, the corresponding subset of the indicator attributes is $X_i = \{X_{i1} \ldots X_{ij} \ldots X_{ij_0}\}$, a naive Bayes classifier N_i is trained on the subset of attributes from the training data, where $1 \leq i \leq \lambda$, $1 \leq j \leq j_0$, $1 \leq j_0 \leq m$. Using the test data to verify Bayes classifier N_i. Estimating the status $\hat{y}$ for the smart meter based on the formula (1).

$$\hat{y} = \arg \max_y P(y) \prod_{k=i1}^{ij_0} P(x_k|y) \tag{1}$$

where y is the performance status. According to the operating value result value of the smart electric meter determined by the λ binary string individual in the initial population Q_k, and the smart meters in the test data. The actual operating state result value of the table calculates the test accuracy of the Bayesian classifier N_i for the test data, and the accuracy is taken as the fitness $F[S_i]$ of the individual S_i.

According to the value of the fitness $F[S_i]$, in each iteration the roulette method is used to select one individual from the current population $\{S_1, S_2 \ldots S_i \ldots S_\lambda\}$, a total of λ repeatable individuals are selected as the parent population. Then the individuals in the parent group are performed by crossover operation, where each crossover operation is random. It is assumed that the gene position of both population is 1 is the dominant gene position, and only one parent is 1 is the non-dominant gene position, and the dominant gene position will be all retained, and the non-dominant gene will be retained. The child-individual will randomly select the gene retaining 1. Then the mutation operation is performed on the individual, changing some gene values in the individual binary string to form a new individual, using the binary string individual after the mutation operation as the new population Q_k, and let $k = k + 1$.

Repeat above procedure until the difference between the fitness value of the individual with the highest fitness value and the lowest fitness value is less than a threshold or the number of iterations k reaches a certain threshold. At this time, the individual with the highest fitness value in the population Q_k corresponds to the subset of attributes most relevant to the assessment of performance of the smart meter.

2.2 Learning Max-Margin Bayesian Classifiers Based on Bagging Ensemble

It is assumed that there are n indicators (attributes) for assessment of the smart meter, which are expressed as $X_1...X_n$. In this paper, the max margin Bayesian classifier (MMBC) [3] is adopted, and it takes advantage of both the Bayesian classifier and the maximum margin classifier, and has high precision. The max margin Bayesian classifier B_t we use contains $n + 1$ variables, where $X_1...X_n$ represent n indicators for evaluating the operating state of the smart meter, and X_{n+1} represents the operating state of the meter, there are k possible values, such as $k = 2$ (Qualified, Unqualified), $k = 4$ (Excellent, Good, Medium, Poor).

Assuming that the data set contains m data samples, in the max margin Bayesian classifier, the classification interval [4] of the i-th data sample m can be expressed as formula [3]:

$$d^i = \frac{P(x^i_{n+1}|x^i_1, \ldots, x^i_n)}{\max_{c \neq x^i_{n+1}} P(c|x^i_1, \ldots, x^i_n)} = \frac{P(x^i_1, \ldots, x^i_{n+1})}{\max_{c \neq x^i_{n+1}} P(x^i_1, \ldots, x^i_n, c)} \tag{2}$$

where x^i_k represents the value of the variable x_k in the i-th sample, and c is the value of all values of x_{n+1} except x^i_{n+1}. When d > 1, the data sample i is correctly classified, and the max operator is replaced by the softmax function [5] $\max_x f(x) \approx \ln\left(\sum_x e^{\eta f(x)}\right)^{\frac{1}{\eta}}$, and make the alogarithm [3]:

$$\ln d^i = \ln P(x^i_1, \ldots, x^i_{n+1}) - \frac{1}{\eta} \ln \sum_{c \neq x^i_{n+1}} \left(P(x^j_1, \ldots, x^j_n, c)\right)^{\eta} \tag{2}$$

In order to make the objective function differentiable and facilitate the parameter learning of the Bayesian classifier, and the data set is divided into three parts based on the smoothing function, D_1 contains $\lambda \ln d \leq 1 - 2\kappa$ samples, and D_2 contains $1 - 2\kappa < \lambda \ln d < 1$ samples, D_3 contains $\lambda \ln d \geq 1$ samples. $\lambda > 0$ and $0.01 \leq \kappa \leq 0.5$.

The scoring function [3] is defined as,

$$Score(B, D) = \sum_{i \in D_1} (\lambda \ln d^i + K) + \sum_{i \in D_2} \left(1 - \frac{(\lambda \ln d^i - 1)^2}{4K}\right) + |D_3| \tag{3}$$

The machine learning process of the max margin Bayesian classifier is to train the classifier that maximizes the value of the aforementioned scoring function from the data. In this paper we do not train a single max margin Bayesian classifier, but to train a set of classifiers with the ideas of ensemble learning. The Bagging strategy is used to train and generate a set F containing T max margin Bayesian classifiers from the data of the smart meter. The specific steps are as follows:

For $i = 1, 2..., T$:

Step 1: Perform the i-th random sampling on the historical data of the smart meters, and collect the m groups of data samples for each sampling to obtain a data set D_t containing m samples.

Step 2: Training the i-th max margin Bayesian classifier B_t with the sampling set D_t, B_t includes $n + 1$ variables, where $X_1...X_n$ indicate the attribute of the smart electric meter, X_{n+1} indicates the status of the electric meter, and the training method is described aforementioned.

Step 3: Determine the set F according to the formula $F = F \cup \{B_t\}$, and if $t < T$, let $t = t + 1$, go to step 1, where the initial value of the set F is empty set.

2.3 Selecting Max Margin Bayesian Classifiers Subset Based on Clustering Algorithm

We select K max margin Bayesian classifiers from the generated T max margin Bayesian classifiers to form a new set ϕ. This strategy can obtain better results than using single classifiers [6]. The steps can be described as follows:

Step 1: Construct a $T \times T$ similarity matrix S, $S(i,j)$ represents the similarity of max margin Bayesian classifier B_i and B_j, assuming the assessment result of B_i and B_j for the sample set are y_k and y'_k, the formula for calculating the similarity of B_i and B_j is:

$$S(i,j) = -\sqrt{\frac{1}{m}\sum_{k=1}^{m}(y'_k - y_k)^2} \tag{4}$$

where m represents the number of samples.

Step 2: randomly select K max margin Bayesian classifiers from set F as the initial centroid max margin Bayesian classifiers B_p^l, where $1 \le l \le L$, $1 \le p \le K$, L is the maximum number of iteration.

Step 3: For each max margin Bayesian classifier in the set F, according to the similarity matrix S, the similarity of B_i and the centroid is investigated. Assigning B_i to the cluster which has the centroid with the highest similarity to B_i, where the initial cluster is empty.

Step 4: For each cluster, recalculate the centroid to make the centroid max margin Bayesian classifier is most similar to the other max margin Bayesian classifier in the cluster, and let $l = l + 1$ to generate a new centroid B_p^l.

Step 5: When the clusters do not change or reach the maximum number L of iterations in two consecutive iterations, go to Step 6, otherwise go to Step 3.

Step 6: Select the max margin Bayesian classifier with the highest precision from each cluster to obtain a new max margin Bayesian classifier set $\phi = \{\phi_1, \phi_2, \ldots, \varphi_K\}$.

Then the traditional inference method is used to obtain each assessment performance results of smart meters of each max margin Bayesian classifier in ϕ. And the K results are combined to generate final result. The frame of the whole procedure is shown as Fig. 1.

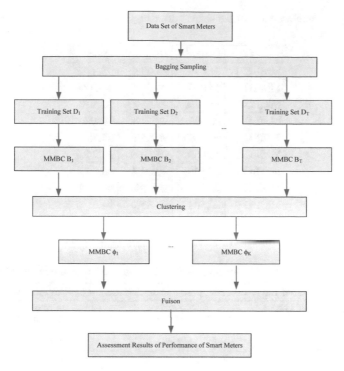

Fig. 1. The procedure of the proposed method

3   Experimental Results and Analysis

The experimental data is collected from the provincial branches of the State Grid, and there are three sets of data. "Dataset One" contains 748 smart electric meter samples, "Data Set 2" contains 645 samples, and "Data Set 3" contains 5000 samples. Select the pre-operation verification error (rated load point error, minimum load point error, maximum load point error), operation error (measurement error one, measurement error two, measurement error three), running time, operating environment, family defect, online monitoring exception and online monitoring clock exception. A total of 11 attributes are used as the indicators to assess the performance of smart electric meter.

The experiment is implemented in Matlab 2014 under 64-bit Win7 system. We first test the performance without ensemble strategy. Ten max margin Bayesian classifiers are trained from the three data sets with ten-fold cross-validation was tested. Table 1 shows the assessment accuracy for the three data sets.

Table 2 shows the results of naive Bayes classifier, support vector machine and the proposed method, it can be observed that the proposed method can achieve higher accuracy.

Figure 2 shows the comparison of the value of k, i.e., number of clusters. It can be observed that for this experimental data, 4 is the ideal value of k.

Table 1. Assessment accuracy of the single max margin Bayesian classifier without ensemble on three data sets

	Dataset 1	Dataset 3	Dataset 3
MMBC 1	81.67%	95.09%	95.87%
MMBC 2	85.00%	96.05%	97.50%
MMBC 3	81.67%	96.05%	97.50%
MMBC 4	90.00%	93.19%	96.1%
MMBC 5	80.00%	96.05%	95.35%
MMBC 6	80.00%	95.10%	94.75%
MMBC 7	87.50%	90.29%	93.50%
MMBC 8	83.33%	96.05%	94.12%
MMBC 9	82.50%	90.33%	94.61%
MMBC 10	80.83%	94.14%	94.75%

Table 2. Comparison of the proposed method with other methods on accuracy

	Naive Bayes classifier	Support vector machines	Proposed method
Dataset 1	80.31%	89.55%	94.67%
Dataset 2	92.13%	95.85%	97.55%
Dataset 3	91.75%	96.39%	98.6%

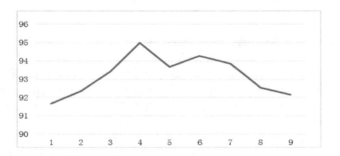

Fig. 2. Accuracy for different values of number of clusters

4   Conclusions and Future Works

In this paper, an operating performance assessment method for smart electric meters is proposed based on selective ensemble of max margin Bayesian classifiers. We present the feature selection, max margin Bayesian classifier training, and clustering based classifiers ensemble. The experimental results illustrate that the accuracy of the proposed method is higher than the strategy without ensemble and other state of the art methods. Therefore, the proposed method can accurately and efficiently assess the performance of smart electric meters.

In future, we plan to make the proposed method adapt to streaming and non-stationary data [7, 8], and convex evidence theory [9], etc.

References

1. Zhu, Y., Qing, X., Liu, J., Tian, Z., Zhou, C.: Data mining application in smart electric meter fault analysis. Jiangsu Electr. Eng. **35**(05), 19–23 (2016)
2. Xu, D., Luan, W., Wang, P., Zhang, Z.: Analysis and applications of smart electric meter data. Distrib. Utilization **32**(8), 25–30 (2015)
3. Pernkopf, F., Wohlmayr, M., Tschiatschek, S.: Maximum margin Bayesian network classifiers. IEEE Trans. Pattern Anal. Mach. Intell. **34**(3), 521–535 (2012)
4. Guo, Y., Wilkinson, D., Schuurmans, D.: Maximum margin Bayesian networks. In: Proceedings International Conference on Uncertainty in Artificial Intelligence, pp. 233–242 (2005)
5. Sha, F., Saul, L.: Comparison of large margin training to other discriminative methods for phonetic recognition by hidden Markov models. In: Proceedings IEEE International Conference on Acoustics, Speech and Signal Processing, pp. 313–316 (2007)
6. Zhou, Z.H., Wu, J., Tang, W.: Ensembling neural networks: many could be better than all. Artif. Intell. **137**(1–2), 239–263 (2002)
7. Zhu, Y., Liu, D., Chen, G., Jia, H., Yu, H.: Mathematical modeling for active and dynamic diagnosis of crop diseases based on Bayesian networks and incremental learning. Math. Comput. Model. **58**(3–4), 514–523 (2013)
8. Zhu, Y., Liu, D., Li, Y., Wang, X.: Selective and incremental fusion for fuzzy and uncertain data based on probabilistic graphical model. J. Intell. Fuzzy Syst. **29**(6), 2397–2403 (2015)
9. Liu, D., Zhu, Y., Ni, N., Liu, J.: Ordered proposition fusion based on consistency and uncertainty measurements. Sci. China Inf. Sci. **60**(8), 1–19 (2017). 082103

Operating Performance Assessment of Smart Meters Based on Bayesian Networks and Convex Evidence Theory

Haibo Yu[1], Helong Li[1], Zehao Zheng[2], and Yungang Zhu[2(✉)]

[1] China Electric Power Research Institute, Beijing, China
[2] College of Computer Science and Technology, Jilin University, Changchun, China
zhuyungang@jlu.edu.cn

Abstract. Smart electricity meters have been widely installed in recent years. In order to automatically assess the operating performance of smart meters, in this paper we propose a method based on Bayesian network. Bayesian network is adopted to represent the casual relationship among attributes and operating performance of smart meter. Multiple Bayesian networks are trained from data with genetic algorithm and bagging sampling, then a subset of Bayesian networks are selected for assessment. The evidence theory is used to fuse the multiple results from Bayesian networks and generate final assessment results for smart meters. The experimental results show the effectiveness and efficiency of the proposed method.

Keywords: Smart electricity meters · Bayesian network · Convex evidence theory

1 Introduction

Nowadays, smart electricity meters are gradually entering every household to achieve complete coverage of power information collection. It is crucial in the construction of smart grid [1]. If artificial intelligence can be utilized to automatically assess the operating performance of smart meters and obtain the information of the smart meter in time, it can greatly facilitate the safe operation, status monitoring, data reading of the power grid. Moreover, It will effectively save the cost of recycling and checking for the smart meters, and provide technical support for replacement of the smart meter, which has important significance [2, 3].

In this paper, we propose a method for assessing the operating performance of smart electric meters. In this method, we utilize Bayesian network to represent the dependent relationship among attributes and operating performance of smart meter. We train multiple Bayesian networks with genetic algorithm and bagging sampling, then select a subset of Bayesian networks for assessment, at last we use convex evidence theory to fuse the multiple results from Bayesian networks and generate final assessment results for smart meters.

© Springer Nature Switzerland AG 2020
F. Xhafa et al. (Eds.): IISA 2019, AISC 1084, pp. 96–102, 2020.
https://doi.org/10.1007/978-3-030-34387-3_12

The remainder of this paper is organized as follows: Sect. 2 presents the method we proposed. Section 3 presents the experiments to verity the proposed method. Finally, conclusions are summarized in Sect. 4.

2 The Proposed Methodology

2.1 Discretization

Since the Bayesian network mainly deals with discrete variables, and the evaluation results of the electric meter are expressed in discrete values, this paper first discretizes the continuous variables of the smart electric meter, based on the idea of [4], the method is describes as follows:

For the index X with continuous values, when there are k non-repeating continuous attribute values x_t in the historical data, calculate the probability $P(x_i)$ of each non-repeating continuous attribute value x_i of the index x, where $1 \leq i \leq k$, the initial value of k is k_0. Calculating the information entropy $H(k) = - \sum_{x_i \in x} P(x_i) \log P(x_i)$ of the index x based on the probability $P(x_i)$. The non-repeating continuous attribute values of the index x are divided into k intervals, wherein each interval corresponds to a non-repeating value of the index x, and the dividing points are recorded.

Selecting two adjacent interval combinations arbitrarily and calculating the probability of occurrence of each of the merged $k - 1$ intervals, and then calculating the information entropy $H(k - 1)$ of the index x according to the probability of each interval, minimizing the difference in entropy between $H(k) - H(k - 1)$ before and after the merger. If two or more sets of adjacent intervals satisfy the condition at the same time, a group is randomly selected; $S_j = (k_0 - 1) \times H(k - 1) - (k - 2) \times H(k_0)$ is calculated according to the information entropy $H(k - 1)$ of the combined interval, where the initial value of j is 1; Each interval corresponds to a discrete value of the index x when $S_j \leq S_{j-1}$. Otherwise let $k = k - 1, j = j + 1$, repeat the above operation.

2.2 Training of Bayesian Networks from Historical Data of Smart Electricity Meters

Bayesian network is an important model to deal with the uncertainty problem in the field of artificial intelligence [5]. The set of indicators and attributes for evaluating the operation state of smart electricity meters after discretization, including basic error, error consistency, operation error, operation time, operation failure rate, full inspection return rate, monitoring anomaly, performance degradation, installation environment, user reputation and family defect, are represented as variable set $X = \{X_1, X_2, \ldots X_n\}$ ($n = 11$). Based on the historical evaluation data of electricity meters' operation state collected in advance and combined with the idea of integrated learning, the Bayesian network representing causality between electricity meter attributes and performance is generated through training by genetic algorithm.

A random sampling is adopted to train and generate a set F of T Bayesian networks from the data. The method is detailed as follows:

Step 1: Carry out the t-th random sampling on the data set, collect m times in total, and get the $D_t = \{C_1C_2...C_m\}$ of the sampling set containing m samples. m is the data quantity, C_i is the data of i group, each group of data is a vector with the length of $n + 1$, $\{X_1...X_i...X_nX_{n+1}\}$ $n = 11$, which is composed of 11 indicator attributes $X = \{X_1...X_i...X_n\}$ for assessing the operating status of the smart electric meter (corresponding to basic error, error consistency, running error, running time, operating failure rate, full inspection return rate, monitoring abnormality, performance degradation, installation environment, user reputation, family defect) and operating status of the smart electric meter X_{n+1}; where the initial value of t is 1, $1 \leq t \leq T$;

Step 2: Train the t-th Bayesian network with the sample set D_t. The Bayesian network contains $n + 1$ variables, where $X_1...X_n$ represent the attribute data of the electric meter as described above, and X_{n+1} represents the state of the electric meter. There are d kinds of possible values, such as $d = 3$ (stability, attention, alert). The method of training Bayesian network is as follows [6]:

Several Bayesian networks are randomly generated based on the training data set $D_t = \{C_1C_2...C_m\}$, as the initial population Pop_t of the optimal Bayesian network which is determined by genetic algorithm, where the initial value of t is 0, and the set number of genetic algorithm iterations is t_0, each The Bayesian network serves as an individual in the initial population Q_t.

Calculate the fitness of each individual S_t^j in the population Pop_t by following formula (1):

$$Score(S_t^j) = F[S_t^j] = \sum_i \left(\sum_j \sum_k N_{ijk} \log \frac{N_{ijk}}{N_{ij}} - \frac{\log m}{2} |\pi_i|(|X_i| - 1) \right) \quad (1)$$

where N_{ijk} is the number of cases in D in which $X_i = x_i^k$ and $\pi_i = \pi_i^j$, $N_{ij} = \sum_{k=1}^{r_i} N_{ijk}$; N'_{ijk} is the number of cases in D' in which $X_i = x_i^k$ and $\pi_i = \pi_i^j$, $N'_{ij} = \sum_{k=1}^{r_i} N'_{ijk}$.

The individual with the largest fitness function $F[S_t^j]$ in the initial population is selected as the optimal individual of the initial population Pop_t, and the Bayesian network S_{tmax}^j is optimal.

Performing a selection operation on the initial population Pop_t, selecting a plurality of Bayesian networks to form a parent group in descending order of the fitness function $F[S_t^j]$; performing a crossover operation on the parent group, Exchange a local substructure of each node and its parent node in any two Bayesian networks of the parent population; Perform mutation operations on cross-processed groups, and the mutation operation includes adding a directed edge to any two nodes in the Bayesian network, delete a directed edge to any two nodes, reverse the direction of the directed edge of any two nodes, and generate the intermediate group Pop_{son}. The fitness is calculated for each individual S_t^j in Pop_t by the same method as calculating the fitness of each individual S_{son}^j of the initial population Pop_{son}; a plurality of individuals are selected from Pop_t and Pop_{son} in order of fitness to form a new generation group Pop_t, let $t = t + 1$.

The above procedures are repeated until $1 \leq t \leq t_0$ and $F\left[S_{tmax}^{j}\right] \leq F\left[S_{(t-1)max}^{j}\right]$, then regarding $S_{(t-1)max}^{j}$ as the optimal Bayesian network for assessment for smart electric meter; or if $t = t_0$ and $F\left[S_{tmax}^{j}\right] > F\left[S_{(t-1)max}^{j}\right]$, then S_{tmax}^{j} is used as the optimal Bayesian network for assessment.

Step 3: Add the Bayesian network B_t generated from Step 2 to the set F, $F \leftarrow F \cup \{B_t\}$, $t = t + 1$.

2.3 Using the Convex Evidence Theory to Integrate the Assessment Results of Bayesian Networks

We use k-means clustering algorithm to make the T Bayesian networks into K clusters, and select the Bayesian network with highest accuracy from each cluster, so as to generate K Bayesian networks. The convex evidence theory [8] is used to integrate the assessment results of K Bayesian networks, and then obtain the final assessment result. Assume that $X_1, X_2,...,X_n$ represent the attributes of smart electric meter, and Y represents operating performance. The integration method is described as below:

Step 1: For the Bayesian network ϕ_1 in the set ϕ, Suppose $x_1, x_2,...,x_n$ represent the values of n attributes for assessing the operating status of the smart electric meter, and Y_l represents the lth of the d operating status of the smart electric meter. The formula [7] for calculating the probability of the operating state Y_l of the smart electric meter is as follows:

$$\lambda_1(s_l) = P(Y_l|x_1,\ldots,x_n) = P(Y_l|\pi_Y) \cdot \prod_{X_i \in Children(Y)} P(X_i|\pi_i) \tag{10}$$

where $1 \leq l \leq d$, $1 \leq i \leq n + 1$, $X_i \in Children(Y)$ indicates X_i is a child variable of Y in the Bayesian network, and π_Y is the parent node of Y;

Step 2: For the Bayesian network ϕ_j in the set ϕ, calculating the probability $\lambda_j(s_l)$ by formula (11):

$$\lambda_j(s_l) = P(Y_l|x_1,\ldots,x_n) = P(Y_l|\pi_Y) \cdot \prod_{X_i \in Children(Y)} P(X_i|\pi_i) \tag{11}$$

where $2 \leq j \leq K$, the initial value of j is 2;

Step 3: Combine $\lambda_1(s_j)$ and $\lambda_j(s_l)$ by formula (12) [8] to generate $\lambda_1(s_l)$:

$$\lambda_1(s_l) = \begin{cases} \sum\limits_{1 \leq k \leq l}[\lambda_1(s_k) + \lambda_j(s_k)]/(g-k+1), & \text{if } l < g \\ \sum\limits_{1 \leq k \leq g}[\lambda_1(s_k) + \lambda_j(s_k)]/(g-k+1) + \sum\limits_{g+1 \leq k \leq d}[\lambda_1(s_k) + \lambda_j(s_k)]/(k-g+1), & \text{if } l = g \\ \sum\limits_{l \leq k \leq d}[\lambda_1(s_k) + \lambda_j(s_k)]/(k-g+1), & \text{if } l > g \end{cases}$$

$$\tag{12}$$

where $g = \begin{cases} \lceil gd \rceil & \text{if } gd - \lfloor gd \rfloor \geq \Delta_2 \\ \lfloor gd \rfloor & \text{if } gd - \lfloor gd \rfloor \leq \Delta_1 \end{cases}$, $gd = \sum_{1 \leq l \leq d} \{\lambda_1(s_l) + \lambda_2(s_l)\} \times l$, Δ_1

and Δ_2 are preset constants;

Step 4: If $j < K$, let $j = j + 1$, and go to Step 2, if $j = K$, go to Step 5;

Step 5: The resulting $\lambda_1(s_l)$ is the probability of each operating status of smart electric meter.

3 Experimental Results and Analysis

The experimental data is obtained from the branches of State Grid, corresponding to data sets 1–3. The operating performance status is {Stable, Attention, Alert}. Table 1 shows the comparison between the assessment results from the proposed method and the actual results. Figure 1 shows the result comparison of the proposed method with the actual results in a graphical way. Due to space limitations, we only select 30% of the samples in each data set for display. In the figure, * indicates the actual result, and + indicates the assessment result from the proposed method. For each sample, if + and * coincide, it means consistency, and if + is above *, it means strict-judgment, if + is below *, it means misjudgment. Table 2 shows the consistency, strict-judgment, misjudgment rate of the proposed method on three datasets.

Table 1. Comparison between the results from the proposed method and actual results on Dataset 1

Performance status	Stable			Attention			Alert		
Proposed method	14763			7192			44		
Actual results	Stable	Attention	Alert	Stable	Attention	Alert	Stable	Attention	Alert
	14763	0	0	36	7156	0	0	27	17

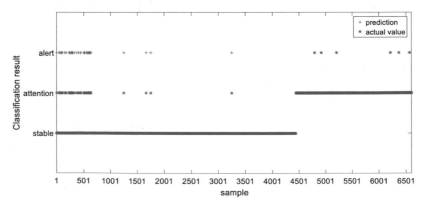

Fig. 1. Comparison between the results from the proposed method and actual results on Dataset 1

Table 2. Consistency, strict-judgment, misjudgment rate of the proposed method on Dataset 1

Status	Stable	Attention	Alert	Overall
Consistency	100.00%	99.50%	38.64%	99.71%
Strict-judgment	0.00%	0.50%	61.36%	0.29%
Misjudgment	0%	0%	0%	0%

It can be observed from the experimental results that, the proposed method can effectively assess the operating performance of the smart electric meter. Although a few samples are strictly judged, the misjudgment rate is zero.

4   Conclusions and Future Works

In this paper, we propose a method for assessing the operating performance of smart electric meters. In this method, we utilize Bayesian network to represent the dependent relationship among attributes and operating performance of smart meter. We train multiple Bayesian networks with genetic algorithm and bagging sampling, then select a subset of Bayesian networks for assessment, at last we use convex evidence theory to fuse the multiple results from Bayesian networks and generate final assessment results for smart meters. The experimental results show the effectiveness and efficiency of the proposed method. In future, we will use novel convex evidence theory [9] and handle the situation of incremental data [10].

Acknowledgments. This paper is supported by the State Grid Corporation of Science and Technology Project "The Research and Application of Smart Meter Operation Status Evaluation Technology Based on Multi-source Data Fusion (Project No. JL71-16-006)".

References

1. Ju, H., Yuan, R., Ding, H., Tian, H., Zhong, W., Pang, F., Xu, S., Li, S.: Study on the whole life cycle quality assessment method of smart meter. Electr. Meas. Instrum. **52**(S1), 55–58 (2015)
2. Zhou, F., Cheng, Y., Xiao, W., Jin, Z.: Method and application of electric meter status assessment based on integrated security domain. J. Autom. Instrum. **07**, 29–33 (2016)
3. Chang, Q., Yan, X., Tao, X., Fu, F.: Research on operating status analysis system for smart meters based on big data technology. Autom. Instrum. (12), 4–6 (2015)
4. He, Y., Zheng, J.-J., Zhu, L.: An entropy-based algorithm for discretization of continuous variables. Comput. Appl. **25**(3), 637–639 (2005)
5. Zhu, Y., Liu, D., Jia, H., Trinugroho, D.: Incremental learning of Bayesian networks based on chaotic dual-population evolution strategies and its application to nanoelectronics. J. Nanoelectronics Optoelectron. **7**(2), 113–118 (2012)
6. Liu, D., Wang, F., Yinan, L., Xue, W., Wang, S.: Structure learning of Bayesian network based on genetic algorithm. Comput. Res. Dev. **38**(8), 916–922 (2001)

7. Zhu, Y., Liu, D., Chen, G., Jia, H., Yu, H.: Mathematical modeling for active and dynamic diagnosis of crop diseases based on Bayesian networks and incremental learning. Math. Comput. Model. **58**(3–4), 514–523 (2013)
8. Liu, D., Yang, B., Zhu, Y., Sun, C.: Fundamental Theories and Methods for Processing Uncertain Knowledge. Science Press (2016)
9. Zhu, Y., Liu, D., Li, Y., Wang, X.: Selective and incremental fusion for fuzzy and uncertain data based on probabilistic graphical model. J. Intell. Fuzzy Syst. **29**(6), 2397–2403 (2015)
10. Liu, D., Zhu, Y., Ni, N., Liu, J.: Ordered proposition fusion based on consistency and uncertainty measurements. Sci. China Inf. Sci. **60**(8), 1–19 (2017). 082103

Research on Object-Oriented Design Defect Detection Method Based on Machine Learning

Yiming Wang[1], Tie Feng[2(✉)], Yanfang Cheng[1], and Haiyan Che[2]

[1] Department of Software, Jilin University, Changchun 130012, Jilin, China
[2] Department of Computer Science and Technology, Jilin University, Changchun 130012, Jilin, China
fengtie@jlu.edu.cn

Abstract. Design defects are one of the main reasons for the decline of software design quality. Effective detection of design defects plays an important role in improving software maintainability and scalability. On the basis of defining software design defects, according to C&K design metrics and heuristics, this paper extracts the relevant features of design defects. Based on classical machine learning methods, classifiers are trained for design defect, and candidate designs are classified by classifiers, so as to identify whether there is a design defect in the design. Experiments show that the method has high accuracy and recall rate in identifying design defects.

Keywords: Design defect detection · Object-oriented metrics · Feature extraction · Machine learning · Classifier

1 Introduction

In software development, due to various technical and non-technical reasons, there will always be more or less design defects to be solved [1–3]. The performance of design defects in software quality characteristics can be summarized as follows: (1) inconsistency between clearly defined functions and user requirements; (2) inconsistency between implicit requirements of user expectations; (3) inconsistency between development standards and design standards specified explicitly. The design defects mentioned in this paper are: Design defects are occurring errors in the design of the software that came from the absence of design patterns or the bad implementation of designing patterns [4].

Software design defects will reduce the comprehensibility, maintainability and scalability of the software. In order to solve this problem, a design defect detection method based on machine learning is proposed in this paper. This paper mainly completes the following tasks: (1) Definition of software design defect features in many design experiences and C&K metrics; (2) Application of five typical supervised machine learning algorithms to train the original data sets. (3) Collect representative design defect source codes, extract their features, and use the extracted feature set as the test set in machine learning algorithm to improve the accuracy of training results.

F. Xhafa et al. (Eds.): IISA 2019, AISC 1084, pp. 103–111, 2020.
https://doi.org/10.1007/978-3-030-34387-3_13

1.1   Motivation

In the design [5] shown in Fig. 1, an example of poor maintainability is given. This example is a typical design defect. The main purpose of this paper is to identify similar examples. Software maintainability refers to the degree of difficulty for maintainers to maintain the software, including the degree of difficulty in understanding, correcting, modifying and improving the software. In the example, class StaffManagement is associated with class Student, Teacher and Dushman through association relations. The method NewTerm calls the semantically similar behavior Register, Enroll and CheckIn of these three classes by judging the identity of the object being processed. Assuming that users require a specific registration process for principals, such as adding class Principal and operating Login, the maintainer must analyze the class StaffManagement method Newterm and find the appropriate place to add code in dozens or even hundreds of lines of code to complete the call to Principal's method Login.

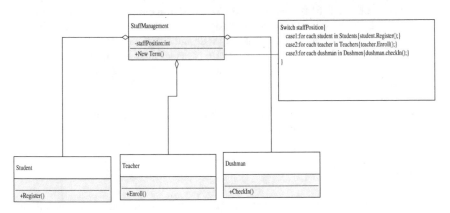

Fig. 1. Examples of poor maintainability

1.2   Related Work

Martin Fowler et al. [6] used Bad Smell to describe problems that need to be refactored in code design. They proposed 22 code design problems and corresponding refactoring methods. In 2000, Robert [7] put forward the principle of object-oriented design, which further improved the quality of software products. Kerievsky et al. [8] expanded Fowler's "bad taste of design" by introducing design patterns into heritage systems through reconstruction. Tahvildari et al. [9] use object-oriented metrics to detect potential design defects and generate corresponding error correction mechanisms through meta-schema transformation. ISO 9126 [10] stipulates that the quality characteristics of software products are composed of six aspects: functionality, reliability, availability, effectiveness, maintainability and portability. Therefore, in the process of software design, if the design results produced make the software products have low functionality, reliability, availability, validity, maintainability or portability, it is said that the design has design defects.

2   Software Design Defects

2.1   Related Concepts

This paper focuses on the design of object-oriented class level, that is, the design of class, attribute, method, inheritance and aggregation between classes and classes. According to heuristics and C&K metrics, the characteristics of software design defects are proposed. Five typical software design defects are focused on:

(1)   Complex Class (CC): It means that this class has most of the functions of the system, but handed over some small work to other classes, which will make this class appear too many instance variables, and the code will become huge. This class violate the principle of single responsibility, and their structure is too complex to understand, modify and maintain. In object-oriented design, we expect the system functions to be evenly distributed to all classes.
(2)   Divergent Class (DC): Each operation of this class will cause changes in several other classes. This class has high coupling, which will seriously affect the maintainability of the system.
(3)   Duplicate Work (DW): Two classes contain the same or similar programs or methods. At this time, the same parts should be extracted to make the code more concise.
(4)   Over-inherited (O2): Classes with deep nested form make the program difficult to understand and maintain.
(5)   Inheritance Error (IE): Subclasses refuse to inherit the interface provided by the parent class or modify the inheritance system.

The metrics involved in this paper refer to object-oriented metrics, which is a method to measure the quality of object-oriented software code. It mainly measures the abstraction, encapsulation, inheritance and polymorphism in object-oriented technology. Common measurement methods include C&K metric [11], Hitz and Montazeri metric [12], Lorenz metric [13] and MOOD metric [14]. The measurement method involved in this paper comes from the C&K measurement used by Chidamber-Kemerer. The complete list of the measures selected in this paper is shown in Table 1.

Table 1. The list of metrics selected in this paper

Weight Methods/Class (WMC)	Response for a Class (RFC)
Number of Children (NOC)	Coupling Between Objects (CBO)
Depth of Inheritance Tree (DIT)	Lack of Cohesion in Methods (LCOM)

The six metrics in Table 1 are selected in this paper because they are representative and widely used in design defect detection. Generally, 90% of design defect representative can be detected by using the six metrics mentioned in the table above. In the method used in this paper, for the input source codes with typical design defects collected from the network, six measures of each design defect are calculated by POM

Tests tool [15]. The verification set in the neural network is formed by combining heuristics, which is prepared for the next machine learning training model.

2.2   Related Definition

Design Defect, denoted as DD, will be used in subsequent algorithms.

Definition 2.2.1 Class Map (CM): class mapping is the corresponding relationship between classes, which can be expressed as a triple, and can be expressed as: CM = (Class1, Class2, Relation).

There are six kinds of mappings between classes as shown in the Fig. 2 above (Bill, Warning, Del), (Bill, Purchase, Agg), (Purchase, Customer, Del), (Warning, Letters, Inh), (Periodic, Messages, Inh), (Message, Customer, Del).

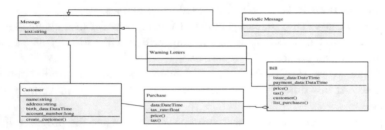

Fig. 2. C&K metrics applied to Royal Service Station's system design [16]

Definition 2.2.2 Design Defect Feature (DDF), a design defect feature is recorded as a five-tuple, expressed as (CM, WMC, NOC, DIT, CBO).

Definition 2.2.3 Design Defect Feature Set (DDFS), the design defect feature set proposed in this paper is all DDFs contained in the source code to be detected.

3   Design Defect Detection Method

3.1   Design Defect Detection

The purpose of design defect detection is to detect the problems existing in the design from the existing design codes. The detection of design defects has been ignored in recent years. To solve this problem, the block diagram of the proposed detection method is shown in Fig. 3.

This method is divided into two stages: training stage and detection stage. In the training stage, we collect the samples needed by the experiment, then generate the design defect feature set DDFS by combining the C&K measure and heuristic, use the feature extraction algorithm in machine learning to create the input feature vector, generate the training data set of the design defect classifier, and finally train and design the defect classifier on the training set. In the detection stage, the sample information in the pre-processed data set is generated into candidate instances, and the feature vectors

Fig. 3. Design defect detection framework

of candidate instances are calculated by C&K metric. The results are input into the defect classifier to get the final source code classification.

3.2 Feature Extraction

We selected CM, WMC, NOC, DIT and CBO as design defect features in this paper, because these features can detect five typical software design defects mentioned in the second section of this paper.

This paper is an experiment on Weka platform. Combining the C&K metrics and heuristics, using Weka Attribute Selected Classifier with feature selection classifier and Greedy Stepwise search class, the classifier continuously searches and designs defect feature set based on Greedy Stepwise using greedy algorithm, evaluates and feeds back the optimal solution, and combines CfsSubsetEval algorithm and Greedy Stepwise to select feature from input source code data. CfsSubsetEval feature selection algorithm evaluates each feature in the defect feature set according to its predictive ability and their correlation. The strong predictive ability of a single feature is considered good by the algorithm, and the defect feature with high correlation with the class is recommended for selection.

3.3 Design Defect Classifier Algorithms

The case base used in this paper is the design defect feature set extracted from all the source codes in the classical MARPLE project by feature extraction algorithm. According to the ratio of 6:2:2, the whole case base is divided into training set, verification set and test set. The design defect feature is vectorized by one-hot coding, and the experimental data set is expanded by ten fold cross-validation and the training result is improved, confirmation generates a design defect classifier with increasing accuracy and recall. Experiments show that when the number of initial iterations of the neural network is 120 and the learning rate $\eta = 0.25$, a part of the results are randomly extracted, after manual intervention and expert knowledge verification, the effect of avoiding over-fitting and under-fitting phenomena in the neural network is the best, and the accuracy of classification results is relatively high.

3.4 Design Defect Detection Algorithms

The main task of the design defect detection stage is to use the trained design defect classifier to identify the design defect in any input system source code. In this paper, five classical two-class supervised learning algorithms are used: Decision Tree (DT), Naive Bayesian Model (NBM), Logic Regression (LR), Support Vector Machine (SVM), K-Nearest Neighbor (KNN) to train the proposed model and classify all subsequent inputs, in order to increase the accuracy and recall rate of the results.

In this paper, the design defect detection algorithm can be described as follows: The design defect detection algorithm takes all the source code to be detected, the design defect feature set and the design defect classifier as input. Firstly, it extracts the source code related features by machine learning correlation algorithm, C&K metric and heuristics, forms the design defect feature set and generates candidate instances. Then, it trains the candidate instances according to the design defect features by using the design defect classifier. Finally, the results of five machine learning algorithms training are used to determine whether the candidate case has design defects after comprehensive consideration of the deviation-variance equilibrium calculation method. The algorithm is shown in Table 2 as follows:

Table 2. Design defect detection algorithm for every instance

Input: Source code to be detected (source), Design defect classifier (ddclassifier), Design defect feature set (DDFS), Design defect feature (DDF), C&K metric (CKm)
Output: Design defect instance collection in the source code to be detected(DDSset)
1.   CaM := CalMe (source, CKm)
2.   While(!end):
3.   foreach CaM ∉ DDFS:
4.   CaM := DDFS ∪ {MeI}
5.   candins := GenFeaIns (source, CaM)
6.   DDSset := φ
7.   foreach instance ∈ candins do
8.   X := φ
9.   foreach DDF ∈ DDFS do
10.   Xi := CMeF (candins, DDFS)
11.   X := X ∪ {Xi}
12.   end
13.   mep := classify (ddclassifier, X)
14.   X := X ∪ {mep}
15.   if label = ture then
16.   DDSset := DDSset ∪ {source}
17.   end
18.   end
19.   return DDSset

4 Experiment and Evaluation

4.1 Experiment

In order to evaluate the effectiveness of the proposed method, we used five representative two-classification supervised machine learning methods to select feature from
experimental data, and finally trained the most accurate design defect classifier model.
Secondly, in order to make our experimental results more illustrative, this paper carries
out experimental validation on three open source software which are most widely used
in recent years and easily accessible on the Internet: JHotDraw for developing GUI
open source framework for graphics applications; Junit for commonly used Java language unit testing framework; Refactory, a refactory tool for Java programming
language.

4.2 Result Analysis

Two widely used metrics, Precision and Recall, are used to evaluate the final results of
the experiment. Table 3 shows the specific representation of the "confusion matrix" of
the classification results:

Table 3. "Confusion matrix" of classification results

Real situation	Prediction results	
	Positive example	Counter-example
Positive example	TP (Real positive example)	FN (False counter example)
Counter-example	FP (False positive example)	TN (True counter example)

$$\text{Precision P} = \frac{TP}{TP + FP} \tag{1}$$

$$\text{Recall R} = \frac{TP}{TP + FN} \tag{2}$$

TP represents the number of original instances with five typical design defects
selected in this input, FP represents the number of original instances without design
defects selected in this input, and FN represents the number of all original instances
entered in this input.

Table 4 compares the experimental results obtained by different supervised learning
methods. P represents the accuracy and R represents the recall rate. The following
conclusions can be drawn from the data analysis in the table:

(1) Comparing comprehensively, the recall rate and accuracy rate of SVM are the best
among the five methods, and the performance is the most stable.

Table 4. Experimental results of this method

Software name	DT		SVM		NBM		LR		KNN	
	P	R	P	R	P	R	P	R	P	R
JHotDraw	75	89	92	88	85	90	77	82	72	85
Junit	72	89	85	90	88	82	80	86	74	82
JRefactory	70	88	84	91	86	84	79	84	78	83

(2) The five methods in this paper can achieve high recall rate, but the accuracy still needs to be improved, and the algorithm needs to be further improved later.

To sum up, there is still a big gap between recall rate and accuracy rate in all open source projects and transmission mode detection method. We need to constantly add new methods to try and improve them. However, there is no particularly good method for design defect detection in academia at present, so the research of our method is worth looking forward to.

Acknowledgments. At the end of this paper, I would like to thank the teachers and classmates who have contributed to this paper, and secondly to those who came to help me.

References

1. D'Ambros, M., Lanza, M., Robbes, R.: Evaluating defect prediction approaches: a benchmark and an extensive comparison. Empirical Softw. Eng. **17**(4–5), 531–577 (2012)
2. Brown, W.H., Malveau, R.C.: "Skip" McCormick III HW, Mowbray TJ, Antipatterns: Refactoring Software, Architectures, and Projects in Crisis. Wiley Computer Publishing, New York (1998)
3. Zhiqiang, L., Xiao-Yuan, J., Xiaoke, Z.: Progress on approaches to software defect prediction. IET Softw. **12**(3), 161–175 (2018)
4. Moha, N.: Detection and correction of design defects in object-oriented designs. In: Companion to the ACM Sigplan Conference on Object-Oriented Programming Systems & Applications Companion. ACM (2007)
5. Feng, T.: An approach to automated software design improvement. J. Softw. **17**(4), 703–712 (2006)
6. Fowler, M.: Refactoring: Improving the Design of Existing Programs. Addison-Wesley, Boston (1999)
7. Robert, M.: Design principle and design patterns (2000)
8. Kerievsky, J.: Refactoring to Patterns. Addison-Wesley, Boston (2004)
9. Tahvildare, L., Kontogiannis, K.: Improving design quality using meta-pattern transformations: a metric-based approach. J. Softw. Maint. Evol. Res. Pract. **16**(4–5), 331–361 (2004)
10. ISO 9126: Software Product Quality Characteristics. http://www.cse.dcu.ie/essiscope/
11. Chidamber, S.R., Kemerer, C.F.: A metrics suite for object oriented design. IEEE Trans. Software Eng. **20**(6), 476–493 (1994)
12. Hitz, M., Montazeri, B.: Measuring coupling and cohesion in object-oriented systems (1995)
13. Lorenz, M., Kidd, J.: Object-oriented software metrics: a practical guide. Prentice-Hall Inc, Englewood Cliffs (1994)

14. Abreu, F.: MOOD-metrics for object-oriented design. In: Proceedings of Oopsla 94 Workshop Paper Presentation (1994)
15. Guéhéneuc, Y.G., Sahraoui, H., Zaidi, F.: Fingerprinting design patterns. In: 11th Working Conference on Reverse Engineering, pp. 172–181. IEEE (2004)
16. Shari, L.P.: Software Engineering Theory and Practice (2003)

Design of Electromagnetic Energy-Saving Intelligent Vehicle

Sujun Zhao, Feng Wang[✉], and Chaoqun Han

College of Mathematics, Physics and Electronic Information Engineering,
Wenzhou University, Chashan University Town, Wenzhou 325035, China
wang_fengwf@163.com

Abstract. This paper designs a set of electromagnetic energy saving intelligent vehicle which can effectively save energy and ensure high speed running. The design includes the design of the mechanical structure of the vehicle model, the design of the circuit module, the design of the sensor signal processing, the design of the supercapacitor module, the preparation of the control algorithm five modules. The design of the mechanical structure of the car model is mainly about the design of the overall structure of the car and the selection of the car model materials, which can reduce the resistance and gravity of the car as much as possible, reduce energy consumption, improve the stability of the car and obtain a faster speed. The key point of circuit module design is to coordinate the work between each module, effectively reduce the total power consumption, and achieve energy saving. The sensor signal processing design focuses on the track information acquired according to the acquisition, so that the car can accurately and quickly adjust the attitude and smoothly pass through various tracks. The ultracapacitor module is the source of energy and power for the whole car. Giving full play to each joule is the key to energy saving. Control algorithm is the soul of the car, this paper is the incremental PID, to deal with a variety of tracks.

Keywords: Energy saving · Super capacitor · Incremental PID

1 Introduction

With the gradual popularization of smart cars in China, more and more students are engaged in the study of smart cars. This paper is about the electromagnetic energy saving intelligent car design, mainly involving the electromagnetic and energy saving these two aspects. The design requirements are based on the rules of the 12th "NXP" cup national college students' intelligent car competition, that is, it is expected to obtain a greater speed with a lower energy consumption, so that the car can complete the next lap on the track, to achieve a perfect match between energy and speed. Nowadays, energy saving and high speed have always been the pursuit direction of intelligent cars. Through the design and test of electromagnetic energy saving cars, it will provide some reference for the development of cars in the future.

© Springer Nature Switzerland AG 2020
F. Xhafa et al. (Eds.): IISA 2019, AISC 1084, pp. 112–118, 2020.
https://doi.org/10.1007/978-3-030-34387-3_14

2   Principles of Operation

Suitable hardware selection and circuit design are the basis of high speed and energy saving of electromagnetic energy saving intelligent vehicle. This paper designs the electromagnetic energy saving intelligent vehicle is from the main hardware circuit and the software design these two aspects carry on the improvement and the adjustment, causes the intelligent vehicle to achieve the energy saving, the high speed goal. Hardware circuit including motherboard, driver board and sensor module, software design focuses on PID control and sensor acquisition and processing.

In the selection of hardware, a number of products were tested and compared, and the most energy-saving components were selected. Motor is the main power output of the car, a German von haber hollow cup motor was selected. The motor has low power consumption and high speed [1]. Given the same voltage input, the motor consumes very low current, which is very suitable for energy saving needs. Encoder is to achieve closed-loop control, so that the car to maintain a constant speed, to avoid sharp turns and rapid deceleration bring a lot of energy consumption. The mini encoder adopted this time is 512 lines incremental rotary speed measurement compatible orthogonal decoder. The encoder has the characteristics of small size, light weight and high stability, which just meets the requirement of energy saving. Ultracapacitor is the energy source of the whole car. If the lifting pressure design scheme is adopted to supply power to the car, the ultracapacitor group with intermediate or larger farad value should be selected. If the step-down method is adopted, the capacitance with low farad value must be selected if it is not needed below 5 V, that is, to reduce C and increase U. After comparison and calculation, the boost efficiency is lower than the step-down efficiency, so the step-down scheme is adopted. Finally, the capacitor group of 4F/12 V is selected. The motherboard includes MCU, voltage regulator circuit and supercapacitor module. The voltage stabilizing circuit is to obtain two-circuit voltage, and the input 12 V voltage is converted into 3 V and 5 V respectively through lm29503-3 and lm2950-5 chip. Then, they are transmitted to the drive circuit, steering gear, KL26 MCU and sensor module for power supply respectively. MCU is to process the feedback information from the sensor and encoder to indicate the next operation of the car. The pin diagram of the main board is shown in Fig. 1 below.

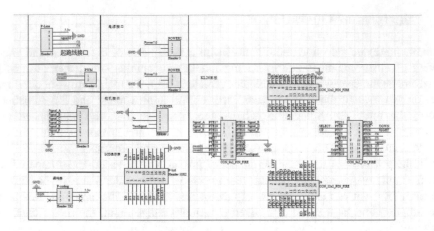

Fig. 1. Pin diagram of main board

The controller, motor drive module and motor are the main components of intelligent vehicle drive system. An excellent intelligent vehicle driving system, not only needs a wide range of speed regulation, but also needs to be able to withstand a large current, to provide a strong power for the car. Of course, the most important is the high energy conversion efficiency, to achieve high speed while saving energy. H bridge drive just meets the demand, it can not only realize the positive and negative drive of dc motor, but also select the right device can reduce energy consumption. The H bridge is shown in Fig. 2 below.

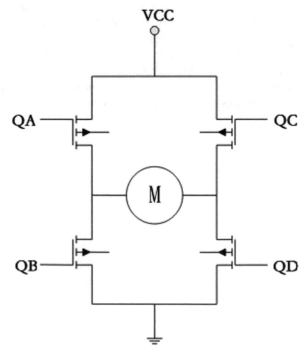

Fig. 2. H bridge drive

Of course, in order to obtain the large current driving ability, other auxiliary circuits are needed. The drive circuit designed in this paper firstly increases the input voltage to 12 V through MC34063 chip, and then provides IR2104S chip, which will obtain a large current to meet the requirements of high-speed operation of the car. Of course, the operation of large load will have an impact on the main control chip, so the 74LVE245 chip is also used for the isolation protection of MCU and MOS tube. The entire drive module circuit is shown in Fig. 3 below.

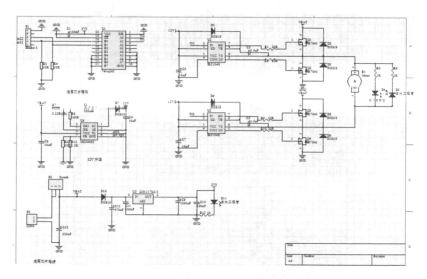

Fig. 3. Drive module circuit

The electromagnetic sensor is composed of LC oscillation circuit P3, TLV2462 op amplifier chip and RC filter circuit [2–4]. Through the principle of electromagnetic induction, to detect the car off the track of the electromagnetic signal line distance.

The tv2462 chip is a rail - to - rail op - amp chip. The chip has high output drive capability and can be close to the power supply voltage output. Moreover, its voltage swing rate of 1.6 V/us and power supply current of only 500 uA make it have good AC performance and low power consumption. Is the first choice of energy-saving car.

In the software, the car motor control using incremental PID control. [5] By calculating the next last error and the PID incremental error of the previous one, the control effect can be achieved by continuously accumulative loop. Incremental PID motor control is more stable than open-loop control, and the car is more responsive [6]. Because the car may encounter the problem of uneven track when driving on the track, the data collected by the sensor must be brushed and removed. The first step of the program is to store the data collected on the track for the first time. The purpose of this is to make the car adapt to the track quickly, that is to say, make it start faster. The second step then uses the Sensor_Get function to get the new data information, that is, to update the data. Then the third step uses Sensor_Filter function to find out the maximum and minimum values in the data and remove them, which can effectively overcome the impulse interference caused by accidental factors and ensure the relia-bility of the data. The final step is to normalize the filtered data between 0 and 100 through the function Sensor_To_One, avoiding floating point Numbers so that the data can be displayed on a small LCD screen [7]. After the above three steps, the sensor data can be assured, for the control of the rear steering gear laid the foundation [8–10]. The following is the flow chart of the acquisition and processing algorithm, as shown in Fig. 4 [11].

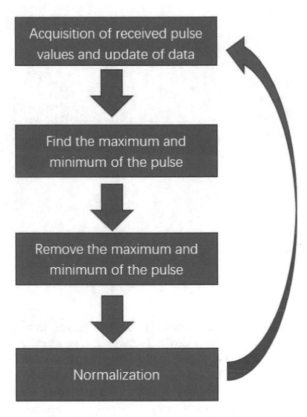

Fig. 4. Sensor sampling processing diagram

3   Test Results

Put the debugged car into the track of 33.6 m and the track of 22.4 m for testing.

The test was not smooth. The car could not enter the small ring effectively and accurately after the continuous small S curve, and sometimes it would lose the line and run out of the track. Through the comprehensive regulation of P and D and the use of different PD combinations, it was finally possible to run into the ring more completely. However, with the continuous improvement of car speed, the previous PD combination had been unable to adapt, and when the speed reached more than 1.8 m/s, adjusting PD could no longer effectively control the car's entry into the ring. Finally, the vehicle speed was set at 1.8 m/s under comprehensive consideration.

Then the first choice of charge was 450 J, which was enough energy for the car to make three steady laps of 33.6 m. But 450 J was actually a lot of wasted energy, leaving enough energy for the microcontroller to work for another 15 s or so. Obviously, this was a huge waste. Later I tried 350 J. The 350 J could cover 33.6 m for three laps, but it was very difficult. The car needed to run a good track in every lap. However, after a wave of thinking and improvement, when the car was in the last lap, it would not limit the duty cycle of the motor, so that it could run with all its strength, which was just able to run perfectly and met the demand.

In the end, the car used 350 J of energy to cover a distance of 33.6 m at a speed of 1.8 m/s.

4 Conclusion

In this paper, an electromagnetic energy saving intelligent vehicle is designed, which is based on the general electromagnetic intelligent vehicle to improve, to achieve high speed, energy saving purposes. Through calculation and comparison, the scheme that can achieve the goal of energy saving and high speed is selected. This electromagnetic energy saving intelligent car has shown good results in the hardware and software, to meet the requirements of the intelligent car race.

The design of the electromagnetic energy saving intelligent vehicle includes the design of the mechanical structure of the vehicle model, the design of the circuit module, the design of the sensor signal processing, the design of the supercapacitor module, and the preparation of the control algorithm.

Acknowledgments. This work was funded by the Wenzhou university open laboratory project (18SK26).

References

1. Sirui, W., Qing, Z.: Performance test method for dc motor of intelligent vehicle model. World Eelectron. Prod. **19**(09), 70–72 (2012)
2. Li, K.Y.: Principles of sensors. Science press **19**(05), 89–95 (2007). (in Chinese)
3. Xiangjie, H., Jiacai, L.: Design of LC resonant amplifier. Electron. Des. Eng. **21**(05), 172–175 (2013)
4. Qian, J., Hui, W., Gao, Y., Zhang, C., Sun, Q.: Experimental design and research on RC filter circuit. Univ. Phys. Exper. **30**(05), 58–62 (2017)
5. Li, F.M.: Research on digital PID control algorithm. J. Liaoning Univ. (Nat. Sci. Edn.) (04), 85–88 (2005)
6. Zhen, M., Xilin, Z.: Design of intelligent vehicle speed regulation system based on incremental PID. J. Hubei Univ. Technol. **30**(02), 72–76 (2015)
7. Wang, F., Zheng, X.: Simulation of the fixed-point number arithmetic. Key Eng. Mater. **467–469**, 2097–2102 (2011)

8. Wang, F., Zhang, Y.: The Comprehensive defence against railway system-wide disaster by computer interlocking system. In: Proceedings of 2nd International Conference on Intelligent Information Management Systems and Technology (IIMST 2007), pp. 95–98. World Academic Press (2007)
9. Wang, F., Zhang, Y., Ye, C.: HBACA-based railway yard route searching. In: The 3rd International Symposium on Intelligent Information Technology Application, pp. 496–499 (2009)
10. Wang, F., Zhang, Y., Chen, F.: Research on WangZhai station route optimization. In: 2009 International Symposium on Image Analysis and Signal Processing, pp. 183–187 (2009)
11. Wang, F., Wang, B.-l., Zhang, Y.: Asymmetry DFT algorithm and its application. J. Chongqing Univ. **31**(2), 162–165 (2008)

New Energy Vehicle Reliability Research by Large Data

Xiao Juan Yang$^{(\boxtimes)}$, Shu Quan Xv, and Fu Jia Liu

Research Institute of Highway, Ministry of Transport, Beijing, China
xj.y@rioh.cn

Abstract. This paper presents a theory of large data analysis of vehicle reliability based on vehicle fault data. By collecting fault data of vehicles with
different energy types, the theory of vehicle fault statistics and reliability analysis suitable for large data analysis is proposed, and the reliability level of
different types of vehicles is evaluated comprehensively by weighted analysis
method. Through comparative analysis, this paper reveals the fault change rule
and reliability level of new energy vehicles, and provides basic support for
improving the reliability of new energy vehicles, reducing maintenance costs,
expanding consumers' purchase choices, and improving the sales volume of
new energy vehicles.

Keywords: New energy vehicle · Large data · Reliability · Failure rate

1 Introduction

According to statistics, the sales of new energy vehicles in China reached 1.256 million
in 2018. China shows obvious and comprehensive advantages in research, industry,
policy innovation and infrastructure construction of new energy vehicles, though the
research on reliability of new energy vehicles just start and is not perfect enough. This
paper focuses on the usage and maintenance of new energy vehicle owners. By collecting fault data of new energy vehicle, it forms a statistical analysis theory of vehicle
failure law, and monitors and evaluates the reliability level of new energy vehicle. By
comparing the reliability level of vehicles with different energy types, it means to
improve the reliability and the research technology of new energy vehicle. Furthermore, we wish the research can make people more ease to buy vehicles, and popularize
the purchase and use of new energy vehicle in our country.

2 Reliability Analysis Theory

Reliability is the ability of a product to perform specified functions under specified
conditions and within specified time [1]. The reliability of automobile products is an
important index to measure the safety quality and fatigue life of automobile products
[2]. The reliability of automobile products is closely related to personal safety and
economic benefits. Improving the reliability of automobile products is the basis of
improving their safety performance [3]. Big data mining is the process of discovering

© Springer Nature Switzerland AG 2020
F. Xhafa et al. (Eds.): IISA 2019, AISC 1084, pp. 119–127, 2020.
https://doi.org/10.1007/978-3-030-34387-3_15

and extracting information or knowledge hidden in a large amount of data, and it is a way to acquire knowledge automatically [4].

3   Acquisition of Automobile Fault Data

In order to compare and analyze the reliability of new energy vehicles, we collects maintenance data of pure electric vehicle, hybrid vehicle and fuel vehicle. The specific information of each vehicle type is in Table 1.

Table 1. Maintenance volume and number of different types of vehicles

Type	Pure electric vehicles		Hybrid electric vehicle		Fuel vehicle	
Vehicle code	A_1	A_2	B_1	B_2	C_1	C_2
Maintenance volume	7807	3102	29626	12309	160005	492084
Number of vehicles	4999	2450	15899	4312	73158	267187

Collection of maintenance information includes license plate number, VIN code, repair date, repair mileage, fault description, repair items, accessories name, number of accessories, etc. The maintenance records of a pure electric vehicle code A_1 is shown in Table 2.

Table 2. Maintenance data of pure electric vehicles A_1

Maintenance information	Number			
	1	2	......	7797
License plate	YU A*0*9*	JING *Q9*3*	......	YU *H*05*
VIN	LS*A2*JX*HA*00*38	LS*A2*JX*GA*03*50	......	LS*A2*JX*GA*01*55
Date of repair	2017-12-21	2017-12-18	......	2018-01-23
Repair mileage	5712 km	6637 km	......	128975 km
Fault description	The acceleration pedal of the car is heavy	The car can't hang up and charges slowly	......	The right front wheels buzz when driving
Maintenance item	Replacement of Electronic Accelerator Pedal Assembly	Updating Charger Procedure and Replacing Shift Control Assembly	......	Replacement of right front brake assembly
Accessories name	Electronic Accelerator Pedal Assembly	Shift Control Assembly	......	Front Brake Corner Assembly (right)
Accessories quantity	1	2	......	1

4   Calculation of Equivalent Fault Number Change Curve of Automobile in Different Mileage Ranges

In this paper, we discuss reliability analysis under large data volume, so the concept of equivalent number of faults in different mileage intervals is proposed. Based on the collected fault data of different types of vehicles, the equivalent fault number curves of different types of vehicles with different mileage intervals are calculated.

4.1   Establishing the Curve of Vehicle Type Changing with the Number of Mileage Faults

Taking pure electric vehicle A_1 as an example, 7807 faults of 4999 vehicle are collected and their distribution with mileage is in Fig. 1.

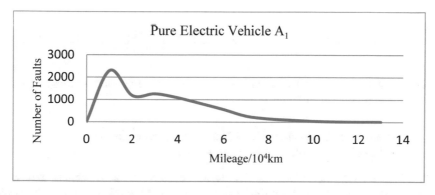

Fig. 1.   Fault distribution diagram of pure electric vehicle A_1

The proportion of the total data is the highest within 20,000 km, and then increases slightly at 30,000 km and then decreases gradually.

4.2   Classification of Fault Hazard Degree

We can see that only the number of faults is calculated in Fig. 1, and there is no distinction between fatal faults, serious faults and general faults. In the actual evaluation of vehicle reliability, we need to consider the number of different levels of failure [5]. The classification principle of specific fault hazard degree [6] is shown in Table 3.

Table 3.   Principles for classification of fault hazard degree

Fault type	Classification principle
Severe fault	Cause major assembly scrap or significant performance degradation
Serious fault	Cause the performance of an assembly to decline or cause damage to the main parts
General fault	Parts are damaged and need to be replaced
Minor fault	It can be repaired without replacing parts

In this paper, the general fault equivalent fault coefficient is set as 1 according to the degree of fault hazard, and the equivalent fault coefficient of other levels is determined accordingly. The number of faults corresponding to the coefficients of different grades and the calculated coefficients of six types of vehicles is in Table 4.

Table 4. Equivalent fault coefficient grade and number of faults corresponding to different vehicle types

Fault type	Severe fault	Serious fault	General fault	Minor fault
Equivalent fault coefficient	$\varepsilon_1 = 10$	$\varepsilon_2 = 5$	$\varepsilon_3 = 1$	$\varepsilon_4 = 0.2$
A_1	132	185	1191	1594
A_2	17	39	342	7409
B_1	799	2181	13986	12660
B_2	383	2133	4295	5498
C_1	7589	24824	66705	60887
C_2	20591	87004	243975	140514

The reliability performance of vehicles with the same fuel type is similar in general distribution and slightly different in details.

4.3 Calculation of Equivalent Fault Number Change Curve of Automobile in Different Mileage Ranges

All the fault data of the vehicle model are aggregated into a coordinate axis whose abscissa is the mileage and the ordinate is the equivalent number of faults. Assuming that the total mileage range of the collected vehicle fault data distribution is 0 to L 10,000 km, the total mileage L is divided into 10,000 km intervals. Calculate the equivalent number of faults in each mileage interval $d(I), (i = 1,2,\ldots\ldots, L)$ and the number of maintenance vehicles in this area $n(i), (i = 1, 2, \ldots\ldots, L)$ which is shown in Fig. 2.

Fig. 2. Equivalent number of faults and number of faulty vehicles with mileage

Let the number of faults with different grades in each mileage interval be minor fault, general fault, serious fault and severe fault respectively like $m_1(i), m_2(i), m_3(i), m_4(i), (i = 1, 2, \ldots\ldots, L)$, then the equivalent number of faults in each mileage interval $d(i), (i = 1, 2, \ldots\ldots, L)$ is as follows.

$$d(i) = 10 * m_4(i) + 5 * m_3(i) + 1 * m_2(i) + 0.2 * m_1(i) \tag{1}$$

Then the equivalent number of each vehicle failures $P(i)$ in each interval is as follows.

$$p(i) = \frac{d(i)}{n(i)}, (i = 1, 2, \ldots \ldots, L) \tag{2}$$

Then the average interval equivalent failure number P of the vehicle in the total mileage L is as follows.

$$P = \left(\sum_{i=1}^{L} \frac{d(i)}{n(i)} \right) / L \tag{3}$$

The variation curves of equivalent number of faults $P(i)$ of six types of vehicles in different sections are shown in Fig. 3. We can see from the figure that the average equivalent number of faults of pure electric vehicle, hybrid vehicle and fuel vehicle decreases in turn.

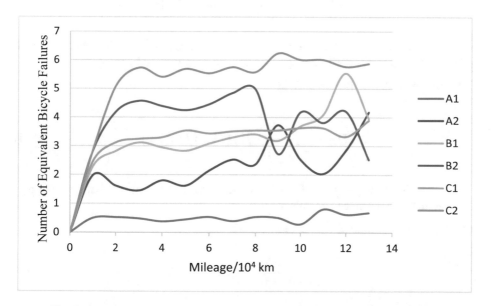

Fig. 3. Variation curve of equivalent fault number P(i) of six types of vehicles

5   Calculation of Reliability Evaluation Index of Vehicle Type

There are many evaluation indexes affecting automobile reliability. Traditionally, reliability indicators include MTBF, equivalent failure rate D and MTTFF. For large data analysis, MTTFF analysis and calculation of average first fault mileage will not be applicable. The corresponding fault interval mileage of vehicle A, B and C in Fig. 4 is Δh_1, Δh_2, Δh_3.

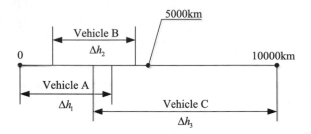

Fig. 4. Distribution of mileage intervals for different vehicles

Specific indicators are calculated as follows:

(1) Mean fault interval mileage MTBF is

$$\mathrm{MTBF} = L/\left(\sum\nolimits_{i=1}^{L} \frac{m(i)}{n(i)}\right), (10^4 \text{ km}) \tag{4}$$

Among them, $m(i)$ is the total number of faults in each mileage interval, $n(i)$ is the number of vehicles in each mileage interval, and L is the total mileage.

(2) The equivalent failure rate D is:

$$D = \frac{\left(\sum_{i=1}^{L} d(i)\right)}{\left(\sum_{i=1}^{L} n(i)\right)}/L, \left(\mathrm{Times}/10^4 \text{ km}\right) \tag{5}$$

Among them, $d(i)$ is the equivalent number of faults per mileage interval, and $n(i)$ is the number of vehicles per mileage interval.

(3) The average interval equivalent fault number P is

$$P = \left(\sum\nolimits_{i=1}^{L} \frac{d(i)}{n(i)}\right)/L, \left(\mathrm{Times}/10^4 \text{ km}\right) \tag{6}$$

Among them, $d(i)$ is the equivalent number of faults per mileage interval, and $n(i)$ is the number of vehicles per mileage interval.

In order to compare the reliability level of different types of vehicles, the six types of vehicles' fault data of the first 130,000 km are selected to calculate, and the specific reliability index values are obtained as shown in Table 5.

Table 5. Calculated value of reliability indicators for six types of vehicles

Vehicle type	Reliability index		
	MTBF (10^4 km)	D (Times/10^4 km)	P (Times/10^4 km)
A_1	0.712	0.0123	2.316
A_2	0.690	0.0037	0.461
B_1	0.492	0.0184	2.929
B_2	0.493	0.0236	3.085
C_1	0.514	0.0267	3.778
C_2	0.569	0.0270	4.023

6 Establishment of Comprehensive Evaluation Model of Vehicle Type Reliability

Different evaluation indicators have different influence on the reliability evaluation of automobiles. After referring to the relevant reliability theory and inviting relevant technical experts to discuss, the weight of the three indicators is determined as follows:

$$A = (\text{Average fault} - \text{free interval mileage, equivalent failure rate, average interval} \\ \text{equivalent number of faults}) = (0.2, 0.3, 0.5)$$

$$(7)$$

The calculation result vectors of vehicle reliability index are as follows:

$$B = (\text{Average fault} - \text{free interval mileage, equivalent failure rate, average interval} \\ \text{equivalent number of faults})T = (b1, b2, b3)T$$

$$(8)$$

Then the reliability level score formula of the vehicle is obtained as follows:

$$S = A \cdot B = 0.2 * b1 + 0.3 * b2 + 0.5 * b3 \qquad (9)$$

This analysis uses Min-max standardized method to process data. Let min (A) and Max (B) be the minimum and maximum values of attribute A respectively. A primitive value x of attribute A is mapped to the value x' in the interval [0, 1] by min-max standardization. The formula is as follows:

$$x' = (x - \min(A))/(\max(B) - \min(A)) \qquad (10)$$

Because MTBF, D and F represent the opposite meaning, so D, F will change the direction.

$$x' = 1 - (x - \min(A))/(\max(B) - \min(A)) \tag{11}$$

By standardizing the values of each index, the values are obtained as shown in Table 6.

Table 6. Calculated value of reliability indicators for six types of vehicles

Vehicle type	Reliability index		
	MTBF (10^4 km)	D (Times/10^4 km)	P (Times/10^4 km)
A_1	1	0.6395	0.4788
A_2	0.8938	1	1
B_1	0	0.3610	0.3074
B_2	0.0281	0.1635	0.2647
C_1	0.1058	0.0382	0.0686
C_2	0.3503	0	0

Then BYD E5 300 has a comprehensive reliability score of

$$Q = 0.2 * 1 + 0.3 * 0.6395 + 0.5 * 0.4788 = 0.63125 \tag{12}$$

In this way, the overall reliability level scores of six models is in Table 7.

Table 7. Reliability level score table

Vehicle type	Reliability level score
A_1	0.63125
A_2	0.97876
B_1	0.26200
B_2	0.18702
C_1	0.06692
C_2	0.07006

We can see from the table that, the overall reliability level of pure electric vehicles is much higher than that of hybrid electric vehicles and fuel-fired vehicles, while that of hybrid electric vehicles is higher than that of fuel-fired vehicles. This calculation is a strong proof that the overall performance of new energy vehicles is better than that of traditional fuel vehicles in terms of reliability, failure rate and damage level. It provides important data support for owners to buy new energy vehicles and maintain them well.

7 Conclusion

In this paper, a large data mining method is proposed to study the performance characteristics and fault rules of automobiles in use, and to form a method for evaluating the reliability level of automobiles based on weighted analysis. By analyzing and comparing the reliability level of different types of vehicles, it is concluded that the reliability performance of new energy vehicles is better than that of traditional fuel vehicles. This will provide support for improving the quality of new energy automobile products, reducing maintenance costs and enhancing the attractiveness of new energy automobile market.

References

1. Sun, X., Fu, Q., Yang, X.: Application research of automobile reliability in automotive system engineering. J. Shenyang Inst. Aeronautical. Ind. **21**(5), 28–30 (2004)
2. Teng, Y.: Research on reliability evaluation of domestic vehicles. Master's thesis of Chang'an University (2010)
3. Wan, Z.: Reliability assessment and failure law of domestic electric vehicles. Master's thesis of Wuhan University of Technology (2008)
4. Huang, Y., Miao, K.: Deep understanding of big data: big data processing and programming practice. Chemical Industry Publishing House (2014)
5. Yan, Y., Dai, R.: Application and reliability evaluation of ST5100TCZ road barrier removal vehicle. Des. Comput. **5**, 33–35 (2006)
6. Standards for Automobile Industry of the People's Republic of China. Quality Assessment Method for Automobile Products (QCT900-1997) (1997)

One-Step Prediction for GM(1,1) with a Sliding Data Window

Zengxi Feng[1,2(✉)], Mengdi Gao[1,3], Bo Zha[4], Peiyan Ni[1],
Xueyan Hou[1], Jiawei Liu[1], and Maijun Gao[1]

[1] School of Building Services Science and Engineering, Xi'an University of
Architecture and Technology, Xi'an, China
fengzengxi2000@163.com
[2] Anhui Key Laboratory of Intelligent Building and Building Energy
Conservation, Anhui Jianzhu University, Hefei, Anhui, China
[3] Xi'an Architecture Design and Research Institute Co. LTD., Xi'an, China
[4] China Northwest Architecture Design and Research Institute Co. LTD.,
Xi'an, China

Abstract. Most studies on grey models (GMs) focus on modeling and optimizing the construction and parameters of GMs, but not on the data and one-
step prediction. Generally, GM(1,1) ignores the function of new data and is
employed for multistep prediction. In multistep prediction, the worse the prediction precision, the larger the number of prediction steps will be. From the
viewpoint of a sliding data window, the current one-step prediction will correspond to the past nth step prediction that was done by the fixed data window.
In addition, the value of the first step prediction, sliding over time, is widely
used in practice and is expected to be very accurate. Therefore, the sliding data
window removes the old data and uses the new data. This is introduced to
improve the past nth step prediction precision that corresponds to the current
one-step prediction precision. Through the example of forecasts of Chinese
energy consumption, the prediction precision can be enhanced effectively.

Keywords: Grey model · Sliding data window · One-step prediction

1 Introduction

Since grey models were proposed by Deng as a prediction tool, they have been widely
applied due to the advantages of forecasting with fewer samples. The GM(1,1) model is
the most generally used grey model [1]. To increase the precision of the GM(1,1)
model, many studies have been reported, such as [2–20]. Ji et al. [2, 3] discussed in
depth the characteristics and properties of GM(1,1). Dang et al. [4, 5] optimized the
initial values of model parameters. Cui et al. [6, 7] optimized the parameters of GMs.
Tan et al. [8, 9] improved the model simulation accuracy by the conformation of
background values. Song et al. [10, 11] analyzed the optimum GM methods. Yao
proposed the discrete GM [12]. Luo et al. [13, 14] developed the non-isometric GM
and its optimization. Wang et al. [15, 16] defined the range suitable for different GMs.

© Springer Nature Switzerland AG 2020
F. Xhafa et al. (Eds.): IISA 2019, AISC 1084, pp. 128–136, 2020.
https://doi.org/10.1007/978-3-030-34387-3_16

Salmeron et al. improved the precision of the model by combining GMs and soft-computing methods [17, 18].

Those reports mainly focus on combining a GM(1,1) model with other modeling technologies, modeling the GM and optimizing its construction and parameters, not on the data and one-step prediction. They do not explore how to improve the prediction precision by combining new data. There are very few reports that study the relation between the one-step prediction precision and the data. Liu et al. proposed the buffer operator used to deal with the original data to enhance the prediction precision [19]. Yin et al. used a grey model of equal dimensional innovation to track objects [20, 21].

GM(1,1) is usually employed for multi-step prediction in many fields based on a fixed width window without the use of new data. For example, Wu et al. [22] forecasted Chinese energy consumption, logistics demand in Jiangsu province and syphilis incidence. Yang et al. forecasted the Yangzi river water level [23]. In addition, the precise one-step prediction is necessary for operators to run ice storage equipment in ice storage air-conditioning systems and for managers to retrofit their energy schemes. For example, operators wish to know the required refrigerating capacity precisely for the following day when running ice storage air-conditioning systems ahead of time. They are not concerned with the capacity for other days. For another example, a manager will want to know exactly the required electric energy production of the next month and does not care about the electric energy production of other times. From the viewpoint of the sliding data window, the current one-step prediction corresponds to the past nth step prediction that was performed by the fixed data window. Therefore, to take full advantage of the value of the updated data and increase the accuracy of the nth step prediction of the GM(1,1), the sliding data window is used in GM(1,1).

2   The GM(1,1) Model

Grey system discovers the inherent regularity of a given data sequence by mining and collating the data. GM(1,1) is a very effective prediction method applied for dealing with the issues with imperfect and uncertain information. When GM(1,1) is used to forecast, GM(1,1) can be built with at least four data samples.

If $X^{(0)}$ is the original sequence and it is non-negative, $X^{(0)}(k)$ is the value of time k.

$$X^{(0)} = \left\{ X^{(0)}(1), X^{(0)}(2), \cdots, X^{(0)}(n) \right\} \tag{1}$$

After 1-AGO (Accumulated Generating Operator) is used for $X^{(0)}$, $X^{(1)}$ can be obtained:

$$X^{(1)} = \left\{ X^{(1)}(1), X^{(1)}(2), \cdots, X^{(1)}(n) \right\} \tag{2}$$

where $X^{(1)}(k) = \sum_{i=1}^{k} X^{(0)}, k = 1, 2, \cdots, n.$

According to the new sequence $X^{(1)}$, the whitenization equation can be built and is defined as GM(1,1):

$$\frac{dX^{(1)}}{dt} + aX^{(1)} = u \tag{3}$$

Where u and a are the unknown parameters. Generally, they are called the grey action and developing coefficient respectively, and can usually be calculated by the method of least squares:

$\hat{a} = [a, u]^T = (B^T B)^{-1} B^T Y_N$ where

$$B = \begin{bmatrix} -0.5(X^{(1)}(2) + X^{(1)}(1)) & 1 \\ -0.5(X^{(1)}(3) + X^{(1)}(2)) & 1 \\ \vdots \\ -0.5(X^{(1)}(n) + X^{(1)}(n-1)) & 1 \end{bmatrix} \quad Y_N = \begin{bmatrix} X^{(0)}(2) \\ X^{(0)}(3) \\ \vdots \\ X^{(0)}(n) \end{bmatrix} \tag{4}$$

The whitenization equation can be solved to obtain

$$\hat{X}^{(1)}(t+1) = (X^{(1)}(1) - \frac{u}{a})e^{-at} + \frac{u}{a} \tag{5}$$

where $\hat{X}^{(1)}(t)$ is the estimated value of $X^{(1)}(t)$.

The inverse accumulated generating operator is as follows:

$$\hat{X}^{(0)}(k+1) = \hat{X}^{(1)}(k+1) - \hat{X}^{(1)}(k) \qquad k = 1, 2, \cdots, n-1. \tag{6}$$

3   Introducing the Sliding Data Window to GM(1,1)

When GM(1,1) is employed to predict, its data window, i.e. the original data sequence, would be selected to form the parameters, B and Y_N. The size of the data window is set to be variable or not by the user. The window is also set to be fixed or sliding by the user. Wu proved the relation between the size of the data window and the MAPE (Mean absolute percentage error) [22]. Liu proved that GM(1,1) can be built with at least four data samples when GM(1,1) is used for forecasting. However, there are no reports on the relation between the sliding data window and one-step prediction precision.

The sliding windows is one significant concept of the streaming model, it only uses the most "recent" elements and discards the rest [24]. The concept of sliding windows shows the importance of recent data. In contrast with sliding windows, the traditional way to deal with data in GM(1,1) does not employ new data, which ignores the function of the new data.

It is apparent that the data sequence $X^{(0)}$ consisting of n elements must be used to form the parameters, B and Y_N, when GM(1,1) is used to forecast. However, the element $X^{(0)}(n+1)$, $X^{(0)}(n+2)$, etc., will be generated constantly by the system or the forecasted object as time goes on. The principle of updating data is shown in Fig. 1. They may have new values, but these are ignored.

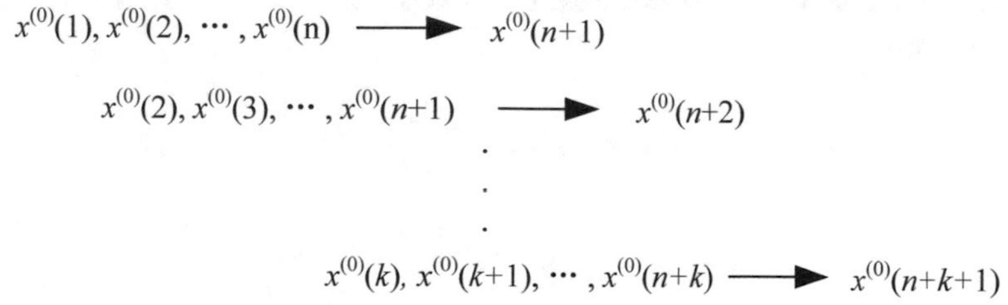

$$x^{(0)}(1), x^{(0)}(2), \cdots, x^{(0)}(n) \longrightarrow x^{(0)}(n+1)$$

$$x^{(0)}(2), x^{(0)}(3), \cdots, x^{(0)}(n+1) \longrightarrow x^{(0)}(n+2)$$

.

.

.

$$x^{(0)}(k), x^{(0)}(k+1), \cdots, x^{(0)}(n+k) \longrightarrow x^{(0)}(n+k+1)$$

Fig. 1. The data update process of the sliding data window

4   The Requirements of Data Samples for GM(1,1)

According to grey theory, satisfying quasi-exponential and quasi-smooth checking conditions is the basic requirements for ensuring X a non-negative sequence [22].

Definition 1 [1]. If a sequence X, $X = \{X(1), X(2), \cdots X(n)\}$, satisfies $\frac{\rho(k+1)}{\rho(k)} < 1 (k = 2, 3, \cdots, n-1)$ and $\rho(k) = \frac{X(k)}{\sum_{i=1}^{k} X(i)} \in [0, \varepsilon] (k = 3, 4, \cdots, n, \varepsilon < 0.5)$, then X is a quasi-smooth sequence.

Definition 2 [1]. For the sequence $X = \{X(1), X(2), \cdots X(n)\}$, if $\forall k, \frac{X(k)}{X(k+1)} \in [a, b], b - a < 0.5, k = 2, 3, \cdots, n$, then X is the quasi-exponent sequence.

To ensure that the study can show accurately the relation between the one-step prediction precision and the data sequence, the original data sequence should be non-negative and satisfy the quasi-exponential and quasi-smooth checking conditions. If the sequence cannot satisfy these requirements, the sequence should be processed with the moving average method. The three-point moving average method is adopted in this paper. The specific formula for its calculation is showed as the follows:

$$x'^{(0)}(1) = [3x^{(0)}(1) + x^{(0)}(2)]/4 \tag{7}$$

$$x'^{(0)}(k) = [x^{(0)}(k-1) + 2 * x^{(0)}(k) + x^{(0)}(k+1)]/4 \quad k = 2, \cdots, n-1 \tag{8}$$

$$x'^{(0)}(n) = [x^{(0)}(n-1) + 3x^{(0)}(n)]/4 \tag{9}$$

5   Experimental Results

The real case of forecasting Chinese energy consumption demonstrated that the prediction precision of GM(1,1) is enhanced effectively by one-step prediction with a sliding data window. Let $GM_m(1,1)$ be the GM(1,1) model where m ($m = 1, 2, \ldots, n$)

is the number of sample. Because there is only one prediction value when the one-step prediction of GM(1,1) is employed at each time, the APE (Absolute Percentage Error) is used to compare the real and forecasted values. In order to analyze the overall prediction precision, MAPE (Mean Absolute Percentage Error) is used to compare the forecasted and real values to evaluate the precision. APE and MAPE are defined respectively as:

$$APE = 100\% \left| \frac{x(k) - \hat{x}(k)}{x(k)} \right| \tag{10}$$

$$MAPE = 100\% \frac{1}{n} \sum_{k=1}^{n} \left| \frac{x(k) - \hat{x}(k)}{x(k)} \right| \tag{11}$$

The data from paper [22] is employed. The actual values and the fitted values or prediction values calculated by different models are listed in Table 1. In one prediction, the fitted values calculated by GM(1,1) correspond to the existing values in the original sequence. The prediction values are those unknown values that are forecasted by GM (1,1). The data from 1990 to 2003 is employed to construct the $GM_{14}(1,1)$ and the data from 1998 to 2003 is employed to construct the $GM_6(1,1)$.

It is clear that the prediction precision of $GM_{14}(1,1)$ without the sliding data window decreases gradually. The APEs of $GM_{14}(1,1)$ at the first, second, third and fourth step prediction are 18.02%, 23.37%, 27.76% and 30.77%, respectively. The MAPE of the fitted values from 1991 to 2003 is 4.13%. The MAPE of the prediction values from 2004 to 2007 is 24.89%.

Considering that the given non-negative sequence of Chinese energy consumption data should meet the quasi-exponential and quasi-smooth checking conditions, the sequence is checked. The result shows that the sequence of Chinese energy consumption data from 1990 to 2003 cannot satisfy the quasi-smooth checking conditions because of $\frac{\rho(14)}{\rho(13)} = 1.047 > 1$ and $\frac{\rho(15)}{\rho(14)} = 1.050 > 1$. To ensure that the study can show accurately the relation between the prediction precision of $GM_{14}(1,1)$ and the data sequence, the sequence is processed with the three-point moving average method. The results of data processing and prediction are listed in Table 1. The forecasting precision of $GM_{14}(1,1)$ without the sliding data window also decreases gradually. The APEs at the first, second, third and fourth step prediction are 18.34%, 23.71%, 28.11% and 31.14%, respectively. The MAPE of the fitted values from 1991 to 2003 is 4.43%. The MAPE of the prediction values from 2004 to 2007 is 25.33%.

Based on the sliding window, the one-step prediction precision of $GM_{14}(1,1)$ increases gradually whether the original sequence is processed or not. The four APEs of one-step prediction with the original sequence are 18.02%, 17.52%, 15.08% and 10.61%, respectively. After the actual values are processed using the three-point moving average method, the four APEs of the one-step prediction of the processed sequence are 13.89%, 9.55%, 8.77% and 7.27%, respectively. It is obvious that the MAPEs decrease from 24.89% and 25.33% to 15.31% and 9.87% respectively.

Table 1. The prediction results of $GM_{14}(1,1)$ with different data windows (unit: 10^4 tons of SCE).

Year	Fixed data window						Sliding data window			
	Actual value	Fitted value & prediction value	APE	The processed value with the three-point moving average method	Fitted value & prediction value	APE	One-step prediction value based on actual value	APE	One-step prediction value based on the processed value	APE
1990	98703			99973.00						
1991	103783	108706.11	4.74%	103859.75	108975.87	5.00%				
1992	109170	112335.53	2.90%	109529.00	112559.23	3.10%				
1993	115993	116086.14	0.08%	115973.25	116260.43	0.23%				
1994	122737	123967.21	1.00%	123160.75	120083.33	2.16%				
1995	131176	128106.16	2.34%	131009.25	124031.93	5.44%				
1996	138948	132383.31	4.72%	136717.50	128110.38	7.80%				
1997	137798	136803.27	0.72%	136689.50	132322.93	3.97%				
1998	132214	141370.79	6.92%	134014.25	136674.00	3.37%				
1999	133831	141370.79	5.63%	134607.25	141168.14	5.48%				
2000	138553	146090.8	5.44%	138534.00	145810.06	5.24%				
2001	143199	150968.42	5.42%	144187.00	150604.61	5.17%				
2002	151797	156008.89	2.77%	155445.75	155556.82	2.48%				
2003	174990	161217.64	7.87%	169191.75	160671.87	8.18%				
MAPE			4.13%			4.43%				
2004	203227	166600.20	18.02%		165955.12	18.34%	166600.30	18.02%	174990.00	13.89%
2005	224682	172162.60	23.37%		171412.09	23.71%	185320.05	17.52%	203227.00	9.55%
2006	246270	177910.70	27.76%		177048.50	28.11%	209135.08	15.08%	224682.00	8.77%
2007	265583	183850.70	30.77%		182870.24	31.14%	237393.28	10.61%	246270.00	7.27%
MAPE			24.98%			25.33%		15.31%		9.87%

To explore the forecasting precision of GM(1,1) with different sequence sizes, the $GM_6(1,1)$ model is selected at random. The prediction results based on the two methods are given in Table 2.

Table 2. The prediction results of $GM_6(1, 1)$ with different data windows (unit: 10^4 tons of SCE).

Year	Fixed data window			Sliding data window	
	Actual value	Fitted value & prediction value	APE	One-step prediction value	APE
1998	132214				
1999	133831	129436.59	3.28%		
2000	138553	138279.61	0.20%		
2001	143199	147726.79	3.16%		
2002	151797	157819.39	3.97%		
2003	174990	168601.51	3.65%		
MAPE			2.85%		
2004	203227	180120.26	11.37%	180120.26	11.37%
2005	224682	192425.96	14.36%	218229.64	2.87%
2006	246270	205572.38	16.53%	253790.89	3.05%
2007	265583	219616.95	17.31%	281030.88	5.82%
MAPE			14.89%		5.78%

It is obvious that the prediction precision of $GM_6(1,1)$ without the sliding data window also decreases gradually. The APEs at the first, second, third and fourth step prediction are 11.37%, 14.36%, 16.65% and 17.31%, respectively. Based on the sliding data window, the four APEs of one-step prediction are 11.37%, 2.87%, 3.05% and 5.82%, respectively. The APEs are improved effectively. The MAPE also decreases from 14.89% to 5.78%.

6   Conclusions

The one-step prediction precision of GM(1,1) with a sliding data window is explored in this paper. Sliding data windows with a fixed size of 14 and 6 are used to forecast the Chinese energy consumption of the following year. The results show that the one-step prediction of GM(1,1) with a sliding data window improves effectively the precision of APE and MAPE. This study demonstrates that the past nth step prediction performed by the fixed data window will correspond to the current one-step prediction with sliding data window, and the prediction precision will be improved effectively by the one-step prediction of GM(1,1). It is helpful for operators to run ice storage equipment in ice storage air-conditioning systems and for managers to retrofit their energy schemes.

Acknowledgement. This study was supported by the Project of Anhui key laboratory of intelligent building and building energy conservation (IBES2018KF08), the Xi'an University of Architecture and Technology Foundation Fund Project (JC1706) and the Special Research Project of Shaanxi Science and Technology Department (2017JM6106).

References

1. Liu, S.F., Lin, Y.: Grey Information: Theory and Practical Applications. Springer, London (2006)
2. Ji, P., Huang, W.S., Hu, X.Y.: Study on the characteristic of grey prediction model. Syst. Eng. Theory Pract. **21**, 105–109 (2001)
3. Zhang, Q.S.: On entropy criterion for grey exponential law. Syst. Eng. Theory Pract. **22**, 93–97 (2002)
4. Dang, Y.G., Liu, S.F., Liu, B.: The GM models that x (1,n) be taken as initial value. Chin. J. Manag. Sci. **1**, 132–134 (2005)
5. Zhang, H., Hu, S.G.: Analysis of boundary condition for GM (1,1) model. J. Huazhong Univ. Sci. Technol. **10**, 110–111 (2001)
6. Cui, J., Zeng, B.: Study on parameters characteristics of NGM (1,1,k) prediction model with multiplication transformation. Grey Syst. Theory Appl. **2**, 24–35 (2012)
7. He, W.Z., Song, G.X., Wu, A.D.: On a gens arithmetic for parameters estimating of model GM (1,1). Syst. Eng. Theory Pract. **25**, 69–75 (2005)
8. Tan, G.J.: The structure method and application of background value in grey system GM (1,1) model. Syst. Eng. Theory Pract. **25**, 98–103 (2005)
9. Li, J.F., Dai, W.Z.: A new approach of background value building and its application based on data interpolation and newton-cores formula. Syst. Eng. Theory Pract. **24**, 122–126 (2004)
10. Song, Z.M., Xiao, X.P., Deng, J.L.: The character of opposite direction AGO and class ratio. J. Grey Syst. **14**, 9–14 (2002)
11. Lin, Y.H., Lee, P.C.: Novel high-precision grey forecasting model. Autom. Constr. **6**, 771–777 (2007)
12. Yao, T.X., Forrest, J., Gong, Z.W.: Generalized discrete GM (1,1) model. Grey Syst. Theory Appl. **2**, 4–12 (2012)
13. Luo, Y.X.: Non-equidistant step by step optimum new information GM and its application (1,1). Syst. Eng. Theory Pract. **30**, 2010–2258 (2010)
14. Wang, Y.M., Dang, Y.G., Wang, Z.X.: The optimization of back-ground value in non-equidistant GM (1,1) model. Chin. J. Manag. Sci. **4**, 159–162 (2008)
15. Wang, Z.X., Dang, Y.G., Liu, S.F.: An optimal GM based on the discrete function with exponential law (1,1). Syst. Eng. Theory Pract. **28**, 61–67 (2008)
16. Xie, N.M., Liu, S.F.: Research on extension of discrete grey model and its optimize formula. Syst. Eng. Theory Pract. **26**, 108–112 (2006)
17. Salmeron, J.L.: Modeling uncertainty with fuzzy grey cognitive maps. Expert Syst. Appl. **5**, 2010–7588 (2010)
18. Zhang, K., Liu, S.F.: Indirect generating grey model based on particle swarm optimization algorithm. Syst. Eng. Electron. **32**, 1437–1440 (2010)
19. Liu, S.F., Zeng, B., Liu, J.F., Xie, N.: Margaret spelling several basic models of GM and their applicable bound (1,1). Syst. Eng. Electron. **4**, 501–508 (2014)

20. Yin, H.L., Wang, L., Nong, J.: Object optical tracking method based on grey model of equal dimensional innovation. Comput. Eng. Appl. **44**, 24–25 (2008)
21. Liu, S.F.: The three axioms of buffer operator and their application. J. Grey Syst. **3**, 39–48 (1991)
22. Wu, L.F., Liu, S.F., Yao, L.G., Yan, S.L.: The effect of sample size on the grey system model. Appl. Math. Modeling **37**, 6577–6583 (2013)
23. Yang, W., Kong, W.J.: Research of the forecast of the yangtze river water level based on improved GM(1, 1) model. J. Wuhan Univ. Technol. **38**, 938–2014 (2014)
24. Vladimir, B., Rafail, O., Carlo, Z.: Optimal sampling from sliding windows. J. Comput. Syst. Sci. **78**, 260–272 (2012)

Evaluation on Well-Canal Combined Irrigation Modes for Cotton Fields in the South of Xinjiang, China Based on an Algorithm of Fuzzy Neutral Network

Zhigang Li[1], Jiaming Lu[2], Mengjie Hao[2], Li Zhang[2], Youbo Guo[2], and Qingsong Jiang[2(✉)]

[1] College of Water Resources and Architecture Engineering, Tarim University, Alar 843300, Xinjiang, China
[2] College of Information Engineering, Tarim University, Alar 843300, Xinjiang, China
Qingsongjiang0827@126.com

Abstract. The research aims to improve water shortage in irrigation of cotton fields and reduce salinization damages in irrigation areas of Tarim River (in the south of Xinjiang Uygur Autonomous Region, China). For this purpose, this study determined soil electric conductivity, ion concentration and key indexes for describing crop growth conditions under different irrigation modes. Based on this, 10 groups of sample data under five irrigation modes were obtained. Moreover, this research analyzed sample data by using an algorithm of fuzzy neutral network (FNN) and discussed the optimal irrigation mode suiting for cotton fields in the region. The results demonstrate that the mode 4 obtains the best overall effects among the five irrigation modes, that is, mixed irrigation mode of 1/4 well water and 3/4 canal water shows the optimal irrigation effects. The relevant research results have practical significance and popularization value for guiding the selection of irrigation modes for cotton fields in the research region.

Keywords: Fuzzy neutral network · Well-canal combined · Irrigation mode · Evaluation

1 Background

The south of Xinjiang Uygur Autonomous Region is located in the northwest border of China, where shows long sunshine time, large temperature differences, dry weather with little rainfall and serious salinization. For this reason, salt-tolerant cotton is main economic crop and Tarim River Basin with relatively abundant water resources becomes the main cotton-producing area. In recent years, due to the lack of scientific

F. Xhafa et al. (Eds.): IISA 2019, AISC 1084, pp. 137–143, 2020.
https://doi.org/10.1007/978-3-030-34387-3_17

planning of reclamation and mining modes as well as unreasonable flooding irrigation and forcing salt down in soil, water resources are greatly wasted, so that water for irrigating farmlands on both banks of Tarim River Basin is extremely scarce at present.

How to alleviate the shortage of irrigation water without excessive salinity in cotton fields has become a key research topic in Southern Xinjiang. A well-canal combined irrigation mode has become an effective method for solving this problem. Some scholars in relevant fields have carried out some researches. For example, by studying changes of groundwater level in Baojixia irrigation area, Shaanxi province, China, Du Wei obtained a scheme for groundwater extraction under well-canal combined conditions [1]. In view of the problems existing in paddy fields in irrigation area of Sanjiang plain in the northeast of China under a single irrigation mode, Huo et al. [2] analyzed high efficiency and energy saving of well-canal combined irrigation. Biswadip et al. [3] studied the optimal allocation policies of land and water resources in sustainable irrigated agriculture. Although some scholars have conducted similar researches, most of the existing research results target at specific problems in specific regions, lacking universality. Because the south of Xinjiang Uygur Autonomous Region, belonging to an extremely arid region, is constrained by extremely arid climate and complicated geographical environment, evaluation on irrigation effects of cotton fields in the region is limited by factors in multiple aspects. In addition, many indexes and variables need to be considered, a complex system is involved, and the constraint boundary of control variables is fuzzy, so it is difficult to describe the irrigation effect using a unified model and paradigm. This study attempted to adopt the method combining fuzzy system with neural network prediction to evaluate the well-canal combined irrigation mode used in cotton fields in south of Xinjiang Uygur Autonomous Region, aiming to select the optimal irrigation mode. The research is of practical significance for water-saving irrigation and improvement of ecological environment in the region.

2 Flow and Steps of the Algorithm

2.1 Flow of Algorithm for Fuzzy Neural Network

See Fig. 1.

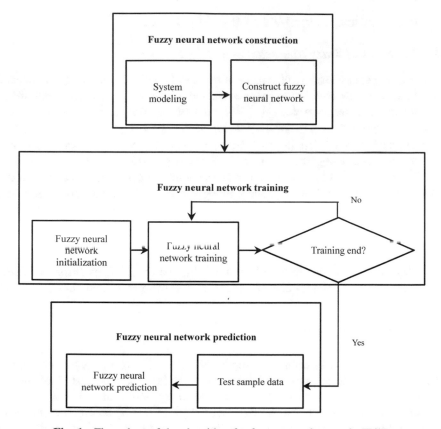

Fig. 1. Flow chart of the algorithm for fuzzy neutral network (FNN)

2.2   Steps of the Algorithm

Step 1: After constructing the fuzzy neutral network (FNN), input and output nodes as well as iterations in the network structure are determined.

Step 2: Parameters of FNN and fuzzy membership are initialized and training data are normalized.

Step 3: During training process of the FNN, training samples are used for training neutral network, so as to correct coefficient and membership parameter.

Step 4: During prediction of the FNN, inverse normalization is continuously performed on the predicted values of the network using the normalized data predicted through the network.

Step 5: The optimal irrigation mode for solving actual problems is determined by the predicted value of the neural network.

3   Mathematical Principles of the Algorithm

3.1   Principles of Fuzzy Mathematics and FNN

The basic concept in fuzzy mathematics is membership degree and fuzzy membership function. Of them, the membership degree refers to the degree of element u belonging to a fuzzy subset f and is expressed as $\mu_f(u)$, a figure in the range of $[0,1]$. The closer the $\mu_f(u)$ to zero, the smaller the degree of u belonging to the fuzzy subset f; the closer to 1, the larger the degree of u belonging to f. Fuzzy membership function is the function for quantitatively calculating the membership degrees of elements. FNN is established according to principle of fuzzy system and each node and parameter in the network has corresponding meaning. During network initialization, the initial value of these parameters can be determined according to the fuzzy or actual situations of the system [4].

3.2   Fuzzy Function and FNN

For Fuzzy function, please refer to literatures [5], and for FNN, its Error calculation, Coefficient correction and Parameter correction, please refer to literatures [5] and [6].

4   Empirical Analysis

In the test, cotton, a major economic crop in the south of Xinjiang Uygur Autonomous Region was used as the research object. By carrying out field test, measures for planting and managing cotton fields, such as weeding, fertilization, chemical control and forced ripening in each test site were consistent. Five irrigation modes were set in the test and they were mode 1 (well irrigation), mode 2 (mixed irrigation of 3/4 well water and 1/4 canal water), mode 3 (mixed irrigation of 1/2 well water and 1/2 canal water), mode 4 (mixed irrigation of 1/4 well water and 3/4 canal water) and mode 5 (canal irrigation). Each irrigation mode was repeated for three times. In the test site with the area of 2 m × 3 m, the groups were randomly arranged, and the cotton was planted in four rows with spacing of 30 cm × 60 cm × 30 cm with two pipes for drip irrigation under one sheet of plastic film mulching. The width of the film was 120 cm and the film spacing was 20 cm. This study mainly discussed the influences of the five different irrigation modes for farmland on growth conditions of cotton crop. The soil electric conductivity, iron concentration and growth conditions of the crop under different irrigation modes were determined, so as to seek the optimal irrigation mode.

4.1   Prediction Image of Training Data Based on FNN Algorithm

To explore the effects of different irrigation modes for farmland on growth conditions of the cotton crop, 10 groups of data were separately obtained under five irrigation modes by determining soil electric conductivity, ion concentration and growth conditions of the crop under different irrigation modes. Moreover, this research analyzed the data by using FNN algorithm and discussed the optimal irrigation mode suiting for cotton growth.

Firstly, the basis for determining network structure is the dimension of training samples and parameters and coefficients of membership function of FNN were initialized and relevant training samples were normalized. Secondly, FNN was trained for 100 times based on training samples, so as to correct coefficient and parameter of membership degree and update each changing value. Moreover, by predicting training data, the prediction image of the training data is obtained and demonstrated in Fig. 2.

As shown in Fig. 2, the actual output and predicted output of the training data are well fitted and an error fluctuates slightly. This indicates that the network structure of the algorithm is established completely and the model is accurate in prediction after parameter correction.

FNN training was performed on the test data that are similar to experimental data and inverse normalization was conducted on prediction results, so as to analyze whether this algorithm has the same adaptability to similar data. The prediction image is obtained after training test data through the FNN, as shown in Fig. 3.

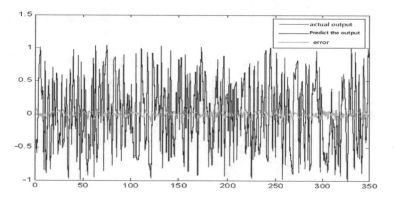

Fig. 2. Prediction of training data

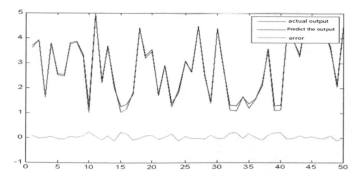

Fig. 3. Prediction image of training data based on FNN

It can be seen from Fig. 3 that predicted output obtained using test data can well fit actual output with error fluctuating in a small range. In view of the experimental data, the above five irrigation modes were evaluated through the well-trained FNN. According to the predicted value of the network, the evaluation indexes for the advantages and disadvantages of the five irrigation modes were obtained and 10 times of predictions were performed on the experimental data in total. Furthermore, by comparing five groups of predicted data at each prediction point, prediction results and the level of advantages and disadvantages of each prediction point can be obtained.

4.2   Results

The five groups of experimental data were normalized and calculated through the well-trained FNN. Moreover, the results were combined on a single graph for analysis and comparison to obtain the qualities of the five irrigation modes, thus obtaining the optimal irrigation mode.

Ten groups of predicted data were obtained based on the FNN and evaluation grades of the five irrigation modes were sorted out and demonstrated in Table 1.

Table 1. Evaluation grades of the five irrigation modes

	1	2	3	4	5	6	7	8	9	10
Mode 1	2	3	1	2	3	3	3	4	3	4
Mode 2	5	2	4	4	5	4	1	5	4	5
Mode 3	4	4	5	5	2	1	4	2	1	3
Mode 4	3	1	2	1	1	2	2	1	2	1
Mode 5	1	5	3	3	4	5	5	3	5	2

By comparing and analyzing Fig. 4 and Table 1, the mode 4 shows the best overall effects among the five irrigation modes, that is, the mixed irrigation mode of 1/4 well water and 3/4 canal water can reach the best irrigation effects.

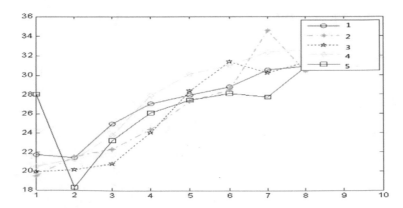

Fig. 4. Comparison and evaluation of the five irrigation modes

5   Conclusions

By establishing the algorithm of FNN for evaluating multiple well-canal combined irrigation modes in the cotton fields, this study performed FNN training on the training data obtained from the experiment by using this evaluation algorithm. Base on this, the actual output and predicted output of training data were well fitted with a small error fluctuation. This indicates that the network structure of the algorithm is well established and prediction using the model is accurate after parameter correction. Then, FNN training was conducted on the test data similar to the experimental data and inverse normalization was carried out on prediction results. The analysis showed that this algorithm had good adaptability to process the same type of data. Finally, data processed under five different irrigation modes in this experiment were predicted by using the evaluation algorithm and the prediction results were graded. Based on the figure for evaluating quality and table of evaluation grades of the five modes, irrigation mode 4 (mixture of 1/4 well water and 3/4 canal water) for cotton reached the optimal level among the different irrigation modes, so the treatment 4 is the optimal.

Acknowledgements. This research was supported by the Chinese Universities Scientific Fund (No. 2019TC158), the President Foundation Youth Project of Tarim University (TDZKQN-201610) and the National Natural Science Foundation of China (No. 61563046).

References

1. Du, W.: Study on the well-canal combined regulation mode in irrigation area based on efficient and safe water use. Northwest Agriculture and Forestry University (2014)
2. Huo, H.Y., Bai, K., Sun, S.Y.: Discussion on well-canal combined irrigation mode in paddy field irrigation area of Sanjiang plain. China Rural Water Hydropower **8**, 515–516 (2006)
3. Biswadip, D., Ajay, S., Sudhindra, N.P., Hiroshi, Y.: Optimal land and water resources allocation policies for sustainable irrigated agriculture. Land Use Policy **42**, 527–537 (2015)
4. Xie, J.J., Liu, C.P.: Fuzzy mathematical method and its application. Huazhong University of Science and Technology Press (2000)
5. Wang, X.C., Shi, F., Yu, L., Li, Y.: Analysis of Fourty-Three Cases of Matlab Neural Network. Beihang University Press (2013)
6. Hao, T.: Classification of traditional Chinese medicine syndromes on clinical symptoms based on artificial neural network algorithm. Lanzhou University (2017)

Grape Downy Mildew On-line Detection Based on Accurate Image Processing Method Preformed on Embedded Ambient Intelligence System

Peifeng Xu[1,2], Qiyou Jiang[2], Zhongying Zhao[2], Nıng Yang[1(✉)], and Rongbiao Zhang[1]

[1] School of Electrical and Information Engineering, Jiangsu University, Zhenjiang 212013, China
505657969@qq.com
[2] Jiangsu Vocational College of Agriculture and Forestry, Jurong 212400, China

Abstract. In this paper, an accurate and intelligent Grape Downy Mildew (GDM) detection method based on common image processing and artificial neural network (ANNs) is proposed. In view of the structure of grape leaves and the distribution characteristics of GDM on leaves, firstly, image processing is mainly to solve the extraction of downy mildew areas on leaves, which is mainly composed of gray-scale processing, gray-scale linear transformation and binarization, and the effects of different processing algorithms and parameters on the results are discussed. Secondly, an ANNs is used to reduce the interference of grape leaf vein on the spot area statistics. Finally, the hardware and software platform based on ARM9 integrated processor, Linux and QT is used to realize the system integration. Compared with the traditional detection method, the accuracy of this detection method can reach 97%, which is close to the accuracy of the human eye, which can be used as an ambient intelligence system in the vineyard inspection, automatic completion of grape GDM detection and grading.

Keywords: Grape Downy Mildew (GDM) · Image processing · Artificial Neural Network (ANNs) · Embedded ambient intelligence system

1 Introduction

Grape is one of the most widely grown fruit crops in the world. It can be directly consumed fresh or processed into raisins and is the main raw material for wine, brandy or other non-fermented beverages. So the quantity and quality of grapes are crucial to the development of the national economy. But grapes are affected by downy mildew, a serious fungal disease caused by Plasmoparaviticola [1]. Incidence and severity of GDM were quantified weekly from the first symptoms appearance on leaves. It has the ability to destroy entire grape plantation and cause great loss to the farmers. In 2010, 1500 hectares' grapes are infected by GDM in Jianou, China, causing a loss up to fifty percent.

© Springer Nature Switzerland AG 2020
F. Xhafa et al. (Eds.): IISA 2019, AISC 1084, pp. 144–154, 2020.
https://doi.org/10.1007/978-3-030-34387-3_18

Traditional GDM detection methods mainly depends on naked-eye observation. To detect the infection in a timely way, experienced farmers have to patrol day and night, which consumes a large number of man-hours [2]. Apart from the naked-eye detection method, there are many other GDM detection methods, most of which are biological methods. Application of PCR for GDM detection is a biological method based on the differences among sequenced and reported cox2 gene sequences of GDM. Loop-mediated isothermal amplification (LAMP) is a novel molecular biological detection method. It uses four specific primers for six regions of the target gene, analysis follows rapid amplification with Bst DNA polymerase [3]. Although these biological methods may have achieved precise results, they are not real-time and on-line.

With the development of precision agriculture, image processing technology is increasingly applied to crop disease identification and degradation [4, 5]. Zou used multiple color cameras to scan the surface of bi-colored apples, and employed thresholding to automatically grade the fruits as normal or defective [6]. Baum managed to separate barley's plaque from its back-ground image with Sobel edge detection [7]. Hence, applying image processing technology to crop disease identification and degradation has many advantages over traditional methods, such as accuracy, and immediacy [8–12]. Image processing technology to GDM detection has not been implemented because downy mildew spot is parallel to grape leaf vein, making it hard to distinguish infected leaves with the naked eyes. This problem is overcome with the application of ANNs that can distinguish the leaf vein.

A highly efficient, timely, accurate and intelligent on-line detection method based on image processing method preformed on embedded ambient intelligence system is proposed in this paper. Using the proposed method, a robot can make an inspection tour in the grape field and can complete the grape GDM detection, diagnosis and grading automatically.

2 Intelligent Image Processing

2.1 Holistic Methods Organization Chart

Image detection methods generally have six steps: preprocessing, window sliding, feature extraction, feature selection, feature classification and post-processing. For different detection objects, the steps can be adjusted or require some special treatment [13, 14]. The chart of system's holistic methods organization includes the processes of image captured, image processing and identification, diagnosis and grading of disease. The logical structure of the model is shown in Fig. 1. Because the veins of grape leaves are thicker and the color is similar to the GDM disease spots, the two mixed together are difficult to distinguish after image processing, which affects the accuracy of detection results. Therefore, this paper adopts the ANNs method to solve this problem.

2.2 Image Processing

This method uses an ARM microcontroller to manipulate the USB camera to capture images. The images were taken from Jiangsu University's experimental greenhouse on

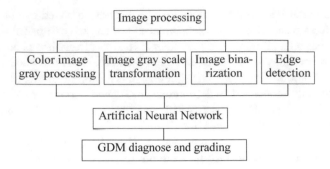

Fig. 1. Holistic methods organization chart

5, May, 2016. Figure 2 is one of the partial enlarged image of the diseased leaf suffering GDM. Image processing is based on mature technology, but the detailed process is omitted due to limited length.

Fig. 2. The partial enlarged image

The image was pretreated by graying firstly, by comparing four gray processing methods according to literature [17], and R component method is selected finally, Fig. 3 shows the resulted.

After gray processing, the image is still has noise and the contrast can't live up to expectations, Fig. 4 shows GDM's gray level histogram. Thus, the method of gray scale linear transformation was applied to increase the contrast [22], according to the pixel distribution in the figure, Fig. 5 shows the image after liner transformation, Fig. 6 shows the image's gray level histogram after liner transformation.

Using 0 or 1 to represent the gray image's all pixels is called image binarization. After using this method, the data to be processed is greatly reduced and it's easier to process this simpler image. In this research, we take the OTSU method after comparison [23]. Figure 11 shows the result after image binarization in the OTSU method.

In the spatial domain of the image, the number of pixels that the region contains is used to represent the region's area. So, the area of GDM in this image can be calculated and recorded as M.

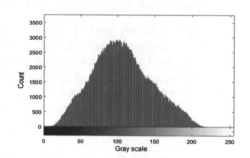

Fig. 3. R component method

Fig. 4. GDM's gray level histogram transformation

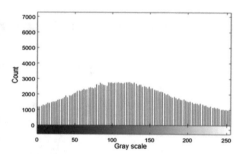

Fig. 5. The image after liner transformation

Fig. 6. The image's gray level histogram after linear transformation

In addition, the initial area of GDM can be calculated with the method of edge detection and recorded as S.

Fig. 7. The image after binarization

2.3   Decrease of Disturbance of Leaf Vein

From Fig. 7, it is clear that leaf vein is the main disturbance in the detection of disease area after three image processing steps. We have found that leaf vein is of relatively regular shape, while the shape of disease area is very irregular. The key to decreasing the disturbance of leaf vein is getting the diseased leaf's vein area. However, it is difficult to get the diseased leaf's vein area directly. The leaf vein area is relevant to leaf's whole area, but the relationship between them is unknown. Thus, a statistical approach based on Artificial Neural Networks (ANNs) is taken to establish the relationship.

In machine learning and cognitive science, ANNs are used to estimate or approximate functions that can depend on a large number of inputs and outputs are generally unknown. ANNs are typically specified using three things: architecture, activity rule and learning rule. As is shown in Fig. 8, an artificial neural network is an interconnected group of nodes, akin to the vast network of neurons in a brain. Here, each circular node represents an artificial neuron and an arrow represents a connection from the output of one neuron to the input of another [15, 16].

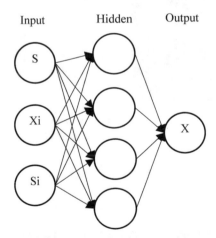

Fig. 8. The ANN structure chart

In this paper, flow diagram Fig. 9 demonstrates the method's process to remove the disturbance of the leaf vein. The number of pixels that the image contains were used to represent the image's area. To avoid the disturbances coming from external environment and different processing methods, the method's process to remove the disturbance of the leaf vein was operated in the designed system, in which system GDM image was processed. Firstly, 100 healthy leaf samples whose size was different from each other were taken. Secondly, every sample was processed by the method same includes four steps as that proposed in this paper. After these steps, the area of the samples and the samples' vein can be calculated and respectively recorded as S_i and X_i ($i = 1, 2, \ldots, 100$), whose unit is pt. According to the foregoing, the area of the diseased leaf whose vein area was needed to be got was measured and recorded as S. Finally, S, Xi and Si were as the input parameter of ANNs, and the area of vein X of diseased leaf was obtained.

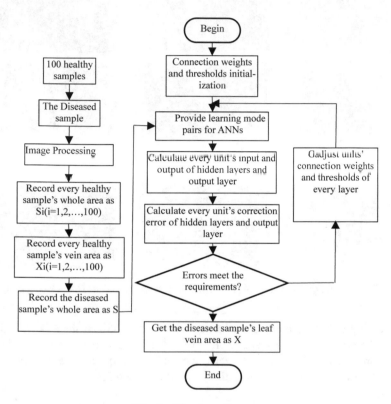

Fig. 9. The flow diagram

Here, images of three samples' contour extraction and leaf vein extraction are shown in Fig. 10. The number of pixels that samples' S and X contain is shown in the statistical Table 1.

Table 1. The number of pixels that samples' S and X contains.

Sample	S(pt)	X(pt)	X/S
1	634195	3863	6.09×10^{-3}
2	521326	3534	6.78×10^{-3}
3	481533	2106	43.7×10^{-3}

Fig. 10. Images of three samples' contour extraction and leaf vein extraction

According to the large number of statistics, we can find that there is a relationship between the area of a grape leaf and the grape leaf vein's area. Figure 11 shows the relationship between the area of a grape leaf and the grape leaf vein's area. The R-square which represents the curve fitting degree is 0.75563.

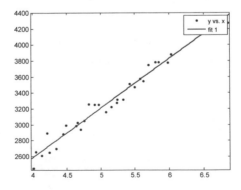

Fig. 11. The relationship between the area of a grape leaf and the grape leaf vein's area.

Among those leafs who contain pixels ranging from 4×10^5 to 6×10^5, the relationship between the area of a grape leaf and the grape leaf vein's area can be expressed by a calculation Eq. (1).

$$X \approx S \times 0.0063 + 92.1478 \tag{1}$$

Based on image binarization, the more accurate area of GDM is M minus X.

3   System Design

In this research, for hardware, SAMSUNG's S3C2440A is chosen as the microprocessor, the web camera connected to ARM microcontroller through USB. The entire process and results are displayed on the LCD. For software, embedded Linux is chosen as the operating system and Qt is used to write application software programs [18–21, 24, 25]. Figure 12 shows the system's block diagram, and Fig. 13 shows the graphics import interface.

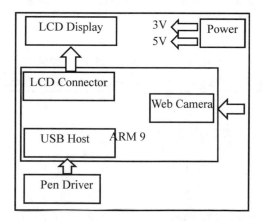

Fig. 12. The system block diagram

Fig. 13. The graphics of system interface

4   GDM Diagnose, Grading and Preventive Measures

GDM grading is an important parameter to measure the prevalence of crop diseases. The description of plant disease from slight to severe is usually expressed as a percentage of the lesion area of the leaf and the entire leaf area [26–28]. In the spatial domain of the image, it is also measured by the proportion of pixels in the plant image matrix, that is, the ratio between the total pixels of the lesion area and the leaf area. The formula for calculating area percentage (s) occupied by diseased region was as follow [29]:

$$
S = \frac{A_d}{A_l} = \frac{P\sum_{(x,y)\in R_d} 1}{P\sum_{(x,y)\in R_l} 1} \times 100\% = \frac{\sum_{(x,y)\in R_d} 1}{\sum_{(x,y)\in R_l} 1} \times 100\% \tag{2}
$$

Here A_d is the area of GDM, A_l is the whole leaf area, P is the area of an unit pixel, R_d is the region of GDM, R_l is the region of the whole leaf. s is the percentage of GDM's area in the whole leaf area.

We divided GDM into three grades according to severity, on this basis, the relationship model of leaf damage size, application concentration and application type was established [30], as is shown in Table 2.

Table 2. Grade classification of GDM.

GDM grade	The percentage of lesion size in the leaf's total area
T_0	$0 \leq K \leq 0.1$
T_1	$0.1 \leq K \leq 0.3$
T_2	$0.3 \leq K \leq 1$

5   Conclusion

In this paper, an efficient automated GDM detection method based on accurate intelligent image processing techniques preformed on embedded ambient intelligence system is proposed. The proposed method adopts image processing techniques and applies ANNs to decrease the disturbance of leaf vein. The whole detection process is performed on the embedded Linux system using ARM9.

Compared with human eye vision, machine vision has stronger grayscale resolution, better spatial resolution, larger photographic range, higher precision and stronger environmental adaptation. Machine vision has a grayscale resolution of 256 grayscales while human eye vision only has a grayscale resolution of 64 grayscales. Human eye vision can't capture small target while machine vision can capture target from micron grade to celestial body using different kinds of optical lens. Machine vision's photographic range is from ultraviolet to infrared while human eye vision's photographic range is only visible light whose wave length is from 400 nm to 750 nm.

Machine vision precision can reach micron grade and can be quantified but human eye vision has poor precision and can't be quantified. Human eye vision has poor adaptation to environment temperature and humidity but machine vision overcomes this shortcoming.

By comparing with the judgment and diagnosis of experts, the proposed method has an accuracy rate up to 97%, which is close to human eyes' judgment accuracy. Because of its intelligence, this proposed system can be used as a robot being able to make an inspection tour in the grape field, which can complete the grape GDM detection, diagnosis and grading automatically.

Acknowledgments. This work was financially supported by the Chinese National Natural Science Foundation (Grant No. 61673195), The University "Blue_Cyanine project" training plan of Jiangsu, Chinese National Natural Science Foundation (Grant No. 31701324), Science and Technology projects plan of Jiangsu Vocational College of Agriculture and Forootry (Grant No. 2018kj11 and 2018kj12).

References

1. Carisse, O.: Development of grape downy mildew (Plasmopara viticola) under northern viticulture conditions: influence of fall disease incidence. Eur. J. Plant Pathol. **144**(4), 773–783 (2016)
2. Bem, B.P.D., Bogo, A., Everhart, S., Casa, R.T., Gonçalves, M.J., Filho, J.L.M.: Effect of y-trellis and vertical shoot positioning training systems on downy mildew and botrytis bunch rot of grape in highlands of southern Brazil. Sci. Hortic. **185**, 162–166 (2015)
3. Kong, X., Qin, W., Huang, X., Kong, F., Schoen, C.D., Jie, F.: Development and application of loop-mediated isothermal amplification (lamp) for detection of plasmopara viticola. Sci. Rep. **6**, 28935 (2016)
4. Rupil, C.: Development of image processing and its applications on cryptography. Int. J. Eng. Sci. Res. Technol. **4**(7), 1160–1162 (2015)
5. Dutta, M.K., Sengar, N., Minhas, N., Sarkar, B., Goon, A., Banerjee, K.: Image processing based classification of grapes after pesticide exposure. LWT-Food Sci. Technol. **72**, 368–376 (2016)
6. Zou, X., Zhao, J., Li, Y., Mel, H.: In-line detection of apple defects using three color cameras system. Comput. Electron. Agric. **70**(1), 129–134 (2010)
7. Baum, T., Navarro-Quezada, A., Knogge, W., Douchkov, D., Seiffert, U.: HyphArea—automated analysis of spatiotemporal fungal patterns. J. Plant Physiol. **168**(1), 72–78 (2011)
8. Woebbecke, D.M., Meyer, G.E., Von Bargen, K., Mortensen, D.A.: Shape features for identifying young weeds using image analysis. Trans. ASAE **38**(1), 271–281 (1995)
9. Jagadeesh, D.P., Yakkundimath, R., Byadgi, A.S.: Image processing based detection of fungal diseases in plants. Proc. Comput. Sci. **46**, 1802–1808 (2015)
10. Koumpouros, Y., Mahaman, B.D., Maliappis, M., Passam, H.C., Sideridis, A.B., Zorkadis, V.: Image processing for distance diagnosis in pest management. Comput. Electron. Agric. **44**(2), 121–131 (2004)
11. Manisha, B., Hingoliwalab, H.A.: Smart farming: pomegranate disease detection using image processing. Proc. Comput. Sci. **58**, 280–288 (2015)
12. Meng, T., Shyu, M.L.: Biological image temporal stage classification via multi-layer model collaboration. In: IEEE International Symposium on Multimedia, pp. 30–37 (2013)

13. Yasutake, T.: U.S. Patent No. 5,483,261. U.S. Patent and Trademark Office, Washington, DC (1996)
14. Fathy, M., Siyal, M.Y.: An image detection technique based on morphological edge detection and background differencing for real-time traffic analysis. Pattern Recogn. Lett. **16** (12), 1321–1330 (1995)
15. Masson, E., Wang, Y.J.: Introduction to computation and learning in artificial neural networks. Eur. J. Oper. Res. **47**(1), 1–28 (1990)
16. Khadse, C.B., Chaudhari, M.A., Borghate, V.B.: Conjugate gradient back-propagation based artificial neural network for real time power quality assessment. Int. J. Electr. Power Energy Syst. **82**, 197–206 (2016)
17. Schowengerdt, R.A.: Techniques for Image Processing and Classifications in Remote Sensing. Academic Press (2012)
18. Manifavas, C., Hatzivasilis, G., Fysarakis, K., Yannis, P.: A survey of lightweight stream ciphers for embedded systems. Secur. Commun. Netw. **9**(10), 1226–1246 (2016)
19. Malinowski, A., Yu, H.: Comparison of embedded system design for industrial applications. IEEE Trans. Industr. Inf. **7**(2), 244–254 (2011)
20. Goodacre, J., Sloss, A.N.: Parallelism and the ARM instruction set architecture. Computer **38**(7), 42–50 (2005)
21. Vijay Babu, M.: Real-time object detection based on ARM9. Int. J. Eng. Trends Technol. **4** (9), 4080–4083 (2013)
22. Munteanu, C., Rosa, A.: Gray-scale image enhancement as an automatic process driven by evolution. IEEE Trans. Syst. Man Cybern. Part B Cybern. **34**(2), 1292–1298 (2004)
23. Liu, J.Z., Li, W.Q., Tian, Y.P.: Automatic thresholding of gray-level pictures using two-dimension Otsu method. In: 1991 International Conference on Circuits and Systems, China. IEEE (1991)
24. Zang, C.Q., Gao, M.Y., He, Z.W.: The transplantion and realization of Qt4. 7.0 based on ARM9 and Linux. Appl. Mech. Mater. **719–720**, 527–533 (2015)
25. Patel, K.S., Kalpesh, R.J.: Implementation of embedded ARM9 platform using Qt and openCV for human upper body detection. IOSR J. Electron. Commun. Eng. **9**(2), 73–79 (2014)
26. Zhu, S.P., Xu, H.R., Ying, Y.B., Jiang, H.Y.: Imaging processing technique to measure plant infection severity. In: Proceedings of the SPIE - The International Society for Optical Engineering, vol. 6381 (2006)
27. Lydia, B., Jumel, S., Picault, H., Domin, C., Lebreton, L., Ribulé, A., Delourme, R.: An easy, rapid and accurate method to quantify plant disease severity: application to phoma stem canker leaf spots. Eur. J. Plant Pathol. **145**(3), 1–13 (2015)
28. Kole, D.K., Ghosh, A., Mitra, S.: Detection of downy mildew disease present in the grape leaves based on fuzzy set theory. In: Advanced Computing, Networking and Informatics, vol. 1, pp. 377–384. Springer (2014)
29. Chen, Z.L., Zhang, C.L., Shen, W.Z., Chen, X.X.: Grading method of leaf spot disease based on image processing. J. Agric. Mech. Res. **11**, 73–75 (2008)
30. Tian, Y.W., Chen, X.: The crops leaf disease grade system based on embedded system. J. Shenyang Agric. Univ. **45**(6), 756–760 (2014)

An Improved Bisecting K-Means Text Clustering Method

Ye Zi, Liang Kun$^{(\boxtimes)}$, Zhiyuan Zhang, Chunfeng Wang,
and Zhe Peng

College of Computer Science and Information Engineering, Tianjin University of
Science and Technology, Tianjin, China
liangkun@tust.edu.cn

Abstract. Bisecting K-means clustering method belongs to the hierarchical algorithm in text clustering, in which the selection of K value and initial center of mass will affect the final result of clustering. Chinese word segmentation has the characteristics of vague word and word boundary, etc. We transformed the corpus into word vector by word2vec, reduced the dimension of data by ontology modeling, and cleaned the data by jieba word segmentation and TF-IDF to improve the accuracy of the data. We propose an improved algorithm based on hierarchical clustering and Bisecting K-means clustering to cluster the data many times until it converges. Through experiments, it is proved that the clustering result of this method is better than that of K-means clustering algorithm and Bisecting K-means clustering algorithm.

Keywords: Text clustering · Bisecting K-means · Ontology theory · Hierarchical clustering

1 Introduction

With the explosive growth of current network information, how to obtain useful information quickly, accurately and effectively in a large amount of text information has become a research hotspot [1]. Text clustering is not only a branch of data mining, but also a hot spot in the field of data mining. Text clustering can summarize and classify different document information in the network and preprocess it, so that the results are more perfect, and the content on the Internet is more structured and clearer. Chinese word segmentation [2] refers to the segmentation of Chinese characters into a single word. This paper discusses the application of text clustering in this field by crawling real estate corpus by crawler program, and uses word2vec, ontology theory and improved Bisecting K-means clustering algorithm to make use of text clustering. It makes the content after clustering more structured and clearer.

© Springer Nature Switzerland AG 2020
F. Xhafa et al. (Eds.): IISA 2019, AISC 1084, pp. 155–162, 2020.
https://doi.org/10.1007/978-3-030-34387-3_19

2   Related Work

In the early days, the research on text clustering adopted a rule-based method, which could solve some simple problems, but could not be put into use fundamentally. Later, with the development of the Internet and big data, the method based on statistics appeared. Text clustering has made a breakthrough.

In [3], the author proposes an improved K-means text clustering algorithm. This method has better clustering quality than the traditional K-means clustering algorithm. Peng et al. [4] improved K-means clustering algorithm to deal with large-scale data with insufficient scalability. This method has good clustering effect and scalability in large-scale text clustering. In [5], the Bisecting K-means clustering algorithm is improved. The clustering performance of this algorithm is better than that of K-means algorithm and Bisecting K-means clustering algorithm. In [6], the algorithm is parallelized by the idea of data parallelism and the strategy of uniform partitioning. The improved algorithm has better speedup and efficiency. In [7], the combination of keywords, K-means clustering algorithm and synonyms is used to improve the accuracy of topic classification. The algorithm not only can effectively classify topics, but also can mine the potential meaning of words.

3   Data Processing

3.1   Corpus Acquisition

There are many methods to obtain text corpus, such as through standard open test data set (Sogou corpus, Wikidata, encyclopedia website, etc.), or through web crawler to obtain web text. In this paper, we choose to obtain corpus from Wikidata and crawlers.

The steps of using Wikidata to obtain corpus are as follows:

> Step 1: download the Chinese Wiki Dump, containing the text, title, and other data.
> Step 2: use Wikipedia Extractor to extract text.
> Step 3: get the text corpus in .txt format, convert it to simple and complicated, and use the open source OpenCV project.

3.2   Data Extraction Using Word2Vec

The word vector is trained by jieba word segmentation and neural network language model in word2vec [8]. Word2vec has two models: the CBOW model and the Skip_Gram model. CBOW is used to predict the current value through context, with low computational complexity and high accuracy, so it is suitable for processing text with large amount of data. Skip_Gram is based on the current value to predict the context, high accuracy, but the calculation is more complex, suitable for processing small data text. In this paper, Skip_Gram is selected to train the word vector. Through the corpus obtained above, the words are mapped to the high-dimensional vector space, which not only solves the dimension problem of the word vector, but also takes into account the language relationship between the words, so that the word vector has better semantic representation ability. The principle of the Skip_Gram principle explanation diagram is shown in Fig. 1.

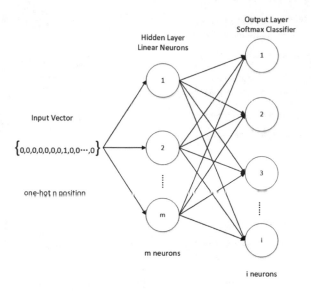

Fig. 1. Skip_Gram principle explanation diagram.

In this paper, the precise pattern of jieba word segmentation is used to segment the corpus obtained above. And calculate the number of words, define the word list length as *num*, to construct the one-hot word vector, and then use the model of Skip_Gram in word2vec to train the word vector, and use the sliding window to obtain the training sample. The generation of the training sample is shown in Fig. 2, we set the size of the sliding window is 2.

3.3   Using TF-IDF to Select Feature Word Vector

TF-IDF is used to assess the importance of words in a text set or corpus [9, 10]. We use TF-IDF to reduce the dimension of data. First of all, the precise pattern of jieba word segmentation is used to segment the corpus, and the TF-IDF value of each word is sorted, and the top N words in the TF-IDF value are selected as the feature word vector of the knowledge cluster [11]. The TF-IDF algorithm process is shown in Fig. 3.

4   The Improved Bisecting K-Means Text Clustering Method

The Bisecting K-means algorithm is based on the K-means algorithm. To a certain extent, it improved the K-means clustering algorithm and obtained good clustering results. However, in some aspects, Bisecting K-means clustering algorithm and K-means clustering algorithm have something in common. For example, clustering

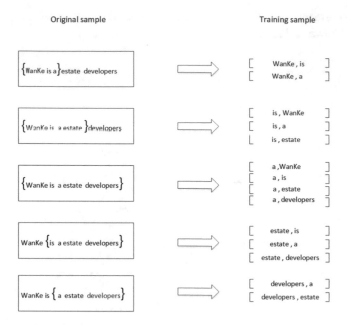

Fig. 2. Original sample to training samples by Chinese word segmentation.

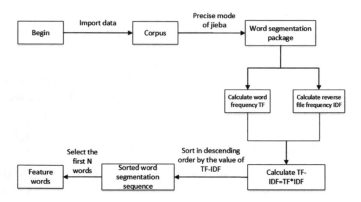

Fig. 3. Flowchart of TF-IDF algorithm.

requires random selection of centroids, and the final clustering results are usually affected by the number of clustering K values.

4.1   Bisecting K-Means Clustering

The idea of Bisecting K-means clustering algorithm: there is a data set $X = \{x_1, x_2, x_3, \ldots, x_n\}$, and the number of clusters is artificially specified as K before clustering. Then, each time a cluster C_i, is selected from the cluster table, the C_i is classified by K-means clustering algorithm, two clusters with the minimum total SSE

are selected, the two clusters are placed in the cluster table S, the centroids of the divided clusters are deleted, and the cluster table S is updated. Repeat the above process until there is a K cluster in the cluster table. Among them, SSE is the sum of error squares, also known as divergence, which is used as the criterion to evaluate the quality of clustering. The definition formula of SSE is as follows (1). Where c_i is the center of mass of cluster C_i and x is a sample point in cluster C_i.

$$SSE = \sum_i^k \sum_{x \in C_i} dist(c_i, x)^2 \tag{1}$$

4.2   Improved Bisecting K-Means Algorithm

The Bisecting K-means algorithm needs multiple K-means clustering to select the cluster of the minimum total SSE as the final clustering result, but still uses the K-means algorithm, and the selection of the number of clusters and the random selection of initial centroids will affect the final clustering effect. In order to solve this problem, this paper uses bottom-up hierarchical clustering to improve Bisecting K-means clustering. The improved algorithm does not need to specify the number of K values, and the discriminant condition can be used to make it converge automatically. As a result, a good clustering effect is obtained.

The process of the algorithm is as follows:

First of all, let the sample set in the cluster C_i be $\{x_{i1}, x_{i2}, \ldots, x_{im}\}$, c_i as the center of mass. Define a measurement function J, which is defined as follows (2):

$$J = \sqrt{\sum_{i=1}^k \sum_{j=1}^m (x_{ij} - c_i)^2} \tag{2}$$

Step 1: set the sample set of the data object to $X = \{x_1, x_2, \ldots, x_n\}$. Set the center of the initial sample set to the initial center of mass, add it to the cluster table S, calculate J^1, and set the number of cycles $P = 1$.

Step 2: each cluster is cyclically selected from the cluster table S, and the cluster is bisected by bottom-up hierarchical clustering, the current SSE value is recorded, and the optimal partition is selected by comparing the values of the SSE. Then calculate the center of the divided cluster as its centroid c_a, c_b.

Step 3: add the center of mass of the new cluster to the cluster table S, delete the center of mass of the decomposed cluster, increase the number of cycles P by 1, and calculate the J^P at this time.

Step 4: set $\varepsilon = \left| \frac{J^P - J^{P-1}}{J^P} \right|$, if $\varepsilon \leq \Delta$, then end the algorithm and output the clustering result; if $\varepsilon > \Delta$, return to step 2 to continue the classification until ε is small enough.

The flow chart of the improved Bisecting K-means clustering algorithm is shown in Fig. 4:

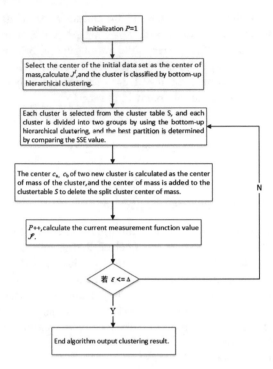

Fig. 4. Flowchart of algorithm optimization.

5 Experimental Results Analysis

5.1 Experimental Process

This paper obtains the corpus of 31.69 MB about real estate from Wikidata. Then we use the model Skip_Gram to process the corpus and convert it into word vector. The experiment sets the dimension of the word vector to 128, sets the sliding size of the model window to 5, and draws the dimension of the trained word vector. Figure 5 is a graph of the training results of a part of the word vector. Then use TF-IDF and ontology theory to clean the trained word vector, and reduce the dimension of the word vector. Figure 6 is a part of the data cleaning result diagram. Finally, the cleaned word vectors are clustered by the Bisecting K-means clustering algorithm and the improved Bisecting K-means clustering algorithm.

5.2 Experimental Results

The Bisecting K-means clustering algorithm and the improved Bisecting K-means clustering algorithm have the same optimal number of clusters on the data. The comparison of the experimental results is shown in Fig. 7. The experimental results show that the improved Bisecting K-means algorithm has a better effect on text clustering. The comparison chart is shown in Fig. 7. The x-axis represents the number of clusters and the y-axis represents the metric of the cluster.

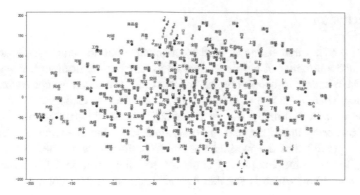

Fig. 5. Training results

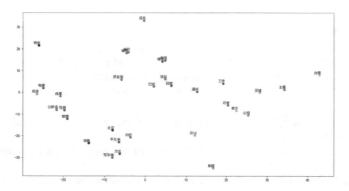

Fig. 6. Data cleaning results

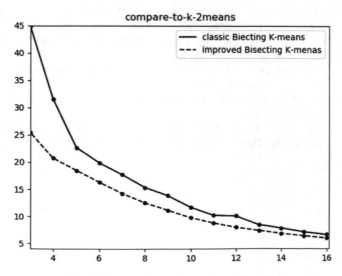

Fig. 7. Comparison of experimental results

6 Conclusion

In view of the fact that Chinese word segmentation has the characteristics of vague boundary and ambiguity of words and words in Chinese environment, this paper uses crawler and Wikidata to obtain corpus, and then uses word2vec to convert corpus into word vector. Then TF-IDF and ontology theory are used to clean the word vector and reduce the dimension. Finally, the word vector is clustered by the improved Bisecting K-means clustering algorithm and the traditional Bisecting K-means clustering algorithm. It is proved that the method has better performance in the accuracy of text clustering, but the time complexity is higher than the traditional Bisecting K-means clustering algorithm, which is suitable for data with small data volume.

Acknowledgements. This work was partially supported by NSFC (No. 61807024).

References

1. Zhang, Y., Huang, T., Lin, K., Zhang, Q.: An improved K-means text clustering algorithm. J. Guilin Univ. Electron. Sci. Technol. **36**(04), 311–314 (2016)
2. Wang, Q.: Chinese word segmentation and word vector. China New Commun. **20**(23), 19–23 (2018)
3. An, J., Gao, G., Shi, Z., Sun, L.: An improved K-means text clustering algorithm. Sens. Microsyst. **34**(05), 130–133 (2015)
4. Liu, P., Lu, J.: Improved K-means text clustering algorithm based on MapReduce. Inf. Technol. (11), 201–205 (2016)
5. Zou, H., Li, M.: An improved bisecting K-means algorithm for text clustering. Microcomput. Appl. **29**(12), 64–67 (2010)
6. Zhang, J., Wang, N., Huang, S., Li, S.: Research on optimization and parallelization of bisecting K-means clustering algorithm. Comput. Eng. **37**(17), 23–25 (2011)
7. Hui, Y., Xia, Y., Chen, Z., Tong, X.: Short text clustering algorithm based on synonyms and K-means. Comput. Knowl. Technol. **15**(01), 5–6 (2019)
8. Tang, X., Zhai, X.: Semantic indexing of text knowledge fragments based on ontology and Word2Vec. Inf. Sci. **37**(04), 97–102 (2019)
9. Dai, Y., Xu, L.: An improved TF-IDF algorithm based on semantic analysis. J. Southwest Univ. Sci. Technol. **34**(01), 6773 (2019)
10. Kui, Z.: Improvement of TF-IDF weight calculation method in text classification. Softw. Guide **17**(12), 39–42 (2018)
11. Liang, K., Wang, C., Zhang, Y., Zou, W.: Knowledge aggregation and intelligent guidance for fragmented learning. Procedia Comput. Sci. **131**, 656–664 (2018)

PLC-Based Control System Design of Intelligent and Efficient Multiple Parking Device

Xiangping Liao$^{(\boxtimes)}$, Wenlong Tang, Zhiguang Fu, Tong Zhang, Zizhen Ou, and Chuangwu Ming

Hunan University of Humanities, Science and Technology, Loudi 417000, Hunan, China
520joff@163.com

Abstract. The development of three-dimensional garage is one of the important ways to alleviate the pressure of modern urban traffic and parking difficulties. This intelligent and efficient multiple parking device mainly relies on the translation movement of grabbing mechanism in three-dimensional direction, which is controlled by PLC motor. Some photoelectric sensors and travel switches are used to ensure the accuracy and stability of parking. By judging the size of the car and the distance between the parking space and the grabbing mechanism and the loading board, the parking device can park the car in the nearest suitable parking space to the loading board, avoid wasting of extra time, which can greatly improve the efficiency. The device has the characteristics of high cost performance ratio, easy to realize large-scale and commercialization.

Keywords: Programmable logic controller · Intelligent and efficient · Three-dimensional garage

1 Introduction

With the continuous development of economy and the improvement of people's living standards, the car ownership has increased rapidly. In the early urban planning, the parking facilities were not considered enough, and most of the parking lot are planar [1]. It is unable to meet the growing demand for vehicles, since the space utilization rate of Parking lot is low, too much space resources is wasted. On the other hand, imported products from abroad are expensive and are not always favored by the market. The design of three-dimensional parking lot which is suitable for the general public in China has become an urgent task. The establishment of three-dimensional and intelligent parking lot is the fundamental solution to the current problem [2]. In this work, an intelligent and efficient multiple parking device which integrates electricity, optics and computer science is proposed. It uses PLC as the core control, which is convenient and reliable. It can be applied to various environments [3]. It can extend indefinitely from front to back and make full use of space resources to cope with parking difficulties. It can be used not only in residential parking space, but also in large open-air parking lots, underground garages and other occasions. The three-dimensional garage with mechatronics technology has broad market prospects, and it will become an effective means to solve the urban traffic difficulties [4].

© Springer Nature Switzerland AG 2020
F. Xhafa et al. (Eds.): IISA 2019, AISC 1084, pp. 163–170, 2020.
https://doi.org/10.1007/978-3-030-34387-3_20

2 Structure and Principle of Intelligent and High Efficiency Multiple Parking Device

The three-dimensional garage can be divided into vertical lifting type, lifting and transverse moving type, roadway stacking type, vertical circulation type, circular horizontal circulation type, box horizontal circulation type, etc. [5].

Roadway stacking garage is a kind of three-dimensional garage which combines machine, light, electricity and automatic control to move the car horizontally and vertically to the predetermined parking position or to take out the car on the contrary. It can parking or picking up car safely, and is suitable for large-scale public lifting buildings and underground three-dimensional garages [6]. Roadway stacked garage equipment has complex structure, no perfect locking and monitoring system, relatively high failure rate.

Vertical circular garage adopts vertical circular motion to access vehicles. Motor drives transmission mechanism through reducer. On the chain of traction components, one loading plate is installed every certain distance. When the motor starts, the loading plate circulates with the chain, so as to achieve the purpose of accessing vehicles. It is easy to be controlled and it only occupies a small area [7]. However, the fastest loading board usually takes two minutes to pick up a car at a time, and the peak period takes too long to pick up the car in turn, which lacks practicability.

In view of the problems of the above three-dimensional vehicle inventory, this paper designs an intelligent and efficient multiple parking device, which belongs to the lifting and transverse three-dimensional garage. There are many types of lifting and transverse three-dimensional garages, which can be large or small in scale, and have strong terrain adaptability, so they are widely used. The parking device has two grabbing mechanisms and three loading board, as shown in Fig. 1, which can solve the problem that the pickup time is too long in turn during the peak period. The device is divided into upper and lower layers, with a total of 27 loading boards. Three of them are used as loading boards. They can be rotated to adapt to different directions, and can be lifted up and down to grasp the mechanism, as shown in Fig. 2. The other 24 loading boards occupy half of the upper and lower floors. Six of the 12 loading boards on the upper floor are used for parking large cars, and the other six are used for parking small cars. The equipment is simple and the failure rate is low.

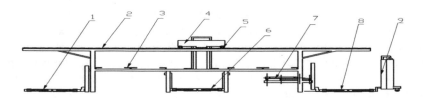

1、6、8- loading board. 2- rack. 3- comb like parking space. 4- grabbing mechanism. 5-gear. 7- linkage mechanism. 9- hydraulic lifting device.

Fig. 1. Intelligent and efficient multiple parking device frame diagram

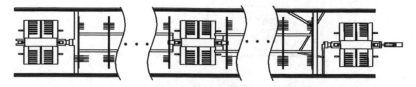

Fig. 2. Top view of intelligent and efficient multiple parking system

When accessing vehicles at the bottom loading board, the car can enter and exit directly without moving the carrier plate. When parking in the upper loading board, it is necessary to determine which plate needs to be parked, which parking space is nearest to the location by photoelectric sensor, and then select the nearest grabbing mechanism to move horizontally. The grabbing mechanism parks the vehicle to the parking space by lifting operation. When picking up the car, the switch is pressed to determine which loading board needs picking up, and then the car is placed on the loading board through a series of operations of the grabbing mechanism. The only difference between parking and picking up is that picking up a car does not need to use photoelectric sensors to determine which loading board is nearest to the loading board, and other operations are generally similar.

3 Design of Control System

The system uses Mitsubishi FX2N-128MT PLC. Its structure is shown in Fig. 3. The input part is mainly composed of photoelectric sensor, travel switch and control switch. The control part is mainly composed of various control motor circuits. When the user needs to access the car, the corresponding start switch is need to be pressed down. Corresponding detection signals are sent to PLC, which is analyzed by the PLC analysis system. The signals are sent to the motor, which drives the gear to rotate, and then transforms them into the lateral movement of the grabbing mechanism through the rack and pinion transmission. The photoelectric sensor is used to judge whether there is a car in the loading board, and the travel switch is used to judge whether the grabbing mechanism reaches the designated position.

The displacement of the loading board in the control system of the three-dimensional three-dimensional garage is completed by PLC. The specific control requirements are as follows.

(1) Automation: When accessing a vehicle, the control system can automatically select a scheduling scheme according to the current situation of the vehicle in stock by pressing the button of selecting the motor parking space, so as to complete the vehicle's entry and exit in the shortest time.

(2) Safety interlock control: Stereo garage is not allowed to cause vehicle damage due to wrong action, so it must have photoelectric detection and other safety protection facilities for timely detection of garage information to avoid accidents and to ensure smooth start and brake. In the aspect of safety control, the mechanical limit switch is used to locate the left and right movements of each parking space.

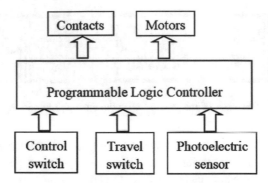

Fig. 3. Control system structure diagram

4   Design of System Hardware

The working theory is that the rotating movement of the loading board can adapt different directions of vehicles, and the parking space is accessed through the translation and lifting of the grabbing mechanism. The rotation of the grabbing mechanism motor is controlled by PLC. There are 44 input signals and 4 output signals of the device (seen in Fig. 4). The input equipment of this device includes control switch, photoelectric sensor, etc. Whether the vehicle need to be accessed is decide by the access vehicle switch, the photoelectric sensor is used to judge whether there is a vehicle on the board, and the travel switch is used to judge whether parked above the board accurately for the grabbing mechanism. The output equipment of the device is used to judge which grabbing mechanism is nearest to the loading board by PLC, and then control the rotation of the motor.

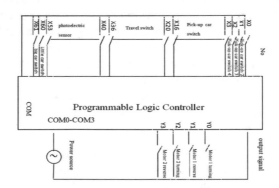

Fig. 4. Structure of control system

5 Design of System Software

The software of GX-Works2 is used by Intelligent and efficient multiple parking device and the program of the system is compiled by the ladder diagram. a programming flow char is in Fig. 5. Including initialization procedures, car storage procedures, car pick-up procedures. The vehicle storage procedure includes judging the size of the vehicle, judging which loading board is available, judging which grabbing mechanism is nearest to the loading board, and judging whether there is a vacancy. The procedure includes judging the position of car which need to be taken, judging which grabbing mechanism is closest to the board, and judging which parking space is closest to the loading board, and put it on the corresponding loading board. Since the program is large, there are many input and output units and many similar parts in between them, the author only lists a small part of the program.

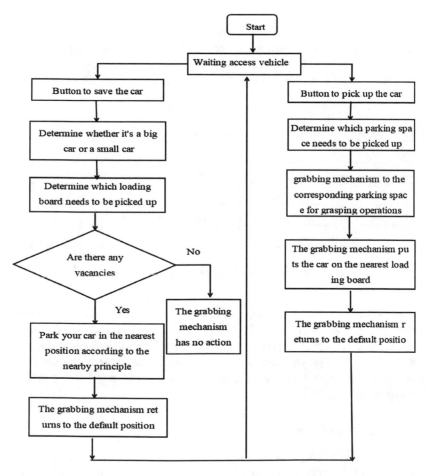

Fig. 5. Parking and pick-up car flow diagram

5.1   The Initialization of System

As shown in Fig. 6, the program initializes the intermediate relay, the counter, and the output. The intermediate relay is aim to prevent program error, the counter is design to delay the program, and the part of output is two motors positive and negative rotation.

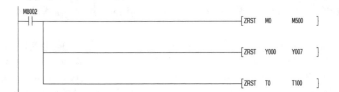

Fig. 6. Initialization program of PLC

5.2   The Procedure of Storage Car

As is depicted in the Fig. 7, procedure about how to determine which store switch is pressed is shown. X0–X2 is the store switch, X3–X16 is the pick-up switch. In order to prevent two or more keys from being pressed at the same time, all switches must be interlocked.

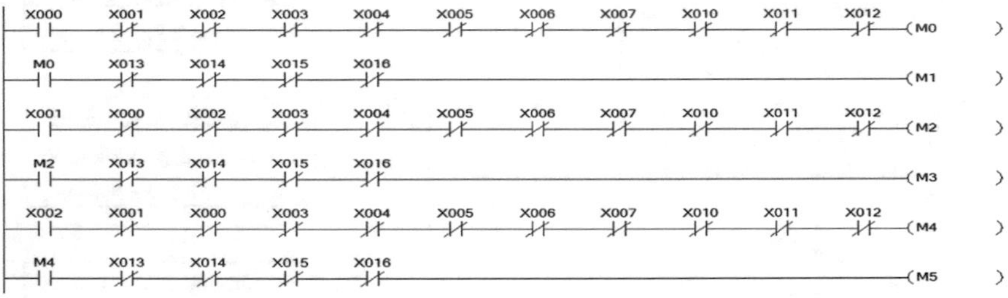

Fig. 7. Determine procedure for pressing the save switch

Figure 8 is a program to park a car in a parking space. M15 is a signal from a car-free photoelectric sensor in a parking space. X20 is a travel switch in the limited position and X21 is a travel switch in that parking space. In software programming, according to the front parking and picking-up process, the most important thing is to consider the sequence and coordination of each component action in the parking and picking-up process.

5.3   The Rotation of Motor

Figure 9 shows a PLC program, by which can judging whether the large parking space is available. M5 is the signal generated by the switch on the rightmost loading plate that needs to be pressed for storage. X40–X45 are photoelectric sensors on the large parking space, X60 is the switch for judging whether the car is a big one. Through the signal

Fig. 8. Part procedures of PLC

Fig. 9. Procedure for judging whether large parking spaces are available or not

transmitted by photoelectric sensor to PLC, judging whether the parking space is available or not.

Figure 10 shows a PLC program of motor inversion on the right grabbing mechanism, M50–M55 are various intermediate relay for judging which parking space is available. X36 is the travel switch at the right end position (in order to ensure that the grabbing mechanism is accurately parked above the loading board, as well as a safety device). T60–T65 are various delay program for the grabbing mechanism to have sufficient time for lifting motion. As long as there is a vacancy among parking space, one of the M50–M55 will be close. When the delay program is finished, the motor will reverse and stop turning until it meets the stroke switch at the most right limit position.

Fig. 10. Motor rotation procedures

6 Conclusion

This article take the Intelligent and Efficient Multiple Parking Device as a object, carry out a study on the design of software and hardware and the functions of the three-dimensional garage control system. Compared with the existing roadway stacking garage and vertical circular garage, the system has the advantages of high efficiency, simple structure and low production cost. Moreover, take the PLC product of Mitsubishi FX2N-128MT is used as the controller to realize the optimal control of intelligent and efficient multiple parking device. PLC makes the operation of three-dimensional garage more simple and convenient, and the system is more safe and reliable.

The advantages are as follows:

(1) The parking space on the first floor of the parking device allows vehicles to enter and exit directly, and the parking space in the second floor depends on various loading boards and grabbing mechanism to access vehicles intelligently according to the principle of proximity, by which the efficiency has been greatly improved.
(2) The comb like parking space can save a lot of steel, which is cheap, easy to process and manufacture, and has a high cost performance ratio.
(3) The device can be extended horizontally and can be applied to various types of parking lots, which is easy to realize large-scale and commercialization.

Acknowledgement. This work was supported by the grant from the Natural Science Foundation of Hunan Province of China (2017JJ3057) and the Research Foundation of Education Bureau of Hunan Province, China (17C0473).

References

1. Zhang, J.: A study of the planning method and its application for the urban parking. Urban Transp. China **1**(1), 23–27 (2003)
2. Chen, S., Zheng, Q.: Design of simple parking control system based on PLC. J. Shaoguan Univ. Nat. Sci. **38**(12), 49–51 (2017)
3. Gilles, M.: Programmable Logic Controllers: Architecture and Application. Wiley, New York (1990)
4. Zhang, G., Geng, C.: PLC-based automatic control for lifting and transfering 3D garage. Process Autom. Instrum. **34**(7), 35–37 (2013)
5. Yi, Q., Ye, Y., Zheng, Z.: Research of the control system based on PLC for a new vertical lift three-dimensional garage. J. Mech. Electr. Eng. **29**(4), 409–412 (2012)
6. Bin, L., Zhang, Z.: The virtual design for new stacking type of three-dimensional carport. J. Henan Inst. Eng. **24**(4), 54–56 (2012)
7. Ma, H.: PLC control of vertical circulating stereoscopic garage. Lift. Transp. Mach. (10), 85–87 (2009)
8. Liang, M., Li, C., Zhang, H.: Discussion on the design scheme of lifting and horizontal moving three-dimensional garage. Lift. Transp. Mach. (1), 25–28 (2013)

Sparse Acceleration Algorithm Based on Likelihood Upper Bound

Pan Liu[✉] and Fenglei Wang

National University of Defence Technology, Changsha, China
liupannudt@foxmail.com

Abstract. It is a great challenge for intelligent analysis and discovery of useful information from a great massive video resources, and the target tracking become more and more popular as the important part of the intelligent video surveillance. The feature extracted by the traditional method cannot keep the invariance of the target when the target surfers occlusion, illumination and pose variation in the moving process, which leads to the failure of the tracker. To handle the problems mentioned above, we conduct a survey on the target tracking based on sparse representation. In view of the problem that a large amount of time is needed to solve the problem of sparse expression optimization, we propose an acceleration algorithm based on the observation likelihood upper bound. The effectiveness of the proposed algorithm is verified by experiments on a number of standard test sets.

Keywords: Sparse acceleration algorithm · Sparse representation · Likelihood upper bound

1 Introduction

Tracking is one of the most important parts in a wide range of applications in computer vision, such as surveillance, human computer interaction, and medical imaging. Sparse expression provides an effective solution to the problem of illumination, noise and local occlusion. However, sparse expression tracking needs to solve a large number of optimization calculations, and the operation speed is slow, which affects the application range of sparse expression tracking. Sparse expression tracking time consumption is mainly concentrated in the process of solving sparse coefficients. Since the algorithm uses more than 600 particles to estimate the target state, it is necessary to perform more than 600 sparse solutions for each frame. Based on the combination of theoretical analysis and experimental verification, this paper studies the method based on likelihood upper bound filtering, also called Bounded Particle Resampling (BPR). Theoretical analysis shows that the BPR method can speed up the tracking speed without losing the tracking effect. The acceleration effect is based on the target background relationship. If the target and the background are different, the acceleration effect is obvious. If the target and background are similar, then the acceleration effect is limited.

© Springer Nature Switzerland AG 2020
F. Xhafa et al. (Eds.): IISA 2019, AISC 1084, pp. 171–179, 2020.
https://doi.org/10.1007/978-3-030-34387-3_21

2 Acceleration Based on Likelihood Upper Bound Filtering

2.1 Sparse Expression Target Tracking

In the sparse expression tracking algorithm, in order to obtain the observational likelihood $p(z_t|x_t)$ of the system, the image vector y represented by the candidate particle is sparsely expressed by the sparse dictionary D, and the problem of solving the sparse coefficient can be expressed as the problem to minimum of the reconstruction error constrained by the non-negative coefficient minimum l_1 norm:

$$\min_b ||Db - y||_2^2 + \lambda ||b||_1, \ s.t. \ c \geq 0 \tag{1}$$

Where, $D = [T, I, -I]$ consists of a set of target templates T and a set of singular value templates I, in which each column of T represents a target template, I was used to represent noise and occlusion. Each column has only one value of 1 and the remaining value is 0, and the columns of I are linearly independent. $b = [a^T, e^T]^T$ consists of a target expression coefficient and a singular value coefficient e.

Finally, the observation likelihood is estimated by the reconstruction error on the target template set, as shown in (2).

$$p(z_t|x_t) = \frac{1}{\Gamma} exp\left\{-\alpha ||Ta - y||^2\right\} \tag{2}$$

Among them, a is the expression coefficient of the target, and α is the parameter that controls the shape of the Gaussian kernel, and Γ is a normalization factor.

For the tracking of time t, the particle that obtains the maximum observation likelihood will be selected as the tracking result, and will be used to update the weight of the particle and perform particle resampling. The method of tracking in this way is called L1 tracking [9].

2.2 Observation Likelihood Upper Bound

It is known from Eq. (2) that the observation likelihood of the L1 tracking algorithm is estimated by the reconstruction error $||Ta - y||^2$ measured by the 2-norm, which a is obtained by solving (1) minimization, so the reconstruction error has a natural lower bound:

$$||Ta - y||^2 \geq ||T\hat{a} - y||^2 \tag{3}$$

Where $\hat{a}$ is the least square approximation of y on T:

$$\hat{a} = arg ||Ta - y||^2 \geq ||T\hat{a} - y||^2 \tag{4}$$

Thus, the observation likelihood $p(z_t|x_t)$ has an upper bound that can be represented by $\hat{a}$:

$$q(z_t|x_t) = \frac{1}{\Gamma} exp\left\{-\alpha||T\hat{a} - y||^2\right\}$$ (5)

Call $q(z_t|x_t)$ the likelihood upper bound. The least squares problem (4) can be solved by Cholesky decomposition or QR decomposition. For a dense matrix, the complexity of Cholesky decomposition is $dn^2 + (1/3)n^3$, and the complexity of QR decomposition is $2dn^2$, where d is the length of the image vector and n is the number of templates. If $d \gg n$, the QR method is twice as slow as the Cholesky method, this is the case with the algorithm in this paper. For a small or medium size problem, the two methods have the same accuracy. In order to prove the effectiveness of the algorithm, the algorithm in this chapter chooses the QR decomposition method.

2.3   Bounded Particle Resampling

Since the calculation of the least square error has a huge computational advantage over the calculation of the l_1-norm, the researchers attempted to use the observation likelihood upper bound $q(z_t|x_t)$ to aid in estimating the observed likelihood [6]. Perhaps the likelihood upper bound does not give a definitive conclusion, but it can rule out some deviations from larger particles. After excluding some particles, the retained particles also need to be solved by l_1 norm minimization to estimate the observation likelihood and update the particle weights. The experimental results show that in the case of the majority, after the likelihood upper bound filtering, only a small number of reliable particles can be retained.

Inspired by this, a two-stage restricted particle resampling method is proposed to calculate the probability of candidate particles. The set of particles at time t is $\chi(t) = \{x_1, x_2, \ldots, x_N\}$, the observed likelihood is $p_i = p(z_t|x_t^i)$, and the upper bound of the observed likelihood is $q_i = q(z_t|x_t^i)$. In the first stage, the upper bound of the observed likelihood q_i of all particles is directly calculated and arranged in descending order $q_1 \geq q_2 \geq \cdots \geq q_N$.

In the second phase, an observation likelihood estimate p_i of the resampled particles needs to be calculated, and the non-related particles are excluded by the dynamically updated threshold τ. Defined τ by the following lemma:

Lemma 1. *If the i-th particle x_i appears at least once after resampling, its likelihood upper bound q_i is not less than the threshold τ_i.*

$$\tau_i = \frac{1}{2N - 1}\sum_{j=1}^{i-1} p_j$$ (6)

Proof of Lemma 1. Since the resampling of the particles is satisfied at the end of each frame of tracking [10]: If a particle appears at least once in the resampled particle set, its observation probability must satisfy the condition:

$$p_i \geq \frac{1}{2N} \sum_{j=1}^{N} p_j \tag{7}$$

Since the total number of particles before and after resampling has not changed, the Eq. (7) is obvious. Then

$$2Np_i \geq \sum_{j=1}^{N} p_j \geq \sum_{j=1}^{i} p_j \tag{8}$$

Both sides subtract p_i, then divide by $2N - 1$, we get

$$p_i \geq \frac{1}{2N - 1} \sum_{j=1}^{i-1} p_j = \tau_i \tag{9}$$

Also, since q_i is the upper bound of p_i, there is $q_i \geq p_i \geq \tau_i$, and the lemma is proved.

By definition, τ_i is non-decreasing, that is $0 = \tau_1 \leq \tau_2 \leq \cdots \leq \tau_N$, and thresholds can be used to dynamically update by

$$\tau_{i+1} = \tau_i + \frac{p_i}{2N - 1} \tag{10}$$

Thus, in the second stage, for a particle whose maximum likelihood is q_i, calculate p_i according to formula (2) and update τ_i according to formula (10), compare q_i and τ_i. For the i-th particle, if $q_i < \tau_i$, the calculation is stopped. According to Lemma 1, particle $x_i, x_{i+1}, \ldots, x_N$ will not appear in the resampled particle set. Table 1 shows the specific calculation flow of the BPR algorithm.

Table 1. Calculation flow of the BPR algorithm.

Input: Particle set of previous frame $\chi_{t-1} = \{x_{t-1}^k\}_{k=1}^N$

Output: Particle set $\chi_t = \{x_t^k\}_{k=1}^N$

---

1. / First stage /

2. for $i = 1 : N$ do

3.   Collect particles x_t^i according to x_{t-1}^i

4.   Build the appearance vector y_t^i of the position x_t^i

5.   Solve the linear minimum problem of y_t^i according to equation (4)

6.   Find q_i according to equation (5)

7. end for

8. Arrange a in descending order, get $q_1 \geq q_2 \geq \cdots \geq q_N$

9. / Second stage /

10. $i \leftarrow 1, \tau_i \leftarrow 0$

11.while $q_i \geq \tau_i$ and $i \leq N$ do

12.   Solve the minimum l_1 norm coefficient of y_t^i according to equation (1)

13.   Calculate p_i according to equation (2)

14.   Update $\tau_{i+1} = \tau_i + \frac{p_i}{2N-1}$

15. $i = i + 1$

16. end while

17. $p_j \leftarrow 0, \ \forall j \geq i$

18. / Resampling /

19. Resampling $\chi_t \leftarrow \{x_t^k\}_{k=1}^N$ according to $\{p_k\}_{k=1}^N$

---

3   Experimental Results and Analysis

In this section, the above proposed methods are verified by experiments, and detailed analysis is made from the aspects of experimental data selection, algorithm description and evaluation indicators. The performance of the algorithm is compared and analyzed from both qualitative and quantitative perspectives.

The target tracking algorithm in this chapter is implemented by Matlab 2010. The running frequency of the computer is 3.1 GHZ, intel (R) core (TM) i5-2400, and the memory is 8 GB. The optimization algorithm is implemented by the well-known sparse optimization toolbox SPAMS, while the training and use of the SVM classifier is implemented by the SVM-Gun toolbox.

3.1   Experimental Data and Algorithm Description

In order to verify the feasibility of the two algorithms, five publicly released tracking data sets were selected for testing, namely David2, Car4, Woman, Subway, Singer1. These video sequences are collected from different scenes, including a variety of target motions, target deformation, lighting changes, target occlusion and many other challenges.

In this paper, four sets of experiments were designed. The first group accelerated the traditional L1 tracking by restrictive particle resampling, and the L1 tracking acceleration method was recorded as L1-BPR. The second group accelerates the target tracking algorithm based on structured sparse features proposed by another article in our paper by restrictive particle resampling. Since the algorithm adopts the idea of blocking, the acceleration is applied to each block separately. BPR and BPR acceleration experiments were applied to the global template. The two methods were recorded as P-BPR and U-BPR, respectively.

3.2   Evaluation Criteria

In order to illustrate the tracking effect of the algorithm after the speed reduction, the results are presented and analyzed from the quantitative and qualitative perspectives. Qualitative evaluation is to evaluate the experimental results from a subjective perspective, and compare the effectiveness and advantages and disadvantages of various algorithms. Quantitative analysis is the analysis of the performance of an algorithm from experimental data. The two most effective evaluation criteria in the field of tracking—CLE (center location error) and TSR (tracking success rate) were introduced to analyze the results.

3.3   Analysis of Results

Experiment 1: The L1-BPR algorithm accelerates validation. The algorithm tracks on the David2 dataset, calculates the particle likelihood upper bound and its likelihood probability for each frame, and counts the time consumed per frame and the calculated number of particles during the entire tracking process.

Result Analysis
Figure 1 shows the likelihood and likelihood upper bounds of the particles in frame 28 and the probability of the threshold in the L1-BPR algorithm tracking process. The horizontal axis is the particle label and the vertical axis is the logarithm of each probability (for convenience). It is found that with the gradual decrease of the likelihood upper bound, the likelihood probability is gradually reduced and the threshold is gradually increased. When the 23rd particle is calculated, the threshold is equal to the likelihood upper bound. According to the BPR acceleration principle, it is not necessary to calculate the residual particle. In the experiment, only 23 times of sparse expression calculation are required. The experiment shows that the algorithm has obvious acceleration effect on the traditional L1 algorithm.

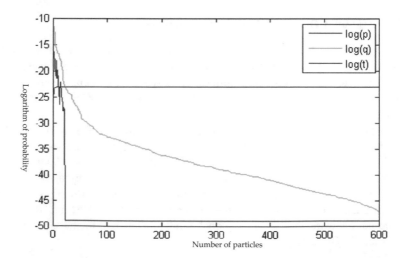

Fig. 1. Probability comparison of all particles in a frame during BPR algorithm tracking

Figure 2 shows the time consumption of tracking each frame during the tracking of the L1-BPR algorithm on the David2 dataset and the comparison of the number of particles that need to be solved. The horizontal axis represents the number of frames, and the vertical axis represents the effective number of particles and the elapsed time (the time unit is adjusted for display convenience). As can be seen from the figure, the more particles that need to be calculated, the more time is consumed for the corresponding frame.

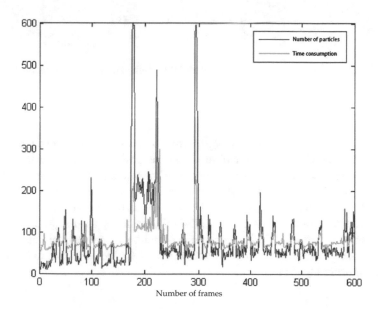

Fig. 2. BPR algorithm tracking process consumes particles and time comparison

Experiment 2: Accelerated verification of P-BPR and U-BPR algorithms. The algorithm tracks the David2 data set, calculates the likelihood upper bound and likelihood probability of each frame, and counts the number of particles that need to be calculated for each frame.

Result Analysis

Figure 3 shows the likelihood and likelihood upper bound of all particles and the probability of the threshold for the U-BPR algorithm and P-BPR algorithm. The horizontal axis is the particle label and the vertical axis is the logarithm of each probability. It can be seen from the results that the BPR acceleration effect depends on the difference of background. If the background and the target are more distinguishable, the acceleration effect is obvious. Otherwise, the acceleration effect is limited.

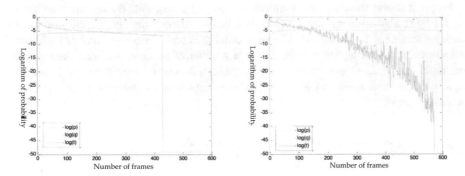

Fig. 3. Comparison of U-BPR (left) and P-BPR (right) algorithm in tracking process

4 Summary

In this paper, we propose to use two algorithms to accelerate the sparse representation of the large number of particles. We observe that he least squares reconstruction error is the natural lower limit of the sparse representation reconstruction error, and the least squares is more than two orders of magnitude faster than the sparse representation velocity. The least square error estimation is introduced to observe the likelihood upper bound. Based on this observation, the BPR algorithm is used to remove the very different particles. Since the BPR algorithm is based on sparse reconstruction error and its performance is related to the background, it's hard for the algorithm to be adapted to structural tracking algorithms.

References

1. Li, X., et al.: A survey of appearance models in visual object tracking. ACM Trans. Intell. Syst. Technol. (TIST) **4**(4), 58 (2013)
2. Ellis, L., et al.: Linear regression and adaptive appearance models for fast simultaneous modelling and tracking. Int. J. Comput. Vis. **95**(2), 154–179 (2011)
3. Kalman, R.E.: A new approach to linear filtering and prediction problems. J. Fluids Eng. **82**(1), 35–45 (1960)
4. Isard, M., Blake, A.: Condensation – conditional density propagation for visual tracking. Int. J. Comput. Vision **29**(1), 5–28 (1998)
5. Arulampalam, M.S., et al.: A tutorial on particle filters for online nonlinear/non-Gaussian Bayesian tracking. IEEE Trans. Sig. Process. **50**(2), 174–188 (2002)
6. Mei, X., et al.: Minimum error bounded efficient $\ell 1$ tracker with occlusion detection. In: 2011 IEEE Conference on Computer Vision and Pattern Recognition (CVPR), pp. 1257–1264. IEEE (2011)
7. Felzenszwalb, P.F., et al.: Object detection with discriminatively trained part-based models. IEEE Trans. Pattern Anal. Mach. Intell. **32**(9), 1627–1645 (2010)

8. Jia, X., Lu, H., Yang, M.-H.: Visual tracking via adaptive structural local sparse appearance model. In: 2012 IEEE Conference on Computer Vision and Pattern Recognition (CVPR). IEEE (2012)

9. Mei, X., Ling, H.: Robust visual tracking and vehicle classification via sparse representation. IEEE Trans. Pattern Anal. Mach. Intell. **33**, 2259–2272 (2011). https://doi.org/10.1109/TPAMI.2011.66

10. Kim, S.J., Koh, K., Lustig, M., et al.: An interior-point method for large-scale $\ell 1$-regularized least squares. IEEE J. Sel. Top. Signal. Process. **1**, 606–617 (2007)

Optimization Design of Quad-Rotor Flight Controller Based on Improved Particle Swarm Optimization Algorithm

Yinliang Yang$^{(\boxtimes)}$, Xiangju Jiang, and Zhen Tong

School of Automation and Electrical Engineering, Lanzhou Jiaotong University,
Lanzhou, China
yang93102@163.com

Abstract. A sliding mode controller with exponential reaching rate is designed for posture control of Quad-rotor aircraft. In order to advance the control quality of the sliding mode controller and reduce the chattering of the system while speeding up the system's arrival at the sliding mode surface, a mean particle swarm optimization algorithm with dynamic adjustment of inertia weight (DIMPSO) is proposed to optimize the gain coefficient and exponential reaching rate parameters of the sliding mode surface. By combining the mean particle swarm optimization algorithm with the method of dynamic adjustment of inertia weight by cosine function, the migration speed and global search ability of particles are improved. The simulation results show that compared with particle swarm optimization (PSO), the DIMPSO algorithm has faster convergence rate and higher optimization accuracy, which makes the attitude control of Quad-rotor aircraft achieve stable tracking control effect.

Keywords: Sliding mode control · Particle swarm optimization · Parameter optimization

1 Introduction

According to the characteristics of four-rotor, there are several common control methods: PID control, sliding mode variable structure control, fuzzy control, backstepping control and so on. These methods have their own advantages and disadvantages in attitude control and position control of aircraft [1].

How to weaken the chattering of the aircraft system in the process of approaching the switching sliding surface is still the core issue of sliding mode control. In Ref. [2], an improved double power reaching rate method is used to improve the chattering problem in sliding mode control. In Ref. [3], an algorithm combining fuzzy and sliding mode control is proposed for the design of attitude angle controller to solve the problem of poor anti-jamming performance of the system. In Ref. [4], backstepping sliding mode control is used in Quad-rotor full-drive system, and underactuation is a cascade sliding mode controller. In Ref. [5], high-order sliding mode is used to reduce buffeting, thus achieving the effect of stable flight of Quad-rotor aircraft.

© Springer Nature Switzerland AG 2020
F. Xhafa et al. (Eds.): IISA 2019, AISC 1084, pp. 180–188, 2020.
https://doi.org/10.1007/978-3-030-34387-3_22

From this paper, an improved particle swarm optimization algorithm is used for tuning the parameters of sliding mode surface gain coefficient and exponential reaching rate, so that the improved system converges faster and has better anti-jamming performance.

2 Dynamics Modeling of Quad-Rotor Aircraft

Quad-rotor aircraft is a four-input-six-output underactuated and strongly coupled system. In order to facilitate the establishment of its mathematical model, the geographic coordinate system: E (O-XYZ), the body coordinate system: B (o-xyz) are introduced (Fig. 1):

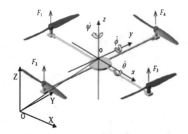

Fig. 1. Quad-rotor aircraft in coordinate system

The transformation matrix of two coordinate systems is shown in Formula (1):

$$R_B^E = \begin{bmatrix} \cos\psi\cos\theta & \cos\psi\sin\theta\sin\phi - \sin\psi\sin\phi & \cos\psi\sin\theta\cos\phi + \sin\psi\sin\phi \\ \sin\psi\cos\theta & \sin\psi\sin\theta\sin\phi - \cos\psi\cos\phi & \sin\psi\sin\theta\cos\phi - \sin\phi\cos\psi \\ -\sin\theta & \cos\theta\sin\phi & \cos\theta\cos\phi \end{bmatrix}$$

$$(1)$$

Euler attitude angles are pitch ϕ, roll θ and yaw ψ. The motion equation can be established according to Lagrange equation. When the influence of air and frictional drag is neglected in flight, the dynamic mathematical model of Quad-rotor aircraft is:

$$\begin{cases} \ddot{x} = (\cos\phi\sin\theta\cos\psi + \sin\phi\sin\psi)\,u_1 - K_1\dot{x}/m \\ \ddot{y} = (\cos\phi\sin\theta\sin\psi - \sin\phi\cos\psi)\,u_1 - K_2\dot{y}/m \\ \ddot{z} = (\cos\phi\cos\theta)\,u_1 - K_3\dot{z}/m - g \\ \ddot{\phi} = u_2 - LK_4\dot{\phi}/I_1 + J_r/I_1\dot{\phi}\Omega_r \\ \ddot{\theta} = u_3 - LK_5\dot{\theta}/I_2 + J_r/I_2\dot{\theta}\Omega_r \\ \ddot{\psi} = u_4 - LK_6\dot{\psi}/I_3 \end{cases}$$

$$(2)$$

Among them, Ω_r is the gyroscopic effect component produced by propeller; Control input (u_1, u_2, u_3, u_4); Rotational inertia of I_1, I_2 and I_3 axes X, Y and Z respectively; J_r is the rotational inertia; L is the distance between the center of mass and the rotor of an aircraft; K_i $(i = 1, \cdots, 6)$ is drag coefficient.

3   Design of Sliding Mode Controller for Quad-Rotor Aircraft

Attitude control of aircraft is realized in the simulation model. A simple virtual prototype of Quad rotor aircraft is drawn by SolidWorks three-dimensional software, as shown in the Fig. 2.

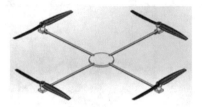

Fig. 2. The structural model of Quad-rotor aircraft.

Then the virtual prototype built in Solidworks is imported into ADAMS for dynamic simulation. The actual parameters of the aircraft model are shown in the Table 1.

Table 1. Actual parameters of aircraft

$I_1/(\text{kg m}^2)$	0.116
$I_2/(\text{kg m}^2)$	0.116
$I_3/(\text{kg m}^2)$	0.312
L/m	0.23
m/kg	1.25

Then, the input and output variables of the system are named and function files are generated. In MATLAB/Simulink, the calculation results of the control system are imported into ADAMS through input variables, and the output variables are displacement and attitude angle (Fig. 3).

ADAMS software can input lift to each rotor in the virtual prototype, and calculate the corresponding attitude angles of the four rotors of the virtual prototype under the action of force according to its own formula library. The attitude angles are input to the sliding mode control system built in MATLAB/Simulink. The control rate is edited by

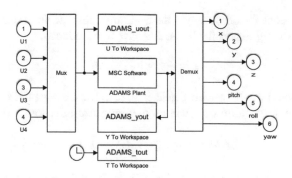

Fig. 3. ADAMS model of Quad-rotor aircraft

S function, and then embedded in the Simulink simulation, the simulation model of sliding mode control is obtained as follows (Fig. 4):

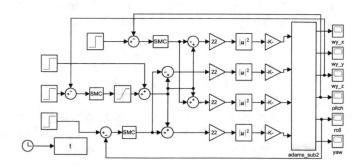

Fig. 4. Simulation diagram of sliding mode attitude control

For attitude control of aircraft, taking pitch channel of Quad-rotor aircraft as an example, the gyroscopic effect component Ω_r is neglected. The dynamic model of pitch angle is as follows.

$$\ddot{\phi} = u_2 - LK_4\dot{\phi}/I_1 \tag{3}$$

The pitch angle tracking error is $e = \phi - \phi_d$. Define the switching sliding surface:

$$s = k_1\dot{e} + k_2e \tag{4}$$

formula (5) can be obtained

$$\begin{aligned} \dot{s} &= k_1\ddot{e} + k_2\dot{e} \\ &= k_1(u_2 - LK_4\dot{\phi}/I_1 - \ddot{\phi}_d) + k_2(\dot{\phi} - \dot{\phi}_d) \end{aligned} \tag{5}$$

In order to reduce the chattering of the system while reaching the switching surface quickly, the following exponential reaching law is introduced.

$$\dot{s} = -\varepsilon \, \text{sgn}(s) - ks \tag{6}$$

In the formula, switch switching gain is ε, k is the exponential coefficient. $\varepsilon > 0$, $k > 0$. The following control rates can be obtained from the formulas (5) and (6).

$$u_2 = k_1^{-1}[k_1 L K_4 \dot{\phi} I_1 + k_1 \ddot{\phi}_d - k_2(\dot{\phi} - \dot{\phi}_d) - \varepsilon \, \text{sgn}(s) - ks] \tag{7}$$

Define the Lyapunov function as $V = 1/2s^2$; Then formula (8) can be obtained.

$$\dot{V} = s(-\varepsilon \, \text{sgn}(s) - ks) \leq -\varepsilon|s| - ks^2 \leq -\varepsilon|s| \leq 0 \tag{8}$$

According to Lyapunov stability criterion, the system state can converge to the switched slide's surface in a limited time.

In order to reduce the chattering in the control process, the saturated function sat(s) is instead of the sign function sgn(s) in the control rate. The expression is as follows.

$$\text{sat}(s) = \begin{cases} 1 & s > \delta \\ ks & |s| \leq \delta, (\text{k} = 1/\delta) \\ -1 & s < -\delta \end{cases} \tag{9}$$

In the process of approaching the sliding surface, switching gain ε has a great impact on the system performance. So as to ensure fast approaching and reduce chattering, choosing appropriate k and ε is the focus of research.

4   Improved Particle Swarm Optimization Algorithm

4.1   Particle Swarm Optimization (PSO)

In 1998, Shi and Eberhart proposed standard particle swarm optimization to improve the search ability of particles. The equation of particle position and velocity are shown in Formulas (10) and (11):

$$V_i^{t+1} = \omega V_i^t + c_1 r_1(P_i^t - x_i^t) + c_2 r_2(P_g^t - x_i^t) \tag{10}$$

$$X_i^{t+1} = X_i^t + V_i^{t+1} \tag{11}$$

Among them, ω is inertia weight; Learning factors are c_1 and c_2; t is the current iteration number of particles.

4.2   Mean Particle Swarm Optimization Algorithm with Dynamic Adjustment of Inertia Weight (DIMPSO)

So as to accelerate the convergence speed of particle swarm optimization and avoid premature phenomena, the mean particle swarm optimization makes full use of the useful information of the particle itself and the global position to better adjust the flight direction of the particle and make it shift to the best position at present, so as to find the global optimal position more quickly. By using the sum of linear combination $(P_i + P_g)/2$ and $(P_i - P_g)/2$ simplified formula (10) and removing the velocity term in formula (12):

$$X_i^{t+1} = \omega X_i^t + c_1 r_1 \left(\frac{P_i + P_g}{2} - X_i^t\right) + c_2 r_2 \left(\frac{P_i - P_g}{2} - X_i^t\right) \tag{12}$$

Although the linear decreasing inertia weight strategy in standard PSO algorithm will improve the search performance of the algorithm, in the later stage of the search, the particles in the population are close to the optimal solution, the diversity of the population will gradually be lost, and the convergence rate of the algorithm will obviously slow down in the later stage of the search. The cosine function is used to control the change of inertia weights to solve the above problems. In the later stage of the search, the larger inertia weights can be obtained, which makes up for the search efficiency in the later stage of the algorithm. The specific expressions are as follows.

$$\omega = \omega_{min} + (\omega_{max} - \omega_{min}) \cos(\pi t / 2 T_{max}) + \sigma \tag{13}$$

Among them, the inertia adjustment factor σ is used to adjust the deviation degree of inertia weight; The maximum and minimum inertia weights are ω_{max} and ω_{min}; The maximum number of iterations of particles is T_{max}.

For the sliding mode controller in this paper, the position of particles in space is composed of four-dimensional vectors: $X_D = (k_1, k_2, \varepsilon, k)$. In order to improve the precision of the parameters of optimized controller, the range of the experiential parameters of the controller is selected before optimization: $\varepsilon \in (0, 1.5), k \in (0, 50)$. Through DIMPSO algorithm optimization, the parameters obtained after optimization are the optimal solution.

4.3 Parameter Optimization Steps

Step 1: The quantity of particles and the maximum number of iterations are initialized.
Step 2: In order to limit the system error in the later stage of transition process, the fitness function selects ITAE error performance index.

$$J = \int_0^{t_s} t|e(t)|dt \tag{14}$$

Among them, $e(t)$ is the system error, t_s is the adjustment time.
Step 3: The fitness value obtained is compared with the individual optimal value P_i. If it is better than P_i, the individual optimal solution is chosen, and then compared with the global optimal value P_g. If it is better than P_g, the globally optimal solution is chosen.
Step 4: The position of particles is updated by formula (12) and the inertia weight is updated by formula (13).

Step 5: Determine whether the optimal fitness or the maximum number of iterations have achieved. If it is reached, the calculation is finished and the particle optimum value is output; otherwise, return to Step 2 to continue execution.

5 Simulation Experiment and Analysis

Based on the above control strategy, the attitude control simulation experiment is carried out. Drag coefficient: K4 = K5 = K6 = 0.012. Four parameters of pitch channel controller are tuned by using MATLAB software simulation. The initial parameters of the DIMPSO algorithm are set to (Table 2):

Table 2. DIMPSO algorithm initialization parameters

Parameter	Numerical value
c_1, c_2	2
Particles	60
Test dimension	4
T_{max}	80
σ	0.1
ω_{max}	0.8
ω_{min}	0.4

The inertia weight of PSO algorithm is 0.6. The fitness comparison curve between PSO and DIMPSO is obtained by simulation as shown in Fig. 5.

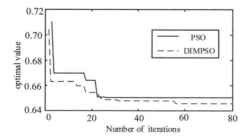

Fig. 5. Fitness value contrast curve

The optimized parameters of the two algorithms are shown in Table 3.

In Fig. 6, the tracking response curves of attitude angle control optimization using two algorithms are given.

The observation chart shows that the improved PSO algorithm converges faster and more accurately than before. The controller can make the aircraft reach the expected

Table 3. Optimized controller parameters

Algorithm	k_1	k_2	ε	k
PSO	9.7001	5.4238	0.8110	11.7167
DIMPSO	8.9952	5.7145	0.9036	13.1509

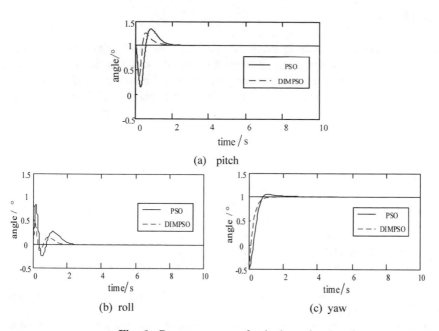

Fig. 6. Response curve of attitude angle control

value within 2 s, and the response rate is faster, which weakens the chattering of the system, and the tracking control effect is better.

When the system reaches a stable state, the disturbance moment with amplitude of 0.25 N m and pulse width of 2 s is added to the model of yaw channel at t = 5 s. In order to verify the stability of DIMPSO algorithm, two algorithms are used to simulate the controller respectively (Fig. 7).

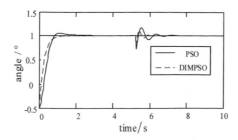

Fig. 7. Comparison of response curves under interference

From the figure, it can be seen that the disturbance fluctuation of the improved DIMPSO algorithm is within 0.3° and restores stability in about 2 s. The convergence speed of the optimized algorithm is faster, and the effect of disturbance suppression is better than that of the improved DIMPSO algorithm.

6   Conclusion

A sliding mode controller is designed for the posture control of a Quad-rotor aircraft. A mean particle swarm optimization algorithm based on dynamic inertia weight adjustment is to optimize the parameters of the controller. Simulation results show that the improved DIMPSO algorithm converges faster and more accurately. When disturbance is added, the algorithm also has good anti-jamming and robustness, and the flight control effect is better.

References

1. Shi, J., Wang, J., Liu, S.: Optimum design of attitude sliding mode controller parameters for vehicle. Aerosp. Shanghai **27**(6), 26–31 (2014)
2. Yang, X., Li, W.: Four rotor aircraft control based on sliding mode controller. J. Hefei Univ. Technol. (Sci. Technol.) **39**(7), 924–928 (2016)
3. Li, Z.: Quad-robot aircraft attitude control system optimization design. Comput. Simul. **34**(5), 58–62 (2017)
4. Wang, C., Chen, Z., Sun, M.: Sliding mode control of a quadrotor helicopter. J. Cent. S. Univ. (Sci. Technol.) **48**(4), 1006–1011 (2017)
5. Cao, L., Zhang, D., Tang, S., et al.: A practical parameter determination strategy based on improved hybrid PSO algorithm for higher-order sliding mode control of air-breathing hypersonic vehicles. Aerosp. Sci. Technol. **59**(3), 1–10 (2016)
6. Huang, Y., Lu, H., Xu, K., et al.: Simplified mean particle swarm optimization algorithm with dynamic adjustment of inertia weight. J. Chin. Comput. Syst. **39**(12), 2590–2595 (2018)

Autonomous Systems

Diversified Development Path of Cultural Exhibition Industry: Perspective of Service Value Chain

Xing Zeng[1,2(✉)] and Ningqiang Huo[2]

[1] School of Economics, Sichuan University, Chengdu 610065, China
gkittyzx@163.com
[2] Research Center of Sichuan Cultural Industry Development,
Chengdu 610213, China

Abstract. Cultural exhibition industry has become an important booster and leader for the development of cultural industry. Rather than a concrete place for transaction, cultural exhibition can provide attendees with opportunities for knowledge, culture education, experience leisure and the like. In other words, culture exhibition is a service and experience-generating place which can foster a sense of cultural community between participants, which makes the study of culture exhibition value chain uniquely meaningful to academia and industry. On the perspective of service value chain, this research analyzes key driving forces, characteristics and operating environment of the cultural exhibition value chain. It further put forward diversified improving path for the cultural exhibition industry.

Keywords: Service value chain · Cultural exhibition industry · Developing path

1 Introduction

The continuous improvement of exhibition service, and the innovation in the cultural creativity industry, will maintain the cultural added value for exhibition, create exhibition consumption growth, and promote the upgrading of exhibition industry. As it is an interrelated system which consists of a series of basic activities and auxiliary activities, how to continuously innovate, optimize and coordinate services value chain has become the key to the cultural exhibition industry development.

This article conducts research on the value chain of culture exhibition industry. Drawing upon the description of definition, classification and characteristics on cultural exhibition industry, it analyzes key driving forces, characteristics and operating environment of the cultural exhibition value chain. With the detailed framework, four types of culture exhibition value chain are elaborated and tested. Investigating the internal mechanism of culture exhibition value chain, this research proposes valuable results and discussions regarding the diversified development path of cultural exhibition industry.

© Springer Nature Switzerland AG 2020
F. Xhafa et al. (Eds.): IISA 2019, AISC 1084, pp. 191–198, 2020.
https://doi.org/10.1007/978-3-030-34387-3_23

2 Theoretical Analysis

2.1 A Framework for Exhibition Value Chain

Firstly, Exhibition value chain is a general term for a series of value-added links which involves major participants (exhibition organizer, exhibitors, service providers, visitors) and related parties (advertising, transportation, construction, tourism, hotel, catering service provider, etc.). They participate in one or more links in the exhibition value chain and gain the maximum value-added interdependently.

Secondly, the exhibition value chain consists of internal value chain and external value chain. The internal value chain mainly refers to a series of activities completed by exhibition organizer who will obtain the main value-added. The external value chain is extended by the participation of relevant parties which are coordinated by exhibition organizer in the exhibition activities.

Thirdly, by providing "differentiated" exhibition services, exhibition enterprises can cultivate customer satisfaction and loyalty. Such core competitiveness should be a priority for exhibition marketers. Therefore, it is of strategic importance to understand the mechanism of exhibition value chain. Referring to the framework for service value chain, we construct the framework for exhibition value chain [1–3] (Fig. 1).

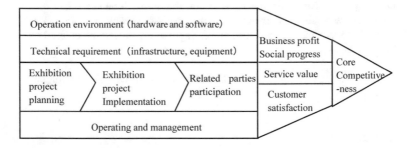

Fig. 1. The framework for exhibition value chain

2.2 The Classification of Cultural Exhibition

The definition of cultural exhibition has been considered as the starting point of the analysis. This study defines cultural exhibition as how the organizer establish the attachment between culture and exhibition to maintain customer satisfaction and gain value-added. In other words, how the cultural function and value can be tapped into exhibition. Based on the definition, the study provides the following four categories:

The first type, the cultural-theme exhibition activities, focuses on displaying, representing and reflecting culture-related themes, such as painting and intangible cultural heritage exhibition.

The second type is defined as business exhibition activities with cultural identity. Comparing to cultural-themes context, this type makes an insight in how the cultural function and value can be tapped into business exhibition.

Nowadays, enterprises emphasize on the culture recognize and culture identity in marketing, which make sense of promoting the culture value of product and service. Therefore, many exhibitors have transferred their goal in the exhibition from concluding on-site transactions to displaying their corporate culture and brand image.

The fourth type, mega-festival activities, involves considerable economic benefit, positive social benefit, large numbers of participants and rich experience, which is based on culture-experience mechanism operating and emotion-exhibition link establishing (Table 1).

3 Description of the Cultural Exhibition Value Chain

On the basis of the classification of cultural exhibition, this article elaborates driving forces, characteristics and operating environment of the cultural exhibition value chain and explores the formation mechanism (Fig. 2).

3.1 Value-Chain of the Cultural-Theme Exhibition Activities

Its primary value-added process is based on the planning and innovation. The chief value-added are explored and obtained by exhibition organizer. Moreover, Ensuring value-added should not be limited to culture creativity itself, but should also include considerations of favorable culture experience and functional conditions, including culture theme, culture atmosphere, culture surroundings, culture service, culture decoration, which provide the important operating environment for cultural-theme exhibition activities, and "cultural creativity" is the key feature to promote the innovation.

3.2 Value-Chain of the Business Exhibition with Cultural Identity

Consumer demand stimulates the upgrading of business Exhibition. Exhibition organizer start to understand the important role which culture attachments plays in achieving competitiveness and maintaining priority. Creating ties between culture and business exhibition should be highlighted in the process of planning. Nevertheless, tough competition urges exhibition marketer to be aware of that business marketing alone fails to play a positive role in promotion and customer-satisfaction, which means that the cultural and emotional connections with attendees must be reinforced in order to induce exhibition satisfaction and loyalty. Thus, the main value-added originates from "culture-ties and satisfaction", which is completed and obtained by exhibition organizer on planning and implementation stages.

3.3 Value-Chain of the Exhibitor's Culture Marketing Activities

Creating cultural ties with the commodities will gain favorable response from customers. The exhibitor's gradually recognize the fact that ensuring a successful promotion on exhibition depends on whether the attendees can be attracted and immersed in a variety of interactive experiences. Such links can be created through incorporating culture marketing activities, which will generate value-added and customer-loyalty.

Table 1. The classification of cultural exhibition

Categories	The cultural-theme exhibition activities	Business exhibition with cultural identity	Exhibitor's culture marketing activities	Mega-festival activities
Description	Professional	Comprehensive/professional	Professional	Comprehensive
Goals	Display/communication	Display/transaction	Marketing	Experience/communication
Competitiveness	Culture creativity	Culture cultivating	Cultural marketing	Cultural experience
Demand-driven	Consumption-demand	Production-demand	Production-demand	Social-demand
Operation-environment	Developing culture-creativity	Creating ties between culture and business exhibition	Promoting culture value of product and service	Exploring culture-experience mechanism

Such requirements provide opportunities with exhibitions to benefit from culture-identity mechanism of marketing. New opportunities are available for profit-seeking companies to create and capture cultural value by servicing previously potential markets. The key to transforming and upgrading the value chain is incorporating "cultural service value". The main value-added is completed and obtained by related parties.

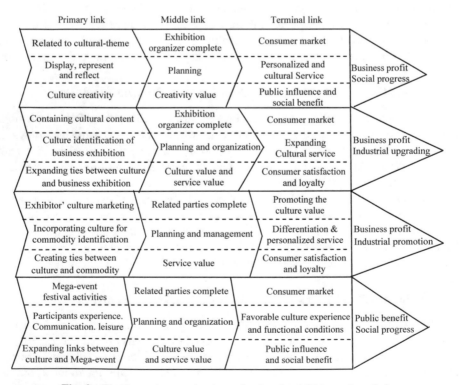

Fig. 2. The formation mechanism of cultural exhibition value chain

3.4 Value-Chain of the Exhibition Activities Referring to Mega-Cultural Event

Mega-festival activities are regarded as mega-cultural events that involve many functions, including cultural communication, cultural entertainment, knowledge popularization and education. Organizers should be aware of that focusing on increasing needs of culture experience and communication will promote economic and social benefits. Ensuring value-added of mega-event should not be limited to the increasing culture experience and strengthening the cultural-identity mechanism, but should also include favorable functional conditions. Therefore, the key to transforming and upgrading the value chain of mega-event is establishing both functional connection and cultural connection to participants. The competitiveness of mega-event mainly originates from the external value chain and the related parties gain the main value-added.

4 The Diversified Development Path for the Cultural Exhibition Industry

4.1 The Cultural Exhibition Driven by Culture Creativity

Cultural creativity should be a priority for exhibition organizer, creating a bond between attendees and exhibition that results in value-added. The key to seek profit in the value chain is cultivating corresponding point of cultural creativity and consumption needs in the social life. Rather than a concrete landscape, exhibitions related to culture theme can foster a sense of community between the individuals and the group. It will transform from niche programming to mass entertainment. In order to reinforce the publicity and sociality of culture exhibition and maintain artistry and creation, organizer should develop various frameworks on culture creative and exhibition (Fig. 3).

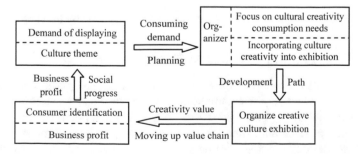

Fig. 3. The development path for the culture exhibition driven by culture creativity

4.2   The Culture Exhibition Driven by Business Exhibition Updating

In the traditional commercial age, many exhibitions were born with "business goal". How to maintain the profitability and achieve customer loyalty is vital to the development of traditional exhibition. Nowadays, consumers pay more attention to personalization, customized, value orientation, group identity, and emotional bonds on exhibition. Therefore, the contents of traditional exhibition service are constantly expanded, which consist of four components: opportunities for information, enterprise promotion, social network, and marketing. Exhibition gradually becomes a kind of cultural forms. The organizer can benefit from knowing how to properly use culture related mobilization techniques to create emotional connections that enrich the attendee experience. Such emotional mobilization can be cultivated through the use of culture, histories, themes, symbols, events, and images. Culture contents can encourage exhibition identification which will cultivate more satisfaction. Only by strengthening the bond between culture resources and traditional exhibition, developing the corresponding point of cultural connotation and group identity, forming the emotional connection on the exhibition, and creating unique and characteristic IP symbols for exhibition, will more exhibitors and visitors be attracted, and will the service value be realized (Fig. 4).

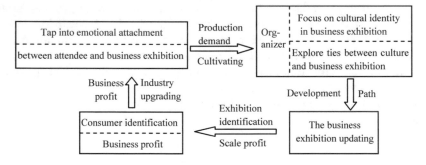

Fig. 4. The development path for the culture exhibition driven by business updating

4.3   The Culture Exhibition Driven by Exhibitor Marketing

Since exhibition can tap into the emotional aspects, culture plays a major role in identity-loyalty mechanism for enterprise marketing. In modern exhibition activities, attendee's demands are not restricted to designing exhibition booth but are extended to publicizing enterprise culture. Meanwhile, since a mega-exhibition or event can provide attendees with opportunities for image-building, social networking, and the like, it can be considered as an important link between marketing and a community of culture communication. How and in what way to develop diverse functions of the exhibition and apply various displaying means is the starting point of moving up the value chain, which plays a significant role in exhibitor promotion and customer satisfaction achievement. Product and brand promotion along fails to play a positive role in marketing field, which means the cultural and social connections with the market must be reinforced on exhibition in order to induce customer-loyalty. Creating and strengthening the interrelationship among marketing and culture identity will provide exhibitor more opportunities for culture value-added which lends important support to the upgrading of value chain and development of the exhibition industry (Fig. 5).

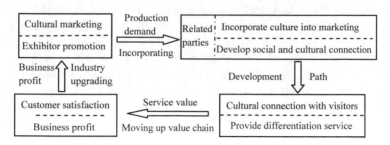

Fig. 5. The development path for the culture exhibition driven by exhibitor marketing

4.4   The Culture Exhibition Driven by Mega-Cultural Experience

Events and exhibitions can also be viewed as cultural tourist attractions. Attendees experience is an important element of exhibition identity and satisfaction for exhibition, organizers and marketers. Particularly for mega-festival events, which last for several days with such a large area of exhibition squares, cultural atmosphere, various experiences activities, and complimentary services is a must. Meanwhile, since the culture experience is important for attendee's satisfaction, organizers need to pay attention to the unique needs of culture and potential attendees instead of basic functional needs. Event organizers should have a better understanding of the contents of features of culture event, which includes experience, public, entertaining, practicing, leisure, and emotion. Moreover, ensuring proper conditions should not be limited to the event itself, but should also include considerations and cultivations of emotional identity which should be originated from culture. Event organizer should make efforts to create more culture communication space for visitor interaction in order to maintain attendee's stay in exhibition and ultimately enhance their satisfaction. Therefore, the role that emotional and cultural identification with the event plays in ensuring value-added should be highlighted [4–7] (Fig. 6).

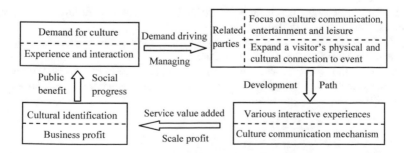

Fig. 6. The development path for the culture exhibition driven by culture experience

5 Conclusion

The innovation of this paper lies in the fact that maximizing the cultural added value for exhibition, will promote the transformation and upgrading of exhibition industry. On the basis of value chain theory, the framework for culture exhibition value chain is modeled and the diversified development paths are put forward. First, based on the classification of four types of culture exhibition, this study explores and tests four internal mechanism of culture exhibition which can provide insights into exhibition market segmentation. Second, we emphasize the importance role of culture creativity plays in the value-added link. Knowing this, the exhibition organizer and manager will pay more attention to enhance culture contents to maintain the interaction between exhibitions and visitors. In particular, the implications of this study are useful for marketing and culture communications strategies that target culture exhibition market. Our findings show that culture plays a major role in the satisfaction-loyalty mechanism, which urge the exhibition organizer to explore culture exhibition segments. Attendee satisfaction and identification depend on whether the participants can be attracted in a variety of experiences which can be achieved by providing culture contents and create more culture community-building space for attendee interaction. Last but not least, by cultivating culture-interaction, will ensure exhibition maintaining sustained value-added and inducing loyalty from attendees.

Acknowledgments. This research was partially supported by Center for Trans-Himalaya Tourism and Culture Studies (LYC15-22) and Research Center of Sichuan County economy development (xyzx1516).

References

1. Getz, D.: Event Management and Event Tourism, p. 21. Cognizant Communication Corporation, New York (2005)
2. Fu, Y., Zheng, X.: Industrial merging of tourism industry and MICE industry: analysis of industrial value chain, approaches and countermeasures. J. Northwest A&F Univ. (Soc. Sci. Ed.) **2**, 151 (2014)
3. Liang, X.: Diversified development path of service industry: perspective of service value chain. China Soft Sci. **6**, 172–178 (2016)
4. Li, H., Pan, J.: Analysis on the culture influence of MICE. J. Guangxi Univ. (Philos. Soc. Sci.) **11**, 88 (2015)
5. Li, Y., Liu, H., Huang, B.: A study on MICE industry value chain and its convergence. Commer. Res. **1**, 10–11 (2016)
6. Xu, Y., Zeng, G.: Exploration of innovation path of China's creative industries under the big data strategy. Theor. Invest. **6**, 108–114 (2018)
7. Fu, X., Yi, X., Fevzi, O., Jin, W.: Linking the internal mechanism of exhibition attachment to exhibition satisfaction: a comparison of first-time and repeat attendees. Tour. Manag. **72**, 92–104 (2019)

Research on Text Classification Method Based on Word2vec and Improved TF-IDF

Tao Zhang and LuYao Wang[✉]

School of Software, Beijing University of Technology, Beijing, China
358226756@qq.com

Abstract. TF-IDF is widely used as the most common feature weight calculation method. The traditional TF-IDF feature extraction method lacks the representation of the distribution difference between classes in the text classification task and the feature matrix generated by the TF-IDF is huge and sparse. Based on this situation, this paper proposes a method of using the feature extraction algorithm of chi-square statistics to compensate for the distribution difference between classes and generating a fixed-dimensional real matrix through word2vec. The experimental results show that the new method is significantly better than the traditional feature extraction methods in the evaluation results such as precision, recall, F1 and ROC_AUC.

Keywords: TF-IDF · Chi-square statistics · Word2vec · Text classification

1 Introduction

In recent years, text classification tasks mainly include text preprocessing, feature selection, model training, and effect evaluation [1]. The text preprocessing stage, especially the feature word extraction part, greatly affects the accuracy of classification. Therefore, researchers at home and abroad are committed to improving the extraction of feature words. In 1988, Salton [2] proposed a combination of word frequency weighting and anti-document frequency calculation, namely TF-IDF. The traditional TF-IDF weight calculation method has the following two problems: Firstly: TF only considers the text information from the perspective of word frequency, lacks the processing of the feature word context, and ignores the structure of the text; Secondly: IDF ignores the deficiency of the distribution of feature items between categories.

This paper proposes to use the chi-square statistical method to describe the distribution information and classification ability of feature words between classes, and use word2vec to generate real-matrix matrices of fixed dimensions. Taking agricultural product public opinion information as an example, the correctness and effectiveness of the improved feature extraction method are verified by comparing the evaluation indexes of TF-IDF before and after improvement on the classification of neutral, neutral and derogatory.

© Springer Nature Switzerland AG 2020
F. Xhafa et al. (Eds.): IISA 2019, AISC 1084, pp. 199–205, 2020.
https://doi.org/10.1007/978-3-030-34387-3_24

2   Introduction of Traditional TF-IDF

2.1   Traditional TF-IDF Calculation Method

In the traditional TF-IDF calculation method, the TF word frequency statistical method is used to describe the high frequency features, and these high frequency features are often noise words that do not help the text classification. Some low frequency words can well represent the text information but because of the low frequency of occurrence, it may be ignored. IDF enhances the weight of feature words with low frequency of occurrence, and to some extent compensates for the lack of TF. The TF-IDF weighting method combines the two methods, the formula is as follows (1):

$$W = TF^* \log \left(\frac{N}{n_k} + 1 \right) \tag{1}$$

In formula (1), N represents the total number of texts in the entire training sample, and nk represents the number of texts containing the feature words.

2.2   The Deficiency of Traditional TF-IDF

The traditional TF-IDF ignores the distribution of feature words between categories [2]. For example, the number of occurrences of the feature word a in the positive class text is m, and the number of occurrences in the negative class text is n, and the total number of occurrences of the feature word in the text collection is recorded as l = m + n. When the m value is large, that is, the feature word appears in the text of the positive class many times, the feature word a can reflect the positive class text very well. However, in TF-IDF, it can be seen from formula (1) that when m and l are larger, the IDF is smaller, so the feature word a is no longer representative.

The traditional TF-IDF generates a model under the premise that the words of the text are completely independent, resulting in the generated model not representing the context of the text and the semantic information of the text; and the dimension of the weight vector matrix generated by TF-IDF is very huge and very, causing wasted storage space.

Moreover, when the feature word a is evenly distributed among each category, it does not have a category representative, so the feature word a should be given a lower weight. However, as can be seen from formula (1), the value of IDF is large at this time, which is not in line with the actual situation. The reason for this result is that the TF-IDF is not considered between categories.

3   Based on the Chi-Square Improved TF-IDF

3.1   Chi-Square Calculation Method

The Chi-square statistical (CHI Square, CHI) is used to measure the degree of deviation between two variables, where the theoretical and actual values are independent of each other. If the deviation is large, then the two variables are independent of each other. The relationship between its characteristics and categories is shown in Table 1.

Table 1. Feature and category relationship table

	In class C_i	Not in class C_i	Total
Text containing the feature word t	A	B	A + B
Text without the feature word t	C	D	C + D
Total	A + C	B + D	N = A+B + C+D

The calculation method based on the feature word t is as shown in the formulas (2), (3), (4), and (5):

$$E_{11} = \left(\frac{A+B}{N}\right) * (A+C) \tag{2}$$

$$E_{12} = \left(\frac{A+B}{N}\right) * (B+D) \tag{3}$$

$$E_{21} = \left(\frac{C+D}{N}\right) * (A+C) \tag{4}$$

$$E_{22} = \left(\frac{C+D}{N}\right) * (B+D) \tag{5}$$

The deviation of the category C_i including the feature word t is calculated according to the deviation calculation method as shown in the formula (6).

$$D_{11} = \frac{(A - E_{11})^2}{E_{11}} \tag{6}$$

Similarly, D12, D21, and D22 can be obtained, and the CHI value division method of bringing in and simplifying the characteristic vocabulary t and the category Ci is as shown in the formula (7).

$$CHI_{(t,Ci)} = \frac{(A*D - C*B)^2 * N}{(A+C) * (B+D) * (A+B) * (C+D)} \tag{7}$$

It can be known from the formula (7) that when the feature vocabulary t and the category Ci are independent of each other, A * D – C * B is equal to 0, and CHI is also equal to zero. The more the feature vocabulary t and the category Ci are related, the larger the value of CHI.

Yang's [3] research shows that chi-square statistics is one of the best feature selection methods in existing text categorization. The larger the chi-square value of a feature item in a category, the more distinguishing the feature word is. However, there are also deficiencies in the statistics of the Chi-square. The shortcomings will be analyzed in detail below.

(1) Traditional Chi-square statistical methods ignore the consideration of low-frequency words. Specifically, if the number of feature words appearing in a text is relatively small and it is repeated in multiple documents, such words cannot distinguish the text categories, so the classification performed is not accurate.
(2) The traditional chi-square does not consider the situation in which the feature words are evenly distributed within the category. It is embodied that the chi-square statistical formula does not reflect the high weight that the feature words should be given when they appear uniformly in a certain type of document [4].
(3) The traditional chi-square statistical method prefers to select feature words that are negatively related to the category. It can be known from formula (7) that the feature vocabulary appears more in other categories, and appears less in the specified class, that is, B and C are larger, and A and D are smaller, and BC AD can be obtained. The CHI value calculated in this way is relatively high, and at this time, it is unreliable to use words feature words which just can distinguish other categories, so the performance of text classification cannot be improved.

3.2   TF-IDF Based on Chi-Square Improvement

This section improves on the traditional TF-IDF, ignoring the distribution of feature words between categories.

As can be seen from the above section, the traditional TF-IDF weight calculation method ignores the distribution difference of feature words between categories. The chi-square statistical value of feature words can well describe the distribution information of feature words between categories. It is embodied that the feature vocabulary classification ability is proportional to its chi-square value. Based on this situation, we introduce the chi-square statistical method to improve the classification ability of feature words between classes. The feature weight improvement method based on the chi-square statistics is shown in formula (8).

$$\text{TFCHI} = \text{TF} * \log\left(\frac{N}{n_k} + 1\right) * \frac{(A * D - C * B)^2 * N}{(A + C) * (B + D) * (A + B) * (C + D)} \tag{8}$$

It can be known from formula (8) that the TF-IDF weight calculation method based on the chi-square statistics makes up for the defect that the feature vocabulary is not considered to be distributed between classes. Words that are evenly distributed across classes but do not have the ability to distinguish between categories can be given lower weights. According to the improved TF-IDF, the traditional calculation method has been improved to some extent. Improved TF-IDF effectively improve the accuracy of weight calculation and the correctness of text classification.

4 Based on the Word2vec Improved TF-IDF

4.1 Improved TF-IDF Combined with Word2vec

In this paper, word2vec uses the CBOW model to use the context information to train the feature words as fixed-size and lower-dimensional real-number vectors. The similar words have similar distance calculation values in vector space. The matrix dimensions generated by TF-IDF are sparse and the text context is ignored. The problem of relationship. Using a fixed dimension word vector effectively solves this problem. The word2vec model ignores the impact of words on document categories due to the impact of different words on different categories of documents. The improved TF-IDF algorithm is used to calculate the weight of document feature items, enhance the distinction between different documents, and improve the combination of word2vec and improved TF-IDF, which not only takes advantage of the fixed dimension of word vector but also add the influence of feature words on document categories. The specific steps are as follows:

(1) Perform the above preprocessing work on the input text, and turn each statement into a form of word collection.
(2) Using the TF-IDF algorithm to calculate the feature value of each word in the sentence, each statement becomes a feature vector form of the TF-IDF value set, and the TF-IDF value representing the i-th word is represented by WiTF-IDF.
(3) According to the above-described chi-square calculation formula, the chi-square value W^i CHI of each word in the sentence is calculated, and the weight $weight^i$ of the word-to-text discrimination ability is calculated according to the improved TFCHI calculation method. Its calculation formula is as follows (9).

$$weight^i = W^i_{TF-IDF} * W^i_{CHI} \tag{9}$$

(4) The CBOW model is used to train the input text sequence to obtain a fixed-dimensional word vector real matrix. The formula is as shown in (10).

$$W^i_{word} = \left[W^{i_1}_{word}, W^{i_2}_{word}, \ldots W^{i_n}_{word} \right] \tag{10}$$

(5) The context semantic word vector word2vec is combined with the weight value TF-IDF of the distinguishing document category. The formula is as follows (11).

$$W^i = \left[weight^i * W^{i_1}_{word}, weight^i * W^{i_2}_{word}, \ldots weight^i * W^{i_n}_{word} \right] \tag{11}$$

5 Experimental Results

5.1 Source of Experimental Data

This paper takes the classification information of agricultural products as an example, and obtains the public opinion information of agricultural news from the Internet through crawling technology. The public opinion information is divided into three categories according to the meaning of commendatory, derogatory and neutral. Each takes 20,000 as the training set and 5000 as the prediction set.

5.2   Evaluation Indicators

In this experiment, the precision rate, recall rate, F1-Score and ROC_AUC were taken as the main indicators for evaluating the text classification effect. These indicators can not only reflect the classification effect of the text classification model, but also help to analyze the influence of the feature extraction method on the classification task while maintaining the same function modules [5].

5.3   Experimental Results

Based on the traditional TF-IDF and improved TF-IDF feature extraction methods, the experimental data are extracted separately, and then the classification algorithm is used to train and verify the classification results.

The comparison effect data of the text classification precision rate is shown in Table 2. The contrast effect data of the text classification recall rate is shown in Table 3. The contrast effect data of the text classification F1 measurement values is shown in Table 4.

Table 2. Comparison of precision

Method	Comendatory sense	Neutral sense	Derogatory sense
Previous TF-IDF	78.9%	77.4%	75.7%
Improved TF-IDF	88.6%	82.8%	86.9%

Table 3. Comparison of recall

Method	Comendatory sense	Neutral sense	Derogatory sense
Previous TF-IDF	72.3%	77.5%	78.7%
Improved TF-IDF	87.6%	92.8%	86.5%

Table 4. Comparison of F1-score

Method	Comendatory sense	Neutral sense	Derogatory sense
Previous TF-IDF	78.3%	75.4%	73.2%
Improved TF-IDF	89.8%	88.3%	87.8%

It can be seen from the data in the table that the improved TF-IDF algorithm has significant improvement in text classification precision, recall rate, and F1 measurement value compared with the traditional TF-IDF algorithm. The specific embodiment is as follows: the improved text algorithm will increase by at least 8% in terms of the precision of the comendatory, neutral, and derogatory effects, at least 11% in the recall rate, and at least 10% in the F1.

6 Conclusion

Based on the traditional TF-IDF, this paper proposes a new feature extraction algorithm for Chi-square statistics, which is evaluated from four aspects: precision rate, recall rate, F1 measurement value and ROC_AUC. The new algorithm makes up for the defect that the feature vocabulary is not distributed between classes. It can assign lower weights to words that are evenly distributed among classes but do not have class distinguishing ability, which effectively improves the accuracy of weight calculation and text classification. Correctness.

Acknowledgments. During the writing of the thesis, I would like to express my gratitude to the teachers, classmates and family members who have given me help and care. First of all, I would like to thank my teacher, Teacher Zhang Tao. Mr. Zhang is passionate and acquainted with enthusiasm and rigor, and has profound professional knowledge. When I encountered academic problems, I gave me a lot of patient guidance and advice. Secondly, every student in the lab family, we exchange and study together in academic research, and the laboratory research atmosphere is strong. Finally, I sincerely thank all the professors and experts who reviewed the paper.

References

1. Jian, J.: Research on Feature Extraction and Feature Weighting in Text Categorization. Chongqing University, Chongqing (2010)
2. Salton, G., McGill, M.J.: Introduction to modern information retrieval. Commun. ACM (1983)
3. Chu, L., Gao, H., Chang, W.: A new feature weighting method based on probability distribution in imbalanced text classification. In: 2010 Seventh International Conference on Fuzzy Systems and Knowledge Discovery, pp. 2335–2339 (2010)
4. Yang, Y., Pedersen, J.O.: A comparative study on feature selection in text categorization. In: Proceedings of the 14th International Conference on Machine Learning (ICML), pp. 412–420 (1997)
5. Sebastiani, F.: Text categorization. In: Encyclopedia of Database Technologies and Applications, no. 6, pp. 109–129 (2005)

Research on Automatic Text Summarization Method Based on TF-IDF

Tao Zhang and Cai Chen[(✉)]

School of Software, Beijing University of Technology, Beijing, China
690799780@qq.com

Abstract. In order to quickly obtain the main information contained in news documents, reduce redundant information and improve the efficiency of finding news with specific content. A Chinese text summarization method based on TF-IDF is proposed. This method uses TF-IDF to calculate the importance of each word in the article, and calculates the TF-IDF of each sentence based on the TF-IDF value of the word. In order to avoid the effect of sentence length on the calculation of sentence TF-IDF value. The sliding window is used to calculate the mean of all words TF-IDF in each sliding window using the given sliding window size. Use the value of the sliding window with the largest mean value in each sentence as the TF-IDF value of the sentence. Combined with other feature of the sentence, the importance of the sentence is calculated. The sentences with the specified length or number of words are intercepted and arranged according to the order of the articles to form a summarization of the article. After comparison experiments, the method is superior to the text summarization scheme based on TextRank method in terms of efficiency and effect.

Keywords: TF-IDF · Text summarization · NLP

1 Introduction

With the coming of the information age, all kinds of information on the Internet are emerging one after another. We can easily obtain the information that we want through search engines and major news release platforms. Most search engines sort by keyword and return the final result. When we search for a keyword, the search engine will return a list of data containing the keywords. When we read the details with the expected click link, we often find that we are not what we want, then we need to return to the search page. If the search results are very large, it may take a lot of time. If there is a tool that helps us read the results of the search and extract the key sentences, This may take a lot of time to get the information you need.

2 Research Status

In the study of automatic text summarization, there are two main types of abstract generation, Extraction-based and Abstraction-based summarization.

F. Xhafa et al. (Eds.): IISA 2019, AISC 1084, pp. 206–212, 2020.
https://doi.org/10.1007/978-3-030-34387-3_25

2.1   Extraction-Based Text Summarization

The Extraction-based text summarization method refers to the evaluation of the importance of the sentence by some method on the basis of the original text, and finds one or more sentences that are closest to the original meaning as the summarization according to the importance of the sentence. At this stage, the research on the method of Extraction-based text summarization is relatively mature. The Extraction-based text summarization assumes that an article can express its meaning through the more important sentences in the article, so the summary task becomes the most important sentence in the article. The core issue is to sort the sentences in the article.

There are two major types of existing sorting methods.

(1) Graph-based sorting, the method treats each sentence as a node, and the similarity of the sentence is used as the weight of the edge. The TextRank method is used to obtain the score of the sentence. The TextRank algorithm is a graph-based sorting algorithm for text. Divide the text into several constituent units (sentences) and build a node connection graph. Using the similarity between sentences as the weight of the edge, the TextRank value of the sentence is calculated by loop iteration, and finally the sentence with high ranking is combined into a text summary.

(2) Feature-based sorting method, the sentences are sorted based on some features of the sentence itself. The commonly used features are sentence length; the sentence position, the first few sentences of the article in the central sentence of the article, according to the sentence in the article Score; the sentence contains the number of keywords, the text is processed after the word segmentation, and the corresponding score is given according to the number of keywords in the sentence. This paper is based on the improvement of the method.

2.2   Abstraction-Based Summarization

Abstraction-based summarization system refers to the computer outputting a summary based on the understood content after understanding the original text. It is similar to summarizing the article after reading comprehension. The main model of the Abstraction-based summarization method is usually the seq2seq model in deep learning, on which the attention mechanism is added. The seq2seq model consists of two parts, (1) the encoder, which is mainly responsible for encoding the original text. (2) The decoder is responsible for decoding the digest. Models with sequence learning capabilities of the RNN series commonly used by encoders and decoders, such as LSTM, GRU, BiLSTM, and variants thereof. Google's latest public textsum model, is a text summary using seq2seq+attention, which can get better results. But training RNN takes a long time and requires huge computing resources.

3 TF-IDF

3.1 Definition

TFIDf is an evaluation metric used to assess how important a word is to a document. The importance of a word is proportional to how often it appears in the article, and inversely proportional to how often it appears in other documents. TF-IDF is divided into two parts, TF word frequency, IDF inverse document frequency, TF indicates the frequency of words appearing in document D, which can be obtained by dividing the number of words by the total number of words in the document. IDF is a general importance of words. The measure can be obtained by dividing the total number of documents by the number of documents containing the word, and then taking the obtained quotient with the base 10 logarithm. In the calculation process, In order to prevent the denominator from being zero in the calculation process, resulting in calculation errors, the denominator is usually added to the original base.

3.2 Method of Calculation

The calculation method of TF-IDF is divided into two parts.

Calculate the TF value for each word of the sentence after the word segmentation is completed.

The formula for calculating TF is:

$$\text{tf}_{i,j} = \frac{n_{i,j}}{\sum_k n_{k,j}} \tag{1}$$

The formula for calculating IDF is:

$$idf_i = \lg \frac{|D|}{|j : t_i \in d_j|} \tag{2}$$

When calculating the idf, if the number of documents containing words is zero, the denominator will be zero. Therefore, the denominator part will be added one by one. The modified idf is calculated as:

$$idf_i = \lg \frac{|D|}{1 + |\{d \in D : t \in d\}|} \tag{3}$$

The formula for calculating TF-IDF:

$$tfidf_{i,j} = tf_{i,j} * idf_j \tag{4}$$

In formula (1), (2), (3), (4): i is the label of the current word in the document, and j is the label of the current document in the corpus |D|: Total number of documents in the corpus.

4 Experiment

4.1 Pretreatment

(1) There are natural separators in English texts. There are no special symbols between Chinese words in Chinese, and there are many double words and multiple words in Chinese. The text segmentation is the first step in Chinese processing. A crucial step, the accuracy of the Chinese word segmentation, it will have an important impact on the next steps. This study used foolnltk as a tokenizer.

(2) The text after stopping the word segmentation often contains a lot of irrelevant words. Adding them to the operation will affect the accuracy of the final result, and more words need more computing performance. This study uses the stop word list to filter the words. Use the stop word table to process the raw data after the word segmentation is completed, and some noise words are deleted.

4.2 Calculation of TF-IDF Value

(1) Construction dictionary
   For the text of the word segmentation, use the stop word table to remove some words, and then add all the filtered words to the dictionary.

(2) Calculate the number of documents containing words
   For each word in the dictionary, first count the number of documents that contain the word. In order to improve the calculation efficiency, the following calculation methods can be adopted. First put all the words that appear in a dictionary, then traverse according to the document, divide each document into words, put all the words into the collection, and then traverse each word in the collection, and count them in the dictionary.

(3) Calculate the TFIDF of each word in each article
   Construct an array of the same length as the document. Each element of the array stores the TFIDF of all the words of an article, and stores it in a dictionary. For each word of each document, use formula (1) to calculate its TF word frequency first, and then use Eq. (3) calculates the IDF value, and finally uses formula (4) to calculate its TF-IDF value.

4.3 Calculate the Importance of a Sentence

There are ways to assess the importance of a sentence. The existing research methods include extracting the keywords of the document by TFIDF or other methods, inverting according to the importance degree, taking the first N keywords as the keywords of the document, and then calculating The number of keywords included in each sentence, divided by the number of keywords divided by the total number of words as the importance of the sentence. This method does not take into account the importance of words. In fact, the contribution of core keywords and general keywords to sentence importance is different. At the same time, the length of sentences will affect the calculation results. If the target core sentence is too long, it will lead to The amount of information contained is relatively sparse and ultimately affects the calculation results.

For the first question, the core word and the general word are distinguished by using the TF-IDF value of the word to represent the importance of the word. For the second problem, two solutions are proposed. Solution 1: Take the mean of the K words with the largest TF-IDF in each sentence as the TF-IDF value of the sentence. Option 2: First we can set the sliding window size W, then calculate the TF-IDF mean of the words in each window, using the largest of them as the TF-IDF value of the sentence. The important formula of the TFIDF of the sentence is as follows:

$$sentence\_tfidf = \max \left(\frac{\sum_{word_j \in w_i} tfidf(word_j)}{|w_i|} \right) \tag{5}$$

In formula (5): Tfidf represents the tfidf value of the acquired word in the document, wi represents the i-th sliding window, word represents the word in the sliding window, and |wi| represents the sliding window size.

After the TF-IDF of the sentence is obtained, it is normalized as the TF-IDF feature of the sentence. According to the position information of the sentence, the position feature of the sentence is extracted. The central sentence of the document tends to appear at the beginning of the document. Therefore, the closer the sentence is to the beginning of the document, the more important it is. The calculation formula of the position feature of the sentence is as follows:

$$index\_feature = \log \left(\frac{|D|}{sentence\_index} \right) \tag{6}$$

The importance of the sentence is calculated by the TFIDF feature and position feature of the comprehensive sentence. The formula for calculating the sentence importance is as follows:

$$weight = sentence\_tfidf * w1 + index\_feature * w2 \tag{7}$$

In formula (7): w1, w2 are the weights of the two features.

After the sentence importance is sorted, the candidate sentences are filtered according to the length of the summarization, but the selected sentences may not guarantee the logic of the original text in reading, so they need to be arranged according to the original order of the sentences and output.

5 Evaluation of Automatic Text Summarization

At present, there is no complete and authoritative evaluation method for the evaluation of text summarization. The existing text abstract evaluation methods are divided into two categories. (1) Internal evaluation methods, which directly evaluate the quality of text abstracts by comparing the consistency of abstracts with documents. Or by matching the generated summary with the manually extracted summary. (2) Indirect evaluation methods to assess the quality of text summary algorithms through changes in indicators of relevant business systems. The evaluation of text summarization is a complex issue that requires further objective evaluation methods.

In this experiment, a total of more than 20,000 news items were extracted from the six news sections of Sina News, including military, finance, technology, and education. First remove the stop words and symbols in the news headline, and then calculate the proportion of words in the news headline that are included in the headline. The calculation formula is:

$$score = \frac{|w \in title : w \in summary\}|}{|title|} \tag{8}$$

The results of the text summarization using TextRank and TFIDF on different datasets using the above method are shown in the Table 1:

Table 1.

	Military	Sports	Education	All
TextRank	0.473	0.451	0.467	0.468
TF-IDF	0.601	0.554	0.589	0.593

It can be seen from the table that the text digest algorithm based on TFIDF has a clear advantage in the production of military, sports, education and all news, which is obviously superior to the TextRank based algorithm.

6 Conclusion

This paper proposes a text automatic summarization method based on TF-IDF, which uses the TFIDF value of the keyword contained in the sentence as the weight, and gives different weights to the core word keyword and the general keyword. At the same time, in order to prevent the inconsistent sentence length from affecting the result, a sliding window is introduced, using the importance of the largest sliding window in the sentence as the sentence importance, and the sentence is sorted according to the sentence length and sentence position, in multiple corpora. It achieved a good result. However, the way to evaluate is to use the captured news headline as the basis for the evaluation. Because the title contains a limited amount of information and there will be a title party, this may affect the results of the evaluation. In the TextRank-based experiment, SnowNLP was used for testing. TextRank runs slower than the TFIDF-based method during the test.

Acknowledgments. During the research, I am very grateful to my tutor. He gave me effective guidance during my research and developed a detailed research plan for me. When I met a problem that I could not solve, he would actively help. I am looking for a solution. I am also very grateful to my school and gave me a good research environment so that I can concentrate on my research work. In addition, I am grateful to those students who have patiently answered me when I have problems.

References

1. Lim, D.: Convolutional attention-based seq2seq neural network for end-to-end ASR (2017)
2. Yong, Z., Wang, Y., Liao, J., et al.: A hierarchical attention seq2seq model with CopyNet for text summarization. In: 2018 International Conference on Robots & Intelligent System (ICRIS) (2018)
3. Jing, L.P., Huang, H.K., Shi, H.B.: Improved feature selection approach TFIDF in text mining. In: International Conference on Machine Learning & Cybernetics (2003)
4. Aizawa, A.: The feature quantity: an information theoretic perspective of Tfidf-like measures (2000)
5. Liu, J., Cheung, J.C.K., Louis, A.: What comes next? Extractive summarization by next-sentence prediction (2019)
6. Templeton, A., Kalita, J.: Exploring sentence vector spaces through automatic summarization (2018)
7. Azhari, M., Kumar, Y.J., Goh, O.S., et al.: Automatic text summarization: soft computing based approaches. Adv. Sci. Lett. **24**(2), 1206–1209 (2018)
8. Ferilli, S., Pazienza, A.: An abstract argumentation-based approach to automatic extractive text summarization. In: Italian Research Conference on Digital Libraries (2018)

Evaluation Methods of English Advanced Pronunciation Skills Based on Speech Recognition

Feng Meng[1], Yanming Jia[2], Wufeng Wang[3], and Xu Ning[4(✉)]

[1] School of Computer Science and Technology,
Wuhan University of Technology, Wuhan 430070, China
[2] Center for Artificial Intelligence and Big Data Research, Global Wisdom Inc.,
Beijing 100085, China
[3] Zhongshan Institute of Advanced Engineering Technology of WUT,
Zhongshan 528437, China
[4] Hubei Key Laboratory of Transportation Internet of Things,
Wuhan University of Technology, Wuhan 430070, China
xuning@whut.edu.cn

Abstract. The assessment of English advanced pronunciation skills evaluates the pronunciation fluency, employing techniques from various fields including artificial neural network, speech recognition and linguistics. The pronunciation fluency includes 4 main parts of pronunciation skills, including speech stress, pause, liaison and plosive. This paper mainly studies liaison and plosive. In the aspect of liaison recognition, GRU (Gated Recurrent Unit) are used to train the classification model. In the field of plosive recognition, words both with 1 plosive label or not are added to the process of language model construction; and acoustic features both with plosive pronunciation skill or not are used to the process of acoustic model construction, with these two parts, static decoding network WFST (Weight Finite-State Transducer) is constructed. The experimental results show that, for different types of liaison, the average accuracy of GRU models are up to 92.34%; for plosive model, the accuracy of the static decoding network model is 72.17%.

Keywords: Liaison · Plosive · GRU · Static decoding network

1 Introduction

At present, foreign language learning in China mainly focuses on English. In the field of English teaching, online education has applied scenarios in all aspects, from words to pronunciation, and the oral English.

For most Chinese English learners, oral English is often their weakness in English learning. Pronunciation fluency is an important index to evaluate the quality of oral pronunciation. Fluency evaluation includes four evaluation points: stress, pause, liaison, plosive. Liaison and plosive are the advanced pronunciation skills.

Liaison is a pronunciation technique that smoothly connects phrases, idioms or two or more words belonging to the same meaning group to pronounce. It sounds like a long

© Springer Nature Switzerland AG 2020
F. Xhafa et al. (Eds.): IISA 2019, AISC 1084, pp. 213–221, 2020.
https://doi.org/10.1007/978-3-030-34387-3_26

word. Liaison can even connect a complete sentence to a super-long sound [1]. Plosive can be divided into complete plosive and fricative plosive. When two plosives are in contact there's a complete plosive of the first sound, and a plosive is heard only after the second consonant, this phenomenon is known as plosive. When a plosive consonant precedes a fricative consonant in a word or at a junction of words it has its release during the pronunciation of the fricative [2]. This phenomenon is called fricative plosive. For the detection method of plosive, this paper proposes an improvement based on the method of Huang et al. Training the plosive and non-plosive parallel audio corpus to the acoustic model while training the language model with the plosive mark vocabulary in [3]. In terms of evaluation, Huang proposed the criteria based on expert scoring, and the evaluation method of this paper is based on English linguistic rules.

For the detection method of liaison, because the speech recognition part of this paper is based on WFST static decoding network, after the acoustic model decodes the phoneme sequence, it is impossible to distinguish whether it is continuous pronunciation from the language model. Therefore, this paper proposes a method based on RNN (Recurrent Neural Network).

In this paper, we propose a method based on speech recognition for the evaluation of spoken English advanced pronunciation skills. Main contributions: 1. For the phoneme sequence in the phonetic sequence obtained by speech recognition system, the method of reconstructing the static decoding network model and the neural network classification method are used to evaluate the plosive pronunciation; For the liaison of speech, RNN is applied to the classification of speech data.

2 Corpus Preparation

2.1 Liaison Corpus Preparation

The study of liaison and plosive needs to include both two of these pronunciation skills and those without them. In this paper, the text corpus containing 300,000 sentences is used to synthesize parallel speech corpus through Amazon Polly speech synthesis interface. Amazon Polly uses Deep Learning to simulate human pronunciation and translate text into audio.

For liaison recognition, this paper analyzes the phoneme level and acoustic characteristics. At the phoneme level, for a liaison phrase, for example: [work out], after acoustic model decoding, a sequence containing 5 phonemes is obtained: /W RT K AW T/, combined with the dictionary, [work] corresponds to phoneme sequence is /W ER K/, and the phoneme sequence corresponding to [out] is /AW T/. Combined with liaison pronunciation rules in English, the last phoneme /K/ of the word [work] is assigned to the first pronunciation of the word [out], which is the liaison pronunciation skill. In the WFST-based static decoding network, it is impossible to accurately assign the phoneme /K/ to [work] or [out] according to the phoneme sequence for liaison and non-liaison pronunciation. In terms of acoustic characteristics, the degree of association between /K/ corresponding features and /AW/ corresponding features also reflects whether there is liaison. Therefore, this paper proposes a sequence classification-based method to distinguish between continuous pronunciation and non-continuous pronunciation.

The corpus preparation task is divided into liaison rules extraction, liaison text extraction including liaison pronunciation phrases, and audio files corresponding to liaison text (including the liaison mode and the non-liaison mode).

2.1.1 Liaison Rules Extraction

Liaison needs some certain preconditions, that is, Liaison can only occur between two or two words within the same meaning group, between the last pronunciation of the previous word and the first pronunciation of the next word. These two pronunciations are generally combined into one syllable when liaison pronunciation. Common liaison pronunciation situations include "consonant + vowel", "/r/ or /re/+ vowel", "consonant + semi-vowel", "vowel + vowel".

2.1.2 Liaison Corpus

Using liaison pronunciation rules extracted in Sect. 2.1.1, combined with the Libir-Speech corpus, the 73 most common types of liaison pronunciations were counted. And the corresponding corpus of each liaison pronunciation is synthesized into corresponding voices.

2.2 Plosive Corpus Preparation

For the plosive, at the phoneme level, for the phrase [second day], the last pronunciation of the first word /d/ and the first pronunciation of the last word /d/ constitute the plosive pronunciation skill, the first /d/ is not pronounced. In the process of decoding the speech recognition model, this phoneme disappears. In order to identify whether or not the plosive pronunciation technique is included, it is necessary to construct a speech recognition model both including the plosive and non- plosive pronunciation skills.

2.2.1 Plosive Extraction

Plosive is generally divided into five cases: plosive and plosive, plosive and nasal consonants, plosive and lingual, plosive and fricative, plosive and squeaky. When a plosive occurs, the previous plosive is generally not emitted or is not completely emitted.

2.2.2 Plosive Corpus

Using the plosive rules extracted in Sect. 2.2.1, combined with the LibirSpeech corpus, the 114 most common types were counted. The words in the collected corpus with the detonation technique are marked, such as [third], marked as [third0], and the language model is trained both with the original text and marked text. The corresponding speech is then synthesized for each corpus.

3 Evaluation

3.1 Evaluation of Liaison

3.1.1 Feature Engineering

Different from the traditional 13-dimensional MFCC (Mel Frequency Cepstrum Coefficient) feature [4], this paper performs the mean variance normalization (CMVN) [5] on the basis of this feature, and the for each frame of feature, we do a frame extension of length of 3. The speech feature engineering is shown in Fig. 1. The feature engineering method introduces the context-related information of the linguistic features, and introduces speaker-independent features through the feature transformation matrix.

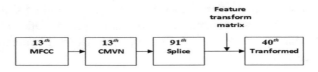

Fig. 1. Speech feature engineering

This paper is based on the Kaldi speech recognition framework. For the speech to be evaluated, according to the liaison rule, extract the possible liaison phrase from the recognition result. And then get the combination of its phonetic symbols and map the phonetic symbols to phonemes. The index of the corresponding state of the phoneme is found in the alignment result, and finally the speech feature corresponding to the phoneme is obtained by the index. The liaison feature extraction process is shown in Fig. 2.

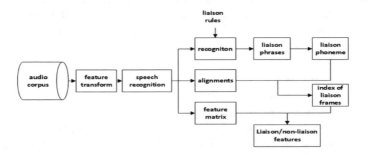

Fig. 2. Liaison feature extraction process

3.1.2 Proposed System for Liaison Recognition

For speech data, it is a frame-by-frame related data, and there is correlation between each frame of data. In processing speech data, it is not meaningful to analyze the characteristics of each frame separately. For speech data, the whole sequence needs to be analyzed. RNN has its unique advantages in dealing with sequential data, and its internal state can display dynamic sequential behavior well [6].

GRU is an excellent variant of RNN. It inherits the characteristics of most RNN models and solves the problem of gradient disappearance caused by gradient shrinkage in the process of reverse propagation [7]. Although CNN still has performance advantages in classification problems, it is relatively weak in the long-term and more complex tasks. The proposed system is shown in Fig. 3.

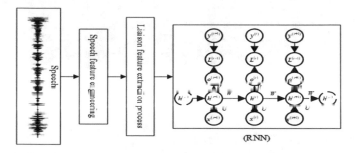

Fig. 3. The proposed liaison recognition system

3.2 Evaluation of Plosive

The plosive speech recognition system is based on the Kaldi speech recognition framework. The system is constructed by WFST [8], including H.fst (transfer phone state to tri-phoneme), C.fst (transfer tri-phoneme to single phoneme), L.fst (transfer phoneme to word), and G.fst (transfer words to sentence). The speech recognition process requires composition of the acoustic model and language model. With these 4 WFSTs above, the static decoding network HCLG.fst is constructed. The process is $G \rightarrow L \rightarrow C \rightarrow H$.

3.2.1 Improvement of Lexicon (L.fst)
The lexicon file contains a one-to-one mapping of words and phoneme sequences, each word has a specific phoneme sequence, similar to the phonetic symbols of words. Since the last pronunciation of the word does not pronounce when the plosive occurs, that is, the last pronunciation of the word is missing, and the last phoneme of the word is omitted corresponding to the lexicon. The method of this paper is to add all the plosive words in the lexicon to the lexicon, and then omit the last phoneme of these words with plosive labels and add them to the lexicon. such as [stop] **/S T AA1 P/**, [stop0] **/S T AA1/** which is shown in Fig. 4.

$$\left[\begin{array}{l} \text{STOP: S T AA1 P} \\ \text{STOP0: S T AA1} \end{array} \right.$$

Fig. 4. Example of plosive word

3.2.2　Improvement of Language Model (G.fst)

N-gram is a statistical language model, which assumes that the *n-th* word is only related to the first *n-1* word and not related to any other word [9]. The probability of the whole sentence is the product of the occurrence probability of each word. Using this language model, there will be a new path with plosive word in the process of decoding to word combination into sentences. For example, [The third chair is broken], if the word [third] contains a plosive, the decoded word is [third0], and the final decoded output is [The third0 chair is broken], as shown in Fig. 5.

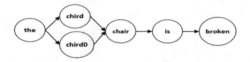

Fig. 5. Example of plosive decoding

3.2.3　Improvement of HMM (H.fst)

Hidden Markov Chain (HMC) model is used to solve practical problems. This model is called HMM (Hidden Markov Model). Hidden Markov model has two randomness: the underlying hidden Markov chain which changes with time (i.e. hidden state); and the observation sequence which has a mapping relationship with hidden state [10]. Such as word [stop], when a phoneme sequence **/S T AA1/** or a phoneme sequence **/S T AA1 P/**, is decoded from HMM, it is no longer just judging it as word [stop], [stop0] can also be decoded, shown in Fig. 6.

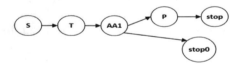

Fig. 6. Example of HMM decoding process

4　Experiment Results and Analysis

4.1　Liaison

In the evaluation of model performance, ROC and AUC are used as comprehensive evaluation criteria [11]. To ensure the best sequence length of the liaison features, 4 type of length of experiments are designed in this paper, including length of 7, 9, 11, 13. And the RNNs in this paper is GRU (Gated Recurrent Unit). The experiment result of liaison type of consonant **/d/** with **/æ/** is shown in Fig. 7.

The Fig. 7 shows the with the sequence length of 11, the experiment result reach the best result from three aspects of accuracy on training set (model_acc), accuracy (test_acc) and AUC (test_auc) on testing set.

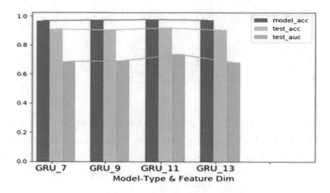

Fig. 7. /d æ/ liaison result

The Table 1 shows 8 kinds of liaison in including **/b ɪ/, /d ɑ/, /d æ/, /d d/, /ð ð/, /d ə/, /d ɪ/, /f ə/**. From the table, we can see that the accuracy of a few models in training set is about 0.9, and most of them are about 0.98. The model can give the correct prediction results. The AUC on the testing set is between 0.6 and 0.8. The average accuracy of the model is over 92% and the average AUC is over 69%.

Table 1. 8 types of liaison results

Liaison type	ACC on testing set (%)	AUC on testing set (%)
/b ɪ/	96.08	64.15
/d ɑ/	86.49	64.22
/d æ/	91.81	73.55
/d d/	96.07	64.17
/ð ð/	97.29	67.19
/d ə/	83.98	78.52
/d ɪ/	91.19	74.79
/f ə/	96.50	68.80

4.2 Plosive

The testing corpus contains 900 speeches, including 450 speeches with 1 plosive pronunciation skill recoding as P_{cnt} and 450 without recoding as N_{cnt}. Speeches with plosive tag (0) after decoding by the improved model in chapter 3.2 is counted as DP_{cnt} and the other is counted as DN_{cnt}. Recall Rate (**R**) and Precision Rate (**P**) is used as evaluation indicator in this chapter. The confusion matrix is shown in Table 2.

From the Table 2, we can easily calculate the R and P. As shown in Formula 4.1.

$$R = 376/450 = 83.56\%$$
$$P = 376/(376 + 145) = 72.1\%$$

(4.1)

Table 2. Confusion matrix for plosive experiment

	P_{cnt}	N_{cnt}
DP_{cnt}	376	145
DN_{cnt}	74	305

The results show that more than 80% of speech with plosive pronunciation skill can be correctly recognized by this model, and the recognition precision is 72.17%. The low recognition precision is due to the insufficient number of audio corpus. For the experiment is based on a small corpus, the low accuracy of speech recognition has a great impact on the identification of plosive.

5 Conclusion

In liaison pronunciation skill recognition method, GRU has a good experimental effect on capturing sequential features. In feature engineering, the features used in this paper extend the information associated with the frames forward and backward the MFCC. Such features are more consistent with the temporality of speech. For the 73 types of liaison mentioned in this paper, the following clustering experiments will be carried out from the perspective of phoneme pronunciation, and the similar liaison will be classified into one category to reduce the number of categories. In plosive pronunciation recognition method, the model with plosive elements was reconstructed, and the phenomenon of missing plosive in the previous experiments was augmented. After the experiment, the method was verified. The next step is to improve the accuracy. From the adaptive direction of the acoustic model, a small audio corpus is added to the large audio corpus to build a large decoding network, which improves the accuracy of speech recognition and improves the recognition rate of the word with plosive pronunciation skill.

Acknowledgments. Supported by the Foundation of Zhongshan Institute of Advanced Engineering Technology of WUT (Grant No. WUT-2017-01).

References

1. Chen, Q.C., Su, P.F., Huang, H.J., et al.: Evaluation for liaison of spoken English: a Sugeno integration approach. In: International Conference on Machine Learning & Cybernetics (2009)
2. Kkese, E., Petinou, K.: Perception abilities of L1 Cypriot Greek listeners - types of errors involving plosive consonants in L2 English. J. Psycholinguist. Res. **46**(1), 1–25 (2016)
3. Huang, S., Li, H., Wang, S., et al.: A method of automatic grading scoring in auxiliary speech scoring system. J. Tsinghua Univ. **1**(49), 20–31 (2009). (in Chinese)
4. Hossan, M.A., Memon, S., Gregory, M.A.: A novel approach for MFCC feature extraction. In: International Conference on Signal Processing & Communication Systems (2010)

5. Hanilçi, C., Kinnunen, T.: Source cell-phone recognition from recorded speech using non-speech segments. Digit. Signal Proc. **35**(C), 75–85 (2014)
6. Miao, Y., Gowayyed, M., Metze, F.: EESEN: end-to-end speech recognition using deep RNN models and WFST-based decoding. arXiv preprint arXiv:1507.08240v3 (2015)
7. Buczkowski, P.: Predicting stock trends based on expert recommendations using GRU/LSTM neural networks. In: Kryszkiewicz, M., Appice, A., Ślęzak, D., Rybinski, H., Skowron, A., Raś, Z. (eds.) Foundations of Intelligent Systems, pp. 708–717. Springer, Cham (2017)
8. Mohri, M., Pereira, F., Riley, M.: Speech recognition with weighted finite-state transducers. In: Benesty, J., Sondhi, M.M., Huang, Y.A. (eds.) Springer Handbook on Speech Processing and Speech Communication, pp. 559–584. Springer, Berlin Heidelberg (2008)
9. Bahrani, M., Sameti, H., Hafezi, N., et al.: A new word clustering method for building n-gram language models in continuous speech recognition systems. In: Nguyen, N.T., Borzemski, L., Grzech, A., Ali, M. (eds.) New Frontiers in Applied Artificial Intelligence, pp. 286–293. Springer, Heidelberg (2008)
10. Rabiner, L.R., Juang, B.H.: An introduction to hidden Markov models. IEEE ASSP Mag. **3** (1), 4–16 (1986)
11. Fawcett, T.: An introduction to ROC analysis. Pattern Recogn. Lett. **25**(8), 861–874 (2006)

An Agent-Based Simulation for Coupling Carbon Trading Behaviors with Distributed Logistics System

Yan Sun[1], Cevin Zhang[2(✉)], and Xia Liang[1]

[1] School of Management Science and Engineering,
Shandong University of Finance and Economics, Jinan 250014, China
[2] School of Engineering Sciences in Chemistry, Biotechnology and Health,
Kungliga Tekniska Högskolan, 14156 Huddinge, Sweden
chenzh@kth.se

Abstract. Access to a timely physical distribution is an important determinant of urban freight transportation. In the context of Internet of Things, emergent sustainability performances of logistics operations are affected by the components of the supply chain, including containers, information protocols, open hubs and marketplaces. The future freight transportation, inspired by the digital internet, can evolve into a physical-internet-enabled scenario, subject to many promises of addressing global logistics sustainability grand challenge. Meanwhile, carbon trading system should be continuously adopted to elucidate production and emission. Among this interaction between infrastructural and behavioral aspects, logistics system design and management must be handled in marketplace-oriented logistics networks, provided that transportation re-sources, if not properly managed, might in practice jeopardizes both the mobility of goods, profits of enterprises as well as the eventual efficiency of the public sector. The deliveries are in a sensitive nature to decreased quality of service when they experience excessive waiting times for being transported to downstream plants. Integrating the carbon trading system into the physical internet simulation of a multi-agent orientation may help technologically forecast logistics benefits. However, we admitted few simulation modelling efforts have been established for addressing sustainability issues in the era of supply chain digitalization. System design and management of physical-internet-enabled transportation understood from emergent behaviors (from decision making autonomy) and metrics (as outputs of operation), especially enabled by the coordination, collaboration and cooperation of many market actors, are rarely present. This study aims at the development and implementation of a virtual carbon trading agenda in the physical-internet-enabled transportation scenario, using multiagent modelling and simulation. Agent-based simulation is considered suitable in this context, for creating a risk-free environment to explore the system contingencies and autonomous decision-makings. Simulation experiments confirm that there is a need to identify a tolerable number of actors presenting in a marketplace as well as the variation of logistical flows in case the paradigm is the Clean Development Mechanism (CDM).

Keywords: Agent-based simulation · Clean Development Mechanism · Carbon trading · Logistics

F. Xhafa et al. (Eds.): IISA 2019, AISC 1084, pp. 222–229, 2020.
https://doi.org/10.1007/978-3-030-34387-3_27

1 Introduction

There is success of introducing carbon trading mechanism. Chicago Climate Exchange, as the second largest carbon trading market place, is based on e-commerce platform and inclusive of handling at least six types of greenhouse gas emissions [1]. This work adds presently implemented a local carbon trading architecture into the agent-simulation modelling that has been built for enumerate physical-internet logistics operations [2]. The new simulation model performs the validation process based on observation and aggregated data. We first of all communicate the carbon trading behaviors in the language of agent-based modeling. The composition of the conceptual model is of four entities, including the enterprise, the public sector agency, the information node and shippers. After this, the suitability of using agents and process simulation is discussed. Agent-based model of enterprises that only deliver freight services is created. Since freight transportation management is not having the same business data intelligence as that of traffic engineering, we synthesized a few data sources available to populate the simulation model. The data sources include operational fact data regarding logistical loads period-wise, geographical coverage, and locations. Results expect to benefit planners, decision making and public services in digitalization.

2 The Simulation Model

We will first of all introduce the architecture of the simulation model. This introduction is accompanied by reflecting how the different actors are represented in the delivery model. Simulation models have been widely used in logistics researches [3–5]. The carbon trading framework is based on a previous work. Readers can refer to Reference [6].

The simulated transportation system forms n logistics enterprises participating carbon trade behaviors, an information node and a public sector agency. Only firms that are granted the license of carbon production can emit greenhouse gas. The information node is responsible for matching requests and replies of carbon trades. The public sector agency monitors the carbon production levels for each enterprise each simulated period. Shippers adapt their procurement strategies in line with market prices. The entities' definitions and assumptions for trading behavior are drafted as below:

(1) This paper preassumes that enterprise k, at simulate time period t, the economic size is Q_{kt} with the very initial value of Q_0. Q_{kt} is updating in line with technological innovations. Meanwhile, $GDP_t = \sum_{k=1}^{n} Q_{kt}$, where GDP_t is the system's GDP at the period t.

(2) The logistics enterprise can invest in CDM project, contributing GDP growth. Suppose that the GDP contribution is linear to the CDM project value. This is expressed as $GDP_t^{CDM} = mC_t^{CDM}$ and $C_t^{CDM} = np_t$, where GDP_t^{CDM} represents the GDP growth thanks to CDM project, and C_t^{CDM} is the CDM project value at period t. m and n are constants. p_t is the carbon price at period t.

(3) As long as the carbon production for enterprise k at period t is E_t^k, the initially granted carbon assignment is A_t^k, in order to implement the task, the needed

carbon assignment reduction via technological developments or procurement from the secondary market should be $(E_t^k - A_t^k)$.

(4) The cost of reducing carbon production for enterprise k is subject to the cost function $c(a_t^k)$, where a_t^k is reduction amount.

Suppose that the carbon trading price for period t is p_t, therefore, the optimization model for enterprise k at period t is as Eq. (1).

$$\min \lfloor c(a_t^k) + p_t(E_t^k - A_t^k - a_t^k) \rfloor \ s.t. 0 < a_t^k < E_t^k \tag{1}$$

(5) Suppose that the enterprise's cost of carbon reduction in the marginal state is linearly related to the reduction amount, in other words, $f(a_t^k) = a_t^k * \partial_t, \partial > 0$, and $f(a) = c'(a_t^k)$, where ∂_t is the parameter of reduction and indicates the technical strength of the enterprise for green logistics.

2.1   The Enterprise Agent

The enterprise is set to profit-oriented. At the beginning of each round, the public sector agency releases freely or through biding to allocate carbon assignments to enterprises. According to situations of the market place, the enterprises can forecast their throughput of logistics for the next round, and a relatively accurate prediction of carbon production is possible based on expected transportation orders and carbon reductions. The enterprise makes final decision making on whether the assignment would be a surplus or deficit, and accordingly decides to procure more or trade out its assignment on hold. The expected profit for enterprise i is as Eq. (2).

$$ep_k = \frac{\left[ica_t^k * p_t + ft_t * tp_t - ft_t * r_t * cr_t - ft_t * cp_t * p_t - oc_t + sub_t - iia_k \right]}{iia_k} \tag{2}$$

where iia_k is the initial investment amount for enterprise k, ica_t^k is the initial carbon assignment obtained by enterprise i, p_t is the carbon trading price, tp_t is the throughputs unit price, ft_t is the throughput, r_t is the required resources for the throughput of one unit, cr_t is the price for resources, cp_t is the carbon production amount for the throughput of one unit, oc_t is the fixed operational cost per period, and sub_t is the public sector subsidy for environmental-friendly enterprises.

2.2   The Public Sector Agency Agent

The main objective of the public sector agency is to facilitate sustainable logistics and reduce carbon production. The responsibility of the agency is twofold: optimize the carbon assignment, and encourages the development of environmental-friendly enterprises through technological support, tax deduction and so forth. The assignment is based on the last round's distribution of carbon production among the enterprises, which is indicated by Eq. (3).

$$ica_t^k = e_{t-1}^k * \frac{total\_cap_t}{total\_cap_{t-1}} * \emptyset \tag{3}$$

where e_{t-1}^k is the carbon production of enterprise k at the period $t - 1$. $total\_cap_t$ is the entire carbon production of the system at period t. $\emptyset$ is the deterioration rate.

2.3 The Shipper Agent

We suppose that the shipper's transportation requirement is in line with the GDP and the dynamics of the product price. The shippers' decision making model of demand is as Eq. (4).

$$D_t = D_{t-1}\left[1 + \left(\frac{GDP_t}{GDP_{t-1}} - 1\right) * a\right] * \left[1 + \left(\frac{P_t}{P_{t-1}} - 1\right) * \partial\right] \tag{4}$$

where GDP_t is the regional GDP value at period t, P_t is the product price at period t, a is the elasticity of demand towards GDP and, ∂ is the pricing elasticity.

2.4 The Information Node Agent

Information nodes are trustworthy organizations or individuals. Their job is informing carbon trading price and constantly importing and exporting carbon assignments to maintain the balance of the marketplace. The accumulated, remained carbon assignment is as Eq. (5).

$$pool(t - 1) = pool(t - 1) + car\_d_{(t)} - car_{s_{(t)}} \tag{5}$$

where $car\_d_{(t)}$ is the total demand of carbon assignment, and $car_{s_{(t)}}$ is the total supply of carbon assignment.

3 Simulation Environment and Experiments

To guarantee the simulator is a reliable and precise representation of logistical flows, an elementary validation is performed. The validation is via checking simulated results with empirical data available. By elementary validation means only certain aspects, e.g., logistical flow characteristics, are validated. This empirical data is a synthesis of different data sources, including infrastructure plans, and the yearly order volumes for all enterprises. The model is then viewed as input-output transformations. Many methods are available for statistical testing. In this work, however, confidence interval is used for the evaluation of the accuracy of simulation [7]. The model presented in this work has a substantial minority of processes. Confidence intervals and the relative differences are calculated. We select different normal hours of the day as the reference. It turns out all ε is less than 5%, as Table 1 presents. This means that the model captures the patient flow characteristics with a less-than-5 disparity to observation. We expect that the patient volume is suitable for validation purposes, because care streams are usually the determinant of coordination in logistics and resource attachment.

Table 1. Validation of logistical flow characteristics

	Observation	Simulation	Confidence interval	$|a - \mu 0|$	$|\varepsilon|$
Rate of arrival (per hour)	1.46	1.46	[−0.95, +0.95]	0.02	2%
	2.22	2.20	[−1.35, +1.35]	0.10	4%
	4.90	4.91	[−2.36, +2.36]	−0.03	1%
	8.67	8.70	[−2.13, +2.13]	−0.09	1%
	9.59	9.72	[−2.79, +2.79]	−0.32	3%

4 Results

The sensitivity analysis is carried out for understanding if the freight transportation service performance would be substantially different to the optimal productive efficiency. Next, we suggest the variations of parameters. Different sets of parameters are simulated to examine if the performance would vary considerably. Increasing the number of present actors is an option in order to achieve a suitable size of market, especially those delayed by a shortage of vehicles. This is, however, associated with higher investments and maintenance efforts that might offset the gains. Planning on the provision of service that attempts to reduce waiting times of delivery in the region might in the end lead to higher operational cost of the system. The enterprise size in favor of cost of the system is a trade-off to be examined. By running scenario simulations, this trade-off is mapped out in accordance of different options, which is illustrated by Fig. 1. The total carbon reduction cost will increase much faster if many market agents are added.

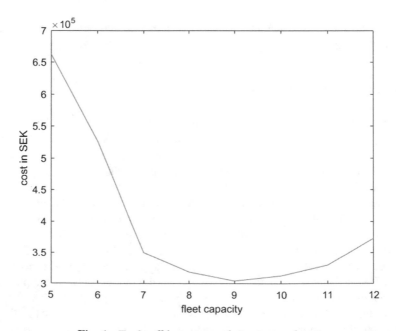

Fig. 1. Trade-off between market actors and cost

Since each enterprise is connected by several facilities in the short range, logistical flows can be intensive on the daily basis. Meanwhile, maximizing throughput is expected. The rate modifier manipulates the overall logistical load towards the system. As the volume increases, its relationship with carbon reduction cost is presented in Fig. 2. A flow even lower than 80% of the current value is not recommended because the resources and physical infrastructure will not be utilized sufficiently and therefore point to economic punishments.

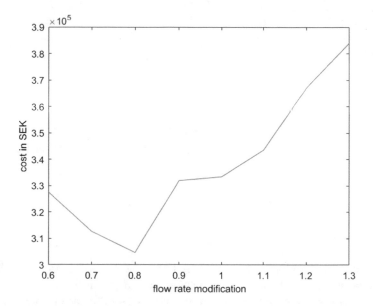

Fig. 2. Trade-off between logistical flow and cost

When the flow comes larger than the current value, adding again market players is essential. This is the conclusion we could acquire by combining the minimum solutions. Both mappings can be simply combined to identify the productivity under different choices of parameters. Carbon reduction cost was mapped regarding reserved margin and rate of arrival. Following this, by exploration of combinations, which aims for the minimal carbon reduction cost for enterprises and consumers, a contour presents the situations besides the optimal. This mapping means that the optimal network size of 9 enterprises would offset the stress of the logistical flow increase, which can be seen in Fig. 3. However, a size smaller than 6, with a rate modifier in the interval of 0.95 and 1.1 will most probably result in a very carbon reduction cost, as indicated by peaks in this contour.

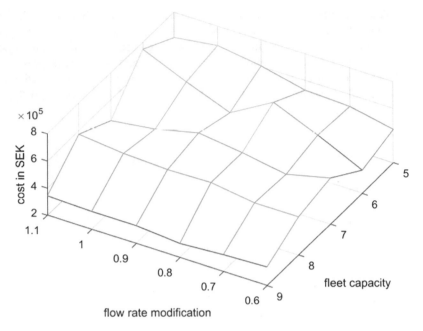

Fig. 3. Contour of objective efficiencies

5 Discussion

In this work, we developed an agent-based simulation model, to assess logistical strategies on a regional transportation network in the physical internet scenario [6], enabling the improved designs regarding objectives on different operation infrastructures. A formulation has been developed, which allows the implementation of such a model for minimization of cost step-wise. Even though it has been represented at a relatively high level, the elementary validation agrees with those in the real operation of the systems. Front of solutions have been computed and presented, which enables the study of trade-off between different alternatives. As the work progress, more operational fact data shall be collected, to improve the details of the model. The development process includes a validation step, however, certain aspects are still not covered, for instance the rational of crew resources, and the resource utilizations of different units. Since those data are usually not directly available, more follow-up work shall be established, to thoroughly validate the simulation as the test bed of various operation strategies. We have explored a couple of operation strategies here, but the interests of different actors of the system need to be further discussed. This will help formulate the objective function.

Acknowledgements. This research was funded by the Project for Humanities and Social Sciences Research of Ministry of Education of China under Grant No. 19YJC630149, the Shandong Provincial Natural Science Foundation of China under Grant No. ZR2019BG006, the National Natural Science Foundation of China under Grant No. 71801142, and the Shandong Provincial Higher Educational Social Science Program of China under Grant No. J18RA053.

References

1. Gans, W., Hintermann, B.: Market effects of voluntary climate action by firms: evidence from the Chicago Climate Exchange. Environ. Resource Econ. **55**(2), 291–308 (2013)
2. Sun, Y., Zhang, C., Dong, K., Lang, M.: Multiagent modelling and simulation of a physical internet enabled rail-road intermodal transport system. Urban Rail Transit **4**(3), 141–154 (2018)
3. Abo-Hamad, W., Arisha, A.: Simulation-based framework to improve patient experience in an emergency department. Eur. J. Oper. Res. **224**(1), 154–166 (2013)
4. Dangerfield, B.C.: System dynamics applications to European health care issues. J. Oper. Res. Soc. **50**(4), 345–353 (1999)
5. Al-Araidah, O., Boran, A., Wahsheh, A.: Reducing delay in healthcare delivery at outpatients clinics using discrete event simulation. Int. J. Simul. Model. **11**(4), 185–195 (2012)
6. Zhu, Q., Wu, J., Wang, Z.: Research on carbon trading market: an agent-based simulation. Geogr. Res. **31**(9), 1547–1558 (2012)
7. Banks, J., Carson II, J.S., Nelson, B.L., Nicol, D.M.: Discrete-Event System Simulation (International Edition). Pearson Prentice-Hall, Upper Saddle River (2005)

The β Coefficient and Chow Stability Test of Chinese Foreign Trade Stocks Under the Sino-US Trade War

Wenting Cao[1,2(✉)]

[1] School of Economics and Management, Yunnan Normal University,
Kunming 650500, China
caowt@163.com
[2] School of Economics, Sichuan University, Chengdu 610065, China

Abstract. This paper selects the daily trading data of foreign trade stocks in China's Shanghai stock market from January 4, 2012 to June 29, 2018. The β Coefficient of each stock is obtained by regression of the single index model of CAPM theory. Taking the outbreak time of the Sino-US Trade War as a structural mutation point, the Chow test method is used to investigate the stability of the β Coefficients of foreign trade stocks. The conclusion is as follows: the systematic risk of China's foreign trade listed companies is generally higher than or equal to the market risk level. Most of Chinese foreign trade stocks have stable β Coefficients (accounting for about 73%), and the rest of the stocks' β Coefficients have undergone structural changes (about 27%) before and after the break-point of the Sino-US Trade War. This means that the Sino-US Trade War has an impact on the expected return model of some Chinese trade-listed companies. In the long run, this effect may expand or spread, and needs to be concerned.

Keywords: CAPM · β Coefficient · Chow stability · Sino-US Trade War

1 Introduction

Sharpe [1], Lintner [2], and Mossin [3] have established asset pricing model (CAPM) based on risk-return trade off principle and portfolio diversification principle, which can determine the price of portfolio or single security. The β coefficient is the most important parameter in the CAPM model and an important indicator for measuring the systematic risk of security. Systematic risk is that affects all assets and cannot be eliminated through asset portfolios. These risks are caused by risk factors that affect the entire market. The universally accepted notion for the measure of a stock's or a portfolio's relative sensitivity to the market based upon its past record is the Greek letter beta (β) (Lee, Finnerty and Wort [4]). In the real economic environment, the β coefficient is estimated from past data, so it can only reflect the past systematic risk situation. Only the β coefficient is stable, it can reflect the current or future risk. Therefore, when analyzing the systematic risk of an industry or a stock, its β coefficient stability test is necessary.

F. Xhafa et al. (Eds.): IISA 2019, AISC 1084, pp. 230–237, 2020.
https://doi.org/10.1007/978-3-030-34387-3_28

The analysis of the β coefficient and its stability have accumulated a lot of results. Huy [5] points out that the β coefficient is the sensitivity of financial assets to market returns. Traditional financial theory considers that if the β coefficient of individual stock is unstable, the diversification effect makes the portfolio of individual stocks have a more stable β coefficient. While Brooks [6] argues that the β coefficient instability of a randomly selected stock combination is higher than that of a single stock. Yuan and Lee [7] suggest that the β coefficient is generally unstable. Alagidede [8] find that the β coefficient is stable after the financial crisis, but not stable before the crisis. This paper focuses on the systematic risk of Chinese trade companies and the stability of the β coefficient.

2 Model and Method

2.1 CAPM Model

In 1952, Markowitz [9] published a paper "Portfolio Selection", which elaborated on quantitative methods for measuring returns and risk levels, and established an analytical framework for the "mean-variance" model to solve the problem of how to optimally allocate funds in the investment decision-making process. Based on Markowitz's portfolio theory, the Capital Asset Pricing Model (CAPM) was respectively derived by Sharpe, Lintner and Mossin [1–3]. The core issue was how the equilibrium price of an asset was formed in the trade-off between income and risk when investors used Markowitz's portfolio theory to allocate assets in the capital market. In order to explore the Capital Asset Pricing Model (CAPM), we must first define the Capital Market Line (CML) and the Security Market Line (SML). CML describes a linear relationship between the expected return and standard deviation of an effective portfolio. It can be expressed as: $E(r_p) - r_f = \frac{E(r_M) - r_f}{\sigma_M} \times \sigma_p$. CML can determine the expected rate of return for a portfolio, but it cannot characterize the expected rate of return for a single security, so it leads to the SML.

SML reveals the correspondence between the risks and rewards of the securities, and the Security Market Line is the CAPM model. The expectation of the expected return of any security i can be expressed as: $E(r_i) = r_f + \beta_i \times (E(r_M) - r_f)$, $\beta_i = Cov(r_i, r_M)/Var(r_M) = \delta_{iM} / \delta_M^2$. CAPM model describes that the expected return of security comes from risk-free return (r_f) and risk compensation ($\beta_i \times (E(r_M) - r_f)$). The higher the risk, the higher the interest rate compensation and the higher the expected return of securities.

2.2 β Coefficient

According to the CAPM model, the β coefficient is obtained from $\beta_i = Cov(r_i, r_M)/ Var(r_M) = \delta_{iM} / \delta_M^2$. The β coefficient is expressed as the risk metric of a certain security relative to the market portfolio, that is, to measure the change degree of a single security return with the market return, which is called Systematic risk. In the CAPM model, β is a very important variable, which is a measure of a security risk. It can also estimate the expected return of a security based on β. When $\beta > 0$, it indicates that the yield of the security changes in the same direction as the market return. Among them, when $\beta = 1$, the risk situation of the security is the same as the risk situation of the market, and the price fluctuation is equivalent to the average market price fluctuation; when $\beta > 1$, the risk of the security is higher than the market risk, and the risk compensation is higher than the risk compensation of the market portfolio, and the price fluctuation is also greater than the average market price fluctuation. This means that if the market yield rises, the yield of the security rises above the average level of rising market returns, and vice versa. When $0 < \beta < 1$, the risk of the security is lower than the market risk, and its price fluctuation is also less than the average market price fluctuation. Explain that if the market yield increases, the rate of increase in the yield of the security is lower than the average level of rising market returns, and vice versa. When $\beta < 0$, it shows that the security return and the market return change in the opposite direction.

2.3   Chow Stability Test

In reality, the introduction of major policies or accidental events can lead to changes in economic operating mechanisms or economic behavior. From the econometric perspective, the parameters of the econometric model have changed. If the above situation occurs in the time window of the research object, the stability of the model parameters needs to be tested. The famous economist Chow [10] uses the method of F to test the stability of the model. According to the understanding of the turning time, all the samples are divided into two parts to construct the F statistic. H_0: There is no significant change in the regression coefficients of the two sub-samples.

$$
\begin{aligned}
F &= \frac{[RSS_N - (RSS_{n1} + RSS_{n2})]/[(N - k - 1) - (n_1 - k - 1 + n_2 - k - 1)]}{(RSS_{n1} + RSS_{n2})/(n_1 - k - 1 + n_2 - k - 1)} \\
&= \frac{[RSS_N - (RSS_{n1} + RSS_{n2})]/(k - 1)}{(RSS_{n1} + RSS_{n2})/(N - 2k - 2)} \sim F(k - 1, N - 2k - 2)
\end{aligned}
$$

RSS_N is the regression of all the data to obtain the sum of squared residuals. RSS_{n1} and RSS_{n2} are respectively obtained by regression using two parts of data to obtain the sum of squared residuals. N is the total sample size. n_1, n_2 are the number of observations of each of the two sub-samples, and k is the number of explanatory variables. α is the test level, if $F \leq F_\alpha(k - 1, N - 2k - 2)$, or the probability of the F is greater than the critical value of the test level, accepting the null hypothesis H_0, indicating that the model structure is stable.

3   Empirical Test and Results

3.1   Data and Variable

According to the industry classification standard of Shenyin Wanguo, this paper selects the stocks in SW Trade III under SW Commercial Trade Classification as the analysis sample. There are 24 stocks under SW Trade III, including 19 Shanghai stocks and 5

Shenzhen stocks. Since the calculation methods of the stock indices of the Shanghai and Shenzhen Stock Exchanges are different and it is difficult to convert between the indices, five Shenzhen stocks are abandoned here, and only the Shanghai stock data is selected for analysis. There is a B share in SW Trade III (900927, Material Trade B share), which has less impact on the sample as a whole, so we exclude it from the analysis sample, and we exclude stocks that continue to ST or *ST status. The stability test of β coefficient requires a long observation time. We select the daily transaction data from January 4, 2012 to June 29, 2018 as a sample, and exclude the transaction data during the stock suspension period. The sample observations of a single stock reached more than 1,296, and the total number of observations of 15 stocks reached 23,521. The data in this article comes from the Wind database and the RESSET database. The sample data is shown in Table 1.

The rate of return is measured by the daily yield of individual stocks, and the formula is $R_i = (P_{t+1} - P_t)/P_t$, P_t is the closing price of a share on the t day. The market rate of return is equal weight average market rate of return, which is calculated as:

$$R_m = \sum_{i=1}^{n} w_i(t) \, r_i(t) / \sum_{i=1}^{n} w_i(t), \quad w_i(t) = 1,$$ Market income is the weighted average return

of all A shares in Shanghai. The risk-free daily rate of return (R_f) is the three-month inter-bank offering rate in Shanghai, and it is converted into the daily rate of return.

3.2 Construction of β Coefficient Model

The expression of the CAPM model is $E(r_i) = r_f + \beta_i \times (E(r_M) - r_f)$, the key factor here is to estimate β. The CAPM model shows that the total risk of a single asset can be divided into two parts: one is the change in the return of the asset i due to the change in the market portfolio M, which is the β_i value (systematic risk); the other is that residual risk called non-systematic risk. The price of a single asset is only related to the size of systematic risk, regardless of its non-systematic risk. In order to facilitate the regression analysis, a single index model of the CAPM model is used here. Sharp assumes that the main factor affecting asset price volatility is the overall price level of the market. He uses the market index to represent systematic factors (macro factors). The expression is $R_{if} = \alpha_i + \beta_i \times R_{Mf} + \varepsilon_i$, It shows that the return on each asset is affected by two risks: one is Systematic risk, which is expressed in R_{Mf}; the other is enterprise-specific risk (non-systematic risk), which is expressed in ε_i. $R_{Mf} = R_M - R_f$ represents the excess return of the market; $R_{if} = R_i - R_f$ represents the excess return of individual stock. By regressing the daily excess returns of individual stocks to the daily excess returns of the market, the estimator of β coefficient is obtained, that is, the β coefficient of individual stocks to be tested for stability.

Table 1. Status of 15 foreign trade stocks in Shanghai Stock Exchange

Stock code	Stock name	Effective sample	Stock code	Stock name	Effective sample
600058	WKFZ	1589	600710	SMD	1296
600120	ZJDF	1595	600735	XHJ	1557
600241	SDWH	1560	600739	LNCD	1590

(*continued*)

Table 1. (*continued*)

Stock code	Stock name	Effective sample	Stock code	Stock name	Effective sample
600250	NFGF	1593	600755	XMGM	1587
600278	DFCY	1605	600822	SHWM	1592
600287	JSST	1588	600826	LSGF	1603
600605	HTNY	1590	600981	HHJT	1585
600626	SDGF	1591			

3.3   Estimation of β Coefficient

According to the linear regression of 15 foreign trade stocks, the β coefficients and the accompanying probabilities of 15 stocks of China's foreign trade are shown in Table 2. The results in Table 2 show that the accompanying probabilities of β coefficients of all stocks in the study sample are all 0, indicating that the estimated β coefficients of 15 foreign trade stocks are significant. The β coefficients of 15 foreign trade stocks in Table 2 are all greater than 0, indicating that the changes in the return of these 15 foreign trade stocks are in the same direction as the changes in market return. Among them, 8 stocks have higher returns than the market return, and the risk of these 8 stocks is also higher than the market risk level. In addition, the yields of four stocks are lower than the market yield, and their respective risks are also lower than the risk level of the market portfolio. Finally, the β coefficients of NFGF(600250), SDWH(600241) and WKFZ(600058) are approximately equal to 1, indicating that the yields of these three stocks fluctuate synchronously with the market yield, and their respective risks are roughly the same as market risks.

Table 2. 15 foreign trade stocks' β coefficients in Shanghai Stock Exchange

Stock name	Stock code	β	P	Stock name	Stock code	β	P
JSST	600287	1.1998	0.0000	NFGF	600250	1.0097	0.0000
DFCY	600278	1.1646	0.0000	SDWH	600241	1.0035	0.0000
LSGF	600826	1.1328	0.0000	WKFZ	600058	0.9161	0.0000
HTNY	600605	1.0833	0.0000	ZJDF	600120	0.8986	0.0000
SHWM	600822	1.0820	0.0000	XHJ	600735	0.7191	0.0000
SDGF	600626	1.0791	0.0000	LNCD	600739	0.6537	0.0000
XMGM	600755	1.0732	0.0000	SMD	600710	0.5357	0.0000
HHJT	600981	1.0162	0.0000				

3.4   Chow Stability Test of β Coefficient

The analysis of the Chow test is to consider whether the sample contains different quality groups, that is, to check whether the parameters between different groups are the same. Since the outbreak of the Financial Crisis and the European Debt Crisis, the world economy is still in the stage of deep adjustment. The sustained downward pressure of the economy has resulted in weak international market demand and slow global economic recovery. At the same time, trade protectionism continues to heat up, and the trend of "anti-globalization" is obvious. Many countries have set up multiple trade barriers to protect domestic real industries, and Chinese companies have experienced an increase in the frequency of trade frictions in the export process. It is worth noting that the Trump administration of the United States has provoked Sino-US trade disputes, and its influence has spread to the global economic market. The United States has adopted a series of trade protectionist measures against China, such as high tariffs, export controls and investment restrictions. The negative impact of the Sino-US trade war on China cannot be underestimated.

Since the time span of the research sample is from January 4, 2012 to June 29, 2018, major changes in the domestic and international economic situation during this period will have an impact on the stability of the foreign trade enterprises' β coefficients. In the context of the Sino-US trade war, the Sino-US trade war began on April 20, 2017. At the request of Trump, the US Department of Commerce initiated a 232 investigation of imported steel products in accordance with Section 232 of the Trade Expansion Act of 1962. At this point, the Sino-US trade war was partially opened. Therefore, we use April 20, 2017 as the mutation time point of the Chow test, and then divide the sample data into two sets. The first set is the sample data from January 4, 2012 to April 19, 2017. The second collection is sample data from April 20, 2017 to June 29, 2018.

Table 3. 15 foreign trade stocks' β coefficients stability test results of China SSE

Stock name	Stock code	Chow stability P	Is it stable?	Stock name	Stock code	Chow stability P	Is it stable?
JSST	600287	0.8783	YES	SDWH	600241	0.4703	YES
DFCY	600278	0.3317	YES	ZJDF	600120	0.3954	YES
LSGF	600826	0.6758	YES	LNCD	600739	0.1057	YES
HTNY	600605	0.5894	YES	SHWM	600822	0.0004	NO
SDGF	600626	0.0504	YES	WKFZ	600058	0.0094	NO
XMGM	600755	0.8834	YES	XHJ	600735	0.0444	NO
HHJT	600981	0.1515	YES	SMD	600710	0.0058	NO
NFGF	600250	0.0879	YES				

The outbreak of the Sino-US trade war may lead to structural changes in the β coefficients of foreign trade enterprises. The results of the Chow stability test of the β coefficients are shown in Table 3. At a significant level of 5%, in order not to reject the null hypothesis (H_0: there is no significant change in the regression coefficients of the two sub-samples). The Chow Test F value has a accompanying probability greater than 5%. Among them, the probabilities of β coefficients stability test of 11 foreign trade stocks are all greater than 5%, and the null hypothesis of Chow stability test is not rejected, indicating that the β coefficients of the 11 listed companies have not changed before and after the Sino-US trade war broke out. Their β coefficients are stable.

While the other four stocks' β coefficients stability tests have probabilities of less than the critical value of 5%, thus rejecting the null hypothesis of the Chow stability test. This means that the β coefficients of these four listed companies have changed before and after the Sino-US trade war broke out, and their β coefficients are unstable.

4 Conclusion

First, considering the β coefficients of foreign trade stocks, the above empirical results can be seen that risks and benefits are accompanied by each other. In general, foreign trade companies with systematic risk higher than market risk account for about 53%, and more than half of these companies have higher returns than the market portfolio. About 20% of foreign trade enterprises have the same systematic risk as market risk. Foreign trade companies with systematic risk below market risk account for about 27%. The shares of these companies are defensive stocks, and their yields are lower than the market average. It shows that the systematic risk of China's foreign trade listed companies is generally higher than or equal to the market risk level. Secondly, considering the stability test of the β coefficients of foreign trade stocks, in general, the β coefficients of most Chinese foreign trade stocks is stable (accounting for about 73%), indicating that the expected return model of China's foreign trade capital market is generally stable. The rest of the stocks have undergone structural changes in the β coefficients before and after the sudden break of the Sino-US trade war (accounting for about 27%), indicating that the Sino-US trade war has an impact on the expected return model of some Chinese trade listed companies. In the long run, this effect may expand or spread, and needs to be taken seriously.

References

1. Sharpe, W.F.: Capital asset prices: a theory of market equilibrium under conditions of risk. J. Finan. **19**(3), 425–442 (1964)
2. Lintner, J.: The valuation of risk assets and the selection of risky investments in stock portfolios and capital budgets. Rev. Econ. Stat. **47**(1), 13–37 (1965)
3. Mossin, J.: Equilibrium in a capital asset market. Econom. J. Econom. Soc. **34**(4), 768–783 (1966)
4. Lee, C.F.J., Finnerty, E., Wort, D.H.: Capital asset pricing model and beta forecasting. In: Lee, C.F., Lee, A.C., Lee, J. (eds.) Handbook of Quantitative Finance and Risk Management, pp. 93–109. Springer, Boston (2010)
5. Huy, D.T.N.: Estimating beta of Viet Nam listed public utilities, natural gas and oil company groups during and after the financial crisis 2007–2011. Econ. Bus. Rev. Cent. South-East. Europe **15**(1), 57–71 (2013)
6. Brooks, R.D.: Alternative point-optimal tests for regression coefficient stability. J. Econom. **57**(1–3), 365–376 (1993)

7. Yuan, F.C., Lee, C.H.: Using least square support vector regression with genetic algorithm to forecast beta systematic risk. J. Comput. Sci. **8**(11), 26–33 (2015)
8. Alagidede, P., Koutounidis, N., Panagiotidis, T.: On the stability of the CAPM before and after the financial crisis: panel evidence from the Johannesburg Securities Exchange. Afr. Rev. Econ. Finan. **9**(1), 180–189 (2017)
9. Markowitz, H.: Portfolio selection. J. Finan. **7**(1), 77–91 (1952)
10. Chow, G.C.: Tests of equality between sets of coefficients in two linear regressions. Econom. J. Econom. Soc. **28**(3), 591–605 (1960)

Dynamical Behavior of an SVIR Epidemiological Model with Two Stage Characteristics of Vaccine Effectiveness and Numerical Simulation

Xiuchao Song[✉], Miaohua Liu, Hao Song, and Jinshen Ren

Department of Basic Sciences, Air Force Engineering University,
Xi'an 710051, China
xiuchaosong@163.com

Abstract. An SVIR epidemiological model with two stage characteristics of vaccine effectiveness is formulated. By constructing the appropriate Lyapunov functionals, it is proved that the disease free equilibrium of the system is globally stable when the basic reproduction number is less than or equal to one, and that the unique endemic equilibrium of the system is globally stable when the basic reproduction number is greater than one.

Keywords: Globally stability · Vaccine effectiveness

1 Introduction

In human history, infectious diseases have repeatedly brought great disaster to human survival. In recent years, the outbreak of some new infectious diseases (SARS, influenza A (H1N1), influenza A (H7N9), etc.) has caused a great impact on people's lives. Vaccines are biological agents made from bacteria, viruses, tumor cells and so on, which enable antibodies to produce specific immunity. Vaccination can provide immunity to those who are vaccinated, can eliminate the spread of some diseases (such as smallpox) [1].

In recent years, more and more authors study the epidemiological models with vaccination [2–5]. Some authors assume that vaccine recipients will not be infected [2, 3]; some other authors assume that vaccine recipients may still be infected [4, 5], but the probability of being infected is smaller than before vaccination. In fact, for some infectious diseases, the vaccinated individuals would not be infected for some time after vaccination. However, bacteria or viruses mutate as time goes by, and the efficacy of the vaccine is correspondingly affected, which makes it is possible for the vaccinated individuals to be infected. For example, the new H7N9 influenza virus mutates more quickly, and the effectiveness of the vaccine depends largely on the extent of the virus mutation [6]. Based on the above facts, we assume that vaccine effectiveness has two stage characteristics: in the first stage, the vaccinated individuals will not be infected; in the second stage, the vaccinated individuals will be infected, but the probability of infection will be smaller than before vaccination. Therefore, this

© Springer Nature Switzerland AG 2020
F. Xhafa et al. (Eds.): IISA 2019, AISC 1084, pp. 238–242, 2020.
https://doi.org/10.1007/978-3-030-34387-3_29

paper studies the epidemiological model with two stage characteristics of vaccine effectiveness, On the basis of getting the basic reproductive number, by using appropriate functionals, the stability of the model is proved by the algebraic approach provided by the reference [8].

In this work, we study the following epidemiological model:

$$\begin{cases} S' = \mu(1-q)A - \beta SI - (\mu+p)S + kV_2 \\ V_1' = \mu qA + pS - \mu V_1 - \varepsilon V_1 \\ V_2' = \varepsilon V_1 - \beta\sigma V_2 I - \mu V_2 - kV_2 \\ I' = \beta SI + \beta\sigma V_2 I - (\mu+\alpha+\beta)I \\ R' = \gamma I - \mu R \end{cases} \tag{1}$$

The model (1) has the same dynamic behavior with the following system

$$\begin{cases} I' = \beta SI + \beta\sigma V_2 I - c_4 I \\ V_1' = \mu qA + pS - c_2 V_1 \\ V_2' = \varepsilon V_1 - \beta\sigma V_2 I - c_3 V_2 \\ S' = \mu(1-q)A - \beta SI - c_1 S + kV_2 \end{cases} \tag{2}$$

2   Existence of Equilibria

Obviously, system (2) has a disease free equilibrium $P_0(S_0, V_{10}, V_{20}, 0)$, where

$$S_0 = \frac{[c_2 c_3 - \mu q(c_3+\varepsilon)]A}{(c_3+\varepsilon)p + c_2 c_3}, \quad V_{10} = \frac{(p+\mu q)c_3 A}{(c_3+\varepsilon)p + c_2 c_3}, \quad V_{20} = \frac{(p+\mu q)\varepsilon A}{(c_3+\varepsilon)p + c_2 c_3}$$

Using [9], we have

$$R_0 = \frac{\beta A[c_2 c_3 - \mu q(c_3+\varepsilon)] + \beta\sigma\varepsilon(p+\mu q)A}{c_4[(c_2+p)c_3 + p\varepsilon]}$$

It can be found the unique endemic equilibrium $P^*(S^*, V_1^*, V_2^*, I^*)$ from the following equations,

$$\begin{cases} \mu(1-q)A - \beta SI - c_1 S + kV_2 = 0 \\ \mu qA + pS - c_2 V_1 = 0 \\ \varepsilon V_1 - \beta\sigma V_2 I - c_3 V_2 = 0 \\ \beta SI + \beta\sigma V_2 I - c_4 I = 0 \end{cases}$$

where

$$S^* = \frac{\mu A c_2(1-q)(c_3+\beta\sigma I^*) + k\varepsilon\mu qA}{c_2(c_3+\beta\sigma I^*)(c_1+\beta I^*) - k\varepsilon p}, \quad V_2^* = \frac{[(c_1+\beta I^*)q + p(1-q)]\varepsilon\mu A}{c_2(c_3+\beta\sigma I^*)(c_1+\beta I^*) - k\varepsilon p},$$

$$V_1^* = \frac{[(c_1 + \beta I^*)q + p(1-q)]\mu A(\beta\sigma I^* + c_3)}{c_2(c_3 + \beta\sigma I^*)(c_1 + \beta I^*) - k\varepsilon p},$$

and I^* satisfies the following equation

$$J(I) = \frac{\beta\mu Ac_2(1-q)(c_3 + \beta\sigma I) + k\beta\varepsilon\mu qA}{-k\varepsilon p + c_2(c_3 + \beta\sigma I)(c_1 + \beta I)} - c_4 + \frac{[(c_1 + \beta I)q + p(1-q)]\beta\sigma\varepsilon\mu A}{-k\varepsilon p + (c_3 + \beta\sigma I)(c_1 + \beta I)c_2}$$
$$= 0$$

3 Stability of Equilibria

Theorem. When $R_0 \leq 1$ the $P_0(S_0, V_{10}, V_{20}, 0)$ is global stable. And P^* is global stable when $R_0 > 1$.

Proof. The global stability of P_0 is firstly proved.
Consider the following Lyapunov functional

$$L_1 = (S - S_0 - S_0\ln\frac{S}{S_0}) + (V_2 - V_{20} - V_{20}\ln\frac{V_2}{V_{20}}) + \frac{\varepsilon}{c_2}(V_1 - V_{10} - V_{10}\ln\frac{V_1}{V_{10}}) + I.$$

so

$$L_1' = \left(1 - \frac{S_0}{S}\right)[c_1(S_0 - S) - \beta SI + k(V_2 - V_{20})] + \left(1 - \frac{V_{20}}{V_2}\right)(\varepsilon v_1 - \beta\sigma V_2 I - c_3 V_2)$$
$$+ \frac{\varepsilon}{c_2}\left(1 - \frac{V_{10}}{V_1}\right)(\mu qA + pS - c_2 V_1) + \beta SI + \beta\sigma V_2 I - c_4 I$$
$$= H(V_2, S, V_1) + c_4(R_0 - 1)I$$

where

$$H(V_2, S, V_1) = c_1(S_0 - S)\left(1 - \frac{S_0}{S}\right) + k\left(1 - \frac{S_0}{S}\right)(V_2 - V_{20}) - c_3 V_2 - \varepsilon V_{20}\frac{V_1}{V_2}$$
$$+ c_3 V_{20} + \frac{p\varepsilon}{c_2}S - \frac{p\varepsilon V_{10}S}{c_2 V_1} + \frac{\varepsilon\mu qA}{c_2} - \frac{\varepsilon V_{10}\mu qA}{c_2 V_1} + \varepsilon V_{10}.$$

For simplicity, denote $x_1 = \frac{S}{S_0}$, $x_2 = \frac{V_1}{V_{10}}$, $x_3 = \frac{V_2}{V_{20}}$, then

$$H(x_1, x_2, x_3) = (2S_0 c_1 + \mu V_{20} + 2\varepsilon V_{10} - \frac{p\varepsilon S_0}{c_2}) - (c_1 - \frac{p\varepsilon}{c_2})S_0 x_1 - (c_1 S_0 - kV_{20})\frac{1}{x_1}$$
$$- kV_{20}\frac{x_3}{x_1} - \mu V_{20}x_3 - \frac{\varepsilon V_{10}x_2}{x_3} - (\varepsilon V_{10} - \frac{p\varepsilon}{c_2}S_0)\frac{1}{x_2} - \frac{p\varepsilon S_0 x_1}{c_2 x_2}.$$

Using the algebraic approach provided by the reference [8], we will prove the function $H(x_1, x_2, x_3) \leq 0$. Firstly, we can get five groups

$$\{x_1, \frac{1}{x_1}\}, \{x_3, \frac{x_2}{x_3}, \frac{1}{x_2}\}, \{\frac{1}{x_1}, x_3, \frac{x_1}{x_2}, \frac{x_2}{x_3}\}, \{x_1, \frac{x_3}{x_1}, \frac{1}{x_2}, \frac{x_2}{x_3}\}, \{\frac{x_3}{x_1}, \frac{x_1}{x_2}, \frac{x_2}{x_3}\}$$

and the product of all functions within each group is one, then we have

$$H_1(x_1, x_2, x_3) = b_1\left(2 - x_1 - \frac{1}{x_1}\right) + b_2\left(3 - x_3 - \frac{x_2}{x_3} - \frac{1}{x_2}\right) + b_3\left(3 - \frac{x_3}{x_1} - \frac{x_1}{x_2} - \frac{x_2}{x_3}\right)$$
$$+ b_4\left(4 - \frac{1}{x_1} - x_3 - \frac{x_1}{x_2} - \frac{x_2}{x_3}\right) + b_5\left(4 - x_1 - \frac{x_3}{x_1} - \frac{1}{x_2} - \frac{x_2}{x_3}\right)$$

Since $H(x_1, x_2, x_3) = H_1(x_1, x_2, x_3)$, we can get

$$\begin{cases} b_1 = c_1 S_0 - kV_{20} - b_4 \\ b_2 = \mu V_{20} - b_4 \\ b_5 = kV_{20} - \frac{p\varepsilon S_0}{c_2} + b_4 \\ b_3 = \frac{p\varepsilon S_0}{c_2} - b_4 \end{cases}$$

As the nonnegativity of $b_i(i = 1, 2 \ldots 5), b_4$ must satisfy the following condition

$$\max\{0, \frac{p\varepsilon S_0}{c_2} - kV_{20}\} \leq b_4 \leq \min\{\mu(1-q)A, \mu V_{20}, \frac{p\varepsilon S_0}{c_2}\},$$

It is easy to prove the existence of the positive number b_4. So $H(x_1, x_2, x_3) \leq 0$ and $H(x_1, x_2, x_3) = 0$ if and only if $x_1 = x_2 = x_3 = 1$. In summary, when $R_0 < 1$ we have $L_1' < 0$, and when $R_0 = 1$, we get $L_1' \leq 0$, and $L_1' = 0$ if and only if $S = S_0 V_1 = V_{10} V_2 = V_{20}$. The largest invariant set for (2) on the set $\{(S, V_1, V_2, I) \in \Omega : S = S_0, V_1 = V_{10}, V_2 = V_{20}\}$ is $\{P_0\}$. Using the literature [10], we can prove the theorem.

4   Numerical Simulation

The numerical simulations on system (2) were carried out. We can see that if $R_0 \leq 1$, then $P_0(S_0, V_{10}, V_{20}, 0)$ is global stable (Fig. 1) and P^* is globally stable when $R_0 > 1$ (Fig. 2).

242   X. Song et al.

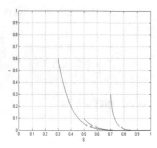

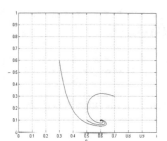

Fig. 1.   Fig. 2.

Acknowledgements. This paper is supported by the Research Fund of Department of Basic Sciences at Air Force Engineering University (2019107).

References

1. World Health Organization: Immunization against diseases of public health importance (2005). http://www.who.int/mediacentre/factsheets/fs288/en/index.html/
2. Pei, Y., Liu, S., Chen, L., et al.: Two different vaccination strategies in an SIR epidemic model with saturated infectious force. Int. J. Biomath. **1**(1), 147–160 (2008)
3. Li, J., Ma, Z.: Stability analysis for SIS epidemic models with vaccination and constant population size. DCDSB **4**(3), 637–644 (2004)
4. Li, J., Yang, Y., Zhou, Y.: Global stability of an epidemic model with latent stage and vaccination. Non. Anal. Real World Appl. **12**(4), 2163–2173 (2011)
5. Yang, W., Sun, C., Arino, J.: Global analysis for a general epidemiological model with vaccination and varying population. J. Math. Anal. Appl. **372**(1), 208–223 (2010)
6. Watanabe, T., Kiso, M., Fukuyama, S., et al.: Characterization of H7N9 influenza A viruses isolated from humans. Nature **501**(7468), 551–555 (2013)
7. Gabbuti, A., Romano, L., Salinelli, P., et al.: Long-term immunogenicity of hepatitis B vaccination in a cohort of Italian healthy adolescents. Vaccine **25**(16), 3129–3132 (2007)
8. Li, J., Xiao, Y., Zhang, F., Yang, Y.: An algebraic approach to proving the global stability of a class of epidemic models. Non. Anal. Real World Appl. **13**(5), 2006–2016 (2012)
9. van den Driessche, P., Watmough, J.: Reproduction numbers and subthreshold endemic equilibria for compartmental models of disease transmission. Math. Biosci. **180**(1), 29–48 (2002)
10. LaSalle, J.P.: The stability of dynamical systems. In: Regional Conference Series in Applied Mathematics, SIAM, Philadelphia (1976)

On Designs and Implementations of Service Request Composition Systems Using a Cooperative Multi-agent Approach

Nguyen Thi Mai Phuong[(✉)] and Nguyen Duy Hung

School of ICT, Sirindhorn International Institute of Technology,
Thammasat University, Rangsit Campus, Pathumthani, Thailand
maiphuong.0112@gmail.com

Abstract. Though much attention has been paid on composing new web services, a closely related but distinct problem of composing requests to existing services has been relatively unexplored. Using existing geo-spatial web services, we illustrated that the problem has become complex that demands support systems of their own kind. We first formalized such systems as logic program, abstracting away from specific implementation languages. We then developed a communication protocol to combine them into a cooperative multi-agent service environment for composing service requests. Advantages that future web service composition could take from our approach was discussed also.

Keywords: Multi-agent · Web services · Agents · Request composition

1 Introduction

Service composition benefits in obtaining answer for user's information demand by sending only one request to the composite service instead of sending multiple requests to different service providers. However, there are two main costs for users that catch our attention the most are discovering service provider, and composing service request. To illustrate our opinion, let's start with an example system in the domain of free and open-access GIS web services.

Example 1: Considering a system which includes three independent services:

– HW_SOS_1 is a web service providing historical weather data of the geographic area foi1 (foi - stands for feature of interest).
– HW_SOS_2 is a web service providing historical weather data of foi2.
– WG_1 is a web service generating future weather scenarios from historical weather under normal climate forecast condition.

Even though HW_SOS_1 and HW_SOS_2 provide the same service, their capabilities are different. Users have to select the correct service provider and construct service request to send out. Sample service requests are represented in Fig. 1. For HW_SOS_1 and HW_SOS_2, the basic form of service request are used. On the other hand, since WG_1 requires input data (e.g. from HW_SOS_1 as in Example 1) to

© Springer Nature Switzerland AG 2020
F. Xhafa et al. (Eds.): IISA 2019, AISC 1084, pp. 243–251, 2020.
https://doi.org/10.1007/978-3-030-34387-3_30

generate answer, a composite form request (which is more complicated) is constructed instead of a basic form request.

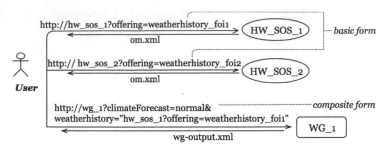

Fig. 1. Sample web service request (basic/composite form)

With the development of technology, the number of free and open-access web services is increasing rapidly that it is beyond the capability of manually composing service request. However, although much attention has been paid on composing new web services, the problem of composing requests to the existing services has been relatively unexplored. To release such burdens for users, we came up with the idea of Service Request Composition system (SRC system) that receives information demand from users as the input and returns as output a service requests. However, equipping a big knowledge for a SRC system to handle both communication and computations is infeasible. Thus, it makes sense to build multiple smaller SRC systems. Since each of them has different capabilities, they need to cooperate to "help" each other to answer the user's queries. In this paper, by inheriting the approach from cooperative information sharing multi-agent model represented in [1], we propose a model where agents are specifically cooperative for composing web service requests.

The rest of the paper is organized as follows. The brief review on Definite Logic Program (DLP) [2] and multi-layered geo-spatial service systems that are related to our work will be presented in Sect. 2. Section 3 represents the knowledge base of SRC system as logic program. Full definition and communication protocol are in Sect. 4. Section 5 is Conclusion and Discussion.

2   Background

2.1   Definite Clause Logic Program

Syntax
To introduce definite logic, let's consider the following logic program:

```
(c1): capabilities(hw1_sep, wh(foi1)).
(c2): comReq(wh(FOI),serReq(SEP,[offering(wh_+FOI)])):-
capabilities(SEP, wh(FOI)).
```

The program consists of two clauses used for composing web service requests for the weather history of the location FOI (feature of interest-foi). The symbol ':-' could be read as "if". Uppercase letters stand for universally quantified **variables** (e.g., SEP, FOI). "hw1_sep" is a **constant** representing a service end point. wh(FOI) is called a **functor**. They are all **terms**. While constants and variables are simple terms, functors are structured terms providing the way to name a complex object. "comReq, capabilities" are **predicates**. A predicate followed by number of terms is called **atoms**.

(c1) and (c2) are samples of **definite clause** which is of the form: $A :- B_1, ..., B_n$ where $A, B_1, ..., B_n$ are atoms. A is called the head of the clause and $B_1, ..., B_n$ is the body of the clause. A definite clause could be considered as three types:

- A clause with head and body is called a rule (e.g. (c1)).
- A clause with an empty head is called a goal.
- A clause with an empty body is called a fact (e,g, (c2)).

A set of definite clauses is called **definite logic program** (DLP).

Semantics

The Herbrand universe of a DLP program P is a set of ground terms constructed from the constants and functors in P. The Herbrand base of P, denoted as hb(P), is the set of all ground atoms constructed using predicates in P and the ground terms in the Herbrand universe. The least Herbrand model, LH(P) consists of all facts $f \in hb(P)$ such that P logically entails f, denoted as $P => f$. All ground atoms in LH(P) are provable.

Proof Theory

Suppose a query q "compose a service request to obtain weather history data of the location foi1" is formalized as: *?-comReq(wh(foi1), SR)*, where prefix '?-' indicates that it is a query. It could be replaced by ':-' to turn a query to a goal. With this goal, a reasoning process is applied to find an answer. Such process could be described as:

For a goal :- $P_1, P_2, ..., P_n$, find a clause $A :- B_1, ...,B_n$ such that A and P_1 are matched, a new goal is generated as :- $B_1, ..., B_n, P_2, ..., P_n$.

Matching a goal with a clause yields one or more substitutions such that by applying those, A and P1 are identical.

A **substitution** $\theta = \{V_1|t_1, ..., V_n|t_n\}$ is a finite mapping of variables V_i to terms t_i (V|t means variable V is replaced by term t). All occurrences of the variables V_i are simultaneously replaced by the term t_i. The process of applying substitution is called **unification**, and the substitution is called unifier. Two or more expressions that have unifier are **unifiable**.

Let $\theta = \{V_1|t_1, ..., V_n|t_n\}$ and $\lambda = \{y_1|u_1, ..., y_m|u_m\}$ be two substitutions. The composition of θ and λ, denoted as $\theta \circ \lambda$, is obtained by:

- Step 1: generating a set $\{V_1|t_1\lambda, ..., V_n|t_n\lambda, y_1|u_1, ..., y_m|u_m\}$.
- Step 2: deleting any element $V_i|t_i\lambda$ such that $V_i = t_i\lambda$.
- Step 3: deleting any element $y_j|u_j$ such that $y_j \in \{V_1, ..., V_n\}$.

A unifier θ is a **most general unifier** (*mgu*) of two or more expressions such that by applying the mgu, every other unifiers can be satisfied.

Proof for a goal is constructed and represented using **SLD backward deduction**. Given a definite logic program P and an original goal G_0, SLD backward deduction for G_0 could be represented as a sequence of pairs (S_i, θ_i) where S_i is a frontier of the proof tree such that:

- $S_0 = G_0, \theta_0 = \varnothing$
- For a selected sentence L in S_i, assume that there is a clause c in P defined L (i.e. L and the head of c are unifiable). S_{i+1} is derived from S_i such that:
  - $S_{i+1} = (S_i \setminus \{L\} \cup body(c)) \circ mgu$
  - $\theta_{i+1} = \theta_i \circ mgu$
- $S_m = \varnothing$ (for successful proof).

2.2   Multi-layered Geospatial Service Systems

Free and open-access service has been ubiquitous in recent years. Weather generation service system (WGS) recently introduced for generating seasonal weather scenarios from historical weather records is a sample [3]. WGS consists of three layers Client, Web Service Server, and Data Source. Clients could be application programs or human analysts. They are not required to self-manage the model. The data sources providing weather observations are local sensors. All collected data needs to be in a web-accessible way and follows SOS standard [4]. The main development focuses on Web Service Server. This layer mainly provides service and manages internal wiring among service components which are able to invoke HTTP methods. Each component provides one or more services whose data obtained is for internal or external use. Clients directly call those services through SOS API by constructing and passing HTTP GET requests to service providers. The responses are encoded in XML file.

Although the system presented in [3] is a very good application, users still need to pay so much effort on self-constructing service request. If the system is scaled up and get more complicated, this burden could be considered as an important issue.

3   The Knowledge of SRC System as a Logic Program

To compose service requests, a SRC system should first discover the capabilities of different service providers in the service environment, and then apply certain composition rules, to translate information demands into service requests. To handle both communication and execution, the required knowledge is considered to consist of two main components as in Fig. 2, and could be formalized as following logic programs.

```
<serReq>:== serReq(<serviceEndpoint>,[<key-value>?{,<key-
value>}*])
<key-value>:== <key>(<value>)
<key>:== <string>
<value>:== <string> | <serReq>
<serviceEndpoint>:== <string>
```

By applying the above grammar, service request to obtain historical weather of location FOI from the service end point SEP could be parsed as:

$$\texttt{serReq(SEP, [offering(wh\_ + FOI)])},\quad \text{denoted as WH\_SR} \qquad (1)$$

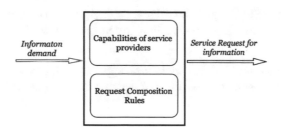

Fig. 2. Context diagram of the service request composition system

Definition 1 (Knowledge of SRC System). *The knowledge of a SRC system could be represented using definite logic program $P = CB \cup CR$, where CB includes ground facts that describe the capabilities of service providers and CR consists of composition rules written in the form of definite clauses.*

$$A :- B_1, B_2, \ldots, B_n$$

Given service end point SEP and the term t which represents information provided by SEP, the fact "*capabilities(SEP, t).*" means SEP is capable to provide t, while "¬ca-pabilities(SEP, t)." means SEP is incapable to provide t.

Query sent to a SRC system has form of "*?-comReq(F, SR)*" where F is a functor representing the demand and information provided by query's sender, and SR is a variable representing the output (service request). If the query matches with head of some rule in CR, a SDL backward deduction is constructed to find answer for it. However, if in the last pair of backward deduction, the multi-set $S_m \neq \varnothing$, that means there is some input which is unavailable in CR. This information demand requires a solution to continue constructing the deduction until $S_m = \varnothing$.

Given a multi-set S such that $S = S_m$, "S." is a set of facts corresponding to S where S. = $\{s. \mid s \in S\}$. S is called a solution for query G if S. together with CR logically entails a successful SLD backward deduction.

$$(CR \cup S.) \Rightarrow \varnothing$$

For a standalone SRC system, $S_m. \subseteq CB$.

Consider two following systems, notation in (1) is used in the following example.

Example 3: System SRC1 with knowledge $P_1 = (CB_1, CR_1)$ composes service requests for historical weather data.

CB_1 consists of a fact: *capabilities(hw1_sep, wh(foi1)).*
CR_1 consists of one composition rule:

$$comReq(wh(FOI), WH\_SR) :- capabilities(SEP, wh(FOI)).$$

If SRC1 receives a composition query: *?-comReq(wh(foi1)), SR)*, denoted as ?-G
By applying knowledge P_1, SRC1 will generate output as

$G\{(SR|serReq(hw1\_sep, [offering(wh\_foi1)])), FOI|foi1, SEP|hw1\_sep\}$

Example 4: System SRC2 with knowledge $P_2 = (CB_2, CR_2)$ composes service requests for weather scenarios of any location FOI given the historical weather data of that location.

CB_2 consists of a fact: $\forall FOI, capabilities(wg1\_sep, ws(FOI))$.
CR_2 consists of a composition rule that accepts user's input as location FOI.

(cr1):   *comReq(ws(WH_SR), serReq(WGSEP, [wh(WH_SR)])) :-*
                                         capabilities(WGSEP, ws(FOI)).

(cr2):   *comReq(ws(FOI), serReq(WGSEP, [wh(WH_SR)])) :-*
                     comReq(wh(FOI), WH_SR), capabilities(WGSEP, ws(FOI)).

Users could parse a web service request to obtain the historical weather data of the interested FOI as in the rule (cr1) or just FOI value as in the rule (cr2). If the system receives a query which includes a web service request as

?-comReq(ws(serReq(hw1_sep, [offering(wh_foi1)])), SR) , denoted as ?-G'

By applying (cr1), SRC2 will generate output as

$G'\{(SR|serReq(wg1\_sep, [wh(serReq(hw1\_sep, [offering(wh\_foi1)]))])), FOI|foi1,$
                 $WGSEP|wg1\_sep, SEP|hw1\_sep\}$

4 SRC System as an Agent

4.1 SRC System Definition

Let consider system SRC3 with knowledge $P_3 = (CB_3, CR_3)$ where $CB_3 = CB_1 \cup CB_2$, and $CR_3 = CR_1 \cup CR_2$. Equipping a powerful knowledge as P3 is infeasible. In practice, smaller scale systems such as SRC1, SRC2 are developed instead of SRC3. However, serving for user's query could be failed if they work independently. To be cooperative in performance, besides the knowledge for computation, an SRC system needs to have a communication protocol to control external interactions. Each SRC system is considered as an agent in a multi-agent environment where agents are cooperative to compose web service request. A SCR system is defined fully as below.

Definition 2 (SRC System). *A SRC system is represented as SRC = (P, I, Δ) where*

- *P = CB ∪ CR is the knowledge of the SRC system as described in Definition 1.*
- *I is information demand which is a set of atoms such that each atom "a" should appear in the body of some rule in CR, but not in the head of any rule in CR and the fact "a." is not in CB.*
- *Δ is the initial state of the SRC system.*

4.2 Communication of SRC Agents

To keep track of communication state, a SRC agent would need some databases. It might require a database to store the input information that comes from other. In addition, the agent needs some database to record the received/sending-out queries also.

Definition 3 (SRC Agent State). *State of SRC system is defined as α = (ACB, RDB, SDB, IDB, t) where*

- *ACB is a set of facts about capabilities of the current active service providers, ACB ⊆ CB.*
- *RDB is a database containing records of received queries (from users or other agents). The records are of the form (**sender, query, S, id**) where*
  - *sender is the query's sender.*
  - *query is in the form of "?- comReq(F, SR)".*
  - *S is solutions for the query which is the multiset S_m that we mentioned before.*
  - *id is non-negative integer to identify the query. A greater id means the more recent query.*
- *SDB is a database to store records of queries that agent sends out when it requires information from other. This record will be deleted when agent receives the reply.*
- *IDB is a database to store replies received from other SRC agents.*
- *t is time stamp of agent's current state. t is updated only when agent receives a query or a reply.*

The state α is changed when the agent receives/sends a query/reply from/to another agent. In the initial state of SRC agent, t = 0, RDB = SDB = IDB = ∅. The communication protocol is illustrated in Example 5.

Example 5 (Continue Example 4). Consider a service environment that includes SRC1 and SRC2 as described in Examples 3 and 4. User E sends a composition query as below to SRC2 for weather scenarios of foi1 and parses only FOI value as *foi1*.

$$?\text{-}comReq(ws(foi1), SR) \text{ , denoted as } ?\text{-}Q$$

Since *?-Q* is matched with the head of rule (cr2) in CR_2, the computation is done by the knowledge base of SRC2. In the end of computation:

The solution S_m = {*comReq(wh(foi1), serReq(SEP, [offering(wh_foi1)])*)}
The substitution $θ_m$ = {*FOI|foi1, WGSEP|wg1_sep*}

S_m is not empty, which means SCR2 requires additional input which is unavailable in its own knowledge. The communication with other agents in the service environment will start to collect the missing information (Table 1).

Table 1. Changes in states of agents during communication

	Event	SRC2	SRC1
0	–	$ACB_2 - CB_2,$ $RDB_2 = SDB_2 = IDB_2$ $= \varnothing$	$ACB_1 = CB_1,$ $RDB_1 = SDB_1 = IDB_1 = \varnothing$
1	E → SRC2	$RDB_2 = (E, \text{?-}Q, S_m,$ $0)$	–
2	SRC2 → SRC1	$SDB_2 = S_m$	$RDB_1 = (SRC2, \text{?-}S_m, \text{capabilities}(hw1\_sep,$ $wh(foi1)), 0)$
3	SRC1 → SRC2	$SDB_2 = \varnothing,$ $IDB_2 = \{SEP\|$ $hw1\_sep\}$	$RDB_1 = \varnothing$
4	SRC2 → E	$RDB_2 = \varnothing$	–

After the communication is successful, the reply that SRC2 sends back to user E would be

$$Q\{(SR\|serReq(wg1\_sep, [wh(serReq(hw1\_sep, [offering(wh\_foi1)]))])), FOI\|foi1,$$
$$WGSEP\|wg1\_sep, SEP\|hw1\_sep\}$$

5   Conclusion and Further Discussion

A SRC system presented in this paper is considered as an instance of information sharing agent (IS agent). They are designed to be cooperative in composing web service request. For such purpose, instead of using normal logic program as in IS agent, we use definite logic program to formalize the knowledge of the SRC system since user's query contains variables. We also defined the communication protocol so that multiple SRC systems could work cooperatively. Several examples were given for the demonstration. The results showed that the concept of the proposed system is feasible. Considering that users might not know exactly the capability of each SRC system in the service environment, we drop the assumption that a query is only sent to the agent which is capable of answering. The SRC system with definitions of both knowledge base and communication protocol provides a very basic design idea for further implementation of web service composition in any different programming languages.

However, in the limited scope of research, there are some scenarios that we were not be able to cover such as when receiving two same requests, whether the system should perform information catching to increase the communication efficiency.

In addition, even though we assume that if at least one SRC agent is capable of providing, users will receive answer eventually, there is a case that user's query is parsed among agents circularly which results in the problem of wasting resources.

Acknowledgments. This research is financially supported under the Thammasat University's research fund, Center of Excellence in Intelligent Informatics, Speech and Language Technology and Service Innovation (CILS), and Intelligent Informatics and Service Innovation (IISI) Research Center.

References

1. Dung, P.M., Hanh, D.D., Thang, P.M.: Stabilization of information sharing for queries answering in multi-agent systems. In: Hill, P.M., Warren, D.S. (eds.) ICLP 2009. LNCS, vol. 5649, pp. 84–98. Springer, Heidelberg (2009)
2. Flach, P.: Simply Logical: Intelligent Reasoning by Example. Tilburg University, The Netherlands (1998). ISBN 0471 94152 2, Chap. 1–2, pp. 3–41
3. Chinnachodteeranun, R., Hung, N.D., Honda, K., Ines, A.V.M., Han, E.: Designing and implementing weather generators as web services. Future Internet **8**, 55 (2016)
4. OGC 12-006 document: OGC® Sensor Observation Service Interface Standard, Published on 2012

A Method of Pointer Instrument Reading for Automatic Inspection

Ning Jiang[1], Jialiang Tang[1], Zhiqiang Zhang[1], Wenxin Yu[1(✉)],
Bin Hu[2], Gang He[3], and Zhuo Yang[4]

[1] Southwest University of Science and Technology, Mianyang, China
yuwenxin@swust.edu.cn
[2] Nantong University, Nantong, China
[3] Xidian University, Xi'an, China
[4] Guangdong University of Technology, Guangzhou, China

Abstract. In automatic inspection, it is required to automatically read the indicator value of the pointer instrument from images. However, in the cases of different shooting positions and focal lengths, the images will be blurred and deformed, which bring difficulties to indicator value reading. This paper proposes a method for reading the indicator values of pointer instruments automatically. First, we make clear templates for each type of dashboard. Then, we use KAZE to extract the features for the input image, use KNN to match the features from this image and the templates, use RANSAC to reduce the wrong matchings, and get the type and area of instrument based of matching results. Finally, we calculate the indicator value by edge information, pointer line information from probabilistic Hough transform, and labeled prior knowledge. The experimental results show the output values within a degree scale is 96.4% in 560 samples, which meets the requirements of automatic inspection.

Keywords: Automatic inspection · Pointer instrument · KAZE

1 Introduction

Pointer instrument is widely used in electric power measurement industry because of its simple structure, convenient maintenance, anti-electromagnetic explosion and other advantages. At present, the reading and inspection of instruments are mostly completed by manpower. This work is low automation and consumes a lot of manpower. With the increase of the number of instruments and the development of information automation in factories, manual detection and instrument reading can't meet the needs of factories. Automatic reading of pointer dashboards has become an urgent problem to be solved.

Many researchers have researched the technology of automatic instrument recognition and reading, mainly focusing on location detection, segmentation, type recognition and pointer extraction [1–5]. Because of pointer dashboard's complexity, there are still many problems in its practicability. In [1], Zheng et al. use Haar-Like Feature and Polar Expansion to realize automatic recognition of single pointer dashboard. In [2], Zheng et al. use edge detection and Hough Transform to realize automatic recognition and reading system in a template case without different shooting positions

F. Xhafa et al. (Eds.): IISA 2019, AISC 1084, pp. 252–259, 2020.
https://doi.org/10.1007/978-3-030-34387-3_31

and focal lengths; The location of the rotation center of the pointer is used in recognition and reading in [3, 4]. In [5], Li uses color information to realize a pointer segmentation system. But all these methods are only suitable for the case of constant illumination and indoor conditions. And all these methods require the dial to have higher clarity and better forehead angle, that is, the camera should not be too far from the dial. However, in practical application, the distance between dials and cameras is often very long, which make above methods difficult in application.

In long-distance, multi-angle photography cases, there are three problems as follows: (1) The proportion of dials in the whole picture is very small, usually only 3%–8%; (2) Because of the long distance or wrong focal length, the number and degree scale of the dial are blurred, and difficult to be recognized; (3) Different inclination and elevation may lead to serious deformation of dial, as shown in Fig. 1. These three problems increase the difficulty of automatic reading.

Fig. 1. Different blurred and deformation cases in automatic inspection

In order to solve these problems, this paper proposes the solutions as follows: (1) Take a template photo for each pointer instrument, manually calibrate the center and radius of the dial, and the angle value corresponding to this instrument. This information will be used as prior knowledge. And the photograph location is provided to automatic inspection system, for taking the images to be recognized by automatic patrol robots. (2) In the process of recognition, extract KAZE features [6] of the image to be recognized and the template image. (3) Estimate the pointer angle by segmentation and Hough transform, and estimate the indicator value by pointer angle and prior knowledge. The experimental results show that this method can get effectively results in long-distance and multi-angle shooting cases.

2 Proposal

Because the instruments are too small in the image, the first step is to locate the dashboard. Detection algorithms based on deep learning, such as RCNN [7–9], YOLO [10], require a large number of samples, especially in deep learning. Therefore, this paper intends to recognized and locate instrumentation through feature matching by instrumentation images and template images. The samples of taken instrumentation images and template images are show in Fig. 2.

(a)   (b)

(c)   (d)   (e)

Fig. 2. Samples of taken images (a–d) and template images (e)

Because of the deviation in photography location and angle, we may get different images as Fig. 2(a)–(d). In these 4 images, the size of instrumentation is same, but the location is different. Figure 2(e) is a clearly template image for this instrumentation.

2.1 Feature Extraction

Alcantarilla et al. proposed a new feature detection algorithm, KAZE algorithm [6], which uses the additive operator splitting algorithm (AOS) to search the local maxima value in the non-linear scale space. In this paper, the experiments are done in rotational scaling, perspective transformation, noise interference, blurred and compressed cases. Compared with SURF [9] and SIFT, KAZE is better in scaling and rotation invariance. This method is stable and repeatable, and achieves better performance in feature detection and matching. The main steps of KAZE algorithm include:

(1) Using AOS algorithm to construct non-linear scale space;
(2) Find the normalized Hessian local maxima of different scales. The calculation of Hessian matrix as follows:

$$L_{Hessian} = \sigma^2 (L_{xx}L_{yy} - L_{xy}^2) \tag{1}$$

Here, σ is the integer value of scale parameter, L is the image filtered by Gauss filter, L_{xx}, L_{yy} and L_{xy} are the second-order differential of L.

(3) At sub-pixel level, the location and scale of feature points are accurately located by spatial scale function using Taylor expansion:

$$L(x) = L + \left(\frac{\partial L}{\partial x}\right)^{T} x + \frac{1}{2} x^{T} \frac{\partial^2 L}{\partial x^2} x \tag{2}$$

Here, $L(x)$ is the spatial scale function, x is a feature point coordinate, and its sub-pixel coordinate is:

$$x = -\left(\frac{\partial^2 L}{\partial x^2}\right)^{-1} \frac{\partial L}{\partial x} \tag{3}$$

(4) Feature vectors construction

In order to achieve image rotation invariance, we need to determine the main direction according to feature points structure in local image. Similar to SURF features, the main direction is obtained by searching in the circular regions. Then, G-SURF is used to describe the feature points. For the feature points with scale parameter σ_i, a rectangular region is taken from the gradient map, with this feature points as center and with width $24\sigma_i$. Then this rectangular region is divided into 4 * 4 sub-windows. Each sub-window is weighted by a Gaussian kernel. Then a sub-region description vector with length 4 is calculated by following formula:

$$d_v = \left(\sum L_x : \sum L_y : \sum |L_x| : \sum |L_y|\right) \tag{4}$$

Each sub-window's feature vector d_v is weighted and normalized by a 4 * 4 Gauss window, then a 64-dimensional feature vector is obtained.

2.2 Key Point Matching

After obtaining the description feature vectors of the key feature points of the image to be recognized and the template image, the next step is to match the feature points between these two images. In order to select the best matching points, we use RANSAC algorithm to compute the transformation $H = [h_{ij}]_{i,j}$ between the source plane and the template plane as following formula:

$$s_i \begin{bmatrix} x'_i \\ y'_i \\ 1 \end{bmatrix} \approx H \begin{bmatrix} x_i \\ y_i \\ 1 \end{bmatrix} \tag{5}$$

The optimal matching point achieves the minimum error in following formula:

$$\sum_i ((x_i' - \frac{h_{11}x_i + h_{12}y_i + h_{13}}{h_{31}x_i + h_{32}y_i + h_{33}})^2 + (y_i' - \frac{h_{21}x_i + h_{22}y_i + h_{23}}{h_{31}x_i + h_{32}y_i + h_{33}})^2) \qquad (6)$$

The position of the dial in the picture to be recognized is determined according to the optimal matching point, and then the dial pointer can be extracted.

2.3   Pointer Extraction

According to the characteristics of the dial, black pointer and white background, the dial area is grayed and binarized. In order to reduce the interference in edge region, the center is taken as the origin and one third of the edge length of dial area is taken as the radius. A histogram distribution of the area is calculated to get two thresholds for binarization.

3   Experiments

Firstly, each dial is calibrated with a template image. As shown in Fig. 2, Fig. 2(a–d) are the images to be recognized, Fig. 3 is the template image of this instrument. The center position and radius of the dial in template image are manually calibrated. At the same time, the pointer angle corresponding to each large degree scale is marked in Fig. 4.

Fig. 3. Template image

Next, we extract the features of the image to be recognized and the template image by KAZE method, and match the key points with KAZE feature by KNN method. In the results, the matching accuracy of KAZE feature points is over 99% in 560 samples. As shown in Fig. 5(a), the matching results of key points by KNN are shown. In Fig. 5(b), the matching results with wrong matching reduced by RANSAC are shown.

After key point matching, we get the dial area in the image. Then, binary segmentation and morphological operation are used to get a new image for pointer recognition and degree value reading. The processing results are shown in Fig. 6.

```
</LinePostions>
<LineAngle>  2.1822135925292969e+002
      1.9932539367675781e+002 1.7703817749023437e+002
      1.4810659790039062e+002 1.2277226257324219e+002
      1.0101653289794922e+002 8.0065742492675781e+001
      5.9001972198486328e+001 3.2622066497802734e+001
      2.4831392765045166e+000 -3.1073944091796875e+001
      -5.6251129150390625e+001
      -7.4456237792968750e+001</LineAngle>
<LineValue>  0. 5. 10. 15. 20. 25. 30. 35. 40. 45. 50. 55.
      60.</LineValue>
```

Fig. 4. Marked pointer angles

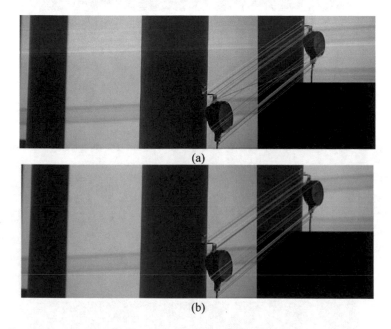

(a)

(b)

Fig. 5. (a) Extracted by KAZE and matched by KNN, (b) Error reduced by RANSAC

Then merges the small lines to a bigger one, with included angle less than 3°, and distance less than a certain threshold. The selection of threshold is related to the size of the dial. The final result of pointer detection is shown in Fig. 7.

Assuming that the pointer angle is α. According to the prior knowledge of calibration as Fig. 3(b), the corresponding calibration line angle $\theta = \{\theta_1, \theta_2, \ldots, \theta_n\}$, and line value $\varpi = \{\varpi_1, \varpi_2, \ldots, \varpi_n\}$. Assuming $\theta_i \le \alpha \le \theta_j$, the dial reading is estimated to be $(\varpi_i + \beta\varpi_j)/(1 + \beta)$, of which $\beta = (\alpha - \theta_i)/(\theta_j - \alpha)$.

In this paper, the result of error control within a degree scale is about 96.4% in 560 sample, which meets the practical requirements.

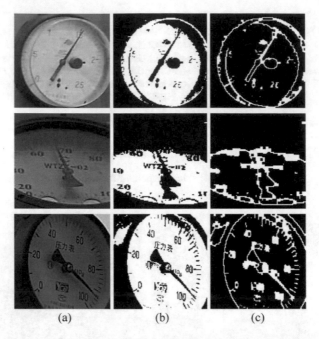

(a)   (b)   (c)

Fig. 6. (a) Original dashboard (b) binary segmentation (c) morphological result

Fig. 7. Final result with pointer angle

4   Conclusion

This paper proposes an automatic reading method of pointer instrument based on KAZE feature for automatic inspection with long distance and multi-angle cases. It consists of three steps: feature extraction by KAZE and matching by KNN to obtain the instrument location, binarization processing and probability Hough transform to estimate the point angle, use prior knowledge to estimate the degree scale. This method effectively solves the difficulty of automatic reading caused by the blurred calibration and serious deformation of the instrument panel taken in long distance and multiple angles cases. The result of reading error within a degree scale is 96.4%, which satisfies the practical requirement.

Acknowledgments. This research was supported by (2018GZ0517) (2019YFS0146) (2019YFS0155) (2019YFS0167) which supported by Sichuan Provincial Science and Technology Department, (2018KF003) Supported by State Key Laboratory of ASIC & System, Science and Technology Planning Project of Guangdong Province (2017B010110007).

References

1. Zheng, X., Chen, X., Zhou, X., Mou, X.: Pointer instrument recognition algorithm based on Haar-like feature and polar expansion. In: IEEE 3rd International Conference on Image, Vision and Computing (ICIVC), pp. 188–193 (2018)
2. Zheng, W., Yin, H., Wang, A., Fu, P., Liu, B.: Development of an automatic reading method and software for pointer instruments. In: First International Conference on Electronics Instrumentation & Information Systems (EIIS), pp. 1–6 (2017)
3. Ma, Y., Jiang, Q., Wang, J., Tian, G.: An automatic reading method of pointer instruments. In: Chinese Automation Congress (CAC), pp. 1448–1453 (2017)
4. Gao, J., Guo, L., Lv, Y., Wu, Q., Mu, D.: Research on algorithm of pointer instrument recognition and reading based on the location of the rotation center of the pointer. In: IEEE International Conference on Mechatronics and Automation (ICMA), pp. 1405–1410 (2018)
5. Li, P., Wu, Q., Xiao, Y., Zhang, Y.: An efficient algorithm for recognition of pointer scale value on dashboard. In: 10th International Congress on Image and Signal Processing, BioMedical Engineering and Informatics (CISP-BMEI), pp. 1–6 (2017)
6. Alcantarilla, P.F., Bartoli, A., et al.: KAZE features. In: European Conference on Computer Vision, pp. 214–227. Springer, Heidelberg (2012)
7. Girshick, R., Donahue, J., Darrell, T., et al.: Rich feature hierarchies for accurate object detection and semantic segmentation. In: Computer Vision and Pattern Recognition, pp. 580–587. IEEE, New York (2013)
8. Girshick, R.: Fast R-CNN. In: IEEE International Conference on Computer Vision, pp. 1440–1448. IEEE, New York (2015)
9. Ren, S., He, K., Girshick, R., et al.: Faster R-CNN: towards real-time object detection with region proposal networks. In: International Conference on Neural Information Processing Systems, pp. 91–99. MIT Press, Cambridge (2015)
10. Redmon, J., Divvala, S., Girshick, R., et al.: You only look once: unified, real-time object detection. In: Proceedings of the IEEE Conference on Computer Vision and Pattern Recognition, pp. 779–788. IEEE, New York (2016)

Research on Data Processing Technology for Heterogeneous Materials Model

Zhenhua Jia[1(✉)], Yuhong Cao[1], Gang Liu[1], and Weidong Yang[2]

[1] North China Institute of Aerospace Engineering,
Langfang 065200, Hebei Province, China
jiazhenhualf@126.com
[2] Hebei University of Technology, Tianjin 300000, China

Abstract. With the rapid prototyping technology becoming more and more mature, the range of products produced by this technology are becoming wider and wider nowadays. In some specific occasions, higher standard is put forward for product performance, so it is difficult for uniform material parts to meet the demand, so the concept of "heterogeneous material parts" appears. At present, the research in the field of rapid prototyping technology for heterogeneous material parts is in its infancy. This paper will make relevant research on data processing technology of heterogeneous material model. Firstly, this paper introduces the background of rapid prototyping technology, the research achievements and research status of foreign and domestic scholars in heterogeneous material parts. Secondly, the data interpolation method of STL file is proposed to realize the data processing of heterogeneous material model. At last, the development of rapid prototyping technology is prospected at the end of this paper.

Keywords: Rapid prototyping · Heterogeneous material parts · STL file · Data processing

1 Introduction

The traditional manufacturing industry is based on the production technology, which requires senior technicians with rich experience to complete the complex process planning tasks, and the limitations of the production technology lead to the inability to produce products with overly complex shapes. The emergence of rapid prototyping (RP) technology makes the manufacturing industry break through the traditional constraints and effectively solve the existing contradictions.

At present, the rapid prototyping technology for uniform material parts has been basically mature. However, with the continuous improvement of product demand, it is difficult for uniform material parts to meet the expected demand, at this time, rapid prototyping in heterogeneous material parts molding manufacturing has obviously become a development trend. At present, there are few researches on the rapid prototyping of heterogeneous material parts. Based on previous researches, this paper will further explore and do research on the technical problems existing in the rapid prototyping of heterogeneous material parts.

© Springer Nature Switzerland AG 2020
F. Xhafa et al. (Eds.): IISA 2019, AISC 1084, pp. 260–267, 2020.
https://doi.org/10.1007/978-3-030-34387-3_32

2   Background of Rapid Prototyping Technology

In the middle and late 1980s, rapid prototyping technology, which started with the Stereo Lithography Apparatus (SLA) technology [1], has developed rapidly. In the subsequent development of rapid prototyping technology, Selective Laser Sintering (SLS), Laser Near Net Shaping (LENS), Fused Deposition Manufacturing (FDM), Laminated Object Manufacturing (LOM) and Three-Dimensional Printing technology (3DP) have appeared successively [2]. Rapid prototyping technology is developing rapidly and has made remarkable achievements in machinery, automobile, aerospace, electronics and other industries [3]. In some cases, higher standard has been put forward for parts. So a large number of scholars began to study the manufacture of heterogeneous material parts.

There are two main factors affecting the molding quality of heterogeneous material parts, one is the CAD modelling of heterogeneous material parts, the other is the processing of model data. The CAD model of the uniform material part can only store the coordinate information, but cannot store the material information. For this reason, some scholars have done a lot of research on how to reflect material information through CAD model. Pinghai et al. proposed a modeling method based on B spline [4]; Kou et al. proposed the modeling method of b-rep [5]; Jackson proposed a modeling method that used the finite element grid to describe the geometric information of parts, and used the distance between the node and the boundary of the internal finite element to represent the material information as a variable [6]; Patil et al. proposed R-function modeling method for describing material structure. They used RM target model to describe heterogeneous entity model [7]; Siu et al. proposed a modeling method based on "gradient source" [8]; Biswas et al. proposed a field modeling method based on geometric domain [9]; Wu Xiaojun et al. proposed the CAD modeling method under the definition of distance field [10]. Unfortunately, because the modeling methods are complex, they can not be applied to the current CAD modeling software. In the current study, most of them are theoretical research on modeling methods of heterogeneous materials parts, and few of them are research on model data processing [11]. Therefore, there is still a big gap in the heterogeneous material model data processing, which deserves further study.

3   Data Processing Method for Heterogeneous Materials Model

Data processing of heterogeneous material model is different from that of uniform material model. It is necessary to add material information into the model and obtain printing path including coordinates information and material information. According to the analysis of the file format and internal data structure, in this paper, OpenGL is adopted as the 3d graphics interface. The material information of STL files is added by coloring the model, and the printing path including position information and material information is obtained by slicing the color model.

3.1 STL Format File

STL format file is an interface between CAD model and data processing system [12]. It has been regarded as the standard format of description file in rapid prototyping field by industry. Data processing software needs to read the STL file information accurately, and then processes the information obtained.

STL file is composed of a large number of disordered planar triangles, that is, differentiating the three-dimensional model, so that the triangular surface can approach the model shape infinitely. A complete STL file should contain the three vertex coordinates of each triangular surface and the normal vector coordinates of each triangular surface, from which the shape of the three-dimensional model can be uniquely determined.

3.2 Representation of Material Information in STL Files

STL files used for data processing of uniform material model are all monochrome STL files. The traditional monochrome STL files only describe coordinate information and have no description of material information, that is to say, they can only store one kind of material. In order to realize the rapid prototyping of heterogeneous material model, it is necessary to add material information into STL file and obtain printing path according to coordinate information and material information, which is the key to solve the problem of heterogeneous material parts rapid prototyping. In this paper, a method of using color STL file to represent material information of heterogeneous material model is proposed. The printing path is obtained by slicing the color model.

A one-to-one relationship is formed between the color information and the material information [13]. One color only represents one kind of material. Assuming that a part is composed of n materials, the STL model of the part can be represented by n different colors. It is assumed that the nth material is represented by m_n and the nth color is represented by c_n. That is,color $c_1, c_2, c_3 \ldots \ldots c_n$ can be used to represent material $m_1, m_2, m_3 \ldots \ldots m_n$.

3.3 Model Coloring

OpenGL is the interface for current 3d graphics programming. In the OpenGL function library, methods glBegin (GL_TRIANGLES) and glEnd () are provided to render triangular facets directly [14].

Through the pre-established color index table, each triangle is colored with OpenGL technology. When using OpenGL technology, it is required to process each vertex as a graphics unit, integrate the color data with the vertex coordinate data, and complete the coloring of all triangular faces in a cycle. The coloring algorithm is as follows:

```
for(int i=0; i<N; i++)

{                                    //N is the total number of triangular patches
    glbegin(GL_TRIANGLES) ;//Color              the          ith          triangle
    glVertex3f(m_triangle[i],V[0]) ;//Definition of the first vertex of the ith triangle
        glVertex3f(m_triangle[i],V[1]) ;//Definition of the second vertex of the ith trian-
gle
        glVertex3f(m_triangle[i],V[2]) ;//Definition of the third vertex of the ith triangle
        glend() ;//The ith triangle coloring is completed
}
```

3.4   Model Visualization

The coordinate transformation function of OpenGL function library can realize the rotation, movement and scaling of the model, so that users can observe the model more directly and truly [15].

(1)   The coordinate translation function glTranslate (x, y, z) can translate local or global coordinates. The concrete operation method is: multiply the coordinate matrix of the model that needs to be translated by the coordinate translation action matrix to get the translated coordinate matrix.

(2)   The coordinate scaling function glScale (x, y, z) causes local or global coordinates to scale. The concrete operation method is: multiply the current model coordinate matrix that needs to be scaled by the coordinate scaling action matrix to obtain the scaled coordinate matrix.

(3)   The glRotate (n, x, y, z) function rotates global or local coordinates by n degrees around an extension line from the origin to the point (x, y, z). The concrete operation method is: multiply the current model coordinate matrix that needs to be rotated by the coordinate rotation action matrix to obtain the rotated coordinate matrix.

The coordinate translation action matrix, the coordinate scaling action matrix and the coordinate rotation action matrix are respectively:

$$
\begin{bmatrix} x & 0 & 0 & 0 \\ 0 & y & 0 & 0 \\ 0 & 0 & z & 0 \\ 0 & 0 & 0 & 1 \end{bmatrix}
\begin{bmatrix} 1 & 0 & 0 & x \\ 0 & 1 & 0 & y \\ 0 & 0 & 1 & z \\ 0 & 0 & 0 & 1 \end{bmatrix}
\begin{bmatrix} x^2(1-c)+c & xy(1-c)-zs & xz(1-c)+ys & 0 \\ yx(1-c)+zs & y^2(1-c)+c & yz(1-c)-xs & 0 \\ xz(1-c)-ys & yz(1-c)+xs & z^2(1-c)+c & 0 \\ 0 & 0 & 0 & 1 \end{bmatrix}
$$

In the coordinate rotation matrix: $C = \cos(n)$, $s = \sin(n)$.

3.5   Slices of Heterogeneous Material Model

Using a set of horizontal section segmentation models, calculate the closed contour lines formed by all triangular faces intersecting the section.

For non-gradient heterogeneous material parts, we obtain the color of a region of the model directly through the color index table, and then distribute the material accurately according to the pre-defined color material mapping table. However, for gradient heterogeneous material parts with no clear limit between materials, we can not divide materials by blocks, and we need to obtain the color information of each point to realize the gradual change of materials. For this reason, we adopt bilinear data interpolation method, which is based on the position coordinate information and color information of the slices, to realize the gradual change between the materials.

In the slicing process, coordinate information and color information of each layer were interpolated to obtain the printing path containing coordinate information and color information.

Taking Fig. 1 as a model, the bilinear interpolation algorithm is described. It is assumed that the four vertices of the heterogeneous material model are V_1, V_2, V_3 and V_4, and the coordinates and colors of the four vertices are known. Make the plane segmentation model parallel to x-y, the thickness of the cutting layer is Z_i, the plane intersects the model boundary line V_1V_2 at the point V_{12}, and the boundary line V_4V_2 at the point V_{42}.

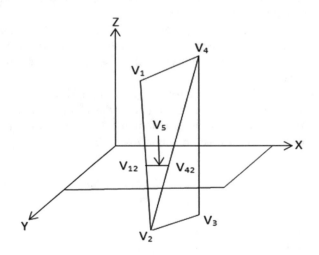

Fig. 1. Bilinear interpolation model

Carry out linear interpolation for V_{12} and V_{42} and calculate the coordinate information of any point V_5 (x_i, y_i).

$$x_i = \frac{(z_i - z_1) \times (x_2 - x_1)}{(z_2 - z_1)} + x_1 \tag{1}$$

$$y_i = \frac{(z_i - z_1) \times (y_2 - y_1)}{(z_2 - z_1)} + y_1 \tag{2}$$

The color information is represented by three primary colors. It is assumed that, given $V_1(R_1, G_1, B_1)$, $V_2(R_2, G_2, B_2)$, $V_3(R_3, G_3, B_3)$, $V_4(R_4, G_4, B_4)$, calculate $V_{12}(R_{12}, G_{12}, B_{12})$ and $V_{42}(R_{42}, G_{42}, B_{42})$.

$$R_{12} = \frac{(Z_{12} - Z_1) \times (R_2 - R_1)}{(Z_2 - Z_1)} + R_1 \tag{3}$$

$$G_{12} = \frac{(Z_{12} - Z_1) \times (G_2 - G_1)}{(Z_2 - Z_1)} + G_1 \tag{4}$$

$$B_{12} = \frac{(Z_{12} - Z_1) \times (B_2 - B_1)}{(Z_2 - Z_1)} + B_1 \tag{5}$$

In the same way, the three primary colors $V_{42}(R_{42}, G_{42}, B_{42})$ are obtained. The color C of the point is the mixture of three primary colors in a certain proportion. Linear interpolation between $V_1 V_2$ points to calculate the color C_{12} and C_{42} of V_{12} and V_{42} points.

$$C_{12} = (1 - \partial)C_1 + \partial C_2 \tag{6}$$

$$C_{42} = (1 - \partial)C_4 + \partial C_2 \tag{7}$$

By linear interpolation between V_{12} and V_{42}, color of any point C_5 can be obtained.

$$C_5 = C_1 + uC_2 + vC_4 \tag{8}$$

In formula (9), u and v can be obtained through coordinates of V_1, V_2, V_3, V_4 and V_5 points.

$$v = \frac{(y_2 - y_1) \times (x_5 - x_1) - (x_2 - x_1) \times (y_i - y_1)}{(y_2 - y_1) \times (x_4 - x_1) - (x_2 - x_1) \times (y_4 - y_1)} \tag{9}$$

$$u = \frac{(x_5 - x_1) - (x_4 - x_1)v}{(x_2 - x_1)} \tag{10}$$

4 Conclusion

(1) In this paper, the development background and research status of rapid prototyping technology are studied and summarized in depth, and the problem to be solved in the molding of heterogeneous material parts is proposed: how to deal with heterogeneous material model data.

(2) This paper briefly introduces the STL file, and puts forward the method of using color STL file to represent the heterogeneous material model.
(3) This paper describes the coloring of monochrome STL files, the visualization of model, and the slicing of color STL files by data interpolation algorithm in detail.

5   Concluding Remarks

In view of the current production and living needs of people, the use of 3D printing technology to achieve the molding and manufacturing of heterogeneous material parts is bound to become a vane in the field of rapid prototyping manufacturing. At present, the research on Rapid Prototyping of heterogeneous material parts is still immature. The existing problems, such as the improvement of hardware details, the selection of materials, the planning of modeling process and the comparison and optimization of data processing algorithms, directly affect the performance and quality of prototyping parts. The problems mentioned above involve several disciplines such as machinery, materials, process planning, mathematics and computer technology. Therefore, it is necessary to integrate the wisdom of researchers in various fields to develop the rapid prototyping technology of heterogeneous material parts into a truly mature advanced technology without short plates.

In addition, the future development direction of rapid prototyping technology will not only be limited to the static model, the design and processing of dynamic model will become the future development trend. The time variable is added into the model, so that the model can represent the material distribution of a certain coordinate at different time points, supporting the constant modification and update of the model. For example, in biomedicine, a growing cancer cell can use a dynamic model to represent the real-time state of the cell.

Based on the rapid prototyping technology in homogeneous material parts molding has reached a mature stage, we need to stand on the basis of previous research to further climb the mountain of heterogeneous material parts rapid prototyping and even dynamic model rapid prototyping.

Acknowledgments. This paper is supported by Natural Science Foundation of Hebei Province (Grant No. E2016202297) and the Support Project of Science and Technology of Langfang (Grant No. 2018011051).

References

1. Kou, X.Y., Tan, S.T.: Heterogeneous object modeling: a review. Comput. Aided Des. **39**, 284–301 (2007)
2. Ji, H.: Analysis and comparison of 3D printing technology and traditional manufacturing technology. Farm Mach. Using Maintenance 3–4 (2018)
3. Guan, Y.: Application of 3D printing technology in three-dimensional modeling course. Qual. Educ. W. China, pp. 169–178 (2018)

4. Zhu, Y., Yang, J., Wang, C.: Research on integrated design and manufacturing of heterogeneous material parts. Mach. Des. 10–17 (2012)
5. Ji, F., Chen, L., Li, Z.: Model and application based on STL file. J. Chang'an University (Nat. Sci. Ed), 36–38 (2006)
6. Li, Z., Xie, C., Yang, J.: 3D Printing layered algorithms based on stl files and fast extraction of adjacent topological information. Comput. Eng. Appl. 32–35 (2002)
7. Yang, G., Liu, W., Wang, W., Tian, F.: Research on topology reconstruction and fast slicing algorithms of STL format files. Modern Manuf. Eng. 75–78 (2009)
8. Zhou, H., Wu, J.: Research on contour adaptive layering method based on STL model. Mach. Electron. 14–17 (2015)
9. Wu, J., Yang, J., Chu, H., Feng, C., Yang, J.: Research on interpolation of slice data of heterogeneous material parts. J. Nanjing Normal Univ. (Eng. Technol. Edn.), 32–36 (2015)
10. Zhu, Y., Yang, J., Wang, C.: Modeling method of heterogeneous material parts based on spatial micro-tetrahedron. J. Mech. Eng. 150–155 (2012)
11. Guan, Y.: Application of 3D printing technology in three dimensional modelling course. Qual. Educ. W. China 169–178 (2018)
12. Hou, B., Liu, X.: Topological reconstruction and defect repair of STL entity model. Comput. Eng. 213–214 (2005)
13. Dai, N., Liao, W., Chen, C.: Key algorithms for fast topological reconstruction of STL data. J. Comput. Aided Des. Graph. 67–72 (2005)
14. Siu, Y.K., Tan, S.T.: Source_based heterogeneous solid modeling. Comput. Aided Des. 41–45 (2002)
15. Xia, Q.: 3D printing technology. J. Changsha Univ. 94–98 (2005)

Trajectory Planning for the Cantilevered Road-Header in Path Correction Under the Shaft

Yuanyuan Qu[1(✉)], Xiaodong Ji[1], and Fuyan Lv[2]

[1] China University of Mining and Technology (Beijing), 100083 Beijing, China
qyy2014@cumtb.edu.cn
[2] Shandong University of Science and Technology, Qingdao 266590, China

Abstract. For path correction of the road-header under the shaft, trajectory planning for the road-header is carried out following different considerations amongst which excavation space is highly concerned. In addition, for the benefit of the mechanical system and also for cutting slipping of the tracks on the road-header, the variation of the required rotation speeds of the two driving wheels are also concerned. Three typical planned trajectories are proposed and compared. Evaluation over these trajectories is also carried out using a simple path tracking algorithm. The calculated results explain that trajectory planning for the road-header is necessary since different trajectories correspond to different scenarios about undesirable excavation space, slipping level and probably power consumption.

Keywords: Road-header · Trajectory planning · Path tracking

1 Introduction

Cantilevered road-header ('road-header' for short in the context) is used for roadway shaping under the shaft in mining industry, by drilling and cutting the coal and rocks following a pre-designed geological blueprint. It is mainly composed of framework, cutting arm, shovel plate, belt conveyor, walking mechanism, hydraulic suits and control system [1], as shown in Fig. 1.

The road-header is always expected to walk along the center line of the desired roadway. The ideal excavation procedure is as following: judge the yawing angle, namely the orientation of the vehicle body at first; drive the road-header with certain steering speed and forward velocity, coal mixed with rocks is cut down and conveyed to the rear while displacement of the vehicle completes; the position and posture information of the vehicle are then updated by estimation or measurements and are used to plan the yawing angle, turning angular speed and forward velocity for the next excavation step [2].

Once off-track happens, namely any of position and orientation fails to follow the pre-designed roadway, path correction is required. Unlike a wheeled robot or an autonomous vehicle to track a certain curve by changing its kinematic parameters neatly and quickly on the ground [3–5], the road-header locates in a sealed space, it can

© Springer Nature Switzerland AG 2020
F. Xhafa et al. (Eds.): IISA 2019, AISC 1084, pp. 268–276, 2020.
https://doi.org/10.1007/978-3-030-34387-3_33

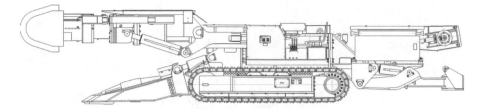

Fig. 1. The structure of a boom-type road-header

hardly reset its orientation in situ and move back to the aimed track immediately [6]. More likely, it may need to retreat, adjust yawing angle, then move forward again, which is a set of inefficient operation. Alternatively, it can keep moving forward instead of retreat, but with well-planned turning angular speed and forward velocity and walk to the aimed track regularly in steps. That is what trajectory planning does in this paper.

2 Trajectory Planning and Movements Scheduling for the Road-Header

For the road-header, the time cost of each piece of displacement is usually longer than that of most caterpillar vehicles on ground. It is more likely to use "adjustment" or "scheduling" instead of "tracking" when illustrating the procedure of path correction in detail.

For safety, the displacement in each adjusting step is limited to be less than 80 cm. Except the natural kinematic restraints of the walking mechanism, considering the sealed space where the road-header locates, the orientation of the road-header can hardly vary randomly because space cost should be concerned since the accompanying space shoring is expensive.

From the road-header at wrong position to the aimed roadway, there could be several trajectories in between, but not all of them are recommended considering excavation cost. Firstly, more excavation means more expending; therefore limited excavation step and excavation space is preferred.

Secondly, since the body size of the road-header is huge, it is inevitable to produce derived space, i.e. the unexpected excavation space, around the main roadway.

Lastly, to shape a roadway close to the pre-designed one as much as possible, the trajectory is expected to be smoothly converging to the main track.

To judge the amount of excavation space during displacement, the top view of the road-header is simplified as a rectangular frame with size $l \times b$, and then judge the area swept by the rectangular during movements. The concerned area in each step is the shadow area illustrated in Fig. 2.

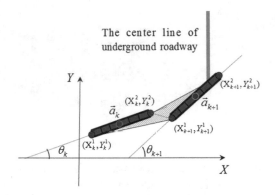

Fig. 2. The area swept by one side of the road-header during one movement

Suppose the feasible trajectory is composed of n steps of displacement, and each step is resulted by running certain turning angular speed ($\dot{\varphi}$) and moving velocity (v). Use A_k to denote the shadow area in Fig. 2, then the resulted over-all area corresponding to the under checked trajectory is then given by:

$$A = \sum_{k=1}^{n} A_k \tag{1}$$

Set the location state of the road-header at step No. k as $s_k(X_k^1, Y_k^1, \theta_k, X_k^2, Y_k^2)$, in which (X_k^2, Y_k^2) and (X_k^1, Y_k^1) are the coordinates of the head and rear point of the road-header, respectively, in the geodetic coordinate system XOY; θ_k is the orientation angle. From state s_k to s_{k+1}, the produced shadow area in Fig. 3 is computed by:

$$A_k = \frac{1}{2}\left|\vec{a}_k \times \vec{d}_{k1}\right| + \frac{1}{2}\left|\vec{a}_{k+1} \times \vec{d}_{k2}\right| \tag{2}$$

where $\vec{a}_k$, $\vec{d}_{k1}$ are vectors $(X_k^2 - X_k^1, Y_k^2 - Y_k^1)$, $(X_{k+1}^1 - X_k^1, Y_{k+1}^1 - Y_k^1)$, respectively; and $\vec{a}_{k+1}$, $\vec{d}_{k2}$ are vectors $(X_{k+1}^2 - X_{k+1}^1, Y_{k+1}^2 - Y_{k+1}^1)$ and $(X_{k+1}^2 - X_k^2, Y_{k+1}^2 - Y_k^2)$, respectively. Then:

$$X_{k+1}^1 - X_k^1 = \int_{t_k}^{t_{k+1}} v\cos\theta dt = \int_0^{t_{k+1}-t_k} v\cos(\theta_k + \omega t)dt \tag{3}$$

t_k and t_{k+n} is the time that excavation step k begins and ends, respectively. Considering $\theta_{k+1} = \omega(t_{k+1} - t_k) + \theta_k$, the following equation is obtained:

$$X_{k+1}^1 - X_k^1 = \frac{v}{\omega}(\sin\theta_{k+1} - \sin\theta_k) \tag{4}$$

Similarly, $Y^1_{k+1} - Y^1_k = \frac{v}{\omega}(\cos\theta_k - \cos\theta_{k+1})$ is   obtained.   Take
$\begin{cases} X^2_k - X^1_k = l\cos\theta_k \\ Y^2_k - Y^1_k = l\sin\theta_k \end{cases}$, the following equations are obtained:

$$\frac{1}{2}\left|\vec{a}_k \times \vec{d}_{k1}\right| = \frac{1}{2}\left|(X^2_k - X^1_k)(Y^1_{k+1} - Y^1_k) - (X^1_{k+1} - X^1_k)(Y^2_k - Y^1_k)\right|$$

$$= \frac{1}{2}\left|l\cos\theta_k \cdot \frac{v}{\omega}(\cos\theta_k - \cos\theta_{k+1}) - \frac{v}{\omega}(\sin\theta_{k+1} - \sin\theta_k) \cdot l\sin\theta_k\right|$$

$$= \frac{1}{2}\frac{vl}{\omega}(1 - \cos(\theta_{k+1} - \theta_k))$$

$$(5)$$

$$\frac{1}{2}\left|\vec{a}_{k+1} \times \vec{d}_{k2}\right| = \frac{1}{2}\left|(X^2_{k+1} - X^1_{k+1})(Y^2_{k+1} - Y^2_k) - (X^2_{k+1} - X^2_k)(Y^2_{k+1} - Y^1_{k+1})\right|$$

$$= \frac{1}{2}\left|\frac{vl}{\omega}(\cos(\theta_{k+1} - \theta_k) - 1) + l^2\sin(\theta_{k+1} - \theta_k)\right|$$

$$(6)$$

Equations (2), (5) and (6) show that A_k is related to v and ω. Therefore, the trajectory planning is somehow a kind of position and orientation scheduling by the configuration of a set of suggested v and ω that comply with the following conditions:

$$\begin{cases} \sum_{k=0}^{n} \frac{v_k}{\omega_k}(\sin\theta_{k+1} - \sin\theta_k) = X_d - X_0 \\ \sum_{k=0}^{n} \omega_k \cdot \Delta t = \theta_d - \theta_0 \\ \min \sum_{k=0}^{n} A_k \end{cases} \tag{7}$$

where X_d and X_0 are the destination X coordinate and current X coordinate, respectively, of a certain particle on the vehicle, saying the centroid; θ_d and θ_0 are the destination and current orientation angle, respectively. The explanation about Eq. (7) is as following:

(1) After n steps, the centroid of the road-header goes onto the center line of the aimed roadway;
(2) After n steps, the orientation angle of the road-header becomes identical with the trend of the aimed track;
(3) The sum of the extra area produced during the n steps is local minimum.

Each step with certain pair of v and ω produces the location state of the road-header $s_k(X^1_k, Y^1_k, \theta_k, X^2_k, Y^2_k)$, then the scheduled trajectory is obtained.

Giving only Eq. (7) cannot figure out a trajectory for practical, some operational requirements are need to be considered commonly. To name one or two:

① Whether if the head of the road-header being allowed to go over the center line in the process?

② Is the value of n constrained to be less than a certain number?

③ Is the Y coordinate of the final point on the planned trajectory required to be less or further than a certain value?

④ The value of v and ω are constant or not? Are they limited numerically or not?

Take the scenario shown in Fig. 2 for example, suppose the center line of the aimed roadway locates along with the line y = 2; and the road-header is originally at a wrong position and orientation denoted by: $s_0(-1.3, 0.25, \frac{\pi}{6}, 1.3, 1.75)$. Figure 3(a) and (b) present three feasible trajectories for the road-header to follow, with different considerations that are listed in Table 1. The values of some necessary parameters are also given in the table.

Table 1. Comparison of the planned Trajectory 1 and Trajectory 2.

The length of the road-header $l = 3$ m; driving speed $v < 0.8$ m/Δt			
Requirements	Trajectory #1	Trajectory #2	Trajectory #3
considerations ①	Yes	No	Yes
considerations ②	<=6	<=6	<=6
considerations ③	None	None	None
considerations ④	$\omega = 0.1$ rad/s $v = 0.75$ m/s	$v = 0.2$ m/s $\Delta\omega \leq 0.02$ rad/s	v and ω are both variable, $v \in [0.3, 0.8]$ m/s; $\omega \in [0.05, 0.2]$ rad/s; $\Delta\omega \leq 0.05$ rad/s

Trajectory #1 and #2 are obtained with different considerations from which the most obvious one is that the head of the road-header is allowed to go over the center line in the process for Trajectory #1 but not for #2. This consideration is prominent because the corresponding excavation space produced in case "yes" is easy to go beyond the value that in case "No". Also, this consideration results to different values of v and ω when they are set to be partially constant.

Trajectory #3 is the one that gives the least level about excavation space with the considerations list in Table 1.

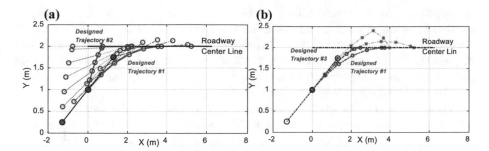

Fig. 3. (a) The planned trajectory #1 and #2; (b) The planned trajectory #1 and #3

In Fig. 3 the road-header is simplified as a line segment composed of three points: the centroid, the front and rear points. The black dotted line denotes the center line of the aimed roadway: $y = 2$; In Fig. 3(a) the red curve that strings the centroids of the segment at different poses is the planned trajectory #1, while the blue curve is the other trajectory #2. The blue points above the aimed roadway are the locus of the front point of the road-header when it moves following trajectory #1. It shows that the head of the vehicle is allowed to be over the center line of the aimed roadway, which is also explained in Table 1. In Fig. 3(b) the planned trajectory #1 and #3 are illustrated by the blue and red curves, respectively. The points marked by stars and above the aimed roadway are the locus of the front point of the road-header, it shows that maximum offset over the aimed roadway produced by trajectory #3 is bigger than that by trajectory #1.

All the presented trajectory plans can guide the vehicle back on to the aimed roadway with different considerations, but which one or two are recommended depends on further evaluation.

3 Trajectory Evaluation

A planned trajectory gives a schedule about position and orientation for the road-header, as well as a set of moving forward velocity v and turning angular speed ω. However, the road-header can hardly follow the trajectory exactly only by carrying out the referred v and ω step by step. An alternative practice is to track only the position and orientation in the schedule, but adjust the turning angular speed ω and moving velocity v accordingly.

The turning angular speed ω is produced by adjusting the rotating speeds of the left and right driving wheels of the tracks by which the road-header moves. If no slips along the tracks, the ω and v are expected to be:

$$v = \frac{1}{2}(r\omega_L + r\omega_R) \; ; \; \omega = \frac{r}{b}|\omega_R - \omega_L| \qquad (8)$$

where r is the radius of the driving wheels; ω_L and ω_R are the rotating speeds of the driving wheels; b is the distance between the left and right tracks.

In practice, for the benefit of the mechanical system and also to prevent slipping of the tracks to some extent, the variation of the rotational speed of the two driving wheels are expected to be smaller, and the differences in between is also expected to be as less as possible during driving. From this point of view, an evaluation method for the planned trajectories is proposed based on a simple path tracking control model.

The used tracking control model is as following [7, 8]:

Take θ as the orientation of the road-header, (x, y), (X, Y) are the coordinates in the vehicle coordinate system xoy and geodetic coordinate system XOY, respectively. Denote the current pose of the centroid of the road-header as $p_c = (X_c, Y_c, \theta_c)^T$, and the destined pose as $p_d = (X_d, Y_d, \theta_d)^T$, and then their difference in the vehicle coordinate system xoy is:

$$
p_e = \begin{bmatrix} x_e \\ y_e \\ \theta_e \end{bmatrix} = \begin{bmatrix} \cos\theta_c & \sin\theta_c & 0 \\ -\sin\theta_c & \cos\theta_c & 0 \\ 0 & 0 & 1 \end{bmatrix} \begin{bmatrix} X_d - X_c \\ Y_d - Y_c \\ \theta_d - \theta_c \end{bmatrix} \tag{9}
$$

Take the differential of p_e:

$$
\dot{p}_e = \begin{bmatrix} \dot{x}_e \\ \dot{y}_e \\ \dot{\theta}_e \end{bmatrix} = \begin{bmatrix} v_d\cos\theta_e - v_c + y_e\omega_c \\ -x_e\omega_c + v_d\sin\theta_e \\ \dot{\theta}_d - \dot{\theta}_c \end{bmatrix} = \begin{bmatrix} v_d\cos\theta_e - v_c + y_e\omega_c \\ -x_e\omega_c + v_d\sin\theta_e \\ \omega_d - \omega_c \end{bmatrix} \tag{10}
$$

To drive $\lim_{n\to\infty} p_e = 0$ within limited steps, i.e. to achieve path tracking, a control law updating v_c and ω_c is given:

$$
\begin{aligned}
v_c &= k_1\dot{x}_e + k_2(a_{max}|x_e|)^{1/2}\text{sgn}(x_e) \\
\omega_c &= k_3\dot{\theta}_e + k_4(\alpha_{max}|\theta_e|)^{1/2}\text{sgn}(\theta_e)
\end{aligned} \tag{11}
$$

k_1, k_2, k_3 and k_4 are random positive control coefficients, a_{max} and α_{max} are the maximum accelerations of the moving velocity v and the turning angular speed ω, respectively. $\text{sgn}(\cdot)$ is a symbolic function.

By carrying out the path tracking through Eqs. (9)–(11) using the planned trajectories presented in previous section. The referred v and ω are obtained and recorded. Then, the referred rotating speeds of the driving wheels on two sides, i.e. the values of ω_L and ω_R, are calculated by Eq. (8), and are shown in Fig. 4.

Figure 4 gives a view about the variance of the rotation speeds. Obviously, trajectory #1 gives the smallest variance during the path tracking and recall that trajectory #1 is not the one gives the least excavation space.

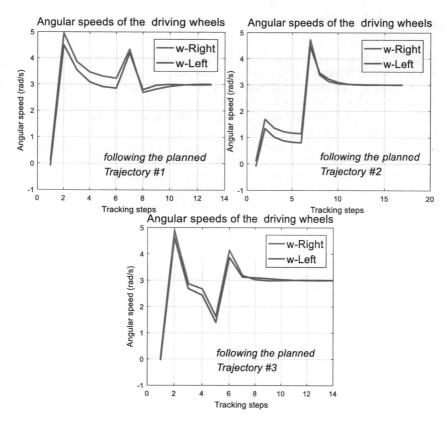

Fig. 4. The angular speeds of the two driving wheels using trajectory #1, #2 and #3 respectively

4   Conclusion

This paper presents a study on trajectory planning for the cantilevered road-header in path correction under the shaft. Excavation space is concerned during trajectory planning since the accompanying space shoring costs in both time and economy. Besides, considering the benefit of the mechanical system and also for cutting slipping of the tracks, the variation of the rotational speeds of the two driving wheels are expected to be smaller, and the differences in between is also expected to be as less as possible during path tracking. Three typical planned trajectories are discussed in this paper and are used based on a simple path tracking algorithm. The calculated results show that trajectory planning is necessary since different trajectories correspond to different scenarios about undesirable excavation space, slipping level and probably power consumption et al.

Acknowledgments. Special thanks to the National Natural Science Foundation of China, for the financial supports to this work through projects No. 61803374 and No. 51874308.

References

1. Yang, Y., Hu, Y.L., Yang, J., et al.: Vibration signal test and analysis of horizontal axial hard rock road-header. Measur. Control Technol. **34**(10), 58–60 (2015)
2. Zhang, M., et al.: Research on roadheader auto rectification in limited roadway space based on regional grid. Chin. J. Sci. Instrum. **39**(3), 62–70 (2018)
3. Grand, C., et al.: Stability and traction optimization of a reconfigurable wheel-legged robot. Int. J. Robot. Res. **23**, 1041–1058 (2004)
4. Chen, G., Jin, B., Chen, Y.: Accurate position and posture control of a redundant hexapod robot. Arab. J. Sci. Eng. **42**(5), 2031–2042 (2017)
5. Qi, R., Zhou, W., Liu, J., et al.: Obstacle avoidance trajectory planning for gaussian motion of robot based on probability theory. J. Mech. Eng. **49**(6), 89–95 (2016)
6. Fu, S.C., Li, Y., Zhang, M., et al.: Ultra-wideband pose detection system for boom-type road-header based on Caffery transform and Taylor series expansion. Measur. Sci. Technol. **29**, 015101 (2018)
7. Han, Q., Liu, S.: Slip control of deep sea tracked miner based on dynamic analysis. J. Central South Univ. (Sci. Technol.) **8**, 3166–3172 (2013)
8. Li, L., Zou, X.: Seafloor robots control on tracking automatically planning mining paths. Chin. J. Mech. Eng. **43**(1), 152–157 (2007)

Development of Intelligent Water Resources System Combined with Artificial Intelligence in Flood Forecasting

Yilin Wang, Jie Wen, Guangru Sun, and Weihua Zhang(✉)

College of Resources and Environment, Southwest University,
Chongqing 400715, China
swuwater@126.com

Abstract. In order to study the application of intelligent water resources system combined with artificial intelligence in flood forecasting, this paper mainly introduces the model of combining BP neural network with genetic algorithm and the expert system. BP neural network is a nonlinear dynamic system composed of simple information processing units. Its flood forecasting process is the process of obtaining the nonlinear function F_{BP} by using historical hydrological data. On this basis, genetic algorithm is added, the initial weight is further optimized, and the flood forecast is more accurate. By collecting and absorbing the local water conservancy and hydrological situation, the expert system matches the case facts with the global facts and rules, and abstracts the wisdom of the experts into the system as a knowledge base or database for flood forecasting. These two methods combine artificial intelligence with flood forecast, and create conditions for the further development of intelligent water resources system.

Keywords: Artificial intelligence · Intelligent water resources system · Flood forecast · Neural network · Expert system

1 Introduction

Along with the concept of the global information tide, intelligent earth, intelligent city and so on, water resources work is gradually approaching to information, automation and intelligence, and the concept of "intelligent water resources system" has gradually begun to enter people's field of view [1–3]. Intelligent water resources system is based on the idea of intelligent earth, which combines the development of computer technology, communication technology, remote sensing and telemetry, network, artificial intelligence and other technologies on the basis of traditional water resources system [4, 5]. According to the chronological order, the development of water resources construction has gone through the following three stages [6, 7]:

In the initial stage, the water resources sector began to be informatized. Its work mainly focuses on the informationization of the collection and organization of hydrological information.

© Springer Nature Switzerland AG 2020
F. Xhafa et al. (Eds.): IISA 2019, AISC 1084, pp. 277–283, 2020.
https://doi.org/10.1007/978-3-030-34387-3_34

The next stage was the rapid development stage of Intelligent water resources, which gradually established a flood control command system to improve the timeliness and accuracy of water transport.

At present, the construction of water resources is in the third stage. Many new development concepts have been proposed at this stage. Among them, the concept of intelligent water resources system is the theoretical result of this stage.

The following is a summary of the development of intelligence water resources system in the field of flood forecasting, focusing on the process of artificial intelligence.

2   Traditional Flow Calculus Methods for Flood Forecasting

2.1   Introduction of Traditional Flow Calculus

Flood forecasting is to predict the flood process which will occur in a certain control section by collecting, transmitting, storing, processing and a series of scientific calculation according to the rainstorm in the basin or the incoming water from the upper reaches of the river, through the collection, transmission, storage, processing and a series of scientific calculations.

The traditional river flow calculation is based on the Saint-Venant equation, which is the basic equation to describe the one-dimensional unsteady flow. When no one enters, the form is as follows:

$$\frac{\partial A}{\partial t} + \frac{\partial Q}{\partial L} = 0 \tag{1}$$

$$-\frac{\partial Z}{\partial L} = S_f + \frac{1}{g}\frac{\partial v}{\partial t} + \frac{v}{g}\frac{\partial v}{\partial L} \tag{2}$$

where A is the wetted cross-sectional area (m^2), Q is the flow rate of cross-section (m^3/s), L is the distance along a river course (m), Z is the level of water(m), v is the mean velocity (m^3/s), g is the acceleration of gravity (m/s^2), S_f is the friction-drag ratio.

The formula (1) and (2) is simplified to obtain the water balance equation and the trough storage equation respectively:

$$\text{Water balance equation}: \frac{I_1 + I_2}{2}\Delta t - \frac{Q_1 + Q_2}{2}\Delta t = \Delta W = W_2 - W_1 \tag{3}$$

$$\text{trough storage equation respectively}: W = f(I, Q) \tag{4}$$

The commonly used hydrographic flow calculation algorithm is the solution to the formula (3), (4). The most typical methods are the Masking method and the characteristic river length method, taking the characteristic river length method as an example below.

The basic principle of the Characteristic River Length Method is that on a particular section of a river of length L (as shown in Fig. 1), its lower bound is the flow measurement section, and at the middle point p of the river section is the basic water gauge section, whether in a steady or unstable flow state. The water level at the p-section (which is the average water level in the reach) and the cross-section discharge of the side-flow can be maintained as a function of a single value [8 and9].

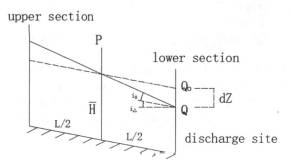

Fig. 1. Characteristic River diagram

2.2 Disadvantages of Traditional Flow Calculus

(1) Because the traditional flow calculation method involves complex theoretical knowledge, the technical personnel's professional request is higher.
(2) The flood forecast results are not accurate. Due to the uncontrollability of human and environmental factors, the traditional manpower calculation results have a large error [10].
(3) When some factors (such as climate, soil, vegetation, etc.) change, the traditional flow calculation method can not automatically change the corresponding parameters, but can only rely on manual adjustment.
(4) The related information of water resources is very scattered, which has brought influence to the normal development of water resources management.

3 Application of Artificial Intelligence in Flood Forecasting

3.1 Genetic Algorithm and Neural Network Model

Artificial neural network [11], genetic algorithm and other artificial intelligence technology has excellent self-organization, self-learning function and can map complex nonlinear functions very well, it can effectively improve the accuracy of hydrological model. With the deepening and development of artificial intelligence technology in this field, researchers take the need of flood forecasting as the goal and draw on the advantages of real-time forecasting technology. In order to establish the unique algorithm and model of flood forecasting, a broader application research is carried out.

3.1.1 Combination of BP Neural Network and Genetic Algorithm

At present, BP Neural Network, which combines BP Algorithm [12] with artificial neural network, is one of the most active methods in the application of artificial intelligence technology in flood forecasting. It is a nonlinear dynamic system composed of simple information processing units, which is suitable for describing the complex relationship between input and output of the system. It can be regarded as a highly nonlinear mapping function from input to output, the Hydrological Model network structure is [13]:

$$Q(t) = F_{BF}[P_r(t), \ldots, Pr(t - X\Delta t), Q(t - Y\Delta t)]$$

where t is the current moment, Q is the site traffic, P_r is the watershed rainfall, Δt is the time interval, X and Y are integers and greater than or equal to 1, F_{BF} is a nonlinear mapping function. In short, the process of forecasting flood by BP Neural Network is the process of obtaining the nonlinear function F_{BP} by using historical hydrological data [14].

On the basis of BP Neural Network model, genetic algorithm can be introduced to improve the accuracy of the prediction model [15]. Genetic algorithm is used to optimize the initial weight of artificial intelligence network, and then use BP algorithm and sample information to train and study the network weight until the optimal weight is obtained. Essentially, the entire network training is divided into two parts: use genetic algorithm to obtain a better set of initial weights, and use algorithms to complete network training. The specific process is as shown in Fig. 2:

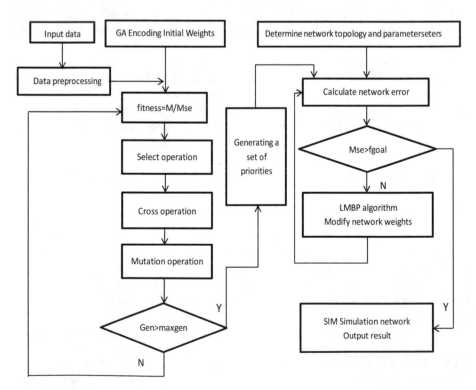

Fig. 2. BP Artificial Neural Network diagram (modified from Chunping Wang, 2005)

3.1.2   Advantages and Development of BP Neural Network and Genetic Algorithm

On the basis of classical hydrological model, artificial intelligence technology is introduced to realize the great improvement of the structure of the original model or the remarkable improvement of prediction accuracy. For example, the sensitive parameters are selected for the Xin'anjiang model [16], and the objective optimization of the parameters is realized; in the estimating Muskingum equation, flow is introduced as a state variable, and the equation can be transformed into a state equation; and use Kalman filter to correct the forecast results in real time [17].

Although BP Neural Network and genetic algorithm simplify the complex operation process of traditional flood calculation methods, it has higher requirements for the distribution range of actual data, and the prediction effect of the test data beyond its range is general, so the reliability of the input data has higher requirements: to ensure that the data entering the network is the measured value as far as possible, try to avoid obtaining input data by applying linear interpolation calculations of measured values [18].

3.2   Expert System

Expert system refers to the computer program which distributes the expert knowledge in a certain field to the computer program, so that the computer program can solve the difficult problems in this field. The establishment of traditional flood forecasting system is mainly aimed at the whole country, but for a specific area, its forecasting accuracy is reduced, resulting in unnecessary economic and even loss of life [18]. Therefore, the introduction of expert system is very necessary [19].

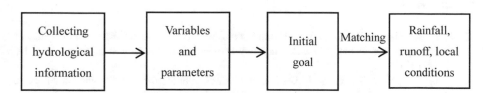

Fig. 3. LFFWS Model flowchart of target establishment

By collecting and absorbing the local water conservancy and hydrological situation, the expert system matches the case facts with the global facts and rules, and abstracts the wisdom of the experts into the system as a knowledge base or database for flood forecasting [20] (Fig. 3).

The LFFWS model [21] is a model based on artificial intelligence and expert system, and serves the expert system. It establishes the information that users need to obtain from flood warning systems, that is, data related to rainfall, including the amount of rainfall and the duration of rainfall; data related to runoff, including the seepage rate and the amount of water stored in the river; and information related to local hydrology and water conditions.

In the model established by LFFWS, by minimizing the boundary of the region of interest, the use of the numerical model is reduced, and the consumption of computing resources is reduced, which makes flood forecasting more cost-effective.

In the coding process of an expert system, experts can work independently or in cooperation with knowledge engineers to develop expert systems in the field of hydrology. At the same time, the professional requirements for technical personnel have been reduced accordingly. Expert system can operate according to changing targets entered, with greater flexibility, can meet the changing situation of local flood management information.

4   Conclusion

The introduction of artificial intelligence technology is the main direction of intelligent water resources development in the future. At present, the artificial intelligence construction of water resources has just started, and the related technical means and development system still need to be improved. Therefore, in the future, artificial intelligence algorithm and expert system will be further improved, which will push the water resources information management business to a higher level.

Acknowledgments. This work was financially supported by Chongqing key research and development projects of social and livelihood (No. cstc2018jscx-mszdX0052). The supports are gratefully acknowledged.

References

1. Ma, F.X., Zhao, H.Y., Xu, Y.X.: Exploration and practice of Internet Water Conservancy-A case study of intellectual Water Conservancy Project in Xuzhou City. Inf. Technol. Inf. **12**, 25–28 (2018)
2. Jiang, Y.Z., Ye, Y.T., Wang, H.: Discussion on intelligent regulation technology architecture for river basin based on interact of things. Water Conservancy Inf. **4**, 1–10 (2010)
3. Jiang, Y.Z., Ye, Y.T., Wang, H.: Smart basin and its prospects for application. Syst. Eng. Theory Pract. **31**(6), 1174–1181 (2011)
4. Wang, Z.J., Wang, G.Q., Wang, J.H., et al.: Developing the internet of water to prompt water utilization efficiency. Water Conservancy Hydropower Eng. **44**(1), 1–6 (2013)
5. Rui, X.L., Wu, Y.F.: Intelligent system of water conservancy based on internet of things. Comput. Syst. Appl. **21**(6), 161–163 (2012)
6. Feng, J., Xu, X., Tang, Z.X., et al.: Research on key technology of water big data and resource utilization. Water Conservancy Inf. **8**, 6–9 (2013)
7. Huang, L.M., Zhang, K., et al.: Management of quality risk of water conservancy projects from perspective of big data. J. Econ. Water Conservancy **35**(6), 66–70 (2017)
8. Bao, W.M.: Hydrological Forecast, 5th edn., pp. 46–50. China Water Conservancy and Hydropower Publishing House, Beijing (2007)
9. Rui, X.F.: Research methods and theoretical innovation of hydrology. Adv. Sci. Technol. Water Conservancy Hydropower **25**(2), 46–50 (2003)
10. Lu, F., Jiang, Y.Z., Wang, H., et al.: Application of Multi-agent genetic algorithm to Parameter estimation of Musingen Model. J. Hydraulic Eng. **38**(3), 289–294 (2007)

11. Liu, P., Gu, O., Xiong, L.H., et al.: Deriving reservoir refill operating rules by using the proposed DPNS model. Water Conservancy Manag. **20**, 337–357 (2006)
12. Liu, F.H., Xie, N.M.: Civilian aircraft cost estimation model and algorithm based on small sample and poor information. J. Syst. Simul. **3**, 687–691 (2014)
13. Dong, C.J., Liu, Z.Y.: Multi-layer neural network Involving chaos neurons and its application to traffic-flow prediction. J. Syst. Simul. **19**, 101–104 (2007)
14. Behling, R., Fischer, A., Herrich, M., et al.: A Levenberg-marquardt method with approximate projections. Comput. Optim. Appl. **S0926–6003**(59), 5–26 (2014)
15. Sao, K.Y., Li, X., Qiu, Y.F., et al.: Improved GA based on imitating diploidic reproduction. J. Syst. Simul. **4**, 816–820 (2012)
16. Wang, J., Shi, P., Jiang, P., et al.: Application of BP neural network algorithm in traditional hydrological model for flood forecasting. Water **9**, 48 (2017)
17. Nazeer, S., Ali, L., Malik, K.N.: Water Quality Assessment of river Soan and source apportionment of pollution sources through receptor modeling. Arch. Environ. Contam. Toxicol. **71**(1), 97–112 (2016)
18. Fang, S., Xu, L., Pei, H., et al.: An integrated approach to snowmelt flood forecasting in water resource management. IEEE Trans. Industr. Inf. **10**, 548–558 (2014)
19. Zhang, X., Moynihan, G.P., et al.: Evaluation of the benefits of using a backward chaining decision support expert system for local flood forecasting and warning. Wiley Online Library **35**(4), 12–16 (2018)
20. Comas, J., Llorens, E., Marit, E., et al.: Knowledge acquisition in the STREAMES project: the key process in the environmental decision support system development. AI Commun. **16**, 253–265 (2003)
21. Ghalkhani, H., Golian, S., Saghafian, B., et al.: Application of surrogate artificial intelligent models for real-time flood routing. Water Environ. J. **27**(4), 535–548 (2012)

Message Transmission-Driven Infectious Disease Propagation Model

Jiefan Zhu, Wenjie Tang$^{(\boxtimes)}$, and Yiping Yao

School of Systems Engineering, National University of Defense Technology,
Changsha 410073, People's Republic of China
tangwenjie@nudt.edu.cn

Abstract. During the spread of infectious diseases such as H5N1 flu, unin-fected individuals may change their behaviors to avoid getting infected after they receive information or warnings about the epidemic. In the study of the spread of infectious diseases, changes of people's behaviors due to the spread of epidemiological messages among the population cannot be ignored. The exist-ing infectious disease model lacks considerations for the spread of epidemic information in the population, which in reality may cause chaos, and thus change the behaviors of the group. The epidemic information is not reflected in these models through the interaction between individuals. For this problem, this article proposes an infectious disease propagation model based and driven by the transfers of information. Objects in the model will send two kinds of information messages during the whole simulation process: one is called triggering message, objects which receive this kind of messages will be triggered to send out messages to other objects, meanwhile the location type object will summarize the overall situation and the individual type object will calculate its own disease situation; The other is the epidemic message. Having been added the parameter 'fear' to measure the degree of panic for the disease, Individual objects who receive this kind of messages will be triggered to increase the value of their parameter 'fear', and once the value of fear reaches a preset number, the indi-vidual will be in panic and choose to stay at home to reduce the contact with public to prevent themselves from getting infected. Test results show that compared with other infectious disease transmission models, the simulation results for the number of infected people are closer to reality, which proves that this model can be used to conduct in-depth research related to the dissemination of epidemic information. Besides, this model can also provide reference for the formulation of public policies to prevent and control infectious diseases.

Keywords: Infectious disease · Human behavior · Information transmission

1 Introduction

Human behaviors will change during the spread of infectious disease due to the changes of individual's own healthy state, or the updated information about the disease known from public media, Internet, or other individuals. This may lead those who are still susceptible to actively avoid going to public places where there may be more infected people to prevent themselves from being infected.

© Springer Nature Switzerland AG 2020
F. Xhafa et al. (Eds.): IISA 2019, AISC 1084, pp. 284–290, 2020.
https://doi.org/10.1007/978-3-030-34387-3_35

Therefore, in the study of the spread of infectious diseases in the population, the dynamic behavior of the population should be in consideration, and when modeling people's behavior, shared information about the disease from other individuals, or in other methods such as public warning, social media is also a key factor that should not be ignored to make the model closer to reality.

The speed of the spread of epidemic information may increase when the number of infected people becomes larger, and the transmission of these information will play a certain role in regulating and decelerating the spread of the disease. For example, a message about the disease published on the internet can be received by a large number of individuals in a very short period of time, thereby those individuals' tendency to change their normal behaviors to stay at home will increase. Nowadays, some active online celebrities have a lot of fans. If they post messages about the disease which is right now spreading in the city, the impact they make to some extent may be as much as the public information warning by the government. Epidemiological information has different effects on different individuals, and individuals may miss a certain piece of information for various reasons. However, as the amount of information about the ongoing spreading disease people receive accumulates among the crowds, it may cause serious panic and thus affect the behaviors of the population. This might affect the actual operation of society. In addition to that, many individuals during the spread of an infectious disease may choose to stay at home because of the information they have known about the spreading disease instead of going to other places where they may get infected. Therefore, in order to make the infectious disease model closer to reality, it is necessary to consider the dynamic changes of people's behavior based on information transmission.

2   Related Works

The susceptible-infective-recovered model(SIR model) by Kermack and McKendrick ① is an early model and also the most classic one that models the infectious disease. But the nodes in the SIR model are fixed, so that the model can not take the dynamic changes of people's behavior into account, and thus the results of SIR models cannot reflect the reality enough.

Epidemic model developed on the basis of the SIR model began to take people's changing behavior into consideration. For example, ZHANG LIN ② pointed out that media reports about the disease can reduce the contact rate between people,which has been confirmed in the 2013 H7N9 ③ infectious disease. As a new type of infectious disease, H7N9 first appeared in Shanghai in 2013, it can spread rapidly in the crowd in a very short period of time. Later, people knew from the TV report that the disease was transmitted just because the contact with infected people. Therefore, people tried to minimize their activities outside to reduce the contact rate with others, and this has indeed reduced the spread of the disease to some extent. However, their further research only pays attention to the public media as the only method in which the disease information spread, totally ignoring the information transmission between individual and individual, or the internet, etc. This leads their conclusions more accurate than the original SIR model, but still not accurate enough to reflect the real dynamic changes of people's behaviors under multiple ways of disease information transmission.

EpiSimdemics ④ is an effective and well-designed algorithm for studying the spread of infectious diseases in large-scale social networks. However, in the algorithmic structure of EpiSimdemics, individuals interact with the location through a message broker only, there is no direct interaction available between individual and individual. This makes it difficult for the algorithm to take the spread of disease information between individual and individual into account. And there for in Keith R. Bisset, Madhav Marathe, Xizhou Feng's research ⑤ based on the EpiSimdemics algorithm to study the interaction of people's behaviors during a infectious disease still fails to consider the disease information spreading directly between individual and individual, making the result still not accurate enough.

In short, the existing infectious disease transmission models and algorithms lack considerations for the spread of the epidemic information among the population which can cause panic, and thus change people's behaviors and the result of the disease.

3 Message Transmission-Driven Infectious Disease Propagation Model

Existing infectious disease models lacked direct interaction between individuals and individuals, which led to the inability of these models to simulate the spread of various information among individuals, which was not consistent with the reality. The algorithm of this information transmission driven infectious disease model derived from the SIR model, people in this model are divided into four categories: susceptible, infected, feared, and recovered (SIFR). These correspond to people who are uninfected (and can be infected), infected (having symptoms of the disease and can infect others), people who fear the disease (choose to stay at home to avoid getting sick because of receiving too much information about the spread and danger of the disease), and those who have fully recovered (individuals who have recovered from the disease are thought to be immune). Among them, regardless of whether the individual is in the susceptible, infected, or recovered stage, every individual is possible to enter the feared stage, depending on the type and amount of the messages with disease information they have received from outside. Objects in the model will send out two types of messages during the simulation process. One is called triggering message. Objects receiving this kind of messages will be triggered to send out messages to other objects and therefore drive the simulation moving on. The location type object which receive this message will summarize the overall situation in it and the individual type object will calculate its own disease situation.

The other is the epidemic message. Having been added the parameter 'fear' to measure the degree of panic for the disease, Individual objects who receive this kind of messages will be triggered to increase the amount of their parameter fear, and once the value of fear reaches a specific number, the individual will be in panic and choose to stay at home to reduce the contact with public to prevent themselves from getting infected.

There are also two types of objects in the model: the individual object and location object. During the simulation, individual objects will retrieve the ID of next location they are going to according to the simulation clock, and sends triggering messages containing its basic information to the location. After summarizing all the messages and the information in a round from individuals, the location objects will send the integrated data back to the individuals. And individuals use the data such as the number of infected people in the location to calculate whether they are infected in this round simulation, and run the stage of disease development function, update their own status, and retrieve the next location to go to, and continue sending massages. Through the constant of triggering information transfer between individual objects and location objects, the simulation clock is driven forward, and the algorithm of the model is also driven by the information.

In the algorithm setting of this model, individuals receive various types of epidemic messages from methods such as other individuals, public media, Internet, etc. Those epidemic messages will change the value a parameter called fear which is added to each individual. And different types of epidemic information will increase the value of the individual's fear parameter to varying degrees. The larger an individual's fear value is, the larger probability individuals choose to stay at home to avoid contact with other who may infect him.

Cases where information transmits between relatives and friends exist in reality, and it is an aspect that other disease models does not considered enough. In this model, each individual will preset their own friends and relatives in the initialization stage, and when they are in panic, they will send out a type of epidemic messages to their relatives and friends, increasing their fear value. Besides, when someone has fully recovered from the disease, they will also send another kind of epidemic messages to reduce their friends' and relatives' fear value.

Media warning is an important part of the prevention and control of infectious disease, it has the characteristics of widely dissemination and credibility of data sources. In the algorithm of this model, the number of individuals in serious symptom is the data that is directly monitored. Once the number is bigger than the preset threshold, media warning will be triggered by sending a kind of epidemic messages to almost all the individuals and therefore increase the public's fear value. And when the number goes down to another preset threshold, the media warning will be cancelled, by sending out another kind of epidemic message to the public to reduce their fear values.

Besides, when too many people who are in the same workplace have been infected and had symptoms, the other staff in the place may get frightened by the disease. To simulate this situation, the algorithm sets: if the number of infected employees in one place in a time reaches a specific number, then all the remaining individuals will receive an epidemic message from the location and get their fear value increased.

4   Test and Evaluation

In order to verify and prove the accuracy of this model, a total of 100,000 individuals and 1000 locations were designed to do the test. And in total 30 people are set to be in latent period (in which they can infect others) and they are divided into two group. They rest of the people are all susceptible. The age ratio of individuals is established according to the Statistical Bureau of the People's Republic of China, <the China Statistical Yearbook>, and select the data for 2016, the ratio of Adolescents, adults, and the elderly (susceptibility varies with age), is as follows: (Fig. 1)

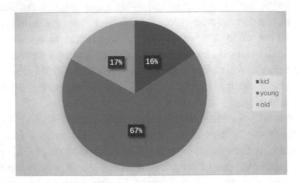

Fig. 1. Age ratio

For each round of simulation, record the number of each types of people. The test is performed on a Lenovo yoga 720 laptop with an inter core i5 processor, 8G memory, and a Windows 10 operating system. The disease set in this test is H5N1 avian influenza, the mechanism of the disease was partially referred from ⑤. And the result of the number of infected people is as follows (Fig. 2) (until the number reaches the peak):

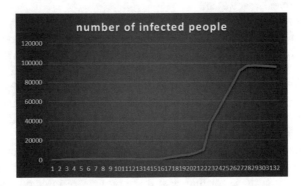

Fig. 2. Number of infected people when reaching the first peak, result of the model test.

And ⑥ shows a real data from 07/12/2009 to 28/01/2010 and the data simulated by SIR models about a SIR type infectious disease: (Fig. 3).

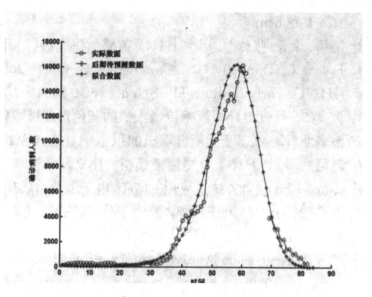

Fig. 3. Data from ⑥, The line with hollow dots is the real data.

Compare results shown above to the result from ⑤: (Fig. 4).

Bisset, Feng, Marathe, and Yardi

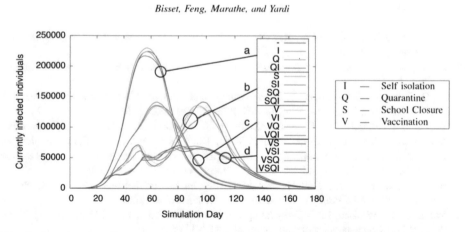

Fig. 4. Results from ⑤, where Currently infected individuals by day for all combinations of interventions. The interventions can be divided into four groups based on the shape of the epidemic curves.

By comparison, it can be seen that even if the algorithm only considers a small number of test objects (only 100,000 individual) and just several spreading ways of disease messages (individual to individual, public warning, small scale Internet spreading), it suits the real data pretty well, no less (even better) than the simulation results in ⑤. This not only proves that the spread of disease messages among the population should not be ignored, but also the correctness of this model.

5 Conclusion and Future Works

In this model, by analyzing the existing infectious disease model, we propose a message transmission-driven infectious disease transmission model. The innovation of the model lies in: the individual can receive disease messages from other individuals or locations, and therefore the disease messages can spread among the population in various methods and finally change the behaviors of the population. It can be used to support the research on the impact of single-channel epidemic message dissemination on disease transmission. Test shows that the algorithm is more feasible and effective than SIR model due to the consideration of the messages spread between individual and individuals. However, because of the limitation of the performance of the test plat form, the test in this model only considers the situation of 100,000 individuals and 1000 locations, And the model lacks considerations of many other situations in people's interaction with others, public, etc. So, in future work, these shortcomings will be fixed and completed, and the aim of the development of this model is to make a model that can use real data to simulate and support the research of infectious disease transmission.

References

1. Kermack, W.O., McKendrick, A.G.: Contributions to the mathematical theory of epidemics. Proc. Roy. Soc. A **115**(772), 700–721 (1927)
2. Zhang, L., Li, C., Guo, W.: An SIRS infectious disease model based on media reports. J. Math. (2018)
3. Li, Q., Zhou, L., Zhou, M., et al.: Epidemiology of human infections with avian influenza A (H7N9) virus in China. New Engl. J. Med. **370**(6), 520–532 (2014)
4. Barrett, C.L., Bisset, K.R., Eubank, S.G., Feng, X., Marathe, M.V.: Network Dynamics and Simulation Science Laboratory, Virginia Tech, Blacksburg, VA 24061 EpiSimdemics: an Efficient Algorithm for Simulating the Spread of Infectious Disease over Large Realistic Social Network. {cbarrett, kbisset, seubank, fengx, mmarathe}@vbi.vt.edu
5. Bisset, K.R., Feng, X., Marathe, M.: Modeling Interaction between Individuals, Social Networks and Public Policy to Support Public Health Epidemiology
6. Yang, Y., Li, W., Zhu, L.: Differential epidemic model and analysis of influenza a (H1N1). Practice and Understanding of Mathematics, 11 (2011)

Robot Path Planning Method Based on Improved Double Deep Q-Network

Wei Xiao[1(✉)] and Qian Gao[2]

[1] College of Mechatronics and Control Engineering, Shenzhen University,
Shenzhen, China
569384164@qq.com
[2] Graduate School, Hohai University, Nanjing, China

Abstract. Considering the slow convergence speed of common deep Q-learning algorithm used in robot exploration strategy, an improved double deep Q-network method based on dueling network is proposed. The use of dueling network structure contributes to the more accurate function estimation, which makes the robot can identify the optimal behavior more quickly, and the network model is more stable. The combination of Boltzmann strategy and ε-greedy strategy improves the speed of robot exploring environment and avoids the constant in-situ movement of robot in local environment. And the use of resampling optimization mechanism solves the problem that samples can't be recovered and the over-estimation of training caused by lack of diversity. The experimental results show that the robot trained with proposed method can adapt to complex environment faster, the convergence speed of the network is improved, and the success rate of reaching target point is increased, which means proposed method can help the robot to obtain the optimal path better, compared with the double deep Q-network.

Keywords: Reinforcement Learning · Deep Q-network · Dueling network · Resampling mechanism · *Boltzmann* strategy · ε-greedy strategy

1 Introduction

Robot path planning technology is one of the core contents of intelligent mobile robot research [1]. Mobile robot path planning refers to that the robot perceives the environment according to the information obtained by the sensors or cameras and plans a route to reach the target independently. With the continuous development of science and technology, mobile robot faces more and more complex environment, traditional path planning can't meet the needs of mobile robots, such as ant colony algorithm [2] and artificial potential field method [3]. In view of this situation, people put forward Deep Reinforcement Learning (DRL) [4], which integrates the Deep Learning (DL) and the Reinforcement Learning (RL) [5]. Where DL is mainly responsible for extracting features from unknown input environment state by using the perception function of neural network, and realizing the fitting of environment state to state action value function. while the main functions of RL is the decision-making according to the

© Springer Nature Switzerland AG 2020
F. Xhafa et al. (Eds.): IISA 2019, AISC 1084, pp. 291–302, 2020.
https://doi.org/10.1007/978-3-030-34387-3_36

output of the deep neural network and certain exploration strategies, thus realizing the mapping from state to action, which can meet the mobile needs of the robot preferably.

Manih et al. [6] combined convolutional neural network with Q-learning algorithm of traditional RL, and proposed Deep Q-Network (DQN), which was well verified in Atari 2600 game. *Van Hasselt* et al. [7] proposed the Deep Double Q-Network (DDQN), which successfully solved the problem of over-optimistic estimation function. *Schaul* et al. [8] used priority-based experience playback mechanism to replace equal probability sampling method in training DQN to improve the utilization rate of valuable samples, but it needs more storage space. *Xin* et al. [9] proposed a DQN-based path planning method for mobile robots for the first time. *Tai* et al. [10] proposed the DQN algorithm for simulation of mobile robot path planning, but DQN algorithm has the disadvantage of overestimating the action value. Therefore, this paper proposes the improved dueling double deep Q-network (IDDDQN) as a new path planning method, where the addition of dueling network makes the estimation of value function more precise and improves the performance of model, the combination of Boltzmann and ε-greedy strategy balances the exploration and utilization of robot action selection strategy, and the data collected by the robot will be stored in the cache memory unit using the improved resampling optimization mechanism, which makes only small batch data can train robot well. The simulation results show the effectiveness of the proposed method in avoiding local optimum, overestimation and reduction of spent time.

2   Method Description

DRL is an end-to-end perception and control system, which is widely used and has strong generalization. The learning process can be described as follows: (1) DL method perceives the high-dimensional information obtained by the interaction between robot and environment at each moment to get the state feature representation from abstract to concrete. (2) The value function of various motions is evaluated based on DL, and the current state is mapped to the corresponding action through a certain strategy. (3) The Robot responds to this action and gets the next visual information. Through the continuous cycle of the above process, the optimal strategy can finally be obtained.

2.1   DQN Algorithm

Manih [12] combined the deep convolution neural network and Q-learning algorithm, and proposed the DQN algorithm, which can improve the instability of the algorithm when non-linear function approximator is used for approximation of value function. Firstly, the DQN algorithm uses two convolutional neural networks, the estimation network and the target network to represent the state value function and the target value function respectively. Where the output of the estimated network is expressed by $Q(s, a|\theta_t)$, which is used to evaluate the value function corresponding to current state action, and the output of target network is expressed by $Q(s, a|\theta_t^-)$, the optimal target of the value function (i.e. objective valve function) is expressed by Y_t approximately as shown in formula (1). The weight θ of estimated network are updated in real time, the weights are assigned to the target network after N times iteration. Initially, $\theta = \theta^-$. DQN updates the weight of the network through TD error, as shown in formula (2).

$$Y_t = r + \gamma max_{a_{t+1}} Q\left(s_{t+1}, a_{t+1}|\theta_t^-\right) \tag{1}$$

$$L_t(\theta_t) = E\left[(Y_t - Q(s, a|\theta_t))^2\right] \tag{2}$$

Secondly, DQN uses experience replay to process the transfer samples obtained during training online. At each time step t, the transfer sample $e_t - (s_t, a_t, r_{t+1})$ of the interaction between robot and environment is stored in the cache memory unit D. In the process of network training, a fixed amount of transfer samples (mini-batch) are randomly sampled from D every time, and the Stochastic Gradient Descent (SGD) algorithm is used to update network parameters, as shown in formula (3).

$$\nabla_{\theta_t} L(\theta_t) - E_{s,a,r,s_{t+1}}[(Y_t - Q(s, a|\theta_t))\nabla_{\theta_t} Q(s, a|\theta_t)] \tag{3}$$

2.2 Deep Q-Network

Hasselt proposed DDQN algorithm based on double Q-learning algorithm [13], which solved the problem of overestimating Q value caused by selecting the next action in DQN algorithm. DDQN algorithm uses two different parameters μ and μ^-, where the parameter μ is used to select the action of Q value and the parameter μ^- is used to evaluate the Q value of optimal action. The two parameters separate the action selection form strategy evaluation, which reduces the risk of overestimating Q value. Where objective value function is shown in formula (4).

$$Y_t^{DDQN} = r + \gamma Q\left(argmaxQ(s_{t+1}, a_{t+1}|\mu_t)\mu_t^-\right) \tag{4}$$

3 Proposed Algorithm

Path planning of mobile robot is the task of interaction between robot and environment, this paper proposes a path planning method for mobile robot based on IDDDQN. Firstly, the robot uses laser ranging sensors to acquire data that could train the network as input, then the system selects the optimal action according to the exploration strategy to reach the next visual observation, and stores the data in the cache memory unit D according to the resampling optimization mechanism, and iterates until the training is completed. The training model of mobile robot path planning is shown in Fig. 1.

3.1 Improved Deep Q Network

Wang [14] designed a DQN algorithm with dueling network structure, where the value function V is updated with the update of Q value, which achieves a better approximation of value function V. In view of this, IDDDQN method combines the dueling

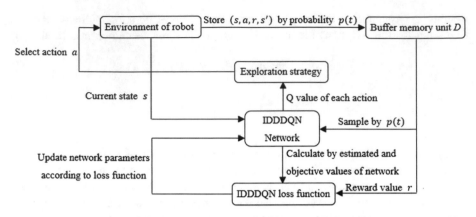

Fig. 1. Training model of mobile robot path planning

network structure with the network structure of DDQN algorithm. The network model is shown in Fig. 2.

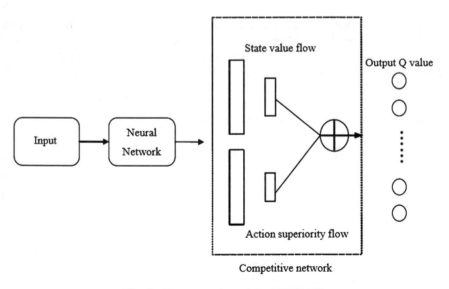

Fig. 2. The network model of IDDDQN

Robot obtains data in observable environment and takes these data as input of improved network. Firstly, the data are analyzed by basic neural network, the maximum Q value is output in real network, and the Q value of optimal action is output in objective network. Then, the dueling network is used to balance the effect of action on Q value in real network and objective network respectively, and finally the Q value of each action is output. The calculation of Q value in dueling networks is shown in formula (5).

$$Q(s,a|\theta) = V(s,a) + \left(A(s,a|\theta) - avg_{a(A(s,a|\theta))}\right) \tag{5}$$

Where $V(s,a)$ represents the state value function, which is the long-term judgment of the current state, $A(s,a|\theta)$ represents the action superiority flow, which is an evaluation for different behaviors under the current state. In this paper, the action superiority flow is set as the value of the single action superiority function minus the average value of all the action superiority functions under the current state.

3.2 Exploration Strategy

The common exploration strategies in DRL are ε greedy, Boltzmann and Gauss exploration strategy. IDDDQN adopts an exploration strategy which combinates ε greedy and Boltzmann exploration strategy. Where ε-greedy is the global selection strategy and Boltzmann is the local selection strategy to solve the problem that ε-greedy strategy will make mobile robots easily fall into local optimum. ε-greedy takes both exploration and utilization into account, The parameter ε represents the exploration factor. When the robot is close to the obstacle, a value p will be generated randomly, If p is less than parameter ε, the mobile robot will select an action randomly. If p is larger than parameter ε, the improved network will learn, and judge the Q value of all the actions (forward, left or right) according to Boltzmann strategy. Finally, the action with maximum probability $p(s, a_i)$ (as shown in formula (6)) will be selected as the optimal action, so that the robot can transfer the current state s_t to the next state s_{t+1}. The specific flow chart is shown in Table 1.

$$p(s, a_t) = \frac{e^{Q(s,a_i)}}{\sum_{a_i \in A} e^{Q(s,a_i)}} \tag{6}$$

Where $a_i (i = 1, 2, 3)$ represents all the optional actions under the current state s_t.

Table 1. Exploration strategy

Steps	Exploration strategy
1	Randomly initialize p, $p \in (0, 1)$
2	If $p < \varepsilon$
3	$a = rand(A)$
4	Otherwise
5	$a = Boltzmann(A)$
6	End

3.3 Resampling Optimization Mechanism

With the continuous training of mobile robot, the number of transfer samples will increase, and the utilization rate of each transfer sample will also change. DDQN algorithm uses the mechanism of cache memory unit to store the transfer samples, online stores and uses the historical samples obtained by the interaction between mobile robot and the environment, which eliminates the correlation between the samples [15]. However, DDQN algorithm extracts small batches of samples from the cache memory unit with equal probability for training every time, this sampling method can't distinguish the importance of different samples. More importantly, the storage capacity of the cache memory unit is limited, this sampling method will cause some samples that haven't been fully utilized to be abandoned. In order to solve this problem, this paper proposes an improved resampling optimization mechanism based on *TD* error. When the robot arrives at the next state from the current state, it will calculate the corresponding reward value r at the same time, so that a complete data tuple (s_t, a_t, r_t, s_{t+1}) can be obtained and stored it in the cache memory unit D. Then, the sample weight $W(t)$ can be calculated by *TD* error δ, and the corresponding weight can be stored. When the transfer samples are extracted from D, the weight $W(t)$ is updated first, and then the mini-batch samples are selected by weight $W(t)$ for training. The specific calculation is shown from formula (7) to formula (9).

$$\delta_t = Y_t^{DDQN} - Q(s_{t-1}, a_{t-1}) \tag{7}$$

$$\omega_t = |\delta_t| + \sigma \tag{8}$$

$$W(t) = \lambda \times \frac{\omega_t^k}{\sum_n \omega_n^k} + \beta \tag{9}$$

Where Y_t^{DDQN} represents the objective value function, the specific calculation is shown in formula (4). $W(t)$ is the weight of transfer sample t, exponent k determines the weight of priority. When k is equal to 0, the samples have equal probability. otherwise, the weight of memory unit is updated proportionally according to $W(t)$. The parameter σ is used to prevent the probability of cache sampling memory unit approaching 0 when *TD* error approximates 0, and the coefficient λ and offset β can ensure that the updating probability of samples is different. This mechanism not only ensures that the cache memory units with higher priority are updated with higher probability, but also ensures that the samples with lower priority can be updated with a certain probability.

4   Experimental Results and Analysis

4.1   Simple Environment

In order to verify the effectiveness of IDDDQN method, we first compare the DDQN and IDDDQN in environment of Cartpole-v0 in OpenAI Gym [16]. The experimental results are shown in Figs. 3 and 4 respectively.

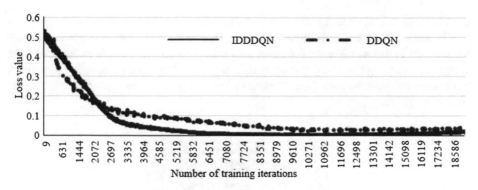

Fig. 3. Comparison of loss function values between IDDDQN and DDQN

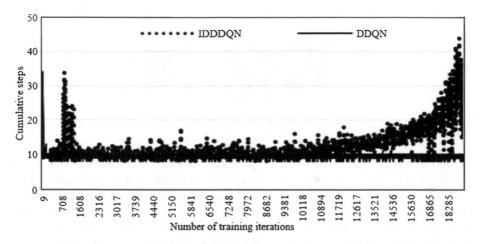

Fig. 4. Comparison of cumulative steps between IDDDQN and DDQN

As can be seen from Fig. 3, the convergence speed of IDDDQN method is obviously faster than that of DDQN algorithm, and the value of loss function is 73.3% lower than that of DDQN algorithm. For cumulative steps, the more cumulative steps, the more target points can be reached by the robot. As can be seen from Fig. 4, the cumulative steps of IDDDQN method are significantly increased than those of DDQN algorithm. Which shows that IDDDQN method has better adaptability and learning ability than DDQN algorithm, and the performance of IDDDQN method is better.

4.2 Complex Environment

In complex environment, robot is trained in the experimental environment of CPU servers, OpenAI gym, Keras, Python 3.5, OpenCV3.4, gazebo 8* and gym-gazebo [17], where mobile robot is equipped with Turtlebot with laser ranging sensors. Turtlebot integrates the popular robotic component (Kinect of Microsoft) into the integrated development platform of ROS applications and builds simulation environment using

Gazebo (Fig. 5). Turtlebot can judge the distance from the obstacles and the target points according to the real-time position relative to its coordinate system during the movement. Turtlebot takes the collected data relative to its coordinate as input and the continuous steering command as output. It does not need to construct a map, and realizes navigation through the laser ranging sensor. The Turtlebot implementation platform is shown in Fig. 6.

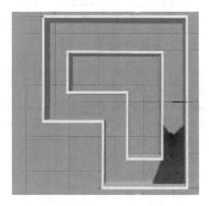

Fig. 5. The simulation environment using Turtlebot robot and laser sensors

In this experiment, the velocity of Turtlebot is set to 3 m/s forward, 0.05 m/s in other directions, and the angular velocity is set to $w = 0.03$ rad/s. The parameters of IDDDQN method are set as shown in Table 2, where the initial value of exploration factor ε is 1, $\varepsilon \in (0.05, 1)$, and the value of ε is linear decreasing with the increase of iteration times of robots. The network applies RMS Random Gradient Decline (RMSProp) to update the parameters, where the momentum coefficient is set to 0.95. Each time, the network is updated with samples size 64 extracted by probability from cache memory unit D through resampling optimization mechanism, the reward value is set as shown in formula (10).

$$r(s_t, a_t) = \begin{cases} 5, & \text{Reach the target point} \\ -200, & \text{Bump into an obstacle} \\ \tau(d_{t-1} - d_t) \end{cases} \tag{10}$$

If the mobile robot detects and arrives at the target position through distance threshold, the reward value is set to 5; if the robot finds collision with obstacles through minimum distance, the reward value is set to -200, and the robot will stop training under both above conditions. Otherwise, the difference between the reward value as the distance from the target and the previous time step will be multiplied by a hyper parameter τ, which enables the robot to be closer to the target position (parameter d_t represents the distance from the current state to the target point).

At the beginning of training, Turtlebot store the transfer samples in cache memory unit D by exploration strategy. When the number of storage in the cache memory unit D reaches 100, the network will be trained by the samples in the cache memory unit D. During the training of each episode, the parameters of the target network will be updated every 100 steps.

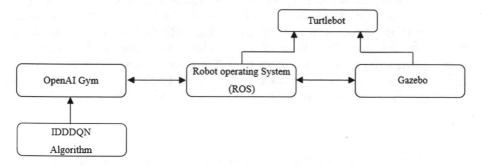

Fig. 6. Implementation Platform of Robot

Table 2. Parameter setting

Parameter	Value
Learning rate α	0.0001
Exploration factor ε	$0.99 \times \varepsilon$
Discount factor γ	0.99
Hyper parameter λ	0.5
Hyper parameter β	0.5
Random number σ	0.2
Parameter K	10000

After training 8000 times, IDDDQN method and DDQN algorithm are compared in terms of average cumulative reward value. The results are shown in Table 3.

Table 3. Comparison of Average Reward Value between DDQN and IDDDQN

Episodes	DDQN	IDDDQN	Episodes	DDQN	IDDDQN
1–500	−133.59	−120.16	4001–4500	−64.60	74.20
501–1000	−108.51	−118.54	4501–5000	−100.92	115.12
1001–1500	−84.45	−65.85	5001–5500	−67.13	46.24
1501–2000	−33.34	−21.36	5501–6000	6.50	34.33
2001–2500	76.15	574.12	6001–6500	−87.02	28.48
2501–3000	29.55	495.44	6501–7000	−66.06	36.17
3001–3500	−56.77	256.45	7001–7500	−9.04	81.23
3501–4000	−75.64	105.05	7501–8000	232.35	66.12

From Table 3, it can be seen that the average cumulative reward value of IDDDQN algorithm reaches the highest value in 2000–2500 times of iteration, which is 574.12, and has been positive in the subsequent iterations. Although the average cumulative reward value of DDQN algorithm appears positive when the times of iteration is small, there are still many negative values with the increase of iteration times, which means that mobile robots are not well trained, and the maximum value was 232.35 at 7500–8000 times. The highest average cumulative reward value of IDDDQN method is 2.5 times that of DDQN algorithm, which means that IDDDQN method can use fewer iterations to complete the training of mobile robots and obtain higher cumulative reward value. Using this method, mobile robot can better adapt to complex environments. Considering the times of iteration needed to well train the mobile robot by the two algorithms and the time needed to train the mobile robot under the same times of iteration, the IDDDQN method can obviously improve the learning speed of the mobile robot.

In the process of training, the more positive reward value, the more correct Turtlebot's judgment is. And the higher the cumulative reward value means the more successful Turtlebot is in avoiding obstacles and reaching more target points, as well as the closer the path is to the optimum. In terms of Turtlebot's success rate p_s as shown in formula (11), IDDDQN method is much higher than DDQN algorithm, which is shown in Figs. 7 and 8, IDDDQN method successfully reaches the target point 2668 times in 8000 times of iteration, while DDQN algorithm only 621 times. Which means the success rate of IDDDQN method to get the optimal path is more than three times higher than DDQN algorithm.

$$P_s = \frac{I_s}{I} \tag{11}$$

Where I_s represents the times of successful arrivals at the target point during training, and I represents the total times of iteration during training.

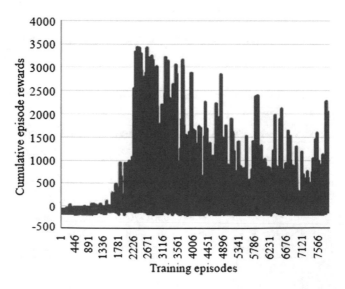

Fig. 7. Cumulative reward value of IDDDQN

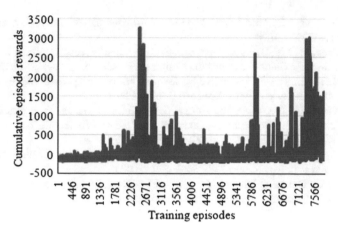

Fig. 8. Cumulative reward value of DDQN

5 Conclusion

The mobile robot estimates the value function of its three actions through the improved DDQN network structure, updates the network parameters, and obtains the corresponding Q value through the training network. On top of that, the mobile robot adopts the exploratory strategy of combining Boltzmann with ε-greedy, and selects an optimal action to reach the next observation. The improved resampling optimization mechanism gives priority to each transfer sample, makes full use of the transfer sample, which improves the convergence speed of the network. In summary, IDDDQN method can get larger cumulative reward value in fewer times of iteration, so that mobile robot can get the optimal path in a shorter time.

References

1. Koubaa, A., Bennaceur, H., Chaari, I., et al.: Introduction to Mobile Robot Path Planning. Robot Path Planning and Cooperation, pp. 3–12. Springer, Cham (2018)
2. Reshamwala, A., Vinchurkar, D.P.: Robot path planning using an ant colony optimization approach: a survey. Int. J. Adv. Res. Artif. Intell. 2(3), 65–71 (2013)
3. Mandava, R.K., Bondada, S., Vundavilli, P.R.: An optimized path planning for the mobile robot using potential field method and PSO algorithm. Soft Computing for Problem Solving, pp. 139–150. Springer, Singapore (2019)
4. Huang, S.H., Zambelli, M., Kay, J., et al.: Learning gentle object manipulation with curiosity-driven deep reinforcement learning. arXiv preprint arXiv:1903.08542 (2019)
5. Sutton, R.S., Barto, A.G.: Reinforcement Learning: An Introduction, pp. 3–11. MIT Press (2018)
6. Mnih, V., Kavukcuoglu, K., Silver, D., et al.: Playing atari with deep reinforcement learning. arXiv preprint arXiv:1312.5602 (2013)
7. Van Hasselt, H., Guez, A., Silver, D.: Deep reinforcement learning with double q-learning. In: AAAI, vol. 2, p. 5 (2016)

8. Schaul, T., Quan, J., Antonoglou, I., et al.: Prioritized experience replay. arXiv preprint arXiv:1511.05952 (2015)
9. Xin, J., Zhao, H., Liu, D., et al.: Application of deep reinforcement learning in mobile robot path planning. In: Chinese Automation Congress (CAC), pp. 7112–7116. IEEE (2017)
10. Tai, L., Liu, M.: Towards cognitive exploration through deep reinforcement learning for mobile robots. arXiv preprint arXiv:1610.01733 (2016)
11. Tai, L., Paolo, G., Liu, M.: Virtual-to-real deep reinforcement learning: continuous control of mobile robots for mapless navigation. In: 2017 IEEE/RSJ International Conference on Intelligent Robots and Systems (IROS), pp. 31–36. IEEE (2017)
12. Mnih, V., Kavukcuoglu, K., Silver, D., et al.: Human-level control through deep reinforcement learning. Nature **518**(7540), 529 (2015)
13. Hasselt, H.V.: Double Q-learning. In: Advances in Neural Information Processing Systems, pp. 2613–2621 (2010)
14. Wang, Z., Schaul, T., Hessel, M., et al.: Dueling network architectures for deep reinforcement learning. arXiv preprint arXiv:1511.06581 (2015)
15. Xiliang, C., Lei, C., Chenxi, L., et al.: Control Dec.-Making **33**(4), 600–606 (2018)
16. Brockman, G., Cheung, V., Pettersson, L., et al.: Openaigym. arXiv preprint arXiv:1606.01540 (2016)
17. Zamora, I., Lopez, N.G., Vilches, V.M., et al.: Extending the OpenAI Gym for robotics: a toolkit for reinforcement learning using ROS and Gazebo. arXiv preprint arXiv:1608.05742 (2016)

Characteristics and Evolution of Citation Distance Based on LDA Method

Benji Li[1], Yan Wang[2], Xiaomeng Li[1], Qinghua Chen[1(✉)],
Jianzhang Bao[1], and Tao Zheng[1]

[1] School of Systems Science, Beijing Normal University, Beijing 100875,
People's Republic of China
qinghuachen@bnu.edu.cn
[2] Department of Mathematics, University of California, Los Angeles 90095,
USA

Abstract. The scientific research behavior of scholars is the core issue of scientific research. The research ideas and methods of complex networks provide a new perspective for the study of science. The scientific citation network and the scientist cooperation network are widely used to study the citation behavior of scholars and the dissemination of scientific ideas, and so far, some results have been obtained. However, due to the lack of information on the content of the article, the research based solely on the network topology has limitations and deficiencies. Combining the textual content analysis through LDA, this paper studies the distribution characteristics of content correlation between articles with citation relations and its evolution with time. It found that the distribution of citation distance has normal characteristics, but the reference distance is visible to be short. Authors have citation preferences for documents at a distance.

Keywords: Scientific reference · Citation distance · Scientist's behavior · LDA

1 Introduction

"Science of science" is a subject devoted to the study of the entities and their relations during the development of science. As an essential subject in the research field, the behavior of scholars has been the focus of attention [1]. However, there has been a lack of systematic quantitative analysis of this research for a long time. Early discussions on the characteristics of scientific development were primarily involved in the fields of philosophy and naturology. In the later period, some quantitative tools and tools introduced to this field and that scientometrics [2] had been proposed; however, the results are not in-depth due to the conventional statistical analysis.

In the near twenty years, the rapid development of complex networks provides new perspectives and methods for research on the science of science [3]. From the perspective of a complex network, each system can be interpreted by the relationship between different entities. Entities can be abstracted into nodes, and the relationships between nodes are recognized as edges. Based on scholars' cooperation network [4] and science citation network [5], many studies are carried. For example, Radicchi et al.

© Springer Nature Switzerland AG 2020
F. Xhafa et al. (Eds.): IISA 2019, AISC 1084, pp. 303–311, 2020.
https://doi.org/10.1007/978-3-030-34387-3_37

measured the influence of a scholar based on the cited diffusion process on the network [6]. Yang et al. discussed on important institutions' collaboration preferences and the effect on publication advantages and citation advantages [7]. Zhou established a bipartite network. This bipartite network contains both academic collaboration and citation information between articles. By discussion on this diffusion process, scholars and articles can be evaluated simultaneously [8]. An and Ding give an new insights into the landscape of the realm of causal inference and may be seen as a case study for analyzing citation networks in a multidisciplinary field [9]. Gualdi evaluated the quality of the paper through random walk processes on science citation networks [10]. Other studies focused on the allocation of credit between co-authors [11] and the selection of representation articles [12].

These studies require an extensive database to build networks, often focusing on static analysis, and are somewhat weak in revealing the dynamics of the science process. Also, this network structure lacks the real relationships between entities, such as the similarity of the content of the two articles, which is restricted in practical applications. The network can simplify a large number of literatures into homogeneous nodes without information, resulting in the complex network method to the application is still lacking in recommended [13] and forecast [14].

Fortunately, Natural Language Processing (NLP) technology has made significant breakthroughs at this stage. With the continuous accumulation of big data, relevant sub-fields of natural language processing, including text categorization [15], automatic abstract technology [16] have made significant progress. This technology allows us to conduct a more in-depth analysis of a large number of academic papers from the perspective of the content of the literature [17].

In this paper, we introduce academic space that is a new perspective to solve those problems. In that space, we discuss the transfer or development of scholars' research fields and the selection of reference materials. This space is created through latent Dirichlet allocation (LDA) [18, 19] which is an extensively used natural language processing technology. This article is organized as follows: in the first part, we give a brief introduction of the data and method. Then we give a precise detail about the data information and how we apply LDA to this dataset. In the next part, we analyze and give a conclusion on the characteristics of the scholars' research field and the selection of citations based on the distance between articles. Finally, we conclude.

2   Materials and Methods

2.1   Data Source

The data is provided by the American Physical Society (APS), includes many physics journals, such as PHYSICAL REVIEW LETTERS, PHYSICAL REVIEW X, PHYSICAL REVIEW (A-E). We used 555915 articles from the year 1919 to 2016. The data example is shown in Table 1.

Table 1. An example of the data records

Field Name	Example
title	Femtosecond planar electron beam source for micron-scale …
doi	https://doi.org/10.1103/physrevstab.4.121301
author	T. C. Marshall; Changbiao Wang; J. L. Hirshfield
affiliation	Department of Applied Physics. Columbia University. …
received_time	2001/10/02
citing_doi	https://doi.org/10.1103/physreve.62.1266; https://doi.org/10.1103/physreve.56.4647;…
pacscode	41.75.Jv;41.75.Lx;41.75.Ht;96.50.Pw
abstract	A new accelerator, LACARA (laser-driven cyclotron…)

2.2   Method

The abstract is the quintessence of the author's research paper so it can approximately represent each article's content. Before we apply the topic vector representation for each article with the LDA method, we have to clean it first.

Convert Case. Excluding some names and place names that have been collected, we change the letters in all words to lowercase. For example, "Cat" and "cat" indicate the same object; if we treat them as different characters, it is not proper and may cause errors. However, "China" and "china" is different.

Remove Stop Words and Punctuations. Delete stop words such as "the", "an", "of", "a", "and" and punctuations because they have little meaning on any topic.

Normalize Text. We changed the plural form of each word into the singular form. We also turned all words' tense to the present. After that step, the revised abstract usually contains only nouns, verb phrases, and other words which are essential to the articles' topic or content.

Get the Topic Vectors. We inputted the abstract texts to the latent Dirichlet allocation (LDA) model [18] to acquire the topic vector for each article. LDA is a generative model. In this model, each document owns multiple topics, and each word of a document supports a certain topic. All the meaningful words in each document in the corpus are used, and the topic model is trained to infer the thematic structure hidden behind the observations. The idea is that

$$P(\text{word}|\text{document}) = \sum_{topic} P(\text{word}|\text{topic}) \times P(\text{topic}|\text{document}) \tag{1}$$

After the training, we got the topic vector for each article that is a high-dimensional representation. In our experiment, we set the topic number to 50. Through the LDA model, we got a topic vector θ_i with 50 dimensions for the i-th article, with each dimension indicates the probability under k-th topic. The sum of the vector θ_i is equal to 1 or $\sum_k \theta_{i,k} = 1$ (Fig. 1).

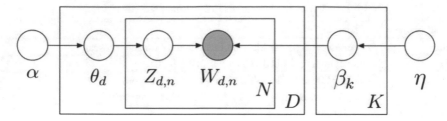

Fig. 1. Graphic model of LDA

Compute the Citation Distance. We then measured the citation distance between a pair of citing-cited papers through Euclidean distance between their vectors. The distance between articles i and j is

$$dis_{i,j} = \left[\sum_k \left(\theta_{i,k} - \theta_{j,k} \right)^2 \right]^{0.5}. \tag{2}$$

Here $\theta_{i,k}$ is the k-th member in the vector θ_i. Obviously, $dis_{i,j} = dis_{j,i}$. The distance between the two articles indicates that there is a big difference in the distribution of the topics, indicating that there is a significant difference between the contents of the two articles. Conversely, if the distance is not too much, the two articles are very consistent in the choice of subject, and the content is likely to be similar.

3   Analysis and Results

3.1   Research Framework

After the LDA process, each article gets a vector representation which can span as an academic space. So that each article can regard as a point in that space, and each author's academic development is a trajectory which corresponds to one of his published articles continually. Figure 2(a) shows the career track of one author whose number of papers is about thirty. Figure 2(b) shows the positions of these articles in the academic space and the topologies of one article between its citations and articles that reference itself. Green circle presents the selected paper. Red squares present the citations of this article. Blue diamonds present the articles that cited this article.

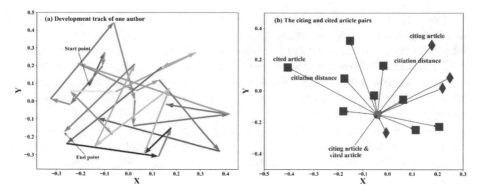

Fig. 2. The academic space frame. (a) Track maps of one author; (b) topologies of one article between its citations and articles that reference itself.

3.2 Trend of Average Citation Distance

Citation means the inheritance and development of scientific knowledge. We can discuss the changes in citation over time, as shown in Fig. 3. It is obvious that the average citation distance from 1990 to 2015 is decreasing overall. The average citation distance in 1990 is above 0.370, but it was near to 0.335 in 2015. The average annual reduction is 0.391%, and the rate of decline is further accelerating. We think there are two main reasons. One is that more and more research work has been carried out over time. The vigorous development of scientific publishing publishes more article than before. There will be more articles in the same academic space unit and more and more similar work. As far as the APS data we use, there were 8,748 articles published in 1990, and the amount of paper published raised to 17,172 in 2015. The average annual growth rate is 2.85%.

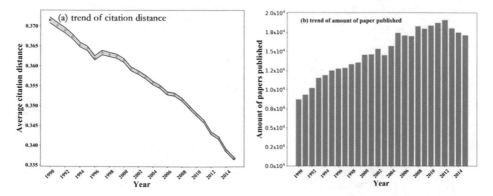

Fig. 3. Trend of citation distance and amount of paper published from 1990 to 2015. (a) the trend of citation distance. The shaded area is the 95% confidence interval; (b) the growth of the amount of paper published.

Also, due to the development of information technology, especially the birth of search engines, scholars can access literature much easier. As an instance, Google Scholar is an accessible web search engine that indexes the full text or metadata of academic literature through a series of publishing formats and subjects. Google Scholar is released in beta in November 2004 and has developed a lot of features over time. In 2006, the function of citation importing is implemented to support bibliography managers (such as RefWorks, RefMan, EndNote, and BibTeX). In 2007, Acharya announced that Google Scholar launched a program to digitize and host journal articles under a deal with their publishers, an effort separate from Google Books, whose scans of older journals do not include the metadata needed to identifying specific articles in specific issues.

3.3   Distribution of the Distances Between Citing and Cited Papers

We separate articles that are from different ears and try to discuss the distances between them and their citations or references. Figure 4 shows the distance distribution for each 5000 randomly chosen articles.

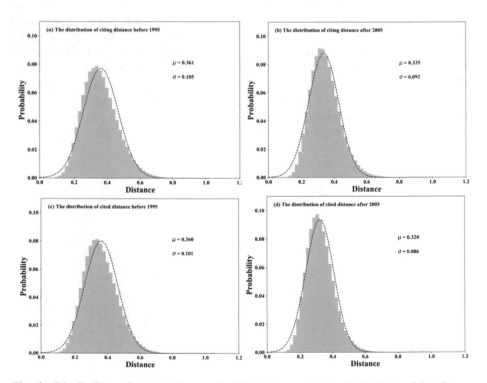

Fig. 4. Distributions of articles' distance in different cases. (a) the distributions of the distance between article published before 1995 and its citations. (b) the distributions of the distance between article published after 2005 and its citations. (c) shows the distributions of the distance between a paper published before 1995 and articles which cited it. (d) shows the distance distributions of between a paper published after 2005 and articles which cited it.

We found that these distributions are approximately follow the normal distribution with u = 0.361 and σ = 0.105 for case (a); u = 0.335 and σ = 0.092 for case (b); u = 0.360 and σ = 0.101 for case (c); u = 0.320 and σ = 0.086 for case (d) respectively. In addition, we calculated the skewness and kurtosis of each profile and found that for case (a) skewness = 0.546 and kurtosis = 0.459; skewness = 0.653 and kurtosis = 0.868 for case (b) skewness = 0.537 and kurtosis = 0.500 for case (c); skewness = 0.658 and kurtosis = 1.089 for case (d). These statistics are little skewed (the skewness is not 0), and more concentrated in the interval with smaller relative distance.

Compared with the normal distribution, all the distributions of citing and cited distances have a certain degree of left-bias and bias toward the smaller distance, which implies that multiple independent random factors do not completely decide the selection of references. From the perspective of time development, both the reference distance and the cited distance have a significant decrease in the mean and standard deviation, indicating that the scholars cited articles are more similar in content and more consistent. We also found that the average citing distance is always slightly larger than the distance of cited before 1995 or after 2005, indicating that the author does not rigidly adhere to the consistency of the article when organizing an article. Besides that, they should satisfy the logic requires, background, the methods used, and so on.

4 Conclusions and Discussions

Understanding the fundamental laws of scientific development and applying them is an important mission of interdisciplinary science, the essence of which is the modeling of complex social systems. In recent years, due to the rapid development of complex network research and natural language processing technology, research on the science of science has been greatly improved.

In this article, we embedded the article's content into a high dimensional space through the LDA model. Different from the citation network and cooperation network which mainly focus on network structure analysis, this paper puts forward a new method based on the content of the articles. This article creatively implements the content analysis based on an article's abstract to get in-depth information behind the article. Thus, the results based on content are more reliable.

We found some interesting results. For instance, the average citation distance from 1990 to 2015 is decreasing overall. Besides, we found that all the distributions of citing and cited have a certain degree of left-bias and bias toward the smaller distance, we can conclude that multiple completely independent random factors do not decide the scholar's citation behavior. Furthermore, both the reference distance and the cited distance showed a significant decrease in the mean and standard deviation with time, indicating that the scholars cited articles are more similar in content and more consistent. Plus, we also found that no matter in which era, the average citing distance is always slightly larger than the distance of cited. We can conclude that the author is not completely confined to the consistency of the article but also need to consider the logic requires when they select a citation.

Our work is a creative attempt, and this practical framework combining network and semantic analysis is a good starting point. According to this idea, we can study the significant differences in behavioral patterns between successful scholars and general scholars through clustering or classifying the scholars' development tracks.

Acknowledgments. We appreciate comments and helpful suggestions from Prof. Zengru Di, Prof. Chensheng Wu, Ms. Weiwei Gu. This work was supported by Chinese National Natural Science Foundation (71701018, 61673070 and 71671017).

References

1. Jia, T., Wang, D., Szymanski, B.K.: Quantifying patterns of research-interest evolution. Nature Human Behaviour **1**(4), 0078 (2017)
2. Leydesdorff, L.: The Challenge of Scientometrics: the Development, Measurement, and Self-Organization of Scientific Communications. Universal-Publishers (2001)
3. Zeng, A., Shen, Z., Zhou, J., Wu, J., Fan, Y., Wang, Y., Stanley, H.E.: The science of science: from the perspective of complex systems. Phys. Rep. **714**, 1–73 (2017)
4. Newman, M.E.J.: The structure of scientific collaboration networks. Proc. Natl. Acad. Sci. **98**(2), 404–409 (2001)
5. Shibata, N., Kajikawa, Y., Takeda, Y., Matsushima, K.: Detecting emerging research fronts based on topological measures in citation networks of scientific publications. Technovation **28**(11), 758–775 (2008)
6. Radicchi, F., Fortunato, S., Markines, B., Vespignani, A.: Diffusion of scientific credits and the ranking of scientists. Phys. Rev. E **80**(5), 056103 (2009)
7. Li, Y., Li, H., Liu, N., Liu, X.: Important institutions of interinstitutional scientific collaboration networks in materials science. Scientometrics **117**(1), 85–103 (2018)
8. Zhou, Y.B., Lü, L., Li, M.: Quantifying the influence of scientists and their publications: distinguishing between prestige and popularity. New J. Phys. **14**(3), 033033 (2012)
9. An, W., Ding, Y.: The landscape of causal inference: perspective from citation network analysis. Am. Stat. **72**(3), 265–277 (2018)
10. Gualdi, S., Medo, M., Zhang, Y.C.: Influence, originality and similarity in directed acyclic graphs. Europhys. Lett. **96**(1), 18004 (2011)
11. Shen, H.W., Barabási, A.L.: Collective credit allocation in science. Proc. Natl. Acad. Sci. **111**(34), 12325–12330 (2014)
12. Niu, Q., Zhou, J., Zeng, A., Fan, Y., Di, Z.R.: Which publication is your representative work? J. Inf. **10**(3), 842–853 (2016)
13. Son, J., Kim, S.B.: Academic paper recommender system using multilevel simultaneous citation networks. Decis. Support Syst. **105**, 24–33 (2018)
14. Acuna, D.E., Allesina, S., Kording, K.P.: Future impact: predicting scientific success. Nature **489**(7415), 201 (2012)
15. Sebastiani, F.: Machine learning in automated text categorization. ACM Comput. Surv. (CSUR) **34**(1), 1–47 (2002)
16. Saggion, H., Poibeau, T.: Automatic text summarization: past, present and future. In: Multi-source, Multilingual Information Extraction and Summarization, pp. 3–21. Springer Berlin Heidelberg (2013)
17. Kim, S.N., Medelyan, O., Kan, M.Y., Baldwin, T.: Automatic keyphrase extraction from scientific articles. Lang. Resour. Eval. **47**(3), 723–742 (2013)

18. Blei, D.M., Ng, A., Jordan, M.: Latent Dirichlet allocation. J. Mach. Learn. Res. **3**, 993–1022 (2013)
19. Hantzsche, A., Kara, A., Young, G., Bates, J.M., Granger, C.W., Geweke, J., Amisano, G., Rossi, B., Elliott, G., Timmermann, A.: Latent Dirichlet allocation. Natl. Inst. Econ. Rev. **246**(1), F4–F35 (2018)

The Design Analysis of Campus Guide Sign System Based on the Comparison of Sino-Korea Colleges

Hongbo Shan[1], Yixuan Xue[1,2], Jinsheng Lu[1], Zhiyong Hou[1],
and Shuxia Li[3(✉)]

[1] College of Mechanical Engineering, Donghua University,
Shanghai 201620, China
[2] Department of Industrial Design, Hytera Communications Corporation,
Nanjing 210012, China
[3] College of Business, East China University of Science and Technology,
Shanghai 200237, China
sxli@ecust.edu.cn

Abstract. The design of the guide sign system is essential to ensure a good interaction mechanism between the user, the space and the sign guidance system, which helps the user complete the spatial positioning and path finding behavior through systematic information transfer. Based on the concept of campus guide sign system, the factors affecting system performance are summarized in the paper. The good strategies of Korean university sign guidance systems are analyzed from function, vision and culture. It is proposed that the college sign guidance system should be designed from the three dimensions of information interaction, visual expression and cultural inheritance considering the relationship between people and campus space, which can become a card to convey the charm of the campus.

Keywords: Guide sign system · University space · Information interaction · Vision expression · Cultural inheritance

1 Introduction

With the development and optimization of higher education in South Korea, the spatial structure of universities is increasingly complex because of the its expansion, combination or reorganization [1]. However, the university sign guidance system has not been developed accordingly. All kinds of buildings bring convenience and comfort to the study and life of teachers and students. At the same time, they also bring some trouble of way finding. Therefore, it is necessary to research the influencing factors, design strategies and corresponding design methods of sign and guide system, which can guide people to complete the way finding quickly and accurately.

F. Xhafa et al. (Eds.): IISA 2019, AISC 1084, pp. 312–318, 2020.
https://doi.org/10.1007/978-3-030-34387-3_38

2   College Sign Guidance System

The university guide sign system is a subsystem of the urban signage guidance system, which exists in public space. However, it has certain particularity due to different service objects and emphases. The conceptual framework is shown in Fig. 1. College sign guidance system is the card of the university charm, leaving the first impression of the campus. Taking the principle of serving the students and teachers essentially, it combines the campus space and guide sign system, and integrates the information of different spaces, such as teaching, life and entertainment space by using the visual forms of symbols, graphics, words, colors and materials. In this way, it can meet the spatial information needs of teachers and students, to standardize the order of college and guarantee the benign operation of the university.

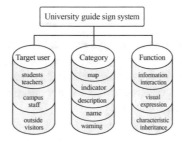

Fig. 1.   University guide sign system conceptual frame diagram

3   Influencing Factors Analysis of Korean University Guide Sign System

3.1   Household Information Quality

The quality of information is based on the number, location and information content, to estimate people's satisfaction with the guide sign system and the information inter-action degree between people and campus space [2]. The problem of information quality mainly includes three aspects: at first, information set density is not balanced in universities, there is too much information in some areas and too little information in other areas. For example, there is an excessive sign at the roundabout of the road, once it is far from the intersection, the number of signs is sharply reduced, and people quickly enter the information vacuum state. Secondly, the information of one guide device is too large, which can make people lose visual focus and reduce the information identification. At last, poor information content means to convey unwanted informa-tion. For example, there are too many road signs and building guides at the university gate, but lack of campus map, which can't satisfy people's need for overall college information.

3.2   Integrated Planning

The overall planning of the campus sign guidance system is mainly reflected in the proportion design and location planning of the guidance equipment, which should accord with ergonomics. On the one hand, if the installation height of guiding equipment is too low, there is a lack of comprehensive consideration of road, pedestrians and driving factors, which not only ignores the needs of driving personnel, but also leads to reduced readability of information [3]. On the other hand, many single guide boards are set at the intersection of campus roads, lacking information integration and reasonable location planning, which adds a narrow sense of space and affects the overall visual effect.

3.3   Cultural Characteristics

Signage with only functional characteristics, such as simple regional guide boards and location boards, can no longer meet the needs of the development of the campus. Nowadays, the university sign guidance system tends to homogeneity because of same color, similar material or shape. This phenomenon not only obliterates the personality and characteristics of universities, but also leads to the loss of university characteristics, which reduces the recognition degree of the society and gradually goes away from the goal of building an innovative university.

4   Design Strategy Analysis of Korean University Guide Sign System

4.1   Reasonable Information Classification Strategy

Taking Inha university as an example, representative buildings are the main building, the 60th anniversary museum, the entrepreneurship center, and the static stone academic library. According to the structure and characteristics of the buildings, its university sign guidance system integrates modern design elements, which is shown in Fig. 2.

Fig. 2. Sign and guide in Inha university

The sign guidance system of Inha university is divided into three levels. The first information is set at the campus gate, the map adopts the perspective method to display the overall layout of the campus, so that people can quickly acquire the whole and local spatial structure information of the college, then master the main route. The secondary information is set at the intersection of the road where the traffic is large, helping people to determine their location and direction of destination, mainly including road diversion signs, road guidance signs, etc. The third information is set in front of campus scenic spots or buildings, and explains the role of explanation, such as habitat lake.

As shown in Fig. 3, because the campus walking route is mainly based on the "dormitory – teaching building – restaurant – playground", guide devices are set on two main road nodes according to the characteristics of population distribution, which can coordinate the behavior of the crowd. The reasonable information grading strategy of Inha university makes the information of the guide consistent, which guarantees the continuity of people's wayfinding.

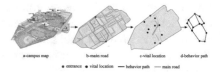

a-campus map   b-main road   c-vital location   d-behavior path

● entrance   ● vital location   — behavior path   — main road

Fig. 3. Guide position distribution on main roads

4.2   Gestures Difficulty Level Test

From the perspective of "man - machine - environment", the distinct visual expression can not only attract people's attention to better realize the information interaction, but also integrate the guide with campus buildings and surrounding environment to enhance the overall visual effect [4].

The south Korean university guide sign system breaks the restriction of traditional expression form, setting position and material, the university is endowed with a strong visual spatial context through distinct visual expression. As shown in Fig. 4, guide devices are hung on the surface of the building with highlighted text, arrows information in Korea University, at the meantime of catering to the structural stability and three-dimensionality, it has improved the recognition degree of guide sign system. Ewha Womans University is known as the "university built on the mountain", there are many rugged roads and forks. In order to enhance the visual impact of road orientation, the vivid colors are used to print the guide information on grey roads in a large area. Its novel and interesting design has become a decoration of the college space, creating a harmonious and warm campus atmosphere.

4.3   Careful Characteristic Inheritance Strategy

Sungkyunkwan university, with its motto of "benevolence, righteousness, propriety and wisdom", continues the Confucian school philosophy since the "Korean times".

Fig. 4. Distinct visual expression   **Fig. 5.** Signs of Sungkyunkwan university

As shown in Fig. 5, taking the classical beauty as the design idea, its guide sign system also integrates the characteristics of universities and Confucian traditional architecture. Through the introverted lines and the steady form, it creates the implicit and quietly elegant university atmosphere, and adds the intrinsic humanistic charm of the university. In addition, with prevalence of "Korean Wave" and "Han Wave", Sungkyunkwan University, as a guide for Confucius commemorative activities, has received a sharply increased number of international students. Considering the language cognition of the people, international standard graphics and symbols are adopted in Chinese, English and Korean to avoid ambiguity, which highlights the unique humanistic care of the university and shapes a humanistic brand of the school.

5   Design Dimensions Analysis of Korean University Guide Sign System

5.1   Design University Guide Sign System from the Perspective of Information Interaction

As shown in Fig. 6, university guide sign system, as the organizer of campus space and the link between different spaces, is designed to realize information interaction between people and university space under the stimulation of people's information needs. Good information interaction could bring enjoyable wayfinding experience for people, which indicates that under the instruction of the signage and orientation system, people could quickly obtain the information for further use and processing to realize the seamless connection between people and campus space.

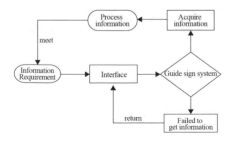

Fig. 6. Information interaction

Good information interaction of guide sign system mostly depends on the information quality, referring to the information grading strategy of Inha university, the university guide sign system can be divided into three levels from high to low. The first level information provides the overall spatial layout to form the images of the university. The second information offers the information of specific zones and important nodes. The third information is the basic level that indicating precise and clear location.

5.2 Design University Guide Sign System from the Perspective of Visual Expression

The visual expression of the guide sign system is very important for shaping the image of colleges. Excellent visual communication is an effective way to enlarge the popularity of universities and enhance campus recognition. The visual elements of the signage and orientation system include information and information carriers.

On the one hand, it is the three-dimensional reflection of the information itself, which forms the three-dimensional effect by stretching, bending or bending the characters and symbols, which can increase the diversity of sign guidance. On the other hand, it is the three-dimensional embodiment of the information carrier, which transfers the plane information to multiple directions through the three-dimensional multidimensionality, and innovates the plane information carrier into the positive and negative information transmission, or the three-sided information transmission of the three-prism vertebra, the multi-dimensional information transmission of the four-prism cone and the cylinder, etc., so as to increase the visibility of the sign guidance.

Rich visual expression of sign guidance can give people a strong visual stimulation, and strengthen people's cognition of the campus, which is conducive to the overall image of the university building.

5.3 Design University Guide Sign System from the Perspective of Characteristic Inheritance

In addition to paying attention to good information interaction and visual expression of the carrier, the establishment of a complete university guide sign system also need to pay attention to inherit the characteristics of colleges.

Guide sign system is a visual expression of the characteristics of the university, the characteristics of universities are specific, which increases people's emotional interaction and brings a strong sense of belonging and identity. As shown in Fig. 7, the elements of university features could be extracted to realize the characteristic inheritance of guide sign system, mainly centered around the shape design of the signs: the first step is to analyze the material, spirit and culture of universities, and extract characteristic elements with strong representativeness and high public recognition to ensure the effective inheritance of characteristics. The second step is to redesign the characteristic elements to enhance the inheritance of characteristics. The third step is to incorporate the designed elements into the shape, color, material and other external forms of the guide sign system to present the most intuitive expression.

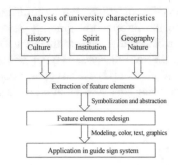

Fig. 7. The method of characteristic inheritance

6 Conclusion

A complete guide sign system design needs to be considered from three aspects: information interaction, visual communication and characteristic inheritance, which is in line with the humanized design concept. The guide sign system truly becomes the charm card, which can not only satisfy the information needs of the people, but also convey the characteristics and charm of universities.

Acknowledgement. This paper is sponsored by Natural Science Foundation of Shanghai (No. 18ZR1409400) and Ministry of Education Humanities and Social Science Research Planning Fund Project (No. 18YJAZH046).

References

1. Xun, S.: Research on college students sports associations and campus culture design based on big data analysis. Bol. Tecnico/Techn. Bull. **55**(10), 272–277 (2017)
2. Shao, H.-P., Mu, W., Ye, Y.-X.: A quantitative model for the validity of guide sign information at intersection. J. Transp. Syst. Eng. Inf. Technol. **17**(4), 70–75 (2017)
3. Beyrouthy, C., Burke, E.K., McCollum, B.: University space planning and space-type profiles. J. Sched. **13**(4), 363–374 (2010)
4. Vilar, E., Rebelo, F., Noriega, P.: Indoor Human Wayfinding Performance Using Vertical and Horizontal Signage in Virtual Reality. Hum. Factors Ergon. Manuf. Serv. Ind. **24**(6), 601–615 (2014)

Design and Implementation of Integrated Operation System for Business and Tourism of Super Large Power Companies

Zhiyong Yu[1]([✉]), Linlin Liu[2], Chen Chen[2], Jia Lai[2], Weitao Zhang[2], and Xuedong Wang[2]

[1] Grid Corporation of China, No. 86, West Chang'an Street, Xicheng District, Beijing, China
zhiyong-yu@sgcc.com.cn
[2] Tianjin Yingdajincai Travel Services Co., Ltd., Lekai Building, No. 7 Guangyuan Street, Xicheng District, Beijing, China

Abstract. The travel needs of large companies are becoming larger and larger, leading to huge difficulties in the management and reimbursement of travel at this stage, such as excessive costs, high cost, low efficiency and poor experience. Some of the existing business travel platforms are dedicated to integrating travel resources to save travel expenses. The purpose of travel is convenient. However, there are still obvious deficiencies in the company's personalized service requirements, paper invoice image recognition technology, and network platform supply and demand. This paper designs a one-stop travel management, which transforms the travel business from offline to online to implement automatic control of the whole process and design a new travel management platform architecture. Practice shows that the travel platform of this paper realizes the unified operation and management of travel business, achieves the purpose of safety, efficiency and convenience, and the platform further intellectualizes the company's travel management. The platform facilitates employee travel, reduces travel costs, improves travel reimbursement efficiency, and optimizes service experience.

Keywords: Travel · Reimbursement · Convenient travel · One-stop

1 Introduction

Business travel management is an important part of company management, especially for large-scale companies. Due to the large variety of business and geographical coverage of large companies, the travel demand is large, and there are differences in time, place and method. Therefore, the management of travel is complicated, especially the management of travel expenses is very difficult. For example, large-scale communication companies and power companies have different travel needs at different times and places because of their respective reasons. From group headquarters to molecular companies, the demand is huge, and it often requires 24 h of uninterrupted service. At the same time, travel management involves a wide range of issues, cumbersome matters, and lack of uniform platform control, usually only control the total

© Springer Nature Switzerland AG 2020
F. Xhafa et al. (Eds.): IISA 2019, AISC 1084, pp. 319–324, 2020.
https://doi.org/10.1007/978-3-030-34387-3_39

amount of travel expenses. Therefore, most enterprise travel management still has the following problems: individual enterprises have problems such as travel expenses exceeding standards and even false reimbursement; failing to give full play to the group scale effect and high cost; travel expenses management still adopts the traditional reimbursement model, and the experience is lower and the experience is poor.

We have great difficulties in dealing with the management and reimbursement of travel for the huge travel needs of large companies. The rapid development of the Internet, Internet of Things, big data, and artificial intelligence technologies provides technical support for building a convenient, intelligent, and reliable service platform. In order to further standardize the difference travel, convenient staff travel, reduce travel costs.

Based on the study of other theories, this paper designs and implements a one-stop business travel service platform. For various industries, travel needs of enterprises of different scales, with the basis of system demand analysis, further improve service efficiency. For the problem that paper invoices are difficult to identify, image recognition and image compression technology are used to enhance the edge and details of the invoice, which improves the recognition of invoices. For the problem of network capacity and network security, the front-end linkage technology and network security technology are adopted to enhance the security of the platform, and the privacy of customers is better protected. The one-stop business travel service platform designed in this paper builds a fast and efficient financial intensification management system based on informationization conditions, steadily promotes the company's financial lean management ability, enhances financial management decision support ability, and enhances the company's core competitiveness.

2   System Design

2.1   System Architecture

Implement online travel control throughout the journey to ensure strict implementation of the system. The travel business will be transferred from offline to online to implement automatic control of the whole process, and the whole process of business trip application, business approval, differential control, ticket reservation, expense reimbursement, batch reconciliation and financial settlement will be automatically completed. The system architecture is shown in Fig. 1.

At the user level, the platform implements the scenario design of unified advancement of funds, unified invoicing, and unified reconciliation settlement, so as to realize the advance payment, invoice, and reimbursement of employees of all units. At the supply level, the platform implements the Group's centralized procurement model to promote cost reduction and efficiency. Relying on the platform and major airlines, hotel groups to carry out centralized procurement of system-wide business travel services, effectively play the scale effect of the platform, and promote cost reduction and efficiency.

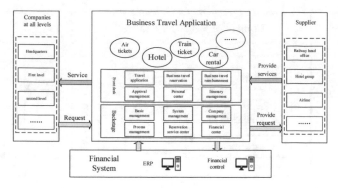

Fig. 1. Travel application system architecture diagram

3 Main Functions and Core Technologies

The platform realizes employees' self-quisition fast booking, one-click reimbursement, realizes travel-free padding, invoice-free, free reimbursement, online processing of travel business, paperless mobile approval, effectively improving the efficiency and user experience of travel management. The platform uses image recognition and image compression technology to enhance the image of the edge and details of the invoice, improve the recognition of the invoice, and lay the foundation for the paperless travel of the platform.

3.1 Travel Service Module

In this module, we use the front-end linkage technology to achieve high expansion and high availability of the platform [1]. The front-end and back-end linkages simultaneously call the technical resources of the back-end department and the customer requirements of the front-end department to establish a regular communication mechanism between the service provider, the demand side and the management party: the mobile Internet itself adaptively adjusts according to the configuration content, the front-end department's customer travel needs are predicted, with strong technical capabilities to ensure the quality of the service is completed. According to the rational use of network diversion, the market department provides an accurate basis for the development of terminal development strategies and reduces coordination costs. The front-end linkage model is shown in Fig. 2.

3.2 Reimbursement Service Module

Take the staff's independent fast booking, one-click reimbursement, to achieve business travel-free, invoice, free reimbursement, travel business online processing, paperless mobile approval. Aiming at the problem of paper voucher confusion and false ticket risk, the platform adopts image compression technology and uses the self-optimized edge-oriented interpolation algorithm to perform high-fidelity compression of invoice images [2]. And adaptive block and adaptive sampling exploration around

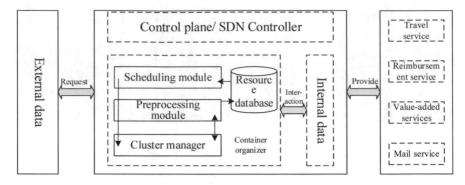

Fig. 2. Front-end linkage model

the complexity of the texture, but there is still a lot of information redundancy compared to the human visual mechanism [3].

This sharpening method is suitable for the purpose of highlighting isolated points, isolated lines, or line endpoints in an image, and the model of image sharpening is shown in Fig. 3. Enhanced technology makes it difficult to determine the position of the edge line for steep edges and slowly changing edges.

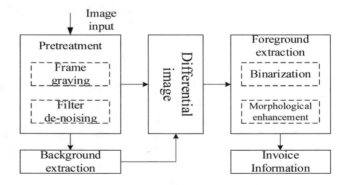

Fig. 3. Image sharpening model

3.3   Intelligent Value-Added Service Module

The platform gradually established a perfect big data cloud system, through the establishment of a travel frequent information database, analysis of employee travel habits, regular products, etc., automatically matching employee differences, targeted push and recommendation of travel products services, improve user experience and satisfaction degree. Simply put, if a certain word appears more frequently, the word can be selected as a feature word [4].

Through semantic association analysis, the business travel situation situational awareness, model construction, event clustering and other functions can be used to build a travel management policy that conforms to the company's current development status, tailor-made for employees to suit their own preferences, and a single user portrait study. By extracting user features and refining user tags for each user in a scenario to construct a user profile, the behavior, needs, interests, preferences, etc. of the specific user can be directly reflected [5]. The user portrait structure based on semantic association analysis is shown in Fig. 4. It ensures the intelligent control of corporate travel and the accurate prediction of user travel.

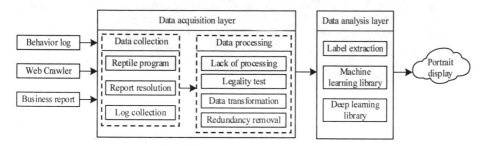

Fig. 4. User portrait structure based on semantic association analysis

3.4 Security Defense Service Module

Due to many factors, information security issues are frequently encountered in computer network technology applications. Take appropriate precautions to address the risks of privacy and sensitive data disclosure [6]. Firewall is a commonly used security isolation technology and data access control mechanism [7]. Firewall is an effective means to prevent hacking and virus attacks, and it is conducive to maintaining the security of computer networks. It's also a security protection barrier widely used in computers [8].

4 Conclusion

As the main body of construction and operation of business travel, E-commerce Company applies current product performance, software and hardware resources, service support capability, business resources on-line, combined with deep pilot application experience and current team size. By relying on the State Grid Corporation and major airlines and hotel groups to conduct centralized procurement of system-wide business travel services, the Group's scale effect will be effectively utilized to promote cost reduction and efficiency. Thereby, further enhance the core competitiveness of the product, enhance the user experience, and make the business travel application a powerful guarantee for employees to travel.

References

1. Chen, K.: Research on the construction of mobile data traffic prediction model based on front-end linkage. Commun. Inf. Technol. (06), 64–66+48 (2012). (in Chinese)
2. Wei, Z.: Histogram based watermarking algorithm robust to geometric attack with high embedding capacity. IEEE Beijing Section, Global Union Academy of Science and Technology, Chongqing Global Union Academy of Science and Technology
3. Wang, W., Dang, J., Li, Y., Yang, H., Yang, W.: A review of research on image compressed sensing theory. Mach. Manuf. Autom. **48**(01), 112–116 (2019). (in Chinese)
4. Wright, J., Yang, A.Y., Ganesh, A., et al.: Robust face recognition via sparse representation. IEEE Trans. Pattern Anal. Mach. Intell. **31**(2), 210–227 (2008)
5. Song, M., Chen, W., Zhang, R.: A review of user portrait research. Inf. Sci. **37**(04), 171–177 (2019). (in Chinese)
6. Dawn. Personal privacy and data security protection for overseas travel. Modern World Police, (11), 72–73 (2018). (in Chinese)
7. Gen, L.: Research on effective application of linkage technology of firewall and IDS in network security management
8. Qi, B.: Research on computer network information security in the era of big data. Sci. Technol. Innovation Appl. (14), 52–53 (2019). (in Chinese)

Simulated Annealing in Talent Scheduling Problem

Cheng Chen, Dan Wang, Gongzhao Zhang, and Mao Chen[✉]

National Engineering Research Center for E-learning,
Central China Normal University, Wuhan 430079, China
chenmao@mail.ccnu.edu.cn

Abstract. Talent Scheduling is about scheduling independent activities (such as scenes in a movie) to minimize the extra cost. Simulated annealing has played an important role in combinatorial optimization since it was proposed. This article explores how the simulated annealing algorithm is applied to solve the Talent Scheduling Problem and aims to generate a sequence as optimal as possible. We also discuss the influence of some factors on the performance of the proposed algorithm and give a suggestion about how SA can be improved.

Keywords: Simulated annealing · Talent Scheduling Problem · Meta-heuristic

1 Introduction

Talent Scheduling Problem (TSP) is an old model built from filming [3]. In fact, one film has lots of scenes and the shooting order normally is not the same as showing on film. To simplify the model, talent scheduling problem assumes that the director hires the actor since the first scene which needs the actor until the last scene with the actor. Each scene needs different actors and the actors should be paid even though he is not need in the current scene. It aims to find the optimal sequence of scenes to get a minimum cost which evaluated by a specific function.

Some researchers have proposed some solutions since the problem was proposed. Genetic Algorithm (GA) [1] was first applied to solve this problem in 1994. The article used the simple genetic algorithm and it was compared with GA combined with pairwise interchange, which could improve the performance of the simple GA. For a small-sized problem instance, the algorithm can find a solution fast, but once the problem size gets bigger, the time is pretty long – more than 20 min. In recent years, dynamic programming [2] was applied and it had a chance to get the best solution for some instances. But the disadvantages are also obvious: it can only process small-sized problem instances, and each choice of the next scene is based on several greedy strategies that are easy to get stuck in local optimum. In the latest research, branch and bound [3] was developed based on dynamic programming which used a memorization technique to enlarge the solvable problem size.

As a heuristic algorithm, Simulated Annealing (SA) [5] can handle big-sized problem instances within an acceptable time limit and it also can jump out the local optimum and convergence to the global optimum by the probability mechanism.

F. Xhafa et al. (Eds.): IISA 2019, AISC 1084, pp. 325–332, 2020.
https://doi.org/10.1007/978-3-030-34387-3_40

The experiment results were compared with Branch & Bound which shows that SA has better performance for almost all instances.

The rest of the paper is presented as follows. Section 2 introduces the talent scheduling problem detailed with the complexity analysis. Section 3 introduces the plan to solve the problem with SA. The experiment results were shown and compared in Sect. 4. Conclusion is presented in the final Section.

2   Talent Scheduling Problem

The Talent Scheduling Problem is a NP problem, which is came up from filming in the real world. During filming, the director would separate the film into many scenes and hire as many actors as needed. But not all the actors show on every scene, the cost to keep hiring actors when they are not needed at some scenes is called extra cost. If the director gets a really awful sequence of scenes, the director would spend a lot of extra money. Since each scene has a start date and an end date, and different scene needs different actors, the salary for each actor is calculated by:

$$S_i = (E_i - S_f) * A_i \tag{1}$$

In formula (1), S_i is the salary of actor$_i$, E_l is the end date of the last scene actor$_i$ attends, S_f is the start date of the first scene actor$_i$ attends. A_i indicates the salary for actor$_i$ per day. To simplify the problem and build a model, this article assumes that the next scene starts once the current scene finished (no interval is needed between two continuous scenes), so that the duration of a scene becomes

$$D_j = S_{j+1} - S_j \tag{2}$$

In formula (2), Dj is the duration of scene$_j$, S_{j+1} is the start date of scene$_{j+1}$ and S_j is the start date of scene$_j$. So we generate a sample table that indicates a film with 10 scenes and 6 actors below.

Table 1. A film schedule with 10 scenes and 6 actors

Actor	Scene										
	S1	S2	S3	S4	S5	S6	S7	S8	S9	S10	Salaries
A1	X	X	0	X	0	0	X	0	X	0	8
A2	0	X	X	X	X	0	X	X	0	0	2
A3	X	0	0	0	0	X	0	X	X	X	10
A4	0	X	0	X	X	X	0	0	X	0	1
A5	0	0	X	0	0	0	0	X	X	X	7
A6	0	0	0	X	X	X	0	0	X	X	3
Duration	2	3	1	5	1	2	1	3	4	1	

In Table 1, X at row i column j means actor$_i$ is needed in scene$_j$. According to the sample table, now we can generate an evaluation function to determine whether it is a good sequence or not. The evaluation function shown below

$$T = \sum_{i=1}^{n} \left(S_i - \sum_{s=1}^{m} (D_s * A_i) \right) \qquad (3)$$

T in formula (3) means the total extra cost of the sequence, sum of D_s means the number of total days that actor$_i$ is needed. According to this evaluation function, we can determine whether the result filming sequence is better or not. The talent scheduling problem is known to be NP-hard [4] even if each actor appears in only two scenes where all actor costs are identical and all durations are identical.

3 Project Solving Plan

3.1 Simulated Annealing

SA was first introduced by Kirkpatrick Scott in 1983 [5], which simulates the annealing process that heats the material to a high temperature and then slowly lowers the temperature to generate crystal with the minimum system energy. Kirkpatrick found the similarities between combinatorial optimization and statistical mechanics and then used the Metropolis algorithm [6] as the approximate numerical algorithm. Instead of accepting only the better solution, SA would accept the worse results with a certain probability to get out of a local optimum.

The problem that needs to be solved is regarded as a system, and each solution indicates a state of the system. All the solutions make up a discrete function. SA algorithm walks through the whole solution space and try to stop at a global optimum. Different from traditional SA, this paper explored several factors that would affect the performance of the SA algorithm, such as how to produce more stable results or produce better but less stable results.

$$P = \frac{1}{1 + e^{\frac{\Delta E}{T}}}, \Delta E = E_{next} - E_{current} \qquad (4)$$

Metropolis acceptance function (formula (3)) can determine whether a new state is acceptable or not. According to the metropolis function ex, at the same temperature, the bigger ΔE is, the smaller the P will be. When T is lowering down, the system would be more willing to accept a better state. Finally, when T is small enough, system would only accept better state and reach a peak. Therefore, based on probability statistics and law of large numbers, SA would approach global optimal result at last. The whole process can be separated into the following steps:

Step 1. Generate a random solution, and set the *start_temperature*, *annealing_rate*, *end_temperature* and *iteration_times* per temperature. Calculate the current energy as $E_{current}$.

Step 2. Find a neighbor of current solution and calculate the new system energy as E_{next}, then calculate $\Delta E = E_{next} - E_{current}$.

Step 3. To find minimum global optimum, use the Metropolis acceptance function to determine whether the new state is accepted or not.

Step 4. If the program reaches the *iteration_times*, go to step 5, otherwise go back to step 2.

Step 5. The program terminates if the temperature reaches *end_temperature*, otherwise lowers *start_temperature* and go back to step 2.

Since its introduction in 1983, SA has been applied for solving many real-world problems and has many variants such as Fast Simulated Annealing in Rd [7]. SA can also combine with other combinatorial optimization methods. For example, SA was combined with particle swarm, genetic algorithm and parallel computing [8–11].

3.2 Neighbors and Steps

There are so many methods to find the neighbors of a solution, we only use two most used, i.e., swapping and insertion. Swapping means that swap the positions of two random elements to form a new solution and the new solution calls the neighbor of the original solution. Different from swapping, insertion would randomly pick an element and insert it into a random position to form a neighbor. In fact, all the methods to find a neighbor is to change the original solution a bit to form a new solution. Steps in three-dimensional space is the distance between two points. The longer the distance is, the bigger the steps would be. In TSP, assuming one swapping or insertion as one step, so that the size of steps is the number of swapping or insertion in one finding neighbor action. The procedure shows as following:

Algorithm 1. Neighborhood based search
1. **Input:** sequence[][]   /*two dimensional array which stored the salaries*/
2. **Output:** sequence[][]   /* neighbor shooting sequence of the original*/
3. steps ← large in high temperature
4. **for** i←0 to steps **do**
5. col_1←random column index
6. col_2←another random column index
7.   swap two columns in the sequence or insert col_1 of sequence to position col_2
8. **end for**
9. return sequence

3.3 Annealing Procedure

To solve the problem with SA, firstly we define all the different shooting sequences as the solution space. Starting with a random shooting sequence and calculate the extra cost with formula (3). The system starts annealing from a high temperature where the atom is highly active, which means we move bigger steps at the beginning. At each iteration, use the neighbor procedure to find the neighbor and re-calculate the extra cost. Metropolis acceptance formula [6] offers a way to decide whether the new state is acceptable. Assuming the current temperature is T, start energy (extra cost) of the system is e_s, the energy of the neighbor is e_t, record the change in energy as $\Delta E = e_t - e_s$. The metropolis probability formula (4) allows the system to be more willing to move to a lower energy state.

Enough iterations are ensured at each temperature to reach a steady state. Then slowly reduce the temperature until the system is terminated at a given freeze point (end temperature). More importantly, it allows the system to accept the higher energy state so that the system would not stick in a local optimum. The whole procedure shows the following:

Algorithm 2. The proposed SA algorithm

Input

1. sequence[][]   /* two dimensional array which stored the salaries*/
2. T_s /* start temperature of the system*/
3. T_t /* temperature that terminate the annealing procedure*/
4. Rate   /* temperature reduce rate*/
5. Times   /* iteration times per each temperature*/

Output

6. Cost   /* minimum extra cost sequence after termination of the program*/
7. E_s ← extra cost of the start shooting sequence
8. **while** $T_s \geq T_t$ **do**
9.   **for** i ← 0 to Times **do**
10.   sequence ← neighbor(sequence)
11. E_t ← extra cost of the current shooting sequence
12. $\Delta E \leftarrow E_t - E_s$
13.   **if** random(0, 1) > P
14.   accept new state
15.   **else**
16.   Refuse the new state and return to initial state
17.   **end if**
18.   **end for**
19. $T_t \leftarrow T_s \times$ Rate
20. **end while**
21. return sequence

Other adaptive approaches such as Thermodynamic Simulated Annealing [12], automatically adjusts the temperature at each step based on the energy difference between the two states, according to the laws of thermodynamics.

4   Experiment Results

4.1   Results Comparison

The problem instances randomly generated by Qin [3] are used in our experiments. There are 200 combinations of different numbers of actors and scenes, and 100 different situations for each combination, which results in 20000 different data in total. This paper picks the most difficult 40 combinations with 5 situations for each combination (only these five situations have reference results provided by Qin). For each situation, the SA procedure is run 100 times and all the results are recorded. The results and comparison with Branch and Bound [3] are shown in Table 2. Note that letter "U" in Table 2 means unknown.

Table 2. Results comparison with Branch & Bound

	No. of scenes								
	56			58			60		
	BB	SA	BEST	BB	SA	BEST	BB	SA	BEST
14-1	12398	11881	U	10797	10278	U	7108	6969	6969
14-2	25525	25136	U	10653	10452	10429	15869	15109	U
14-3	14643	14006	13892	21869	19909	U	13008	12474	U
14-4	8705	8333	U	16197	15170	15166	11762	10868	U
14-5	10738	9703	9701	14573	13929	13929	11856	10899	10899
16-1	17689	17159	U	10916	10815	10813	13906	13439	U
16-2	19977	19466	U	16478	14963	14963	24452	23731	U
16-3	21637	20138	19932	17484	16163	16139	13857	13374	U
16-4	19453	19113	U	15802	15618	15613	17093	16647	16593
16-5	13002	12582	12545	14710	14510	U	14231	13216	13216
18-1	15752	14979	14894	13930	13319	13098	20796	20012	U
18-2	15261	13180	13168	12877	11316	U	13994	13147	U
18-3	15875	15791	15791	18965	17313	17313	16374	15348	15263
18-4	20578	18027	18027	23965	22655	U	26013	23594	23164
18-5	7481	6736	6562	19259	18711	U	19505	18446	U
20-1	21731	21265	U	27050	25744	U	12167	9783	9783
20-2	16402	15679	U	12870	12224	U	23636	21134	U
20-3	28586	27757	U	20562	20274	U	22626	22134	U
20-4	30534	28529	U	17460	15661	U	20920	20296	U
20-5	23817	23099	U	23140	18472	U	24272	23286	U
22-1	22376	21268	U	33826	32558	U	17224	16189	U
22-2	17136	16444	16332	21206	19177	U	18926	18364	U
22-3	19541	18884	U	18579	17131	U	30533	28596	U
22-4	17107	16872	U	23512	23338	U	31529	30030	U
22-5	24290	22713	U	18598	16970	U	17636	16965	U

In Table 2, for each column, BB means the upper bound of Branch & Bound, SA means the best result of SA, and BEST is the known global optimum. For each row, 'i-j': i means the number of actors, j means the different random situations generated with i actors. According to the comparison, apparently, SA can almost totally reach better results than Branch and Bound. In some of the known cases, the SA can even reach the best results. So that SA apparently has advantages compared to Branch and Bound in the Talent Scheduling Problem.

4.2   Influence Parameters

The most important part of the SA algorithm adjusts the parameter to maintain a balance between performance, CPU time and stability. The size of the steps is quite important to the performance. Generally, the influence of the step size is shown in Fig. 1. However, in experiments of SA, the metropolis function already offers a way to jump out of local optimum. At the beginning with high temperature, the program accepts all new states (new shooting sequences) and finally, it refuses all new states. The effect of metropolis function with small steps can get better results but with less stability. Decreasing step size based on temperature can reach more stable results but has a little higher extra costs.

The annealing rate is the most important factor in time cost. It is generally accepted that time can be exchanged for better performance. But the experiments show that more execution time does not necessarily result in a better outcome.

The choice of start temperature and terminate temperature varies for different situations. The start temperature should ensure that the system is quite active (accept almost all new states) while the terminate temperature ensures that the system denies almost all worse states. It can let the system have time to jump out of local optimum and finally reach a peak.

| i. small steps | ii. big steps | iii. decreasing steps |

Fig. 1. The influence of different size of steps. (i) Small steps can reach a peak, but easy to trap into local optimum. (ii) Big steps can easily jump out of local optimum, but meanwhile, it's hard to reach a peak. (iii) Decreasing steps can jump out of local optimum at the beginning and approach a peak at the end.

5   Conclusion

The paper applies a clear, specific and easy-to-understand SA algorithm to solve the talent scheduling problem. The comparison with the latest work which used Branch and bound shows that SA has a better performance in less time. A discussion of how different factors affect the performance of SA algorithm is presented.

Acknowledgement. The work was financially supported by the National Key R&D Program of China (2017YFB1401300, 2017YFB1401302) and the self-determined research funds of CCNU from the colleges' basic research and operation of MOE (Grant No. CCNU19ZN011).

References

1. Nordström, A.L., Tufekci, S.: A genetic algorithm for the talent scheduling problem. Comput. Oper. Res. **21**(8), 927–940 (1994)
2. Garcia de la Banda, M., Stuckey, P.J., Chu, G.: Solving talent scheduling with dynamic programming. INFORMS J. Comput. **23**(1), 120–137 (2011)
3. Qin, H., Zhang, Z., Lim, A., et al.: An enhanced branch-and-bound algorithm for the talent scheduling problem. Eur. J. Oper. Res. **250**(2), 412–426 (2016)
4. Cheng, T.C.E., Diamond, J.E., Lin, B.M.T.: Optimal scheduling in film production to minimize talent hold cost. J. Optim. Theory Appl. **79**(3), 479–492 (1993)
5. Kirkpatrick, S., Gelatt, C.D., Vecchi, M.P.: Optimization by simulated annealing. Science **220**(4598), 671–680 (1983)
6. Metropolis, N., Rosenbluth, A.W., Rosenbluth, M.N., et al.: Equation of state calculations by fast computing machines. The J. Chem. Phys. **21**(6), 1087–1092 (1953)
7. Rubenthaler, S., Rydén, T., Wiktorsson, M.: Fast simulated annealing in Rd with an application to maximum likelihood estimation in state-space models[J]. Stoch. Process. Appl. **119**(6), 1912–1931 (2009)
8. Fang, L., Chen, P., Liu, S.: Particle swarm optimization with simulated annealing for TSP. In: Proceedings of the 6th WSEAS International Conference on Artificial Intelligence, Knowledge Engineering and Data Bases (AIKED 2007), pp. 206–210 (2007)
9. Rere, L.M.R., Fanany, M.I., Arymurthy, A.M.: Simulated annealing algorithm for deep learning. Procedia Comput. Sci. **72**, 137–144 (2015)
10. Gu, X., Huang, M., Liang, X.: The improved simulated annealing genetic algorithm for flexible job-shop scheduling problem. In: 2017 6th International Conference on Computer Science and Network Technology (ICCSNT), pp. 22–27. IEEE (2017)
11. Ye, Z., Xiao, K., Ge, Y., et al.: Applying Simulated Annealing and Parallel Computing to the Mobile Sequential Recommendation. IEEE Trans. Knowl. Data Eng. **31**(2), 243–256 (2019)
12. De Vicente, J., Lanchares, J., Hermida, R.: Placement by thermodynamic simulated annealing. Phys. Lett. A **317**(5–6), 415–423 (2003)

Is He Angry or Happy? Recognizing Emotionally-Accented English Speech

Hui Peng[1], Siliu Liu[2], Hongyan Wang[2(✉)], and Jeroen van de Weijer[2]

[1] Shenzhen Autumn Harvest School, Shenzhen University, Shenzhen, China
[2] Shenzhen University, Shenzhen, Guangdong, People's Republic of China
wanghongyan0069@hotmail.com

Abstract. In this study, we investigate how listeners recognize emotional speech and if they are better recognizing some emotions than others. Chinese listeners with relatively basic and relatively more advanced English skills were asked to recognize three kinds of emotional speech (expressing anger, joy, and sadness), as well as neutral speech, produced by native English and Chinese English speakers. The Chinese listeners with a more advanced English level showed significantly better skills of speaker identification, while sadness any joy were better recognized than anger. This research has implications for cross-cultural communication and speaker identification in general.

Keywords: Speaker identification · Emotional speech · Accented-English

1 Introduction

1.1 Research Background

Speech is used to convey emotion, which forms an important part of human communication. The detection of emotions in speech is therefore of considerable importance in many areas, including inter-personal communication and human-to-machine communication. Computational systems are able to infer much information from a person's spoken input based on acoustic, lexical and syntactic features [1], but human speakers appear to have even fuller access to emotional (re)cognition.

Not only do listeners recognize emotions in speech—they recognize the speaker as well. Based on characteristics of the speaker' voice, factors such as the shape of their vocal tract, throat size, structure of other vocal organs etc. help to identify speakers as male or female, young or old, their socio-economic status, and if they are an L1 or an L2 speaker (e.g. [2]). Things like speech rhythm, intonation, manner of articulation, vocabulary selection also determine a speaker's characteristics. Speaker identification research is currently still limited to text-dependent, environment-dependent, normal-mood-dependent and language dependent factors, i.e. it does not generally consider the contribution of emotional speech in speaker identification.

This paper investigates speaker identification of accented-English speech with different emotions. It focuses on Chinese listeners with different levels of knowledge of English. Both groups recognize the accented English speech with different emotions produced by native English speakers and Chinese English speakers.

F. Xhafa et al. (Eds.): IISA 2019, AISC 1084, pp. 333–338, 2020.
https://doi.org/10.1007/978-3-030-34387-3_41

According to the results of an acoustic analysis of related studies [3–5], the emotions manifested in English read-out speech can be described by the following characteristics (see Table 1):

Table 1. Acoustic characteristics of English read-out speech

Emotions	Acoustic characteristics		
	Pitch	Intensity	Rate
Neutral	average	high	average
Anger	low	average	high
Sadness	average	low	low
Joy	high	average	average

Neutral speech has an average pitch, intensity and speech rate. Angry speech has a low pitch, average intensity and high speech rate. Sadness has average pitch, low intensity and low speech rate. Joy has a high pitch, average intensity and average speech rate. This means that the three emotions (and neutral speech) can be arranged on three acoustic continua as in Fig. 1:

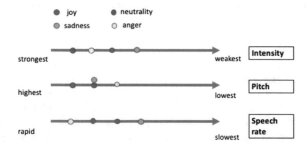

Fig. 1. Parameter dynamics of acoustic features (joy, sadness, neutrality and anger)

One question is whether these parameters are universal or language-specific. [6] investigated the influence of language and culture on the understanding of vocal emotions. Their results indicated that similarity of cultural values and language background were not advantages in recognizing vocal emotions expressed in a different culture and language. We therefore decided to investigate how well these different emotions are recognized in English sentences by speakers of Chinese.

1.2 Research Questions

Our specific research questions are the following:

1. How well do Chinese listeners recognize the emotions produced by native English speakers and by Chinese English speakers?
2. How does emotional speech affect speaker identification?

3. Is there a difference in recognition between lower-level and more advanced speakers of English?
4. Which emotions are best recognized?

2 Research Methods and Data Collection

2.1 Subjects

We recorded eight Chinese (five females and three males; mean age = 22.6 years, range = 20–24 years) and eight English native speakers (four females and four males; mean age = 23.6, range = 22–28). They were asked to record sentences using different, acted, emotions (anger, joy, sadness) as well as in neutral speech. All speakers were born and raised in their respective countries. The Chinese speakers were recruited at a university in China. The English native speakers were students at universities in the United States. None of the Chinese participants was bilingual in English, but all reported being fluent in English and regularly using the language as students at a university where the medium of instruction is English.

The test data therefore consist of 1,280 utterances (16 speakers × 20 sentences × 4 moods). We selected 96 utterances (24 neutral, 24 sad, 24 joyful and 24 angry) with an emotion identification rate of no less than 65% (i.e. more than 2.5 times better than chance) as listening materials.

Two groups of adult subjects were recruited as listeners: the first group consisted of five males and five females from China with an age from 20 to 50 (average age M = 28.1) who spoke Chinese as their mother tongue and had never learned English or had only learned English before at a low level (at most: high school). The second group consisted of ten Chinese listeners (average age M = 23.0) as well, with an age from 20 to 27 whose first language (L1) was Chinese and spoke English (L2) at an advanced level (university level). These subjects were chosen from those over 20 years old. One of the reasons was a recommendation made in an earlier study that very young listeners are not ideal subjects for this kind of research [7].

2.2 Stimuli

The test corpus consisted of 20 sentences read by two groups of speakers. These sentences were taken from casual conversation and consisted of 8-10 words (e.g. "I have received a letter from my cousin". The content was neutral, but the utterances were spoken in a way that communicated each of the three intended emotions, as well as a neutral statement. These sentences were used for the listening test.

2.3 Procedure

The listening test was carried out using a web environment. Participants had computer access and used either speakers or headphones. They were provided with the listening material and an answer sheet. Before the test, the participants were asked to complete a short language questionnaire, which confirmed their personal data given above.

The listening test required the two groups of Chinese listeners to listen to 24 separate context-free English sentences produced with different emotions and they were asked to identify the speakers. The Chinese listeners would hear neutral speech at first, and then three sentences with the particular emotions (joy, sadness, anger) would be played. The listeners were asked to choose which of these three sentences was produced by the speaker of the neutral-speech sentence. After 24 items, the answer sheets were collected for data analysis.

3   Results

3.1   Presentation of Experiment

The study required 20 Chinese listeners to recognize emotional sentences presented by native English speakers and Chinese English speakers. These 20 people are averagely divided into two ordinal groups according to the level of English. The present education level of Group 1 (five males and five females) ranks from primary school to high school and the accuracy is 47.92% (104 correct responses in all 240 responses) of emotional identification tests. People (five males and five females) in Group 2 all have bachelor degree and the accuracy is 64.58% (156 correct responses in all 240 responses) of emotional identification tests.

3.2   Data Analysis and Results

All the test groups identified the target speakers of the emotional utterances with an accuracy of more than chance probability. Yet the second group of listeners (Chinese listeners at the advanced level) performed better than the first group (see Table 2). The accuracy rate of speaker identification of Group 1 was 47.92% and of Group 2 was 64.58% (see Table 2).

Table 2. The accuracy of speaker identification

Groups	Correct responses	Total responses	Accuracy rate
Group 1	103	240	47.92%
Group 2	155	240	64.58%

Secondly, the data demonstrate that in group 1 sadness was best recognized, followed by joy and anger. For group 2, joy was best recognized, followed by sadness and anger. Combining the two groups, the conclusion seems warranted that speakers are easier to identify on the basis of utterances produced with sadness and joy than with anger. We can relate this finding to the acoustic features of emotions (see Fig. 2) because sadness and joy have similar acoustic features to neutrality, specifically pitch and speech rate. This may be the reason why joyous and sad speech are judged correctly more easily.

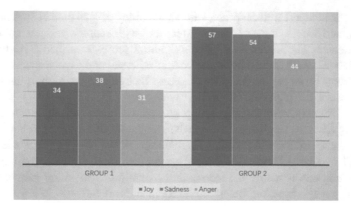

Fig. 2. The correct responses of speaker recognition in joy, anger and sadness. *Note:* Emotions: Blue represents joy; orange represents sadness; grey represents anger.

Table 3 shows that Chinese listeners at a lower level of English did a *better* job in recognizing utterances produced by Chinese English speakers than by English native speakers. The Chinese listeners at an advanced English level recognized utterances produced by English native speakers better. This may be related to an interlanguage speech intelligibility benefit [8–11], which might play a greater role at relatively lower levels.

Table 3. The accuracy rate of speaker recognition with utterances produced by Chinese English speakers and Native English speakers (%)

Groups	Utterances produced by English native speakers	Utterances produced by Chinese English Speakers
1	38.75	45.63
2	66.86	62.5

To summarize, both groups of Chinese listeners recognized the emotional utterances produced by both groups of speakers well, and listeners from group 2 performed better than from group 1. Moreover, listeners from group 1 performed better when recognizing emotional sentences produced by Chinese English speakers, but listeners from group 2 performed better with native English speakers. Utterances simulated by sadness and joy were better recognized, and angry utterances scored the lowest recognition rate among all participants.

4   Conclusions

We tested the recognition of emotional speech by two groups of Chinese listeners: one at a relatively lower level and one at a relatively higher level. The listeners at a higher level of English recognized the speakers in accented-English speech with an accuracy of over 67%, while the listeners at a lower English level recognized speakers in emotionally accented-English speech with an accuracy of less than 50%. The advanced listeners gained scores of over 65% for utterances with sadness and joy and of 55% for angry utterances. The lower-level listeners gained scores less than 50% for all emotional utterances. Both groups performed better with sad and joyous speech, and scored lower accuracy for anger. Sadness was the best recognized emotion in the lower-level group, while for the higher-level group it was joy.

Finally, listeners from the first group performed better when the utterances were produced by Chinese English speakers, whereas the second group did better when the utterances were produced by native English speakers. This indicates that language background is also a factor that affects speaker recognition in a foreign language. Future research can focus on the recognition of emotions for other language pairs, as well investigate the acoustic factors involved in more detail.

References

1. Oudeyer, P.-Y.: The production and recognition of emotions in speech: features and algorithms. Int. J. Hum.-Comput. Stud. **59**(1–2), 157–183 (2003)
2. Shriberg, E., Ferrer, L., Kajarekar, S.S., Scheffer, N., Stolcke, A., Akbacak, M.: Detecting nonnative speech using speaker recognition approaches, pp. 26–32 (2008). (in Odyssey)
3. Tamuri, K.: "Fundamental frequency in Estonian emotional read-out speech," Eesti ja soome-ugri keeleteaduse ajakiri. J. Est. Finno-Ugric Linguist. **6**(1), 9–21 (2014)
4. Tamuri, K., Mihkla, M.: Emotions and Speech Temporal Structure. Linguistica Uralica **48** (3), 209–217 (2012)
5. Tamuri, K., Mihkla, M.: "Expression of basic emotions in Estonian parametric text-to-speech synthesis," Eesti ja soome-ugri keeleteaduse ajakiri. J. Estonian Finno-Ugric Linguist. **6**(3), 145–168 (2015)
6. Altrov, R., Pajupuu, H.: "The influence of language and culture on the understanding of vocal emotions," Eesti ja soome-ugri keeleteaduse ajakiri. J. Estonian Finno-Ugric Linguist. **6**(3), 11–48 (2015)
7. Altrov, R., Pajupuu, H.: Estonian emotional speech corpus: culture and age in selecting corpus testers. In: Skadina, I., Vasiljevs, A. (eds.) Human Language Technologies – The Baltic Perspective: Proceedings of the Fourth International Conference Baltic HLT 2010. IOS Press, Amsterdam, pp. 25–32 (2010)
8. Algethami, G., Ingram, J., Nguyen, T.: The interlanguage speech intelligibility benefit: The case of Arabic-accented English. In: Levis, J., LeVelle, K. (eds.) Proceedings of the 2nd Pronunciation in Second Language Learning and Teaching Conference, Iowa State University, Ames, pp. 30–42 (2010)
9. Wang, H., van Heuven, V.J.: The interlanguage speech intelligibility benefit as bias towards native-language phonology. i-Perception **6**, 1–13 (2015)
10. Bent, T., Bradlow, A.R.: The interlanguage speech intelligibility benefit. J. Acoust. Soc. Am. **114**(3), 1600–1610 (2003)
11. Xie, X., Fowler, C.A.: Listening with a foreign-accent: the interlanguage speech intelligibility benefit in Mandarin speakers of English. J. Phonetics **41**(5), 369–378 (2013)

Research on Combinatorial Feature of Total Ionizing Dose (TID) Effect on Interconnect

Xiaofei Xu[1,3(✉)], Denghua Li[1,3], and Shuhui Yang[2]

[1] School of Electronic and Information Engineering, Beijing Jiaotong University, Beijing 100044, China
xu18910782910@126.com
[2] School of Information Engineering, Communication University of China, Beijing 100024, China
[3] School of Automation, Beijing Information Science and Technology University, Beijing 100192, China

Abstract. In mixed-mode simulation of this paper, the copper interconnect line structure in VDSM (very deep submicron) was connected to a circuit with the silica-based device (on-chip or inter-chip). A physical model of the mass transport of Cu interconnect is found based on the vacancy exchange mechanism based on a continuity equation on Cu diffusion including different TID ionizing dose settings of test. The TID failure tests are operated on interconnect and device samples, and the comparison between the model and literature shows that the simulation result is in accordance with the experiment data excellently (deviation <10%) at high electric field (beyond 2.5 MV/cm), small line space (90 nm) on the condition of different temperature, current density and line spaces. The electrical characters of interconnect lines with different width are simulated with the ISE. The numerical result of this model was in accordance with the data showed in literature and the experimental data. A number of research data contribute greatly to the study of sensitivity of the copper interconnect line and device, and find the method to enhance the anti-TID ability.

Keywords: Mixed-Mode · Cu interconnect · Irradiation · TID

1 Introduction

With the development of modern integrated circuits, it was necessary to do a further research about the interconnect liability, and the parameters model and circuit performance of interconnect was distributed with different size, which had been studied in some domestic and foreign papers [1–3].

According to the different sensitivity of measurement, a whole set of radiation tests normally would need to be performed by the Total Ionizing Dose (TID) radiation simulation using an proton beams [4, 5]; Electrical properties entry of large size device were susceptible to ionizing radiation in some papers [6]; The smaller size devices with thinner oxide layer were more impressionable to the stronger total effect of ionizing radiation. Research and analysis had been done on how to reduce the total dose effect

© Springer Nature Switzerland AG 2020
F. Xhafa et al. (Eds.): IISA 2019, AISC 1084, pp. 339–349, 2020.
https://doi.org/10.1007/978-3-030-34387-3_42

of parasitic effects and improve the anti-radiation ability of the device in interconnect nanometer level, so as to meet the needs of military [7–9].

The interconnect model had become better and reasonable for the further research, but the calculation of anti-radiation effectiveness was always difficult to evaluate. In order to make the model closer to the practical conditions, the on-chip or inter-chip interconnect should be considered to monitor currently during the irradiation. What's more, based on the combination of devices total dose effect and damage mechanism, Visual-TCAD construct copper interconncct model could be observed the electric field distribution with changing line width, and the reliability of copper interconnect system. Hence, it could provide a precise method to model the equivalent circuit and analyze the radiation damage reinforcement technology of interconnect [10–12].

2 Theory Analysis

The total dose effect of the ionizing radiation was the cumulative statistical results of a large number of individual photons and particles. Its energy, damage deposit and the effects of radiation had been distributed averagely throughout the circuit. Radiation-induced trapping electric charge was produced by ionizing radiation in sio_2 films, and interface stated at the interface of si/sio_2, which leaded to the variation of electric field intensity and threshold voltage shift of CMOS device, and the depression of channel carrier migration rate brought about the increase of leakage current. The ionizing radiation was effected on interconnect, and the components were concentrated in the drift of the threshold voltage of the device, which could be expressed by using the VT, and referred with Eq. 1:

$$\Delta V_T = -e \cdot \Delta Q_{ot}/C_{ox} + e \cdot \Delta Q_{it}/C_{ox} = \Delta V_{ot} + \Delta V_{it} \tag{1}$$

Here, the first term of the formula was the space charge generated by the ionizing radiation in the unit area and the effect of the threshold voltage. The second term was the interface state charge and the effect of the threshold voltage.

2.1 TID Model

The total dose radiation effect model based on the defect trapping theory considers the two mechanisms of defect trapping and emission, and there was only one dominant defect level in the body defect. There was a total dose irradiation effect model for the charge transfer efficiency decay based on the charge transfer efficiency, and was referred with Eq. 2:

$$CET = 1 - 3\frac{N_t}{n_s}(\frac{\tau_s}{\tau_c} - r_f(0))(1 - \exp(-\frac{t_{sh}}{\tau_s})) \tag{2}$$

Here, N_t and n_s were defect density and the signal electron density, respectively; while τ_s was defect trapping time constant, and τ_c was defect capture time constant and defect emission time constants related to the numerical. $r_f(0)$ was initial value of defect filling density, and t_{sh} was signal charge transfer time.

The radiation model of semiconductor in software TCAD ISE was referred with Eq. 3:

$$G = g * D * \left(\frac{E + E_0}{E + E_1}\right)^n \tag{3}$$

Where, E, G and D were the electric field intensity, the electron hole pair generation rate and the radiation doses were defined in the input file severally. Among them constant, $E_0 = 0.1$, $m = 0.9$, $E_1 = 1.35 \times 10^6$ v/cm

2.2 Interconnect Preprocessing

The ionizing irradiation had made metal atoms moving, which had made rise to mass accumulation or loss and the short circuit or increased resistance, so the following were the main reasons of interconnect failure: thermal stress, mutual diffusion, electro-migration and other factors. Electrical current analysis of interconnect was based on the continuity equation for conducting media, model equations or the Poisson's equation, the electron and hole continuous equation, electron and hole lose transport equation, and the order was stated as follows, referred with Eq. 4:

$$\begin{aligned}
\nabla \cdot \varepsilon \nabla \phi &= -q(p - n + N) \\
\nabla \cdot J_n &= qR + q\frac{\partial n}{\partial t} \\
-\nabla \cdot J_p &= qR + q\frac{\partial p}{\partial t} \\
J_n &= -qnU_n\nabla\phi_n \\
J_p &= -qpU_p\nabla\phi_p
\end{aligned} \tag{4}$$

Where, ε was the dielectric constant, φ was electrostatic potential, q was the total charge, n was electrons, p was holes, N was impurity concentration respectively, R was the lattice and electron hole recombination probability, and J was the electron and hole current density.

For efficiency reasons, there were some physical models to simulate radiation effects, such as electrostatic model, boundary condition per electrode, and electron scattering model in software. Mutual connection and device simulation results were completely transparent in terms of semiconductor technology to save the tape costs, design time and tape out cycle. The reliability of the electro-migration of the metal copper interconnects was related to the properties of the coating, barrier layer, and the grain boundary between the metal and copper, and its equivalent coefficient diffusion was simple, referred with Eq. 5:

$$D = D_g\left(\frac{\delta_g}{d}\right) + D_{c1}\delta_{c1}\left(\frac{2}{w} + \frac{1}{h}\right) + D_{c2}\left(\frac{\delta_{c2}}{h}\right) \tag{5}$$

Here, D_g, D_{c1}, and D_{c2} were respective for copper grain boundaries, the copper layer and the copper diffusion coefficient of the interface layer coverage; d was grain diameter, w and h were the width and height of a interconnect; δ was three dimensions.

This decomposition was used in subsequent equations to discuss the constitutive equation for the dilatational and deviatoric parts independently. The dilatational part, which corresponds to the trace of the tensor, described the material behavior in the case of a pure volume change; the deviatoric part described an arbitrary deformation changing the volume, referred with Eq. 6:

$$\varepsilon = \varepsilon^e + \varepsilon^p = \frac{\sigma}{E} + \alpha(\frac{\sigma_y}{E})(\frac{\sigma}{\sigma_y})^n \qquad (6)$$

Here, ε was the elastic strains and the plastic strains, while α and n were material parameters, σ, ε and E were the stress in one dimension, the total strain in one dimension and Young's modulus respectively. No extending the formula to three dimensions.

2.3   Material Model and Setting

On the equivalent circuit model of interconnect structure in nanoscale integrated circuits, CMOS (low resistivity silicon complementary metal oxide semiconductor) process was used in the top layer of metal copper for the signal line, with silicon as the signal loop, whose mutual connection and silicon-based devices were oxide and were based on the environment temperature 300 K. UseISE software to set material parameters and model.

Firstly, the relative dielectric constant of copper conductive metal was 0, where the lattice heat capacity and lattice thermal conductivity severally were set as constants $3.42 \, j/k * cm^3$ and $3.85 \, w/k * cm$. When the band gap width was 0, Fermi level kept 11.7 eV, resistivity stayed 1.56×10^{-6} Omega. Secondly, the relative permittivity of the oxide was 3.9, the optical refractive index perpendicular to the direction remained 1.46, lattice heat capacity was set as constant $1.67 \, j/k * cm^3$, lattice thermal conductivity was maintained $0.014 \, w/k * cm$, and the band gap width of the Fermi level was of 0.09 eV.

3   Experimental

The radiation response of nm bulk silicon MOS devices after exposure to TID was experimentally investigated. Due to the uncertainty of the incident location, different performance degradation behavior was irradiated by TID.

As shown in Fig. 1, the silicon-based devices of mixed mode circuit interconnects model can be seen. According to the model, the boundary condition schematic of interconnect structure was shown in Fig. 2. Due to lower interconnect dimensions, interface diffusion was most important influence on the electro-migration characteristics, which could precisely reflect the vacancy concentration with time or position parameter of structural dynamic characteristics.

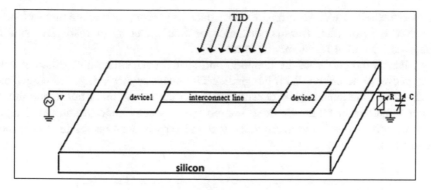

Fig. 1. An Interconnects structure and devices was connected in mixed-mode.

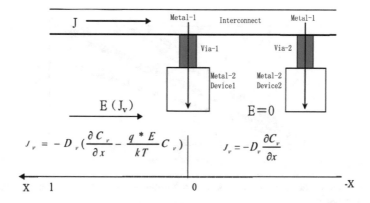

Fig. 2. The boundary condition schematic of interconnects line structure.

As shown in Fig. 2, providing appropriate initial conditions and boundary conditions change is necessary, which could be calculated by the continuity Eq. 7 and obtain vacancy concentration with time or location [13]. Prior to the start of the experiment, the initial vacancy concentration profile was found to be uniform, equal to the concentration of the equilibrium without stress conditions.

$$\frac{\partial U}{\partial \zeta} = \frac{\partial^2 U}{\partial x^2} - \alpha \frac{\partial U}{\partial x} + \beta \cdot U \tag{7}$$

Here, $\alpha = \frac{q^* E L}{kT}$, $\beta = A \cdot \frac{C'}{C} \cdot \frac{L^2}{D_{v0}}$, α was positive or negative, related to the electric field; U was relative vacancy concentration in interconnects; ζ was relative time; x was relative position; C' was the interfacial defect concentration; C was dislocation concentration.

Based on standard available values for 90-nm IMEC technologies, it was used by the hydrodynamic model with high-field saturation and mobility degradation models including doping-dependence and carrier-carrier scattering. The energy of particles was set up to 5.62099 MeV. According to other data discussion, such as the threshold of the device or the grid-plate trans-conductance, it didn't alter so dramatically with TID change as 1 Mrad, 4 Mrad and 30 Mrad.

In Fig. 3, the curves of the temporal and spatial concentration of different values with time were described. With the greater TID intensity increasing, the bigger mean electric stress, the more current density, the easier the formation of holes, and then the shorter interconnect failure time, so interconnect vacancy concentration was higher. There was distinction between the numerical and experimental result, due to the model difficult to simulate exactly.

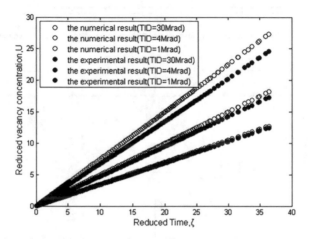

Fig. 3. The relative vacancy concentration in interconnects of both the theoretical calculation and simulation result

As shown in Fig. 4, 2-D structure of interconnect and physical device on silicon was simulated with ISE Visual-TCAD software in Fig. 4(a). The Gate was the metal copper, and the relation position was the oxide layer, then the following was the device of the chip; the effect of the irradiation and its behavior were found by the indirect definition of the interface charge, which helped explain the degradation of series resistance through optimizing the mesh operation and simulating.

The Ids-Vgs curve was shown in semi-logarithmic scale for different amount of negative trapped charge in the LDD spacers in Fig. 4(b); the irradiation dose enhancement effect of oxide layer was produced a considerable number of charged particles and the injection of large amounts of charged particles. So we had learned, the characteristics of the transistor current were increased, where the switching characteristics had become poor due to the dose sensitive.

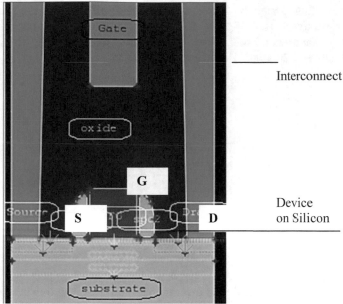

(a) 2-D structure of the interconnect and device model on silicon used in our
     simulation

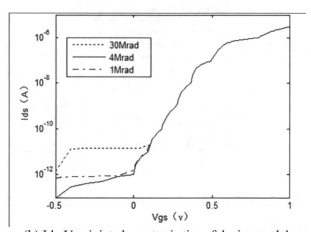

(b) Ids-Vgs joint character is tics of device model

Fig. 4. Ids-Vgs character is tics in semi-logarithmic

4 Results and Discussions

An amount of research was carried on the major TID effects reliability problems of Cu
interconnect system based on the theory, emulation and experiment in the paper. The
experimental condition was the same as Sect. 2. The isolation between two consecutive
metal layers was considered and simulated by using the spacer/silicon interface

shrinking gradually between 90 nm and 180 nm of the interconnect dimension respectively (see Fig. 5); On the condition of the total dose effect, the electrical characters of the copper interconnect were studied with different characteristic factors of 20mrad ~ 360mrad and 10 keV.

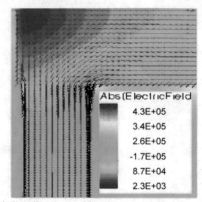

(a)   The electrical field distribution of interconnect line irradiated by TID

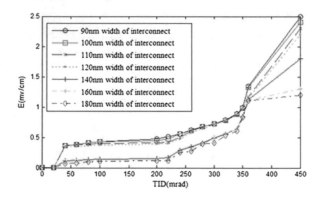

(b)   The electrical field output of interconnect line irradiated by different TID

Fig. 5.   The dependences of relative E variation on different TID for interconnect line

Figure 5 showed the simulated TID-E curves in semi-logarithmic scale for different Cu interconnect of line width 90 nm ~ 180 nm, and the amplitude of the electric field obviously increased on the structure of less than 120 nm interconnect line width, due to the EM mechanism of Cu interconnect changing with the dimension beyond the 90 nm/65 nm. As the size of interconnect spacing decreased, the roughness of the edge of the figure increased obviously, and the electric field distribution in the dielectric layer would become inconsistent, and it would also affect the electric field distribution in the dielectric layer. Breakdown time had an effect. Rough edges of interconnects could lead to many fixed areas of high electric field, and make it easier for electrons to inject into the anode from the cathode.

An amount of electro migration (EM) failure tests were operated on the samples in order to study the EM character of Cu interconnect on the condition of different temperature system. A physical model of the mass transport of Cu interconnect was found based on the vacancy exchange mechanism. A phenomenon was found through the experiment and the model that the voids grow and agglomerate along the interface of Cu and cap dielectric and concentrate at cathode end of the line. The interface diffusion had become an important failure mechanism. The EM characters of the structures were evaluated through the result of the experiments and emulations, which were instructive to the design of the Cu interconnect system.

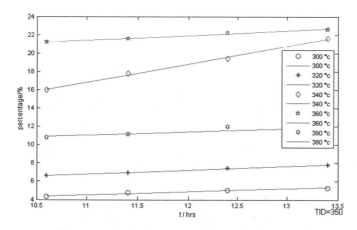

Fig. 6. Theelectro migration failure test based on different temperature

The electro-migration failure tests were operated on interconnect samples, and the experimental conditions was 120 nm width of interconnect and TID = 350, as shown in Fig. 6. Above, there were five curves, which stood for different electro-migration failure percentage. On the same graph consequently, the metal interconnects were with higher temperature and more increasing power density, which made the electro migration become one of the main latent damage modes.

The maximum electric field intensity was distributed at the edge of interconnect, and in Fig. 7, the test showed there was some electric field distribution in the area close in the line. As the interconnect width and feature size was decreasing, the integrated circuits made the electric field strength increasing and enhancing at the corners of the wire continuously for narrow line width of the interconnect. The test data of the mean electric field-TID were collected and adopted a linear configuration as follow. It was helpful to improve the reliability of Cu interconnect system.

Figure 7 showed clearly that with the reduction of the distance between the lines, the maximum electric field and the average electric field were gradually enhanced. The work showed that an improved method for the EM failure mechanism of Cu interconnect was proposed when the line space was beyond the 90 nm/65 nm.

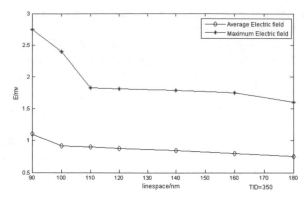

Fig. 7. The electro migration failure test based on different temperature

5   Conclusion

In the paper, Cuinterconnects and devices structure model in VDSM technologies were presented. The degradation mechanism and the electric field characterization of Cu interconnect and devices were studied based on the main components of the electric field acceleration factor as different TID, line space and different temperature. The results showed that the model was in good agreement with the experimental data at high field strength (greater than 2.5 MV/cm) and small spacing (90 nm) (deviation less than 10%).

The electrical characteristics from the simulation and corresponding TID failure test were developed with different spacing of lines and different widths of interconnect structure, and the results showed that with decreasing interconnect width and line spacing, electric field intensity of interconnect was increasing, which would accelerate the degradation of TID interconnects, and shorten its service life. The effect of the interaction between the two lines should not be ignored when the space between the lines were below 180 nm. In mixed-mode simulation, the parameters such as leakage current and electric field were analyzed for Total Ionizing Dose (TID) changing. The simulation results showed that with the TID increasing, the device on silicon was defected. The decreasing width of the line within the size range was 90 nm ~ 180 nm, and E transfer efficiency was sensitive within TID 50mrad or beyond TID 200mrad. The effects of irradiation on the Cu interconnect could change and worsen parasitic circuit performance seriously.

The curve was made by the effect of characteristic parameters on differential thermal and the line space analysis, which could contribute greatly to study sensitivity of the copper interconnect and device on dose enhancement, where the ratio of grain size to interconnect width would be larger and larger when the line size was reduced to a certain extent to find the method to enhance the anti-TID ability. Next step, more serious measurements were proposed to improve the TID failure of Cu interconnect system.

Acknowledgments. The paper was supported by National Natural Science Foundation of China (Grant No 61171039). We also wish to thank KeJingda Electronic Co. Ltd. for preparing software.

References

1. Lifei, J., Lingling, S., Lei, Z.: Analysis the sensitivity of interconnect to process variation for 65 nm technology node. Chin. J. Electron Devices **31**(3), 781 (2008)
2. Ming, D.: The study on the reliability and failure mechanism of copper interconnection in VDSM, Ph.D Thesis, Xi'an Xidian University, Xi'an (2010)
3. Zhiyuan, L., et al.: Analyzing influence of interconnect on performance for CMOS circuits. J. Nat. Sci. Heilongjiang Univ. 31(5) (2014)
4. Lloyd, J., Rodbell, K.P. (eds.): Reliability in Handbook of Semiconductor Interconnection Technology. Elsevier, New York (2006)
5. Hughes, H., Benedetto, J.: Radiation effects and hardening of MOS technology; devices and circuits. IEEE Trans. Nucl. Sci. **50**(3), 500–521 (2003)
6. McLain, M., Barnaby, H., Holbert, K., Schrimpf, R., Shah, H., Amort, A., Baze, M., Wert, J.: Enhanced TID susceptibility in sub-100 nm bulk CMOS I/O transistors and circuits. IEEE Trans. Nucl. Sci. **54**(6), 2210–2217 (2007)
7. Fleetwood, D., Eisen, H.: Total-Dose radiation hardness assurance. IEEE Trans. Nucl. Sci. **50**(6), 552–563 (2003)
8. Ceschia, M., Paccagnella, A., Cester, A., Scarpa, A., Ghidini, G.: Radiation induced leakage current and stress induced leakage currentin ultrathin gate oxides. IEEE Trans. Nucl. Sci. 45 (6), 55, 2375–2382 (1998)
9. Ceschia, M., Paccagnella, A., Sandrin, S., Ghidini, G., Wyss, J., Lavalle, M., Flament, O.: Low field leakage current and soft breakdown in ultrathin gate oxides after heavy ions, electrons or X-ray irradiation. IEEE Trans. Nucl. Sci. 47(6), 552, 566–573 (2000)
10. Barnaby, J.: Total-Ionizing-Dose effects in modern CMOS technologies. IEEE Trans. Nucl. Sci. 53(6), 3103–3121 (2006)
11. Gerardin, S., Bagatin, M., Cester, A., Paccagnella, A., Kaczer, B.: Impact of heavy-ion strikes on minimum-size MOSFETs with ultrathin gate oxide. IEEE Trans. Nucl. Sci. **53**(6), 3675–3680 (2006)
12. Turowski, M., Raman, A., Schrimpf, R.: Nonuniform total-dose induced charge distribution in shallow-trench isolation oxides, IEEE Trans. Nucl. Sci. 5351(6), 3166–3171 (2004)
13. Jeung-Woo, K., Won-Sang, S., Sam-Young, K., Hyun-Soo, K., Hyun-Goo, J., Chae-Bog, L.: Characterization of Cu extrusion failure mode in dual-damascene Cu/Low-k interconnects under electro migration reliability test. In: Proceedings of 8th IPFA, pp. 174–177 (2001)

Research on Modal Analysis of Lifting Mechanism of Traffic Cone Automatic Retracting and Releasing System

Linsen Du[1(✉)], Guosheng Zhang[1], Jiaxin Zhang[2], and Hongli Liu[1]

[1] Research Institute of Highway Ministry of Transport, Beijing 100088, China
ls.du@rioh.cn
[2] Shenyang University of Technology, Shenyang 110870, China

Abstract. The lifting mechanism of the traffic cone automatic retracting and releasing system is installed on the vehicle frame. During normal operation, the vibration of the mechanism will be affected by the engine and road roughness. By using the finite element software ANSYS to carry on the modal analysis to the lifting mechanism, it obtains each order natural frequency and the mode diagram of the mechanism. And compared with the vibration frequency produced by engine excitation and road excitation. Avoid the resonance of lifting mechanism of traffic cone automatic retracting and releasing system because of the similar between the excitation frequency and the natural frequency of the mechanism. It affects the operation safety of traffic cone automatic retracting and releasing system.

Keywords: Traffic cone · Automatic retracting and releasing system · Modal analysis · Excitation

1 Introduction

Traffic cone automatic retracting and releasing system vehicles are usually affected by the excitation of their own engine vibration and road roughness during operation. If the excitation frequency is close to the natural frequency of their own structure, the mechanical structure will resonate, leading to deformation and damage of the mechanism. Based on the basic data of modal analysis of the lifting mechanism of the traffic cone automatic retracting and releasing system, the vibration frequencies of each order are obtained. The vibration frequency caused by engine vibration and road roughness is compared with the natural frequency of the structure in order to avoid mechanical resonance.

© Springer Nature Switzerland AG 2020
F. Xhafa et al. (Eds.): IISA 2019, AISC 1084, pp. 350–356, 2020.
https://doi.org/10.1007/978-3-030-34387-3_43

2 Theory of Modal Analysis

For an undamped free vibration system, the kinetic equation is defined as:

$$[M]\{x''\} + [K]\{x\} = \{0\} \tag{1}$$

Where, $[M]$ is the mass matrix of the structure; $\{x''\}$ and $\{x\}$ respectively represents the acceleration matrix and displacement matrix of the structure; $[K]$ is the total stiffness matrix.

The free vibration of the structure is simple harmonic vibration, that is, the displacement is sine function:

$$x = xsin(\omega t) \tag{2}$$

The Eq. (2) is substituted into Eq. (1) to get:

$$([K] - \omega^2[M])\{x\} = \{0\} \tag{3}$$

The above equation is the characteristic equation of the system, the problem of solving the natural frequency and natural mode of the system is the problem of solving the eigenvalue and eigenvector of the equation. According to the linear algebra theory, the necessary and sufficient condition for the equation to have non-zero solution is:

$$\left|[K] - \omega^2[M]\right| = 0 \tag{4}$$

Because there are n degrees of freedom after the discrete structure, [K] and [M] are square matrices of order n, so The n-order natural frequency and corresponding main mode of the structure can be obtained by solving the equation.

3 Extrinsic Motivators

3.1 Influence of Engine Vibration

The main excitation source of traffic cone automatic retracting and releasing system is engine vibration. The burst pressure generated by the gas in the engine cylinder, the inertia force and inertia moment generated by the motion are transmitted to the frame through the connection support between the engine and the frame. Among them, the reciprocating inertia force and torque caused by the unbalanced motion of crankshaft, connecting rod and piston are easy to cause the resonance of the car chassis, which has a greater impact on the vibration of the car, and the other components have a smaller impact. Therefore, to reduce the vibration, the modal frequency of the whole mechanism should avoid the excitation frequency range of the inertial force of the engine.

The excitation frequency generated by the engine depends on the engine rotation speed and the number of cylinders. The calculation formula is:

$$f_e = 2nZ/(60\tau) \tag{5}$$

Where, n is the engine revolving speed; Z is the number of engine cylinders; τ is the number of engine strokes.

As known, The number of engine cylinders Z is 4, the number of engine strokes τ is 4. In the operating state, the vehicle speed is generally 10–20 km/h, the rotation speed n is generally 1200–1600 r/min, Therefore, the excitation frequency f_e generated by the engine in the working state is between 40 and 53 Hz.

3.2   Influence of Road Roughness

During driving, the chassis frame is stimulated by the uneven road surface from the wheels. Uneven road surface exerts displacement and impact disturbance on the wheels driving on it. The vibration generated by this random excitation may cause fatigue failure of the mechanism. Caused by road roughness excitation vibration, and is associated with the speed of the car, when the car travels at speed V (km/h) for the road roughness space frequency Ω (m^{-1}) road, produced by the road excitation frequency fr is the product of Ω and V, that once the road excitation frequency and modal frequency of the frame phase coupling, car body resonance is going to happen, and the resonance speed as follows:

$$V = 3.6L_{min}f_r \tag{6}$$

Where, L_{min} is the minimum irregularity wavelength of road surface.

It is inversely proportional between the wavelength L and roughness space frequency Ω of road roughness, In our country, the roughness of different road surface spectrum wavelength of the measured results are shown in Table 1.

Table 1. Different pavement roughness wavelength L

Road conditions	Unpaved road	Gravel road	Washboard road	Flat road
Road roughness wavelength /m	0.77–2.5	0.32–6.3	0.74–5.6	1.0–6.3

In order to prevent the resonance of lifting mechanism caused by road excitation frequency, the lifting mechanism frequency f_r shall meet the following requirements:

$$f_r \geq V_{max}/(3.6L_\omega) \tag{7}$$

Where, V_{max} is the maximum travel speed; L_ω is the actual road condition irregularity wavelength.

As known, the off-duty driving speed of the traffic cone automatic retracting and releasing system is 60 km/h. The road surface is flat expressway, so the L_ω is defined as

3.0 m. So f_r is computed to 5.56 Hz. The driving speed of the traffic cone automatic retracting and releasing system is 10–20 km/h, so it is computed that f_r is greater than or equal to 1.851 Hz.

4 Modal Analysis of Lifting Mechanism

4.1 Three-Dimensional Model

The lifting mechanism part in the three-dimensional mechanical design model of traffic road cone automatic retracting and releasing system (see Fig. 1) is selected, and Ignore stiffeners, partitions, holes and accessories, The simplified structure of the lifting mechanism is shown (see Fig. 2).

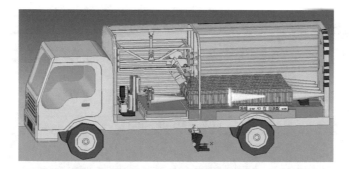

Fig. 1. Road cone automatic pickup and release system vehicle model

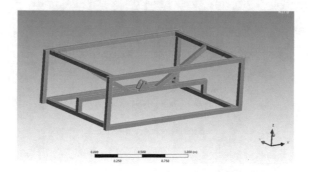

Fig. 2. The lifting mechanism model

4.2 Meshing

The simplified model was imported into the ANSYS workbench analysis software to conduct grid division, and the grid model was shown (see Fig. 3).

4.3   Boundary Conditions for Modal Analysis

In the actual working process, the lifting mechanism is not in a completely free state. The pin seat of the mechanism is connected to the carriage frame by a straight rod. The lifting mechanism can move up and down by a telescopic straight rod. According to the actual working conditions, the lifting mechanism pin base two holes for 6 degrees of freedom constraints, x, y direction limit displacement, x, y, z direction limit rotation.

4.4   The Solution Results of Modal Analysis

By ANSYS workbench analysis and solution, the vibration frequency (Table 2) and mode diagram (see Figs. 4, 5, 6, 7, 8 9) of the lifting mechanism of the 6 order modal analysis can be obtained.

Table 2. Modal vibration frequencies and modes of the lifting mechanism

Order number	Frequency	Vibration mode
1	–	–
2	9.34	The vertical beam bends in the x direction
3	9.55	Beam bending
4	11.96	The vertical beam bends in the Z direction
5	18.55	Center beam bending
6	28.57	Diagonal distortion

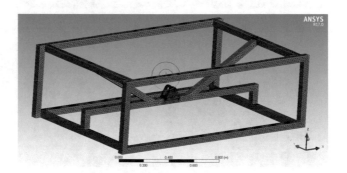

Fig. 3. Grid model

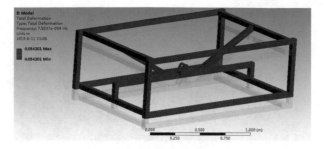

Fig. 4. The first order vibration mode

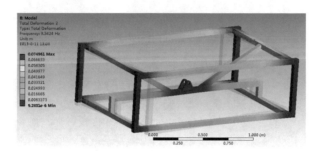

Fig. 5. The second order vibration mode

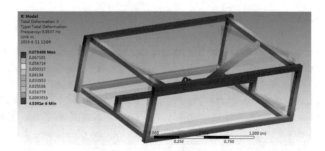

Fig. 6. The third order vibration mode

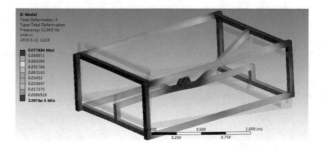

Fig. 7. The fourth order vibration mode

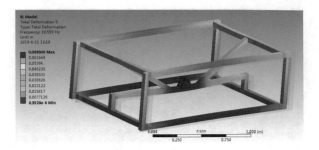

Fig. 8. The fifth order vibration mode

Fig. 9. The sixth order vibration mode

5   Conclusion

From the 6 order mode diagram of the modal analysis of the lifting mechanism, it can be concluded that the natural frequency of the lifting mechanism is roughly in the range of 9.34 Hz–28.57 Hz. According to the theoretical calculation of excitation factors, the excitation frequency of the engine is approximately 40–53 Hz. The excitation frequency of pavement roughness is roughly within the range of 5.56. The natural frequency of the mechanical structure is not close to the excitation frequency, so the structure will not cause resonance of the mechanical structure due to the influence of engine vibration and road roughness.

References

1. Wang, Y., Xie, Y.: Structural modal analysis of four cylinder internal combustion engine block. Trans. Csice **20**(1), 75–78 (2002)
2. Cui, L., Zhang, L., Ma, W., Zhang, X.: The modal analysis and improvement design of the rotary lifting mechanism of a certain road cone retractor. Machine Manufacturing, 589 (2013)

Research on Blockchain-Based Commercial Paper Financing in Supply Chain

Lin Zou, Shiyu Jia, Qiujun Lan, and Zhongding Zhou(✉)

Business School of Hunan University, Changsha, China
acrowise@126.com

Abstract. Blockchain provides a secure and reliable guarantee for information transaction. The trading partners share information and benefits, as well as take risks together, which prompts the integration with blockchain. However, the establishment of trust could be the biggest challenge in this integration. In this article, a solution to address the challenge by building a supply chain financial service system based on the league blockchain is proposed. When small and medium-sized enterprises (SMEs) have financing needs, they can get a loan through blockchain system effectively. In the supply chain, with the help of consensus mechanism and encryption technology, the system provides enterprises, banks and regulators with more convenient financing audit and supervision methods, which improves the financing speed of non-core enterprises. At the same time, the system makes the financing process transparent, safe and traceable. On the one hand, this system changes the way information is stored from SMEs' respective systems to the blockchain. On the other hand, regulators can supervise SMEs' actual use of loan funds effectively. This, to some extent, has greatly solved the financing difficulties of SMEs.

Keywords: Blockchain · Supply chain · Commercial paper financing

1 Introduction

In the supply chain, the cooperation between core enterprise and its upstream and downstream enterprises make it possible to maximize the efficiency of supply chain. The SMEs play an indispensable role in the cooperation which regarded as supporting enterprises. However, due to the relatively small scale, SMEs are often at a disadvantageous position in the supply chain, and the way of returning funds is frequently governed by the payment methods of the core enterprises in supply chain. Therefore, SMEs may face a funding gap in the production process. It is easy to throw the whole supply chain out of balance. At the same time, in the process of lending from banks, SMEs often face the following difficulties: weak anti-risk ability; vulnerable to economic fluctuations; unregulated management; banks have a high cost of loan management for SMEs, thereby they raise loan interest rate to balance the cost; lack of collateral; other loan methods (such as bond, private loans, etc.) have a high threshold or high financing cost, and so on.

© Springer Nature Switzerland AG 2020
F. Xhafa et al. (Eds.): IISA 2019, AISC 1084, pp. 357–364, 2020.
https://doi.org/10.1007/978-3-030-34387-3_44

Traditional financing method makes it difficult for SMEs to obtain funds in time. Commercial paper financing can provide a better financing solution for SMEs. It refers to the financing with commercial paper. SMEs can transfer the commercial paper they hold and then get a loan with the deduction of relevant expenses to carry out their production activities. From the perspective of banks, commercial paper financing improves the stickiness of SMEs to them and expands their customer groups. As for enterprises, on the one hand, it delays the time limit for the buyers to pay for the goods. On the other hand, the suppliers can collect the payment in time to carry out production activities. Hence, commercial paper financing can make the use of funds more efficient, and then promote the circulation of funds to SMEs.

In practice, banks often have insufficient trust in SMEs. To strengthen the management of receivables, SMEs can entrust third parties (factory) to carry out the agency. As a way of financing for SMEs, commercial paper financing helps SMEs to obtain funds. The factoring business involves the factor, banking institution, core enterprise in the supply chain and other SMEs. Information storage is scattered regardless of large information exchange between different entities. Consequently, there are some problems during the financing process.

First, it is difficult to verify the authenticity of commercial paper. It usually requires the holder to provide relevant certificates such as contract and invoice for examination. There is no doubt that this process will take a lot of time. Sometimes they can't screen out the situation that both parties of the transaction make up the trade jointly or apply for financing in different institutions repeatedly with the same commercial paper. At the same time, the paper version is more prone to counterfeit than the electronic version.

Second, it is difficult to assess the credit of enterprises, which issues commercial paper. As a third party, factoring companies can't participate in the real trade process between commercial sides. Insufficient information of the historical transactions may easily lead to deviation in credit assessment, and result in uncollectible debts. When there is a dispute in the financing process, more efforts will be spent on investigating due to the scattered storage of information.

As a result of the difficulties in the commercial paper financing mentioned above, this paper proposes to combine the blockchain technology with supply chain business. Blockchain as a decentralized mechanism constructs a trusted decentralized database through distributed storage, peer-to-peer transmission, consensus mechanism and encryption algorithm etc. Banks and financial institutions can use the data from blockchain to review commercial paper and evaluate credit. After the loans are issued, the flow of funds can be monitored through the blockchain. This combination can provide traceable and trusted ledger to resolve disputes and reduce survey costs. Hence, a system under the scenario of commercial paper financing is designed in this paper.

2 Related Works

Wang et al. pointed out that for a long time, only large and medium-sized enterprises were served by China's financial system. SMEs are restricted by such factors as small scale, limited production technology, imperfect management system and low market competitiveness and so on. They are often faced with strict qualification examination and higher interest rate by the lending Banks. With the rapid development of Internet in China, Internet finance has played an important role in solving the financing difficulties of SMEs [1].

Yi [2] combined the multi-stage credit transfer problem with blockchain. Distributed ledger technology is adopted to solve the problem of information asymmetry and use credible data to carry out systematic risk management. In order to realize the integration of payment and settlement, the participants are monitored by intelligent contracts in supply chain finance which ensures the data consistency through consensus mechanism.

Tang [3] thinks the core function of commercial paper has evolved into a form of financing. Financing through the commercial paper in our country has become an important tool for SMEs. And the continuous innovation of the system promotes the application of commercial papers in supply chain finance.

Beck [4] said the emerging blockchain world is the combination of traditional ways of doing things and those that are enabled by blockchain. The generation and management of economic behavior need intermediary organization as support. It is commonly believed that such organizations and intermediaries will act as expected. But if they don't, the trust deteriorates. It's hard to find trusted intermediaries, and that's where blockchain comes in.

Jessel and DiCaprio [5] discussed that can blockchain make trade finance more inclusive? The inefficient trade finance makes many applications go unfunded. Relying on the credit system built by blockchain, more trade financing will be permitted. More SMEs and emerging markets will receive financial support, and then contribute to a more inclusive trade finance structure.

Wang et al. [6] established a theoretical model to analyze the new financing way of SMEs through blochchain. SMEs with low risk and high quality are more likely to obtain bank loans through information sharing.

3   Design the Blockchain-Based Financing System

According to the degree of openness, blockchain can be divided into three types: public blockchain, private blockchain and league chain. Public blockchain has a high degree of openness, and anyone in the world can read the data, participate in trade and compete the right to charge to an account of the new block. However, it takes 10 min to produce a new block. The low efficiency makes it difficult to meet the demand of real-time transaction. At the same time, the protection of data privacy is poor. Private chains are only open to individuals or entities, and have the lowest degree of decentralization. Therefore, it needs less time to reach consensus, so the transaction speed is faster, and the efficiency is higher. Private chains are more suitable for the internal use of specific institutions but not suitable for the supply chain field with multi-enterprise interaction. League chain is between the public and private chain, which can realize the "part of decentralize". Joining league chain should be authorized. Each node of the chain usually has the corresponding entities or organizations, and each institution controls one or more nodes. The transaction data were recorded collectively, and only these organizations and institutions can read and write the data in the league chain.

Compared with the public chain, the league chain improves the speed of reaching consensus and the degree of privacy protection.

Therefore, in this paper, in order to achieve the security and efficiency of the transaction process, the supply chain financial system based on the blockchain is in the form of league chain. Banks, the factoring company, core and none-core enterprise of supply chain, and the relevant supervisory authority join the league chain through registration and certification.

The blockchain-based supply chain financial service system is designed as the integration of several different business modules. Each module interacts through the interface and can communicate with the external.

Figure 1 shows the sketch map of system module.

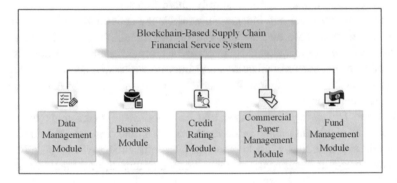

Fig. 1. Sketch map of system module

The data management module is used for collecting, storing and sharing data on the blockchain. Specifically, the data management module can be used to enable factoring companies to obtain credit information of core enterprises from blockchain. It can also be used to record information after financial institutions make loans to suppliers and suppliers supply to core enterprises of the supply chain in accordance with smart contracts.

The business module is used for uploading, confirming and terminating contract. Specifically, based on the trade contract and credit information of core enterprises in the data management module stored in the blockchain, factoring companies can sign factoring contracts with suppliers. Moreover, the trade contract between suppliers and core enterprises can be stored on the blockchain in the form of smart contracts.

The credit rating module is used to assess the credit rating of an enterprise. Managers can set credit rating standards, and only enterprises which are up to standard can sign with the factoring company.

The commercial paper management module is used to issue and transfer the commercial paper. Through this module, the receivables of the supplier to the core enterprise can be transferred to the factoring company. This module also enables financial institutions to verify the bills provided by factoring companies through blockchain, so as to review the loan application made by factoring companies for suppliers. When core enterprises in the supply chain make token repayment to financial institutions, smart contracts will be triggered and financing bills between factoring companies and Banks will be cleared through the commercial paper management module.

The fund management module is used for fund exchange, fund receipt and payment. After approving the bill, financial institutions will make loans to the supplier through this module. Also, core enterprises of the supply chain can pay back to financial institutions in the way of Token.

The detailed functions of the system module are shown in Fig. 2.

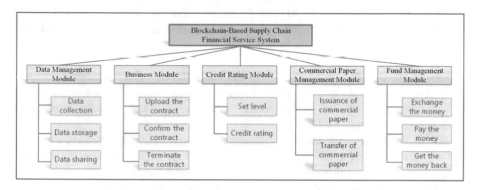

Fig. 2. Function of system module

4 Design the Commercial Paper Financing Process Based on Blockchain

Integrating supply chain finance with blockchain, the core and non-core enterprises of supply chain, banking institutions, factors and financial regulators integrate in the league chain. Therefore, the exchange of information between them can be more smoothly and also improve the financing efficiency of SMEs. At the same time, regulators can track the whereabouts of funds effectively.

Blockchain-based supply chain financial service system can provide efficient and safe financing functionalities. Connecting the capital flow of upstream and downstream enterprises in the supply chain as well as enabling SMEs to obtain financing loans in a more convenient and efficient way. In brief, it can improve the financing efficiency of SMEs.

Non-core enterprises in the supply chain can conduct financing through following steps in the system. And Fig. 3 shows the processes of commercial paper financing.

Step 1: Both parties of the transaction in the supply chain (core enterprises and suppliers) upload the contract in the business module. The contract was stored in the form of smart contract. And it will take effect after the signing and confirming of both parties. Then the receivables of suppliers to the core enterprises will be formed.

Step 2: In blockchain, factoring company evaluates the credit rating of core enterprises in the supply chain. Factoring company can choose rating methods voluntarily, such as factor analysis, comprehensive analysis method, weighted scoring method and so on. According to the assessment results, factoring company can choose whether or not to factor.

Step 3: Capital demanders and the factoring company sign the contract. The contract is stored in the system in the form of intelligent contract.

Step 4: The factoring company sends the confirmation of transfer to the core enterprise and the capital demander (supplier), and waits for the Hash signature confirmation of the two parties.

Step 5: The core enterprise and the capital demander sign Hash signature. And then the transfer of accounts takes effect. The information is recorded in the blockchain. The receivables of the supplier to the core enterprise can be transferred to the factoring company.

Step 6: Financial institutions (commercial banks) have complete blockchain ledger, and they can verify the authenticity of transaction notes (e.g. supply contract, factoring contract and receivables notes) through the blockchain.

Step 7: In the blockchain, factoring companies can finance suppliers through discount, pledge financing, pledge invoice and other means from commercial banks.

Step 8: In blockchain, commercial banks will top up to their account, and obtain the equivalent tokens of funds through fund management module, then pay the tokens to the demand side.

Step 9: In accordance with the requirements of the intelligent contract, the capital demander supplies to the core enterprises and the information is recorded in the blockchain.

Step 10: The core enterprise of supply chain will refund through blockchain. They charge to the blockchain account and exchange for the equivalent token. Commercial banks will withdraw money when they receive the tokens of core enterprises. Then the smart contract will close the account automatically after the repayment.

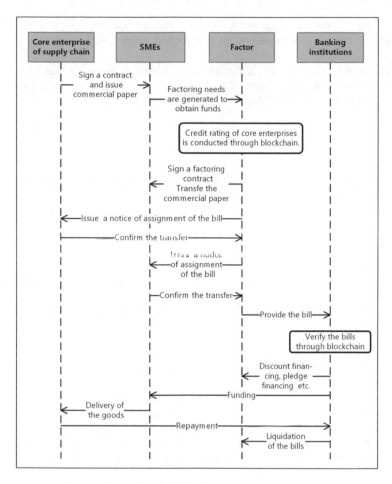

Fig. 3. Financing process

5   Conclusion

The blockchain-based supply chain financial service system constructed in this paper is an attempt of blockchain technology in the field of supply chain finance. Non-core enterprises in the supply chain can conduct commercial paper financing through this system. The probability of information asymmetry can be reduced by consensus mechanism. And it establishes a point-to-point trust mechanism. Complete transaction records are recorded in the blockchain, and the records cannot be tampered, which helps enterprises in the supply chain to establish trust quickly and improve the operation efficiency of the supply chain. However, there still are some deficiencies in the current system, such as privacy protection is not considered. Privacy protection methods can be added in future studies to protect information of enterprises on the chain. For example, the differential privacy method in literature [9] can be used to help enterprises realize privacy protection in the blockchain. And in work [10], asymmetric cryptography, zero-knowledge proof, the design of multi-channel and privacy data collection are the privacy protection measures used to prevent from privacy disclosure.

Acknowledgments. This research is supported by the National Natural Science Fund of China (Project No. 71871090) and National Natural Science Fund of China for Emergency Management Project (Project No. 71850012).

References

1. Wang, L., Liu, H.Z.: Financing of small and medium enterprises based on internet finance. In: IOP Conference Series: Materials Science and Engineering, vol. 394, p. 052015 (2018)
2. Yi, Y.: Research on several problems of multi-level credit transfer based on blockchain. Zhejiang Finance **2018**(9), 3–7 (2018)
3. Tang, Y.W.: Evolution of commercial bill market functions and SMEs' choice of financing methods under a changing credit system. J. Financ. Res. **2018**(05), 37–46 (2018)
4. Beck, R.: Beyond bitcoin: the rise of blockchain world. Computer **51**(2), 54–58 (2018)
5. Jessel, B., Dicaprio, A.: Can blockchain make trade finance more inclusive? J. Financ. Transformation **47**, 35–50 (2018)
6. Wang, R., Lin, Z.X., Luo, H.: Blockchain, bank credit and SME financing. Qual. Quant. **53**, 1–14 (2019)
7. Kshetri, N.: Blockchain's roles in strengthening cybersecurity and protecting privacy. Telecommun. Policy **41**(10), 1027–1038 (2017)
8. Treleaven, P., Brown, R.G., Yang, D.: Blockchain technology in finance. Computer **50**(9), 14–17 (2017)
9. Zhu, T.Q., Li, G., Ren, Y.L., Zhou, W.L., Xiong, P.: Differential privacy for neighborhood-based collaborative filtering. In: 2013 IEEE/ACM International Conference on Advances in Social Networks Analysis and Mining (Asonam), pp. 758–765 (2013)
10. Ma, C.Q., Kong, X.L., Lan, Q.J., Zhou, Z.D.: The privacy protection mechanism of Hyperledger Fabric and its application in supply chain finance. Cybersecurity **2**(1), 1–9 (2019)

Efficient Subspace Clustering Based on Enhancing Local Structure and Global Structure

Qianqian Yu, Yunjie Zhang$^{(\boxtimes)}$, and Chen Sun

Department of Mathematics, Dalian Maritime University, Dalian 116000, China
yjzhang@dlmu.edu.cn

Abstract. Subspace clustering (SC) achieves excellent clustering result via learning an affinity matrix and applying the spectral clustering on the matrix in order to divide data points into several different subspaces. One of the challenges faced by various SC methods is to construct a fitting affinity matrix from the given data. Sparse subspace clustering (SSC) learns the affinity matrix by minimizing l_1 norm. Log-determinant approximation to rank (CLAR) learns the affinity matrix by employing a special function, called Log-determinant (Log-Det), to take the place of rank function. In this paper, we propose an improved SSC algorithm by using LogDet function and Frobenius norm. The improved algorithm could effectively improve the grouping effect and obtain block diagonal solution, especially when the database size is large. Experiments on the Synthetic database, the 'Hopkins155' database and 'Extended Yale B' database, are demonstrations of the proposed approach could get a better clustering result.

Keywords: Sparse subspace clustering · Spectral clustering · Least squares representation · Low-rank representation

1 Introduction

In real world problems, large amounts of datasets that need to be processed are often high-dimensional, such as image clustering, image segmentation, etc. In practice, however, such datasets are often assumed to originate from a union of numerous subspaces, and the subspaces and their membership are unknown. Therefore, we need to partition the data into their potential subspaces by subspace clustering (SC). In the past two decades, various SC methods have been raised, among which the spectral clustering-based techniques attract more attention because of its simplicity. The first step of these techniques is constructing an affinity matrix to measure the similarity between data points. And the second step is segmenting the dataset by implementing spectral clustering for the affinity matrix [1, 2].

The learning of affinity matrices is arguably the key to the SC approaches based on spectral clustering, which largely relies upon the self-representation property of datasets and the penalty form of representation coefficients. In the past few years, numerous algorithms have been developed to learn an effective affinity matrix, and difference primarily lies in choosing the penalty form of representation coefficients. One influential

© Springer Nature Switzerland AG 2020
F. Xhafa et al. (Eds.): IISA 2019, AISC 1084, pp. 365–373, 2020.
https://doi.org/10.1007/978-3-030-34387-3_45

classes of algorithms is sparse subspace clustering (SSC), whose objective is to seek out the sparse self-representation for every point by making use of few points from one subspace [3, 4]. SSC involves solving a minimization problem in which the l_1 norm of representation vectors is used as sparsity-promoting penalty [3]. While, another influential algorithm LRR is suggested penalizing the rank of representation matrix [4]. Ultimately, the goal is the resulting affinity matrix could have the block-diagonal property. However, SSC is suboptimal because it is likely to lose the connection between data points. On the other hand, LRR is difficult to solve because it's much hard to minimize the rank of matrix. Although nuclear norm can provide a good surrogate for matrix rank, it doesn't always work in many situations.

Kang et al. [5] presented an algorithm, called log-determinant approximation to rank (CLAR), to use log-determinant (LogDet) function to take the place of rank function. Lu et al. [6] propose an algorithm, called least squares regression (LSR), to build the penalty form of representation coefficients by using Frobenius norm, and showed that if the subspaces are independent, the solution of affinity matrix can be block-diagonal. Hu et al. [7] also used this method to strengthen the block-diagonal structure of the representation coefficients matrix. However, it is difficult to obtain satisfactory grouping and block-diagonal results for the large amount of data sampling. A new algorithm, called Enhancing Local Structure and Global Structure (LSGS), is proposed in this paper to solve above problems. The proposed algorithm may enhance the grouping effect and block-diagonal property of affinity matrix when the database size is large. Experiments on some datasets testify the validity of LSGS algorithm.

The structure of this paper's rest part is as follows. In Sect. 2, it's a short description of several algorithms. There are the LSGS algorithm and theoretical analysis in Sect. 3. Experimental on several databases are carried out in Sect. 4. In the final Section, there are summary conclusions.

2   Related Work

Assume that a data set $X = [x_1, x_2, \ldots, x_n] \in R^{(d \times n)}$ is high-dimensional, and it can be regarded as a union of several low dimensional subspaces of R^n. Subspace clustering could be implemented into two part: obtaining a sparse affinity matrix and applying spectral clustering on it. First, we have to find a matrix C which can satisfy the condition as:

$$X = XC, c_{ii} = 0 \tag{1}$$

The representation coefficients $W = (|C| + |C^T|)/2$ is used to indicate the similarity between the data points. The next step is using the spectral cluster method to receive the error rate of clustering. If the solution C is freedom from constraint, there are many solutions. Therefore, for getting an accurate and concise solution, we can give it a set of constraints.

2.1 Sparse Subspace Clustering

The core thought behind SSC is getting a spares affinity matrix to represented X as a sparse linear combination of other points in X. Then SSC apply spectral clustering to the spares affinity matrix to obtain a clustering result. If there are noises in the data, the model is formulated as follows:

$$\min\|C\|_1 + \lambda\|E\|_F^2 \text{ s.t. } X = XC + E, diag(C) = 0 \tag{2}$$

where $\|C\|_1$ denotes the l_1 norm of C, SSC use the F norm for the error term and λ is used for balancing the effect of this two terms.

Despite SSC's success, there are still disadvantages. For example, the structure of data could be broken.

2.2 LRR and CLAR

The purpose of LRR is to search a lowest-rank representation for the database. The advantage of LRR is that it could better capture the global structure of data. But calculating the rank of the matrix C is NP-hard, the model use nuclear norm replace it:

$$\min\|C\|_* + \lambda\|E\|_{2,1} \text{ s.t. } X = XC + E \tag{3}$$

where $\|C\|_*$ denotes the nuclear norm of C.

Nevertheless, if there are more noisy data, it's better to calculate the logarithm-determinant function instead of the nuclear norm $\|C\|_*$. The model of CLAR is shown below:

$$\min_C \text{ logdet}(I + C^T C) + \lambda\|E\|_l \quad \text{s.t. } X = XC + E \tag{4}$$

where $\|\cdot\|_l$ denotes a selected norm which is determined by the noise type of data. Compared with LRR, the new model performs better, especially when the singular values of C vary greatly.

2.3 Subspace Segmentation via Least Square Regression

It's known that, when using LRR model, it's easy to lose some important structure of the data, but LSR which has grouping effect could avoid this problem [8, 9]. The grouping effect guarantee the data, which are in a same subspace, could be grouped together. The model is written as:

$$\min_C\|C\|_F^2 + \lambda\|E\|_F^2 \text{ s.t. } X = XC + E, diag(C) = 0 \tag{5}$$

SSC and LSR all aim at obtaining a block-diagonal [9] affinity matrix. In ideal situation, the data in the same lump belong to the same subspace, and the number of lump shows the number of subspace.

3 Proposed Method: LSGS

3.1 Motivation

For a given dataset $X = [x_1, x_2, \ldots, x_n] \in R^{(d \times n)}$, a good clustering algorithm should produce segmentation of X which show a relationship that be spares between subspaces and closely be knitted in subspaces. Thus, both local and global structures of X should be considered in the algorithm model. In other words, the resulting affinity matrix should be as sparse and block-diagonal as possible. Following such a train of thought, we build an improved SSC model, called Enhancing Local Structure and Global Structure (LSGS), by incorporating logarithm-determinant function with Frobenius norm, as following:

$$\min_C \text{logdet}(I + C^T C) + \lambda_1 \|C\|_F + \lambda_2 \|E\|_F \quad \text{s.t.} \, X = XC + E \tag{6}$$

where C denotes the representation matrix, λ_1 and λ_2 are two balanced penalty parameters, E represents the error component, and $\|\cdot\|_l$ denotes a selected norm depending on the type of data noise, such as l_1 norm, $l_{2,1}$ norm, or squared Frobenius norm.

3.2 Optimization and Algorithm

Thanks to the non-convexity of logdet $(I + C^T C)$, the model (6) is not convex. To solve this problem, we set $Z = C$ and then convert model (6) to the following equivalent form:

$$\min_C \text{logdet}(I + Z^T Z) + \lambda_1 \|Z\|_F + \lambda_2 \|E\|_l \, \text{s.t.} \, X = XC + E, C = Z \tag{7}$$

Consequently, we can figure out the model (7) by augmented Lagrange multiplier (ALM) method, and the corresponding augmented Lagrangian function is:

$$\begin{aligned} L(E, Z, Y_1, Y_2, C, \mu) &= \log \det(I + Z^T Z) + \lambda_1 \|Z\|_F + \lambda_2 \|E\|_l \\ &\quad + \text{tr}(Y_1^T (Z - C)) + \text{tr}(Y_2^T (X - XC - E)) \\ &\quad + \frac{\mu}{2} (\|Z - C\|_F^2 + \|X - XC - E\|_F^2), \end{aligned} \tag{8}$$

where $\mu > 0$ represents a penalty parameter, and Y_1 and Y_2 represent two Lagrange multipliers. Considering that the variables contained in the model are interdependent and therefore cannot be solved directly, we apply the idea of alternate minimization to update one variable. The corresponding update scheme is as follows:

$$\begin{aligned} C^{t+1} &= \arg\min_C \text{tr}((Y_1^t)^T (Z^t - C)) + \text{tr}((Y_2^t)^T (X - XC - E^t)) \\ &\quad + \frac{\mu^t}{2} \|Z^t - C\|_F^2 + \frac{\mu^t}{2} \|X - XC - E^t\|_F^2 \end{aligned} \tag{9}$$

$$Z^{t+1} = \arg\min_{Z} \log \det(I + Z^T Z) + \lambda_1 \|Z\|_F$$
$$+ \frac{\mu^t}{2} \left(\left\| Z - (C^{t+1} - \frac{Y_1^t}{\mu^t}) \right\|_F^2 \right) \tag{10}$$

$$E^{t+1} = \arg\min_{E} \lambda_2 \|E\|_l + \mathrm{tr}((Y_2^t)^T (X - XC - E))$$
$$+ \frac{\mu^t}{2} (\|X - XC^{t+1} - E\|_F^2) \tag{11}$$

Updating C. After taking the derivative of (9), it's easy to obtain a solution of C:

$$C^{t+1} \quad (I \mid X^T X)^{-1} [X^T (X \quad E') \mid J' \mid \frac{Y_1^t + X^T Y_?^t}{\mu^t}] \tag{12}$$

Updating Z. The most important part is updating Z by using the following theorem [10]:

Theorem 1. If an optimization problem meets the conditions that $F(Z)$ is a unitary invariant function. Meanwhile, the SVD of A is $A = U\Sigma_A V^T$, then the optimization problem

$$\min_{Z} F(Z) + \frac{\beta}{2} \|Z - A\|_F^2 \tag{13}$$

could be solved. The optimal solution of (13) is Z^* with SVD $U\Sigma_Z^* V^T$, where $\Sigma_Z^* = diag(\sigma^*)$; moreover, $F(Z) = f \circ \sigma(Z)$, and $\sigma(Z)$ donates a vector of noincreasing singular values of Z. Then we use the *Moreau-Yosida* proximity operator to receive $\sigma^* = \mathrm{prox}_{f,\beta}(\sigma_A)$, where $\sigma_A := diag(\Sigma_A)$, and

$$\mathrm{prox}_{f,\beta}(\sigma_A) = \arg\min_{\sigma} f(\sigma) + \frac{\beta}{2} \|\sigma - \sigma_A\|_2^2 \tag{14}$$

So we could let $F(Z) = \log\det(I + Z^T Z) + \lambda_1 \|Z\|_F$. After taking the derivative of the function, we have

$$\frac{2\sigma_i}{1 + \sigma_i^2} + \lambda_1 \sigma_i + \beta_k(\sigma_i - \sigma_{i,A}^t) = 0, \text{ s.t. } \sigma_i \geq 0, \text{ for } i = 1, \ldots, n. \tag{15}$$

After rearranging this Eq. (15), we could obtain a cubic equation with three roots. There is a solution $\sigma_i^* \in [0, \sigma_{i,A}^t]$ if $\beta = 3$, therefore the parameter $\mu^0 = 3$ will be used in our experiments.

Then it's simple to get the update of Z:

$$Z^{t+1} = U diag(\sigma_1^{t+1}, \ldots, \sigma_n^{t+1}) V^{\mathrm{T}} \tag{16}$$

Update E. The solutions for E could be obtained in a similar way. When E is for squared Forbenius norm,

$$E^{t+1} = \frac{Y_2^t + \mu^t(X - XC^{t+1})}{\mu^t + 2\lambda} \tag{17}$$

When E is for l_1 norm, let $Q = X - XZ^{t+1} + Y_1^t/\mu^t$, the solution is as follows:

$$E_{ij}^{t+1} = \begin{cases} Q_{ij} - \frac{\lambda_2}{\mu^t} \operatorname{sgn}(Q_{ij}), & \text{if } |Q_{ij}| < \frac{\lambda_2}{\mu^t} \\ 0, & \text{otherwise} \end{cases} \tag{18}$$

When E is for $l_{2,1}$ norm, if we define $Q = X - XZ^{t+1} + Y_1^t/\mu^t$, the solution is

$$[E^{t+1}]_{:,i} = \begin{cases} \frac{\|Q_{:,i}\|_2 - \frac{\lambda_2}{\mu^t}}{\|Q_{:,i}\|_2} Q_{:,i}, & \text{if } \|Q_{:,i}\|_2 < \frac{\lambda_2}{\mu^t} \\ 0, & \text{otherwise.} \end{cases} \tag{19}$$

The rest of multipliers is easy to update:

$$\begin{aligned} Y_1^{t+1} &= Y_1^t + \mu^t(Z^{t+1} - C^{t+1}) \\ Y_2^{t+1} &= Y_2^t + \mu^t(X - XC^{t+1} - E^{t+1}) \end{aligned} \tag{20}$$

Algorithm. To sum up the above argument, there is a complete process to solve the LSGS model (6). It's shown in Algorithm 1.

Algorithm1: LSGS

Input: data X, parameters $\lambda_1, \lambda_2, \alpha$, iter, $\mu^0 > 0, \gamma > 0$

Output: C

1: Initialize: $Y_1 = Y_2 = E = \mathbf{0}, C = I$.

2: For $t = 1, 2, \ldots$, iter.

3: Update C by (12) and update Z by (16).

4: Obtain E by (17), (18) or (19) according the type of noise.

5: Update Y_1 and Y_2 by (20).

6: Update the parameters μ^t by $\mu^{t+1} = \gamma\mu^t$.

Until stopping criterion is met.

---

4   Experimental Results

The effectiveness of LSGS algorithm is illustrated by three experiments on Synthetic Database, images clustering and segmentation problems, along with giving the comparison results with SSC, LSR and CLAR algorithms. The parameters of those experiments are set as follows: $\lambda_1 = 2.4$, $\lambda_2 = 0.003$, $\mu^0 = 3$, $\alpha = 2$, iter $= 100$, $\gamma = 1.1$.

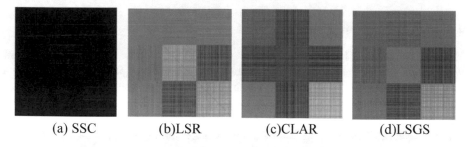

| (a) SSC | (b)LSR | (c)CLAR | (d)LSGS |

Fig. 1. Affinity matrices of 4 algorithms

4.1   Synthetic Data

The synthetic data was constructed by three 1D linear subspaces which intersect at a point, and every subspace contains 300 points. In this experiments, $\lambda_1 = 7.5$ is changed to get a better result. The clustering errors are shown in Table 1. It's clearly that the new method performs better than others. This proves the efficiency of LSGS.

For a great clustering result, the relationship should be spares between subspaces and closely be knitted in subspaces. This phenomenon could be shown in the structure of affinity matrix. Figure 1 shows the affinity matrix of several algorithms. From left to right, they belong to SSC, LRR, CLAR, LSGS separately, and it's clearly that LSGS obtains the most obvious Block-diagonal characteristic.

Table 1. Average clustering errors (%) on synthetic data

Algorithm	SSC	LSR	CLAR	LSGS
Errors	38.33	20.89	31.44	17.78

4.2   Face Clustering

In this experiment, 'Extended Yale B' database is used to prove the effectiveness of LSGS. The data consists of $n = 38$ individuals face images. We change the size of sample images as 48×42 pixels and regard each image as a data point. In Table 2, the misclassification errors of the first 8 and 10 subjects face images are shown. It's clear that the new algorithm performs better than others.

Table 2. Average clustering errors (%) on face clustering

Algorithm	8 subjects	10 subjects
SSC	6.01	7.34
LSR	26.40	27.66
CLAR	3.36	3.85
LSGS	3.30	3.44

4.3 Motion Segmentation

In this experiment, we will verify the effect of LSGS for motion segmentation problem which is increasingly showing its broad usefulness. This sort of problem is essentially a pixel level description of the image, which gives each pixel the meaning of the category. Table 3 shows average clustering errors rate on four kinds of clustering method on the 'Hopkins155' database. In this experiment, except setting $\lambda_2 = 700$, we follow experimental settings in [5] for fair comparison. After taking off several singular points, the error rate of LSGS is lower than others.

Table 3. Average clustering errors (%) on motion segmentation

Algorithm	SSC	LSR	CLAR	LSGS
Errors	2.18	2.84	1.61	1.57

5 Conclusions

The penalty of the representation matrix has always been the topic of attention for spectral clustering-based subspace clustering algorithms. Two representative processing techniques, SSC and LRR, are aimed at finding sparse or low rank representation matrices. Motivated by both, this paper puts forward a modified SSC algorithm (LSGS) which combines the logarithm-determinant function and Frobenius norm together to penalize the representation matrix. By solving a joint optimization problem, the resulting affinity matrix may be more conducive to enhancing the classification effect when the database size is large. The experiments on three datasets show that LSGS can achieve better results.

References

1. Vidal, R.: Subspace clustering. IEEE Trans. Signal Process. Mag. **28**(2), 52–68 (2011)
2. Ng, A., Weiss, Y., Jordan, M.: On spectral clustering: analysis and an algorithm. Neural Inf. Process. Syst. **2**, 849–856 (2002)
3. Elhamifar, E., Vidal, R.: Sparse Subspace Clustering: Algorithm, Theory, and Applications. IEEE Trans. Pattern Anal. Mach. Intell. **35**(11), 2765–2781 (2013)
4. Liu, G., Lin, Z., Yu, Y.: Robust subspace segmentation by low-rank representation. In: International Conference on Machine Learning (ICML), pp. 663–670(2010)

5. Kang, Z., Peng, C., Cheng, Q.: Robust subspace clustering via smoothed rank approximation. IEEE Signal Process. **22**(11), 2088–2092 (2015)
6. Lu, C.Y., Min, H., Zhao, ZQ., Zhu, L., Huang, D.S., Yan, S.: Robust and efficient subspace segmentation via least squares regression. In: Fitzgibbon, A., Lazebnik, S., Perona, P., Sato, Y., Schmid, C. (eds) Computer Vision – ECCV 2012. ECCV 2012. Lecture Notes in Computer Science, vol 7578, pp. 347–360. Springer, Berlin (2012)
7. Wu, Z., Yin, M., Zhou, Y.: Robust Spectral Subspace Clustering Based on Least Square Regression. Neural Process. Lett. **48**, 1359–1372 (2018)
8. Hu, H., Lin, Z., Feng, J., Zhou, J.: Smooth representation clustering. In: IEEE Conference on Computer Vision and Pattern Recognition (CVPR), pp. 3834–3841 (2014)
9. Zhang, S., Li, Y., Cheng, D.: Efficient subspace clustering based on self-representation and grouping effect. Neural Comput. Appl. **29**, 51–59 (2018)
10. Kang, Z., Peng, C., Cheng, J., Cheng, Q.: Logdet rank minimization with application to subspace clustering. Comput. Intell. Neurosci. **2015**, 68 (2015)

E-Enabled Systems

An Efficient Bridge Architecture for NoC Based Systems on FPGA Platform

S. P. Guruprasad[1(✉)] and B. S. Chandrasekar[2]

[1] Department of ECE, Jain University, Bangalore, India
spgpl306@gmail.com
[2] CDEVL, Jain University, Bangalore, India

Abstract. The growing demands at the electronic consumer application areas have lent the requirement of embedded devices to be integrated on the same System on Chip (SoC). The SoC uses on-chip bus interconnection with shared memory communication. However, these buses are not scalable and limited to specific interface protocol. The Network on chip (NoC) provides a better interconnection solution for the SoC with reliable and scalable features. The bridge architecture is necessary to communicate the SoC through the NoC paradigm. Thus, the manuscript introduces an efficient bridge with Ethernet-Media Access Control (MAC) and also presented an interconnection of the bridge architecture having NoC based systems on targeted FPGA. The bridge architecture is consists of FIFO buffers, Serializer, priority based Arbiter, Credit counter Packet formation for Ethernet-MAC Transceiver module followed by packet parser and deserializer. The bridge with a single router and 2X2 NoC based systems are designed by using congestion free adaptive XY-routing. The proposed bridge architecture and the bridge with NoC are implemented over Artix-7 FPGA with prototyping. The performance analysis is considered in terms of Average latency, Maximum throughput at different packet injection rate for a bridge with NoC based systems.

Keywords: Bridge · Ethernet-MAC · FPGA · NoC · Network interface · Router

1 Introduction

The usage of electronic consumer devices in different applications has made complexity and size enhancement in a system on chip (SoC). The design of SoC composed of the number of processing elements as processors, IP cores, memory devices, and other peripherals on a single chip. This SoC design needs to process the synthesis, implementation, verification, validation, and prototyping on chips or Field programmable gate array (FPGA) which induces more complexity and challenges in design analysis. The traditional SoC design solutions are impractical to implement large resources on a single chip. Hence, the partition of the larger SoC designs is needed to be done into smaller modules. Consideration of the interface protocols is necessary for communication of each module and multi-chip interconnection method to communicate further. The existing on-chip interconnects relay on the bus based

© Springer Nature Switzerland AG 2020
F. Xhafa et al. (Eds.): IISA 2019, AISC 1084, pp. 377–383, 2020.
https://doi.org/10.1007/978-3-030-34387-3_46

protocols, and it offers shared memory communication along with fixed memory models. The buses had limited support for data transfer and suited to specific interface protocol. Network on Chip (NoC) is the best interconnect solutions in SoC's designs and address the scalability and reliability features. Thus, to interface any device on Multiprocessing SoC (MPSoC) using NoC paradigm, a bridge module is essential and is connected through the internet or LAN. The internet as a network provides resource information. Hence, the NoC based systems use interface protocols and are different from the Internet. Thus, bridge architecture is needed to be designed for protocol conversions [1–3]. The proposed bridge architecture uses the Ethernet-MAC protocol, and it provides long-distance wired communications with scalable connections, and it is suitable for on-chip and off-chip Bridging.

In this manuscript, introduces an efficient, cost-effective hardware architecture of a bridge with the inclusion of Ethernet-MAC and also bridge interconnection with NoC based systems. The manuscript is categorized with Sect. 2 explaining the review of the existing work on interface interconnects on SoC, interface protocols on NoC and bridging approaches. Section 3 describing the hardware architecture of the bridge. The bridge with NoC architecture is defined in Sect. 4. Section 5 deals with the synthesis results and performance evaluation of the proposed bridge with NoC architecture. Finally, the analysis of overall work with results and conclusions are presented in Sect. 6.

2 Related Work

A review of the relevant existing works is considered in this section which includes different interface protocols for NoC and SoC, bridging approaches with NoC and its applications. Vani et al. [4] describe the Interface Bridge Module using Advanced Microcontroller Bus Architecture (AMBA) on-chip bus based Advanced High-performance Bus (AHB) and Advanced Peripheral Bus (APB). A functional verification was conducted where it is observed that AHB2APB bridge model supports low latency and high bandwidth. Beyranvand et al. [5, 7] introduces the bridging schemes in NoC based systems to improve the quality of service (QoS). The bridging scheme is introduced between the multiple chips like ASIC, FPGA, Digital signal processing (DSP), etc. of each chip contain NoC systems. The QoS provided as Guaranteed Throughput (GT), and Best effort (BE) traffic to improve latency and bandwidth of the multiple connections for input traffic are analyzed on NoC based sub-systems. Michel et al. [6] present the NoC bridging solution for dynamically reconfigurable architectures. The AHB to SoC-wire bride architecture is designed in a single configurable controller (static FPGA), which provides a connection between a SoC-wire and processor bus systems to communicate the data processing Unit (DPU) in reconfigurable FPGA's. The data rate is calculated for a bridge in full duplex mode. The Kyriakakis et al. [8] explains the inter-chip NoC communication bridge on FPGA which supports the globally-asynchronous locally synchronous and Fault-tolerant features. The bridge model incorporates the Forward Error Correction (FEC) technique for fault tolerance and interconnection with NoC, Serial Peripheral Interface (SPI) protocol is used.

It has been noticed from the review of the existing works for bridge based inter-connection for NoC systems that very few hardware-based design approaches are designed for bridge architecture and adopted in the NoC based systems. Most of the existing works are considered bus based protocols for interconnection with NoC and lacks with higher hardware complexities and cost-effective solutions. Also, very less work on On-chip FPGA prototyping for bridge architecture with the inclusion of Ethernet-MAC are exist. Most of the hardware-based bridge architecture uses off-chip Xilinx Ethernet-MAC Wrapper. Thus a cost-effective standalone bridge architecture with the inclusion of Ethernet-MAC for NoC based systems need to be designed to fulfill the above gaps with better outcomes. The next section explains the detailed architecture of the bridge with Ethernet-MAC.

3 Bridge Architecture

The full bridge architecture is presented in Fig. 1 which consists of FIFO Buffers, Serializer, packet formation for Ethernet-MAC Transceiver, packet parser, Deserializer, credit counter, and arbiter modules. The bridge architecture received data from the network interfaces via different ports and stored in individual FIFO. Asynchronous FIFO is designed to receive the data values and write it sequentially into the memory locations. The data values are read out sequentially from the memory locations. The same clock signal can be used for write and read data values. The FIFO data width is fixed to 32-bit. The FIFO memory locations are changed based on the packet injection rate (PIR) for NoC based systems.

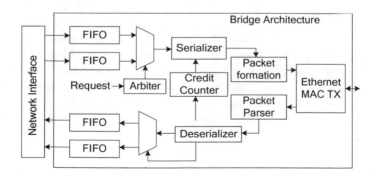

Fig. 1. The Bridge Architecture with Ethernet-MAC

If two or more FIFO wants to communicate in a bridge, the priority based arbitration are considered between the connections. The multiplexor receives the FIFO's data parallelly and generates the 32-bit output data based arbitration permission. The arbitration requests are made through Data Transaction Level (DTL) port with memory-mapped access. The priority-based arbitration acts as a scheduler and provides the QoS requirements in every connection. Based on the request, the Serializer receives data parallelly and generates the output serially, which is used in a payload of an

Ethernet-MAC Frame. The Serializer converts 32-bits data into 8-bit data like Parallel In serial Out (PISO) manner. The deserializer gives the credits as inputs to a credit counter. The 2-bit credit counter provides the counter data, for every count, the parallel data converted to serially and passed to a payload. Later, the packets are framed as per IEEE 802.3 specification and adopted in bridge architecture design and the serialized data stored as a payload for Ethernet–MAC [9]. Figure 2 shows the Ethernet format containing 7-bytes of a preamble, 1-byte of a start of delimiter (SFD), 6-bytes of a source address and destination address, 2-bytes of Type/Length which gives the MAC data information. The 1500-bytes of payload and 4-bytes of Frame check sequence (FCS) detects the errors using Cyclic Redundancy Check (CRC).

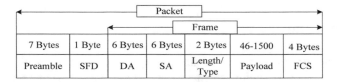

Fig. 2. Ethernet Packet format

The Ethernet MAC Transceiver architecture [9] mainly consists of Transmitter and Receiver FSM, Frame Length Counter (FLC), CRC, Receiver-FIFO and Transmitter-FIFO. The Ethernet MAC check the half or full duplex mode of operations based on carrier sense and collision signals. The receiver-FSM receives the data packets, and if the data packet is valid, then it will process further else discard. The transmitter and receiver-FSM works based on a medium request. Packet parser receives only the payload data and breaks into 8-bit serial data. The deserializer acts a Serial in parallel out (SIPO) and receives the 8-bit data serially till the last payload data. Then the shifting operation is performed with counting for parallel conversion of 32-bit. These counted values are considered as input to the credit counter. The 32-bit deserializer data is regarded as inputs to the demultiplexer. Based on a counter, the demultiplexer generates the two or more data and are considered as inputs to FIFO buffers. The transmitted FIFO's data must be the same as received FIFO's data to validate the bridge architecture.

4   Bridge with NoC

The bridge architecture connected to NoC based systems supports off-chip communications for data, on-chip or off-chip data control, scheduling between the multiple connections, and off-chip various connections of the NoC. The chips are considered as FPGA Devices or ASIC. For prototyping the bridge architecture with NoC, the FPGA devices are considered. The placement of the bridge with 2x2 NoC architecture is represented in Fig. 3. The data packets are received via the internet or LAN to Ethernet-MAC Module. The Ethernet-MAC Module receives the valid packets and transmits to bridge architecture. If the packets are not valid, it will discard the packets. The bridge received the data packets and passed to the router module via the Network interface (NI). In Fig. 3, 2x2 mesh topology based NoC is designed, and It mainly contains 4- Routers R1, R2, R3 and R4 and all are interconnected with link wires. The first Router (R1) receives the data packets from the Bridge and perform the communication based on the destination address. The Router architecture mainly consists of the five –port input registers, followed by packet formation with priority based arbitration and adaptive-XY routing algorithm. To improve the hardware complexities in NoC based systems, an efficient router architecture is designed using adaptive –XY routing algorithm. In this design, the router R4 is considered as the destination location. Based on the routing algorithm, the bridge data packets will be reached to the routing destination.

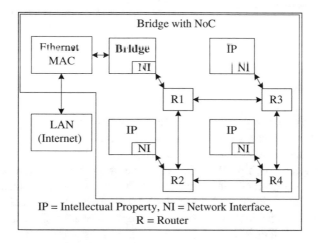

Fig. 3. The Placement of Bridge with NoC

5 Results and Analysis

The bridge with NoC architecture is designed and synthesized using Xilinx Platform and implement on Artix-7 FPGA Development board. The performance analysis of this work is evaluated using average latency and throughput for Packet injection rate (PIR) are presented in Fig. 4(a) and (b) respectively.

The latency of bridge-router uses 740 clock cycles and bridge with 2x2 NoC uses 2961 clock cycles at 0.6 PIR. The maximum throughput of the bridge NoC is calculated by using the number of IP's followed by data packet size (32-bit), PIR and Maximum operating frequency (MHz) obtained. An assumption is made that the PEs are connected to the NoC boundary via network interfaces. The maximum throughput of bridge-router is 4.615 Gbps and bridge with 2X2 NoC is 17.132 Gbps at 0.6 PIR.

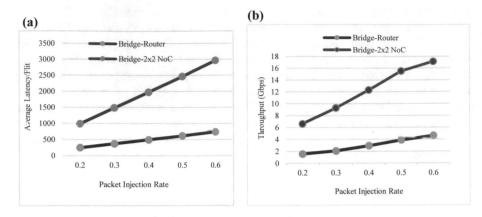

Fig. 4. (a) Average Latency (b) Maximum throughput v/s PIR of Bridge with NoC

6 Conclusion

This manuscript presents the design of standalone bridge architecture with Ethernet-MAC and also bridge with a single router and 2X2 NoC based systems. The proposed bridge architecture increases the robustness with the Ethernet-MAC inclusion and targets efficiently on on-chip FPGA Devices. The bridge with a single router and 2X2 NoC is designed using congestion free adaptive-XY routing. The performance analysis of the bridge with a single router and 2X2 NoC includes the average latency and throughput for different PIR. The maximum throughput of a bridge- with a single router is 4.615 Gbps and bridge with 2X2 NoC is 17.132 Gbps at 0.6 PIR. This architecture can be incorporated in futuristic researches with the security features to bridge and NoC based systems to strengthen the data packets from attacks.

References

1. Jiang, S.: Rapid implementation of the MAC and interface circuits for the wireless LAN cards using FPGA. J. Commun. Netw. **1**(3), 201–212 (1999)
2. Elkeelany, O., Chaudhry, G.: A prototype of wideband/ethernet bridge using WEMAC. In: The 2002 45th Midwest Symposium on Circuits and Systems, 2002, MWSCAS-2002, vol. 1, p. I-595. IEEE (2002)
3. Kommineni, B.P., Srinivasan, R., Holsmark, R., Johansson, A., Kumar, S.: Modeling and Evaluation of a Network on Chip (NoC)–Internet Interface (2005)
4. Roopa, M.: Design of AMBA based AHB2APB bridge. IJCSNS **10**(11), 14 (2010)
5. Nejad, A.B., Martinez, M.E., Goossens, K.: An FPGA bridge preserving traffic quality of service for on-chip network-based systems. In: 2011 Design, Automation & Test in Europe, pp. 1–6. IEEE (2011)
6. Michel, H., et al.: AMBA to SoCWire network on chip bridge as a backbone for a dynamic reconfigurable processing unit. In: 2011 NASA/ESA Conference on Adaptive Hardware and Systems (AHS), San Diego, CA, pp. 227–233 (2011)

7. Nejad, A.B., Molnos, A., Martinez, M.E., Goossens, K.: A hardware/software platform for QoS bridging over multi-chip NoC-based systems. Parallel Comput. **39**(9), 424–441 (2013)
8. Kyriakakis, E., Ngo, K., Öberg, J.: Implementation of a fault-tolerant, globally-asynchronous-locally-synchronous, inter-chip NoC communication bridge on FPGAs. In: 2017 IEEE Nordic Circuits and Systems Conference (NORCAS): NORCHIP and International Symposium of SoC, Linkoping, pp. 1–6 (2017)
9. Guruprasad, S.P., Chandrasekar, B.S.: An optimized packet transceiver design for ethernet-MAC layer based on FPGA. In: International Conference on Intelligent Data Communication Technologies and Internet of Things, pp. 725–732. Springer, Cham (2018)

A Method for Acquiring Experiential Engineering Knowledge from Online Engineering Forums

Zuhua Jiang[1(✉)], Bo Song[2], Xiaoming Sun[1], and Haili Wang[1]

[1] Department of Industrial Engineering and Management,
Shanghai Jiao Tong University, Shanghai, China
zhjiang@sjtu.edu.cn
[2] China Institute of FTZ Supply Chain, Shanghai Maritime University,
Shanghai, China

Abstract. Question and answer (Q&A) is the most primitive and common way of knowledge exchange. There is a lot of empirical knowledge accumulated in the Q&A record of online engineering forums. In order to acquire experiential engineering knowledge from the Q&A record, first an ontology of experiential engineering knowledge is constructed by referring to authoritative and structured domain text, whereby a formal conceptualization of a specific engineering field is obtained. In view of the relatively lower quality of empirical Q&A compared with authoritative knowledge source, the quality of online Q&A in the field of computer-aided engineering is evaluated by referring to the relevant methods in the quality evaluation of community Q&A. Finally, by selecting high-quality Q&A and express it as ontology concepts and attributes, the acquisition of experiential engineering knowledge is achieved.

Keywords: Knowledge acquisition · Experiential engineering knowledge · Online engineering forum · Q&A

1 Introduction

The acquisition and reuse of experiential engineering knowledge can greatly improve the efficiency of knowledge accumulation and knowledge management and thus enhance the competitiveness and creativity of knowledge-intensive enterprises. With the continuous development of Web 2.0 technology, large-scale knowledge collaboration platforms are constantly created and used. The accumulation of knowledge also shows the characteristics of high update frequency, wide range of participation and increased demand for automatic knowledge acquisition. As one of the representatives of the knowledge collaboration platform, online engineering forum hosts a large number of problem discussions, in which the situation and solution of engineering problems have been elaborated through multiple rounds of interaction between the questioner and answers. These engineering problem scenarios and their corresponding solutions constitute experiential engineering knowledge. This paper presents a method to acquire empirical knowledge from online engineering forums.

© Springer Nature Switzerland AG 2020
F. Xhafa et al. (Eds.): IISA 2019, AISC 1084, pp. 384–391, 2020.
https://doi.org/10.1007/978-3-030-34387-3_47

2 Literature Review

In the field of experiential knowledge acquisition, Gavrilova et al. summarized the characteristics and limitations of several commonly used knowledge acquisition methods and tools, such as interview, observation, brainstorming, questionnaire, role playing, roundtable, etc., and concluded that observation and role playing were more suitable for experiential knowledge acquisition since they could help better understand the motivation and inclination of experts [1]. Argote et al. collected targets, technologies, strategies, motivations and other organizational context information in the process of organizational learning, which were then used to describe and share the experiential knowledge generated in the learning process of the organization [2]. Ruiz et al. extracted the situation, event, analysis and solution from empirical cases, and used "attribute-value" pairs to express the experiential knowledge extracted from the cases [3]. Chen used the three elements of question, reason and solution to constitute the representation model of experiential engineering knowledge, and conducted knowledge reasoning by constructing ontology [4]. Liu et al. proposed a two-stage method for acquiring experiential tacit knowledge: with the recorded utterance of experienced engineers, they first obtained the tacit knowledge in the form of natural language and then applied the key graph algorithm to obtain the core content of experiential tacit knowledge [5]. Song et al. proposed to extract Q&A modes from engineering task context and transform them into experiential engineering knowledge by text-to-ontology mapping [6]. In terms of Q&A quality evaluation, Blooma et al. considered accuracy, completeness and relevancy as three important indicators to measure the quality of answers [7]. Shah et al. designed 13 evaluation indicators to score the answers in Yahoo! Q&A community and identified the most effective indicators by comparing the predicted scores with the actual scores of the answers [8]. Nie et al. designed a method including offline learning and online search to select the best answer to a question [9].

In summary, the existing experiential knowledge acquisition study has shortcomings in terms of the quantity and diversity of knowledge source. At the same time, the empirical Q&A contained in online forums as a large and extensive source of knowledge has not been fully utilized. Through quality evaluation and formal representation of the empirical Q&A in online engineering forums, this paper will achieve more efficient experiential engineering knowledge acquisition.

3 Process of Experiential Engineering Knowledge Acquisition

3.1 Construction of Experiential Engineering Knowledge Ontology

Ontology is an explicit description of conceptualization, which includes representative vocabulary for sharing domain knowledge, as well as categories, relationships, functions and other objects defined based on this vocabulary [10]. In order to simplify the process of ontology construction and ensure the authority of ontology concept source, we use the engineering textbook catalog as the input of ontology construction. Figure 1 shows an excerpt from the tutorial of the CAE software ANSYS 10.0, which defines the basic concepts and operations in the domain with concise noun phrases and verb-object phrases. The steps of ontology construction from the catalog shown in Fig. 1 are as follows:

Step 1. Put the root title into the queue to be processed;

Step 2. Get the first title t from the queue;

Step 3. Mark the part of speech of the words, remove stop words, change nouns into singular forms, and change verbs into root forms;

Step 4. Take out a noun phrase $w_i w_{i+1} \ldots w_{j-1} w_j$ which has not been analyzed in t (w_j is a noun and the word following it is not a noun. w_i is a noun, verb or adjective. $w_{i+1} \ldots w_{j-1}$ are nouns). If it has already expressed a concept in the ontology, go to step 6, otherwise step 5 is executed;

Step 5. Create ontology concept $c(w_i w_{i+1} \ldots w_{j-1} w_j)$;

Step 6. Create an instance $v(w_i w_{i+1} \ldots w_{j-1} w_j)$ of the concept $c(w_i w_{i+1} \ldots w_{j-1} w_j)$;

Step 7. For each instance $v'(w_m w_{m+1} \ldots w_{n-1} w_n)$ in the father title of the current:

(1) If w_i and w_m are nouns, and $w_j = w_n$, then connect $c(w_i w_{i+1} \ldots w_{j-1} w_j)$ and $c'(w_m w_{m+1} \ldots w_{n-1} w_n)$ using an Is_a relation;

(2) If $w_j \neq w_n$, then connect $v(w_i w_{i+1} \ldots w_{j-1} w_j)$ and $v'(w_m w_{m+1} \ldots w_{n-1} w_n)$ using a Part_of relation;

Step 8. If all noun phrases in t have been processed, go to the next step; otherwise go back to step 4;

Step 9. Take out all direct subtitles of t and add them to the top of the queue to be processed in their original order;

Step 10. If the queue is empty, end; otherwise return to Step 2.

```
2. Structural Static Analysis .................................................
     2.1. Linear vs. Nonlinear Static Analyses ...............................
     2.2. Performing a Static Analysis .......................................
          2.2.1. Build the Model .............................................
               2.2.1.1. Points to Remember ..................................
          2.2.2. Set Solution Controls .......................................
               2.2.2.1. Access the Solution Controls Dialog Box ........
               2.2.2.2. Using the Basic Tab .................................
```

Fig. 1. Input information for ontology construction

3.2   Quality Evaluation of Empirical Q&a

The quality of empirical Q&A is often in various levels. Low-quality Q&A often contains unrelated or emotional statements which are not suitable for acquiring experiential engineering knowledge and often focuses on very broad or abstract questions, such as "how to do finite element analysis". Thus low-quality Q&A is not helpful to solve specific engineering problems. In order to obtain more useful empirical knowledge, this paper designs the following 13 indicators to evaluate the quality of empirical Q&A.

(1)   Number of first person pronouns (PPR1st) – candidate indicator of answer
(2)   Number of second person pronouns (PPR2nd) – candidate indicator of answer
(3)   Number of verbs (Verb) – candidate indicator of question and answer
(4)   Number of nouns (Noun) – candidate indicator of question and answer
(5)   Number of adjectives (Adj) – candidate indicator of question and answer
(6)   Text length (Length) – candidate indicator of question and answer
(7)   Proportion of overlap of words between Q&A (Overlap) – candidate indicator of answer
(8)   Proportion of new words (words that do not appear in questions) in answer (Inova) – candidate indicator of answer
(9)   Inverse Document Frequency of new words in answer (Idf_inova) – candidate indicator of answer. If the responder introduces unusual concepts, it may mean the responders think profoundly and innovatively. In order to measure whether the concepts are usual or not, the IDF values of new words in an answer are calculated, and the highest five values are averaged to generate the value of this index.

$$IDF(w) = \log \frac{|D|}{|\{d|d \in D, w \in d\}|} \tag{1}$$

In the above formula, w is a word, D is the set of existing empirical Q&A documents, and d is one empirical Q&A document.

(10)   The semantic relatedness between new words in the answer and words in the question (Asc_inova) – candidate indicator of answer. The concepts introduced in high quality answers should be related with the question. Pointwise Mutual Information (PMI) is used here to estimate the semantic relatedness between words, and the mean of the maximum PMI calculated for the new words in an answer and the words in the question is taken as the value of this index.

$$PMI(w_1, w_2) = \log \frac{p(w_1, w_2)}{p(w_1)p(w_2)} \tag{2}$$

where $p(w)$ is the proportion of sentences containing word w in all the sentences in the Q&A library, and $p(w_1, w_2)$ is the proportion of sentences containing both words w_1 and w_2.

(11)   Average concept depth (Avg_depth) – candidate indicator of question and answer. The concepts in the ontology form a hierarchical structure according to the relation Part_of and Is_a. Thus the concepts in Q&A have their corresponding depths in the ontology. As the concepts with greater depths express more specific meanings, they may signify higher quality of questions and answers.

(12) Extended word overlap ratio based on ontology (Ext_overlap) – candidate indicator of answer. To further evaluate the relatedness between question and answer, ontology concepts directly related with the new words in an answer are selected and added to the answer, then the proportion of overlapping words is calculated again for this index.

(13) User experience (User_exp) – candidate indicator of question and answer. Experiential knowledge is highly individualized. Because of the complexity and diversity of engineering questions, high quality Q&A may not have obvious linguistic characteristics, but it may relate with responsible and honorable users. As users with strong sense of responsibility and honor tend to help more people to solve problems, thus the number of users that a user has helped (Indegree) could help measure the quality of Q&A. We use log (1 + Indegree) as a user's experience value.

3.3   Ontology Concept Mapping

Ontology concept mapping completes the transformation from text to knowledge. Its essence is to find corresponding domain ontology concepts for the words in Q&A, and then use ontology concepts to annotate the Q&A text semantically. For the noun phrases in Q&A, we first try to map them to the concepts that match them perfectly in the ontology. If there is no ontology concept that perfectly matches a phrase to be mapped, we adopt a semantic similarity based mapping method and map the phrase to the ontology concept whose semantic similarity with the phrase is the largest and surpasses 0.7 in value. For semantic similarity, it is calculated using the following formula:

$$Sim(c, c') = \frac{1}{|c|} \sum_{w \in c} \max_{w' \in c'} \frac{PMI(w, w')}{PMI(w, w)} \tag{3}$$

where c is a phrase to be mapped and c' is a concept in the ontology, w and w' are words in c and c' respectively, $|c|$ is the number of words in c.

4   Case Study

Totally 16106 Q&A threads in the online CAE forum XANSYS are used as the source of experiential engineering knowledge. The threads involve 54,878 posts and 5,090 users. The IDF value of words, the PMI value between word pairs and the user experience measured by Indegree are calculated based on these threads. The concept depth and the extended word overlapping ratio between question and answer are calculated based on the ontology constructed in Sect. 3.1. In order to identify the real effective indicators for evaluating Q&A quality, a data set containing 45 questions of high quality, 195 questions of low quality, 48 answers of high quality and 271 answers of low quality is selected, and Logistic Regression is performed on this data set. The result of Logistic Regression is shown in Figs. 2 and 3.

It can be seen from Fig. 2 that verb number (Verb) and user experience (User_exp) are effective indicators for judging the quality of CAE questions. The corresponding quality evaluation equation is:

$$Q_{CAE}(q) = e^{-7.686 + 0.204*Verb + 0.428*User\_exp} / \left(1 + e^{-7.686 + 0.204*Verb + 0.428*User\_exp}\right) \quad (4)$$

It can be seen from Fig. 3 that, although the number of first-person pronouns (PPR1st), the proportion of overlap of words between Q&A (Overlap) and the inverse document frequency of new words in the answer (Idf_inova) have a role in judging the quality of CAE answers, they work poorly in identifying high quality answers (18.8%). This means that the essence of high quality answers does not lie in the linguistic features of these answers but in their usefulness in helping people to solve the problem. Based on this conclusion, we use positive comment templates and reference analysis to identify high-quality answers. Positive comment templates are regular expressions such as "that(is|'s) (it|right|correct)", "(that|it) work", "you(are|'re) (right|correct)", "thank(s a lot| you very)". After using them to find the posts containing positive comments, the high quality answers are identified by looking for the posts referred to by these comments.

Classification Table[a]

Observed			Predicted		
			GoodQues		Percentage
			0	1	Correct
Step 4	GoodQues	0	189	6	96.9
		1	14	31	68.9
	Overall Percentage				91.7

a. The cut value is .500

Variables in the Equation

		B	S.E.	Wald	df	Sig.	Exp(B)
Step 4[a]	Verb	.204	.023	36.853	1	.000	1.226
	User_exp	.428	.188	5.161	1	.023	1.534
	Constant	-7.686	1.191	41.622	1	.000	.000

a. Variable(s) entered on step 4: User_exp.

Fig. 2. Effective indicators for quality evaluation of CAE questions

Classification Table[a]

Observed			Predicted		
			GoodAnsw		Percentage Correct
			0	1	
Step 3	GoodAnsw	0	265	6	97.8
		1	39	9	18.8
	Overall Percentage				85.9

a. The cut value is .500

Variables in the Equation

		B	S.E.	Wald	df	Sig.	Exp(B)
Step 3[a]	PPR1st	-.135	.070	3.766	1	.052	.873
	Overlap	13.549	2.865	22.366	1	.000	7.663E5
	Idf_inova	1.146	.452	6.427	1	.011	3.146
	Constant	-6.728	1.566	18.461	1	.000	.001

a. Variable(s) entered on step 3: PPR1st.

Fig. 3. Effective indicators for quality evaluation of CAE answers

Applying the method proposed in this paper, we evaluate the quality of the questions in the online forum XANSYS, select the discussion threads with high quality questions and search for positive comments in the threads to identify high quality answers. By mapping the noun phrases in high quality Q&A to the domain ontology, we acquire experiential engineering knowledge as shown in Table 1.

Table 1. A case of acquired experiential knowledge

<QA title="ANSYS STRUC Nonlinear layered material properties">
  <post author= "Cutter P. ">
    <question> Dear all, I am having problems in modeling a <OWL: ~layered material> layered material </OWL: ~layered material> where in some of the layers the <OWL: ~Young's modulus> Young's modulus </OWL: ~Young's modulus> in the x-direction is considerably (factor of 1e9) larger than that in the y-direction. It is trying to simulate the effects of a panel where some of the layers have been burnt and are just uni-directional fibers. The result is that there is a very large increase in <OWL: ~deflection> deflection </OWL: ~deflection> which is not what I have measured experimentally. Has anyone else experienced anything similar or know of any way to tackle the problem? </question>
  </post>
  <post author="danbohlen">
    <answer> Are you trying to model the individual layers or the <OWL: ~bulk effect> bulk effect </OWL: ~bulk effect> of all the layers? In a bulk sense you may not be able to simulate a <OWL: ~ratio> ratio </OWL: ~ratio> that big. There's a thing about the material being <OWL: ~positive definite> "positive definite" </OWL: ~positive definite>...</answer>
  </post>
</QA>

5 Conclusion

A large number of discussion records have accumulated in online engineering forums for various engineering problems. As a source of empirical knowledge, they make up for the shortcomings of existing knowledge sources in terms of knowledge quantity and diversity. Through quality evaluation and formal representation of Q&A in online engineering forums, this paper achieves more efficient experiential engineering knowledge acquisition. The proposed method uses authoritative and structured domain text for formal concept extraction and ontology construction. By setting up 13 quality evaluation indicators, the quality of forum Q&A in the field of computer-aided engineering is evaluated. High-quality Q&A with its key phrases mapped to ontology concepts and attributes is acquired as experiential engineering knowledge. The proposed method integrates natural language processing, knowledge engineering and machine learning approaches to achieve acquisition of experiential engineering knowledge with a high degree of automation, thereby laying a foundation for the in time reuse of experiential engineering knowledge.

Acknowledgments. This work was supported by National Natural Science Foundation of China (No. 71601113, 71671113).

References

1. Gavrilova, T., Andreeva, T.: Knowledge elicitation techniques in a knowledge management context. J. Knowl. Manage. **16**(4), 523–537 (2012)
2. Argote, L., Miron-Spektor, E.: Organizational learning: from experience to knowledge. Organ. Sci. **22**(5), 1123–1137 (2011)
3. Ruiz, P.P., Foguem, B.K., Grabot, B.: Generating knowledge in maintenance from Experience feedback. Knowl.-Based Syst. **68**, 4–20 (2014)
4. Chen, Y.J.: Development of a method for ontology-based empirical knowledge representation and reasoning. Decis. Support Syst. **50**(1), 1–20 (2010)
5. Liu, L., Jiang, Z., Song, B.: A novel two-stage method for acquiring engineering-oriented empirical tacit knowledge. Int. J. Prod. Res. **52**(20), 5997–6018 (2014)
6. Song, B., Jiang, Z., Liu, L.: Automated experiential engineering knowledge acquisition through Q&A contextualization and transformation. Adv. Eng. Inf. **30**(3), 467–480 (2016)
7. Blooma, M.J., Chua, A.Y.K., Goh, D.H.L.: A predictive framework for retrieving the best answer. In: Proceedings of the 2008 ACM Symposium on Applied Computing, pp. 1107–1111 (2008)
8. Shah, C., Pomerantz, J.: Evaluating and predicting answer quality in community QA. In: Proceedings of the 33rd International ACM SIGIR Conference on Research and Development in Information Retrieval, pp. 411–418 (2010)
9. Nie, L., Wei, X., Zhang, D., et al.: Data-driven answer selection in community QA systems. IEEE Trans. Knowl. Data Eng. **29**(6), 1186–1198 (2017)
10. Gruber, T.R.: A translation approach to portable ontology specifications. Knowl. Acquis. **5**(2), 199–220 (1993)

Concept, Method and Application of Computational BIM

Lushuang Wei[1(✉)], Shangwei Liu[1], Qun Wei[1,2], and Ying Wang[2]

[1] North China University of Water Resources and Electric Power,
Zhengzhou, China
weils@ncwu.edu.cn
[2] University of Chinese Academy of Sciences, Beijing, China

Abstract. Dynamo which is developed based on the basic concept of visual programming has become a platform mainly used in AEC of construction industry. Dynamo has three prominent features: (1) with internal architecture completely independent of any software platform, Dynamo has independent graphics engine nodes, and through node parameterization, it can make general geometric modeling, edit and extend line, surface, shape and entity; (2) Dynamo has an independent, full-featured calculation engine which can process many conventional mathematical operations and logic decisions, and the flexible input and output can export and store data visually; (3) Dynamo's open source code provides convenience for users to develop and compile. Developers, according to their different needs, can import python or other languages into Dynamo to extend it into a more professional and powerful software system. Combined with practical engineering applications, this paper uses Dynamo's enormous computing power and BIM's visualization, coordination, simulation, optimization and drawing to solve problems in engineering design. The emergence of Dynamo has pushed BIM into a new stage of integrated calculation & programming and graphics generation.

Keywords: Digital graphic medium · Computational BIM · Dynamo · CAD · Graphics and image processing

1 Introduction

In terms of construction engineering industry, China is now in a new stage of BIM learning, application and development. Most people simply think that BIM is just an abbreviation of Building Information Modeling. While in fact, BIM includes many connotations: Building Information Management, Building Information Manufacture, Building Information Business and Building Information Built. Building generally refers to the construction of all civil, water and electricity works, and it is also a technology that visual modeling, simulation, coordination, optimization and automatic drawing are realized on a computer. Therefore, the author believes that it may be more appropriate to explain BIM as Building Intelligence Modeling. At present, many scholars have made researches in this regard [1].

© Springer Nature Switzerland AG 2020
F. Xhafa et al. (Eds.): IISA 2019, AISC 1084, pp. 392–398, 2020.
https://doi.org/10.1007/978-3-030-34387-3_48

The emergence and development of BIM benefit from the common advancement and improvement of computer hardware and software technology, graphics and image technology, computer vision technology and computer numerical analysis technology. The essence of BIM is defined as follows: BIM technology is essentially a dynamic correlation system with digital graphics of structural engineering and related information integrated, among which, related information updates with change in graph structure, and digital graphics is the main line of BIM technology. The concept, construction method, description criteria, theoretical basis and connotation of the proposed digital graphic medium form the mapping format and simulation module that the research object in the engineering structure corresponds to in the natural space and computer space, which makes the "graphics" a five-dimensional information carrier that combines geometric features and dynamic changes of non-graphical properties with time. For such engineering structure described by the digital graphic information system, data in the process of design, analysis, production and installation can be directly obtained and fed back from graphics, which presents the characteristics of big data. This is the advantage and motivation that BIM technology keeps developing [2–7].

It can be seen that BIM technology is an information-based product. Among the popular three mainstream software, namely Autodesk CAD, Bently and Catia, Autodesk's CAD series products enjoy a large user base in engineering, such as Revit, Civil 3D, 3Dmax, Maya and Inventor. Autodesk Revit, originally used in building design, has become one of the tools of choice for BIM. Since 2010, Revit has integrated Revit Architecture, Electronic and Piping. It uses five primitives (stereo, component, annotation, fundamental, and view) to depict various complex structure graphs such as buildings, machines and electrical tubes and their corresponding non-graphic properties. It should be noted that the modeling method provided by Revit is quite stiff and the mapping method is also limited; it requires a large number of parameterized family libraries to support the structure construction. But Revit has a prominent advantage in BIM information acquisition and management, making it a popular BIM modeling tool.

2 Concept and Content of Computational BIM

Lan Keough, when working for a company in New York, USA, often struggled with data exchange between Rhino and Revit. Inspired by another graphics software - Grasshopper, he tried to make similar development and experiment in Revit according to the idea of driving the structure of Rhino in a way generated by parametric calculations, and based on the basic concept of visual programming, he began to compile "dynamo", with the purpose of making it a key platform used in AEC of construction industry. Dynamo's core functions are all related to Revit under the BIM environment, but independent of Revit. Actually, Dynamo introduces Revit's related functions as a module. Dynamo paves the way for Revit's close collaboration extensions. With the development and application of BIM technology, Dynamo's function and power have been gradually shown.

2.1 Dynamo Has Three Prominent Features

(1) With internal architecture completely independent of any software platform, Dynamo has independent graphics engine nodes, and through node parameterization, it can make general geometric modeling, create and edit straight lines, curves, curved surfaces, polygons, polyhedral surface shapes and solids, as well as break points, difference sets, union sets and intersections of geometric solids.

(2) Dynamo has an independent, full-featured calculation engine that can meet a variety of common computing needs, such as creation and editing of multiple lists, sorting and interference testing, mathematical operations and logic decisions, flexible input and output, visual export and storage of data.

(3) Dynamo's open source code provides convenience for users to download and compile. Developers, according to their different needs, can import python or other scripting languages, or C# or C++ language into Dynamo to extend it into a more professional and powerful software system.

The said "two independent and one open source" enables most AEC workers to solve engineering design problems with Dynamo's powerful computing capabilities. BIM technology is characterized by visualization, coordination, simulation, optimization and drawing, the solutions of which can be generated using Dynamo's automation, analog computation, scripting and parameterization.

2.2 Dynamo's Computational Design

Dynamo's visual programming language allows engineers to create programs through a node-based graphical interface, and designers have no need to write code. When the predefined functional modules (nodes) are connected, Dynamo's computing power can be given full play, and tedious production processes can be displayed visually in real time, convenient for improving the design process, which provides a new idea for exchange between Revit and Autodesk Vasari. Engineers can freely create design models and use sophisticated data processing, correlation structures, geometric control and other functions. It makes possible the free creation and editing of structure, visualization process exchange, quick acquisition of analytical data, effective optimization of scheme design, increase of computational efficiency, and reduction of engineering costs. In addition to inheriting BIM information from Revit, it also provides real-time processing tools for green design analysis, energy consumption analysis and solar radiation analysis of Vasari cloud, which is not possible in traditional CAD-written software.

2.3 Dynamo's Node

In the Fig. 1 Work Interface, the tree structure listed on the left is Dynamo's node library which has three to four nodes after expanded. The existing Dynamo 2.0 has 8 classes of node libraries, more than 960 basic nodes in total, which are divided into three categories: creation, operation and query according to different functions.

Multiple nodes can be connected in series or in parallel to show the workflow of the nodes through guide lines (Figs. 2, 3 and 4).

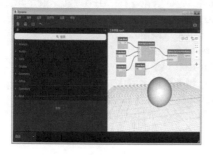

Fig. 1. Work interface

Fig. 2. Classified node library and its hierarchy

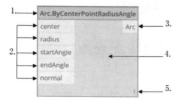

Fig. 3. Node Description Diagram

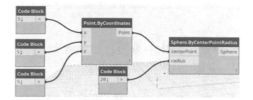

Fig. 4. Program node connection diagram

2.4 Dynamo's Common Nodes and User-Defined Nodes

The "Code Block" node provides node calculation for directly writing Design Script or Python code; values, texts and mathematical formulas can be directly input at the input end, and multiple operation expressions can be input at one node with a line break symbol:

- Create a list randomly: Fig. 5 shows a list of values that are incremented or decremented, a multidimensional list, and a nested list (Fig. 6).

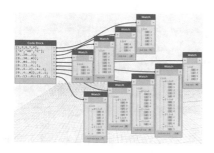

Fig. 5. Function description of code block node

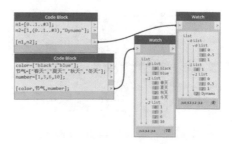

Fig. 6. Edit list of code block node

- Edit any list: Multiple lists can be combined into a multi-dimensional list, and any item herein can be extracted; on the contrary, a multi-dimensional list can also be flattened into a single-item list.
- Control node operations: The operation of a node depends on the target element (executable target). It is controlled according to command method table (executing a node function), and separated by "•" to form a node command.
- Enter a command in Code Block, and a drop-down menu that lists related commands will appear. It is an easy-to-use tool for users in visual programming.
- User-defined nodes: This is another advantage of Code Block; that is, a function command can be created using simple statements, which is similar to the Lisp language in AutoCAD. Methods of defining functions or calculations can also be written with Python language.

User-defined nodes can save Dynamo's finished programs as a file in dyf format, which can be used in current or multiple Dynamo files as a subprogram. Two ways of creating user-defined nodes have been detailed in the software. Users can save them in a given path, and call them in the program like "block" in CAD.

3   Application of Dynamo in the BIM Model of Sunxihe Bridge

3.1   Introduction to the Sunxihe Bridge

As a key control project of Jiangjin-Xishui Expressway, Sunxihe Bridge is an important part of the southwest channel connecting Chongqing and Guizhou, an important inter-provincial channel in the "three rings, twelve radiations and seven links" highway network in Chongqing, and a typical extra-long highway bridge in mountain areas with a super-high pier and a large span. With the full length of 1,578 m, its main bridge is a gravity-anchored steel truss suspension bridge spanning 660 m; the tower is about 200 m high, and the bridge is about 280 m high from the valley. The main girder uses reinforced steel truss girder, with the height of 5.5 m and the width of 28 m. The project features large investment, long construction period, wide coverage, huge data and information processing, and complicated construction management.

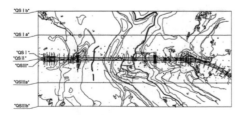

Fig. 7. Geological map of the Sunxihe bridge points and virtual points

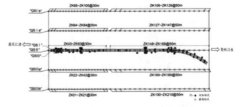

Fig. 8. Location map of geological drilling

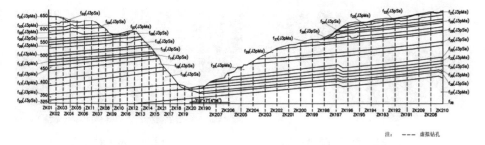

Fig. 9. Sectional view of geological data

3.2 Pre-treatment of Geological Data of the Sunxihe Bridge

According to Figs. 7, 8 and 9, Excel points can be calculated from 5 geological river surfaces, 50 drilling points and 210 virtual drilling points. The lattice data of multi-geological rock stratums is listed in an Excel sheet, and the curved surface forming function of civil 3D is used to generate three-dimensional solid figure. as shown in Fig. 10.

3.3 Formation of 3D Solid

The Dynamo tables at each level can easily generate their own fitting planes. Coupled with upper and lower planes and boundary constraints, the three-dimensional solid graphics of each layer are generated, as shown in Fig. 11. The graphics can be subdivided into numerically calculated grids Fig. 12. Such graphics can also generate physical entities through a 3D printer Fig. 13.

曲面-f22

序号	X	Y	Z
1	3433. 369	1616. 221	509. 113
2	3383. 369	1616. 221	506. 63
3	3333. 369	1616. 221	500. 572
4	3283. 369	1616. 221	487. 813
5	3233. 369	1616. 221	479. 772
6	3210. 627	1616. 221	477. 44
7	3383. 369	1416. 221	502. 149
8	3333. 369	1416. 221	496. 652
9	3439. 913	1416. 221	509. 008
10	3283. 369	1416. 221	484. 072
11	3233. 369	1416. 221	479. 884
12	3183. 369	1416. 221	474. 473
13	3133. 369	1416. 221	469. 088
14	3083. 369	1416. 221	463. 19
15	3033. 369	1416. 221	453. 321
16	2983. 369	1416. 221	430. 229
17	2933. 369	1416. 221	404. 382
18	2923. 375	1416. 221	403. 706
19	3383. 369	1216. 243	501. 544
20	3333. 369	1216. 243	497. 969
21	3421. 604	1216. 243	504. 487
22	3283. 369	1216. 243	486. 795
23	3233. 369	1216. 243	480. 619
24	3183. 369	1216. 243	476. 191
25	3133. 369	1216. 243	468. 052
26	3083. 369	1216. 243	465. 202
27	3033. 369	1216. 243	458. 141
28	3014. 523	1216. 243	455. 013
29	3383. 369	1016. 243	506. 707
30	3333. 369	1016. 243	500. 542
31	3433. 5	1016. 243	509. 092
32	3283. 369	1016. 243	487. 828
33	3233. 369	1016. 243	479. 751
34	3210. 62	1016. 243	477. 418
35	3383. 369	816. 243	502. 133
36	3333. 369	816. 243	496. 642
37	3439. 935	816. 243	508. 986
38	3283. 369	816. 243	483. 495
39	3233. 369	816. 243	479. 859

Fig. 10. Sheet   of   curved surfaces

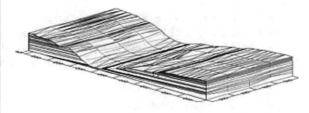

Fig. 11. 3D Solid based on the chart

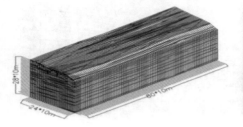

Fig. 12. Typical section of 3D solid

Fig. 13. Printed 3D solid

4 Conclusion

Computational BIM, in the eyes of BIM engineers, is not simply digital modeling with software operations, but using Dynamo, a visual programming and modeling tool to solve complex engineering problems. It can also improve design efficiency, and reduce or avoid design errors. We can make full use of the integrated digital graphics and information, and follow the theory and method of digital graphic media, to make it possible that the digital modeling of complex structures and alien systems and the use of building information are accomplished in a plane environment. We should accumulate and gradually improve the professional node libraries, and develop an English platform that is more suitable for China's engineering needs and has independent property rights by fully using the functions of Excel sheet, rich resources of Matlab, as well as Visual Lisp and Activex development tools, to realize a significant increase in BIM technology and levels.

References

1. Qun, W., Weibo, Y., Shangwei, L.: Research progress of digital graphic information fusion system in BIM technology [EB/OL]. Sciencepaper, Beijing, 19 March 2014. http://www.paper.edu.cn/releasepaper/content/201403-758
2. Shangwei, L., Hao, H., et al.: Steel Structure Inspection Method Based on BIM System, 26 April 2017. CN.ZL201410825237.3
3. Wei Qun. BIM-based Thin-wall Channel Composite Building, 13 October 2017. CN. ZL201510525764.7
4. Qun, W., Guoxin, Z., Zongmin, W., et al.: Digital Graphic Medium Simulation Method for 3D Spatial Structure, 02 July 2014. CN.ZL201210047628.8
5. Qun, W.: Research & development and application of BIM technology in steel structure engineering. In: Academic Report of China Steel Structure Industry Conference, Kunming (2012)
6. Liu, S., Wei, L., Zhu, X.: The applicable study cloudy technology in steel-structured engineering software framework. In: Huang, Y. (ed.) ICCET. Trans Tech Publications, Kunming (2013)
7. Keough, I.: What Revit Wants: Dynamo Revit Test Framework [EB/OL], 16 October 2013. https://www.revitforum.org/blog-feeds/16596-what-revit-wants-dynamo-revit-test-framework-ian-keough.html

Research on Game Strategy of Information Sharing in Cross-Border E-commerce Supply Chain Between Manufacturers and Retailers

Zhao Min[1], Xi Mikai[2(✉)], and Dai Debao[2]

[1] SHU-UTS SILC Business School, Shanghai University,
Shanghai 201800, China
[2] School of Management, Shanghai University, Shanghai 200444, China
2339084755@qq.com

Abstract. In the cross-border e-commerce transaction supply chain, information sharing can coordinate the relationship between node companies, promote long-term cooperation, and maximize the overall benefits of the supply chain. This paper uses evolutionary game to study the information sharing between retailers and manufacturers in the cross-border e-commerce supply chain, builds a related evolutionary game model, analyzes the game equilibrium conditions of retailers and manufacturers, and the influencing factors of both parties' information sharing. The research results show that the initial sharing ratio of information, cost of information sharing, income after information sharing, risk of loss of important information, and trust level all affect the effect of information sharing between the two parties. Measures to maximize the interests of both parties include: increasing mutual trust, rational distribution of income, compliance with laws and regulations, avoidance of loss of important information, and reducing information sharing costs.

Keywords: Information sharing · Cross-border e-commerce · Evolutionary game · Retailer

1 Introduction

Cross-border e-commerce refers to a type of international business activity in which different parties are involved in transactions, through e-commerce platforms to achieve transactions, payment settlement, and through cross-border logistics companies to deliver goods and complete transactions. Under a cross-border e-commerce, after a product is produced from a factory in China, it is sold directly to overseas retailers or terminal consumers by relying on the Internet and international logistics, greatly reducing intermediate links, lowering costs and barriers, and increasing the speed and efficiency of payment collection. Speed up the pace of foreign trade. Information sharing is the key to achieve coordinated operation between the upstream and downstream of the supply chain. It is a profit-driven process. Each member will decide whether or not to continue sharing based on its own benefits. Achieving information sharing will bring great benefits to the entire supply chain, so it is necessary to combine the two for analysis.

© Springer Nature Switzerland AG 2020
F. Xhafa et al. (Eds.): IISA 2019, AISC 1084, pp. 399–406, 2020.
https://doi.org/10.1007/978-3-030-34387-3_49

From the perspective of the existing literature, it mainly focuses on cross-border e-commerce and information sharing. The first is to describe the meaning of cross-border e-commerce and future development prospects. Ma (2017) considered that the application of B2C sales channels has the effect of improving consumer welfare and reducing social welfare. Lu (2016) proposed solutions to the development of cross-border e-commerce platform. Xu (2015) proposed that cross-border e-commerce should become a new channel for China's foreign trade. The second is to study the influencing factors of information sharing. Lu (2017) supposed that only one retailer and one manufacturer, then model and perform numerical analysis to discuss the impact of information sharing on it. Li (2015) used the Stackelberg master-slave game model to analyze the impact of information sharing and non-sharing on the pricing and profits. Mao (2008) conducted exploratory research and quantitative analysis of the behavioral factors that influence the willingness of retailers and suppliers to share information and list the factors.

However, the above literature does not combine the two studies, and it rarely involves the construction of an evolutionary game model. Evolutionary game theory emphasizes that participants are not completely rational, it is impossible to maintain rationality in every decision stage, and it is possible to find breakthroughs for the study of game theory from the point of view of limited rationality. Therefore, the paper will use the evolutionary game model and build the replication dynamic equation to analyze the factors and propose several solutions. The game model diagram shown in Fig. 1.

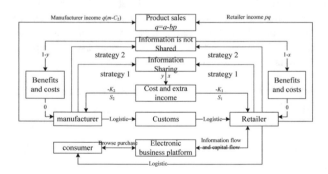

Fig. 1. Cross-border electricity supplier game model

2 Construction of Evolutionary Game Model

2.1 Model Description and Assumptions

When shares information, the retailer can only obtain the manufacturer's judgment of the market demand from the wholesale price, and the manufacturer can obtain the market forecast information from the retailer.

In this game model, both sides of the game are manufacturers and retailers that conduct cross-border transactions, assume that retailers dominate in this supply chain. The strategy sets of the two sides are: information sharing, information not sharing.

Hypothesis 1: In this model, all manufacturers as a group are players 1, all retailers are players 2, and the supply chain is a simple two-tier supply chain with only retailers and manufacturers, also it is a retailer-led type and both manufacturers and retailers trade directly with no other intermediate links.

Hypothesis 2: The two sides are bounded rational assumptions and the risk appetite is neutral. The purpose of both is to pursue the maximization interests.

Hypothesis 3: In the case where both sides of the game choose the strategy that information is not shared, it is stipulated that both sides of the game have zero returns.

Hypothesis 4: The probability that the retailer chooses to share information is x, not share is $1 - x$, for the manufacturer, share information is y, not share is $1 - y$.

Hypothesis 5: In this model, the retailer is dominant and will charge the manufacturer K, the product sales amount q, the retailer sales price p, and the p, q satisfaction relationship $q = (a - bp)$, The retailer's distribution cost is C_1, the manufacturer's manufacturing cost is C_2, the manufacturer sells the retailer's wholesale price is m $(p > m)$, the retailer information sharing cost is K_1 and the manufacturer's cost is K_2, the retailer gains additional revenue when the manufacturer selects the information sharing strategy is S_1, on the contrary, the manufacturer gains additional revenue is S_2 $(S_1 < K_1, K_2 < S_2, S_1 < S_2)$, the cost of manufacturer predicts product demand is t. The revenue matrix can be shown in Table 1.

Table 1. Revenue matrix of retailers and manufacturers

Retailer (game player 1)	Manufacturer (game player 2)	
	Information Sharing (y)	Information is not shared ($1 - y$)
Information Sharing (x)	(Π^1_{11}, Π^2_{11})	(Π^1_{12}, Π^2_{12})
Information is not shared ($1 - x$)	(Π^1_{21}, Π^2_{21})	(Π^1_{22}, Π^2_{22})

Among them, Π^k_{ij} $(k = 1, 2)$ indicates the revenue value that the game player k selects the strategy i (1 indicates information sharing, 2 is not shared) and the player 2 selects strategy j (1 indicates information sharing, 2 is not shared):

$$\Pi^1_{11} = (p - m)(a - bp) + K - C_1 - K_1 + S_1, \quad \Pi^2_{11} = (m - C_2)(a - bp) + S_2 - K_2 - K$$

$$\Pi^1_{12} = (p - m)(a - bp) + K - C_1 - K_1, \quad \Pi^2_{12} = (m - C_2)(a - bp) + S_2 - K$$

$$\Pi^1_{21} = (p - m)(a - bp) + K - C_1 + S_1, \quad \Pi^2_{21} = (m - C_2)(a - bp) - K_2 - K - t, \quad \Pi^1_{22} = \Pi^2_{22} = 0.$$

2.2 Constructing a Replication Dynamic Equation

The most important thing in the evolutionary game is to find the stability strategy (ESS). The evolutionary stability strategy is a stable state of a certain group. By replicating the dynamic equation, the stable state of the evolutionary game can be described. Based on the above-mentioned revenue matrix and game relationship, a replication dynamic equation between retailers and manufacturers can be constructed.

(1) Retailers choose the expected benefits of information sharing

$$T_1 = y \prod_{11}^1 + (1 - y) \prod_{12}^1 \tag{1}$$

(2) Expected income of retailer selection information not shared

$$T_2 = y \prod_{21}^1 + (1 - y) \prod_{22}^1 \tag{2}$$

(3) Manufacturers choose the expected benefits of information sharing

$$U_1 = x \prod_{11}^2 + (1 - x) \prod_{21}^2 \tag{3}$$

(4) Manufacturers choose the expected benefits of information sharing

$$U_2 = x \prod_{12}^2 + (1 - x) \prod_{22}^2 \tag{4}$$

(5) Average Expected Revenue of Retailers

$$\overline{T} = xT_1 + (1 - x)T_2 \tag{5}$$

(6) Average Expected Revenue of Manufacturers

$$\overline{U} = yU_1 + (1 - y)U_2 \tag{6}$$

(7) According to the Malthusian equation (Friedman 1991), the replication dynamic equation of retailers and manufacturers can be

$$F(x) = \frac{dx}{dt} = x(1 - x)[y(C_1 - (p - m)(a - bp) - K) + (p - m)(a - bp) + K - C_1 - K_1] \tag{7}$$

$$G(y) = \frac{dy}{dt} = y(1 - y)[x(t - (m - C_2)(a - bp) + K) + (m - C_2)(a - bp) + K - t - K_2] \tag{8}$$

3 Evolutionary Stability Strategy Analysis

3.1 Analysis of Evolutionary Stability Strategy for Information Sharing

Make Eqs. (7) and (8) equal to zero finding the equilibrium point.

$$\begin{cases} F(x) = \frac{dx}{dt} = 0 \\ G(y) = \frac{dy}{dt} = 0. \end{cases} \tag{9}$$

1. Discuss $F(x) = \frac{dx}{dt} = 0$. Have $y^* = \frac{-(p-m)(a-bp)-K+C_1+K_1}{-(p-m)(a-bp)-K+C_1+1}$, $y = y^*$, $y > y^*$, $y < y^*$. The three-phase copy dynamic phase diagram is shown in Fig. 2.

2. Discuss $G(y) = \frac{dy}{dt} = 0$. Have $x^* = \frac{-(m-C_2)(a-bp)+K+t}{(m-C_1)(a-bp)+K_L+K+t}$, $x = x^*$, $x > x^*$, $x < x^*$, The three-phase copy dynamic phase diagram is shown in Fig. 3.

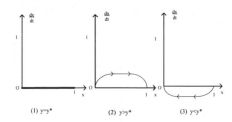

(1) y=y*   (2) y>y*   (3) y<y*

Fig. 2. Copying the dynamic phase diagram

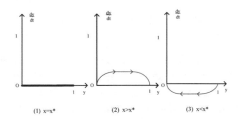

(1) x=x*   (2) x>x*   (3) x<x*

Fig. 3. Copying the dynamic phase diagram

3. Discuss the duplicated dynamic phase diagrams for both as shown in Fig. 4.
① When the initial state falls into zones 1 and 4, it is unstable;
② When the initial state falls into zone 2, ESS is $x^* = 1$, $y^* = 1$;
③ When the initial state falls in zone 3, ESS is $x^* = 0$, $y^* = 0$.

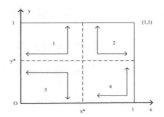

Fig. 4. Dynamic phase diagram

3.2   Impact of Parameter Changes on ESS

The effect of parameter changes on the ESS is shown in Table 2. As is seen from the table that the ESS to eventually stabilize at (1, 1). Both parties of the game will eventually choose to share information.

Table 2. Effect of parameter changes on ESS

Parameters change	District 2 area	ESS
$C_1\uparrow$	$\uparrow$	(1, 1)
$C_2\uparrow$	$\uparrow$	(1, 1)
$K\uparrow$	$\uparrow$	(1, 1)
$K_1\uparrow$	$\downarrow$	(0, 0)
$K_2\uparrow$	$\uparrow$	(1, 1)
$t\uparrow$	$\uparrow$	(1, 1)
$(p-m)(a-bp)\uparrow$	$\downarrow$	(0, 0)
$(m-C_2)(a-bp)\uparrow$	$\uparrow$	(1, 1)

4   Simulation Analysis

This paper uses MATLAB software to simulate the influence of each parameter on Strategy Selection under different initial states of both players. $C_1 = 15$, $K = 30$, $K - C_1 - K_1 = 10$, $a = 0.5$, $p - m = 0.5$, $m - C_2 = 0.2$, $w = 20$, $K - w - K_2 = 5$, so that the initial value is changed equivalently between [0, 1]. In a certain time, other variables remain unchanged, as shown in the left of Fig. 5, when x' > 0.8, the retailer finally chooses the strategy of information sharing, and when x' < 0.8, the retailer finally chooses the strategy of information not sharing; as shown in the right of Fig. 5, no matter how much value y' is taken, the manufacturers' stabilization strategy is information sharing. Therefore, the evolutionary stabilization strategy of both sides is information sharing.

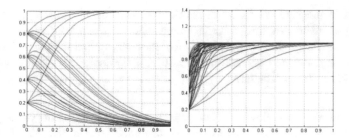

Fig. 5. Evolutionary game simulation diagram

5   Conclusions and Recommendations

5.1   Influence Factors

1. Status relationship between the two. Information sharing allows manufacturers to more accurately predict the retailer's demand during the lead-time, thereby reducing inventory fluctuations. Retailers believe that sharing information with manufacturers will lead to a passive position.
2. Information asymmetry. The retailers in a dominant position can use the terminal market information they hold as a bargaining chip to collect additional benefits from manufacturers. For example, retailers may require manufacturers to provide price protection, cash discounts, delayed payments, and other benefits through possession of information such as sales volume, inventory and so on.
3. Benefits. Retailers generally think that information sharing can bring greater benefits to manufacturers, but they have relatively little benefit from them, which also reduces the enthusiasm of retailers for information sharing.
4. Trust issues. Both parties pay more attention to short-term interests in cooperation. In the absence of trust, the authenticity and integrity of the information have been questioned, making it difficult for both parties to share information.

5.2   Suggestions

According to the influencing factors mentioned above, the following three points are given in this paper.

1. Establish an incentive mechanism to strengthen the cooperation between the two parties and increase trust. Information sharing will increase the overall expected profit of the supply chain, but for retailers, the expected profit does not necessarily increase. Manufacturers can compensate retailers in some extent.
2. Rationally distribute income and enhance communication. Manufacturers and retailers should subjectively have the desire and ability to establish an information sharing mechanism, increase trust, and pay more attention to long-term benefits. They should abide by relevant laws and regulations of the country and conduct business with integrity. In addition, the income should be distributed according to the proportion of initial information sharing.

3. Establish a mechanism. Establish restriction mechanism and integration management mechanism to improve management efficiency and supervision capabilities, promote the development of domestic logistics, strengthen the connection between domestic logistics and international logistics, and enable cross-border e-commerce to develop faster and better.

Acknowledgments. As for this section, the paper is nearing the end. Thank you in particular to my dear graduate tutor, Professor Dai Debao, also thanks for Teacher Zhao Min. From the topic selection, research methods, paper structure framework, literature reading, and later revision of the paper, Teacher Dai Debao carefully guided and strictly demanded that regular discussions and puzzles, helping me solve the problems encountered in the process of writing smoothly.

References

Lu, J., Feng, G., Wang, N., Yungao, M.A.: Research on the influencing factors of bullwhip effect in inventory under information sharing. J. Manage. Sci. **20**(03), 137–148 (2017)

Lu, X., Zhou, M.: Problems and paths in the development of china's cross-border e-commerce platform. Econ. Rev. **03**, 81–84 (2016)

Ma, S., Chen, A.: Cross-border e-commerce: B2B or B2C - based on sales channel perspective. Int. Trade Issues **03**, 75–86 (2017)

Li, B., Sun, P., Li, Q.: Research on information sharing value in dual-channel supply chain. J. Syst. Eng. **30**(04), 530–538 (2015)

Xu, S., Zhang, Y.: Cross-border e-commerce should be built into a new channel of "Made in China" export. Econ. Rev. **02**, 26–30 (2015)

Wei, D.: The impact of forecasting information sharing on the pricing of dual channels and retailers' own products. Tianjin Polytechnic University (2017)

Mao, W., Jiang, L.: Analysis of behavioral factors affecting the willingness of information sharing between retailers and suppliers. J. Hebei Univ. Econ. Trade **01**, 66–72 (2008)

Wang, W.: Status quo and prospect of evolutionary game theory research. Stat. Decis. **03**, 158–161 (2009)

Zhou, M., Dan, B., Yu, H.: Group purchasing and demand information sharing of complementary products manufacturing supply chain. J. Manage. Sci. **20**(08), 63–79 (2017)

Cross-Modal Multimedia Archives Information Retrieval Based on Semantic Matching

Gang Wang[1], Yu Wang[1], and Rujuan Wang[2(✉)]

[1] Northeast Normal University, Changchun, China
[2] College of Humanities & Sciences of Northeast Normal University,
Changchun, China
wangrujuan_1108@sina.com

Abstract. In view of the low recall rate of archives retrieval at this stage, this paper proposes a cross-modal multimedia archives information retrieval model. According to the characteristics of archives, this method carries out feature processing and semantic mapping of archives information, applies the integrated learning method to cross-media archives retrieval, proposes Bagging-SM method for semantic matching of different modes of multimedia archives information, and finally proves that the cross-modal multimedia archives information retrieval model can effectively improve archives information retrieval and retrieval. Full rate is helpful to the development of intelligent retrieval in archives system.

Keywords: Semantic matching · Archives information retrieval

1 Introduction

At present, there are some studies in the field of archives digital information construction at home and abroad [1]. Good Practice Guidelines, funded by the Sixth Framework of the Council of Europe's Information Society Technology (IST), is an excellent example of digital integration of books, museums and archives [2]. The project provides guidance for policy makers and archives professionals in building digital archives. The World Digital Library Project, constructed by UNESCO, provides archival information materials in multilingual cultures around the world on the Internet [3]. Therefore, in order to effectively retrieve digital archives information, it is necessary to analyze the association of archives data [4].

Cross-modal multimedia archives information retrieval (CMAIR) is to solve the problem of heterogeneity and decentralization of digital archives resources. Through the association construction and data fusion of digital multimedia archives, it forms whole information resources, realizes the association links between digital archives information resources and fully excavates the implicit semantic information of digital archives.

© Springer Nature Switzerland AG 2020
F. Xhafa et al. (Eds.): IISA 2019, AISC 1084, pp. 407–413, 2020.
https://doi.org/10.1007/978-3-030-34387-3_50

2 File Feature Extraction

Digitized archives have a variety of styles, carrier forms and expression forms. Therefore, digital archives information has the characteristics of multi-source and heterogeneous. Some feature information can be obtained by cataloguing items, and some feature information needs to be extracted from the contents of archives by artificial intelligence technology such as pattern recognition. These feature information are attribute data of archives. Different attribute data belong to different feature views, and the extracted features are constructed according to different views. In order to describe conveniently, the basic elements of the file association model are defined as follows:

Definition 1: Multi-source heterogeneous archival information data. Archival information data can be described as a collection of archival entities:

$$S = \left\{ A_i^t | t = 1, 2, 3, \ldots, T; \ i = 1, 2, 3, \ldots, Nt; \ \sum_{t=1}^{T} Nt = |A| \right\} \quad (1)$$

Among them, T represents the number of file types and Nt represents the number of file entities under t type.

The feature extraction of archives information is to format the relationship between archives entity A and feature relation f by RDF triple. After feature extraction steps, archives information entity is expressed as a set of multiple features.

Definition 2: Archival entities are described as collections of archival feature information:

$$\forall A_i \in D, A_i = \{ f_{it} | i = 0, 1, 2, 3, \ldots, m_t; \ \sum_{t=1}^{T} m_t = |f| \} \quad (2)$$

(Among them, f denotes the characteristic information of archives and m_t denotes the number of characteristic information of archival entity A_i under type T.

2.1 Feature Extraction of Text Archives

Text archives are one of the most widely used types of archives at present, including various conference documents, telegrams, Leaders' speeches, contracts, agreements, manuscripts of publications, etc. Multimedia type text files are pre-processed by data, and then feature extraction is carried out according to text type files. The text feature extraction process is shown in Fig. 1.

Text files mostly use the method of text information extraction to extract features. Text information extraction is a technology to extract specific information (such as noun phrases, place names, time, person names, etc.) from text data. Archives data need to extract features with strong specificity; the main need is to identify the names of people, places, institutions, time in the archives, suitable for the use of part-of-speech tagging for extraction.

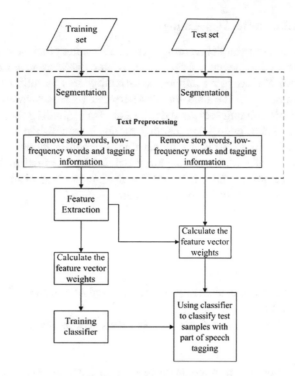

Fig. 1. Text information extraction

2.2 File Feature Extraction of Graphic Images

Picture files generally include two parts: picture and description, which are supplemented artificially in the description part of the file, and Archive Retrieval Based on the content of the description part. It is helpful to fully mine the information value of picture archives to carry out archival features from the content of the picture itself. Through effective training of extracted features, the computer can understand and recognize graphics and images. Commonly used image features include color feature, texture feature, and shape feature, spatial relationship feature and so on.

3 Cross-Modal Multimedia Archives Retrieval Based on Semantic Matching

This section analyses the relationship between the underlying features and the high-level semantics of multimedia information. According to the underlying feature spaces of different modal objects, isomorphic high-level semantics space is constructed. In the cross-media retrieval of the application of ensemble learning method, Bagging-SM method is proposed to match the semantics of different modal multimedia objects.

3.1   Bagging (Bootstrap Aggregating)

Bagging method can run parallel learning, improve learning efficiency, and because the base learners are as independent as possible, it can get a more generalized model. Bagging algorithm obtains many different sub-training sets by uniform sampling with playback of training sets. The data samples between these sub-training sets may be repeated, but the sub-training sets are independent. This sampling method is Bootstrap sampling method. Then, one-to-one learning model is established and trained for the obtained subset of training, so that multiple final prediction models can be obtained. Finally, the data samples of the test set are predicted by the final multiple prediction models, and the results of the voting are the final prediction results, which can be regarded as a probability distribution matrix here.

Pseudo-code description of the algorithm:

Input: Training *Set-TrS*, Test Set-*TS*, Category-*k*, Number of Learning Models-*l*.

Output: Predictive probability distribution M of Bagging learning model.

Algorithmic process:

For $i = 1$ to l

n data samples with feedback is sampled randomly from the training set *TrS* and stored in the sub-training set TrS_i.

TrS_i is used to train the learning model, and the corresponding learning model C_i is established.

End For

Predictive learning is performed on the test set *TS*. Each learning model C_i obtains a corresponding learning result, and l results are obtained. The results l are voted on to the corresponding categories, that is, the probability distribution M of a predicted category is obtained according to the voting results.

3.2   Semantic Matching

Semantic matching is to map different modes of multimedia objects to an isomorphic semantic subspace [5, 6], through semantic learning of extracted underlying features, and then match similarities according to semantic relevance to complete cross-media retrieval tasks.

$$D : R_D^{m \times p} \rightarrow S_D^{m \times k} \tag{3}$$

$$I : R_I^{m \times q} \rightarrow S_I^{m \times k} \tag{4}$$

For text document D and image I, their corresponding feature spaces are marked as $R_D^{m \times p}$ and $R_I^{m \times q}$ respectively. m denotes the number of samples, because text documents and images exist in pairs, so their number of samples is the same. p and q denote the dimensions of text documents and image features respectively, and m and k denote the corresponding isomorphic semantic subspaces of text documents and images after semantic mapping, respectively. k denotes the number of semantic categories, then the semantic category can be expressed as $V = (v_1, \cdots, v_k)$. On the premise of obtaining the underlying feature spaces $R_D^{m \times p}$ and $R_I^{m \times q}$, the text documents and images are semantically mapped, and the corresponding isomorphic semantic subspaces $S_D^{m \times k}$ and $S_I^{m \times k}$ are obtained through training. These two isomorphic semantic subspaces are independently trained from the underlying feature space, and both are generated by the same semantic concept base model.

From the above, we can get the probability distribution vector.

$$P_{V|D}(\upsilon_i|D), i \in (1, \cdots, k) \tag{5}$$

$$P_{V|I}(\upsilon_i|D), l \in (1, \cdots, k) \tag{6}$$

The k-dimensional probability distribution vectors in text document and image isomorphic semantic subspace are obtained respectively. By using the proposed Bagging-SM method, the above semantic matching process can calculate the probability distribution of each sample in a text document and an image which belongs to the corresponding semantic category, and thus the isomorphic semantic subspace of the text document and the image can be obtained [7, 8].

The last step is to carry out cross-media retrieval based on semantic matching.

$$O_{D2I}(D, I) = dist(\pi_D, \pi_I) \tag{7}$$

$$O_{I2D}(I, D) = dist(\pi_I, \pi_D) \tag{8}$$

There are two ways of retrieving images and text documents. By calculating the complete distance matrix O of query data and retrieval data, O is a distance matrix of $m * m$, where $O[a, b]$ represents the distance between the query data a and the retrieval data b. In this paper, normalized correlation coefficient is used as a measurement tool to calculate the similarity between query data and retrieval data.

4   Experiment

4.1   Experimental Data Set and Environment

The data set used in this paper is the public archives data released by an archive on the Internet (not including sensitive information), which contains two parts: feature view and archives entity. Among them, 2000 are archives entity. It contains two types of data files: text and photo. Experimental environment: Windows 10 operating system, Intel Xeon E3-1501 Mv6 processor, 16 G memory, programming language is java.

4.2   Experimental Analysis

In order to judge the accuracy of archival information retrieval, the retrieval results are compared with manual analysis. Specifically, a volume of archives is randomly selected from the experimental samples, and then related samples are selected by manual screening of 1000 archives. Compared with the results of the archives in information retrieval and manual selection, the closer the samples are to the results of manual screening, the more effective the information retrieval is. The experimental

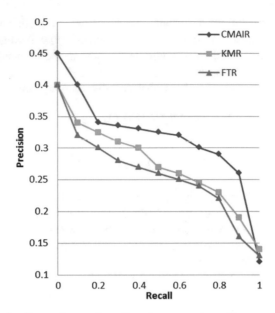

Fig. 2. Comparisons of recall and accuracy in archives retrieval

evaluation index of this paper adopts the precision-recall curve (PR chart). The results show in Fig. 2. By comparing the proposed method with different archives information retrieval methods, it shows that the proposed method has a good performance in query results.

5 Conclusion

In view of the low recall rate of archives retrieval at this stage, this paper proposes a cross-modal multimedia archives information retrieval model. According to the characteristics of archives, this method processes and semantically maps archives information, applies the method of ensemble learning to cross-media archives retrieval, and proposes Bagging-SM method to match different modes of multimedia archives information semantically. Experiments show that this method has greatly improved the recall rate of archives retrieval. It has high reference value and practical significance. In the follow-up work, we will consider how to use the association of archives information, extract the relationship between the characteristics of the object of archives, and optimize the archives retrieval model through the association analysis of archives information.

Acknowledgments. This work was partially supported by the project of Fundamental Research Business Expenses of Central Universities, No. 2412018JC018. The Education Department of Jilin province science and technology research project "13th Five-Year Plan" No. JJKH20181306KJ. Research on the Long-term Digital Preservation Repository Pattern and Service System Construction of College Audiovisual Archives (Social Science Research Project of Jilin Provincial Department of Education "13th Five-year Plan"/No. 2015-547).

References

1. Jin, R., Ruan, N., Dey, S., et al.: SCARAB: scaling reachability computation on large graphs. In: ACM SIGMOD International Conference on Management of Data. ACM (2012)
2. Cui, B., Tung, A.K.H., Zhang, C., et al.: Multiple feature fusion for social media applications. In: ACM SIGMOD International Conference on Management of Data. ACM (2010)
3. Karampatakis, S., Bratsas, C., Antoniou, I.: Library linked data: the case of public library of Veroia. In: 2016 11th International Workshop on Semantic and Social Media Adaptation and Personalization (SMAP). IEEE (2016)
4. Cai, P., Wang, Z., Fu, X.: Semantic-based cross-media information retrieval technology research. Microelectron. Comput. **27**(3), 1–12 (2010)
5. Karampatakis, S., Bratsas, C., Antoniou, I.: Library linked data: the case of public library of Veroia. In: 2016 11th International Workshop on Semantic and Social Media Adaptation and Personalization (SMAP). IEEE (2016)
6. Cai, S.: Research on Probability Based Cross-Media Retrieval Method. Huazhong University of Science and Technology (2013)
7. Wei, Y.: Semantic Classification and Retrieval of Cross-Media Data. Beijing Jiaotong University (2016)
8. Wei, Y., Zhao, Y., Zhu, Z., et al.: Modality-dependent cross-media retrieval. ACM Trans. Intell. Syst. Technol. **7**(4), 1–13 (2016)

Modeling and Simulation Analysis of Semi-active Oil and Gas Spring Characteristics

Chunming Li, Yijie Chen[✉], Qiangshun Lei, Yafeng Zhang,
and Xiaodong Gao

China North Vehicle Research Institute, Beijing, China
chenyijie1206@163.com

Abstract. In order to further improve the average off-road speed of special vehicles, a semi-active oil and gas spring design scheme is proposed, and the multi-stage damping adjustable function can be realized by parallel proportional throttle valve. Based on the physical model of oil and gas springs, the series-parallel relationship of adjustable damping throttling is extracted, and the derivation of analytical equations between flow and differential pressure is carried out; In addition, the flow characteristics of the proportional throttle valve were fitted by least squares fitting, and the exact correspondence between current and flow coefficient was obtained. Combined with the actual gas state equation, the simulation results of the oil and gas springs under different currents are simulated and compared with the experimental data to verify the correctness of the model; Finally, the analysis of the influencing factors of the performance characteristics was carried out, which laid the foundation for the design of oil and gas springs.

Keywords: Semi-active · Damping controlled · Oil gas spring · Proportional valve · Speed characteristics

1 Introduction

After the structural parameters of the passively suspended damping valve are determined, the system output force will not be able to be adjusted over a wide range. The maneuverability and steering stability of the vehicle when driving under different road conditions and vehicle speeds are difficult to achieve optimally, and the driver cannot effectively adjust the vibration damping device according to the vibration state of the vehicle body. According to the structural principle of the semi-active suspension, the electromagnetic proportional control valve is connected in parallel with the oil and gas spring damping valve, through changing the current intensity to drive the position change of the spool, thereby achieving the purpose of adjusting the throttle area of the valve port and outputting different resistance values [1].

© Springer Nature Switzerland AG 2020
F. Xhafa et al. (Eds.): IISA 2019, AISC 1084, pp. 414–421, 2020.
https://doi.org/10.1007/978-3-030-34387-3_51

2 Damping Valve Structure Principle

Figure 1 is schematic diagram of the damping adjustable valve system, the electro-magnetic proportional valve is connected in parallel with the oil and gas spring damping valve, and the damping valve is mounted on the piston. The working oil realize throttling and reciprocating flow through the damping valve and the proportional valve respectively.

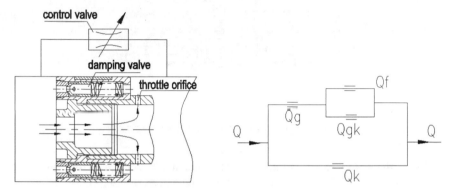

Fig. 1. Adjustable damping valve schematic   **Fig. 2.** Valve system throttling diagram

Figure 2 is the schematic diagram of the damping valve flow, where Q is the total flow of the damping valve, Q_k is proportional valve flow, Q_f is passive damping valve flow, Q_g is constant through hole flow, and Q_{gk} is high pressure pipeline flow.

3 Damping Adjustable Mathematical Modeling

According to the string parallel relationship of the oil road, the following equation can be written [2]:

$$\begin{cases} Q_f = Q_g \\ Q_k = Q_{gk} \\ Q = Q_f + Q_k \\ \Delta p = (\Delta p_f + \Delta p_g) = (\Delta p_k + \Delta p_{gk}) \end{cases} \tag{1}$$

In which, Δp_f is pressure difference at both ends of the gap, Δp_k is pressure difference at both ends of the control valve, Δp_g is pressure difference between the ends of the oil hole, Δp_{gk} is loss of resistance to high pressure lines.

3.1 Proportional Valve Flow Fitting

Figure 3 is the schematic structural diagram of the proportional valve, which mainly adjusts the output force of the electromagnet by changing the current intensity to achieve the purpose of changing the opening degree of the valve core; When the current is 0 A, the throttle channel area is the largest, and the generated damping force is the smallest. When the current is 1.5 A, the throttle channel is closed, and the oil only passes through the passive damping valve, and the damping force generated is the largest.

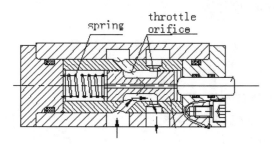

Fig. 3. Proportional valve structure diagram

Due to the complicated structure of the internal oil passage of the proportional valve, it is difficult to obtain accurate analytical calculation results. Therefore, based on the test of the proportional valve flow, the data is polynomial fitted by the least squares method [3]:

$$Q_k = C_k \sqrt{\frac{2\Delta p_k}{\rho}} \tag{2}$$

In which, C_k is control valve flow coefficient.
The flow least squares expression is:

$$\Pi = \sum_{i=1}^{n} (Q_k - Q_{ki})^2 = \sum_{i=1}^{n} \left(C_k \sqrt{\frac{2\Delta p_{ki}}{\rho}} - Q_{ki} \right)^2 \tag{3}$$

In which, Q_{ki} is the proportional valve flow, Δp_{ki} is the test pressure difference, $i(1, 2 \cdots n)$ is a data group.

$$\frac{\partial \Pi}{\partial C_k} = 2 \sum_{i=1}^{n} \sqrt{\frac{2\Delta p_{ki}}{\rho}} \left(C_k \sqrt{\frac{2\Delta p_{ki}}{\rho}} - Q_{ki} \right) = 0 \tag{4}$$

Deriving the flow coefficient formula:

$$C_k = \sqrt{\frac{\rho}{2}} \sum_{i=1}^{n} \frac{Q_{ki}}{\sqrt{\Delta p_{ki}}} \tag{5}$$

Take the maximum throttle as an example, $I = 0$ A, the curve fitting of the test data shows the comparative effect as shown in Fig. 4, It is not difficult to see that the deviation is very small and proves that the selection method and polynomial form are reasonable. Based on this, the flow coefficient of the control valve under different currents is solved separately, as shown in Table 1.

Table 1. Proportional valve fitting flow coefficient of different current

I/A	0	0.3	1.5
C_k/mm^2	12.69	9.92	0

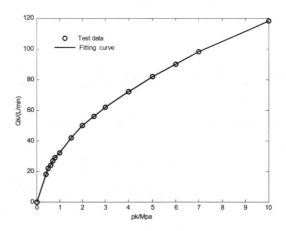

Fig. 4. Proportional valve flow fitting curve

3.2 Damping Characteristic Analytical Calculation

The gap flow formula is:

$$Q_f = \frac{\pi r_n \Delta p \delta^3}{6\mu n h} \tag{6}$$

In which, μ is oil dynamic viscosity, n is number of stacked valve slice, h is single sheet thickness, r_n is inner diameter of the gap, Δp is damping valve pressure difference, δ is ring gap width.

The damping valve throttling equations are:

$$\begin{cases} \Delta p = \Delta p_f = \Delta p_k \\ A_u V_d = \frac{\pi r_n \Delta p \delta^3}{6\mu n h} + C_k \sqrt{\frac{2\Delta p}{\rho}} \end{cases} \tag{7}$$

In which, A_u is the piston area, V_d is reciprocating speed of the piston, ρ is oil density.

Deriving the pressure difference is: [4]

$$\Delta p = \frac{6\mu n h}{\pi r_n \delta^3}(A_u V_d - Z_2) \tag{8}$$

In which,

$$Z_1 - \sqrt{18 C_k^2 \mu^2 n^2 h^2 + 6\pi r_n \rho \delta^3 \Lambda_u V_d \mu n h}$$

$$Z_2 = \frac{C_k\left(-6\sqrt{2}C_k \mu n h + 2Z_1\right)}{\sqrt{2}\pi r_n \rho \delta^3}$$

The slice turbulent flow formula:

$$Q_f = 2\pi r_n \left\{ -127.79v + \left[3 + 2.5\ln\left(\sqrt{\frac{\Delta p \delta^3}{8v^2 n h \rho}}\right)\right]\sqrt{\frac{\Delta p \delta^3}{2nh\rho}}\right\} \tag{9}$$

The total pressure difference of semi active suspension is:

$$\begin{cases} Z_3 = 127.79v + \frac{A_g}{2\pi r_n}\sqrt{\frac{2\Delta p_g}{\xi_g \rho}} \\ Z_3 = \frac{A_u V_d - C_k \sqrt{\frac{2\Delta p_k}{\rho}}}{2\pi r_n} + 127.79v \\ \Delta p_k = \left(\frac{A_{gk}}{C_k}\right)^2 \frac{\Delta p_{gk}}{\xi_{gk}} \\ \Delta p = \Delta p_f + \Delta p_g = \Delta p_k + \Delta p_{gk} \end{cases} \tag{10}$$

In which, A_g is passing oil orifice cross-sectional area, A_{gk} is cross-sectional area of pressure pipeline, ξ_{gk} is the partial pressure loss, ξ_g is the partial pressure loss of passing oil orifice.

The system resistance value under arbitrary displacement can be solved by iterative calculation based on formula $F_d = \Delta p A_u$ programming.

3.3   Elastic Force Value of Semi-active Oil and Gas Spring

From the point of view of accuracy and simplicity, Fandewaer's actual gas state equation was used for modeling:

$$p = \frac{RgTm_q}{V_q - m_q b} - \frac{am_q^2}{V_q^2} \tag{11}$$

In which, L_j is static equilibrium position chamber length, A_{gw} is external section area of piston rod, A_{gn} is intersection area of piston rod, s is oil and gas spring displacement.

The system elastic force at any displacement of oil and gas spring is calculated by the following formula:

$$F_t = pA_{gw} \tag{12}$$

The total output value of the system can be solved:

$$F_z = F_d + F_t \tag{13}$$

4 Comparison of Test and Simulation

Input sinusoidal excitation, $f_z = 1\,\mathrm{Hz}$, $A_z = 0.04\,\mathrm{m}$, and the external characteristic test and mathematical modeling simulation of different current damped adjustable oil and gas spring are carried out (Tables 2, 3, 4 and Figs. 5, 6, 7).

Table 2. Force value comparison when current $I = 1.5\,\mathrm{A}$

The force name	$F_{z\max}$	$F_{z\min}$	F_{zy0}	F_{zf0}
Test data (KN)	56.99	11.29	41.44	17.23
Simulation data (KN)	57.55	11.95	41.63	16.54
Data error (%)	0.97	5.52	0.46	5.6

From the above Figures, it can be seen that the deviation between the simulation results and the test data is very small, the mathematical model of the damped adjustable oil and gas spring and the data fitting method of the control valve are verified. The output force of the current $I = 1.5\,\mathrm{A}$ is similar to that of the passive oil and gas spring, indicating that the control valve is closed, and the oil fluid is mainly reciprocated in the oil cavity through the damper valve on the piston; When the control valve is in a fully open state and $I = 0\,\mathrm{A}$, the local resistance loss of the throttle channel is minimal, from Fig. 8 to compare the test data of the two limit cases, the difference is very obvious, and the change rate of damping force at the displacement zero reaches 53%, which shows that the method of parallel electromagnetic proportional control valve for oil and gas springs can effectively regulate the output of the system and provide a reliable platform for the study of semi-active suspension.

Table 3. Force value comparison when current $I = 0.3\,\mathrm{A}$

The force name	$F_{z\max}$	$F_{z\min}$	F_{zy0}	F_{zf0}
Test data (KN)	53.12	17.37	36.89	21.96
Simulation data (KN)	53.51	17.17	36.93	21.19
Data error (%)	0.7	1.2	0.1	3.5

Table 4. Force value comparison when current $I = 0\,\text{A}$

The force name	$F_{z\max}$	$F_{z\min}$	F_{zy0}	F_{zf0}
Test data (KN)	52.02	17.85	35.17	23.83
Simulation data (KN)	53.51	18.87	34.84	24.71
Data error (%)	0.6	3.5	2.0	1.0

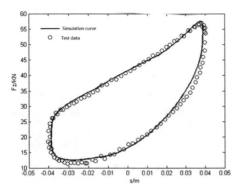

Fig. 5. Indicator characteristics when $I = 1.5\,\text{A}$

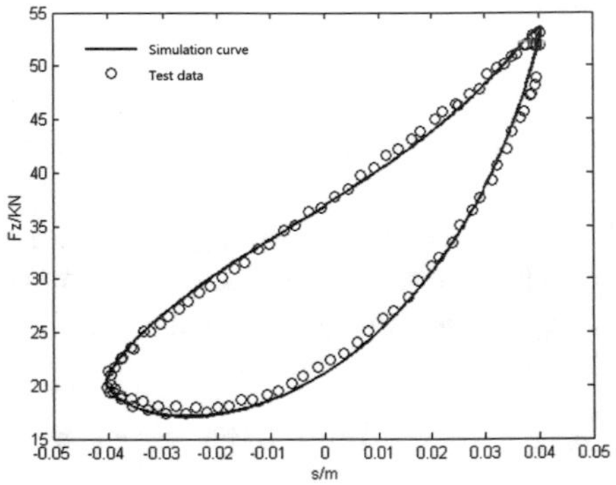

Fig. 6. Indicator characteristics when $I = 0.3\,\text{A}$

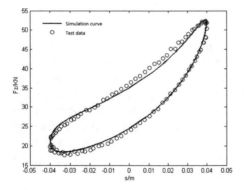

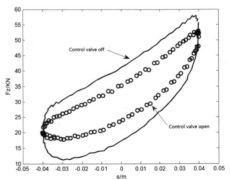

Fig. 7. Indicator characteristics when $I = 0\,\text{A}$

Fig. 8. Comparison of control valve open and closed

5 Analysis of Influencing Factors

The electric and spring characteristics of the oil and gas spring are calculated separately, which lays a foundation for analyzing and studying the output force value of the system at any displacement and speed.

The increase of the diameter of the piston and the decrease of the outer diameter of the piston rod make the effective working area of the piston become larger, and the oil flow through the gap increases, and the damping force increases (Figs. 9 and 10).

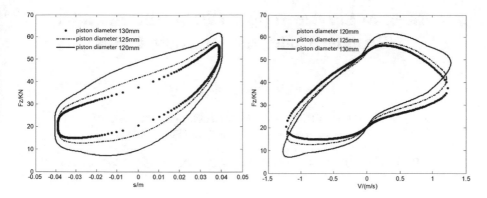

Fig. 9. Spring performance and speed characteristics of different piston diameters

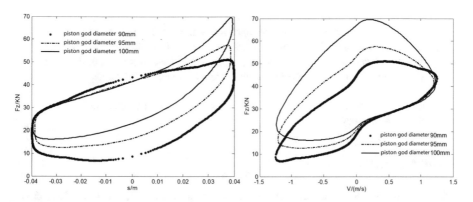

Fig. 10. Spring performance and speed characteristics of different piston rod diameters

References

1. Lee, K.: Numerical modelling for the hydraulic performance prediction of automotive monotube dampers. Veh. Syst. Dyn. **28**, 25–39 (1997)
2. Chen, Y.-J.: Research on analytical computation of valve parameters and design of hydro-pneumatic spring. Beijing Inst. Technol. (2008)
3. Moona, B.Y., Showa, K., Chung, S.W.: Mechanical properties study and design of a suspension system by considering tension force. Mater. Process. Technol. **140**, 385–390 (2003)
4. Chen, Y., Gu, L., Guan, J.F.: Research on the mathematical model and analysis the throttle aperture of hydro-pneumatic spring. Beijing Inst. Technol. **5**, 388–391 (2008). (in Chinese)

Effect of Service Supply Chain with Sensitive Demand Under the Dynamic Game

Jun Hu[1,2(✉)] and Haiying Li[1]

[1] College of Economics and Trade, Guangxi University
of Finance and Economics, Nanning 530003, China
52618265@qq.com
[2] Beijing Oriental Landscape Environment Co., Ltd., Beijing 100015, China

Abstract. In order to discusses the ratchet effects and the reputation effects in service supply chain under the multi-period dynamic game. The paper depicts the model, and builds a multi-period dynamic game programming model. Such decision-making as quality efforts and revenue sharing coefficient are resolved. It analyzes the ratchet effect and the reputation effect due to the dealer's revenue. It summarizes the meaning of all effects with reputation and ratchet. We can find: the hazard risk in service supply chain is reduced in long-term dynamic game. The effect is the same as that under the asymmetric information.

Keywords: Sensitive demand · Dynamic game · Service supply chain · Reputation · Ratchet

1 Introduction

Under international logistics practice, members' behavior usually is dynamic and long-term. The current service affects the next period decision-making, such as the logistics service providers and agents. In multi-period and multi-level supply chain, there are complex interactions among many members in the supply chain network especially. The interaction between decision makers, which own rationality, feedback and time delay, can also lead to compound behavior. Therefore, decision coordination and behavior of members becomes a new research field in the complex supply chain system (Larsen 1999). Based on this, this paper will consider the multi-period dynamic game model of service supply chain, in which demand is the sensitive with price, quantity and member effort, which has the reputation effect and the ratcheting effect in secondary service supply chain.

2 Model Description

There is a kind of international logistics business, which is secondary service supply chain. There are two participants: service provider (set ic. Provider) and the service agent (set ic. Agent). Provider can determine service level θ_t, while incurring certain service costs $c(\theta_t)$ by managing the service. Agent buys the service from provider at wholesale price w_t, and according to market demand D_t then orders from provider.

© Springer Nature Switzerland AG 2020
F. Xhafa et al. (Eds.): IISA 2019, AISC 1084, pp. 422–428, 2020.
https://doi.org/10.1007/978-3-030-34387-3_52

Agent's efforts e_t, accordingly, bring agency costs to the agent. Thereafter, agent provides the service to final consumers at retail price p_t and gains profits. According to the principle of maximizing expected profits, both service provider and agent make decisions. The provider is leader and the agent is follower in the game. Under the revenue sharing contract, wholesale price w_t by the provider's to the agent is generally equal to the provider's cost c ($w_t = c$). Through the whole game, service providers, as the dominant players, guide agent to take the expected effort by providing incentive contract. After observing service provider's decision-making (i.e. the contract given), agent takes it as a constraint and chooses the effort to maximize its utility. Then, according to the performance of agent, the provider updates its beliefs and gives a new contract. Under new contract constraints, the agent chooses its effort, which can maximize its utility. Such reciprocation constitutes a multi-period dynamic game in service supply chain.

Assumptions:

(1) Unilateral information asymmetry in supply chain, i.e. service provider does not know the type of agent's capability γ but only its distribution: $Y \sim N(\bar{y}, \sigma_\gamma^2)$ and both sides' capability λ and γ do not change with time.

(2) Service provider's cost remains unchanged and lead time is zero, that is to say, the agent's demand can be met immediately. Shortage rate is zero.

(3) Provider and agent are independent entities in economic and risk-neutral in decentralized decision-making.

(4) $\varepsilon_t \sim (0, \sigma_\varepsilon^2)$, λ, ε_1 and ε_2 are independent.

(5) Agent's cost function is $c(e_t) = \frac{\eta e_t^2}{2}$, and $c(e_t) > 0$, $c'(e_t) > 0$, $c''(e_t) > 0$. η ($\eta > 0$) is the cost coefficient. The greater the cost coefficient is, the higher the cost of same effort e_t. The provider cost is $c(\theta_t) = \frac{\zeta \theta_t^2}{2}$, when, ζ ($\zeta > 0$) is cost coefficient of the provider. The greater the cost coefficient is, the higher the cost of same service.

(6) Market demand D_t is sensitive to price, quantity and effort, and is linearly correlated in quantity. Agency's price p_t, agency effort e_t, service effort θ_t, wholesale price w_t and market random variables all affect demand. The dynamic market demand is as follows: $D_t = a - p_t + \gamma e_t + \lambda \theta_t + \varepsilon_t$.

(7) Firstly, agent has not yet established a reputation for its ability, and he needs to send a signal about his ability through transactions. When incentive contract offered by the provider satisfies the agent's participation constraints, the agent will accept it.

(8) Provider can provide explicit incentive contract satisfying its participation constraint at each stage (i.e. $E(U_t) \geq \overline{U}$, the agent's reserved utility $\overline{U}$), but can not commit to long-term contracts (i.e., the service providers will make the second stage contracts according to the performance of the agent in the first stage).

The meaning of variables in the model:

a—market capacity; γ—agent ability; t—transaction stage; e_t—agency effort at the t stage, ε_t—random variable at a t stage, probability density of $f(\varepsilon_t)$; π_t—whole supply chain profit at the t stage; δ—discount factor; π_r—agent profit in t stage; $c(e_t)$—agency

Cost; U—agent utility; V—provider utility; W—whole revenue in supply chain; $E(\gamma|\pi_{t-1})$—agent's reputation in the t stage, referring to the belief to γ (i.e. expectation) to provider; b—agent's bargaining power, agent's profit share about its reputation $E(\gamma|\pi_{t-1})$; ϕ_t—agent's profit share from excess performance.

3 Planning Equation

3.1 Incentive Contract

In multi-cycle game, dynamic principal-agent game, incentive contract offered, which is provided by the provider to the agent, includes visible explicit incentive and implicit incentive with good reputation. In revenue-sharing contract, π_t is common knowledge about the provider and the agent, so agent's actual performance is $\pi_t - E(\pi_t|\pi_{t-1})$. Then $\phi_t[\pi_t - E(\pi_t|\pi_{t-1})]$ is explicit incentive in t stage. If the principal-agent is multiple stage, even without explicit incentives, agent is motivated to work more harder, because it will gain its reputation and get more future income. Therefore, $bE(\gamma|\pi_{t-1})$ is implicit incentive provided by the provider in t stage.

In t stage, T, which is service provider pays agent, as follows:

$$T = \phi_t[\pi_t - E(\pi_t|\pi_{t-1})] + bE(\gamma|\pi_{t-1}) \tag{1}$$

T is the incentive provided by service provider in the t stage, in which the first incentive is explicit and the second incentive is implicit.

3.2 Agent's Expected Utility

As the assumptions, the agent expected utility in two-stage is:

$$E(U) = \sum_{t=1}^{2} \delta^{(t-1)} E(U_t) = \sum_{t=1}^{2} \delta^{(t-1)} \int [T - c(e_t)] f(\varepsilon_t) d\varepsilon_t \tag{2}$$

Agent need to choose right effort e_t in T stage under the incentive contract so as to maximize the expected utility in the whole cycle.

$$\max E(U) = \sum_{t=1}^{2} \delta^{(t-1)} \int \{\phi_t[\pi_t - E(\pi_t|\pi_{t-1})] + bE(\gamma|\pi_{t-1}) - c(e_t)\} f(\varepsilon_t) d\varepsilon_t \tag{3}$$

By calculating the upper formula's integral, we can get:

$$\max E(U) = \sum_{t=1}^{2} \delta^{(t-1)} \{\phi_t[p_t \cdot (a - p_t + \gamma e_t + \lambda\theta_t + \varepsilon_t) - \frac{\eta e_t^2}{2} - \frac{\zeta\theta_t^2}{2} - E(\pi_t|\pi_{t-1})] + bE(\gamma|\pi_{t-1}) - \frac{\eta e_t^2}{2}\} \tag{4}$$

When an agent makes its decision at each stage, it must consider the current decision-making on the subsequent stage's utility. Agent's effort is to influence the next stage utility through service provider's belief at each stage.

3.3 Provider's Expected Utility

By providing incentives to the agent, the provider maximizes its expected utility. The provider aims to maximize its expected utility.

$$\max E(V) = \sum_{t=1}^{2} \delta^{(t-1)} \int \{(1 - \phi_t)[\pi_t - E(\pi_t|\pi_{t-1})] - bE(\gamma|\pi_{t-1}) - c(\theta_t)\}f(\varepsilon_t)d\varepsilon_t \tag{5}$$

By the upper formula integral, we can get:

$$\max E(V) = \sum_{t=1}^{2} \delta^{(t-1)}\{(1 - \phi_t)[p_t \cdot (a - p_t + \gamma e_t + \lambda\theta_t + \varepsilon_t) - \frac{\eta e_t^2}{2} - \frac{\zeta\theta_t^2}{2}$$
$$- E(\pi_t|\pi_{t-1})] - bE(\gamma|\pi_{t-1}) - \frac{\eta\theta_t^2}{2}\} \tag{6}$$

Assuming that service provider has rational expectations, in t stage, its effort to act as an agent are recorded as follows:

$$\bar{e}_t = E(e_t|\pi_{t-1}), \ \tau = \frac{\sigma_\gamma^2}{\sigma_\gamma^2 + \sigma_\varepsilon^2}$$

and the larger the service provider τ is, the higher the proportion of the observed agents π_1 will be (Fama 1980). From the rational expectation formula, the following state transfer equation is obtained.

3.4 Planning Equation

From the service provider and agent's expected revenue function in the initial state, we can get the dynamic multi-period programming equation of service provider, which is provided to the agent concluding explicit incentive and implicit incentive (reputation incentive), when agent's capability is unknown.

$$\max_{\theta_t, \phi_t} E(V). \ \text{Constraint condition:} \ (p_t, e_t) \in \max E(U), \ E(U_t) \geq \overline{U}, \ E(\gamma|\pi_0) = \bar{\gamma}.$$

By substituting the above formulas into the programming equation, we can get the results.

$$\max_{\theta_t, \phi_t} E(V) = \sum_{t=1}^{2} \delta^{(t-1)} \{ (1 - \phi_t)[p_t \cdot (a - p_t + \gamma e_t + \lambda \theta_t + \varepsilon_t) - \frac{\eta e_t^2}{2} - \frac{\zeta \theta_t^2}{2}$$

$$- E(\pi_t | \pi_{t-1})] - bE(\gamma | \pi_{t-1}) - \frac{\eta \theta_t^2}{2} \} \tag{8}$$

S.t:

$$(p_t, e_t) \in \sum_{t=1}^{2} \delta^{(t-1)} \{ \phi_t[p_t \cdot (a - p_t + \gamma e_t + \lambda \theta_t + \varepsilon_t) - \frac{\eta e_t^2}{2} - \frac{\zeta \theta_t^2}{2} - E(\pi_t | \pi_{t-1})] + bE(\gamma | \pi_{t-1}) - \frac{\eta e_t^2}{2} \},$$

$$E(U_t) \geq \bar{U}, \ E(\gamma | \pi_0) = \bar{\gamma}, \ E(\pi_1 | \pi_0) = p_1 \cdot (a - \bar{p}_1 + \bar{\gamma} \bar{e}_1 + \lambda \bar{\theta}_1) - \frac{\eta e_1^{-2}}{2} - \frac{\zeta \theta_1^{-2}}{2}.$$

4   Solution and Analysis of Equations

The Bellman equation of the objective function is established, and the service provider's quality effort and revenue sharing coefficient are obtained by solving the Bellman equation. They are:

$$\Phi_t = \frac{U_{-R}}{\frac{\zeta K_t(a-c)}{(\zeta - K_t \lambda^2)} + \delta \tau \cdot \Phi_{t-1} \cdot [b - \cdot \frac{\zeta(\theta_t)^2}{2\eta}]}$$

$$\theta_t = \frac{\lambda K_t(a-c)}{(\zeta - K_t \lambda^2)} + \frac{\delta \tau}{\eta} [b - \theta_{t-1} \cdot \frac{\lambda(K_t - K_{t-1})(a-c)}{(\zeta - K_t \lambda^2)}].$$

Assuming that Φ_0 and θ_0 are known, service provider's service effort θ_t and revenue sharing coefficient Φ_t can be solved.

Substituting above (Φ_t^*, θ_t^*) to $E(V)$, we can obtain:

$$e_t^* = \frac{\gamma}{a\eta} \{ \Phi_t^* \cdot \theta_t^* + \delta \tau [b - \Phi_{t+1}^* \cdot (1 + \frac{\Phi_{t+1}^*}{a\eta})] \}.$$

There are two parts in agent's effort e_t^*: $\frac{\gamma}{a\eta} \cdot \Phi_t^* \cdot \theta_t^*$ is the component belonging to explicit incentive of agent; $\frac{\gamma}{a\eta} \cdot \delta \tau [b - \Phi_{t+1}^* \cdot (1 + \frac{\Phi_{t+1}^*}{a\eta})]$ is the component unrelated to decision-making at this stage, belonging to implicit incentive.

5   Reputation and Ratcheting Effect

Agent's incentive includes not only explicit but also implicit. Implicit incentive is the result of two effects' interaction: ratcheting effect and reputation effect, which are as follow: $\frac{\gamma \delta \tau b}{a\eta}$, $-[\frac{\gamma \delta \tau}{a\eta} \cdot \Phi_{t+1}^* \cdot (1 + \frac{\Phi_{t+1}^*}{a\eta})]$.

5.1 Reputation Effect

It refers to the agent's efforts at a certain stage to bring more benefits to the next stage. It is precisely because of reputation benefits that it will encourage agents to continue to strengthen their agency efforts in order to achieve better reputation in the next stage and generate more revenue.

From the reputation effect $\frac{\gamma\delta\tau b}{a\eta}$, we find when the agent's bargaining power b, business ability γ, discount factor δ and value τ are stronger, the agent will be motivated to improve its effort, because it will gain better reputation and more benefit.

The smaller market capacity a and cost coefficient η are, the more the agency effort can bring reputation and revenue. The more market capacity a is, the greater the agent's influence to market, and the lower the agent's cost with the same effort.

5.2 Ratcheting Effect

Ratcheting effect is that principal's assessment about agent's performance is based on the previous stage performance at a certain stage, so agent will reduce its previous effort stage to avoid undertaking too high performance task at that stage. Ratcheting effect is implicit incentive, which has nothing with decision-making in current stage but is related to decision-making in next stage.

Ratcheting effect in agent's effort is $-[\frac{\gamma\delta\tau}{a\eta} \cdot \Phi^*_{t+1} \cdot (1 + \frac{\Phi^*_{t+1}}{a\eta})]$ that we make conclusion that the greater income distribution coefficient Φ^*_{t+1} in next stage, the lower agent's effort. It is the reserved space for the agent's performance improvement. The larger income distribution coefficient, the higher income distribution in next stage.

In service supply chain, members' behavior is a multi-period dynamic game process. Considering sensitive characteristics of price, quantity, member effort and market demand, this paper studies a kind of service supply chain model under multi-period, dynamic game, which's reputation effect and ratcheting effect are based on dynamic game and a revenue sharing model.

It is found that the agent's effort is the result due to reputation effect and ratcheting effect in multi-cycle, repeated game, which also explains the importance of reputation in multi-cycle supply chain. Through long-term game, moral hazard can be reduced in service supply chain, so that the same results can be achieved almost under asymmetric information.

Acknowledgments. Thank you for the program from College of Economics and Trade, Guangxi University of Finance and Economics (Title: Research on the Quality of Cross-Border E-Commerce Services; No. 2019YB07), and High Level Innovation Team and Excellent Scholars Program of Guangxi University.

References

Radner, R.: Monitoring cooperative agreement in a repeated principal-agent relationship. Econometrica **49**(7), 1127–1198 (1981)

Rubbinstein, A.: Equilibrum in supergames with the overtaking criterion. J. Econ. Theory **21**(2), 16–29 (1979)

Fama, E.F.: Agency problems and the theory of the firm. J. Polit. Econ. **88**(2), 288–307 (1980)

Larsen, E.R., John, D.W., Thomsen, J.S.: Complex behavior in a production-distribution model. Eur. J. Oper. Res. **119**(1), 61–74 (1999)

Agrella, P.J., Lindrothb, R., Norrmanb, A.: Risk, information and incentives in telecom supply chains. Int. J. Prod. Econ. **90**(1), 1–16 (2004)

Monnahan, G.E., Petrruzzin, N.C., Zhao, W.: The dynamic pricing problem from a newsvendor's perspective. Manuf. Serv. Oper. Manag. **6**(1), 73–91 (2004)

Wang, X.: Contractual incentives, information sharing and dynamic coordination of supply chains. Manag. World **4**, 106–115 (2005)

Lu, S.: Coordination of behavior and decision-making in multi-stage and multi-cycle supply chains. Logist. Technol. **29**, 127–129 (2006)

Ma, S., Zhou, J.: Dynamic quantitative elastic contract model with time flexibility. Storage Transp. Maint. Commod. (2), 1–4 (2007)

Xiong, Z., Li, G., Tang, Y., Li, W.: Research on channel coordination considering dynamic pricing in network environment. J. Manag. Eng. **21**(3), 49–55 (2007)

Jin, Y., Ci, X., Ye, Z., Xi, Y.: Multi-stage dynamic signal transmission contract arrangement for venture capital. J. Shanghai Jiaotong Univ. **38**(3), 434–437 (2004)

Weng, M., Liu, N.: Supply chain dynamic incentive contract under information asymmetry. J. Guangxi Univ. Finance Econ. **21**(2), 25–29 (2008)

Nie, J., Xiong, Z.: Two-cycle telecom supply chain ordering strategy under uncertain demand. Ind. Eng. **12**(1), 17–23 (2009)

Survey on Monitoring and Evaluation of Gas Insulation State in GIS

Fanqi Meng, Lei Xia, and Jingdong Wang[✉]

Northeast Electric Power University, Jilin 132012, China
286604374@qq.com

Abstract. Gas Insulated Switchgear (GIS) with advantages such as less occupation, low noise, high reliability and safety has been widely used in substations of various voltage levels. However, gas insulation deterioration can cause failure to the GIS equipment and even shutdown to the whole substation. This consequence seriously threatens the operation safety of the power grid. Therefore, monitoring and evaluating of gas insulation states are necessary. In order to obtain an overview of current research status of this field, so that the further studies could be done efficiently. This paper summarizes the researches in monitoring and evaluating of gas insulation states in recent years. The remote online monitoring technologies and the intelligent evaluation methods are respectively reviewed which including the establishment of index system and the creation of evaluation model. At the end of the paper, according to the challenges faced by the current researches, the future research directions and development trends are given.

Keywords: GIS equipment · Remote online monitoring · Gas insulation state assessment

1 Introduction

Gas Insulated Switchgear (GIS) is the primary devices in substations. Once it breaks down, it will cause power outages, even seriously affecting the operation safety of power grid. The main reason for the malfunction of GIS equipment is the deterioration of gas insulation 7. Therefore, the monitoring and evaluation of gas insulation status of GIS are necessary. In order to have a clear understanding of this research field, this paper overviews the existing remote on-line detection technology and gas insulation state evaluation method in Sects. 2 and 3. Section 4 summarizes the establishment of index system and evaluation model. Section 5 discusses the future research trend.

2 On-line Monitoring of Gas Insulation Status

The traditional method to detect the insulation status was manual, and relied on some specific measuring instruments. The method is susceptible to interference from external conditions and subjective judgment, so it will easily lead to large faults. To address this problem, researchers have developed new online monitoring technologies.

© Springer Nature Switzerland AG 2020
F. Xhafa et al. (Eds.): IISA 2019, AISC 1084, pp. 429–436, 2020.
https://doi.org/10.1007/978-3-030-34387-3_53

2.1   UHF Partial Discharge Monitoring Technology

Partial discharge is an electric phenomenon that may penetrate the insulation medium, then cause insulation deterioration [2]. The Ultra High Frequency (UHF) method has many advantages in monitoring GIS partial discharge, such as highly sensitive and strong immunity to interference. Through UHF sensors, the electromagnetic wave signal produced by partial discharge can be collected. Then signal will be sent to server to analysis the partial discharge degree and estimate the insulation deterioration state [3].

2.2   Gas Component Analysis Technology

Some discharge phenomenon, such as arc discharge, corona discharge and spark discharge, can cause the decomposition of SF6. Thus, the concentration of insulated gas will decrease [4]. Eventually, it will affect the insulation performance. Therefore, gas-component analysis technology evaluates gas insulation state according to the components and proportion of gas in GIS. Since the method monitors the chemical compositions of the equipment, it can be more immunity to the vibration inside the equipment and the external noise compared with the acoustic analysis [5].

2.3   Comprehensive Intelligent Analysis Technology

Comprehensive intelligent analysis technology can automatically measure various gas-electric parameters and estimate the insulation states by using artificial intelligence. At the same time, the operation state information can be shared by using 4G communication technology [6]. With the further construction of smart grid, the intelligent analysis technology will become more mature. Therefore, the intelligent on-line monitoring technology will become the trend. It will be helpful to ensure the safe and stable operation of power grid.

3   Intelligent Evaluation of Insulation Condition

Traditional methods such as Threshold Diagnostic Method and Time domain waveform etc. to evaluate the insulation state only use few parameters. The accuracy is doubtful since the state usually determined by multiple gas-electric factors [7].

3.1   Diagnostic Methods of Expert System

Expert system uses Delphi method to make evaluation. The core of the system is the knowledge base and inference machine, the Knowledge base contains expert knowledge and experience in the field, while the inference machine makes decisions according to the knowledge and experience in the information base through the computer simulation expert decision-making method. This method is simple to operate, but it requires high knowledge coverage and low fault tolerance in the knowledge base, which may cause the forecast result to deviate from the actual [8]. Figure 1 shows the model of expert system.

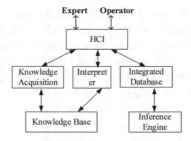

Fig. 1. Expert system model

3.2 Fuzzy Comprehensive Evaluation Method

The Fuzzy comprehensive evaluation method evaluates the insulation state of GIS equipment by combining different interfering factors. The method considers the hierarchy and fuzziness of the evaluation objects. Considering the objectivity of indicators [9], the method discovers the intrinsic relationship of interfering factors through establishing the fuzzy evaluation model to obtain scientific and reliable evaluation results. The factor set is divided into m class according to the proportion of the index, as shown in Formula 1. And all the index sets and index factors of GIS equipment are evaluated by establishing fuzzy comment set, as shown in Formula 2, and on the basis of a large amount of historical data, the Power Index set and fuzzy evaluation matrix are established, as shown in Formula 3:

$$U = \{u_1, u_2, u_3 \ldots u_i \ldots u_m\} \, n = 1, 2, 3 \ldots; \, i = 1, 2, 3 \ldots \tag{1}$$

$$V = \{v_1, v_2, v_3 \ldots v_i \ldots v_k\} \, k = 1, 2, 3 \ldots; \, i = 1, 2, 3 \ldots \tag{2}$$

$$A_i = \{a_{i1}, a_{i2} \ldots a_{ij} \ldots a_{imi}\} \, i = 1, 2, 3 \ldots; \, j = 1, 2, 3 \ldots. \tag{3}$$

The evaluation matrix R_i represents the first indicator set, where r_{jk} represents the weight of u_{ij} in the U_i:

$$R_i = \begin{bmatrix} r_{11} & r_{12} & \cdots & r_{1k} \\ r_{21} & r_{22} & \cdots & r_{2k} \\ \cdots & \cdots & \cdots & \cdots \\ r_{j1} & r_{j2} & \cdots & r_{jk} \\ \cdots & \vdots & \ddots & \vdots \\ r_{m1} & r_{m2} & \cdots & r_{mk} \end{bmatrix}$$

3.3 Artificial Intelligence Based State Assessment Method

There are many types of factors affecting the GIS insulation state, each of which is related. Experts use artificial intelligence algorithm to analyze the GIS insulation state. For instance, the state evaluation method of Fish swarm optimization algorithm based on rough set theory [11], the state evaluation method based on FKNN algorithm [12], and the state evaluation method based on Decision Tree algorithm [13]. Among them, the neural network has powerful self-learning function and data processing capability, and can map highly nonlinear input and output functions.

A state evaluation method combining artificial neural network and information fusion technology is introduced, where the artificial neural networks adopt a radial base function (RBF). The RBF uses the selected indicators as evaluation data, analyzes and evaluates the state of the device by adjusting the weight between the output layer and the hidden layer. The information fusion technology is to give a credible state assessment through rule-based reasoning [14] by comprehensively considering the output of RBF. The transformer state evaluation model of information fusion is simple and flexible. The basic credibility of the initial diagnostic layer output structure is more accurate and reliable. The neural network feature level fusion can obtain more accurate diagnosis conclusions [15]. RBF can solve the difficulty of ordinary BP neural network - the local minimum value problem.

A combination of long-short-time memory network (LSTM) and Bagging integrated learning method is proposed to evaluate the severity of GIS partial discharge. Based on the cyclic neural network, LSTM determines the influence of historical information on the instantaneous information by controlling the gating unit. The method solves the problem that the traditional cyclic neural network is difficult to model the long-term span sequence [16]. As shown in the Fig. 2, the fault-free samples and the fault samples are first sampled to obtain N majority sample sets and a few sample sets, and then N majority sample sets and a few sample sets are combined to form N training sample sets, and the filtered data set is utilized. The established evaluation model was trained, and the influence of different types of partial discharge information on the assessment of partial discharge severity was analyzed. The risk assessment of GIS partial discharge under operating conditions was realized, and the GIS insulation state assessment was realized.

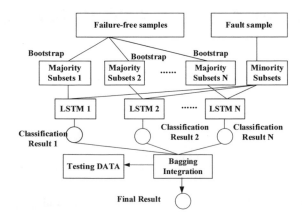

Fig. 2. Partial discharge assessment model based on bagging-LSTM

The Literature proposes a method combining Markov model and Analytic Hierarchy Process (AHP). The device state is divided into nine states from the perspective of GIS equipment failure. Based on the Markov model, the probability of the nine states is calculated and analyzed. The Markov model is to express and calculate state probabilities by mathematical and Bayesian formulas. Finally, the AHP is used to analyze the failure probability of various states. Thus the status of the GIS device is evaluated.

Firstly, Analytic Hierarchy Process (AHP), through the analysis of GIS equipment genus, the GIS state is divided into target layer, criterion layer, index layer, and the analysis structure model is established by 30% times, as shown in Fig. 3. The GIS state is analyzed by the weight value of the index layer for the target layer. After the hierarchical structure of GIS state analysis is established, the eigenvalues and eigenvectors of the judgment matrix are obtained by constructing the Judgment Matrix A, and then the various state weight coefficients of GIS devices are calculated. Finally, the eigenvector group is normalized, that is, the weight index of each state.

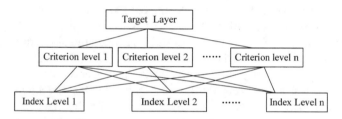

Fig. 3. Hierarchical analysis structure diagram

4   Selection of Evaluation Indicators

4.1   Equipment Operating Environment Indicators

High internal humidity will make the GIS equipment inside damp, and decrease insulation performance. Thus the literature proposes to evaluate the insulation status according to the internal humidity index of GIS equipment [17]. Due to the integration s of the equipment, the heat is hard to disperse. Therefore, the literature proposes to monitor internal temperature of GIS equipment [18].

GIS equipment insulation status depends on insulating gas SF_6. SF_6 gas concentration must be more than 99.8% to ensure the safe and stable operation of equipment. The insulation performance and arc extinguishing ability of SF_6 are closely related to the pressure. It's crucial to maintain the stability of SF_6 gas pressure inside GIS equipment. Therefore, the internal air pressure in GIS equipment is proposed to monitored in the literature [19].

4.2   Gas Composition Index

In the literature [20], the gas decomposition products of 550 kV GIS equipment are detected by gas chromatography combined with chemical detection tube method, and the composition and content of decomposition products are statistically analyzed, which is used as an index to evaluate the insulation status of equipment. Its main products are shown in Table 1:

Table 1. SF_6 decomposition products

Main decomposition	Components of products
SOF_2	$CO_2 + CF_4$
SO_2F_2	$SOF_2 + SO_2F_2 + SO_2$
SOF_4	SO_2/H_2S
SOF_4	$SOF_2 + SO_2F_2 + SO_2$

Most state assessment methods are based on single or limited state parameters, such as humidity, temperature, air pressure, failed to comprehensively consider the related parameters. The accuracy and relevance of evaluation results need to be improved, and the evaluation results are relatively one-sided. The literature proposes to evaluate the GIS insulation state based on multiple parameters [20].

Firstly, the method summarizes the indexes and defines the main factors affecting the insulation state of GIS. Then the composition, temperature, humidity, pressure and purity of gas are introduced as the initial index sets. Then, the analysis method of R clustering is used to classify the indexes preliminarily and use K-W to verify the rationality of classification. Secondly, for each type of index, the amount of information in the factor load matrix is used to filter the index. Finally, the contribution rate of the index is proposed to verify the rationality of the screening results, and establish the multi-parameter index system to evaluate the insulation state of GIS more comprehensively and thoroughly, as shown in Fig. 4.

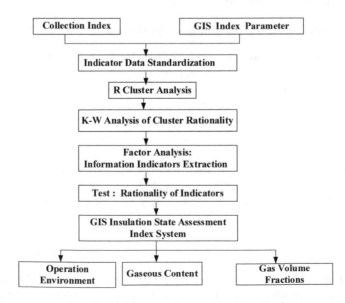

Fig. 4. Multi-parameter assessment method

5   Summary

In recent years, due to the development of artificial intelligence, both the monitoring technologies of GIS equipment and evaluation methods of gas insulation status have made significant progress. Nevertheless, some valuable and realistic problems haven't been solved very well, they are as follows:

(1) Deep fusion of multi parameters. Most existing researches on online monitoring only measure few parameters, such as temperature, humidity, infrared spectroscopy, sound or vibration. So the state of GIS can't be reflected comprehensively and evaluate accurately. Hence, it is worthy to study the multi-sensor information fusion technology to realize deep fusion of multi parameters of GIS gas and electric power.

(2) Intelligent prediction of insulation deterioration. Most evaluation methods use some intelligent algorithm to classify the equipment insulation state or fault. However, the classification results only represent the current state, and this discrete evaluation conclusion can't accurately describe the possible future security risks. Therefore, how to predict its future running state through a comprehensive intelligent analysis of the state of the device is worth studying.

(3) Real-time confirmation and ensure of the system. Some studies have focused on the real-time problem of system, but there is no special detection and optimization of the program execution time in the system. With the combination of various application systems proposed, the objective attributes the system determine that the information part must have predictable and stable real-time performance. Therefore, the monitoring and evaluation of the gas insulation state of GIS equipment must also meet the requirements of real-time. So how to identify and ensure that the overall real-time nature of the system becomes a problem that must be solved.

In conclusion, with the advent of the intelligent era, in order to ensure the safe and stable operation of the power grid, remote on-line real-time monitoring and intelligent evaluation of the GIS gas insulation state is still the focus of future research.

Acknowledgments. The Education Department of Jilin Province (JJKH20190711KJ) and Science and Technology Department of Jilin Province (20190303107SF) funded this work.

References

1. Wu, Y.: Application analysis of SF6 gas decomposition products in fault diagnosis of high voltage electrical equipment. Autom. Appl. (10), 112–113, 115 (2017)
2. Liu, M.: Design and application of partial discharge test system for SF6 circuit breaker based on ultra-high frequency method. North China Electric Power University (2014)
3. Rao, Z.Q., Zheng, S.S., Wang, Z.B., Wu, J., Lu, H.: Phase characteristics of GIS partial discharge UHF signal under negative OLIV. High Volt. Technol. 1–8 (2019)

4. Li, K., Hassan, J., Zhang, G.Q.: Research status and prospect of decomposition gas analysis detection technology for electrical equipment in air discharge. High Volt. Appl. **52**(12), 15–22 (2016)
5. Tang, J., Yang, D., Zeng, F.P., Zhang, X.X.: Research status of insulation fault diagnosis methods and techniques for SF_6 equipment based on decomposition component analysis. J. Electr. Technol. **31**(20), 41–54 (2016)
6. Nie, C.: Design and application of GIS on-line monitoring system for intelligent substation. Hubei University of Technology (2016)
7. Liao, R.J., Wang, Y.Y., Liu, H.: Research status of state evaluation methods for power transmission. High Volt. Technol. **44**(11), 3454–3464 (2018)
8. Bai, S.J., Zeng, L.C., Zhang, Y.Q., Zhang, H.J., Zhang, M., Wang, X.H., Rong, M.Z.: Intelligent GIS. Smart Grid **5**(01), 75–83 (2017)
9. Huang, Y.C., Sun, H.: Dissolved gas analysis of mineral oil for power transformer fault diagnosis using fuzzy logic. IEEE Trans. Dielectr. Electr. Insul. **20**(3), 974–981 (2013)
10. Yuan, L., Du, L., Wu, J.M.: Comprehensive fault diagnosis method for multi-parameter faults of transformers based on fuzzy membership function. High Volt. Appar. **47**(5), 35–42 (2011)
11. Chen, X.Q., Liu, J.M., Huang, Y.W.: Transformer fault diagnosis based on improved artificial fish swarm optimization rough set algorithm. High Volt. Eng. **38**(06), 1403–1409 (2012)
12. Fang, Q., Chen, J.X., Zhang, X.X., Li, X.W.: Research on the evaluation of GIS operation status based on FKNN algorithm. J. Hubei Univ. Technol. **33**(02), 62–66 (2018)
13. Dong, M., Qu, Y.M., Zhou, M.G.: Fault diagnosis of oil-immersed power transformers based on combination decision tree. Proc. CSEE (16), 35–41 (2005)
14. Ruan, L., Xie, Q.J., Gao, S.Y.: Application of artificial neural network and information fusion technology in transformer state assessment. High Volt. Technol. **40**(3), 822–828 (2014)
15. Qi, Z.Z.: Real-time state evaluation of transformer based on multi-information fusion. High Volt. Electr. Appl. **48**(1), 95–100 (2012)
16. Song, H., Dai, J.J., Li, Z., Sheng, G.: Evaluation method of partial discharge severity in GIS under operating conditions. J. Electr. Eng. China
17. Li, Y.C., Wang, H.: Analysis of the influence of moisture on SF6 electrical equipment and the influencing factors of humidity measurement. Shaanxi Electr. Power. (03), 25–28 (2007)
18. Dai, W.W., Gao, K., Ma, L.: Study on GIS temperature rise anomaly state evaluation based on multiphysics coupling. Mech. Eng. **35**(06), 623–626 (2018)
19. Wen, T., Zhang, M., Li, H.: Modeling and analysis of SF_6 gas pressure in GIS equipment of substation. J. Shaanxi Univ. Technol. **34**(02), 12–15 (2018)
20. Wang, B.W., Yang, T.Z., Wang, X.P., Zhao, L.: Insulation fault diagnosis based on short-term SF_6 gas decomposition product data. High Volt. Appar. **54**(11), 55–61 (2018)
21. Song, R.J., Zhao, M., Guan, Y.Z., Du, Y.Q.: Evaluation of GIS gas insulation state based on correlation vector machine. High Volt. Appar. **52**(12), 237–243 (2016)
22. Qu, C.Y., Sun, L.J., Xu, X.Q.: Power big data distributed index based on B+ tree. J. Northeast. Electr. Power Univ. **36**(05), 80–85 (2016)

Advance in Evaluation System Research of Internet Public Opinion

Fanqi Meng, Xixi Xiao, and Jingdong Wang[(✉)]

Northeast Electric Power University, Jilin 132012, Jilin, China
286604374@qq.com

Abstract. Internet public opinion represents netizens' emotions, attitudes and opinions, at the same time, the Internet of openness make some public opinion turn to social public crisis. Therefore, the evaluation of Internet public opinion is essential as well as vital for predicting the influence of public events. A comprehensive indicator system is the key to realize the accuracy evaluation. Thus, many researchers have created their own indicator system from different perspectives. In order to make an overall understanding of this research field and make the further research more efficient, this paper sorts out the existing researches on the evaluation indicator system of Internet public opinion by literature measurement and visual analysis. The results show that the existing indicator systems have diversification and stratification, but some indicators are difficult to measure. Based on the conclusions, this paper predicts the development trend of this research field and provides suggestions for the future research.

Keywords: Internet public opinion · Indicator system · Literature measurement · Visual analysis

1 Introduction

Internet has become a primary and important tool to spread information and express opinion. According to the 43rd report of CCNIC [1], Chinese had 829 million Internet users in December 2018, and 98.6% of them were mobile Internet users. Internet penetration rate was 59.6% and the number is still increasing. Therefore, Internet public opinion has a significant impact on social harmony and political stability. Once malicious rumor or slander about some certain public incident appears on Internet, it is likely to expand rapidly and eventually causes harmful impact on social harmony.

Internet public opinion monitoring and early warning are necessary to precaution the public crisis. The basis and the kernel of the research is an evaluation indicator system [2]. Hence, scholars have constructed different evaluation index system of Internet public opinion from different angles. In order to make an overall understanding of this research field, this paper has analyzed 191 articles selected from CNKI database. Keywords includes "Internet public opinion", "sentimental warning", "sentimental analysis" and "sentimental monitoring", and the journals must be indexed by "CSSCI". The publication time is from 20100101 to 20190114. The analyzing was aiming at analyzing the existing research results, combing the research context, predicting its

© Springer Nature Switzerland AG 2020
F. Xhafa et al. (Eds.): IISA 2019, AISC 1084, pp. 437–444, 2020.
https://doi.org/10.1007/978-3-030-34387-3_54

development trend, and providing reference and suggestions for the sustainable development of Internet public opinion monitoring and early warning research.

2 Quantitative Analysis of Document Form Characteristics

2.1 Research Journals, Subject Analysis

In terms of the distribution of source journals in the sample literature (see Table 1) mainly focuses on 39 intelligence magazines (20.4%), 30 modern intelligence (15.7%), 24 intelligence science (12.6%). It can be seen that the intelligence journals are the mainstream journals to introduce Internet public opinion research.

In order to understand the main research materials, this paper uses the sample literature to classify the disciplines, mainly focuses on 108 Information technology (56.5%), 34 social science II series (17.8%), 27 social science I series (14.1%). Internet public opinion is an interdisciplinary research field, in which information technology is the primary subject of the research.

Table 1. Journal distribution.

Journal	Number of articles	Proportion
Intelligence magazine	39	20.4%
Modern intelligence	30	15.7%
Information science	24	12.6%
Intelligence theory and practice	7	3.7%
Library and information work	6	3.1%
E-government	5	2.6%

2.2 Organization, Author Analysis

According to Price's law, the number of articles issued by the core authors must satisfy (n_{max} represents the number of papers with the most authors, $n_{max} = 16$, m = 2.996). Therefore, the author who published three or more articles are assumed to be the core author (see Table 2).

Table 2. High-yield authors and mechanism.

Serial number	Author	Mechanism	Number of articles (pieces)
1	Lan Xinyue	China People's Police University	16
2	Zhang Yuliang	Henan Polytechnic University	6
3	Xia Yixue	China People's Police University	5
4	Dong Xilin	China People's Police University	5

(*continued*)

Table 2. (*continued*)

Serial number	Author	Mechanism	Number of articles (pieces)
5	Li He	Jilin University	4
6	Zhang Yanfeng	Jilin University	4
7	Chen Yue	Information Engineering University	4
8	Peng Lihui	Jilin University	4
9	Liu Bingyue	Tianjin Transportation Technical College	3
10	Huang Hui	Jilin University	3

Internet public opinion governance research is mainly carried out in the well-known universities. Table 2 shows the high yield authors and core institutions in the field of Internet public opinion evaluation in proper order. The Chinese People's Armed Police Academy has the largest number of papers, and Lan Xinyue is the high-yield author of the unit. The core author has not yet formed, and the relationship between institutions is not close enough.

3   Quantitative Analysis of Document Content Characteristics

Keyword is the core of the article, and also a high summary of the article's theme. In this paper, the literature data are introduced and analyzed by CiteSpace, and the knowledge clustering is carried out to show many hidden complex relations between knowledge individuals and knowledge groups in the research of Internet public opinion detection and early warning, such as network, structure, interaction, intersection and so on.

3.1   Statistical Analysis of Keywords

The keyword with the highest frequency is "Internet public opinion". The keyword "Indicator system" has a frequency of 24, ranking second, and the centrality is 0.06, ranking third (see Table 3). From the frequency and centrality of keywords, the "indicator system" is a research hotspot in this field.

3.2   Analysis of Mutation Words

Mutation words indicate that the dynamic variation of word frequency distribution over a certain period of time is relatively large, reflecting the frontier themes of research and trends in the field [3, 4]. Its basic principle is to use the term that the frequency of statistical occurrence increases abruptly in a relatively short period of time or that the rate of frequency growth increases markedly [5]. Cite Space can determine the mutation words by calculating the frequency change rate (see Fig. 1). They are emergencies (strength 2.6712, 2010–2010), public opinion analysis (strength 2.8651, 2011–2011),

Internet public opinion on emergency (strength 2.9333, 2011–2011). In summary, the "evaluation indicator system" is a research hotspot in 2014. The "social media" has become a hotspot in recent research.

Table 3. Keywords and frequency and centrality.

Freq	Centrality	Keywords
114	1.4	Internet public opinion
9	0.13	Big data
4	0.07	Weibo
4	0.06	Monitoring indicators
8	0.06	Monitoring
24	0.06	Indicator system
11	0.05	Internet opinion monitoring
13	0.03	The emergency
9	0.02	Warning
3	0.01	Public opinion transmission

Top 10 Keywords with the Strongest Citation Bursts

Keywords	Year	Strength	Begin	End	2010 – 2019
emergency	2010	2.6712	2010	2010	
public opinion analysis	2010	2.8651	2011	2011	
Internet public opinion on emergency	2010	2.9333	2011	2011	
Public opinion transmission	2010	2.8257	2013	2013	
Weibo	2010	3.7939	2013	2013	
Internet public opinion monitoring	2010	2.6013	2014	2014	
Public opinion information	2010	2.6013	2014	2014	
Evaluation indicator system	2010	2.6013	2014	2014	
Bp neural network	2010	2.9333	2016	2016	

Fig. 1. Internet public opinion evaluation study of mutation words and their parameters.

Through Fig. 1 and Table 3, we get the research hotspots and evolution of Internet public opinion. The research direction of scholars is from the initial focus on emergencies to the present monitoring, evaluation index system and early warning of Internet public opinion. Therefore, the Internet public opinion needs to be monitored more intelligently and efficiently.

3.3   Analysis of the Indicator of Internet Public Opinion Evaluation

Because of the information is complex and diverse, the Internet public opinion transmission path is diversified. In order to screen out the dangerous sources and factors that can directly affect the security Internet public opinion, we must dig deep into the law of the evolution of public opinion. For example, Li [6] analyzing the evolution of online public opinion topics. Du [7] analyze the similarities and differences between the performances of five typical case events and the propagation paths.

Evaluate the security level of specific Internet public opinion information, select various levels of evaluation indicators through scientific methods, and construct a Internet public opinion evaluation indicator system [8], Dai [9, 10] has successfully proposed various characteristic Internet public opinion indicator systems. Wang [11, 12] sorted and summarized the previous Internet public opinion monitoring indicator system (see Table 4), and constructed a more scientific system of Internet public opinion monitoring and early warned indicator system. Lan [13] based on the research mechanism and evolution law of Internet public opinion in emergencies, constructed an emergency network evaluation indicator system for emergencies in the three dimensions of netizen response, information characteristics of emergencies and event diffusion. Zhang [14] constructed an emergency network risk assessment indicator system including total indicators, classification evaluation indicators and one-way evaluation indicators in three levels and a total of 21 indicators. Then applied the analytic hierarchy process, the weight of the evaluation indication system is determined. The research on the construction of risk assessment indication system for public opinion information flow was based on UML method [15]. Li, Yan and Li [16] proposed an indicator system for quantitative evaluation of Internet public opinion impact degree,

Table 4. Comparison of commonalities between indicator systems [11].

Indicator category	Indicator system	Author	Indicator characteristics
Facing communication	Internet public opinion security evaluation index system	Dai Yuan et al.	Analysis of target topics from a security perspective based on the topic of communication
	Internet word of mouth comprehensive indicator	RI Word of mouth consulting agency	Measuring sentiment influence based on audience engagement based on audience
	5-dimensional public opinion monitoring indicator system	Jin Jianbin	Based on the communication process, focusing on the characteristics of public opinion in a single website/forum, focusing on the characteristics of the process of communication
	Internet public opinion warning level indicator system	Wu Shaozhong, Li Shuhua	Integrating all the key elements of public opinion communication
Content oriented	Standard Internet sentiment indicator	Zhao Xudong	Based on content recognition, determine the influence of the keyword based on the category of the keyword/subject
	"10°" indicator system	Xie Haiguang, Chen Zhongrun	Content-based analysis to identify topics of interest based on content analysis

and a quantitative calculation method for Internet public opinion impact degree that was based on the indicator system. Based on the public opinion information data collected from various Internet public opinion sources, and based on the establishment of evaluation indicator systems for news sources, forums, blogs and videos. The overseas sites were evaluated separately, and six first-level indicators were established (see Table 5). Xing [17] combined with information entropy and analytic hierarchy process to establish an evaluation indicator system, and divided the influencing factors of Internet public opinion information dissemination into five first-level indicators: public opinion, public opinion attention, netizen emotion, public opinion influence and public opinion sensitivity and set 20 indicators released.

Table 5. Quantitative evaluation indicator system for Internet public opinion impact degree [16].

Primary indicator	Secondary indicators	Primary indicator	Secondary indicators
News	Overall amount of information Maximum amount of information per day Important site information volume Netizen comment information volume Mobility clue information	Blog	Overall amount of information Maximum amount of information per day Netizen comment information volume Positive and negative information ratio Mobility clue information
Forum	Overall amount of information Maximum amount of information per day Netizen comment information volume Positive and negative information ratio Mobility clue information	Video	Overall amount of information Maximum amount of information per day Netizen comment information volume Positive and negative information ratio Mobility clue information
Weibo	Overall amount of information Maximum amount of information per day Netizen participation Important netizen indicator Positive and negative information ratio Mobility clue information	Overseas	Overall amount of information Maximum amount of information per day Twitter spreads information Main dissident website information volume Positive and negative information ratio Mobility clue information

Above indicator system reveals the audience, process and other elements of lyric communication at the communication level [18]. At the content level, the lyrical value is deeply explored from the perspectives of identification and analysis, which basically covers most of the Internet public opinion monitoring points from the theme, content, communication process, and public opinion audience. Focusing on the following four points during construction: First, pay attention to the evolution process of Internet public opinion, and different risks at different stages; Second, pay attention to the selection of indicators so that ensure accurate calculation and reliable evaluation; Third, pay attention to the research results of the predecessors, and draw on their valuable results; The research on Internet public opinion evaluation indicator system involves many subject areas, and the construction of Internet public opinion indicator system of each scholar's evaluation indicator is also different.

4 Summary and Outlook

This paper focused on the researches of indicator system to Internet public opinion evaluation, overviewed its research context and research content through the analysis of related articles of the recent decade. The results show that scholars have created various evaluation indicator system from different perspectives such as information technology, social science and economic management science. They also combined the qualitative and quantitative indicators to achieve a more accurate evaluation. However, the research of Internet public opinion indicator system still have some problems and shortcomings. First, how to determine Internet public opinion information collection standards, what information to monitor and how to quantitatively evaluate and how to quantitatively evaluate, lack of data and quantitative research on such information, and lack of empirical research and data analysis. Second, Internet public opinion evaluation indicator system research is not only closely related to social media, but also integrates multidisciplinary knowledge, the relationship between research disciplines is not close enough. Third, the level of danger level is related to many factors, there is currently no clear quantitative standard. This shows that the mature academic community of academic research in this field has not yet formed, which will not be conducive to the further development of in-depth research.

In the all-media era, we can continue to study the indicator system from the following aspects. First, pay attention to the basis of each indicator, determine Internet public opinion information collection standards, ensure the scientific nature of the evaluation system indicators, and make the level of indicators at all levels clearer, the final indicators are easier to quantify and the actual application is more accurate. Second, focusing on the quantitative evaluation of the final indicators and the classification of Internet public opinion evaluation level, there must be a reliable evaluation level. Third, strengthen data capture and interpretation capabilities, discover and grasp the laws that lead to the occurrence and development of Internet public opinion, widely absorb advanced theories, explore theoretical innovations and breakthroughs, empirical research and data analysis. Fourth, connect theory with practice, meet the needs of the all-media era, and conduct cutting-edge evidence. Sex research, put forward a scientific, reliable and easy to implement Internet public opinion evaluation indicator system.

Acknowledgments. The Education Department of Jilin Province (JJKH20190711KJ) and Science and Technology Department of Jilin Province (20190303107SF) funded this work.

References

1. CNNIC released the 43rd Statistical Report on the Development of China's Internet Network in Beijing. http://www.cac.gov.cn/2019-02/28/c_1124175686.htm. Accessed 28 Feb 2019
2. Cao, R.: Full-sample based literature review on the study of network public opinion index systems. J. Intell. **34**(05), 154–158, 28 (2015)
3. Chen, Y., Chen, C.M., Liu, Z.Y.: The methodology function of cite space mapping knowledge domains. Stud. Sci. Sci. **33**(02), 242–253 (2015)
4. Wu, H.J., Zhou, L.P., Xin, Y.: Comparative study about personalized recommendation of home and abroad based on knowledge map. J. Northeast Dianli Univ. **32**(05), 124–128 (2012)
5. Sun, Y.S., Qou, R.R., Deng, X.: Research development of mapping knowledge domains in China—analysis based on CiteSpace II. J. Mod. Inf. **34**(1), 84–88 (2014)
6. Li, Z.W., Xing, Y.F.: Research on the evolution law of public opinions on emergent events in the new media environment—taking the topic of Jiuzhaigou earthquake on Sina Weibo as an example. Inf. Sci. **35**(12), 39–44, 167 (2017)
7. Du, H.T., Wang, J.Z., Li, J.: Research on evolution model for online public opinion of emergent events based on multiple cases. J. China Soc. Sci. Tech. Inf. **36**(10), 1038–1049 (2017)
8. Wang, L.C., Chen, L.F.: A summary of theoretical research on internet public opinion evolution, early warning and coping at home and abroad. Libr. J. **37**(12), 4–13 (2018)
9. Dai, Y., Yan, F.: Research on information mining and evaluation index system based on network public opinion security. Inf. Stud. Pract. **31**(06), 873–876 (2008)
10. Dai, Y., Hao, X.W., Gao, Y., Yu, Z.H.: Research on the construction of China's network public opinion security evaluation index system. Netinfo Secur. **04**, 12–15 (2010)
11. Wang, Q., Chen, Y., Cao, N.P.: Review on index systems of network public opinion for monitoring and early warning. Inf. Sci. **29**(07), 1104–1108 (2011)
12. Wang, Q., Chen, Y., Cao, N.P.: Research on the construction of internet public opinion monitoring and early warning index system. Libr. Inf. Serv. **55**(08), 54–57, 111 (2011)
13. Lan, Y.X.: Construction of emergency network evaluation index system for emergencies. Intell. Mag. **30**(07), 73–76 (2011)
14. Zhang, Y.L.: Network opinion risk evaluation index system based on the cycle of emergency. Inf. Sci. **30**(07), 1034–1037, 1043 (2012)
15. Zhang, Y.L.: Establishment of risk evaluation index system about information flow of network public opinion on emergency based on UML method. Libr. Inf. **33**(11), 100–106 (2016)
16. Li, Y.Q., Yan, L., Li, J.F.: A quantitative evaluation system and calculation method for internet public opinion influence. Netinfo Secur. **09**, 196–200 (2015)
17. Xing, Y.F., Wang, X.W., Wang, D.: Research on the negative network public opinion monitoring index system based on information entropy in new media environment. J. Mod. Inf. **38**(09), 41–47 (2018)
18. Song, R.J., Xu, J.: Establishment and application of evaluation indicator system for relay protection device. J. Northeast Dianli Univ. **35**(06), 70–76 (2015)

The Generalized Inverse of a Class of Block Matrix from Models for Flow in a Dual-Permeability System

Tangwei Liu[✉], Jinfeng Lai, Jingying Zhou, Kangxiu Hu,
and Jianhua Zhang

School of Science, East China University of Technology,
418 Guanglan Road, Nanchang 330013, China
595109035@qq.com

Abstract. The generalized inverse method has grown to become an effective and important method for solving linear mathematical problem as well as applications of linear mathematics, such as computational fluid dynamic, constrained least-squares problems, constrained quadratic programming, and so on. The commonly used generalized inverse methods mainly include minus inverse, plus inverse, least square generalized inverse, reflexive generalized inverse and minimum norm generalized inverse. In this paper, generalized inverse method is studied for a special class of block four-by-four matrix problems discretizing the Dual-Permeability flow in porous media model by mixed multi-scale finite element methods. Firstly, some partial differential equation models of flow in a Dual-Permeability system are given in this paper. Then a class of block matrix from the discrete systems of mathematic models by the multi-scale finite element methods is analyzed. And for both singular and nonsingular cases, the minus inverse of the block two-by-two block matrices are discussed, respectively. Finally, the generalized inverse of the block four-by-four matrix is obtained by transforming the block four-by-four matrix problems into the two-by-two matrix problems, and the algorithm for computing the minus inverse is presented.

Keywords: Generalized inverse · Block matrix · Partial difference equation · Flow model

1 Introduction

The numerical computing methods of the coupling partial differential equation (PDE) system for flow models have been the focus of many studies. In general speaking, mixed multi-scale finite element methods (MMsFEM) are regarded as the general methods to solve the flow models with special boundary conditions [1–4]. We can see that numerical results for PDEs with FEMs require to solve some algebraic equations. There are many kinds of inverse matrices methods for linear algebraic equations. In previous studies, some scholars have given many expressions for MP generalized inverse of partitioned matrix [5–9]. However, it is still difficult to calculate the generalized inverse of a general matrix. In this paper, the generalized inverse of four

© Springer Nature Switzerland AG 2020
F. Xhafa et al. (Eds.): IISA 2019, AISC 1084, pp. 445–452, 2020.
https://doi.org/10.1007/978-3-030-34387-3_55

by four block matrices is used to solve a special class of linear systems derived from the numerical solution of PDEs for dual-permeability flow in porous media.

2   Four by Four Block Matrix from the Dual-Permeability Flow in Porous Media

In this section, we consider the MMsFEM for the following PDE models consisting of four equations for pressure P_1, P_2 and Darcy velocity U_1, U_2 for a Dual-Permeability flow system [3, 4],

$$(\phi_1 C_1)\frac{\partial P_1(x,t)}{\partial t} - div(U_1(x,t)) - \lambda\alpha(P_2(x,t) - P_1(x,t)) = 0, x \in \Omega, t > 0, \quad (1)$$

$$(\phi_2 C_2)\frac{\partial P_2(x,t)}{\partial t} - div(U_2(x,t)) + \lambda\alpha(P_2(x,t) - P_1(x,t)) = 0, x \in \Omega, t > 0, \quad (2)$$

$$U_i(x,t) = -\lambda K_i(x)\nabla P_i(x,t), \quad (i = 1,2), \quad (3)$$

$$P_i(x,0) = P_i^0, \quad (i = 1,2), \quad (4)$$

where Ω is a domain in R^d $(d = 2,3)$, ϕ_j $(j = 1,2)$ are the porosity, C_j $(j = 1,2)$ denote the compressibility, $k_j(x)$ $(j = 1,2)$ denote the permeability tensor. We assume that U_j, P_j $(j = 1,2)$ are unknown. We also assume that the Eqs. (1)–(4) are equipped with the following Neumann boundary conditions:

$$\vec{U}_j \cdot \vec{n} = 0 \ \ on \ \ \partial\Omega \ \ (j = 1,2), \quad (5)$$

where $\vec{n}$ denotes the outward normal of $\partial\Omega$.

By using the MMsFEM, we obtain the approximations of the Laplace transform of the pressure P_1, P_2 and Darcy velocity U_1, U_2:

$$\bar{U}_i^h = \sum_{j=1}^N \tilde{U}_i^j(s) \cdot \psi_j(x) \ \ (i = 1,2), \quad \bar{P}_i^h = \sum_{j=1}^N \tilde{P}_i^j(s) \cdot \varphi_j(x) \ \ (i = 1,2), \quad (6)$$

where ψ_j are velocity basis functions and φ_j are pressure basis functions $(j = 1,2,\cdots N)$.

By solving the following algebra system, the coefficients $\tilde{U}_1^j$, $\tilde{P}_1^j$, $\tilde{P}_2^j$ and $\tilde{U}_2^j$ are obtained.

$$\begin{bmatrix} A_1 & B_1 & 0 & 0 \\ B_2 & D_1 & D_2 & 0 \\ 0 & D_3 & D_4 & B_3 \\ 0 & 0 & B_4 & A_2 \end{bmatrix} \begin{bmatrix} \tilde{U}_1^j \\ \tilde{P}_1^j \\ \tilde{P}_2^j \\ \tilde{U}_2^j \end{bmatrix} = \begin{bmatrix} 0 \\ 0 \\ F_1 \\ F_2 \end{bmatrix}, \quad (7)$$

where

$$A_1 = [\int_\Omega \psi_k \cdot (\lambda K_1)^{-1} \psi_j dx], A_2 = [\int_\Omega \psi_k \cdot (\lambda K_2)^{-1} \psi_j dx], B_1 = B_4 = [-\int_\Omega \varphi_j div(\psi_k) dx], B_2 = B_3 = [\int_\Omega \varphi_k div(\psi_j) dx],$$
$$D_1 = [-\int_\Omega (\Phi_1 C_1 s + \lambda\alpha)\varphi_j\varphi_k dx], D_2 = D_3 = [\int_\Omega \lambda\alpha\varphi_j dx], D_4 = [-\int_\Omega (\Phi_2 C_2 s + \lambda\alpha)\varphi_j\varphi_k dx], F_1 = [-\int_\Omega \Phi_1 C_1 p_1^0 \varphi_k dx],$$
$$F_2 = [-\int_\Omega \Phi_2 C_2 p_2^0 \varphi_k dx], (k,j = 1,2, \cdots N).$$

Then the linear algebra system (7) can be written as:

$$\begin{bmatrix} A_1 & B_1 & 0 & 0 \\ B_2 & D_1 & D_2 & 0 \\ 0 & D_2 & D_4 & B_2 \\ 0 & 0 & B_1 & A_2 \end{bmatrix} \begin{bmatrix} \tilde{U}_1^j \\ \tilde{P}_1^j \\ \tilde{P}_2^j \\ \tilde{U}_2^j \end{bmatrix} = \begin{bmatrix} 0 \\ 0 \\ F_1 \\ F_2 \end{bmatrix}. \tag{8}$$

3 The Generalized Inverse of a Class of Two-by-Two Block Matrix

3.1 The Basic Concept of Inverse Matrices

There are many kinds of inverse matrices for block matrices such as minus inverse, plus inverse, least square generalized inverse, reflexive generalized inverse and minimum norm generalized inverse. In this section, we consider only the minus inverse of four-block matrices.

For the matrix $A \in F^{m\times n}$, we consider the following formulas:

$$\forall A \in F^{m\times n}, \quad (i)\ AXA = A; \quad (ii)\ XAX = X; \quad (iii)\ (AX)^H = AX; \quad (V)\ (XA)^H = XA.$$

For an $m \times n$ matrix A, if there exists a matrix X satisfy (i), then matrix X is called a minus inverse of A, and denoted by A^-. And $\forall A \in F^{m\times n}$, minus inverse A^- exists.

If matrices A, X satisfy formulas (i) and (ii), we call X the reflexive generalized inverse of the matrix A. If matrices A, X satisfy formulas (i) and (iii), we call X the least squares generalized inverse of the matrix A. If matrices A, X satisfy formulas (i) and (V), we call X minimum norm generalized inverse of the matrix A. If matrices A, X satisfy formulas (i), (ii), (iii) and (v), we call X plus inverse of the matrix A.

3.2 Minus Inverse of a Two-by-Two Block Matrix

In this section, we consider the generalized inverse of the following two-by-two block matrix

$$A = \begin{pmatrix} A_{11} & A_{12} \\ A_{21} & A_{22} \end{pmatrix}. \tag{9}$$

We discuss the minus inverse A^- of matrix A when $A_{12} = A_{21} = 0$.
We partition A^- into appropriate blocks as follows:

$$A^- = \begin{pmatrix} B_{11} & B_{12} \\ B_{21} & B_{22} \end{pmatrix}. \tag{10}$$

Since $AA^-A = A$, we have

$$AA^-A = \begin{pmatrix} A_{11} & 0 \\ 0 & A_{22} \end{pmatrix} \begin{pmatrix} B_{11} & B_{12} \\ B_{21} & B_{22} \end{pmatrix} \begin{pmatrix} A_{11} & 0 \\ 0 & A_{22} \end{pmatrix} = \begin{pmatrix} A_{11}B_{11}A_{11} & A_{11}B_{12}A_{22} \\ A_{22}B_{21}A_{11} & A_{22}B_{22}A_{22} \end{pmatrix} = A$$
$$= \begin{pmatrix} A_{11} & 0 \\ 0 & A_{22} \end{pmatrix}.$$

From $A_{11}B_{11}A_{11} = A_{11}$, $A_{22}B_{22}A_{22} = A_{22}$, we obtain $B_{11} = A_{11}^-$, $B_{22} = A_{22}^-$.

The general solution of the matrix equation $AXB = 0$ is $X = Y - AA^-YBB^-$, where Y is an arbitrary matrix of appropriate order. So the general solutions of the matrix equation $A_{11}B_{12}A_{22} = 0$ and $A_{22}B_{21}A_{11} = 0$ are

$$B_{12} = Y - A_{11}^-A_{11}YYA_{22}A_{22}^-, B_{21} = Z - A_{22}^-A_{22}ZA_{11}A_{11}^-,$$

where Y and Z are arbitrary matrices of appropriate order. Then we have

$$A^- = \begin{pmatrix} A_{11} & 0 \\ 0 & A_{22} \end{pmatrix}^- = \begin{pmatrix} A_{11}^- & Y - A_{11}^-A_{11}YYA_{22}A_{22}^- \\ Z - A_{22}^-A_{22}ZA_{11}A_{11}^- & A_{22}^- \end{pmatrix}. \tag{11}$$

As A_{12} and A_{21} are nonzero, we discuss some singular and nonsingular cases about matrix A.

(I) When matrix A is singular, assume that $\det(A_{11}) \neq 0$, by the elementary transformation of the partitioned matrix

$$\begin{pmatrix} E & 0 \\ -A_{21}A_{11}^- & E \end{pmatrix} \begin{pmatrix} A_{11} & A_{12} \\ A_{21} & A_{22} \end{pmatrix} \begin{pmatrix} E & -A_{11}^-A_{12} \\ 0 & E \end{pmatrix} = \begin{pmatrix} A_{11} & 0 \\ 0 & A_{22} - A_{21}A_{11}^-A_{12} \end{pmatrix}.$$

Notes

$$P = \begin{pmatrix} E & 0 \\ -A_{21}A_{11}^- & E \end{pmatrix}, \quad Q = \begin{pmatrix} E & -A_{11}^-A_{12} \\ 0 & E \end{pmatrix},$$
$$D = \begin{pmatrix} A_{11} & 0 \\ 0 & A_{22} - A_{21}A_{11}^-A_{12} \end{pmatrix},$$

such that $PAQ = D$.

According to A singular, $\det(A_{22} - A_{21}A_{11}^{-1}A_{12}) \neq 0$, the following formula can be obtained.

$$
\begin{aligned}
A^- &= \begin{pmatrix} A_{11} & A_{12} \\ A_{21} & A_{22} \end{pmatrix}^- = QD^-P = \begin{pmatrix} E & -A_{11}^-A_{12} \\ 0 & E \end{pmatrix} \begin{pmatrix} A_{11} & 0 \\ 0 & A_{22}-A_{21}A_{11}^-A_{12} \end{pmatrix}^- \begin{pmatrix} E & 0 \\ -A_{21}A_{11}^- & E \end{pmatrix} \\
&= \begin{pmatrix} A_{11}^- + A_{11}^-A_{12}(A_{22}-A_{21}A_{11}^-A_{12})^-A_{21}A_{11}^- & -A_{11}^-A_{12}(A_{22}-A_{21}A_{11}^-A_{12})^- \\ -(A_{22}-A_{21}A_{11}^-A_{12})^-A_{21}A_{11}^- & (A_{22}-A_{21}A_{11}^-A_{12})^- \end{pmatrix}.
\end{aligned}
$$

$$(12)$$

If $\det(A_{22}) \neq 0$, then $\det(A_{11} - A_{12}A_{22}^-A_{21}) \neq 0$. We have

$$
A^- = \begin{pmatrix} (A_{11}-A_{12}A_{22}^-A_{21})^- & -(A_{11}-A_{12}A_{22}^-A_{21})^-A_{12}A_{22}^- \\ -A_{22}^-A_{21}(A_{11}-A_{12}A_{22}^-A_{21})^- & A_{22}^- + A_{22}^-A_{21}(A_{11}-A_{12}A_{22}^-A_{21})^-A_{12}A_{22}^- \end{pmatrix}.
$$

$$(13)$$

(II) When A is nonsingular, if $\det(A_{11}) \neq 0$, the $PAQ = D \Rightarrow A = P^-DQ^-$. Since

$$
(P^-DQ^-)(P^-DQ^-)^-(P^-DQ^-) - P^-DQ^-QD^-PP^-DQ^- = P^-DD^-DQ^- \\
= P^-DQ^-,
$$

D is a quasi-diagonal matrix, the following formula can be obtained

$$
D^- = \begin{pmatrix} A_{11} & 0 \\ 0 & A_{22}-A_{21}A_{11}^-A_{12} \end{pmatrix}^- = \begin{pmatrix} A_{11}^- & 0 \\ 0 & (A_{22}-A_{21}A_{11}^-A_{12})^- \end{pmatrix},
$$

$$
A^- = \begin{pmatrix} A_{11}^- + A_{11}^-A_{12}(A_{22}-A_{21}A_{11}^-A_{12})^-A_{21}A_{11}^- & -A_{11}^-A_{12}(A_{22}-A_{21}A_{11}^-A_{12})^- \\ -(A_{22}-A_{21}A_{11}^-A_{12})^-A_{21}A_{11}^- & (A_{22}-A_{21}A_{11}^-A_{12})^- \end{pmatrix}.
$$

$$(14)$$

In the same way, when $\det(A_{22}) \neq 0$, we can get

$$
A^- = \begin{pmatrix} (A_{11}-A_{12}A_{22}^-A_{21})^- & -(A_{11}-A_{12}A_{22}^-A_{21})^-A_{12}A_{22}^- \\ -A_{22}^-A_{21}(A_{11}-A_{12}A_{22}^-A_{21})^- & A_{22}^- + A_{22}^-A_{21}(A_{11}-A_{12}A_{22}^-A_{21})^-A_{12}A_{22}^- \end{pmatrix}.
$$

$$(15)$$

If A is semi-positive definite Hermite matrix, similar conclusions will be drawn regardless of whether A_{11} or A_{22} is invertible or not.
A is semi-positive definite Hermite matrix, there exists matrix B, such that $A = B^H B$.
Note that

$$
B = (B_1 \quad B_2), A = \begin{pmatrix} A_{11} & A_{12} \\ A_{21} & A_{22} \end{pmatrix},
$$

the following equation can be obtained

$$\begin{pmatrix} A_{11} & A_{12} \\ A_{21} & A_{22} \end{pmatrix} = \begin{pmatrix} B_1^H B_1 & B_1^H B_2 \\ B_2^H B_1 & B_2^H B_2 \end{pmatrix}.$$

The equations $A^H A x = 0$ and $A x = 0$ have the same solution. Since $A^H A (A^H A)^- A^H A = A^H A$, the following formulas can be obtained.

$$A^H(A(A^H A)^- A^H A - A) = 0, A(A^H A)^- A^H A - A = 0, A = A(A^H A)^- A^H A,$$
$$A_{21} A_{11}^- A_{11} = B_2^H B_1 (B_1^H B_1)^- B_1^H B_1 = B_2^H B_1 = A_{21}, \quad A_{11} A_{11}^- A_{12} = B_1^H B_1 (B_1^H B_1)^- B_1^H B_2 = B_1^H B_2 = A_{12}.$$

Notes

$$P = \begin{pmatrix} E & \mathbf{0} \\ -A_{21} A_{11}^- & E \end{pmatrix}, Q = \begin{pmatrix} E & -A_{11}^- A_{12} \\ \mathbf{0} & E \end{pmatrix}, D = \begin{pmatrix} A_{11} & \mathbf{0} \\ \mathbf{0} & A_{22} - A_{21} A_{11}^- A_{12} \end{pmatrix},$$

the following formulas can be obtained

$$D = PAQ = \begin{pmatrix} E & \mathbf{0} \\ -A_{21} A_{11}^- & E \end{pmatrix} \begin{pmatrix} A_{11} & A_{12} \\ A_{21} & A_{22} \end{pmatrix} \begin{pmatrix} E & -A_{11}^- A_{12} \\ \mathbf{0} & E \end{pmatrix}$$

$$= \begin{pmatrix} E & \mathbf{0} \\ -A_{21} A_{11}^- & E \end{pmatrix} \begin{pmatrix} B_1^H B_1 & B_1^H B_2 \\ B_2^H B_1 & B_2^H B_2 \end{pmatrix} \begin{pmatrix} E & -A_{11}^- A_{12} \\ \mathbf{0} & E \end{pmatrix} = \begin{pmatrix} A_{11} & \mathbf{0} \\ \mathbf{0} & A_{22} - A_{21} A_{11}^- A_{12} \end{pmatrix}.$$

Obviously, P and Q are invertible, we can obtain

$$A^- = QD^- P = Q \begin{pmatrix} A_{11} & \mathbf{0} \\ \mathbf{0} & A_{22} - A_{21} A_{11}^- A_{12} \end{pmatrix}^- P = Q \begin{pmatrix} A_{11}^- & \mathbf{0} \\ \mathbf{0} & (A_{22} - A_{21} A_{11}^- A_{12})^- \end{pmatrix} P$$

$$= \begin{pmatrix} E & -A_{11}^- A_{12} \\ \mathbf{0} & E \end{pmatrix} \begin{pmatrix} A_{11}^- & \mathbf{0} \\ \mathbf{0} & (A_{22} - A_{21} A_{11}^- A_{12})^- \end{pmatrix} \begin{pmatrix} E & \mathbf{0} \\ -A_{21} A_{11}^- & E \end{pmatrix}$$

$$= \begin{pmatrix} A_{11}^- + A_{11}^- A_{12}(A_{22} - A_{21} A_{11}^- A_{12})^- A_{21} A_{11}^- & -A_{11}^- A_{12}(A_{22} - A_{21} A_{11}^- A_{12})^- \\ -(A_{22} - A_{21} A_{11}^- A_{12})^- A_{21} A_{11}^- & (A_{22} - A_{21} A_{11}^- A_{12})^- \end{pmatrix}.$$

4   Generalized Inverse Solutions of a Class of Linear System

And for the system (8), we note

$$A_{11} = \begin{pmatrix} A_1 & B_1 \\ B_2 & D_1 \end{pmatrix}, A_{12} = \begin{pmatrix} \mathbf{0} & \mathbf{0} \\ D_2 & \mathbf{0} \end{pmatrix}, A_{21} = \begin{pmatrix} \mathbf{0} & D_2 \\ \mathbf{0} & \mathbf{0} \end{pmatrix}, A_{22} = \begin{pmatrix} D_4 & B_2 \\ B_1 & A_2 \end{pmatrix},$$
$$U = (\tilde{U}_1^j, \tilde{P}_1^j)^T, P = (\tilde{P}_2^j, \tilde{U}_2^j)^T, F = (F_1, F_2)^T, \mathbf{0} = (0, 0)^T$$

then the following equation can be obtained

$$\begin{bmatrix} A_{11} & A_{12} \\ A_{21} & A_{22} \end{bmatrix} \begin{bmatrix} U \\ P \end{bmatrix} = \begin{bmatrix} \mathbf{0} \\ F \end{bmatrix}. \tag{16}$$

According to the above formulas, the solutions can be obtained

$$\begin{bmatrix} U \\ P \end{bmatrix} = \begin{bmatrix} A_{11} & A_{12} \\ A_{21} & A_{22} \end{bmatrix}^{-} \begin{bmatrix} \mathbf{0} \\ F \end{bmatrix}, \tag{17}$$

$$\begin{bmatrix} A_{11} & A_{12} \\ A_{21} & A_{22} \end{bmatrix}^{-} = \begin{bmatrix} (A_{11} - A_{12}A_{22}^{-}A_{21})^{-} & -(A_{11} - A_{12}A_{22}^{-}A_{21})^{-}A_{12}A_{22}^{-} \\ -A_{22}^{-}A_{21}(A_{11} - A_{12}A_{22}^{-}A_{21})^{-} & A_{22}^{-} + A_{22}^{-}A_{21}(A_{11} - A_{12}A_{22}^{-}A_{21})^{-}A_{12}A_{22}^{-} \end{bmatrix}. \tag{18}$$

At last, we can get the following recurrence

$$\begin{bmatrix} U \\ P \end{bmatrix} = \begin{bmatrix} (A_{11} - A_{12}A_{22}^{-}A_{21})^{-} & -(A_{11} - A_{12}A_{22}^{-}A_{21})^{-}A_{12}A_{22}^{-} \\ -A_{22}^{-}A_{21}(A_{11} - A_{12}A_{22}^{-}A_{21})^{-} & A_{22}^{-} + A_{22}^{-}A_{21}(A_{11} - A_{12}A_{22}^{-}A_{21})^{-}A_{12}A_{22}^{-} \end{bmatrix} \begin{bmatrix} \mathbf{0} \\ F \end{bmatrix}, \tag{19}$$

where

$$\begin{aligned} A_{22}^{-} &= \begin{bmatrix} D_4 & B_2 \\ B_1 & A_2 \end{bmatrix}^{-} \\ &= \begin{bmatrix} (D_4 - B_2A_2^{-}B_1)^{-} & -(D_4 - B_2A_2^{-}B_1)^{-}B_2A_2^{-} \\ -A_2^{-}B_1(D_4 - B_2A_2^{-}B_1)^{-} & A_2^{-} + A_2^{-}B_1(D_4 - B_2A_2^{-}B_1)^{-}B_2A_2^{-} \end{bmatrix}. \end{aligned}$$

Using the above results, the algorithm for computing matrix $\begin{bmatrix} A_{11} & A_{12} \\ A_{21} & A_{22} \end{bmatrix}^{-}$ is given as following.

Algorithm 1. Implementing process of computing matrix $\begin{bmatrix} A_{11} & A_{12} \\ A_{21} & A_{22} \end{bmatrix}^{-}$

1. Compute A_2^{-}, B_1^{-};
2. Compute $(D_4 - B_2A_2^{-}B_1)^{-}, (D_4A_2^{-}B_1)^{-}$;
3. Compute A_{22}^{-};
4. Compute $(A_{11} - A_{12}A_{22}^{-}A_{21})^{-}$;
5. Compute $\begin{bmatrix} A_{11} & A_{12} \\ A_{21} & A_{22} \end{bmatrix}^{-}$ according to (18).

Then we can compute the numerical value of $U_1^j(t), U_2^j(t)$ $(j = 1, 2 \ldots N')$ and $P_1^j(t), P_2^j(t)$ $(j = 1, 2 \ldots N')$ by $\tilde{U}_1^j(s), \tilde{U}_2^j(s)$ $(j = 1, 2 \ldots N')$ and $\tilde{P}_1^j(s), \tilde{P}_2^j(s)$ $(j = 1, 2 \ldots N')$ using the numerical formula of Laplace inversion transform.

5 Conclusion

In this paper, we derive the generalized inverse of the block four-by-four linear systems which arise from partial differential equation models of flow in a Dual-Permeability system. Numerical solutions of pressure and Darcy velocity in the mathematic models can be obtained.

Acknowledgments. The authors are grateful to the referees for their valuable comments and suggestions which helped to improve the presentation of the paper.

Funding. This research was funded by the National Natural Science Foundation of China under Grant No. 41766001, 41962019.

References

1. Aarnes, J., Efendiev, Y.: Mixed multiscale finite element methods for stochastic porous media flows. SIAM J. Sci. Comput. **30**, 2319–2339 (2008)
2. Efendiev, Y., Thomas, H.Y.: Multiscale Finite Element Methods: Theory and Applications, pp. 27–33. Springer, New York (2009)
3. Liu, T., Xu, H., Qiu, X.: A combination method of mixed multiscale finite-element and Laplace transform for flow in a dual-permeability system. ISRN Appl. Math. **1**, 1–10 (2012)
4. Vogel, T., Gerke, H.H., Zhang, R., Van Genuchten, M.Th.: Modeling flow and transport in a two-dimensional dual-permeability system with spatially variable hydraulic properties. J. Hydrol. **238**, 78–89 (2000)
5. Yin, Zh., Jia, S.: Solution of generalized inverse matrix and linear equations. Math. Pract. Cogn. 239–244 (2009)
6. Luo, J., Fang, W.: Introduction to Matrix Analysis, pp. 208–215. South China University of Technology Press, Guangzhou (2006)
7. Tian, Y., Takane, Y.: More on generalized inverse of partitioned matrices with Banachicwicz-Schur forms. Linear Algebra Appl. **430**, 1641–1655 (2009)
8. Spellacy, L., Golden, D., Rungger, I.: Performance analysis of a pair wise method for partial inversion of complex block tridiagonal matrices. Concurr. Comput. Pract. Exp. e4918 (2018)
9. Yan, Z.: New representations of the Moore-Penrose inverse of 2×2 block matrices. Linear Algebra Appl. **456**, 3–15 (2012)

An Improved Method of Joint Extraction

Yanfang Cheng[1], Yinan Lu[2(✉)], and Hangyu Pan[2]

[1] Software Institute, Jilin University, Changchun, China
[2] College of Computer Science and Technology,
Jilin University, Changchun, China
luyn@jlu.edu.cn

Abstract. Named entity recognition (NER) and relation extraction (RE) are the basic tasks of Natural Language Processing (NLP). However, previous works always treat them as two separated subtasks, a novel improved method of joint extraction was present in this paper to solve the problem of internal relationship and error propagation in traditional pipeline model, Named entity recognition is regarded as a sequential annotation problem. In relation extraction task, the relationship between two entities and their relationship types are predicted at the same time, the possible multiple pairs of relationships in sentences are identified. Finally, the work innovatively use sequential patterns to correct the results. The experiment on two authoritative datasets verifies the advancement in our model.

Keywords: Joint extraction · Sequential patterns revision · Information extraction · Relation extraction

1 Introduction

The goal of information extraction (IE) is to transform text information into structured information, which is initially used to locate specific information in natural language documents and belongs to a sub-domain of natural language processing. The classic tasks of IE include Named Entity Recognition (NER) [1] and Relation Extraction (RE) [2]. The major task of NER is to identify related named entities, such as person name, place name, etc. from a piece of natural language text and mark the location and type. RE mainly mining and recognize the implicit grammatical or semantic association between entities in text, it is the key link of information extraction, which is also widely used in the field of Data Reduction and Knowledge Graph [3].

The architecture of traditional relation extraction task uses pipeline model, firstly, identifies named entity, then extracts relational entity. The design of system is highly coupled but simplifies the problem too much, splits the named entity recognition and relation extraction, and does not make full use of the information between the two tasks. Moreover, if there are errors in entity recognition, relation extraction task cannot identify and feedback it, which directly affects the accuracy of the pipeline models.

In order to overcome all the influence brought by traditional pipeline models, we propose a novel joint model based on sequential pattern modification. The problem of error propagation is alleviated by combining the two sub-tasks simultaneously. At the same time, lower level parameters in some models are shared by entity recognition and

© Springer Nature Switzerland AG 2020
F. Xhafa et al. (Eds.): IISA 2019, AISC 1084, pp. 453–459, 2020.
https://doi.org/10.1007/978-3-030-34387-3_56

relation classification sub-models in the joint model, which accelerated the model capture the relationship between the two sub-tasks, and improves the accuracy of extraction relation by utilizing sequential pattern modification.

2   Related Work

Traditional joint model of named entity recognition and relation extraction relies on the features of manual annotation. The feature space of joint task is usually composed of sub-task feature space, which has the problem of sparse features. Yang et al. introduced constraints and used integer linear programming (ILP) to optimize the uncertainty of prediction [4]. Li combines incremental beam search with structured perception mechanism to build an end-to-end joint model framework [5].

Some deep neural network-based methods have been applied to the joint model to achieve entity recognition and relationship extraction, alleviating the dependence on expert features in the traditional joint model. Li et al. used CNN to encode word characters as character representations, and input them into the network as text representations along with word embedding and part-of-speech embedding [6]. Joint extraction was accomplished by stacking and parameter sharing of two Bi-LSTMs. Katiyar et al. used a multi-layer Bi-LSTM network, using a model similar to the attention mechanism at each time step, and did not rely on dependency tree information to complete joint extraction [7].

BERT model [3], which is proposed by Devlin, provides a new research direction for natural language processing. Combining the advantages of traditional distributed text representation models GPT and ELMo [8], we use Transformer to extract features and train them. Combining with bidirectional information, randomly hide some words and predict them with uncovered words. We use BERT model with very amazing effect to jointly identify named entities and extract relationships from data in ACE04 and ADE datasets. Finally, the innovative addition sequence model is helpful to improve the accuracy of the results to a certain extent.

3   Model Framework

Transformer can reduce the computational load and improve the parallel efficiency without weakening the final experimental results. In this paper, the BERT representation model is used to process the original data and then as the input of Bi-LSTM. NER is regarded as the problem of sequence annotation. Bi-LSTM is followed by a CRF layer to process its output and obtain the global optimal output sequence. The entity annotation is embedded and concatenated with the output of Bi-LSTM, as the input of relation extraction. Finally, the sequential pattern method is used to correct the model results. The whole process is shown in Fig. 1.

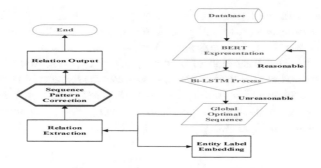

Fig. 1. Process of proposed method

3.1 Text Representation Based on BERT

Traditional word embedding models such as Word2voc and Glove can represent text as corresponding word vectors [9], but they cannot solve the problem of different meanings of polysemous words. In this development process, ELMo—the context-independent static vectors are transformed into context-dependent dynamic vectors. BERT [10]—the downstream specific NLP task calls are more convenient. The potential problems of ELMo model are corrected. Sentence representations or sentence pairs are obtained by negative sampling at sentence level. Transformer model replaces LSTM to improve the efficiency of expression and time.

3.2 Sequence Annotation and Relationship Extraction

The most prominent advantage of LSTM network is that it can solve the problem of contextual incoherence in long-text input by processing long-distance dependent information with memory units. Bi-LSTM takes both the above information and the context prediction information into account, calculates the most probable entity label of each Token at CRF level, obtains the global optimal output sequence, and selects the

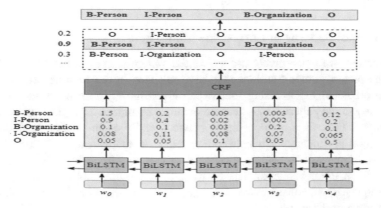

Fig. 2. Sequence annotation and relationship extraction based on BERT-LSTM-CRF

label sequence with the highest prediction score as the final result. The process of sequence annotation is shown in Fig. 2. A set of output results with a score of 0.9 (B-Person, I-Person, O, O-Organization, O) are selected.

In this module, the basic unit of model operation is token. Setting up the token resume relationship between the two entities, indicates the relationship between the two entities. Given sentence $S = (w_0, w_1, \ldots, w_N)$, the task is to find each word $w_i \in \{w_1, w_2, \ldots, w_n\}$ most likely corresponds to the words $w_j \in \{w_0, w_1, \ldots, w_N\}$. The probably calculating method was showed as (1.1):

$$P(w_j|w_j, S) = \frac{\exp(g(a_j, a_i))}{\sum_{K=0}^{N} \exp(g(a_K, a_i))} \tag{1.1}$$

Our task is to identify each token $w_i (i = 0, 1, \ldots, n)$ the most possible corresponding vector $\hat{y}_i \subseteq w$ and the most possible corresponding relationship label $\hat{r}_i \subseteq R$. Given the relationship label r_k, the scores between w_i and w_j are calculated as follows, Where f(.) is the activation function (relu, tanh):

$$s^{(r)}(z_j, z_i, r_K) = V^{(r)} f(U^{(r)} z_j + W^{(r)} z_i + b^{(r)}) \tag{1.2}$$

$V^{(r)} \in R^1, U^{(r)} \in R^{1\times(2d+b)}, W^{(r)} \in R^{1\times(2d+b)}, b^{(r)} \in R^1$, d is the number of hidden units in LSTM, b is the size of tag embedding, L is the layer width. Given token w_i, the probability that the relationship between w_j and w_i is r_k:

$$P_r(\text{head} = w_j, \text{label} = r_K|w_i) = \sigma(s^{(r)}(z_j, z_i, r_k)) \tag{1.3}$$

Where $\sigma(\cdot)$ is sigmoid function. Minimizing cross-entropy loss $\mathcal{L}_{\text{rel}}$ during training. Among them, $\hat{y}_i \subseteq w$ and $\hat{r}_i \subseteq R$ are the truth vectors of the corresponding w_i words and related labels, respectively, and M is the number of w_i. After training, the corresponding probability can be calculated by substituting the real label values $\hat{y}_i, \hat{r}_i$ into (1.3). The threshold value is the lowest probability that all real labels deserve. In prediction, if the probability calculated by (1.3) is greater than the threshold value, the relationship is considered to be valid.

3.3   Sequence Patterns to Correct the Model Results

In this section, we innovatively use sequential pattern to correct the results, mine the sequential patterns among entities that have relations in the training set, and further filter the predicted results using PrefixSpan algorithm [11]. In our task settings, all frequent sequential pattern sets LS can be obtained by mining from the sequential set S. For example, there is a relationship R between entity e_i and entity e_j to determine whether the dependency path between e_i and e_j contains enough frequent pattern sequences. In our task, as long as it contains at least one frequent pattern sequence, in other words, if the dependency path between e_i and e_j does not contain any frequent pattern sequence, we consider two. The relationship between entities is unreliable and is removed from the result.

4 Experimental and Analysis

4.1 Experiment Setting

Our improved method is designed by with Python and TensorFlow, machine learning libraries to develop our joint model. The size of LSTM is set to d = 64, the network layer width L = 64, and the training is carried out by Adam optimizer. The learning rate is a = 10^{-3}, the label embedding size is 25, relu function is used to activate in ACE04, and tanh activation is used in ADE dataset. In addition, in order to avoid the over-fitting problem, Dropout technology is applied in the hidden layer of the two tasks, early stopping technology is used in the verification set, and the optimal Super-parameters are obtained after 60 to 200 iteration cycles, and the final termination period is selected according to the results of the verification set.

4.2 Experimental Results and Comparative Analysis

ACE04: We follow the cross-validation settings of Li [5] and Gupta [12], delete DISC relationships, cross-validate benws and nwire subsets five times, and randomly select 15% of the data from the training set for validation and selection of hyperparameters.

ADE: The ADE (Adverse drug effect) task is designed to identify two entities (drugs and diseases) and extract drug and disease-related relationships [13]. Biologists aim to annotate drug disease entities in ADE-related sentences, so we only use ADE-related sentences to participate in the experiment, similar to Li [6]. 10-fold cross-validation was used to evaluate our model. The final result is that F1 represents the macro average.

Our method can directly infer all entities and relationships of a sentence by modeling the whole sentence at one time, and naturally capture multiple relationships and model them as multi-label problems. Table 1 shows the comparison of our experimental results in two datasets. Through the analysis of the data.

Table 1. Comparison of experimental results on the dataset of ACE04 (left) and ADE (right)

Dataset	Experiment	Feature dependency	NER			RE			Comprehensive F1
			P	R	F1	P	R	F1	
ACE04	Li and Ji [5]	Yes	83.50	76.20	79.70	60.80	36.10	45.30	62.50
	Miwa [14]	Yes	80.80	82.90	81.80	48.70	48.70	48.40	65.10
	Katiyar [7]	No	81.20	78.10	79.60	46.40	45.53	45.70	62.65
	Our model	No	81.21	81.14	81.17	50.19	44.25	47.03	64.10
ADE	Li [15]	Yes	79.50	79.60	79.50	64.00	62.90	63.40	71.45
	Li [6]	Yes	82.70	86.70	84.60	67.50	75.80	71.40	78.00
	Our model	No	84.20	88.12	86.12	70.90	76.85	73.76	79.94

(1) The experimental results of ACE04, the F1 value of our model in both tasks is 2% better than that of Katiyar [7]. The main reason for this result is that we obtained more effective word representation by using BERT language pre-training [9] model.

(2) Compared with the standard of 0.5% entity recognition task and 1% relationship extraction task in Miwa [14], our model performance is also in a reasonable range. Compared with Li [5], although the accuracy is not significantly improved, our entity recognition F1 value has been improved by 2%, and the relationship extraction task F1 value has been improved by 3%, because our model can generate automatic extraction. These features are very good in different contexts, have strong practicability in various data sets, and the generalization ability of the model is good.

In addition to the demonstration and analysis of our normal experimental results, we also performed ablation tests on ACE04 datasets to analyze the effectiveness of some of the designs in our joint model. As shown in Table 2 is our experimental data:

Table 2. Comparison of ablation results

Index	NER			RE			Comprehensive F1-values
	P	R	F1	P	R	F1	F1
Our model	81.21	81.14	81.17	50.19	44.25	47.03	64.10
No result amendment	81.00	81.28	81.14	50.02	44.36	47.02	64.08
No label embedding	80.72	80.47	80.59	49.80	43.67	45.53	63.06
No CRF	80.65	81.04	80.84	48.73	43.06	45.72	63.28

The following conclusions can be drawn from the ablation results in the Table 2:

(1) Although the recall rate decreases, the overall accuracy of output results has been improved, and the F1 value has been slightly improved through the result revision of the sequential model.

(2) After removing the tag embedding, the performance of the relational extraction task decreases (the difference of F1 value is 1%), which indicates that the existence of entity annotation provides meaningful information for the relational extraction task.

(3) The CRF layer can effectively capture the effective dependencies in the data set, so as to obtain the globally optimal output sequence. Removing the CRF layer and replacing it with softmax results in a slight reduction in F1 value of entity recognition tasks and a reduction in performance of relational extraction tasks by about 2%.

Acknowledgments. This work was supported by Jilin Provincial Science & Technology Development (20180101054JC), and Talent Development Fund of Jilin Province (2018).

References

1. Lample, G., Ballesteros, M., Subramanian, S., et al.: Neural architectures for named entity recognition (2016)
2. Kanya, N., Ravi, T.: Modelings and techniques in named entity recognition: an information extraction task. In: IET Chennai 3rd International Conference on Sustainable Energy and Intelligent Systems (SEISCON 2012). IET (2012)
3. Kong, B., Xu, R.F., Wu, D.Y.: Bootstrapping-based relation extraction in financial domain. In: 2015 International Conference on Machine Learning and Cybernetics (ICMLC). IEEE (2015)
4. Yang, B., Cardie, C.: Joint inference for fine-grained opinion extraction. In: Proceedings of the 51st Annual Meeting of the Association for Computational Linguistics (Volume 1: Long Papers), pp. 1640–1649 (2013)
5. Li, Q., Ji, H.: Incremental joint extraction of entity mentions and relations. In: Proceedings of the 52nd Annual Meeting of the Association for Computational Linguistics (Volume 1: Long Papers), pp. 402–412 (2014)
6. Li, F., Zhang, M., Fu, G., et al.: A neural joint model for entity and relation extraction from biomedical text. BMC Bioinform. **18**(1), 198 (2017)
7. Katiyar, A., Cardie, C.: Going out on a limb: joint extraction of entity mentions and relations without dependency trees. In: Proceedings of the 55th Annual Meeting of the Association for Computational Linguistics (Volume 1: Long Papers), pp. 917–928 (2017)
8. Peters, M.E., Neumann, M., Iyyer, M., et al.: Deep contextualized word representations (2018)
9. Devlin, J., Chang, M.W., Lee, K., et al.: BERT: pre-training of deep bidirectional transformers for language understanding (2018). arXiv preprint arXiv:1810.04805
10. Lee, J., Yoon, W., Kim, S., et al.: BioBERT: a pre-trained biomedical language representation model for biomedical text mining (2019)
11. Pei, J., Han, J., Mortazavi-Asl, B., et al.: Prefixspan: mining sequential patterns efficiently by prefix-projected pattern growth. In: Proceedings 17th International Conference on Data Engineering, pp. 215–224. IEEE (2001)
12. Gupta, P., Schütze, H., Andrassy, B.: Table filling multi-task recurrent neural network for joint entity and relation extraction. In: Proceedings of COLING 2016, the 26th International Conference on Computational Linguistics: Technical Papers, pp. 2537–2547 (2016)
13. Gurulingappa, H., Rajput, A.M., Roberts, A., et al.: Development of a benchmark corpus to support the automatic extraction of drug-related adverse effects from medical case reports. J. Biomed. Inform. **45**(5), 885–892 (2012)
14. Miwa, M., Bansal, M.: End-to-end relation extraction using LSTMs on sequences and tree structures (2016). arXiv preprint arXiv:1601.00770
15. Li, F., Zhang, Y., Zhang, M.: Joint models for extracting adverse drug events from biomedical text. IJCAI. AAAI Press (2016)

New Measurement Method of Three-Phase Voltage Unbalance Rate Based on DSOGI-PLL

Congwei Yu[1], Xiao Liang[1], Jiafeng Ding[1(✉)], Xinmei Li[1],
Guan Tong[1], Qin Luo[1], and Ting Lei[2]

[1] School of Physics and Electronics, Central South University,
Changsha 410083, China
csjfding@csu.edu.cn
[2] Electric Power Research Institute of Guangdong Power Grid Corporation,
Guangzhou 510080, Guangdong Province, China

Abstract. Three-phase Voltage Unbalance Rate (TPVUR) is an important indicator to measure power quality, whose accurate measurement requires fast and accurate extraction of fundamental sequence component. Considering harmonic interference and shortcomings that Vector Symmetric Component (VSC) is only suitable for extracting sequence components in steady-state circuits. A new measurement method of TPVUR based on Double Second Order Generalized Integrator-Phase Locked Loop (DSOGI-PLL) and Instantaneous Symmetrical Component (ISC) has been put forward. Firstly, the three-phase voltage signal is converted to the $\alpha\beta$ signal in two-phase stationary coordinate system, and then the DSOGI-PLL and the ISC are combined to filter out the harmonics and extract the instantaneous positive sequence and negative sequence components of fundamental, thereby the real-time measurement of TPVUR are realized. Simulink simulation results show that the new method not only satisfies the real-time detection of TPVUR, but also achieves the same accuracy as the recommended algorithm in GB/T 15543-2008 and IEC-6100-4-27-2015.

Keywords: TPVUR · VSC · ISC · DSOGI-PLL · Sequence components · Simulink

1 Introduction

With the rapid development of the global economy, the demand for power supply is increasing. New energy converters such as solar energy and wind energy, variable frequency speed control devices and high-power uncontrollable rectifiers are impactful, the pollution of the power grid caused by the massive access of no-lineal, harmonics and unbalance equipment lead to a decline in power quality [1, 2]. Among them, unbalance rate is an important index to measure power quality. When the voltage of three-phase power grid is seriously unbalanced, the output of distribution transformer decreases, which affects the normal operation of electrical equipment and increases the loss of power grid [3, 4]. Therefore, it has important theoretical value and practical

© Springer Nature Switzerland AG 2020
F. Xhafa et al. (Eds.): IISA 2019, AISC 1084, pp. 460–468, 2020.
https://doi.org/10.1007/978-3-030-34387-3_57

significance to study the measurement method of TPVUR and improve its accuracy and real-time performance.

Accurate measurement of TPVUR requires both a fast acquisition sequence component and an accurate estimation of the voltage fundamental phasor (amplitude and phase). At present, the VSC is used to extract the fundamental positive, negative and zero sequence components of the three-phase voltage at home and abroad, and then the TPVUR is obtained by the ratio of the Root-Mean-Square (RMS) value of the negative sequence component and the positive sequence component. Accurate fundamental amplitude and phase values are obtained by complex vector operations in the frequency domain, thus affecting the accuracy and real-time performance of TPVUR measurements. The most common method for extracting fundamental phasors is Fast Fourier Transform (FFT) [5]. When the grid frequency deviates from the power frequency, it is difficult to achieve synchronous sampling and integer period truncation when sampling the voltage signal, FFT algorithm will cause spectrum leakage and fence effect, which affects the accuracy of three-phase voltage phasor calculation. Wen [6, 8] and Bing [7] used windowed interpolation FFT algorithm to reduce spectrum leakage and fence effect, but the long data window results in poor real-time performance, which is not conducive to dynamic measurement. Rodriguez [9], Yuan [10] and Zhu [11] used ISC to extract the instantaneous value of sequence component of the voltage. The algorithm expands the application range of the VSC, and does not require vector operations and has good real-time performance, so it can be used to dynamically analyze three-phase circuits.

Comprehensive comparison of the TPVUR algorithm in Table 1, the VSC are used to extract the positive and negative sequence components of the voltage phasor in

Table 1. Common three-phase unbalance rate algorithm

Source	Voltage unbalance rate	Note
GB/T 15543 IEC-6100	$\dfrac{\|\vec{V^-}\|}{\|\vec{V^+}\|} \times 100\%$	$\vec{V^+} = \left(\vec{v_a} + \partial\vec{v_b} + \partial^2\vec{v_c}\right)/3$ $\vec{V^-} = \left(\vec{v_a} + \partial^2\vec{v_b} + \partial\vec{v_c}\right)/3$ $\vec{v_a}, \vec{v_b}, \vec{v_c}$ are voltage phasor $\partial = e^{j2\pi/3}$
GB/T 15543 IEC-6100 GIGRE	$\sqrt{\dfrac{1-\sqrt{3-6\beta}}{1+\sqrt{3-6\beta}}} \times 100\%$	$\beta = \dfrac{\|v_{ab}\|^4 + \|v_{bc}\|^4 + \|v_{ac}\|^4}{\left(\|v_{ab}\|^2 + \|v_{bc}\|^2 + \|v_{ac}\|^2\right)^2}$ v_{ab}, v_{bc}, v_{ac} are line voltage
IEEE Std 936-1987	$\dfrac{max[v_a,v_b,v_c]-min[v_a,v_b,v_c]}{V_{pav}} \times 100\%$	$V_{pav} = (v_a + v_b + v_c)/3$ v_a, v_b, v_c are phase voltage
IEEE Std 112-1991	$\dfrac{max[\|v_a-V_{pav}\|,\|v_b-V_{pav}\|,\|v_c-V_{pav}\|]}{V_{pav}} \times 100\%$	$V_{pav} = (v_a + v_b + v_c)/3$ v_a, v_b, v_c are phase voltage
NEMA	$\dfrac{max[\|v_{ab}-V_{lav}\|,\|v_{bc}-V_{lav}\|,\|v_{ac}-V_{lav}\|]}{V_{lav}} \times 100\%$	$V_{lav} = (v_{ab} + v_{bc} + v_{ac})/3$ v_{ab}, v_{bc}, v_{ac} are line voltage
Ghijselen [12]	$\dfrac{\overline{V^-}}{V^+} \times 100\% = \dfrac{t-js}{r+js} \times 100\%$	$s = \dfrac{v_{ca}^2-v_{bc}^2}{2\sqrt{3}v_{ab}} \quad r = \dfrac{v_{ab}}{2} + \dfrac{4A_s^2}{2\sqrt{3}v_{ab}}$ $t = \dfrac{v_{ab}}{2} - \dfrac{4A_s^2}{2\sqrt{3}v_{ab}} \quad p = \dfrac{v_{ab} + v_{bc} + v_{ca}}{2}$ $A_s^2 = \sqrt{p(p - v_{ab})(p - v_{bc})(p - v_{ca})}$ v_{ab}, v_{bc}, v_{ac} are line voltage

GB/T 15543 and IEC-6100, the size and phase of each phasor must be measured, which is cumbersome vector operation. In order to solve this problem, many international organizations (such as IEEE, GIGRE, NEMA, etc.) began to redefine the unbalance rate, that is, calculate the TPVUR by line voltage, some scholars (such as Ghijselen [12]) deduced the formula defined by IEC, the positive and negative sequence components of the phase voltage are replaced by the positive and negative sequence components of the line voltage, however, since the phase offset is neglected, the voltage unbalance characteristics cannot be accurately grasped, the VSC is mainly used to accurately extract the sequence components at present, but only for steady state analysis of three-phase circuits. Foreign scholar Lyon extended the vector symmetry component (VSC) to the time domain, namely ISC [13].

Rodriguez uses the ISC to extract the Positive Sequence Component (PSC) of the voltage [9], which proves that ISC can track the PSC of the voltage in real time and achieve the same accuracy of the VSC, the flow chart is shown in Fig. 1. The voltage signals can be converted from *abc* to the $\alpha\beta$ reference frames by the Clarke transformation, and then the PSC of voltage is extracted by the Quadrature-Signals Generator (QSG) and ISC.

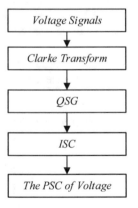

Fig. 1. Positive sequence component extraction process based on ISC

At present, the method in literature [9] is only used in the field of grid-connected control and power compensation, this paper introduces the ISC into the field of dynamic measurement TPVUR, and a TPVUR measurement new method based on DSOGI-PLL is proposed. After filtering harmonics by SOGI, the calculation of positive and negative sequence components of the fundamental is effectively reduced by variable substitution, thereby improving the real-time and accuracy performance of the TPVUR, the PLL makes the method have a frequency adaptive function. This work provides a new solution for dynamically measuring TPVUR.

2   The Design of Second Order Generalized Integrator (SOGI)

According to the conclusion in literature [9], the premise of reducing the calculation of ISC is to construct a quadrature signal, and considering the harmonics in the power grid, if the necessary filtering operations are not considered in the specific calculation, the accurate sequence component cannot be extracted by ISC. Therefore, SOGI is designed in this paper, its structure is shown in Fig. 2. SOGI can not only accurately track the grid voltage and filter, but also generate quadrature voltage signals.

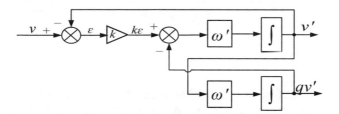

Fig. 2. The structure diagram of SOGI

In Fig. 2, the input v of SOGI is the grid voltage, whose frequency ω is generally fluctuating, ω' is the resonant frequency, k is the damping coefficient, and the transfer function of output v' and qv' to input v is as follows.

$$D(s) = v'(s)/v(s) = k\omega's/\left(s^2 + k\omega's + \omega'^2\right) \tag{1}$$

$$Q(s) = qv'(s)/v(s) = k\omega'^2/\left(s^2 + k\omega's + \omega'^2\right) \tag{2}$$

The Bode diagram of $D(s)$ and $Q(s)$ with different k values is shown in Fig. 3. Combining Eqs. (1), (2) and Fig. 3, when the resonance frequency ω' coincides with the input frequency ω, the amplitude of $D(s)$ and $Q(s)$ are the same and the phases are

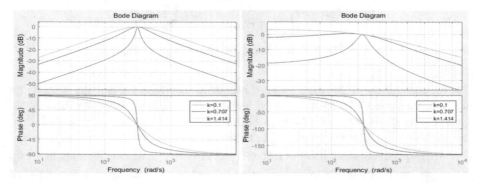

Fig. 3. The Bode diagram of $D(s)$ and $Q(s)$

different by 90°. Therefore, the output signals $v'(s)$ and $qv'(s)$ are quadrature signals of the same magnitude, and v' has the same magnitude and phase as v.

It can be seen from Fig. 3 that $D(s)$ is a second-order band-pass filter whose transfer function can be expressed as follows.

$$D(s) = A_0 \frac{\omega'}{Q_D} s \Big/ \left(s^2 + \frac{\omega'}{Q_D} s + \omega'^2 \right) \tag{3}$$

Where A_0 is the gain of $D(s)$, Q_D is the quality rate, and the Q_D can be obtained by comparing Eqs. (1) and (3) as follows.

$$Q_D = 1/k \tag{4}$$

From Eq. (4), the quality factor of SOGI is not affected by the resonance frequency ω'. According to Fig. 3, the passband of SOGI varies with the value of k, the larger the value of k, the smaller the quality factor, the wider the passband, the lower the selectivity, which facilitates the fast and accurate phase locking of the PLL below. The bandwidth of $D(s)$ is only affected by the k, which is independent of ω', and has significant attenuation at other frequencies besides ω'. Similarly, $Q(s)$ is a low-pass filter whose steady-state gain is also affected only by k, independent of ω'. Adjusting k to set the bandwidth of $Q(s)$, the smaller the value of k, the better the filtering effect of $Q(s)$, but the longer the dynamic response time. In order to balance the filtering effect and response speed of SOGI, k is generally in the vicinity of $\sqrt{2}$, which is consistent with the requirement of the quality factor $Q_D = 0.707$ of the filter.

3   TPVUR Measurement System Based on DSOGI-PLL

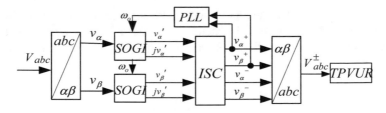

Fig. 4. TPVUR measurement system schematic diagram

The research object of this paper is the three-phase four-wire low-voltage distribution network, combined with the above theoretical analysis, a new TPVUR measurement system is designed, which is shown in Fig. 4. Firstly, the three-phase voltage instantaneous value V_{abc} is transformed by Clarke transformation to obtain the v_α and v_β, and they are respectively input into the DSOGI, and only the fundamental components v_α', jv_α', v_β', jv_β' are retained after DSOGI filtering, v_α' and jv_α' are quadrature components of the same waveform, so are v_β' and jv_β'. Then, the fundamental sequence component

$v_a^+, v_\beta^+, v_\alpha^-, v_\beta^-$ are extracted by ISC. The transformation formula of ISC is shown in Eqs. (5) and (6).

$$V_{abc}^+ = \begin{bmatrix} v_a^+ \\ v_b^+ \\ v_c^+ \end{bmatrix} = [T^+]V_{abc} = \frac{1}{3} \begin{bmatrix} 1 & \partial & \partial^2 \\ \partial^2 & 1 & \partial \\ \partial & \partial^2 & 1 \end{bmatrix} \begin{bmatrix} v_a \\ v_b \\ v_c \end{bmatrix} \tag{5}$$

$$V_{abc}^- = \begin{bmatrix} v_a^- \\ v_b^- \\ v_c^- \end{bmatrix} = [T^-]V_{abc} = \frac{1}{3} \begin{bmatrix} 1 & \partial^2 & \partial \\ \partial & 1 & \partial^2 \\ \partial^2 & \partial & 1 \end{bmatrix} \begin{bmatrix} v_a \\ v_b \\ v_c \end{bmatrix} \tag{6}$$

Where $V_{abc}^\pm$ is the instantaneous positive and negative sequence components represented by complex numbers, $\partial = e^{j2\pi/3}$ is the Fortescue phase shifting operator. The instantaneous positive sequence component in two-phase stationary coordinates is

$$V_{\alpha\beta}^+ = \begin{bmatrix} v_\alpha^+ \\ v_\beta^+ \end{bmatrix} = m[T_{\alpha\beta}][T^+][T_{\alpha\beta}]^T V_{\alpha\beta}' = \frac{1}{2} \begin{bmatrix} 1 & -j \\ j & 1 \end{bmatrix} V_{\alpha\beta}' = \frac{1}{2} \begin{bmatrix} v_a' - jv_\beta' \\ jv_a' + v_\beta' \end{bmatrix} \tag{7}$$

Similarly, the negative sequence component is

$$V_{\alpha\beta}^- = \begin{bmatrix} v_\alpha^- \\ v_\beta^- \end{bmatrix} = m[T_{\alpha\beta}][T^-][T_{\alpha\beta}]^T V_{\alpha\beta}' = \frac{1}{2} \begin{bmatrix} 1 & j \\ -j & 1 \end{bmatrix} V_{\alpha\beta}' = \frac{1}{2} \begin{bmatrix} v_a' + jv_\beta' \\ -jv_a' + v_\beta' \end{bmatrix} \tag{8}$$

Where $[T_{\alpha\beta}] = \begin{bmatrix} 1 & -1/2 & -1/2 \\ 0 & \sqrt{3}/2 & -\sqrt{3}/2 \end{bmatrix}$, $[T_{\alpha\beta}]^T = \begin{bmatrix} 1 & -1/2 & -1/2 \\ 0 & \sqrt{3}/2 & -\sqrt{3}/2 \end{bmatrix}^T$, $j = e^{j\pi/2}$ is the phase shifting operator in time domain.

Since the above mentioned SOGI constructs the quadrature component of the input signal, its resonant frequency needs to be consistent with the input signal. In order to ensure that SOGI can track the input signal with zero steady-state error, which improving the accuracy of the TPVUR measurement, the PLL in literature [14] was introduced when designing a real-time measurement system. The v_a^+ and v_β^+ are input the PLL module that can realize the instantaneous phase lock by combining the Park transform and the PI controller, the whole phase locking process constitutes a feedback, the algorithm achieves phase lock by non-stop closed-loop iteration, so that the voltage difference caused by the output phase and the input phase is zero. Next, the output ω_0 of PLL will be input into SOGI, so that the whole system has a frequency adaptive function. Finally, the positive and negative sequence components of the grid voltage signal are obtained by the Clarke inverse transform as follows,

$$V_{abc}^\pm = \begin{bmatrix} v_a^\pm \\ v_b^\pm \\ v_c^\pm \end{bmatrix} = [T_{\alpha\beta}]^T \begin{bmatrix} v_\alpha^\pm \\ v_\beta^\pm \end{bmatrix} = \begin{bmatrix} v_\alpha^\pm \\ -\frac{1}{2}v_\alpha^\pm + \frac{\sqrt{3}}{2}v_\beta^\pm \\ -\frac{1}{2}v_\alpha^\pm - \frac{\sqrt{3}}{2}v_\beta^\pm \end{bmatrix} \tag{9}$$

According to the positive and negative sequence components of Eq. (9), the TPVUR can be obtained by referring to the definitions in GB/T 15543-2008 and IEC-6100-4-27-2015 as shown in Eq. (10).

$$TPVUR = \frac{|v_{abc}^-|}{|v_{abc}^+|} \times 100\% = \frac{\sqrt{(v_a^-)^2 + (v_b^-)^2 + (v_c^-)^2}}{\sqrt{(v_a^+)^2 + (v_b^+)^2 + (v_c^+)^2}} \times 100\% \qquad (10)$$

Where $|v_{abc}^+|$ and $|v_{abc}^-|$ are the RMS values of the three-phase voltage positive sequence and negative sequence components, respectively.

4   Simulation and Results Analyses

In order to prove the correctness and real-time of the new method proposed in this paper, three sets of three-phase voltage signals (that is, three scenes) are set and simulated. The unbalance rate of three sets of voltage signals is displayed by probe, and the algorithm in GB/T 15543-2008 and IEC-6100-4-27-2015 are used to calculate the unbalance rate (Table 4) for comparison.

The amplitude and phase parameters of the three sets of voltage signals are shown in Table 2. The first set of signals contain harmonic components, and the parameters of the harmonic components are shown in Table 3, where the fundamental frequency ω is equal to 50 Hz, the sampling period T_s is equal to $5e^{-7}$s, the odd harmonic frequency $\omega'' = h \times \omega$ (h = 3, 5, 7, 9), the second and third sets of signals contain only the fundamental wave. The three sets of signals shown in Table 2 are simulated by Matlab/Simulink. Since the GB/T 15543-2008 and IEC-6100-4-27 recommend unbalance rate calculation after 10 V cycles, so the TPVUR acquisition should be after 0.2 s, and the waveform of the TPVUR of three sets of signals is shown by probe in Fig. 5.

Table 2. The parameters of three-phase voltage signals

Scene	v_a		v_b		v_c	
	$|v_a|$/V	φ_a/deg	$|v_b|$/V	φ_b/deg	$|v_c|$/V	φ_c/deg
1	220	0	220	−120	220	120
2	207.76	15.41	218.55	255.71	220.36	134.52
3	220	0	0	0	220	120

Table 3. The parameters of harmonic components

v_3		v_5		v_7		v_9	
$|v_3|$/V	φ_3/rad	$|v_5|$/V	φ_5/rad	$|v_7|$/V	φ_7/rad	$|v_9|$/V	φ_9/rad
80	−5pi/3	70	−4pi/3	60	2pi/3	50	pi/3

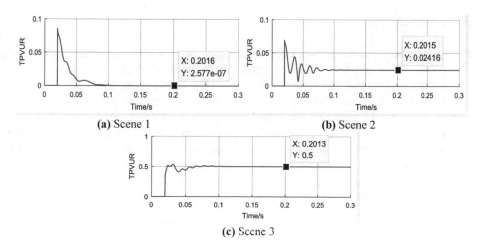

(a) Scene 1   (b) Scene 2

(c) Scene 3

Fig. 5. The waveform of TPVUR of three sets of signal

Combined with the simulation results of Fig. 5, the proposed new method can dynamically measure the TPVUR in real time within one cycle of the input signal, but the TPVUR is unstable at this time. With the phase-locked process of the PLL, the extraction of the sequence components will quickly stabilize from fluctuation. The harmonics shown in Table 3 are added to the first set of signals to verify the filtering effect of SOGI. After filtering and phase locking by DSOGI-PLL, the TPVUR is 0.00002577%, the TPVUR of the second set of signal is 2.416%, the TPVUR of third set of signal is 50.000%. Finally, the TPVUR measured in this paper and the recommended method of GB/T 15543-2008 & IEC-6100-4-27-2015 are shown in Table 4.

Table 4. Comparison of TPVUR (%)

Signals	Scene 1	Scene 2	Scene 3
GB/T	0.00000000	2.415	50.000
IEC	0.00000000	2.415	50.000
This paper	0.00002577	2.416	50.000

Referring to Table 4, the first set of signal is different, the reason is that after adding harmonics, the filtering operation will make the arithmetic components in Simulink generate iterative error, which is negligible within the allowable range, and the TPVUR of the signals of the second and third sets obtained by the method proposed in this paper is the same as that calculated in GB/T 15543-2008 and IEC-6100-4-27-2015. Therefore, the method proposed in this paper has certain accuracy and feasibility.

5 Conclusions

A new measurement method of TPVUR based on DSOGI-PLL has been proposed in this paper. SOGI can filter harmonics and obtain the quadrature components of fundamental, PLL makes SOGI have frequency adaptive function, it can effectively reduce the computational complexity of ISC by a certain variable substitution, and the TPVUR can be dynamically analyzed. The simulation experiments showed that the proposed method can accurately extract the positive and negative sequence components of the fundamental under the premise of filtering out the harmonics in the power grid, and can effectively reduce the calculation amount, the obtained TPVUR can reach the same accuracy as the recommended algorithm in GB/T 15543-2008 and IEC-6100-4-27-2015, so the new method proposed in this paper can be applied in real-time monitoring of grid voltage unbalance system.

References

1. Zhe, W., Yiru, W., et al.: Development status of China's renewable energy power generation. In: IEEE International Conference on Sustainable Power Generation and Supply (2009)
2. Singh, M., Allen, A., Hodge, B.-M.S.: Grid connection and power conditioning of wind farms. In: Handbook of Clean Energy Systems (2004)
3. Yang, C., Meng, H., et al.: The transient response analysis of SRF-PLL under the unbalance grid voltage sag. Trans. China Electro Tech. Soc. S2, 34–44 (2016)
4. Kabiri, R., et al.: Control of active and reactive power ripple to mitigate unbalanced grid voltages. IEEE Trans. Ind. Appl. 52(2), 1660–1668 (2015)
5. Hao, P., Li, D., et al.: An improved algorithm for harmonic analysis of power system using FFT technique. Proc. CSEE 23(6), 50–54 (2003)
6. Wen, H., Teng, Z., et al.: Simple interpolated FFT algorithm based on minimize side lobe windows for power-harmonic analysis. IEEE Trans. Power Electron. 26(9), 2570–2579 (2011)
7. Bing, J., Zhirui, L., Shengsuo, N.: FFT harmonic analysis based on nine terms minimum side-lobe cosine-sum window. Electr. Power Sci. Eng. 28(10) (2012)
8. He, W., Ran, J., et al.: Fast measurement method of voltage unbalance factor based on windowed FFT. Trans. China Electro Tech. Soc. 32(16), 275–283 (2017)
9. Rodriguez, P., Teodorescu, R., et al.: New positive-sequence voltage detector for grid synchronization of power converters under faulty grid conditions. In: IEEE Power Electronics Specialists Conference, pp. 1–7 (2006)
10. Yuan, X.: An improved method of instantaneous symmetrical components and its detection for positive and negative sequence current. Proc. CSEE 28(1) (2008)
11. Zhu, J.: Three-phase four-wire D-STATCOM control based on instantaneous symmetrical component method. Electr. Meas. Instrum. 51(23) (2013)
12. Ghijselen, B.: Exact voltage unbalance assessment without phase measurements. IEEE Trans. Power Syst. 20(1), 519–520 (2005)
13. Lyon. Discussion on "transient conditions in electric machinery". J. Am. Inst. Electr. Eng. 42(10), 1076–1078 (1923)
14. Rasheduzzaman, M., Khorbotly, S.: A modified SRF-PLL for phase and frequency measurement of single-phase systems. In: IEEE Energy Conversion Congress and Exposition (2017)

Native Language Identification by English Stop VOT of Chinese, French and English Speakers

Meichen Wu, Zhehan Dai, Lou Lv, Hongyan Wang[✉],
and Jeroen van de Weijer

Shenzhen University, Shenzhen, Guangdong 518060,
People's Republic of China
wanghongyan@hotmail.com

Abstract. The present study examines the effect of the stop consonants' Voice Onset Time (VOT) on the speaker identification of accented English with Chinese accent, French accent and American native English. The measurement of the degree of intelligibility unfolds by the production and perception experiments; in production experiment, three male adult speakers with different linguistic backgrounds (Chinese, French and American English) are singled out to record the experimental materials containing six stop consonants as targets, and phonetic analysis were conducted. In perception experiment, 30 listeners from three linguistic backgrounds (Chinese, French and American English) are asked to identify the speakers' native linguistic backgrounds based on the sound features they hear. Results indicate that accented English does play a special role on speaker identification. Specifically, listeners score highest when listening to the speakers from same linguistic backgrounds; in addition, two stop consonants /p/ and /g/ are the most special cue for speaker identification.

Keywords: Voice Onset Time (VOT) · Accented English · Speaker identification

1 Introduction

1.1 Research Background

The origins of the study of VOT (Voice Onset Time) could be dated back to the studies of Adjarian in the 19th century, which is characterized by the "relation qui existe entre deux moments: celui où la consonne éclate par l'effet de l'expulsion de l'air hors de la bouche, ou explosion, et celui où le larynx entre en vibration" (relation that exists between two moments: the one when the consonant bursts then the air is released out of mouth, or explosion, and the one when larynx starts vibrating) (Adjarian 1899).

After that, people were more receptive to the definition 'the time interval between the burst that marks release of the stop closure and the onset of quasi-periodicity that reflects laryngeal vibration' which specified in distinguishing the categories of voice and voiceless stops by looking at how VOT functioned to separate the stop categories of 11 kinds of languages in three groups: voiced unaspirated (−125 ms to −75 ms),

© Springer Nature Switzerland AG 2020
F. Xhafa et al. (Eds.): IISA 2019, AISC 1084, pp. 469–474, 2020.
https://doi.org/10.1007/978-3-030-34387-3_58

voiceless unaspirated (0 to +25 ms) and voiceless aspirated (+60 ms to +100 ms) depending on voicing position compared to the time of release [1]. They examined the word-initial stops in 11 languages. Since then, a large number of studies of many languages had been started, including a research on VOT in 51 languages, [2] and another more recent report in 18 languages [3].

With the notion of English as lingua franca (ELF) emerging, people are increasing concerned with accented English. With VOT a cue for speaker identification, there are fruitful achievements concerning this topic. For example, Both English and French possess six plosive phonemes: voiceless and voiced bilabial consonants /p, b/, alveolar consonants /t, d/ and velar consonants /k, g/. However, French and English differ substantially in the manner in which they instantiate the phonological voicing distinction [4]. In word-initial position, English VOICED stops (/b, d, g/) are voiced or voiceless unaspirated, and stops (/p, t, k/) are voiceless aspirated [5]. In citation speech, English VOICED stops are produced with short voicing lag while VOICELESS stops are produced with long voicing lag. Based on the analyses of Lisker & Abramson and Keating, English is described as having short lag and long lag patterns. In comparison with English, French VOICED stops (/b, d, g/) are voiced and that VOICELESS stops (/p, t, k/) are voiceless aspirated. According to investigations conducted by Caramazza, [6] VOT duration for French /b, d, g/ ranges between −150 ms to 0 ms and are produced with voicing lead, while that of French /p, t, k/ ranges between 0 ms and 30 ms and were produced with short voicing lag. Therefore, French is characterized as having voicing lead and short lag patterns

It has also been studied that Mandarin Chinese has a phonetic contrast in two ways in the Chinese stops compared to English: voiceless aspirated and voiced [7]. And the result demonstrated that these two characters could be aligned with English while a short delay in the VOT of Chinese voiceless aspirated stops refer as English voices stops; voiced Chinese plosives go to English voiced likewise.

1.2   Research Questions and Hypotheses

The main objective is to find out whether the VOT could be a special cue in the native language speaker identification in different languages: English, Chinese and French. To be more specific, we would like to know is there advantage for people to recognize the English with the accent of their mother tongue; in addition, which sound can be the most representative and salient one for speaker identification task. Consequently, we hypothesized that (a) correctness of speaker identification task might highly match the nationality of different listeners, which means Chinese Listeners would have a higher correctness in identifying English Stops spoke from Chinese; (b) Considering the literature review and the language itself, French and English bear more distinctions compared with English and Chinese in voiced consonants, ex. /ti/ could show a higher identification score than the other stops.

2   Experiment

2.1   Materials

It is statistically shown that the six English stop consonants are best performed when producing with a high vowel /i/ followed after it [8]. Therefore, the high vowel /i/ was employed to ensure the biggest deviances. In addition, in order to avoid the confounds caused by different vowel contexts and durations, /i/ will be used in every piece of materials. Due to the fact that the syntax level of a language is higher than the lexical level and then the phoneme level, [9] we decided to use sentences as experimental materials instead of single words.

Consequently, we will take a combination of two related studies mentioned above and create a new model of "I say S+/i/ again", and "S" represents English plosive consonants (stops) /b, p, d, t, g, k/. It allows us to have 18 sentences with six English stops in three accented English languages (English with Chinese accent, English with French accent and American English).

2.2   Participants

We singled out three male adult speakers respectively from Chinese, French and American English linguistic backgrounds to do the recordings of our experimental materials. We prepared a pretest in the form of questionnaire to them beforehand for ensuring they are qualified as the speakers in our present study. The questionnaire is mainly to check the nationality and whether the English is their first language.

For listeners, 30 participants were selected, equally 10 persons of whom are from Chinese, French and American English linguistic backgrounds. All those 30 listeners are students from Shenzhen University, China, and share the same language environment with the speakers selected.

2.3   Methods

Speakers were arranged in a soundproof room in a controlled environment of the phonetics and phonology laboratory in School of Foreign Languages, Shenzhen University. All the recordings were made individually by using a Tascam DR-06 solid-state recorder (44.1 kHz, 16 bit) and a Sennheiser HD800 close-talking headset microphone. Each 3 participants was directed to produce 6 sentences consisting six English stops: /b/ /p/ /t/ /d/ /k/ /g/ in a model like "I say S+/i/ again", (e.g. I say /ti/ again.) thus $6 \times 3 = 18$ pieces sounds material were produced for further analysis in the PRAAT [10].

18 trials were decoded into sentences, and then presented to the listeners. The listeners were first required to listen to the sentences recorded by the speakers, and there is a question form for them to mark the one answer that matches the sentences that the listeners heard. Each sentence was played only once. After collecting all the data, they will be analyzed by the method of one-way ANOVA in the SPSS, and later reported their F value to check the statistical confidence.

3 Results

Figures 1, 2, 3, 4, 5 and 6 show the correct rates that the listeners from three linguistic backgrounds got on the decision of the English stops, respectively /b/, /p/, /d/, /t/, /k/, /g/. We can find that when the listeners were presented with the recordings from the speaker from the same linguistic background, they scored highest; This indicates that speaker identification is relatively easier for the listeners whose linguistic background is identical with the speakers.

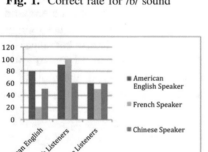

Fig. 1. Correct rate for /b/ sound

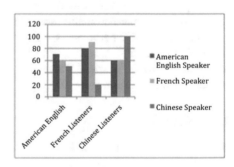

Fig. 2. Correct rate for /p/ sound

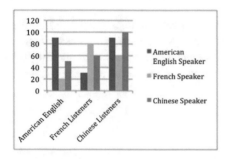

Fig. 3. Correct rate for /d/ sound

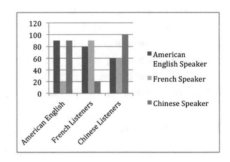

Fig. 4. Correct rate for /t/ sound

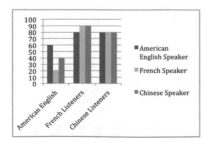

Fig. 5. Correct rate for /k/ sound

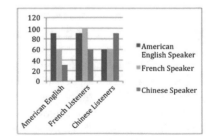

Fig. 6. Correct rate for /g/ sound

Figure 7 shows the mean correction rates of listeners from the three linguistic backgrounds on six single English stops. In addition, the total mean correction rates for identifying the six English stops are further calculated. For /b/ sound, the mean correction rate is 64.3%, 67.7% for /p/ sound, 67.7% for /d/ sound, 65.6 for /t/ sound, 78.9% for /k/ sound and 71.1% for /g/ sound.

However, there are some extremely high scores and low score; for example, the mean correct rates for /p/ sound, the French listeners scored extremely low with only 33% as correct rate, while Chinese listeners scored high with 90%; for the identification of the /g/ sound, American listeners scored extremely high with 90% as the correct rate but Chinese listeners failed to identify it properly with only 50% as correct rate. We would wonder whether the sounds /p/ and /g/ can function as the special cues to identify the linguistic background of a speakers; we ran a one-way ANOVA test to show its statistic confidence. The result of ANOVA shows that /p/ in $F (2,27) = 10.84$, $p < 0.01$, statically significantly while /g/ also has a value of $F(2.27) - 3.248$ $p = 0.054$, which is marginally significant. And the total $F (2,27) = 1.572$, $p = 0.226$ run by the ANOVA.

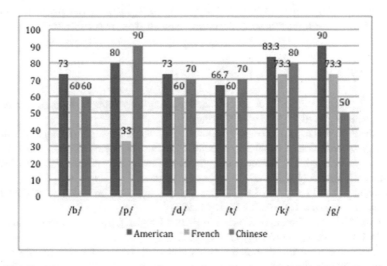

Fig. 7. Mean correct rates for listeners from the three linguistics backgrounds

4 Conclusions

4.1 Research Findings

We can draw the following conclusions based on the data collected and analyzed.

The first finding lies on the phenomenon that each language background listeners respectively have the highest correctness for their own language speaker. The data shown in this study indicate that the task of speaker identification is not based on the chance alone (in a three-alternative forced choice task, the score should at least be 33%). However, the scores are not more than between 12% and 27% points better than change, although the correction rate of the scores for some specific trials were higher than 70%. It is clearly shown that, compared to the cases in which the speakers are from America and France, Chinese listeners tend to be more successful when they listen to Chinese speaker. This effect was predicted by the shared inter-language benefit hypothesis but did not reach statistical significance.

Secondly, special cues do exist for speaker identification. Through the ANOVA analysis, we can see that /p/ and /g/ phonics demonstrate the highest F value in ANOVA, which means these 2 stops could be a special cue for speaker language identification.

4.2   Limitations

For the time and experience limited, in the data collecting phase and encoding part, we couldn't control all the variables to avoid confounds like: the difference in speakers voice characteristics which could possibly affect the efficiency of identification by listeners; the practical effect for listeners while doing the experiment because of the lack of amply participants, which might lead to some data discrepancies and misjudges, as well as the design of the questionnaire and the edition of the recording itself, as possible which gave us a not so convincing data in this field of testing.

References

1. Lisker, L., Abramson, A.S.: A cross-language study of voicing in initial stops: acoustical measurements. Word **20**(3), 384–422 (1964)
2. Keating, P.A., Miko, M.J., Ganong, W.F.: A cross-language study of range of voice onset time in the perception of initial stop voicing. J. Acoust. Soc. Am. **70**(5), 1261–1271 (1981)
3. Cho, T., Ladefoged, P.: Variation and universals in VOT: evidence from 18 languages. J. Phon. **27**(2), 229 (1999)
4. Kessinger, R.H., Blumstein, S.E.: Effects of speaking rate on voice-onset time in Thai, French, and English. J. Phon. **25**(2), 168 (1997)
5. Docherty, G.J.: The Timing of Voicing in British English Obstruents. Introduction, pp. 1–3 (1992). https://doi.org/10.1515/9783110872637
6. Caramazza, A.: The acquisition of a new phonological contrast: the case of stop consonants in French-English bilinguals. J. Acoust. Soc. Am. **54**(2), 421 (1973)
7. Shimizu, K.: A study on VOT of initial stops in English produced by Korean, Thai and Chinese speakers as L2 learners. In: ICPhS, Hong Kong, pp. 1818–1821, August 2011
8. Nearey, T.M., Rochet, B.L.: Effects of place of articulation and vowel context on VOT production and perception for French and English stops. J. Int. Phon. Assoc. **24**(01), 1 (1994)
9. Efstathopoulou, P.N.: VOT productions and accented speech by Greek/English bilinguals. In: Proceedings of the 8th International Conference on Greek Linguistics, vol. 30, pp. 107–120 (2009)
10. Boersma, P., Weenink, D.: Praat: doing phonetics by computer (2017)

Discussion on Enterprise Competitive Capability Based on Radar Chart Analysis—Taking "H V Company" as an Example

Yueming Cheng[1(✉)], Wen Zhou[1], Yi Cheng[2], Ting Cheng[3],
Kang Cheng[4], and Fangyao Wang[5]

[1] School of Economics and Management,
Jiangxi Science and Technology Normal University, Nanchang 330013, China
729209745@qq.com
[2] Jiangxi Building Materials Scientific Research and Design Institute,
Nanchang 330013, China
[3] School of Finance and Economics, Jiangxi University of Technology,
Nanchang 330000, China
[4] School of Management, Jiangxi University of Technology,
Nanchang 330000, China
[5] College of Arts and Sciences, The Ohio State University,
Columbus 43210, USA

Abstract. The competitiveness of enterprises plays a decisive role in the survival and development of enterprises. This requires enterprises to explore their own competitiveness and cultivate and develop their own competitive advantages in order to be invincible in market competition. There are many methods for analyzing the competitiveness of enterprises, and the radar chart analysis method is a very practical analysis method. This paper uses the radar chart analysis method to analyze the competitiveness of the company and propose relevant improvements.

Keywords: Enterprise competitiveness · Radar chart analysis · Competitive advantage

1 Introduction

With the accelerating process of economic globalization and the increasingly fierce competition in the international market, the establishment and development of the company's own competitiveness has become the development goal that all enterprises generally pursue. Enterprise competitiveness has a decisive role in the survival and development of enterprises. Having a strong competitive enterprise means that the company has the ability to maintain a stable competitive advantage for a long time and earn profits that exceed the average profit of the society, which can keep the company in an invincible position. If a company wants to survive and develop, its competitiveness is a key factor. Any company must think about how to develop and develop its

© Springer Nature Switzerland AG 2020
F. Xhafa et al. (Eds.): IISA 2019, AISC 1084, pp. 475–483, 2020.
https://doi.org/10.1007/978-3-030-34387-3_59

own competitiveness to gain its own competitive advantage. From this point of view, attaching importance to the formation and development of the company's own competitiveness is the key to the survival and development of the company in the fierce global competition. The competitiveness of Chinese enterprises is still far from the enterprises of developed countries. The overall development situation is still severe. Therefore, one of the major problems that Chinese enterprises must solve at present is how to cultivate and enhance their competitiveness. After consulting various relevant materials, it is found that although there are many analytical methods for the discussion of the competitiveness of enterprises, after careful comparison, many evaluation ideas have some subjective colors, which do not reflect the real problems of enterprise development. According to incomplete investigations, at least 20 methods have been explored to analyze the competitiveness of enterprises. They are mainly divided into two categories: single-index analysis methods and comprehensive index system analysis methods. The more representative one is factor analysis method, contrast to gap method, comprehensive index evaluation method, etc. Based on the comparative analysis, the author finds that the radar chart analysis method has obvious advantages in evaluating the competitiveness of enterprises. It can comprehensively analyze and evaluate the business situation of the enterprise, and intuitively find and reflect the advantages and disadvantages of the enterprise, thus helping the enterprise to clearly judge its own competitive position. The author believes that using "H V" as an analysis case, using radar chart analysis to explore the competitiveness of enterprises will have certain practical significance and theoretical significance.

2   Related Concepts and Basic Theory

2.1   The Basic Theory of Enterprise Competitiveness

The connotation of enterprise competitiveness has different definitions in different development periods. As time goes by, its development has gone through the following processes: First, it is the advantage theory from the early economics. They think that the competitiveness is the difference in all aspects of the enterprise, such as absolute advantage—comparative advantage—the comparative advantage of the basic production factors possessed by enterprises—the difference in market share. Second, as people's understanding deepens and then gradually transitions to the essence of enterprise competitiveness is the element resources occupied by enterprises, including the theory of enterprise competitiveness (external market resources) - resource potential differences (internal resources). Third, to the present, people believe that the essence of competitiveness is the ability to acquire and use resources. To sum up, we define the competitiveness of enterprises here. It refers to the enterprise as an independent economic entity. Under the conditions of competitive market economy, by cultivating its own resources and capabilities, acquiring externally searchable resources, and comprehensively making full use of it, On the basis of creating value for customers, they can achieve their own value and compete with their competitors in the market.

2.2   Introduction to Radar Chart Analysis

The radar chart analysis method is a graph that comprehensively analyzes and evaluates the development status and financial level of the enterprise from the aspects of production (production, safety, profitability, growth and liquidity). The financial ratios drawn in this way are analyzed by graphs such as radar radiation and have the role of "heading" guidance, hence the name. (As shown in Fig. 1.) This method is a comprehensive analysis tool for economic benefits of enterprises. Therefore, it is very practical to evaluate the competitiveness of enterprises. It is an intuitive reflection of the results of the "five-sex analysis" of enterprises. Due to its intuitive expression, it is now The radar chart has surpassed the financial category and is widely used in various fields. Radar chart mainly analyzes the business development and financial status of enterprises from both dynamic and static aspects. Static analysis is a horizontal analysis of the financial ratios of other similar or similar enterprises or industries belonging to the enterprise and their own indicators, dynamic analysis is to vertically compare the current financial ratio of the enterprise with the financial indicators of the previous accounting cycle. The radar map combines the horizontal and vertical analysis and comparison methods to calculate the five types of indicators of profitability, growth, safety, liquidity and productivity of the integrated enterprise. These five primary indicators also contain many secondary indicators. (See Table 1 for details.) Analysis of profitability indicators can explore the profitability of a company's operations; safety refers to the degree of security of a company's operations, which is mainly used to measure its liabilities to ensure the safety of normal business; Sex indicators can grasp the efficiency of corporate capital use through the operation of their funds; while analyzing growth indicators mainly depends on observing the development trend of the company's operating capacity in a certain period of time. Even if a company has high profitability, growth may not be good,

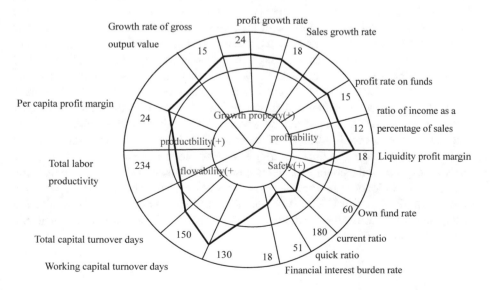

Fig. 1. The radar map

indicating its future profitability will declining. Therefore, analysis of production indicators is mainly to understand the production and operation capabilities of enterprises in a certain period of time. It is possible to clearly understand the advantages and disadvantages of the enterprise, and help the company to develop its strengths, avoid weaknesses and continuously improve its competitiveness.

2.3   Application Steps of Radar Chart Analysis

Establishing an indicator system is the first step in using a radar chart to explore the competitiveness of an enterprise. The indicator system is the best determined according to the analysis purpose we have established. It can refer to the common indicator system used for financial analysis, such as the indicator system in the DuPont analysis system. However, at the same time, it is necessary to consider the characteristics of the competitiveness of each enterprise, and make corresponding changes to the adopted indicators, so as to be able to comprehensively analyze the competitiveness of enterprises. According to this, the author selected the following five types of indicators in the process of designing the indicator system, namely, profitability, operational capability, solvency, marketing capability and development capability. These 5 parts are the concentrated expression of the competitiveness of enterprises and the organic components of enterprise competitiveness as a whole.

The application steps of the radar chart analysis method: 1 establish the index system used in the analysis 2 collect the relevant indicator data of the enterprise and the industry data 3 draw a radar chart based on the data. The method of drawing the radar chart is: 1 draw 1 standard circle (with a radius of 1 unit), and then draw two concentric circles (radius 1.5 and 2 respectively) with the same center and different radii. These concentric circles are divided into five regions at 72° in each region, representing the indicators of profitability, operational capability, solvency, marketing capability and development capability of the enterprise. The standard circle in the concentric circle represents the 1/2 value or worst case of the industry average; the second largest circle represents the average of the same industry level; the largest circle represents 1.5 times or the best state of the industry average. 2 According to the number of secondary indicators, the corresponding number of rays are made from the center of the circle, and the 360° of the concentric circle are divided into a plurality of sectoral regions, respectively representing different indicator regions. 3 Determine the value of each ray according to the specific size of the index value, and mark the corresponding position according to the financial indicators of the enterprise. After the various ratio values of the company are linked by lines, an irregular closed-loop diagram is drawn.

Nowadays, with the continuous development of computer technology, the financial radar chart is no longer the original manual depiction of the above, and the common office software Microsoft Office, WPS, etc. have already had the automatic generation of the radar chart. This article will also use the relevant software of Office to generate radar analysis charts.

3   Application of Radar Chart Analysis Method in Enterprise Competitive Capability Analysis—Taking "H V" as an Example

3.1   Basic Situation of "H V" Company

Known for its ultimate service, the H V brand was founded in 1994. A hot pot restaurant that specializes in Sichuan-style hot pot and integrates hot pot features. After years of development, H V has grown into an internationally renowned catering company. By the end of 2017, H V has operated more than 300 direct-operated stores in more than 100 cities including mainland China, Hong Kong and Taiwan, as well as Singapore, the United States, Australia, South Korea and Japan, with more than 50,000 employees.

H V is known for its warmhearted service and the experience of providing customers with a sense of belonging, such as dolls, birthday surprises, free shoeshine, etc., to satisfy the customer's requirements and to make customers feel welcome and popular. Warm heart, great feeling and a sense of belonging, so it has a high popularity. In addition to free nail art and hand care for female customers, some stores also have children's amusement parks and special care services, but also actively promote the hot pot culture with Chinese characteristics, provide the characteristics of Sichuan's national genius, and integrate Chinese martial arts, noodle show and etc. Their corporate culture gives them a great mission to "create a happy hot pot time with carefully selected products and innovative services, and deliver healthy hot pot food culture to food lovers around the world", and make every effort to provide extraordinary dining for every customer.

In recent years, H V has opened new stores in second- and third-tier cities at a faster rate, but many industry insiders have analyzed that the larger the chain operation of the catering industry, the more likely it is to make mistakes. And Swill face some new challenges, such as controlling food safety and quality stability, increasing the restaurant manager's talent pool, and strengthening supply chain management. Therefore, in view of the recent development of H V, this paper uses radar chart analysis to explore the competitiveness of the company, and gives a glimpse of the author with detailed analysis.

3.2   Discussion on the Competitiveness of "H V" Based on Radar Chart Analysis

The main steps of using the radar chart analysis method to analyze the competitiveness of H V company are as follows: (1) According to the financial indicators related to the financial statements of H V company in 2018, the actual value of each index of the company is determined. (2) Determine the reference index of "H V" company. Comparing the standard level or average level of "H V" companies with similar enterprises, it is also necessary to determine the relevant reference indicators, using the same industry average as a standard control. Mainly based on the "National Economic Industry Classification and Code", "Enterprise Size Division Standards" and relevant data found on the official website of the National Bureau of Statistics. (3) Calculate the contrast value of the indicator. For example, the index value of the industry index value is 0.5 times, the value of the index is 2 times, and the ratio of the index value to the industry average index value. (4) Create a radar chart using Excel. (See Table 2 and Fig. 2 for details).

Table 1. The value of S & its industry indicators.

Primary indicator	Secondary indicator name	Indicator value	Industry average value	0.5 times industry value	2 times industry value	Ratio value/industry refers to
Profitability (profitability)	Gross profit margin (%)	0.5938	0.4955	0.2478	0.991	1.198
	Sales net interest rate (%)	0.0966	0.1176	0.0588	0.2352	0.0113
	Sales profit margin (%)	1.0016	0.688	0.344	1.376	1.455
Safety (solvency)	Current ratio (%)	1.7350	1.6354	0.8177	3.2708	1.06
	Quick ratio (%)	1.6000	2.0128	1.0064	4.0256	0.793
	Assets and liabilities (%)	0.2775	0.4623	0.2312	0.9246	0.6
Fluidity (operating capacity)	Total asset turnover rate (times)	2.18	1.9	0.95	3.8	1.147
	Fixed asset turnover rate (times)	5.61	3.66	1.83	7.32	1.5327
	Current assets turnover rate (times)	4.74	2.55	1.275	5.1	1.859
Growth (marketing ability)	Sales revenue growth rate (%)	0.5915	0.21	0.105	0.42	2.816
	Profit before tax (%)	0.3871	0.1822	0.0911	0.3644	2.094
	Total assets growth rate (%)	2.1973	0.985	0.4925	1.97	2.23
Productive (development ability)	Per capita sales Sales income (million)	31.2727	19.9	9.95	39.8	1.571
	Per capita salary (million)	6.2400	3.6886	1.8443	7.3772	1.692

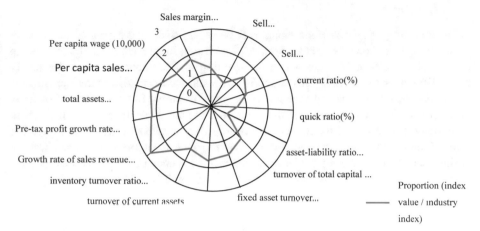

Proportion (index value / industry index)

Fig. 2. The radar chart of H V company

Operation steps: 1. Create related data Table 2, select the data to be used to make the radar chart, that is, select the cell of B1:B18 on the left side, then hold down [Ctrl] and select the G1:G18 cell area. (The column B data is used to identify the ray axis name of the radar chart, and the G column is the data to be used for the actual drawing.) 2. Click the Insert Chart option, select the radar chart in the pop-up chart type box, and select the radar target in the sub-target option. Select the appropriate radar pattern as needed and click on the graphical representation to confirm. 3. After the radar pattern is automatically generated, we can modify and beautify the image according to the needs of the analysis, in order to make the graphic clearer and more beautiful, so as to better reflect the financial status of the enterprise. Operate the icon title, axis, grid lines, data markers, and classification flags separately. Click the corresponding position of the chart to pop up the dialog box, and then select the relevant tabs and options.

First, calculate the area of the standard circle, $S = \pi * R^2 = 3.14$.

Second, calculate the area of the polygon in the graph and convert it to the sum of the areas of the triangles. Use the formula $S = 1/2 * a * b * \sin c$ (where a and b are the adjacent sides of the triangle, and the length is the radar map index value. c is the angle between the adjacent sides. Since 15 indicators are used, the c of each triangle is $360/15 = 24°$, so the calculated polygon area $S' = 4.196$.

Third, the ratio of the calculated polygon area to the standard circle area is $4.196/3.14 * 100\% = 133.6\%$.

This paper refers to the "Enterprise Size Division Standard" and the performance evaluation scale of large and medium-sized enterprises. The results are divided into four grades. The results are better than 150%, 100%–150% is good, 50%–100% is medium, low at 50%, it is considered to be low in competitiveness. According to the above analysis, the competitiveness of "H V" is relatively good. Among them, the operational capacity and marketing ability are superior, and its development ability is obviously insufficient, and the profitability needs to be enhanced.

3.3   Suggestions for the Improvement of the Competitiveness of "H V" Company

After the listing of H V in Hong Kong in September 2018, they expanded the number of newly opened stores as soon as possible. According to relevant data, H V has set up at least 100 stores in 2018, of which the opening in December accounted for newly opened stores last year 27%. This action is also well documented, and their prospectus shows that new stores can achieve a short supply of supply in a short period of time. However, their fast expansion of the store still attracts the attention of all relevant people in the industry. At the same time, considering the split statistics of the operating days of new and old stores and considering the factors of the new store climbing period, the effect and rhythm of the store expansion are uncertain. We find that the full-year performance of 2018 is less than expected, and will be 2019. Faced with greater performance pressure, H V will usher in a real test in 2019, and the economic development trend in 2019 has yet to be considered, or affecting the mid-to-high-end catering industry. The rapid expansion of the industry has high requirements for back-end capability, and the sea fishing has been fast for the past 2 years. The results of the expansion have yet to be tested. How to maintain its competitiveness in the future, and continue to hand over the beautiful business data is worthy of continued attention. Combining the five indicators of the above radar analysis chart and its actual development status, the following suggestions are proposed for the "H V" company to enhance its competitiveness:

(1) Implement the concept of "continuous interest, lock management"
    The labor-intensive nature of the food service industry makes it difficult to ensure that employees comply with the company's system regulations, strengthen management incentives, and be people-oriented. On the one hand, we must highly align the interests of employees and the company, fully stimulate the growth of vitality, and achieve the next big step in sales revenue; on the other hand, we must control food safety and other risk factors. A two-pronged approach to achieve further improvement in development indicators.
(2) Strengthen supply chain management and maintain its competitiveness
    The new entrants in the market have weaker bargaining power with suppliers and lack of management experience. On the contrast, H V has a great competitive advantage. The acquisition of high-quality resources and the grasp of the upstream of the industrial chain can help to control product quality and save costs, in order to increase the value of the profitability indicator.
(3) Take advantage of the initial capital expenditure The large-scale catering industry needs a large amount of funds to ensure normal turnover operations. For new entrants, the initial investment in large-scale operations is too large, which in turn leads to cash flow difficulties. H V can take advantage of its existing advantages and increase its investment in technology, such as optimization, research and development of smart restaurant technology, and strong business systems.
(4) Adhere to the brand first The quality of ingredients is the basis of excellent catering enterprises, and management ability is the catalyst for outstanding catering enterprises to stand out. Continuously improve service capabilities and

provide value-added services to member customers. H V should provide its customers with the ultimate service experience and build a brand moat with a good quality of food and excellent management capabilities. As a leading company in catering, we must give full play to our own advantages.

4 Conclusion

Any company that wants to succeed in development must conform to the times. In the fierce global competition, enterprises know how to cultivate and enhance their own competitiveness, which plays an important role in promoting the development of the world economy. Every enterprise must know how to use the competitive analysis method such as radar chart analysis to analyze self-development trends from horizontal, vertical and even deeper, and to identify the position of their own enterprises in the market competition. Defining the comparative advantages and disadvantages of enterprises and competitors will help enterprises to make strategic adjustments quickly and accurately, and escort the development of enterprises.

References

1. Yuan, J., Cheng, L.: The connotation and characteristics of enterprise competitiveness. Jiangsu Bus. Theory (6), 95–96 (2003)
2. Shi, Q., Mei, Q.: Research on the analysis system of enterprise competitiveness. J. Jiangsu Univ. Soc. Sci. Edn. (03), 54–56 (1993)
3. Su, L., Zhou, M.: Literature review of enterprise competitiveness and its evaluation. Sci. Technol. Econ. Mark. (09) (2017)
4. Zhang, R.: Research on the competitiveness of enterprises. Friends Account. (14) (2007)
5. He, X.: Radar diagram analysis of enterprise competitiveness in competitive intelligence research. Libr. Inf. Work (04), 86–90 (2010)
6. Zhang, C.: The application of radar chart in financial analysis—take a listed company in the construction industry as an example. Manag. Informatiz. China (23), 7–9 (2011)
7. Yu, J.: Analysis of core competitiveness model of Haidilao hot pot. Goods and quality (S7) (2012)
8. Xiao, W., Zhao, C.: Finally going to Hong Kong for listing, the future challenge of Haidilao is not small. Chin. Food (11), 126–127 (2018)
9. Wang, H.: Three main paths of enterprise culture construction—taking Haidilao as an example. Account. Lett. Rep. (B05) (2018)
10. Song, Z., Zhang, X.: The enlightenment of "Haidilao" business model to the service marketing of catering industry. Manag. Sci. Technol. Small Medium-Sized Enterp. (02) (2018)
11. Su, L.: Research on the operating situation of Lao listed companies based on Radar Diagram Analysis. Guangxi University (2013)
12. Guo, H.: Research on the competitiveness of industrial enterprises in the central PRD—taking Heyuan City of Guangdong Province as an example. In: IOP Conference Series: Materials Science and Engineering, vol. 439, no. 3 (2018)

The Construction and Key Technologies Research of the Integration of EMS and DMS with Rush Repair Scheduling in Area Grid

Yinghua Song[1(✉)], Ming He[2], Leifan Li[2], and Wei Yuan[1]

[1] NARI Technology Development Co., Ltd., Nanjing 210006, Jiangsu, China
ssyyhh786@163.com
[2] State Grid Sichuan Electric Power Co., Ltd., Chengdu 610041, Sichuan, China

Abstract. This paper analyzes the power grid in grid current small area power grid dispatching system, distribution automation, repair the present situation of the construction of power grid dispatching system, put forward for the middle and small area deployment rob integration system construction model. For different construction patterns are analyzed, key technology used for the deployment of the construction mode of integration finally rob of the system to the description of the. Regional power grid deployment rob integration system is in power network dispatching small area and repair the trend and direction of scheduling system construction, is the rapid and effective way to promote the distribution automation in small area power grid and rush repair scheduling construction. Provides an important technical means a kind of stability, safety, economy, an important guarantee to improve the quality of power supply service area.

Keywords: EMS · DMS · Rush repair scheduling

1 Introduction

In recent years, under the background of building a smart grid, distribution network construction has been highly valued by power companies at all levels as an important link in smart grids. Since 2008, some large and medium-sized cities in China have successively built distribution automation main stations and distribution network repair systems, which have a positive effect on improving the monitoring level of distribution network operation, improving the efficiency of fault repair and improving the reliability of power supply. However, for the distribution network in small and medium-sized areas, the radiation power supply is mainly composed of long lines and branch lines, low power load density, wide dispersion, poor operating conditions, and difficulty in building communication networks. In addition, the construction of the distribution automation main station system in small and medium-sized cities is still independent of the construction of the main network dispatching system, and the manpower and financial pressure of the automation system construction is relatively large [1].

How to use limited construction funds, operation and maintenance teams, combined with the characteristics of small and medium-sized regional power grids, to build

F. Xhafa et al. (Eds.): IISA 2019, AISC 1084, pp. 484–492, 2020.
https://doi.org/10.1007/978-3-030-34387-3_60

an automation system that can meet its own development needs, is stable, reliable, simple to maintain, economical and practical, and is currently the construction of urban power grid automation system in small and medium-sized regions. There is an urgent need to solve the problem in the process.

2   The Concept of Integrative System

The power grid deployment and grab integration system integrates dispatching automation, distribution automation and distribution network repair scheduling on a unified open software and hardware platform. Under the premise of ensuring stable operation and safe operation of the system, it is implemented in the same system, with two-level power grid scheduling and distribution network repair scheduling application integration. The integrated system models the network of high-, medium-, and low-voltage three-level power grids within the jurisdiction of the regional power grid. At the same time, it integrates the low-voltage users and the power grid model by integrating with the deployment information to construct an integrated distribution network model. To achieve unified equipment modeling, unified data access, unified monitoring and scheduling, and unified management. At the same time, the integrated analysis of the high, medium and low voltages is carried out for the system application function [2] (Fig. 1).

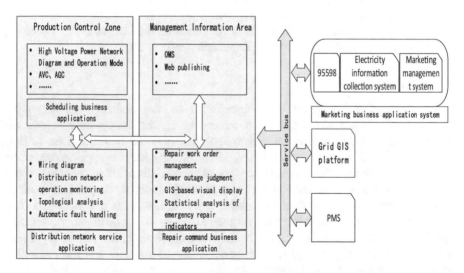

Fig. 1. Block diagram of the overall architecture of the deployment

The construction of regional power grid deployment and integration system is the development trend and direction of power grid automation system construction in small and medium-sized areas. The key technology of integrated system construction is an important technical means to improve the safety, stability and economic operation of power grids in small and medium-sized areas, and promote small and medium-sized enterprises. Intensive operation and management of regional power grids. At the same time, in view of the fact that the dispatching automation system has been basically established in the main network dispatching of domestic power grids, the deployment and grabbing integrated system should be based on the economical application and resource reuse, and should fully draw on the construction results of the previous dispatching automation, taking into account the distribution network operation and maintenance. Demand, adopt various safe and reliable extension technologies on the dispatching automation system, expand the wiring diagram electronic management software based on the existing main network dispatch control system, add new distribution network data collection, distribution network SCADA, distribution network foundation Network analysis, feeder automation processing, etc. One-area distribution network dispatching operation function and three-zone repair scheduling functions such as repair work order management, fault analysis, power outage analysis, GIS-based visual display, etc.

3   Research on the Mode of Deployment and Integration

The construction of distribution automation main station system should be based on the principle of economy and practicability, make full use of existing resources, reasonably select the construction mode and investment scale, and meet the dispatching and operation requirements of distribution network [3].

The specific construction method should be based on the integration of the county automation system, and there are three implementation modes:

3.1   Full Integration

The fully integrated deployment and integration model is based on a unified support platform, maintenance and application according to the responsibility area and authority, and an integrated system with the same human-machine interface. In business applications, it can be divided into grid dispatching service applications, distribution network scheduling service applications, and distribution network repair scheduling service applications; in the system architecture, three parts of business applications are respectively configured with application servers and workstations to implement independent data for their respective business applications. Acquisition, analysis, maintenance and application; in the basic application, the platform provides unified network topology analysis, alarm service, human-machine interface, workflow and other applications; in data processing and storage, it is unified by the database server, data processing server To complete, to achieve unified processing and storage of various business data (Fig. 2).

The fully integrated deployment and robbing integration mode is a mode in which data is centrally processed and stored, and the responsibility area and service authority are divided according to the application service, which is suitable for the area where the unified integration system is unified and the platform is uniformly maintained.

This mode is applicable to the areas where the dispatching automation system, the distribution automation system, and the emergency dispatching system are not put into

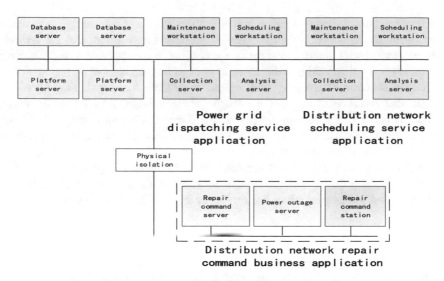

Fig. 2. Structure diagram of the system of deployment and complete integration

operation or are being replaced, and the number of lines covered by the automation planning of the area (including the county) is 800 or less. Real-time information 500,000 points and below.

Due to the new construction mode of such construction mode data, the system after construction has all the business functions of power grid dispatching; in the distribution network scheduling, it has the electronic function of distribution network wiring diagram, distribution data access, distribution network operation monitoring, and distribution. Automatic fault handling of network, distribution network topology application, etc.; repair scheduling has repair worksheet management, power outage analysis, fault diagnosis and judgment, GIS-based visual display, etc.; external interface has integrated communication with 95598 system, PMS, grid GIS platform.

3.2 Integration of Matching and Rescue

The integration mode is the distribution automation and repair command business application based on a unified platform. They are located in the distribution control and repair command of the distribution network in the first and third districts. The dispatch automation system is coupled as an independent system through the service bus and the integrated system. To realize the exchange of basic grid data and form a generalized deployment and integration system [4]. In terms of basic applications, the integrated system for providing and robbing is provided by the platform for unified network topology analysis, alarm service, human-machine interface, workflow, etc. In terms of system architecture, the integrated system with robbing has unified data storage and platform services. The distribution network operation control and the repair command are respectively configured with application servers and workstations to realize analysis, maintenance and application of their respective services.

The integration mode is a mode in which the scheduling automation is relatively independent, the distribution data is centrally processed and stored, and the responsibility area and service authority are divided according to the application service, which is suitable for the dispatching automation system to be put into operation.

The construction of the integrated system is suitable for dispatching automation systems that have been put into operation. There are plans to build distribution automation and repair scheduling areas, and the area (including county) distribution network automation planning coverage lines are more than 200, real-time measurement Point at 100,000 points and above.

Since the regional dispatch automation system has been completed, such systems do not require scheduling control functions. The functions of distribution network dispatching and repair scheduling are similar to those of integrated systems.

3.3   Integration of Allocation and Distribution

The deployment integration mode is that the dispatch automation and distribution automation business applications are implemented based on a unified platform, and have the same jurisdiction. The distribution network repair scheduling is an independent system coupled with the deployment integration system through the service bus to obtain the real-time operation mode of the automation system and Network analysis services, the three form a generalized integration system. In terms of basic applications, the deployment integration system provides data processing services, network analysis services, alarm services, human-machine interfaces, etc. by the platform; in the system architecture, the integrated system has unified data storage and platform services, and distribution. The automation business can configure independent data collection services as needed.

The deployment integrated system mode is a mode in which the distribution network repair scheduling application is relatively independent and deployed based on an integrated platform. It is suitable for the scale of the distribution network relative to the main network, the configuration requirements for the distribution network automation function are low, and the distribution network is repaired. Dispatched areas that are uniformly constructed and maintained by provincial companies.

Since the construction of such a system is based on the existing power grid dispatching automation system, it is suitable for this mode in the existing dispatching automation system and the area where the distribution automation system is planned, and the area (including county) power distribution. The network automation plan covers 800 lines and below, and the real-time information points are 500,000 points and below [5].

In terms of function, it no longer focuses on the main network dispatch control, but only focuses on the functional construction of the distribution automation system, including the electronic function of the distribution network wiring diagram, distribution data access, distribution network operation monitoring, distribution network automatic fault handling, Main functions such as distribution network topology application. At the same time, consider integrating information with other business systems, such as repair scheduling, PMS, GIS, etc.

3.4   Comparison of Construction Models

The table below lists the differences in the three construction models for comparative analysis (Table 1).

Table 1. Construction model difference table

Mode	Scheduling system requirements	System size (number of lines)	Funding needs	Functional range	Construction period
Complete integration	None	<800	More	All	Longer
Matching and integrating	Need	>200 && <800	Normal	50%	Normal
Deployment integration	Need	<800	Less	40%	Shorter

4   Research on Key Technologies

The regional power grid deployment and integration system involves a number of key technologies in the construction process, such as communication message bus, feeder automation integrated with the main distribution network, and topology analysis of the full voltage model. This paper combines several key technologies studied in the construction of Sichuan Mianyang deployment system.

4.1   Integrated Analysis and Application Technology of Main Distribution Network

The grid analysis application function completes the effective analysis of the grid operation status. The main distribution network integration system uses the main distribution network operation data and the result data provided by other application software to analyze and evaluate the operation of the main network and the distribution network, and realize the main distribution network. Optimized operation [6].

4.1.1   Topology Analysis of the Main Distribution Network Integration Network

The network topology analysis based on the main distribution network integration system can be dynamically analyzed according to the connection relationship of the high, medium and low voltage grids and the operating state of the equipment. The analysis results can be applied to the main network monitoring, power distribution monitoring, and security constraints. Through the analysis of the electric island, the charging state of the high-medium and low-level three-level voltage grid equipment is analyzed, and the electric island is divided according to the topological connection relationship and the charging state of the main distribution network equipment.

Through the power point traceability, the power supply path and power supply of the high, medium and low three-level voltage grid equipment are analyzed, which provides a topology basis for related analysis applications [7]. Through the operation of the main distribution network equipment listing, temporary jump connection, etc., analyze the impact of such operations on the high-middle and low-level three-level voltage network topology.

4.1.2 Main Distribution Network Integration Topology Coloring Research

The main distribution network integrated topology coloring can determine the charging status of various electrical equipments of the high, medium and low three-level voltage model in the system according to the real-time status of the main distribution network switch, analyze the power supply point and the power supply path of each point, and the result is in the human machine. The interface is represented by different colors. At the same time, according to the results of the top-level analysis of the high-middle and low-level three-level power grids, the operating states of the grid model components exhibiting different voltage levels, such as live, power-off, grounding, etc., are dynamically displayed, and the power supply areas of different power supply points are dynamically displayed, and all paths of load transfer are displayed. Display the power supply range of different substations. The fault area can be shaded according to the results of the medium and low voltage fault analysis.

4.1.3 Main Distribution Network Integration Load Transfer Analysis

The main distribution network integrated load transfer analysis analyzes the impact load according to the target equipment of different voltage levels, and transfers the affected load safely to the new power supply point, and proposes a load transfer operation scheme including the transfer route and the transfer capacity. At the same time, the load information can be statistically analyzed to analyze the load and basic information of the load equipment affected by the target equipment. Through the transfer route search, the topology analysis method is used to search for all reasonable load transfer paths. Combined with the results of topology analysis and power flow calculation, the transfer load capacity and the transferable capacity of the transfer path are analyzed. It is also possible to analyze the results of dual-supply customer transfer [8].

4.2 Automatic Fault Diagnosis Technology Based on Battalion Data Integration

For an integrated system, when a fault occurs, the integrated system uses multi-party coordination to analyze and locate the fault. Comprehensively dispatch OMS blackout plan information, real-time information of the automation system, the power-off information of the mining system, and the geographic information of the GIS to locate the possible range of faults as comprehensively and accurately as possible.

4.2.1   Fault Diagnosis Based on Scheduling OMS Blackout Plan Information

For the planned power outage information, the research engine analyzes the equipment affected by the planned power outage and records it in the results table as a first-level trusted fault source. The emergency repair dispatch system obtains the next-day planned power outage information from the OMS, analyzes the power outage impact distribution according to the disconnection point information of the planned power outage information, and then records the power failure impact distribution into the judgment result table as a first-level trusted fault source.

4.2.2   Fault Diagnosis Based on Power Distribution Automation Fault Blackout Information

For the switch trip information, the research engine first analyzes the impact device of the trip switch, and then queries the first-level trusted fault source. If it belongs to the first-level trusted fault source, it directly performs the merge operation; if it is not the first-level trusted fault source, According to the route merged, it is recorded in the pre-judgment result table as a secondary trusted fault source. For the secondary trusted fault source, manual verification is supported, and the manually confirmed secondary trusted fault source is recorded in the known fault result table as a primary trusted fault source.

4.2.3   Based on the Fault Diagnosis of the Power Information Collection System

For the distribution change power failure information, the research engine queries the first-level trusted fault source. If it belongs to the first-level trusted fault source, it directly performs the merge operation; if it is not the first-level trusted fault source, it records the pre-judgment result according to the route merged. In the table, as a secondary trusted source of failure. For the secondary trusted fault source, manual verification is supported, and the manually confirmed secondary trusted fault source is recorded in the known fault result table as a primary trusted fault source.

Acknowledgments. The regional power grid deployment and grabbing integrated system construction mode is to unify the three architectures of power grid dispatching, distribution network scheduling, and distribution network repair and dispatching. The software functions are accurately positioned and the hardware resources are highly multiplexed to form a complete integrated system. With the improvement of power grid planning and construction, the society has increased the requirements for power supply reliability and power supply service quality, and promoted the construction and perfection of the deployment and integration system. At the same time, the key technologies related to power grid dispatching, distribution network scheduling, distribution network repair scheduling and system integration are increasing, and the development is becoming more and more mature. There are more and more cases of different modes of deployment and integration systems. Value is gradually emerging.

At present, some typical integrated engineering cases in some parts of the country, such as Mianyang in Sichuan, Langfang in Hebei, and Suzhou in Jiangsu, have been put into operation or are being implemented. The regional power grid deployment and grab integration system is the development trend and direction of distribution network dispatching and repair scheduling in small and medium-sized areas. It is an important technical means to improve the stability, safety and economic operation of small and medium-sized power grids, and is an important guarantee for improving the quality of regional power supply services. Promote the intensive operation and management of the power grid.

National Key R&D Program of China (SGTYHT/13-JS-175); State Grid Sichuan Electric Power Co., Ltd. technical project "Research on Key Technologies of Integrated Scheduling Control and Fault Repair Scheduling System for EMS and DMS".

References

1. Shen, B., Wu, L., Wang, P.: Technical characteristics and application effective-ness analysis of distribution automation pilot project. Power Syst. Autom. **36**(18) (2012)
2. Li, S., Ren, J., Huang, L., Liu, B., Jing, G.: Analysis of the development of urban distribution network automation and its operation management model. Electr. Abstr. (2) (2009)
3. Lu, W., Lin, Q.: Research and analysis of the construction of integrated systems for integration. In: China Electrical Engineering Society Power System Automation Committee 2012 Academic Exchange Conference (2012)
4. Jiang, J., Zhuang, X., Mei, F.: Design and application of distribution network production and repair command platform. Power Inf. **11**(5) (2013)
5. Zhang, D.: Development and application of integrated county-level power grid deployment system. Power Equip. **9**(4) (2008)
6. Guo, J., Qian, J., Chen, G., Zhang, W., Du, P., Cui, L., Shang, X.-w.: Technical scheme for intelligent distribution network dispatching control system. Autom. Electr. Power Syst. **39**(1), 206–212 (2015)
7. Liu, J., Zhao, S., Zhang, X.: Progress in China's distribution automation and several suggestions. Autom. Electr. Power Syst. **36**(19), 6–10 (2012)
8. Zhao, J., Chen, X., Lin, T., et al.: Distribution automation construction based on smart grid. Autom. Electr. Power Syst. **36**(18), 33–36 (2012)

Design of Sleeve Type Fruit Picking Device

Xiangping Liao[✉], Xiong Hu, Miao Jiang, Lin Long, and Yuan Liu

Hunan Institute of Humanities, Science and Technology,
Loudi 417000, Hunan Province, China
520joff@163.com

Abstract. In order to solve the problems of large workload, wide range of operation (uneven fruit distribution) and high requirement of touch force control in fruit (juicy fruits are easily bruised) harvesting for mass production, a sleeve type fruit picking device is proposed. Based on the principle of force parallelogram, a sleeve type fruit picking device is designed by using a special blade mounting method and lightweight materials that meet mechanical properties, and the mechanical analysis and calculation of the fruit picking process are carried out. The results showed that the sleeve type fruit picking device can pick and collect the fruit efficiently with a small force without bruising it.

Keywords: Sleeve type · Parallelogram principle of force · Picking device · High-efficiency

1 Introduction

The sleeve type picking device is designed with various mechanical and artificial assistance devices. The devices are currently used to pick fruits include vibratory picking set [1], scissors picking set [2] and robotic picking set [3]. Even though Vibratory picking set has high working efficiency, it may do damage to fruits and its trees. Scissor picking set is more accurate and less harmful to pick fruits. However, it has the disadvantage of low efficiency because this is a method that needs to be picked by hand continuously. Robotic picking set has high precision in picking fruit, but lower working efficiency and higher cost in fruit picking. Therefore, it is particularly important to improve the efficiency of the fruit picking and reduce the cost, while without damaging fruits. In this work, a new sleeve type fruit picking device was put forward to solve these problems. It can achieve good performance in the process of fruit picking through the mechanical analysis and calculation.

2 Structure and Operating Principle of Fruit Picking Device of Sleeve Type

We can figure out the operating principle of the picking device form its structure as shown in Figs. 1 and 2

© Springer Nature Switzerland AG 2020
F. Xhafa et al. (Eds.): IISA 2019, AISC 1084, pp. 493–500, 2020.
https://doi.org/10.1007/978-3-030-34387-3_61

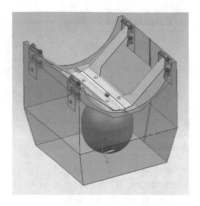

Fig. 1. Picking box 3D model

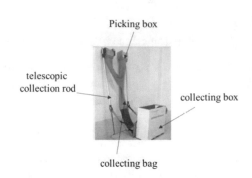

Fig. 2. Physical picture

When the fruit is put into the picking box, the fruit stem are clamped by baffles and blades. Then the fruit is stuck in the picking box under the combined action of two baffles and rubber bands. Afterwards, the telescopic collection rod should be pulled down, and the blade is used to cut the carpopodium with a small force to separate the fruit from the fruit branch, so as to achieve the purpose of picking. Finally, the fruit falls under gravity and it can be buffered and guided by the collecting bag in the process. So it can be transported to the collecting box properly.

As can be seen from Fig. 2, the fruit picking device has two collecting rods and two picking boxes. When the left picking boxes and collecting rods are used for fruit harvesting, the right ones are used to support the branches of fruit; when the right picking boxes and collecting rods are used for fruit harvesting, the left ones are used to support the branches; thus, through the coordination of left and right devices, the efficiency of fruit harvesting can be greatly improved.

The blade installation method is shown in the Fig. 3. We can figure out that the yellow part is the blade. When fruits enter the picking box, the cutter saddle will move and follow the baffle, then the top end of cutter saddle will moves backward relative to the baffle because of different hinged joints, so that the fruit body will not touch the blade and be scratched by the blade.

When the fruits enter the picking box, the picking box will move down. Afterwards, the baffle gradually clamps the carpopodium. Then the top end of the cutter saddle moves forward relative to the baffle plate until the blade cut into the carpopodium. Under the action of parallelogram force, cutting force between the two blades gradually increases, thus cutting the stem and separating the branch from the fruits.

The manufacturing method of the fruit picking device is as follows.

1. The galvanized sheet is processed into a sleeve by using a riveting method.
2. The baffle is connected with the sleeve by rivets, hinges and rubber bands. Meanwhile, it should be inclined to the sleeve at an angle, by which it can be restored to an initial state.
3. The galvanized plate at the end of the sleeve should be made into an inward-contracting pocket shape, which not only can buffer and guide the falling process of fruits, but also can be well connected with the buffer bag.

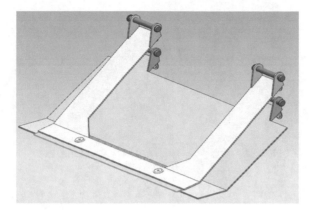

Fig. 3. Blade mounting position

3 Force Analysis and Calculation of Picking Process of Sleeve Type Fruit Picking Device

As shown in Fig. 4, When the fruit enters the picking box, the collecting rod will be put an downward pulling force to pick the fruit. Consequently, the baffles, blades and fruits on the picking box form a parallelogram structure as shown in Fig. 5. Therefore, when a pulling force is applied to the collecting rod, the blade generates a shear force on the fruit stem as shown in Fig. 6.

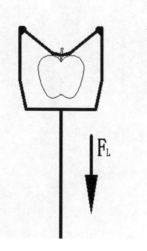

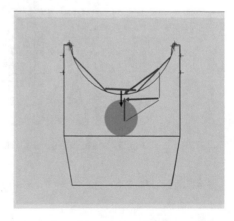

Fig. 4. Diagrammatic sketch of fruit picking

Fig. 5. Forced analysis diagram of parallelogram

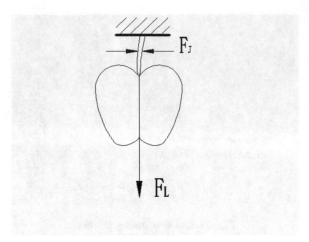

Fig. 6. Stress analysis diagram of fruit stem

Because the fruit stem has a certain elastic deformation, the length and diameter of the fruit stem will be deformed under the action of the pulling force. The deformation analysis of the fruit stem is shown in Fig. 7.

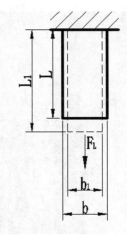

Fig. 7. sketch of the fruit stem tensile deformation

Under axial tension F_L, the stem is axially elongated

$$\Delta l = \frac{F_L l}{EA} \tag{1}$$

Linear strain

$$\varepsilon = \frac{\Delta l}{l} = \frac{l_1 - l}{l} \tag{2}$$

Under axial tension F_L, the transverse strain is

$$\varepsilon' = \frac{\Delta b}{b} = \frac{b_1 - b}{b} \tag{3}$$

Because the tensile stress does not exceed the proportional limit.
μ is Poisson's ratio

$$\left| \frac{\varepsilon'}{\varepsilon} \right| = \mu \tag{4}$$

When there is no tension F_L in the axial direction under the action of

$$\tau = \frac{F_s}{A} \le [\tau] \tag{5}$$

$$A = \pi \left(\frac{b}{2} \right)^2 \tag{6}$$

When $Fs \ge [\sigma]A$, the stem can be cut off.
   Where, F_S is the shear force; A is the cross-sectional area of branches; $[\sigma]$ is the allowable shear stress.
   When there is axial tension F_L

$$v^2 = 2gh\frac{F_s'}{A'} \le [\tau] \tag{7}$$

$$A' = \pi \left(\frac{b_1}{2} \right)^2 = \pi \left(\frac{b - \Delta b}{2} \right)^2 \tag{8}$$

Because $\Delta b > 0$, $F_s' < F_s$. With an axial tension, cutting the same stem will be less laborious than without axial tension.
   The force applied to the collecting rod during the picking process is an sudden applied dynamic load

$$Fd = KdFst \tag{9}$$

Where F_d is the dynamic load, K_d is the dynamic load factor, K_{st} is the static load. The impact is caused by pulling down from high h, then

$$K_d = 1 + \sqrt{1 + \frac{2h}{\Delta st}} \tag{10}$$

Due to the sudden addition of the load to the collecting rod, it is equivalent to the case where the object moves downwards to be $h = 0$. At this time $K_d = 2$. Therefore, under the sudden load, the dynamic load is twice the static load. That is, simply by applying a force greater than or equal to $Fs/2$ on the collecting rod, the fruit stem can be sheared. In other words, it has a good labor-saving effect.

4 Calculation and Analysis of Casing Wall Thickness and Selection of Materials

The idealized analysis of the force on the picking box wall is shown in Fig. 8.

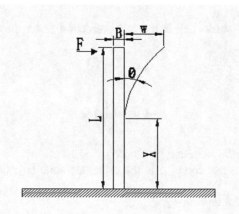

Fig. 8. Analysis of the force on the wall of picking box

The bending moment on any cross section

$$M = -F(L - X) \tag{11}$$

The differential equation of a flexural line

$$EI\omega'' = M = -F(L - X) \tag{12}$$

$$EI\omega' = \frac{F}{2}X^2 - FLX + C \tag{13}$$

$$EI\omega = \frac{F}{6}X^3 - \frac{FL}{2}X^2 + CX + D \tag{14}$$

Where E is the elastic modulus of the material (200×10^3 N/mm^2); F is the force; I is the moment of inertia of material cross section on bending neutral axis ($I = BL^3/12$).

At the fixed end of A

$$X = 0 \, \omega_A = 0 \tag{15}$$

$$\omega_A' = \theta_A = 0 \tag{16}$$

The boundary conditions (16) are substituted into (13), and the Eq. (15) are substituted into (13), we can obtain:

$$C = EI\theta_A = 0 \tag{17}$$

$$D = EI\omega_A = 0 \tag{18}$$

And then the integral constant C, D is substituted into (13) and (14)

$$EI\omega' - \frac{F}{2}X^2 \quad FLX \tag{19}$$

$$EI\omega = \frac{F}{6}X^3 - \frac{FL}{2}X^2 \tag{20}$$

The cross - section B coordinate $X = L$ is substituted into the above two equations

$$\theta_B = \omega_B' = -\frac{FL^2}{2EI} \tag{21}$$

Where θ_B is negative means that the angle of rotation of section B is clockwise

$$\omega_B = -\frac{FL^3}{3EI} \tag{22}$$

Where ωB is negative indicates the downward deflection of point B.

Reference [7] combined with the analysis and calculation under ideal conditions and the specification and performance of related materials, by which 0.7 mm thick galvanized sheet is selected for the picking box. The selected materials are thin and light and accord with the required mechanical properties, which is conducive to efficient fruit picking.

5 Conclusion

1. Compared with the existing fruit picking device, a sleeve type fruit picking device is designed, by which fruits can be located uncertainly and picked efficiently without bruise. At the same time, the advantages of picking for a variety of fruits (more than one fruit at one time) are reflected in practical experiments.
2. The analysis on the theory shows that the sleeve type of fruit picking device requires smaller force during picking. That's to say, the fruit stem can be cut off with a shear force which is greater than or equal to $Fs/2$. The effort of saving effect can be demonstrated well.

3. The analysis on the theory shows that just 0.7 mm thick galvanized sheets can meet the material performance requirements. Using the lightweight material is also good for efficient picking.

Acknowledgement. This work was supported by the grant from the Natural Science Foundation of Hunan Province of China (2017JJ3057) and the Research Foundation of Education Bureau of Hunan Province, China (17C0473).

References

1. Yi, X.: Virtual Prototype Design and Dynamic Simulation of Vibratory Forest Pickers. Central South University of Forestry and Technology (2013)
2. Chen, S., Wang, H., Chen, J., Hu, P., Chen, Y., Zeng, Q., Dong, J.: Design of a rotary fruit picker. Sci. Technol. Innov. **09**, 164–165 (2018)
3. Jin, C.: Application of robot technology in agriculture—fruit picking robot. Sichuan Agric. Mach. (06), 32 (2009)
4. Wang, W., Wu, L., Yang, Z., et al.: Cut straight edge shear mathematical model and experiment. J. South China Agric. Univ. **37**(4), 105–111 (2016)
5. Li, Y., Song, X., Yang, L.: Experimental study on mechanical properties of apple tree branches. Chinese J. Appl. Mech. **33**(04):720–725 + 745 (2016)
6. Wen, B.: Mechanical Design Manual: Single-line. Mechanical Engineering Materials, vol. 12, 5th edn., pp. 387–388. Mechanical Industry Press, Beijing (2014)
7. Cheng, D.: Mechanical Design Manual: Single-line. Commonly Used Mechanical Engineering Materials, vol. 1, 5th edn., pp. 91–92. Chemical Industry Press, Beijing (2010)
8. Liu, H.: Materials Mechanics I, vol. 1, 5th edn., pp. 185–187. Higher Education Press, Beijing (2011)
9. Shaojun, W.: Mechanical Manufacturing Process Design Manual. Mechanical Industry Press, Beijing (1985)

Skill Assessment Plagiarism Screening Based on Access Control System

Runxin Yang and Jimei Li[✉]

School of Information Science,
Beijing Language and Culture University, Beijing, China
ljm@blcu.edu.cn

Abstract. In recent years, teaching, testing, competing and training of practical skills has been developed rapidly. However, there are still some deficiencies in the existing skills assessment system. At present, the automatic assessment system cannot support plagiarism screening based on the answer data. According to the characteristics of access control system, this paper takes the Yonyou ERP-U8 software as an example to design and implement a plagiarism screening method. The method will identify different users by signing the answer data based on the identification information of the users in the assessment system and then identify the plagiarism by plagiarism screening algorithm. The plagiarism screening method proposed in this paper can be applied to any organizational information system with access control system which need to be assessed in practice. It also has certain reference significance for other IT skills assessment that need to identify users' operational identity by plagiarism screening.

Keywords: Plagiarism screening · Skills assessment · Access control system

1 Introduction

With the rapid development of IT technology and the implementation of national informatization policies, the enterprises' demand of enterprise information systems practical and the qualified personnel is increasing, which is crucial for the application education and personnel training of real software.

In addition, the influence of the skills competitions or certifications has been enlarging rapidly as well. For example, the Competition of Accounting Skills (Accounting informatization) held by Ministry of Education and the "Seentao Cup" National Accounting Information Technology Competition organized by Ministry of Industry and Information Technology. Both of them are held once a year with up to a million students.

At present, the mainstream research work of skills assessment can be classified into two categories. The first one is to change operation questions into objective ones. The second one is to ask the users to operate in a real software environment. In the latter case, the automatic assessment is accomplished by an independent assessment system. However, the existing assessment system does not have the plagiarism screening ability to evaluate the operation results of the users in the real software. Basically, the

F. Xhafa et al. (Eds.): IISA 2019, AISC 1084, pp. 501–509, 2020.
https://doi.org/10.1007/978-3-030-34387-3_62

assessment system publishes tasks. According to the user identity required in the task, users log in the software through the same user identity, and then process the business informationally according to the requirement of tasks. That is, different users in the software use the default and the same user information, so that it is impossible to distinguish the answer data of different users, thereby causing the answer data not to be screened.

To sum up, teaching, testing, competing and training of practical skills has been developed rapidly. However, the automatic assessment system cannot support plagiarism screening based on the answer data. Therefore, according to the characteristics of access control system, this paper takes the Yonyou ERP-U8 software as an example to design and implement a plagiarism screening method. The plagiarism screening method can be applied to any organizational information system with access control system which need to be assessed in practice. It also has certain reference significance for other IT skills assessment that need to identify users' operational identity by plagiarism screening.

2 Related Works

The skills assessment studied in this paper belongs to the automatic assessment of skills non-objective questions. It is the key and difficult point in computer-aided evaluation (CAA) [1–8], which is a multi-disciplinary research topic. The research status of this topic includes the following aspects:

Domestic: Xu, Liu and He of Beijing Normal University have developed a complete IT skill assessment system iTAS and tutor system iTutor for IT skill training, and given the automatic scoring algorithm for general IT operation questions [3–7]; Jin, Ma, Luo and Wu, from Zhejiang Normal University, designed and developed the information technology level certificate examination system for primary and secondary schools in Zhejiang Province [10]; Based on the research results, Wang designed and implemented a computer-aided assessment system based on network collaboration [1, 2]; Li and Sun of Beijing Normal University have designed an automatic evaluation system for static web page production problems [11]; In view of the introduction to the SAP ERP course, Guo, from Shanghai Institute of Foreign Trade, and Yang, from Beijing SAP Software System co., ltd., developed an automatic assessment system by turning the operation tests into the fill-in-the-blanks problem [12]; At present, the mainstream assessment system is based on the C/S framework and it is used only for Yonyou ERP accounting skill assessment [13]; Our team has researched ERP skill assessment features, modeled ERP practical skills online intelligent assessment system last year [14, 15] and researched ERP practical skills assessment strategy and intelligent assessment architectures this year [16, 17].

Foreign: Roy Dowsing, Steward and Roman Sleep of the University of East Anglia in the UK developed the Word-Task system and the SpreadTask system for word processing and spreadsheet applications; American Testing Authority Inc., or ATA, has developed a dynamic assessment technology (DST) based on dynamic simulation technology, which provides a virtual interactive test environment and real software environment [19].

Through the analysis of the research work, the existing skills assessment system for enterprise real software meets the basic needs of teaching, testing, competing and training, but it cannot support plagiarism screening in skills assessment. It has always been a key and difficult issue in the field of distance education. This paper proposes a solution to plagiarism and provides its core technology, the plagiarism screening method of skills assessment based on access control system.

3 Design and Application of Plagiarism Screening Method Based on Access Control System

Taking Yonyou ERP-U8 software as an example, this paper designs and implements plagiarism screening method based on access control system. The following describes the requirements and design scheme,

3.1 Requirements Analysis

Through the analysis of the existing skills assessment system, the existing system for enterprise real software meets the basic needs of teaching, testing, competing and training, but does not have the functions of plagiarism screening.

It should be noted that access control system has not been fully utilized in the existing work, but the assessment system releases the task, and the users use the unified identity required in the task to login. That is, the users' answer data has the same operator information, so even if the user U1 submits the answer data of user U2, the system cannot identify the user U1's plagiarism behavior, and it is impossible to distinguish the answer data of different users, thereby causing the answer data not to be screened. Therefore, in view of the problem that the existing skills assessment system for enterprise real software does not have the plagiarism screening function and cannot identify the plagiarism of the user, a method of plagiarism screening based on access control system is proposed.

3.2 Design Scheme

Based on requirements analysis, the design scheme of the plagiarism screening method based on access control system proposed is as follows, the assessment system architecture is shown in Fig. 1 and the specific data processing flow is shown in Fig. 2.

(1) Obtain identification information and perform data signature.

Embedding the identity information of the user in the operation of the user, so that different users' answer paper data and teacher's answer data have different identity information, and different users' answer paper data, answer paper data and answer data can be distinguished.

The data involved in this study includes three data sets, the initial data set D1, the answer data set D2, and the users' answer paper data set D3. The initial data refers to the data before the teacher answers the question; based on the initial data, the teacher answers the data according to the requirements of the task, which is called the answer data; based on the initial data, the user answers the data according to the requirements of the task, which is called the answer paper data. The answer data representing the correct answer by the teacher and the answer paper data after the users' operations are collectively referred to as the answered data.

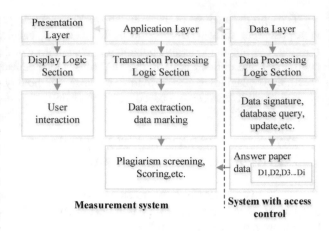

Fig. 1. Assessment system architecture

This paper will use the access control system of Yonyou ERP-U8 and use the operator information field. When the users log in to the assessment system, the identities of the users will be embedded in the user information data table of Yonyou ERP-U8 and then store in the specific data item of the relevant data table in the answer paper data, that is, the identity of the user in the software is the distinguishable personal identity. After the operation is completed, the

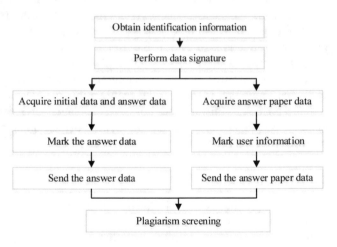

Fig. 2. Data processing flow

assessment system extracts the answer paper data containing the identity information of the user. At this time, the answer paper data set after the data signature is converted into D3i (i = 1, 2, 3..., i is the i-th user). The answer paper data set contains the identity information of the users.

By answer paper data signature, it is possible to distinguish the answer paper data of different users, and then the individualized operations such as plagiarism screening and post skill analysis can be performed on the users. The specific steps are as follows:

① Obtain identification information of the user from the assessment system. The user will log in to the assessment system through the login name and password to view the published tasks, and the system will capture the identification information N1 of the user through the user login information.

② Perform data signature in the answer paper data operated by the users. The assessment system generates the identity information string of the user based on identification information N1 of the user, and then the identities of the users will be embedded in the user information data table of Yonyou ERP-U8 and then store in the specific data item of the relevant data table in the answer paper data, That is, the default operator name N2 in Yonyou ERP-U8 is replaced with the identification information N1.

(2) Acquire data with signature information

Acquiring data with signature information refers to obtaining the initial data set D1, the answer data set D2 and the users' answer paper data set D3. Usually, the initial data is imported on the user's local machine during the assessment preparation, so that the teacher and the user have the same initial data set D1.

In Yonyou ERP-U8, each of the answered data sets D2 and D3 includes one or more data tables, each of which includes one or more records, each of which includes one or more data items. In the case of complete operation, the answer data set D2 corresponds to the data table in the answer data set D3 and the number of records in the table. The following analysis is based on the premise of complete operation, and other special cases are treated similarly.

(3) Signed answer data marking for plagiarism screening.

① Mark the answer data. The assessment system matches the corresponding data of the answer paper data set D3 and the answer data set D2, and calculates the actual scores based on matching results. At this time, both D2 and D3 have D1. If the data of D2 is not marked with operation type, the initial data in D3 will be scored according to the standard of operation data during automatic scoring, thereby affecting the quality of scoring. Therefore, before the automatic scoring, the initial data D1 and the answer data D2 are matched based on the primary key of the data table, and based on the matching result, the records in the answer data D2 are marked and divided into four types of tags: add, delete, change, and consistency. If a certain record r211 of a certain table t1 in D2 has no matching record based on the value of the primary key in the corresponding table t1 of D1, the operation type of r211 is " add"; If a certain record r211 of a certain table t1 in D2 has a matching record r112 in the corresponding table t1 of D1 based on the value of the primary key, then, the value of the field is compared between r211 and r112. If they are equal, the operation type of r211 is "consistency"; if there is a difference, the operation type of r211 is "change", and the modified data item is marked at the same time; If all records of a table t1 in D2 have been identified, but there is still a record r114 in the corresponding table t1 of D1, r114 is added to t1 of D2 (denoted as r215), and the operation type of the label r215 is "delete".

② Mark user information. This study will classify and mark the assessment data items in the software data table from the perspective of scoring importance, and divide it into main identification field F0, basic field F1, key field F2, comprehensive field F3, calculation field F4, extension field F5, operator name field F6, etc., different field categories represent different scoring weights. The "operator name field F6" is used as a scoring category alone, which is helpful for plagiarism screening of users. In Yonyou ERP-U8, auditor, orderer and biller are all operator name fields.

(4) Answer paper data plagiarism screening based on the identification information and data marking.

Answer paper data plagiarism screening refers to the identification of the user's answer paper data, to determine whether the operation is answered by the person rather than submitting the data of others. When scoring, the answer paper data D3 needs to be matched with the answer data D2. For example, a certain record r311 of a certain table t1 in D3 and a record r211 of a certain table t1 in D2 are matched. Wherein, the value V3 of the operator name field marked by r311 in D3 is different from the value V2 of the operator name field of r211 in D2, but the field value comparison cannot be directly performed. The score of this field needs to follow the principle of plagiarism screening. If the value V3 of the operator name field of the record r311 in the answer paper data D3 matches the login information N1 of the assessment system, it means that the user does not have plagiarism, and the data item is scored correctly.

As shown in Fig. 3, the generated results feedback file records the users' scores, operational field errors, and detailed operational information. In Yonyou ERP-U8, auditors, bookkeepers, billers are all operator name fields. Therefore, when the operator

Operator:	Ding Yi		
Scores:	100.0		
Grades:	1496		
Total errors:	(T,1496);(F,0)		
Table name	(Field name,Evaluator answer,Teacher answer,True or False)		
Input table	(Operation type,input,input,T)	(Auditor,Ding Yi,Zhao,T)	(Originator,Ding Yi,Zhao,T)
Input table	(Operation type,input,input,T)	(Auditor,Ding Yi,Zhao,T)	(Originator,Ding Yi,Zhao,T)
Output table	(Operation type,output,output,T)	(Auditor,Ding Yi,Zhao,T)	(Originator,Ding Yi,Zhao,T)
Output table	(Operation type,output,output,T)	(Auditor,Ding Yi,Zhao,T)	(Originator,Ding Yi,Zhao,T)

Fig. 3. Feedback file

name field is inconsistent with the teacher field, the score of this field needs to follow the principle of plagiarism screening. If the value of operator name is consistent with assessment system login information, the field is true.

The skills assessment plagiarism screening method based on access control system provided in this paper mainly has four steps. First, the user's answer paper data is signed. The user's login information is synchronized to the access control system of Yonyou ERP-U8. Through data signature, the phenomenon that operators have the same name when different users operate the software is avoided. After signing the answer data, the data pre-processing is mainly to obtain the initial data set D1, the answer data set D2 and the user's answer paper data set D3 as well as to mark the operation type of the answer data and the users' information. The answer data and the answer paper data after data pre-processing can be used for plagiarism screening. The score of the operator name field needs to follow the plagiarism screening principle. By matching the value of operator's information with the login information N1 of the assessment system, it can check whether the operator of the software is the login of the assessment system, and then distinguish whether the answer data has plagiarism.

4   Summary

With the process of informatization, software application and personnel training of real skills software are extremely important. Therefore, according to the characteristics of access control system, this paper takes the Yonyou ERP-U8 software as an example to design and implement a plagiarism screening method and applies to the actual assessment. Practice shows that plagiarism screening method designed in this paper has the following main features and functions:

(1) The method realizes the answering data signature and plagiarism screening function of the skills assessment system, and can identify the operator's identity and determine whether there is plagiarism.
(2) This method can be applied to any organizational information system with access control system and requiring skill assessment.
(3) The method can provide a technical basis for plagiarism screening for various skills such as examinations, competitions, exercises, trainings, etc. It can also be used as a reference for other IT skills assessment that need to identify the user's operational identity.

The following research can be stated from the two aspects:

(1) The design and implementation of the algorithm related to personalized learning analysis, which includes job skill analysis, personalized feedback generation, behavior pattern mining based on the data edited by users, learning analysis and data mining based on online grade.
(2) The algorithms related to intelligent marking of exam paper including the optimal matching between the records of result that is submitted by examinee and the records of answer that is provided by examiner, calculating the exam paper scores and the grading model selection.

Acknowledgments. This research project is supported by Science Foundation of Beijing Language and Culture University (supported by "the Fundamental Research Funds for the Central Universities") (No: 19YJ040004, 17PT01, MOOC201810, 17ZDJ02), by BLCU supported project for young researchers program (supported by the Fundamental Research Funds for the Central Universities) (19YCX129, 19YCX128), and by Computer Foundation Education Research Institute of National Colleges and Universities (No: 2018-AFCEC-177, 2019-AFCEC-100).

References

1. Wang, Y.: Research on the System of Computer-Assisted Assessment in IT Skills Based on Network Cooperative. Zhejiang Normal University, Jinhua (2011). (in Chinese)
2. Wang, Y.: The design of a network automatic scoring system for IT skills. Comput. Knowl. Technol. **10**(10), 2270–2272 (2014). (in Chinese)
3. He, K., Xu, J.: The new development of research on computer assisted assessment (CAA). Open Educ. Res. **11**(02), 78–83 (2005). (in Chinese)
4. Xu, J., Liu, Q., He, K.: New field of CAA research – IT skills assessment automation (part one). eEduc. Res. (01), 33–37 (2002). (in Chinese)
5. Xu, J., Liu, Q., He, K.: New field of CAA research – IT skills assessment automation (part two). eEduc. Res., (02), 44–48 (2002). (in Chinese)
6. Xu, J., Liu, Q.: IT Skill Automated Testing and Assessment: Theory, Technologies and Applications. Science Press, Beijing (2001). (in Chinese)
7. Xu, J.: The Theory of Automated Skills Testing and Assessment and its Application. Beijing Normal University, Beijing (2001). (in Chinese)
8. Xiong, Y., Peng, D.: Current status and development of computer-assisted chinese characters assessment. China Examinations (09), 20–26 (2013). (in Chinese)
9. Liang, H.: Design and implementation of the computer basic applications course online test system. Sun Yat-sen University, Guangzhou (2011). (in Chinese)
10. Jin, B., Ma, Y., Luo, H., Wu, Z.: The formal deseription of check information and its applications. Comput. Sci. **32**(01), 106–107 (2005). (in Chinese)
11. Li, J., Sun, B.: Study on skill assessment method based on marking criteria and execution engine. Comput. Appl. Software **29**(10), 128–132 (2012)
12. Guo, X., Yang, K.: Design of ERP test automatic marking based on the practice. University Education (01), 71–72 (2013). (in Chinese)
13. SowerPower Technology Co. Ltd. Competition system (national vocational skills contest accounting skills competition) introduction. [EB/OL], 23 April 2019. http://www.sower.com.cn/index.php?list=5. (in Chinese)
14. Li, J., Wang, R., Feng, Y.: Modeling of ERP skill assessment features. In: 2018 13th International Conference on Computer Science & Education (ICCSE 2018). IEEE, Colombo (2018)
15. Wang, R., Li, J.: Modeling of ERP practical skills online intelligent assessment system. In: 2018 International Conference on Information Technology and Management Engineering (ICITME 2018). Atlantis Press, Beijing (2018)
16. Yang, R., Wang, R., Li, J.: Research and development of automatic assessment strategy for ERP practical skills based on MOOC. In: 2019 14th International Conference on Computer Science & Education (ICCSE 2019). IEEE, Toronto (2019)
17. Wang, R., Li, J.: Architectures for intelligent assessment on complex information system skills. In: 2019 14th International Conference on Computer Science & Education (ICCSE 2019). IEEE, Toronto (2019)

18. Seentao Technology Co., Ltd. Introduction. [EB/OL], 23 April 2019. http://www.seentao.com/about/introduce.html. (in Chinese)
19. ATA Company. Testing Service [EB/OL]. 23 April 2019. http://www.ata.net.cn/ETService/
20. Competition of Accounting Skills, 24 April 2019. http://www.chinaskills-jsw.org/
21. Paperless Evaluation System (Computerized Accounting), 24 April 2019. http://www.sower.com.cn/index.php?list=4
22. Wu, Y..: Design and Development of the Network Examination and Evaluation System Based on B/S Framework. Qufu Normal University, Qufu (2005). (in Chinese)

Crowd Density Estimation Based on the Improved Vibe Algorithm and a Noval Perspective Correction Method

Junhang Wu[1,2], Yuanyuan Ren[2], Tiaojun Zeng[2,3(✉)],
and Chuanjian Wang[2]

[1] The Computer School of Wuhan University, Hubei, China
[2] The School of Information Science & Technology,
Shihezi University, Shihezi, China
zengtiaojun@163.com
[3] Key Laboratory of Computer Network and Information Integration
(Southeast University), Ministry of Education, Nanjing, China

Abstract. Unusual crowding is often associated with unusual events. Estimating population density is a good way to control management, maintain crowd safety, and prevent riots or high-risk activities. Therefore, in the paper, an improved method is proposed for crowd density estimation. First, a dynamic controller is introduced to Vibe algorithm, helping error detection and ghost suppression for faster adaptation to scene changes. Second, take perspective correction into account for pixel counting and propose a novel modeling method based on pixel level. It is of great significance to enhance the accuracy of estimation. Next, hybrid features based on global and local texture information are proposed to provide robustness to noise. Finally, the improved method was verified using images selected from the UCSD database. Experimental results demonstrate the performance and potential of the method.

Keywords: Crowd density estimation · Dynamic controller · Perspective correction · Hybrid features

1 Introduction

Estimating the number of people in the image is a practical machine vision task that is increasingly popular in the field of security surveillance. For safety and economic reasons, it is critical to reliably estimate the population density that occurs in public places over time. However, manual monitoring is low in accuracy and time consuming, and cannot meet the requirements of public safety. A lot of work has been done to solve this problem [1–5]. Methods for estimating population size and density using image processing techniques can be divided into two categories: direct methods and indirect methods [6].

For the first category, the direct approach attempts to estimate population density by simultaneously identifying individuals and their locations. For pedestrian detection, a part-based detector is typically used for processing, such as a detector formed by the head and shoulders [7] or a detector only for the head [8]. These detection-based

© Springer Nature Switzerland AG 2020
F. Xhafa et al. (Eds.): IISA 2019, AISC 1084, pp. 510–515, 2020.
https://doi.org/10.1007/978-3-030-34387-3_63

methods can be further divided into two categories: model-based methods and trajectory-based clustering methods. For the former method, the author attempts to segment first and then examine each human body with a mannequin or shape [9]. The latter approach attempts to detect independent motion by gathering points of interest on people being tracked over time [10, 11].

For the latter category, indirect methods attempt to extract several local and global features for estimation, rather than directly detecting people. Based on the detection of different features, the indirect population estimation method is mainly divided into two parts: a pixel-based method and a texture-based method. Pixel-based methods typically extract various features of foreground pixels based on motion of the object, such as foreground regions, number of foreground pixels, histograms of edge directions, or edge counts [12, 13]. Most of these methods indicate the relationship between the foreground area and the total number of people. Near linear. The texture-based approach explores a coarser approach and requires analysis of the plaques of the image to achieve better performance in the presence of high population densities with severe overlap and occlusion.

Compared with the direct method, the indirect method is more difficult to detect a single human target, and it is easier and more efficient to extract features from the image. For pixel-based methods, background removal [14, 15] is the first step, and it directly affects the estimation results. How to improve the effectiveness of foreground detection algorithms for complex scenes has become a research hotspot. At the same time, the original foreground pixels are exposed to the perspective distortion because the camera is always placed on the side, forming an angle with the ground plane. Perspective distortion refers to the fact that far objects in the scene appear smaller than objects in the image. It is of great significance to the accuracy of estimation. So far, there has been little work for perspective correction yet.

This paper proposes an improved population density estimation method. Our contribution has three aspects. First, Vibe algorithm is improved for foreground detection using a dynamic controller, helping error detection and accelerating the ghost suppression for faster adaptability to scene changes. Second, perspective correction is taken into account for pixel size normalization with the new modeling method. Third, the description of the image does not use a single feature, but rather a texture feature with global and local texture information. This representation is more powerful than a single function in dealing with noise.

2   Proposed Method

Foreground pixels may be determined as background pixels during image processing, when the background model is interfered by external factors (noise). It may cause error detection when the next frame comes. If so, it will form an endless loop and make detection error increase. To alleviate the problem, a kind of judgment method based on spatial consistency is applied to updating mechanism. Next, the improved algorithm will be described in details.

For any pixel $x(x_i, x_j)$, determined as the background pixel in frame I, its $k \times k$ neighborhood is defined as follows:

$$N_x = \left\{ y = (y_i, y_j) \in I : |x_i - y_i| \leq k, |x_j - y_j| \leq k, k \in Z^+ \right\} \tag{1}$$

It provides a rigorous mathematic representation for the square neighborhood with the size of $(2k+1) \times (2k+1)$. Ω_x is denoted by the set of pixels matching to their background models in (N_x):

$$\Omega_x = \left\{ y \in N_x : \#\{P(y) \cap S_R(I(y))\} < \#_{min} \right\} \tag{2}$$

Neighborhood Coherence Factor (NCF) [20] is defined as follows:

$$NCF(x) = \frac{|\Omega_x|}{|N_x|} \tag{3}$$

Where $| \bullet |$ refers to the cardinality of a set, i.e. the number of element in a set. $NCF(x)$ is taken as a benchmark to measure the accuracy of its background model for its representation of spatial consistency. It has the values in the close interval of $\left[\frac{1}{(2k+1)(2k+1)}, 1 \right]$.

From the original update policy, the probability for updating, $1/\varphi$, is a fixed parameter throughout the algorithm flow. However, it is not reasonable for some special scenes because the probability, $1/\varphi$, should better be a dynamic parameter to adapt to complex background. So a dynamic controller is introduced for φ. Then we redefine φ as follow:

$$\varphi = \varphi \times \frac{1}{2 * NCF(x)}, NCF(x) \in \left[\frac{1}{(2k+1)(2k+1)}, 1 \right] \tag{4}$$

According to Eq. (4), the larger $NCF(x)$ becomes, the more pixels in its surrounding can be described by their background models and the probability that pixel x is used for background updating should be appropriately increased. As it can be seen from Eq. (4), it makes the adaptive adjustment to φ based on $NCF(x)$, which helps background model improve robustness to complex background.

Meanwhile, shape expansion and corrosion are used to eliminate isolated singularities caused by noise interference. In addition, n-polygon sharp of ROI selection is adopted instead of conventional regular shape to further reduce external interference and bring more flexibility to foreground detection.

2.1   Geometric Correction for Pixel Size Normalization

The original foreground pixels are exposed to the perspective distortion because the camera is always placed on the side, forming an angle with the ground plane. Perspective distortion means that a distant object in the scene looks smaller than a near object in the image, but may actually be a phenomenon of the same group of people. At low and medium population densities, occlusion and overlap have little effect on maintaining a linear relationship between the foreground area and the number of people. This article maps the total foreground pixels directly to the total number of people with low and medium density. The accuracy of the pixel count based method is largely affected by perspective distortion. Therefore, geometric correction of the original pixels is the key to improving the estimation accuracy.

2.2   Description of Our Method

The basic principle of the improved algorithm is to place all objects at different distances in a scene on the same scale. Assume that the vanishing line (horizon) is parallel to the horizontal scan line in the image. As the picture shows, four points $\{Pi, i = 1, 2, 3, 4\}$ are selected along the road and two to either side, forming a quadrilateral. Note that the points should be selected in the way that the extension of the lines parallel to road border lines will intersect at the "vanishing point" Pv.

Notice that, the image consists of one layer of arranged pixels just like a chessboard when it is enlarged. Every pixel has the same size in the image. Meanwhile, the pixel is the smallest unit for pixel counting based methods, so more attention should be paid to the pixels when modeling. Thus a new modeling method is proposed for geometric correction based on pixel level.

Instead of taking reference scale as a line, it is treated as a row of closely spaced pixels and it really does in fact. Then $\overline{P1P2}$ is defined as the reference scale:

$$\overline{P1P2} = \{(x_i, y_1) | i \in Z\} \tag{5}$$

Where every pixel is considered as a point and is also treated as a small square, with a side length of 1 for computation. Meanwhile, S_{side} is defined as the scale for the side having undergone geometric correction. the pixel (x, y), needing geometric correction, is taken form the horizontal scan line and mapped to reference scale for correction. From triangular relation, the scale of square side is derived:

$$\frac{1}{S_{side}} = \frac{\overline{P'_v P_m}}{\overline{P'_v P_n}} = \frac{y - y_v}{y_1 - y_v}, \text{ Side scale for pixel}(x, y) S_{side} = \frac{y_1 - y_v}{y - y_v} \tag{6}$$

The scale is only a function of one side of a square. What we want is the proportional relationship between pixels, i.e. the area ratio of pixel. So, we write:

$$S = S_{side} \times S_{side}, S = \left(\frac{y_1 - y_v}{y - y_v}\right)^2 \tag{7}$$

Equation (10) provides the ratio for pixel counting. In another word, one pixel is counted as S, instead of 1. However, since most of the pixels of the human body are not on the ground, the ratio just exported is only valid for pixels on the ground and not for human subjects. Fortunately, Ma et al. [21] not only derived the ground geometry correction, but also officially proved that it can be directly applied to all foreground pixels. This means that all foreground pixels can be mapped directly to the number of people after geometric correction. This article uses the progressive scan method to count the foreground pixels. The total number of pixels can be written as follows:

$$N_{total} = \sum_{i=1}^{M} N_i * S_i \tag{8}$$

Here, M denotes the number of rows. N_i refers to the number of foreground pixels form horizontal scan line i and S_i is the weight to be applied to each pixel on the line i.

3   Experiments for the Estimation of Crowd Densities

To test and validate our method for crowd density estimation, some experiments and result analysis have been done. In these experiments, test images set, consisting of 1600 frames selected from UCSD database, is classified into 4 levels (400 frames for each level) based on their crowd densities. In order have a deep impression on the definition of the 4 density levels, the examples with low, moderate, high and very high densities. For each level, we select 35% of frames as the training set and the remaining 65% frames as test samples.

Consequently, the crowd densities are classified into 4 level based on the method in Sect. 3. The experimental results are shown in Table 3. It provides the classification accuracy for our improved method. The estimated levels mostly distribute in main diagonals, demonstrating the validity of the method. It can be seen that our method can achieve great classification performance for each density level. Compared to the method, our method can gain higher classification rate of crowd density with effort for the improvement of method. Considering the factor affecting the classification performance, complex scenes and poor resolution should take main responsibility. Yet, at the same time there are still two possible points for improvement. First, we notice that at the lower-left corner of our selected ROI, It presents darker background than other regions because of the shadow, causing much missed detection. The methods based on background compensation can be selected to improve detection performance. Most previous works on crowd density with feature classification hardly take into account of removing outlier samples. So another work we can do is outlier detection and removal in the process of sample training. The behavior is of great significant to the training set with large number of samples.

Acknowledgements. We thank the funding agents for their support and the contributions of all authors. This work was supported by the open project of Key Laboratory of Computer Network and Information Integration (Southeast University), Ministry of Education under Grants K93-9-2018-10; the National Natural Science Foundation of China (NNSF) under Grants 61561041; The National key research and development plan of China under Grant 2017YFB0504203; XJCC Innovation Team of Geospatial Information Technology under Grant 2016AB001.

References

1. Zhang, Y., et al.: Single-image crowd counting via multi-column convolutional neural network. In: IEEE Conference on Computer Vision and Pattern Recognition IEEE Computer Society, pp. 589–597 (2016)
2. Yuan, Y., Wan, J., Wang, Q.: Congested scene classification via efficient unsupervised feature learning and density estimation. Pattern Recogn. **56**, 159–169 (2016)
3. Foroughi, H., Ray, N., Zhang, H.: Robust people counting using sparse representation and random projection. Pattern Recogn. **48**(10), 3038–3052 (2015)
4. Hashemzadeh, M., Farajzadeh, N.: Combining keypoint-based and segment-based features for counting people in crowded scenes. Inf. Sci. **345**, 199–216 (2016)
5. Zhang, Z., Li, M.: Crowd density estimation based on statistical analysis of local intra-crowd motions for public area surveillance. Opt. Eng. **51**(4), 047204 (2012)
6. Saleh, A., Mohsen, S.A., Suandi, S.A., Ibrahim, H.: Recent survey on crowd density estimation and counting for visual surveillance. Eng. Appl. Artif. Intell. **41**, 103–114 (2015)
7. Li, M., et al.: Estimating the number of people in crowded scenes by MID based foreground segmentation and head-shoulder detection. In: International Conference on Pattern Recognition IEEE, pp. 1–4 (2008)
8. Lin, S.-F., Chen, J.-Y., Chao, H.-X.: Estimation of number of people in crowded scenes using perspective transformation. IEEE Trans. Syst. Man Cybernet.-Part A: Syst. Hum. **31**(6), 645–654 (2001)
9. Ma, H., Zeng, C., Ling, C.X.: A reliable people counting system via multiple cameras. ACM Trans. Intell. Syst. Technol. (TIST) **3**(2), 31 (2012)
10. Rabaud, V., Belongie, S.: Counting crowded moving objects. In: IEEE Computer Society Conference on Computer Vision and Pattern Recognition IEEE Computer Society, 2006:705–711. LNCS. http://www.springer.com/lncs. Accessed 21 Nov 2016
11. Cheriyadat, A.M., Bhaduri, B.L., Radke, R.J.: Detecting multiple moving objects in crowded environments with coherent motion regions. In: Computer Vision and Pattern Recognition Workshops. CVPRW 2008. IEEE Computer Society Conference on IEEE, pp. 1–8 (2008)
12. Hou, Y.-L., Pang, G.K.H.: People counting and human detection in a challenging situation. IEEE Trans. Syst. Man Cybernet.-Part A: Syst. Hum. **41**(1), 24–33 (2011)
13. Chan, A.B., John Liang, Z.-S., Vasconcelos, N.: Privacy preserving crowd monitoring: Count- ing people without people models or tracking. In: IEEE Conference on Computer Vision and Pattern Recognition. CVPR 2008, pp. 1–7. IEEE (2008)
14. Lee, S., Lee, C.: Low-complexity background subtraction based on spatial similarity. EURASIP J. Image Video Process. **2014**(1), 30 (2014)
15. Zhao, C., Wang, X., Cham, W.-K.: Background subtraction via robust dictionary learning. EURASIP J. Image Video Process. **2011**(1), 972961 (2011)

Research on Digital Twin Technology for Production Line Design and Simulation

Xiang Li[1,2(✉)], Jinsong Du[1,2], Xiaolong Wang[1,2], Dongmei Yang[3], and Bintao Yang[3]

[1] Shenyang Institute of Automation, Chinese Academy of Sciences, Shenyang 110016, China
lixiang@sia.cn
[2] Institutes for Robotics and Intelligent Manufacturing, Chinese Academy of Sciences, Shenyang 110016, China
[3] China Guizhou Liyang Aero-Engine Co., Ltd., Guiyang 210016, China

Abstract. With the rapid development of big data, artificial intelligence and internet of things, digital twin technology becomes a new research hotspot in the field of intelligent manufacturing. In this paper, the digital twin technology for production line design and simulation is studied. Emphasis is laid on the building and fusion of production line model, virtual-real mapping and real-time interaction technology and virtual production line simulation and verification technology. The research content of this paper provides theoretical and technical reference for the application of digital twins in the design and implementation of manufacturing production line.

Keywords: Digital twin · Modeling simulation · Production line design · Intelligent manufacturing

1 Introduction

With the rapid development of industrial technology and new generation of information technology, automatic production lines in the fields such as aerospace and industrial manufacturing are more and more complicated. The life cycle cost of production lines including design, development, testing, operation and maintenance increases substantially. Meanwhile, the probability of design defect and production difficulty greatly increases because of the complexity of production lines.

In recent years, digital twin technology develops rapidly in the field of industrial manufacturing. It provides a new train of thought to solve the above problems. Gartner, a world-renowned consulting firm, has listed digital twin in the top ten strategic trends in science and technology for three years from 2017 to 2019 [1, 2]. Conceptual model of digital twin appears for the first time in 2003. It is proposed in the course of product lifecycle management (PLM) by professor Grieves [3]. It is defined as "in-formation mirror model" and "digital twin" in the literature [4]. NASA introduces digital twin conception in the space technology roadmap [5] for the first time in 2010. NASA wants to adopt digital twin technology to realize comprehensive diagnosis and pre-diction of flight system. In recent years, there are many research findings of applying digital twin

© Springer Nature Switzerland AG 2020
F. Xhafa et al. (Eds.): IISA 2019, AISC 1084, pp. 516–522, 2020.
https://doi.org/10.1007/978-3-030-34387-3_64

technology in the field of industrial manufacturing. A preliminary study on data fusion is carried out in the literature [6, 7]. In the literature [8], basic theory and key technology of applying digital twin technology to build cyber-physical systems (CPS) are analyzed from four aspects including physical fusion, model fusion, data fusion and service integration. In the literature [9], the definition of digital twin in the fields of aerospace, industry 4.0 and intelligent manufacturing is set forth, and the role of digital twin technology in industry system based on CPS is summarized.

Digital twin collects feedback data, and is assisted by artificial intelligence, machine learning and software analysis, to build a digital simulation on information platform. This simulation can automatically change as physical entity changes. Application of digital twin technology in production line design and simulation can reduce design time of production line, find design defects in time, improve performance of production line and reduce failure rate of production line. Digital twin technology is one of the research hotspots in the field of internet of things (IoT). But as a very new concept, theory and technology of digital twin in digital production line simulation are not complete. Few mature research results have been achieved. In view of this, research on digital twin technology for production line design and simulation is carried out in this paper. Three key technologies involved in the simulation of production line are elaborated emphatically.

2 Building and Fusion of Production Line Model

Digital twin virtual simulation model is the real mapping of actual production line. The modeling process is as follows. Firstly, the production line is modeled from four dimensions including geometry, physics, behavior and rules. And the model is evaluated and validated to ensure the correctness and validity of the model. Then each dimension model is related, combined and integrated so that they can be integrated into a complete virtual production line model with high fidelity in the information space.

The process of model fusion mainly involves the construction of multi-dimensional models, evaluation and validation, association and mapping, fusion and consistency analysis. To realize multi-dimensional model fusion of production line, actual production line should be described and modeled from geometric shape, physical properties, behavior response and rules, and the relevance and mapping relations between each dimension model must be analyzed fully. Geometric model and physical model describe heterogeneous elements of production line. Behavior model de-scribes driving and disturbing factors, which enable each factor to possess behavior-al characteristics, response mechanism and the ability to perform complex behaviors. Rule model characterizes the rules reflected by the actual production line and its model at the geometric, physical and behavioral levels, and maps the rules to the corresponding models, so that each model has the ability of evaluation, evolution, reasoning and so on. By establishing the relationship among the dimension models, multi-dimensional models of production line are integrated and fused to generate the virtual production line model. The virtual production line model adopts a unified three-dimensional representation to support visualization and simulation running.

After determining the model, in order to ensure the validity and correctness of the model, the multi-dimensional model is validated based on verification validation & accreditation (VV&A) in this paper. The model is validated from several aspects including accuracy of input and output, simulation confidence, sensitivity and simulation precision.

3   Virtual-Real Mapping and Real-Time Interaction Technology

In the actual production line, huge amounts of data are generated in the manufacturing process of products. These data can only show the current working status of the production line, and can not predict the future working status. Pure virtual model carries out simulation calculation according to pre-set. Space environment changes, manual intervention and other factors can not be integrated into the pure virtual model. Stated thus, physical entity and pure virtual model are two information is-lands. It is impossible to fully and thoroughly grasp the working status and performance indexes of product manufacturing process in real environment.

For this reason, bidirectional mapping between virtual model and real data should be established. Positive input mapping relation is using real data as simulation input data of virtual model. Reverse feedback mapping relation is using simulation results of virtual model as predicted performance indexes of actual production line. Through this bidirectional mapping relationship, virtual model and actual production line can be effectively integrated and fused. Afterwards, real-time interaction mechanism between virtual and real in product manufacturing process is established. Through digital twin technology of virtual-real fusion, working status and performance indexes in the production and manufacturing process can be mastered.

Working principle of bidirectional mapping between virtual model and real data is shown in Fig. 1.

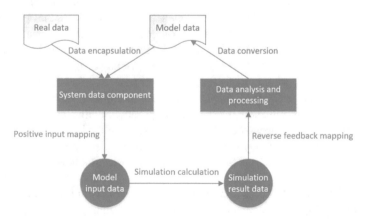

Fig. 1. Working principle of bidirectional mapping between virtual model and real data

(1) Real data is standardized and encapsulated into system data component. Data information is expressed in a unified interface form.
(2) Simulated input data of virtual model is expressed in a unified data structure.
(3) In the process of positive input mapping between real data and virtual model, information interaction is accomplished by data adapter.
(4) Simulation result data generated by virtual model is stored in XML file, which is processed and analyzed to generate feedback information. Thus reverse feedback mapping is established.

Principle of virtual-real interaction based on bidirectional mapping is shown in Fig. 2.

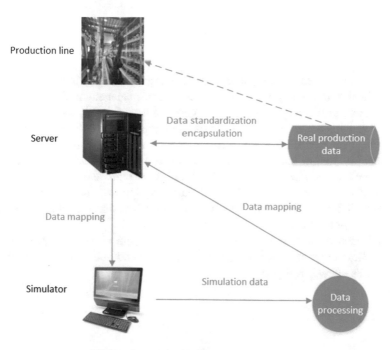

Fig. 2. Principle of virtual-real interaction

(1) Simulation model adopts the one-step solution method. The simulation process is synchronization with physical time.
(2) Based on bidirectional mapping between virtual model and real data, real status and data can be received by each simulation step and be input into virtual model, which is as input for solution of next simulation step. Running state of actual product line can be simulated in real time.
(3) Through the views of curves, animations and data tables, working state and performance indexes of actual product line are displayed dynamically.
(4) Historical data generated by the production process is recorded and analyzed to generate valuable information.

On the basis of realizing real-time interaction between virtual model and physical entity of production line, virtual reality synchronized display technology of production line is studied. With this technology, operators can manipulate virtual model to influence physical entities, observe and monitor digital models of production line from different angles and positions. By arranging a large number of high-precision sensors to collect operation data of production line in real time, virtual reality (VR) technology can display the real-time running state of production line in surreal form. VR technology can provide immersive virtual reality experience from aspects of visual, acoustic and tactile, realize real-time and continuous human-computer inter-action. With VR technology, operators can quickly grasp the principle, structure, characteristics, changing trend and health status of production line.

4 Simulation and Verification Technology of Virtual Production Line

Procedure design scheme of digital production line is researched to simulate every link in the production line and guide the construction and management of entity production line. Take parts processing as an example, virtual simulation of production line carries out virtual running according to the technological route of parts. The virtual simulation involves all main equipment such as machine tool, robot, logistics system and so on. The virtual simulation can optimize layout of equipment, work-place and logistics system, simulate automatic loading and unloading of parts and action of each equipment, and calculate production rhythm of parts, equipment utilization rate and other performance indexes. Thus virtual simulation can provide improvement basis for the design of real production line.

Virtual simulation of production line involves layout simulation, process simulation, robot simulation, production logistics simulation and human-computer interaction simulation, as shown in Fig. 3.

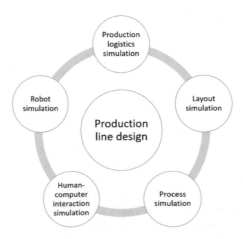

Fig. 3. Virtual simulation of production line

The simulation and verification technology of PLC is a key technology involved in the simulation and verification of production line. PLC simulation is the preview and mirror of real-time production line. PLC simulation for virtual production line is divided into four modules: control module, communication module, equipment library module and simulation module. The overall framework of PLC simulation and verification is shown in Fig. 4.

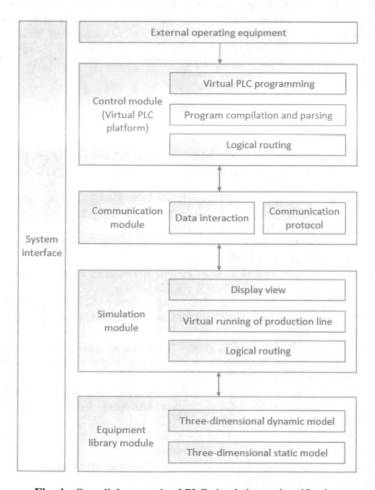

Fig. 4. Overall framework of PLC simulation and verification

5 Conclusion

Digital twin provides new ideas and tools for the innovation and development of current manufacturing industry. It has attracted more and more attention from industry and academia. Digital twin technology is the inevitable choice to cut down the development cycle and cost of automation production line and realize intelligent manufacturing and service in the future. With the continuous conquering of key technologies, digital twin technology will be an important means to realize digital design, manufacturing and service guarantee of production line in the future. Digital twin can make the innovative design, manufacture and reliability of production line rise to a new height.

Acknowledgments. This work was supported by National Key R&D Program of China (No. 2017YFB1303701), AECC Independent Innovation Special Fund Project (ZZCX-2018-035), the Project of Intelligent Manufacturing Integrated Standardization and New Model Application, Science and Technology Service Network Initiative (KFJ-STS-QYZD-107), Made-in-China 2025 Sichuan Action Fund Project (2017ZZ003) and K.C. Wong Education Foundation.

References

1. Beate, B., Vera, H.: Digital twin as enabler for an innovative digital shopfloor management system in the ESB logistics learning factory at Reutlingen-University. Procedia Manuf. **9**(1), 198–205 (2017)
2. Seongjin, Y., Jun, H. P., Won, T. K.: Data-centric middleware based digital twin platform for dependable cyber-physical systems. In: Ninth International Conference on Ubiquitous and Future Networks (ICUFN), pp. 922–926. (2017)
3. Grieves, M.W.: Product lifecycle management: the new paradigm for enterprises. Int. J. Product Dev. **2**(1–2), 71–84 (2005)
4. Grieves, M.W.: Virtually Perfect: Driving Innovative and Lean Products Through Product Lifecycle Management. Space Coast Press, Florida (2011)
5. Piascik, R., Vickers, J., Lowry, D., et al.: Technology Area 12: Materials, Structures, Mechanical Systems, and Manufacturing Road Map. NASA Office of Chief Technologist, Washington, DC (2010)
6. Ricks, T.M., Lacy, T.E., Pineda, E.J., et al: Computationally efficient solution of the high-fidelity generalized method of cells micromechanics relation, 05 Jan 2017. http://www.cavs.msstate.edu/publication/docs/2015/09/140441700_Ricks.pdf
7. Cai, Y., Starly, B., Cohen, P., et al.: Sensor data and information fusion to construct digital-twins virtual machine tools for cyber-physical manufacturing. In: Proceedings of the 45th SME North American Manufacturing Research Conference, vol. 10, pp. 1031–1042. (2017)
8. Tao, F., Cheng, Y., Cheng, J.F., et al.: Theories and technologies for cyber-physical fusion in digital twin shopfloor. Comput. Integr. Manuf. Syst. **23**(8), 1603–1611 (2017)
9. Negri, E., Fumagalli, L., Macchi, M.: A review of the roles of digital twin in CPS-based production systems. Procedia Manuf. **11**(1), 939–948 (2017)

Deployment Optimization Algorithm of Brigade First-Aid Stations for Battlefield Rescue

Zhou Wenming$^{(\boxtimes)}$, Zheng Lin, Wu Yaowu, Yuwen Jingbo,
and Lan Shibin

Joint Service Command and Training Centre, Joint Logistic College, National
Defense University of PLA, Taiping Rd., No. 23, Beijing 100858, China
zwmedu@163.com

Abstract. Save oneself and rescue each other in battlefield are called platinum ten minutes. Position data of brigade first-aid stations and combat units in battlefield are analyzed and studied. Based on the shortest route algorithm of graphics theory, a optimization design algorithm of first-aid station precinct is proposed, and a optimization model of place and number what needs to be added of first-aid station are built. It strongly supports the decision-making of battlefield rescue and Medical evacuation.

Keywords: Battlefield rescue · First-aid station · Shortcut algorithm · Rescue precinct · Optimization design

1 Introduction

It is the only described for battlefield rescue that platinum ten minutes and golden one hour. For wounded soldiers, carrying out cardiopulmonary resuscitation, stop bleeding in six minutes, and carrying them to the nearest first-aid station for professional rescue in another six minutes are efficient save oneself and rescue each other method. It can efficiently reduce casualty rate [1].

Currently, according to the position of combat units, it will play a vital role for confirming the medical evacuation route and carrying out rescue that deploying first-aid station and confirming its precinct. Based on position data analysis of combat units and deployed first-aid station, calculate the distance from each combat unit to the nearest first-aid station, and confirm each first-aid station precinct. To the combat unit that can't carry the wounded to the nearest first-aid station in 6 six minutes, have to optimize the number and position of first-aid stations deployed additionally, assure that the number of added first-aid station is least, and the balanced degree of work of all first-aid station is best. It can effectively help make the wounded rescue plan [2].

F. Xhafa et al. (Eds.): IISA 2019, AISC 1084, pp. 523–533, 2020.
https://doi.org/10.1007/978-3-030-34387-3_65

2 Data Analysis of Combat Units and Brigade First-Aid Stations

The place coordinates of ninety two combat units of one brigade are expressed in Table 1. Deployed twenty first-aid stations based on someone combat units among the brigade. The route between each combat unit and first-aid station is as Fig. 1.

Table 1. Serial number index and coordinates of combat units

Serial number	Abscissa	Ordinate	Serial number	Abscissa	Ordinate
1	414	360	47	326	373
2	404	344	48	321	375
3	384.5	352	49	343	373
4	382	378.5	50	346	383
5	340	377	51	349.5	381.5
6	336	384	52	352	378
7	318	363	53	349	370
8	335.5	354.5	54	371	364
9	334	343	55	372	354
10	283	326	56	355	375
11	248	302	57	364	383.5
12	220	317	58	358	388
13	226	271	59	352	383
14	281	293	60	370	389
15	291	336	61	336	396
16	338	329	62	382	382
17	416	336	63	392	376
18	433	372	64	393	367
19	419	375	65	396	362
20	445	392	66	399	363
21	258	273	67	402	360
22	230	279	68	406	361
23	226	266	69	411	356
24	213	291	70	409	351
25	228	301	71	416	352
26	257	302	72	419	348
27	251.5	307	73	423	355
28	244	329	74	419.5	357
29	247	338	75	406.5	365.5
30	315	368	76	406	369
31	316	352	77	410	371
32	327	356	78	418	365
33	328	351	79	421	371
34	329	343.5	80	425	373

(*continued*)

Table 1. (*continued*)

Serial number	Abscissa	Ordinate	Serial number	Abscissa	Ordinate
35	337	340	81	439	369
36	337	335	82	439.5	374
37	332	336	83	435	377
38	372	331	84	439	386
39	372	334	85	441	393
40	389.5	331.5	86	448	393
41	412	328.5	87	446	379
42	420	345	88	442.5	381
43	412	344	89	439	383
44	395	347	90	438.5	3798.5
45	343	343	91	434	349
46	343	349	92	422	342

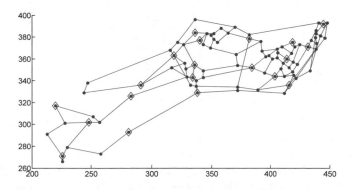

Fig. 1. Plane graph of combat units and brigade first-aid station

Where the diamond logo is the position of first-aid station. According to the relative position and route network among all combat units and first-aid stations, distribute first-aid station precinct, and the purpose is that each combat unit can carry the wounded to the nearest first-aid station in the precinct. This is a goal programming problems. Here adopt the shortest route algorithm of graphics theory to solve it.

3 Floyd Algorithm

The base thinking of the multi-source Floyed algorithm of shortest route [3] is as follows:

It defined two two-dimensional matrix:

Matrix D records the shortest route. for example: D[5][7] = 10, it explains that the shortest route between vertex 5 and vertex 7 is 10.

Matrix P records the intermediate transit point in shortest route between two vertexes. For example: P[5][7] = 1, it explains that the shortest route track between vertex 5 and vertex 7 is as: 0 -> 1 ->3.

Through triple cycle, with k as intermediary point, v as start point, w as terminal point, it compares D[v][w] with D[v][k] + D[k][w] to attain the least value. If D[v][k] + D[k][w] is the smaller value, then overwrite storage it to D[v][w].

Using the graphics to explain Floyd algorithm principle as follows [4, 5]:

Corresponding relations between Vertex name and subscript are as follow:

A B C D E F G
0 1 2 3 4 5 6

The first step: initialize matrix D and P (Fig. 2).

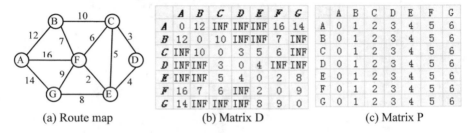

(a) Route map   (b) Matrix D   (c) Matrix P

Fig. 2. Initialization sketch map

The second step: renew matrix D and P with A as intermediary point.

The value D[B][G] is INF, but with A as intermediary point, D[B][A] + D[A][G] = 12 + 14 = 26, less than D[B][G] = INF, thus, D[B][A] + D[A][G] is the least value of B -> G, then it overwrite storage in D[B][G] with 26 (Fig. 3).

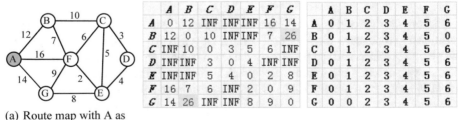

(a) Route map with A as
intermediary point   (b) Matrix D renewed   (c) Matrix P renewed

Fig. 3. Sketch map with A as the intermediary point

The third step: renew matrix D and P with B as intermediary point.

In matrix D After the second step, the value of D[A][C] is INF, but through vertex B, D[A][B] + D[B][C] = 12 + 10 = 22 is less than D[A][C] = INF, thus D[A][B] + D[B][C] is the least value of A -> C, overwrite storage in D[A][C] with 22 (Fig. 4).

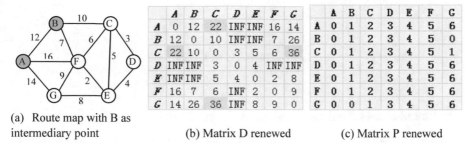

	A	B	C	D	E	F	G
A	0	12	22	INF	INF	16	14
B	12	0	10	INF	INF	7	26
C	22	10	0	3	5	6	36
D	INF	INF	3	0	4	INF	INF
E	INF	INF	5	4	0	2	8
F	16	7	6	INF	2	0	9
G	14	26	36	INF	8	9	0

	A	B	C	D	E	F	G
A	0	1	2	3	4	5	6
B	0	1	2	3	4	5	0
C	0	1	2	3	4	5	1
D	0	1	2	3	4	5	6
E	0	1	2	3	4	5	6
F	0	1	2	3	4	5	6
G	0	0	1	3	4	5	6

(a) Route map with B as intermediary point   (b) Matrix D renewed   (c) Matrix P renewed

Fig. 4. Sketch map with B as the intermediary point

In a similar way, renew matrix D and P with vertex C, D, E, F, G as intermediary point in turn. Through programming calculation, the final result is as follows (Fig. 5):

For example: the least distance value from vertex 0 to 3 is 22, that is D[0][3]. the route: 0 -> 5 -> 4 -> 3.

0	12	22	22	18	16	14
12	0	10	13	19	7	16
22	10	0	3	5	6	13
22	13	3	0	4	6	12
18	9	5	4	0	2	8
16	7	6	6	2	0	9
14	16	13	12	8	9	0

0	1	1	5	5	5	6	14
0	1	2	2	5	5	5	16
1	1	2	3	4	5	4	13
4	2	2	3	4	4	4	12
5	5	2	3	4	5	6	8
0	1	2	4	4	5	6	9
0	5	4	4	4	5	6	0

(a) Matrix D renewed   (b) Matrix P renewed

Fig. 5. Final result chart

4 Modeling of Jurisdiction Distribution of First-Aid Stations

The purpose of modeling is that gains Reasonable division of first-aid jurisdiction of first-aid stations, and assures that each combat unit can carry the wounded to designated first-aid station in 6 min by off-road rescue vehicle (average speed 30 km/h).

4.1 Mathematical Modeling

Supposed control variable x_{ij} $(i = 1, \cdots 20)$, that is

$$x_{ij} = \begin{cases} 1, & Unit\ j\ is\ under\ the\ jurisdiction\ of\ the\ first-aid\ station\ i \\ 0, & Unit\ j\ is\ not\ under\ the\ jurisdiction\ of\ the\ first-aid\ station\ i \end{cases} \qquad (1)$$

Where j is the combat unit number. Adopt Floyd algorithm to calculate the shortest path from j to i, denote as r_{ij}. The mathematical model is as follows:

$$f_j = \min_{i=1}^{20} r_{ij} x_{ij} \ (j = 1, 92)$$

$$s.t. \sum_{i=1}^{20} x_{ij} = 1 (j = 1, 92)$$

(2)

4.2   Solution and Analysis of the Model

Function (2) can be solved through 0–1 integer programming, but here adopt Folyd algorithm to calculate, and deal with the result properly. It is expressed in Table 2.

Table 2. Serial number index and coordinates of combat units

Unit num	First-aid station num	Distance	More than 6 min	Unit num	First-aid station num	Distance	More than 6 min
1	1	0	0	47	7	12.8062	0
2	2	0	0	48	7	15.0505	0
3	3	0	0	49	5	5	0
4	4	0	0	50	5	8.4853	0
5	5	0	0	51	5	12.2932	0
6	6	0	0	52	5	16.5943	0
7	7	0	0	53	5	11.7082	0
8	8	0	0	54	3	22.7089	0
9	9	0	0	55	3	12.659	0
10	10	0	0	56	5	20.837	0
11	11	0	0	57	4	18.6815	0
12	12	0	0	58	5	23.0189	0
13	13	0	0	59	5	15.2086	0
14	14	0	0	60	4	17.3924	0
15	15	0	0	61	7	40.8575	1
16	16	0	0	62	4	3.5	0
17	17	0	0	63	4	10.3078	0
18	18	0	0	64	4	19.3631	0
19	19	0	0	65	3	15.2398	0
20	20	0	0	66	3	18.402	0
21	14	30.4795	1	67	1	16.1942	0
22	13	8.9443	0	68	1	12.0711	0
23	13	5	0	69	1	5	0
24	13	23.8537	0	70	2	8.6023	0
25	12	17.8885	0	71	1	11.4031	0
26	11	9	0	72	2	16.0623	0
27	11	16.433	0	73	1	10.2961	0
28	15	47.5184	1	74	1	6.265	0

(*continued*)

Table 2. (*continued*)

Unit num	First-aid station num	Distance	More than 6 min	Unit num	First-aid station num	Distance	More than 6 min
29	15	57.0053	1	75	1	9.3005	0
30	7	5.831	0	76	1	12.8361	0
31	9	20.5572	0	77	19	9.8489	0
32	7	11.4018	0	78	1	6.4031	0
33	8	8.2765	0	79	19	4.4721	0
34	9	5.0249	0	80	18	7.2801	0
35	9	4.2426	0	81	18	7.2801	0
36	16	6.0828	0	82	18	12.1165	0
37	16	11.1818	0	83	18	6.7082	0
38	16	34.0588	1	84	20	11.4032	0
39	?	36.8219	1	85	20	4.1231	0
40	2	19.1442	0	86	20	3.1623	0
41	17	8.5	0	87	20	17.3044	0
42	17	9.8489	0	88	20	14.8478	0
43	2	8	0	89	20	10.8167	0
44	2	9.4868	0	90	20	14.3522	0
45	9	10.9508	0	91	17	39.1927	1
46	8	9.3005	0	92	17	25.3003	0

The result show that the distance is more than 3000 m from seven units to the nearest first-aid station respectively. That is they can't reach the nearest first-aid station in 6 min. However, they are allowed to be assigned to the nearest first-aid station. In addition, all combat units can reach their first-aid stations within 6 min.

5   Modeling and Analysis of Number and Location of Additional First-Aid Stations

In response to the problem that seven units cannot meet the requirement of reaching the nearest first-aid station respectively within 6 min, consider adding first-aid stations. The goal is to add the least number of it, and the added all meet the requirement. In addition, ensure the best work balance of all first-aid stations.

5.1   Optimization Model of the Number of Newly Added First-Aid Stations

Suppose n is the number of new first-aid stations, Mathematical modeling is as follows:

$$\min f(n) = ax(n) + bn \tag{3}$$

$$s.t. \begin{cases} n \in \{2,3,4,5\} \\ x(n), \textit{the number of units that cann't} \\ \qquad \textit{reach station within 6 minutes when added n newly} \end{cases}$$

Where a and b are weight coefficients. Suppose $a = 2$, $b = 1$, the results of matlab programming are as follows (Figs. 6, 7 and Table 3):

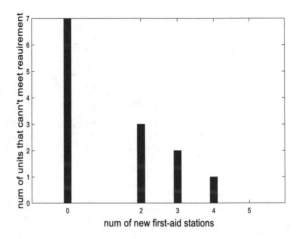

Fig. 6. The num of new first-aid stations and the num of combat units what cann't meet requirement

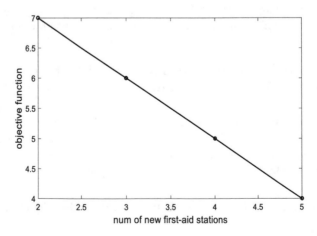

Fig. 7. The relation chart of objective function and num of new first-aid stations

Table 3. The results of the optimization modeling

The nun of new first-aid stations	The num of units that don't meet requirement
0	7
2	3
3	2
4	1
5	0

Visible, the optimal number of newly added first-aid stations is 5. At this time, all combat units can send the wounded to the nearest first-aid station within the stipulated 6 min.

5.2 Optimlzation Model of the Position of Newly Added First-Aid Stations

Ensuring the best work balance of all first-aid stations is very important for rescuing effectively the wounded. The workload balance is expressed by the variance of workload. The smaller the variance, the higher the work balance. Mathematical modeling is as follows:

$$\min f(C) = \sum_{j=1}^{N} (\sum_{i=1}^{M} \omega_i X_{ij} - \frac{\sum_{i=1}^{M} \omega_i}{N})^2 | C \tag{4}$$

$$s.t. \begin{cases} C \in \{C_{\{21,\cdots,N\}}^5\} \\ unsatisfied\_6\text{min } s\_num = 0 \\ \sum_{j=1}^{N} X_{ij} = 1 \\ X_{ij} \in \{0, 1\} \end{cases}$$

Where $i = 1, \cdots, M$ is first-aid stations, $j = 1, \cdots, N$ is combat units, objective function is the least workload variance under condition C. $C \in \{C_{\{21,\cdots,N\}}^5\}$. The model solving steps are as follows:

The first step: take one of all possible combinations;

The second step: calculate the best deployment scheme of first-aid stations under the combination. If some combat units exist that cann't meet requirement, repeat the second step;

The third step: calculate the workload balance under the scheme; if it less than the current least workload balance, then overwrite save it in the optimization projects;

The fourth step: repeat the above steps until all schemes are traversed.

The number of all feasible schemes satisfying the conditions is 48 after programming and calculation. The number of the best schemes is 8 expressed in Table 4, and the workload balance corresponding to optimal scheme is 5.9006 (Fig. 8).

Table 4. The best schemes and workload balance

Num	Workload balance 工作均衡度	Num of combat units				
1	5.9006	22	28	39	48	91
2	5.9006	22	28	39	48	92
3	5.9006	22	28	40	48	91
4	5.9006	22	28	40	48	92
5	5.9006	22	29	39	48	91
6	5.9006	22	29	39	48	92
7	5.9006	22	29	40	48	91
8	5.9006	22	29	40	48	92

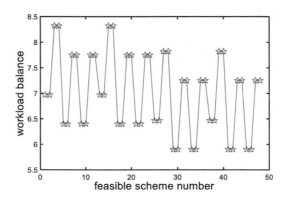

Fig. 8. The degree of workload balance

6   Conclusion

According to deployment information of brigade first-aid stations and combat units in battlefield, optimal jurisdiction distribution algorithm for first-aid stations is proposed, and a optimization model of place and number what needs to be added of first-aid station are built. It strongly supports the decision-making of battlefield rescue and Medical evacuation.

References

1. Zhang, S.H., Guo, S.S., Sun, J.S.: Health Service Tutorial, pp. 20–30. PLA Publishing House, Beijing (2010)
2. Chen, W.L.: Modern Health Service Front Along Theories, pp. 1–20. Military Medicine Science Publishing House, Beijing (2006)
3. Han, Z.G.: Method and Application of Mathematical Modelling, pp. 337–347. Higher Education Press, Beijing (2017)
4. https://blog.csdn.net/jeffleo/article/details/53349825

5. https://blog.csdn.net/qq_35644234/article/details/60875818
6. Jiang, Q.Y., Xie, J.X., Ye, J., et al.: Mathematical Model, pp. 10–12. Higher Education Press, Beijing (2018)
7. Zhuo, J.W., Wang, H.J.: Method and Practice of MATLAB Mathematical Modelling, pp. 219–230. Beihang University Press, Beijing (2018)

Cloud Computing, Mobile and Wireless Networking

Multiple Key Encryption (MKE) Technique: An Encryption-as-a-Service Delivery Model in Cloud Computing

K. Kuppusamy[1]([✉]) and J. Mahalakshmi[2]

[1] Department of Computational Logistics, Alagappa University, Karaikudi, India
kkdiksamy@yahoo.com
[2] Department of Computer Science, Alagappa University, Karaikudi, India

Abstract. Cloud Computing, a form of on-demand service offering resources such as platform, infrastructure, software, database etc., Cryptography, the art of altering the user defined plain data into inarticulate form, and that exists as a prime solution to protect the data on network transmission. This research work specifically focused on, an interesting, Encryption-as-a-service application delivery model, with a symmetric key algorithm based on the conventional encryption mode is developed. Key generation for the encryption is a most important phase of every cryptographic process. Multiple keys are involved for encryption of data to strengthen the encryption quality via the ICBC Process. It is tested with various inputs and the quality of the algorithm is investigated. The Motivation behind this application development is to minimize the execution time for the encryption within limited storage capacity. The proposed application model in this research work, avoids unauthorized attacks such as brute-force, chosen plaintext, CCA and CPA and the results are analyzed using the cryptanalysis process. The Performance of the proposed algorithm is analyzed with statistical measures such as frequency analysis. Comparative analysis is done to reveal the enhancement of the algorithm. The homogeneity of the data from the experimental results shows the efficiency of the data of how well it is encrypted. Evaluation reports state that the developed application service relic's potential for authenticating text files with better satisfaction while out-sourced as an application in cloud computing environment.

Keywords: ICBC · Multiple Key Encryption Algorithm · Matrices · Symmetric key encryption · Substitution operations · Cloud computing environment

1 Introduction

Cloud Computing offers Services on-demand to its users, based on pay-as-you-use scheme. Security-as-a-Service or Encryption-as-a-Service is an emerging paradigm in cloud computing environment, an application model that allows the user to encode their data to articulate format. There are certain limitations in the existing cloud computing security models which are related to the authenticity of the data [6]. This paper deals with a new symmetric key encryption algorithm to enhance the security of the data

© Springer Nature Switzerland AG 2020
F. Xhafa et al. (Eds.): IISA 2019, AISC 1084, pp. 537–545, 2020.
https://doi.org/10.1007/978-3-030-34387-3_66

when the client prefers the security service. The opportunity of hacking the data is more prevalent in the transmission spectrum. Cryptography offers the security of data through the conversion of data into an intelligible format. Public Key encryption and Private Key encryption are the types of encryption scheme, in which both the consigner and consignee share the same secret key that is related in an easily computable way in the first method and the second involve separate keys [11]. Moreover, the multi-tenancy model and the pooled computing resources in computing have introduced new security challenges [7] that require novel techniques. It is mandatory to use them in an optimized manner to avoid overheads [1]. The logical operations employed in key generation method in the proposed work offer high security for keys.

2   Related Research Works

Sastry, Udaya Kumar and Vinaya Babu [12], have developed block ciphers by an iterative method. Sravan Kumar, Suneetha and Chandrasekar, [15], reported on the utilization of the logical operator XOR logical operator and its working algorithm, in block ciphers. According to the authors, key scheduled algorithm is less prone to timing attacks. Srikantaswamy and Phaneendra, [16], demonstrated the use of one time padding that can be used for efficient encryption scheme by involving arithmetic and logical operations. Kameswari, Kumari and Kumar [3], explained the use of block ciphers in encryption and multi code generation based on secret key generation. Paul, Kumar and Mandal, [8], proposed a technique that is very secure and suitable for encryption of large files of any type. Session Based Symmetric Key cryptographic Technique (SBSKCT) considers the plain text as a string with finite number of binary bits. This input binary string is broken down into blocks of various sizes. Roy [14], demonstrates the use of a cryptographic method called UES – III (Ultra Encryption Standard III). It is a Symmetric key Cryptosystem which includes multiple encryptions, bit-wise randomization and a new advanced bit-wise encryption technique with feedback. Ramakrishna Das, Saurabh Dutta [12], extended their views on private key encryption with use of some logical operators.

Anupriya, Agnihotri, Soni and Babelay [2], put forward a novel approach for encryption using XOR based Extended Key. Satyajeet R. Shinge, Rahul Patil, in the year of 2014 [13], explained the encryption algorithm based on ASCII characters. Padhmavathi, B. Ray, Arghya, Anjum, Alisha, Bhat, Santhoshi in 2014 [9], explained the advancement of conventional encryption technique by using the Merkle-Hellman Knapsack Cryptosystem. Vaidehi and Rabi, in the year 2014 [17], experimentally verified the design and analysis of AES-CBC mode for high security applications. Li, Li and Shi [19], reported about tripartite secret key protocol in the private cloud file encryption. An encryption algorithm that is certificate less, being employed in the work by authors, yields better results for large scale environment.

3 Preliminaries

3.1 Access Structure

The basic mathematical rationale for the implementation of this proposed encryption Algorithm is in a matrix form. Each text file with varied input (text, numerical, alphanumeric, special characters) is converted to its corresponding binary bits, each of 8 bits. Let us assume that the input data is to be filled in 8×8 Matrix blocks with 64-bit input elements as binary bits. If M is a matrix, then Mnm is a 8×8 matrix, whose elements are represented using 8-bit binary structure, where m, n indicates the elements of the matrix. The Improved Cipher Block Chaining operation mode is followed after this design rationale to strengthen the key generation process.

3.2 Preprocessing

Block cipher is a form of substitution cipher where the data processed as chunks. Cipher Block Chaining is one of the encryption schemes that acts as a primary model for encrypting data stored in the cloud data centers as chunks. It needs an Initialization Vector (IV), with 8 elements of which the other elements are disjunctive. The key generation is random most of the time. The output of the first block remains as input to the consequent block. Every preceding block element is XORed, with the successor so that the decryption has to be dependent on the previous block. Hence, a single bit error in any block element will lead to decryption failure.

4 Encryption-as-a-Service in Cloud Computing Using New Cryptosystem MKE

4.1 Multiple Symmetric Key Encryption Algorithms

Multiple Symmetric key encryption algorithm is a newly proposed algorithm which employs the same key for both encryption and decryption. Consider the given input as in the text file. The process begins with the separation of input characters into binary bits, which is then filled in the matrix format of 8×8 blocks, with 64 elements. The elements referred here are in binary values and is placed sequentially across every row. The following is the algorithm to encrypt the given text file.

4.2 Algorithm

Step 1: Read the input text file consists of the original plain text to get encrypted.
Step 2: Convert the corresponding character (Including Numeric's, Characters and Alphanumeric) to corresponding Binaryvalues.
Step 3: Fill the converted binary strings in the basic access structure, Matrix form comprises of 8×8 blocks, with 64 elements, till end of file is reached.

4.3 Key Generation

Key generation is the most significant part of the cryptographic process. In this proposed algorithm key generation takes place in two steps. The first key generation is from the input matrix itself. The following procedure describes the key generation process in brief.

Step 1: Consider the 8×8 blocks, with 64 elements, from the input matrix.

Step 2: Set the Initialization vector with 8-bit binary input as 1×8 matrix.

Step 3: Assume that Xi is the initial 8-bit of the block matrix and Yi be the Initialization vector with 1×8 blocks.

Step 4: The first key is generated as a result of the improved cipher block chaining algorithm, which ends from the chaining process.

Step 5: The key thus generated from the input file is XORed with the input file blocks.

Step 6: Second key is generated with the help of pseudorandom number generators.

Step 7: The resultant partially encrypted blocks as a result of first key generation is again XORed with the key generated from the PRNG.

Step 8: Repeat the process, till end of file is reached.

Step 9: Convert the binary strings to corresponding ASCII codes.

Step 10: The evolved matrix blocks is completely ciphered blocks.

4.4 Decryption Algorithm

Step 1: Read the output text file consists of the encrypted text values.

Step 2: Convert the corresponding character (Including Numeric's, Characters and Alphanumeric) to corresponding Binary values.

Step 3: Fill the converted binary strings in the basic access structure, Matrix form comprises of 8×8 blocks, with 64 elements, till end of file is reached.

Step 4: The first key is generated as a result of the improved cipher block chaining algorithm, which ends from the chaining process.

Step 5: The key thus generated from the input file is XORed with the output file blocks.

Step 6: Second key is generated with the help of pseudorandom number generators.

Step 7: The resultant partially decrypted blocks as a result of first key generation is again XORed with the key generated from the PRNG.

Step 8: Repeat the process, till end of file is reached.

Step 9: convert the binary strings to corresponding ASCII codes.

Step 10: The evolved matrix blocks is completely deciphered blocks.

The following Fig. 1, is the block diagram of the proposed Multiple Key Encryption Algorithm.

4.5 Illustration for the Proposed Algorithm

The following example shows the working procedure of the proposed encryption algorithm. Let as consider text file with finite number of characters that includes

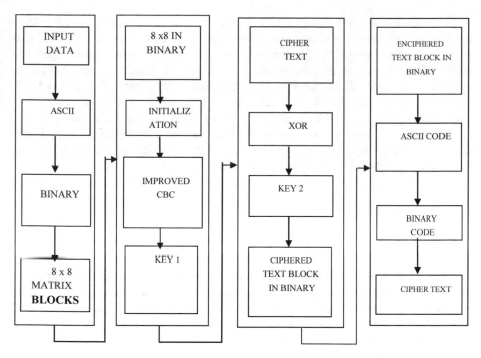

Fig. 1. Block diagram for encryption process with key generation

numbers, letters, and special characters all. The sample text considered here is "IN-PUTVALAlagappaUniversityresearch][.;/'-,". Every input character is converted to 8×8 matrix blocks. From the given input series the first eight characters INPUTVAL is taken and equivalent ASCII code is processed. Now the ASCII code is transformed into their binary values and occupied in the 8×8 matrix block in horizontal manner. Consider the given block as PBXi (Process Block X where I indicates 0–7 bits).

PBXi

0	1	0	0	1	0	0	1
0	1	0	0	1	1	1	0
0	1	0	1	0	0	0	0
0	1	0	1	0	1	0	1
0	1	0	1	0	1	0	0
0	1	0	1	0	1	1	0
0	1	0	0	0	0	0	1
0	1	0	0	1	1	0	0

PBYi

0	1	0	0	1	0	0	1
0	1	0	0	1	1	1	0
0	1	0	1	0	0	0	0
0	1	0	1	0	1	0	1
0	1	0	1	0	1	0	0
0	1	0	1	0	1	1	0
0	1	0	0	0	0	0	1
0	1	0	0	1	1	0	0

The initialization vector be [0 1 0 1 0 1 0 0]. The converted binary bits matrix PBXi now fed to the chaining operation mode. The first input bit values (0–7) is XORed with the initialization vector IV. The output is again XORed with the second input bits (0–7). This is continued until end of the loop is reached. Since, it is operated on the block cipher mode, all blocks were simultaneously transformed. The output is the resultant matrix block PBYi. Now the Block matrix PBYi is XORed with PBXi. The XOR logical operator is employed in every place since it is the self invertible operator. Hence, when the number of blocks increases the computational complexity of the process becomes low. Again, this is converted to binary format. For the decryption, the reversible order of the encryption steps was followed.

5   Results and Discussions

The proposed method is implemented for Windows 7 operating system with Core-i3 and 3 GB RAM, on Visual Studio 2010, with C# language. The experimental results are analyzed for various form of text with various volumes and their performance evaluation is done. The results show the security of the proposed encryption algorithm against various attacks. Efficacy of the proposed algorithm is demonstrated through the statistical measures such as the entropy/Frequency analysis, poker test measure etc. The experimental result is done with various standard input data and their outcomes are analyzed.

5.1   Experimental Results

This section presents the results and outcomes in Table 1, by employing the proposed Multiple Key Encryption algorithm. Another notable feature is that since every data is stored as matrix the computational complexity is also becoming less, the improved cipher block chaining works simultaneously as chunks and made the execution faster.

Table 1. Experimental results of encrypted and decrypted text with time in milliseconds

Plain text	Encrypted text	Decrypted text	Encryption time in (msec)	Decryption time in (msec)
sampleout	♥σ⌐ ⌈*ʇP ₋	sampleout	1340 ms	1345 ms
12345678	6_5♠⊣‖ ◊◘ ◄	12345678	595 ms	590 ms
Alagappa	#aJ_ð↑♠q	Alagappa	1245 ms	1245 ms
(19ma!`)	&t%_@E"	(19ma!`)	1036 ms	1036 ms
&89as + -6	©rwb↔◄Λ♥	&89as + -6	1559 ms	1560 ms

5.2   Performance Analysis

Encryption and Decryption Time
Table 2 depicts the implementation done on various volumes of text file in Kilobytes. The measurement was taken place in seconds.

Table 2. Table with varied volume of contents in file and the time taken

Size of file in kilo- bytes	Time in milli seconds
560 kb	13000 ms
187 kb	5000 ms
16 kb	2000 ms
9345 kb	18904 ms
126789 kb	98754 ms

5.3 Security Analysis

In the scenario of this proposed research work two important general Cipher Block chaining is against this CCA and CPA. Beside this, algorithm works on multiple key chaining feature and again XORed with the PRNG. This makes the system more complex, such that if a single bit error lies then the whole text will lead to collapsed. Moreover, when multiple keys are employed for the encryption, also in binary bits if a single element is misplaced then the intruder will end with wrongkeys.

5.4 Comparative Analysis

A comparative analysis is done between the proposed algorithm to that of various existing algorithms that work on bitwise operations basis. The proposed algorithm is compared with various existing works such as [5, 18]. The following Fig. 2, represents the difference analysis between the Extended MSA method (DJSA), Advanced Encryption Algorithm as well as our proposed scheme.

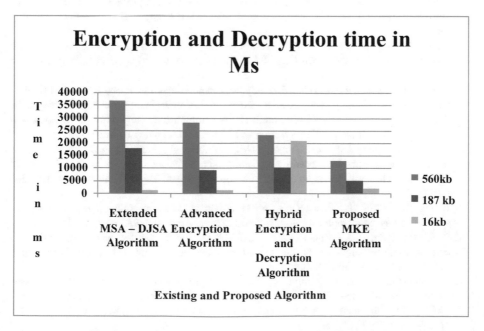

Fig. 2. Chart depicting comparative statistics in milliseconds.

6 Conclusion

In the existing cloud services, the security or the encryption services remain low while comparing the others. The Proposed Multiple Key Encryption Algorithm, presented in this research work basically alters the form of data and works better for data security in cloud paradigm. It is verified with various experiments that improve the encryption strength. The block ciphers are involved for the implementation, which works against the parallclization and the altering of entire data to 0's and 1's, leads to most important devastation if there is an even error in a single bit. Since consistency followed in the design rationale it automatically fastens the processing speed. The experimental results show minimum time for encryption of data and prove this presented algorithm is less prone to timing attacks. Hence, in cloud computing, this hybrid Improved cipher block chaining with multiple key encryption security services yield a better solution for encryption services to the users.

Acknowledgement. The author express deep sense of gratitude to the Alagappa University, Karaikudi, India for the financial assistance through the RUSA Phase 2.0, to carry out this research work.

References

1. Kakkar, A., Singh, M.L., Bansal, P.K.: Mathematical analysis and simulation of multiple keys and S-Boxes in a multinode network for secure transmission. Int. J. Comput. Math. **89** (16), 123–214 (2012)
2. Anupriya, E., Agnihotri, A., Soni, S., Babelay, S.: Encryption using XOR based extended key for information security – a novel approach. Int. J. Comput. Sci. Eng. **3**(1), 146–154 (2011)
3. Pratha, A.K., Kumari, R.C., Kumar, L.P.: Scheme of encryption for block ciphers and multi code generation based on secret key. Int. J. Network Secur. Appl. (IJNSA) **3**(6), 141–148 (2011)
4. Schiestl, C.: Pseudozufallszahlen. Der Kryptographie Klagenfurt Press, Klagenfurt (1999)
5. Chatterjee, D., Nath, J., Dasgupta, S., Nath, A.: A new Symmetric key Cryptography Algorithm using extended MSA method: DJSA symmetric key algorithm. In: International Conference on Communication Systems and Network Technologies. IEEE (2011)
6. Elashry, I.F., Faragallah, O.S., Abbas, A.M., ElRabaie, S., Fathi, E., El Samie, A.: A new method for encrypting images with few details using Rijndael and RC6 block ciphers in the electronic code book mode. Inf. Secur. J. Global Perspect. Taylor Francis **22**(1), 193–205 (2012)
7. Huang, K., Chiu, J., Shen, S.: A novel structure with dynamic operation mode for symmetric-key block ciphers. Int. J. Network Secur. Appl. (IJNSA) **5**(1), 17–36 (2013)
8. Paul, M., Kumar, J., Mandal, A.: Universal session based bit level symmetric key cryptographic technique to enhance the information security. Int. J. Network Secur. Appl. (IJNSA) **4**(4), 123–136 (2012)
9. Padhmavathi, B., Ray, A., Anjum, A., Bhat, S.: Improvement of CBC encryption technique by using the Merkle-Hellman Knapsack Cryptosystem, Intelligent systems and controls (ISCO). In: 7th International Conference 4–5 January, Coimbatore, Tamilnadu, India, pp. 340–344. IEEE (2013)

10. Kuppuswamy, P., Al-Khalidi, S.Q.Y.: Hybrid encryption/decryption technique using new public key and symmetric key algorithm. MIS Rev. **9**(2), 1–13 (2014)
11. Das, R., Dutta, S.: An approach of bitwise Private-key Encryption, technique based on Multiple Operators and numbers of 0 and 1 counted from binary representation of Plain Text's single character. Int. J. Innovative Technol. and Exploring Eng. **2**, 1–6
12. Sastry, V.U.K., Udaya Kumar, S., Vinaya Babu, A.: A large block cipher using modular arithmetic inverse of a key matrix and mixing of the key matrix and the plaintext. J. Comput. Sci. **2**(9), 698–703 (2006)
13. Shinge, S.R., Patil, R.: An encryption algorithm based on ASCII value of data. Int. J. Comput. Sci. Inf. Technol. **5**(6), 7232–7234 (2014)
14. Roy, S., Maitra, N., Agarwal, S., Nath, J., Nath, A.: Ultra Encryption Standard (UES) Version-III: advanced symmetric key cryptosystem with bit-level Encryption algorithm. Int. J. Modern Educ. Comput. Sci. **7**, 50–56 (2012)
15. Kumar, D.S., Suneetha, C.H., Chandrasekhar, A.: A block cipher using rotation and logical XOR operations. Int. J. Comput. Sci. Issues **8**(6), 142–147 (2011)
16. Srikantaswamy, S.G., Phaneendra, H.D.: Enhanced one time pad cipher with more arithmetic and logical operations with flexible key generation algorithm. Int. J. Network Secur. Appl. (IJNSA). **3**(6), 243–248 (2011)
17. Vaidehi, M., Rabi, B.J.: Design and analysis of AES-CBC mode for high security applications. In: Current Trends in Engineering and Technology 2nd International Conference, pp. 499–502 (2014)
18. Gupta, V., Singh, G., Gupta, R.: Advanced cryptographic algorithm to improve data security. Int. J. Adv. Res. Comput. Sci. Software Eng. **2**(1), 1–6 (2012)
19. Li, X., Li, W., Shi, D.: Enterprise private cloud file encryption system based on tripartite secret key protocol. In: International Industrial Informatics and Computer Engineering Conference, pp. 166–169. Atlantis Press (2015)

DNS Proxy Caching Method for False Invoice Detection

Qiurong Zhu[1], Zhichen Cao[2], Haibo Hou[1(✉)], and Jiangbing Yang[1]

[1] Beijing Lanxum New Technology Co., Ltd., Beijing, China
houhaibo@lanxum.com
[2] School of Cyber Engineering, Xidian University, Xian 710126, China

Abstract. With the rapid development of web technology, web applications have grown rapidly, electronic invoices have become increasingly diversified, and false electronic invoices on the Internet have also increased dramatically. How to quickly and accurately detect false electronic invoices is a challenge. However, a large number of electronic invoice query requests may cause the false invoice detection service to be blocked, and the query results cannot be returned in time and accurately. To this end, this paper proposes a false invoice detection method based on DNS proxy cache. The method firstly designs a DNS cache storage and retrieval method based on the access characteristics of the domain name of the high-heat electronic invoice. At the same time, through a reasonable cache replacement method based on table entry probability, to ensure that the hot invoice domain name is replaced with a lower probability. Experiments show that this method reduces the average query delay of electronic invoice query and improves the stability of false invoice detection service.

Keywords: Domain name system · Proxy caching · False invoice detection

1 Introduction

At present, the optimization technology for the DNS proxy cache [10–13] mainly uses the HashMap-based method to store all the attributes in the domain name response record [3–5]. Yan et al. [3] used the HashMap structure to store all the response resource records with the "domain name" as the key, and the "response IP list, current time, and life cycle" as the value; Yang et al. [4] will all the domain names. The response resource record is stored in the cache in the form of a HashMap. Although the hash table-based caching mechanism can speed up domain name retrieval, these solutions still have the following disadvantages:

(a) The hash table size is subjectively selected and chained to handle hash collisions. Subjectively selected hash table sizes can have an impact on cache performance. If the hash table is set too large, it may cause space waste. If the setting is too small, the conflict rate may increase.
(b) The resource record itself is not optimized for storage. When the storage domain name is large, the limited storage space will store fewer entries and the storage utilization is lower.

© Springer Nature Switzerland AG 2020
F. Xhafa et al. (Eds.): IISA 2019, AISC 1084, pp. 546–553, 2020.
https://doi.org/10.1007/978-3-030-34387-3_67

In addition, when performing cache record update replacement, the existing methods are mostly updated based on the lifetime of the response record [1, 2, 6]. The setting of the time to live is a compromise between query accuracy and query latency. Ballani [1] and Ramasubramanian et al. [6] pointed out that although the response records of some domain name resources can be stable for several months, the lifetime is only set to a few hours, which will reduce the hit rate of DNS records in the cache, which is not conducive to shortening the DNS. The query latency will also increase the load on the remote domain name server. In [2], to prevent the cache from being attacked by DDoS, the lifetime of the domain name record is increased, but this may result in an obsolete response record and inaccurate return results. At the same time, the domain name lifetime is specified by the remote server and will not change due to domain name popularity. However, in the actual application environment, domain name access usually has strong locality [7], and only a limited number of domain name request behaviors occur within a certain period of time. However, cache updates based on time-to-live cannot guarantee that hot-hot domain names are often hit. Therefore, the life-time based update strategy is not conducive to the cached domain name server has a higher hit rate and a smaller recursive query load in the case of querying a large number of domain names.

In summary, for the problem of large domain name query and complex domain name/IP correspondence, the storage pressure brought by the proxy cache domain name server needs to design a reasonable proxy cache storage and retrieval method to ensure a large number of domain name queries. When requested, the storage pressure of the cache is small. At the same time, in order to ensure the user's small query delay and accurate response result, it is necessary to implement a reasonable proxy cache record update replacement algorithm for the proxy cache design.

2   Cache Storage and Retrieval Methods

According to the characteristics of high-heat domain name obtained from the analysis of DNS information, this paper designs a Multi-Feature-Based DNS Proxy (MFBDP) based on multi-dimensional features. MFBDP uses the multi-level hash table classification and fast retrieval speed to design DNS proxy cache storage and retrieval technology. It mainly includes two parts: DNS proxy cache storage and retrieval method and cache update replacement method based on table entry probability. The workflow of MFBDP is shown in Fig. 1:

(1) Parsing the domain name, time, message type, etc. from the DNS packet;
(2) Filter the illegally queried domain name. If the query domain name hits the cache, it will directly respond to the result; if it does not, it will enter (3).
(3) Forwarding the missed legal domain name request to the cached domain name server configured by the operator, and then synthesizing the domain name and the response IP address of the response result into the proxy cache, and updating the frequency, time, and the like of each entry information.
(4) In the process of caching the domain name, if the frequency of the response of the secondary or tertiary domain is "$NXDOMAIN" exceeds the threshold within a

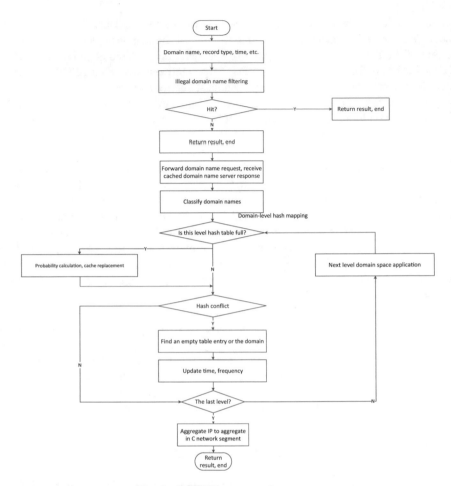

Fig. 1. MFBDP proxy cache process

certain period of time, the maximum public mode extraction of the prefix entry is performed by introducing the wildcard '*'. The corresponding record in the cache is updated, and only the aggregated maximum common mode prefix string is stored.

(5) When a certain level of hash table is full, a cache update replacement algorithm based on the probability of the entry is used to select an appropriate entry from the hash table for cache update.

The design and implementation of the MFBDP storage and retrieval method are described in detail below.

The overall module design of the MFBDP storage and retrieval method is shown in Fig. 2. It mainly includes the interaction module with the client, the DNS packet parsing module, the illegal domain name filtering module, the cache retrieval module, the forwarding domain name request module, the receiving DNS response module, and the query. The result is stored in the cache module and the asynchronous mechanism management module. The solid line in the figure indicates the DNS request process, and the dotted line indicates the DNS response process. Each module cooperates with the corresponding control flow () and the data stream () to complete the storage and retrieval function of the DNS proxy cache. Compared with the existing common DNS proxy cache structure, the MFBDP has the biggest difference in that the query result is stored in the cache module: MFBDP aggregates the domain name and the reply IP address, and uses the probability-based cache replacement method for caching in the storage process. A replacement update for a table entry.

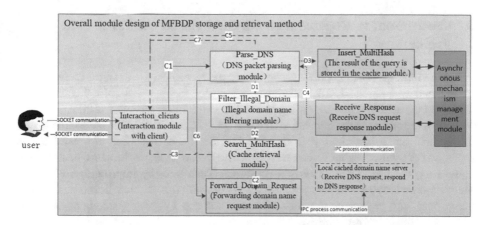

Fig. 2. MFBDP cache storage and retrieval overall module structure

The control flow is described as follows:

- C1: The DNS query packet sent by the user to the domain name request is transmitted to the DNS packet parsing module.
- C2: If there is no hit query domain name in the cache retrieval module, the forwarding domain name requesting module is notified to perform domain name forwarding query.
- C3: If the query domain name is hit in the cache retrieval module, the client interaction module is notified of the domain name response by the query result.
- C4: The receiving DNS request response module transmits the DNS response message of the local cached domain name server to the DNS packet parsing module.
- C5: If the query domain name is hit in the cache retrieval module, the client interaction module is notified of the domain name response.
- C6: Notifying the parsed non-type A domain name query request to the forwarding domain name requesting module, and then forwarding the domain name requesting module to the cached domain name server deployed by the operator.
- C7: The parsed non-A query type response result is transmitted to the client interaction module, and the client interaction module is notified to respond to the user.

The data flow is described as follows:

- D1: The domain name parsed by the DNS packet parsing module from the DNS request packet.
- D2: Domain name that conforms to the standard RFC definition format after filtering by the illegal domain name filtering module.
- D3: The DNS packet parsing module resolves the domain name and all response IP addresses from the DNS response message.

3   Cache Update Replacement Method Based on Table Entry Probability

When the hash table of the proxy cache is full, the records in the existing cache need to be updated and replaced. The existing DNS proxy caching technology updates and replaces the cache records according to the life cycle information of different records. However, according to the analysis of the life cycle of the DNS information, the heat of the domain name has nothing to do with the life cycle. The lifecycle-based update replacement method cannot guarantee that the cache replaces the low-heat domain name from the hash table as early as possible. Therefore, by improving the existing lifecycle-based cache update replacement algorithm, this section combines domain name heat with cache update, and proposes a cache update replacement algorithm based on table entry probability.

3.1   Principle of Probability Calculation

Combine the domain name heat, the cache record hit rate, and the cache record accuracy rate by the entry probability to ensure that the hot domain name is replaced with a lower probability and the accuracy of the response result is guaranteed. The popularity of the domain name is related to the base query base number and the domain name query frequency. Therefore, the substitution probability for each hash table item is shown in formula (1):

$$p_n = p_{n-1} * e^{-a*(t_n - t_{n-1})} \tag{1}$$

The definitions of each parameter in the cache replacement probability formula and the function description are shown in Table 1.

The cooling factor a and time difference $(t_n - t_{n-1})$ parameters are adaptively adjusted as the domain name query behavior changes. For example, the high-heat domain name at a certain time becomes smaller as time passes, and the cooling factor a becomes larger, and the time interval t becomes larger. Therefore, the popularity of the domain name gradually cools down to become a low-heat domain name, and the possibility of replacement becomes correspondingly larger.

Table 1. Cache replacement probability formula parameter description

Probability formula parameter	Definition	Functional description
P_{n-1}	Represents the probability of a hash entry calculated at the time of the last replacement, initially 1	The cache replacement probability is used to determine whether the domain name in the cache needs to be replaced. The larger the p, the higher the heat of the domain name, and the less likely it is to be replaced
P_n	Represents the hash table entry probability calculated at the time of current replacement	
a	Cooling factor is a description of the heat of the domain name	In inverse proportion to the query base x, assuming that the calculation of a satisfies the power law formula $a = e^{-\gamma x}$ (γ is a power law parameter), the larger the query base of the high-heat domain name, the smaller the value of the corresponding a
t_{n-1}	Represents the last query time of the item	$(t_n - t_{n-1})$ can effectively reflect the frequency of querying domain names. The faster the query frequency, the smaller the time interval, the higher the domain name heat, and the larger the cache replacement probability p. The less likely a hot domain name is to be replaced
t_n	Represents the current query time	

3.2 Probability Calculation Parameter Optimization

The probability calculation parameters of each entry are related to the time difference and the cooling factor a. The time difference is an exact parameter, but the power factor relationship parameter γ between the cooling factor and the query frequency is a variable. In this paper, the optimal power rate relationship $a = e^{-\gamma x}$ is selected by the cache hit ratio. According to the trend of different power rate curves, the representative γ values are shown in Fig. 3, which are (1) $a = e^{-0.01x}$, (2) $a = e^{-0.1x}$, (3) $a = e^{-x}$, (4) $a = e^{-10x}$.

Obtain 100,000 domain names in a continuous DNS traffic from a carrier's high-speed network, and then deploy the MFBDP in a LAN with controllable range. The hit rate of the DNS proxy cache is calculated by continuously initiating a domain name request on the client of the local area network on which the MFBDP is deployed.

The results of the hit rate of actively detecting 100,000 domain names are shown in Table 2. (1)–(3) As the long tail of the curve becomes more and more obvious, the cache hit rate gradually increases. When γ is 80.3%, it is optimal. The long tail of (4) is the most obvious, but the cache hit rate is reduced at this time. Therefore, the power rate relationship between the query base and the cooling factor is the best when $a = e^{-\gamma x}$, and the cache hit rate is the highest.

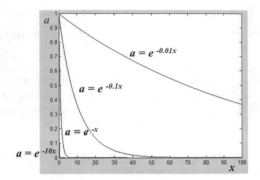

Fig. 3. Different trends of $a = e^{-\gamma x}$

Table 2. Effect of γ change on cache hit ratio

$a = e^{-\gamma x}$	Cache hit ratio
(1) $\gamma = 0.01$	67.78%
(2) $\gamma = 0.1$	77.79%
(3) $\gamma = 1$	80.30%
(4) $\gamma = 10$	73.62%

At the same time, in order to prevent the jitter of the update replacement and ensure the accuracy of the response record, the proxy cache needs to be completely updated in a certain period (such as January).

In order to verify that the probability-based cache replacement algorithm has a good query hit rate, it compares it with the commonly used replacement algorithm for query hit ratio. When the replacement algorithm is (1) based on probability replacement, (2) based on TTL replacement, (3) based on LRU replacement, (4) sending 100,000 identical domain name requests based on FIFO replacement, respectively calculating the query hit ratio. The query hit rate results for different replacement algorithms are shown in Table 3. The probability-based cache replacement algorithm is optimized by 10.81% compared to the commonly used TTL-based replacement algorithm. Therefore, the probability-based cache replacement method has a better optimization effect on reducing the response delay of the proxy cache.

Table 3. Query hit ratios for different replacement algorithms

Replacement algorithm	Cache hit ratio
Probability-based	80.30%
TTL-based	69.49%
LRU-based	63.40%
FIFO-based	22.02%

4   Conclusion

Based on the characteristics of high-heat domain names, this paper proposes a false invoice detection method based on DNS proxy cache. The method reduces the storage pressure of the cache, ensures that the domain name responds faster, and enhances the stability of the domain name query behavior. The experiment proves that the probability-based cache replacement algorithm is optimized by 10.81% compared with the commonly used TTL-based replacement algorithm, which plays an important role in reducing the response delay of the proxy cache.

Acknowledgments. This work is supported by National Key R&D Program of China (No. 2017YFB0802700).

References

1. Ballani, H., Francis, P.: A simple approach to DNS DoS defense. In: Proceedings of HotNets (2006)
2. Pappas, V., Massey, D., Zhang, L.: Enhancing DNS resilience against denial of service attacks. In: Annual EEE/IFIP International Conference on Dependable Systems and Networks, pp. 450–459 (2007)
3. Yan, B., Fang, B., Li, B., Wang, Y.: Detection and defence of DNS spoofing attack. Comput. Eng. **32**(21), 130–132 (2006)
4. Bellis, R.: DNS Proxy Implementation Guidelines (2009)
5. Stolikj, M., Verhoeven, R., Cuijpers, P.J.L., et al.: Proxy support for service discovery using mDNS/DNS-SD in low power networks. In: 15th International Symposium on World of Wireless, Mobile and Multimedia Networks, pp. 1–6 (2014)
6. Ramasubramanian, V., Sirer, E.G.: The design and implementation of a next generation name service for the internet. ACM SIGCOMM Comput. Commun. Rev. **34**(4), 331–342 (2004)
7. Yuchi, X., Li, X., Yan, B., et al.: Internet usage measurements in DNS services. Comput. Eng. Appl. **45**(34), 85–88 (2009)
8. Long, L.: A protection method and device against changing prefix domain name attacks. Chinese Patent: CN201110190540.7, 09 January 2013
9. Zou, D., Long, W.J., Ling, Z.: Winnowing-based similar text positioning method. In: International Conference on Internet Technology and Applications, pp. 1–5 (2010)
10. Klein, A., Shulman, H., Waidner, M.: Internet-wide study of DNS cache injections. In: IEEE INFOCOM 2017-IEEE Conference on Computer Communications, pp. 1–9 (2017)
11. Su, J., Li, Z., Grumbach, S., Salamatian, K., Han, C., Xie, G.: Toward accurate inference of web activities from passive DNS data. In: 2018 IEEE/ACM 26th International Symposium on Quality of Service (IWQoS), pp. 1–6 (2018)
12. Felton, M., Tang, J., Baker, L., Edwards, C.: Evaluating DNS and cache coherence. J. Comput. Sci. Software Eng. **10**(3) (2018)
13. Hao, S., Wang, H.: Exploring domain name based features on the effectiveness of DNS caching. ACM SIGCOMM Comput. Commun. Rev. **41**(1), 36–42 (2017)

Research on Multi-source Data Fusion Technology Under Power Cloud Platform

Xiaomin Zhang[1], Qianjun Wu[2(✉)], Xiaolong Wang[2],
and Yuhang Chen[2]

[1] State Grid Information System Integration Company,
NARI Group Corporation, Nanjing City, Jiangsu Province, China
[2] Information System Integration Company, NARI Group Corporation,
Nanjing City, Jiangsu Province, China
wuqianjun@sgepri.sgcc.com.cn

Abstract. The power cloud platform has a large number and variety of software and hardware resources, and the relationship is complicated. The existing data sensing methods have become more and more difficult to meet the real-time and accuracy requirements of the power information system when processing the massive monitoring data generated by it. In the global sensing process of power cloud platform, the most important thing is the aggregation, processing and analysis of monitoring data. Therefore, this paper studies the multi-source data fusion technology in the global joint sensing technology of power cloud platform and proposes a multi-source data fusion architecture and method suitable for power cloud platform.

Keywords: Data fusion · Global sensing · Power cloud platform

1 Introduction

With the rapid spread of cloud platforms, power companies have entered the era of "large networks, large systems, large concentration, high reliability, and high security" [1]. The power cloud platform system is becoming more complex and larger. At the same time, with the gradual completion of power cloud platform construction, the existing power cloud platform has changed from system construction to system operation and maintenance. System operation and maintenance plays an important role in the development of power cloud platform, and gradually expands into a global role. However, in the actual production environment, due to the complicated relationship between the cloud platform network and the device resources, it is often difficult to quickly locate the system failure point. In most cases, O&M personnel can only manually troubleshoot faults through expert experience [2]. This process usually takes a lot of time, and the reliability of the system is often difficult to guarantee. In order to realize the early warning and quickly positioning of faults and improve the reliability of power information systems, it is necessary to detect the internal operating status of the system in real time by globally sensing the application servers, database servers, storage, network channels and other resources involved in the existing power information systems.

© Springer Nature Switzerland AG 2020
F. Xhafa et al. (Eds.): IISA 2019, AISC 1084, pp. 554–559, 2020.
https://doi.org/10.1007/978-3-030-34387-3_68

To archive the global sensing of the power cloud platform, the most important thing is to aggregate, process and analyze the monitoring data in the power cloud platform. However, the current power cloud platform equipment is large in number and variety, and massive monitoring data is generated every moment [3]. How to effectively integrate such huge monitoring data has become an urgent issue.

Data fusion [4] is to achieve the precise location of the target, identity estimation, and real-time battlefield situation and enemy threat assessment. The process of inter-connecting and synthesizing data of single or multiple information sources can be roughly divided into five steps: Quasi, interconnect, filter, identify, evaluate. In the power cloud platform, multi-source data fusion is an information processing process that automatically analyzes and comprehensively processes different node-aware data obtained in time series under certain rules to complete real-time evaluation of the cloud platform status.

This paper studies the multi-source data fusion technology, discusses its data fusion architecture and data fusion method under the power cloud platform to realize the further integration of multi-source data in the power cloud platform, and comprehensively evaluates the entire information system to improve the reliability of the information system.

2 Multi-source Data Fusion Architecture

According to the different levels of data representation in data fusion, its architecture can be divided into 3 categories [5]:

(a) Raw data fusion. As shown in Fig. 1, this architecture directly combines the raw data of each sensor to form new raw data (the raw data of each sensor must match each other). Then perform feature extraction (usually get a feature vector) and make decisions (or recognition) based on this feature. Finally, new data that containing multiple source information will be generated.

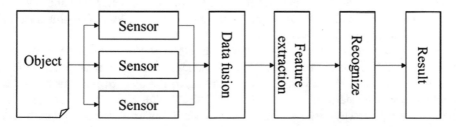

Fig. 1. Raw data fusion

(b) Feature fusion. As shown in Fig. 2, This architecture first extracts raw data separately from each sensor, then fuse these feature vectors to form a new vector (usually vector connection and compression). In this architecture, it only performs data fusion on the extracted features, so the information loss is inevitable, but its occupied bandwidth is relatively small.

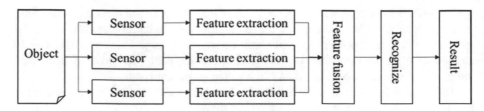

Fig. 2. Feature fusion

(c) Scheme fusion. As shown in Fig. 3, this architecture only performs data fusion on decision of each sensor. Since the information has been transmitted twice before data fusion (feature extraction and recognition), the information loss is serious. But correspondingly, the bandwidth occupied is also minimal on data fusion.

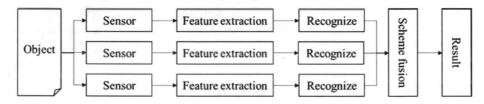

Fig. 3. Scheme fusion

The above three architectures, the raw data fusion is the best in accuracy, and the system can perceive all internal details to make detailed and accurate evaluation of the system operation status. However, considering the numerous sensors in power cloud platform, the processing of massive monitoring data not only consumes a large amount of computing resources, but also blocks the entire information network. The performance of whole system will be greatly reduced and it's also difficult to archive real-time dynamic monitoring of the power cloud platform. The latter two architectures, although reducing the bandwidth requirements, can not avoid loss of information, and the final accuracy is also lower.

Considering the actual operation of the power cloud platform, this paper proposes a new data fusion architecture (a hierarchical architecture) for the power cloud platform based on the above three architectures. According to the modern information system level protection standards (such as TCSEC, ISO/IEC 15408, GB/T 18336 and etc.) and the granularity strategy of power information system, under the consideration of efficiency and accuracy, this paper divides the sensing level of power cloud platform into three levels: (a) node layer,the operating state of device or application of cloud platform bottom, such as switches, MySQL, OS; (b) component layer,the operating state of the complete service, application or operating system which constituted by bottom device and application, such as business system, middleware, distribution service bus; (c) platform layer, the operating state of the whole platform which constituted by all components.

As shown in Fig. 4, sensor data fusion is performed at the node level, feature fusion is performed at the component layer, and scheme fusion is performed at the platform layer. The data fusion of each layer is directly calculated by its layer without directly uploading to the computing center.

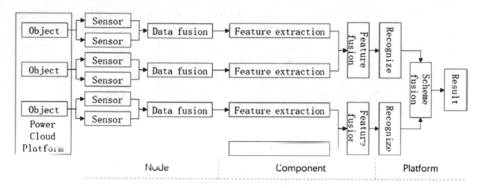

Fig. 4. Hierarchical data fusion

On node layer, it can be aggregated at the first time of monitoring data generation. The cloud platform can maximize the use of its subordinate IoT facilities, all sensor information will be gathered and processed at the end of net (terminal device or application), so that the cloud platform can sense all real-time information in a short time.

On component layer, many underlying devices or applications form a complete cloud platform component. One underlying object may be included by multiple components. Each component can be considered as a minimal system which has its own computing center, it extracts features from each node and try to fuse the features together. All gathered features should be compressed and recoverable with less information loss.

On platform layer, the whole platform is formed with basic components from previous layer. The platform identifies component runtime status by some predefined rules, then the status is gathered and processed to form the overall status of the platform. When scheme fusion is running, there are many methods to process, such as Bayesian reasoning [6], Kalman filtering [7], neural network [8–10].

The above hierarchical architecture realizes the layered sensing and data fusion of the cloud platform, avoiding the impact of massive monitoring data on network system. At the same time, it also improves the real-time nature of the cloud platform.

3 Multi-source Data Fusion Method

In previous section, this paper proposed a hierarchical architecture to realize the layered sensing and data fusion of the cloud platform, but there is no implementation method for data fusion. This section will study the specific methods of data fusion to find a data fusion method suitable for power cloud platform layers (node, component, platform).

On node layer, this paper, which combines the vendor evaluation criteria and expert experience, weights the original data to obtain the final fusion value. Assume that there are n sensors to measure an object, and the data output by the i-th sensor is X_i, where $i = 1, 2, 3, \cdots, n$. The output measurement value of each sensor is weighted and averaged, the weighting coefficient is W_i, the weighted average fusion result is

$$\bar{X} = \sum\nolimits_{i=1}^{n} W_i X_i \tag{1}$$

The above node layer data fusion formula requires that all X_i should be matched. For the unmatched detection data $\bar{X}_j$, the data is directly spliced (usually vector splicing) to obtain the final fusion result $\bar{\chi}$.

$$\bar{\chi} = \sum \bar{X}_j \tag{2}$$

For component-layer feature fusion, this paper uses a deep autoencoder (as shown in Fig. 5) to compress the stitched features, and further reduce the transmission bandwidth under the condition of ensuring less information loss. In order to reduce the performance loss caused by less feature fusion, only a 2-layer deep autoencoder is used to compress the input features. The deep autoencoder consists of two symmetric deep confidence networks, one of which has two shallow layers that form the part responsible for encoding and the other two layers of which are the decoding part.

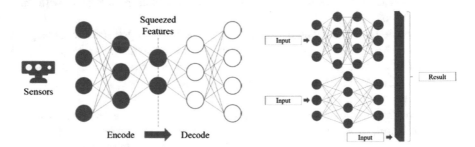

Fig. 5. Deep autoencoder for component fusion

Fig. 6. Deep network for scheme fusion

For platform-layer decision fusion, this paper first normalizes the decision (or scheme), forms a new feature vector by splicing, and then uses a simple deep neural network (as shown in Fig. 6) to identify and classify it for final recognition result. This paper uses a 3-layer neural network (N3), a 2-layer neural network (N2) and a fully-connected network (FC) to form a deep neural network, and directly contact the output of the 3-layer network and the 2-layer network and the original input features, then import them into a fully connected network. The raw data directly input into the fully connected network can be considered as a special 0 hidden layer neural network (N0) whose input is equal to the output. Adding the outputs of N0, N2, and N3 to form an approximate residual network can effectively reduce information loss during neural network transmission.

4 Conclusion

Multi-source data fusion is not only the core of the power cloud platform to achieve global sensing, but also a sufficient condition to improve platform reliability. Based on the traditional data fusion method and the current situation of power cloud platform and machine learning method, this paper proposes a multi-source data fusion architecture and method suitable for the existing power cloud platform to further improve the global sensing accuracy and real-time performance of the power cloud platform. It provides strong support for the safe and efficient operation of the power cloud platform.

Acknowledgment. This research was financially supported by the Science and Technology projects of State Grid Corporation of China (NO. 500409081).

References

1. Zhao-Hui, Y., Qin-Ming, J.: Power management of virtualized cloud computing platform. Chin. J. Comput. **6**, 015 (2012)
2. Ericson, C.A.: Fault tree analysis. In: System Safety Conference, Orlando, Florida, vol. 1, pp. 1–9 (1999)
3. Simmhan, Y., Aman, S., Kumbhare, A., et al.: Cloud-based software platform for big data analytics in smart grids. Comput. Sci. Eng. **15**(4), 38 (2013)
4. Castanedo, F.: A review of data fusion techniques. Sci. World J. **2013** (2013)
5. Salerno, J.: Information fusion: a high-level architecture overview. In: Proceedings of the Fifth International Conference on Information Fusion. FUSION 2002. (IEEE Cat. No. 02EX5997), vol. 1, pp. 680–686. IEEE (2002)
6. Mahler, R.P.S.: Statistical Multisource-Multitarget Information Fusion. Artech House Inc, Norwood (2007)
7. Sun, S.L., Deng, Z.L.: Multi-sensor optimal information fusion Kalman filter. Automatica **40** (6), 1017–1023 (2004)
8. Zhang, J.: Improved on-line process fault diagnosis through information fusion in multiple neural networks. Comput. Chem. Eng. **30**(3), 558–571 (2006)
9. Broussard, R.P., Rogers, S.K., Oxley, M.E., et al.: Physiologically motivated image fusion for object detection using a pulse coupled neural network. IEEE Trans. Neural Networks **10** (3), 554–563 (1999)
10. Carpenter, G.A., Martens, S., Ogas, O.J.: Self-organizing information fusion and hierarchical knowledge discovery: a new framework using ARTMAP neural networks. Neural Networks **18**(3), 287–295 (2005)

Research on Task Allocation and Resource Scheduling Method in Cloud Environment

Pengpeng Wang[1,4], Kaikun Dong[1], Hongri Liu[1], Bailing Wang[1,2], Wei Wang[1(✉)], and Lianhai Wang[3(✉)]

[1] Harbin Institute of Technology, Harbin, China
wpp_hiter@163.com
[2] Harbin University of Technology (Weihai) Innovation Pioneer Park Co. Ltd., Harbin, China
[3] Qilu University of Technology (Shandong Academy of Science), Jinan, China
[4] Luoyang Electronic Equipment Test Centre of China, Luoyang, China

Abstract. Task allocation and resource scheduling capability are important indicators for evaluating cloud environment. Aiming at the problems of low resource utilization, high algorithm time complexity and low task allocation efficiency of existing task allocation strategies, a task allocation and resource scheduling method based on dynamic programming in cloud environment is proposed. Using the idea of dynamic programming, this method regards the matching of tasks and servers as a combination of multi-stage decision-making, and obtains the optimization scheme of task allocation, which reduces the completion time of tasks. The experimental results show that the proposed method can reduce the task completion time and the resource load is relatively balanced, which can effectively improve the task execution efficiency.

Keywords: Cloud environment · Dynamic programming · Task allocation · Resource scheduling

1 Inrtoduction

The cloud environment refers to the Internet or large data environment that can provide computing power, storage capacity or virtual machine services to users or various application systems on demand from the dynamically virtualized resource pool [1]. The scalability and dynamics of hyperscale resources in cloud environment and the scheduling adaptive requirements of super-large computational tasks make task allocation and resource scheduling much more complicated than general algorithms. Task allocation and resource scheduling problems have been proved to be a non-deterministic polynomial (NP) problem [2]. How to make more efficient use of resources in cloud computing systems has been a research hotspot.

Zheng et al. proposed an improved task scheduling algorithm based on NSBBO (Non-dominated Sorting Biogeography-based Optimization, NSBBO). The uniform distribution weighting strategy is used to obtain the weighting coefficient group of the scheduling target according to the qualitative user preference [3]. Shen et al. proposed an improved ant colony-based cloud service composition algorithm, improved the ant

© Springer Nature Switzerland AG 2020
F. Xhafa et al. (Eds.): IISA 2019, AISC 1084, pp. 560–567, 2020.
https://doi.org/10.1007/978-3-030-34387-3_69

optimization path through social cognition optimization, and adopted an optimized ant colony pheromone update strategy to improve the algorithm search efficiency [4]. Shen et al. proposed that the immune evolution algorithm is beneficial to the artificial immune principle to complete the task global optimization, reducing the scheduling time overhead [5]. Wang et al. proposed a cloud computing resource scheduling method based on the dual fitness dynamic genetic algorithm. The weighted method is used to fuse the two goals of energy consumption and time consumption, improve convergence accuracy [6]. Kong et al. proposed a task scheduling strategy based on virtualized fuzzy prediction model, and established a fuzzy rule prediction model to complete task scheduling [7]. Liu proposed a cloud computing task scheduling strategy based on genetic algorithm, which improved the low efficiency of traditional genetic algorithm and improved the convergence speed [8]. The above algorithm unilaterally improves time performance or resource utilization, and the overall performance of the algorithm needs to be further optimized.

Based on the principle of optimality, this paper proposes a dynamic programming scheduling algorithm (DPSA) to solve the task allocation and resource scheduling problems in cloud environment. While ensuring that the task execution time meets the user's requirements, the load balancing of resources is guaranteed. Finally, simulation experiments verify the feasibility of the proposed algorithm.

The remainder of this paper is organized as follows: Sect. 2 describes the DPSA and shows the method of task allocation and resource scheduling based on DPSA. Section 3 verifies the method by experiments and analysis the results. Section 4 concludes this paper.

2 Methods

2.1 Dynamic Programming Scheduling Algorithm (DPSA)

The dynamic programming scheduling algorithm solves the problem by splitting the problem and defining the relationship between the state of the problem and the state, so that the problem can be solved in a recursive (or divide-and-conquer) manner. In general, for a class of problems that are suitable for solving with DPSA, the sub-problems that are decomposed are often interrelated. The problem solved by DPSA is a multi-stage decision problem. The problem to be solved is decomposed into several sub-problems, and the sub-phases are solved in order. The solution of the previous sub-problem provides useful information for solving the latter sub-problem. When solving any sub-problem, various possible local solutions are listed, and those that are likely to reach the optimal local solution are reserved by decision, and other local solutions are discarded. The sub-problems are solved in turn, and the last sub-problem is the solution to the initial problem. These decisions form a sequence of decisions and determine an optimal path of activity to complete the process [9]. The process of multi-stage decision making is shown in Fig. 1.

The optimization problem is the most suitable field for DPSA. In general, it should follow the following steps:

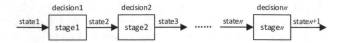

Fig. 1. Multi-stage decision making process

(1) *Dividing stages:* According to the time or space characteristics of the problem, the problem is divided into several stages. In the division phase, note that the stage after the division must be ordered or sortable, otherwise the problem cannot be solved. The variables describing the stage are called stage variables, which are commonly used k.

(2) *Determining the state:* The various objective situations in which the problem develops to each stage are represented by different states. Of course, the choice of state should be met without aftereffect. Variables that describe the state are called state variables, and the state variables of phase k are usually represented by s_k.

(3) *Determining decisions and strategies:* Decisions represent the choice decisions made by a decision maker in several scenarios when a certain stage is in a certain state. The variables that describe the decision are called decision variables, and the decision variables of phase k are commonly by U_k. The value of the decision variable is subject to the state variable and is limited to a certain range, called the allowable decision set, which is denoted as $D_k(s_k)$.

The whole decision process from the beginning of the first phase to the end of the final phase is called the whole process of the problem; and the decision process from the beginning of the k phase to the final phase is called the k-subprocedure [10]. In the whole process, the decision sequence $P_{1,n} = \{u_1, u_2, \ldots, u_n\}$ composed of the decision-making of each stage is called the whole process strategy, which is called the strategy; and the decision sequence $P_{k,n} = \{u_k, u_{k+1}, \ldots, u_n\}$ on the k-subprocedure is called the k-subprocedure strategy, which is called the sub-strategy.

(4) *Writing the state transition equation:* The state transition equation describes the evolution of the state from one phase to the next. If the state variable value of phase k is s_k, when the value of the decision variable u_k is determined, the value of the state variable u_k of the next stage is completely determined. This correspondence is denoted by $s_{k+1} = T_k(s_k, u_k)$ as the state transition equation [11].

(5) *Determining the index function and the optimal value function:* The index function is divided into a stage indicator function and a process indicator function. The stage indicator function is a measure of the target profit and loss value (generated by state and decision) at a certain stage, denoted by $v_k(s_k, u_k)$.

The process indicator function refers to the total target profit and loss value (generated by state and decision) at each stage involved in the process, denoted as $V_{k,n} = V_{k,n}(s_k, u_k, s_{k+1}, u_{k+1}, \ldots, s_n, u_n)$.

The process indicator function required for dynamic programming should be separable and can be expressed as a function of the index function of each stage it contains.

According to the principle of optimality of dynamic programming, the whole process optimal strategy is required, which can be optimized from the optimization of sub-process strategies. For the process index function is the form of the phase index function sum, consider the k-subprocedure optimal value function $f_k(s_k)$.

$$f_k(s_k) = \underset{\{u_k,\ldots,u_n\}}{opt} \left\{ \sum_{j=k}^{n} v_j(s_j, u_j) \right\} = \underset{\{u_k,\ldots,u_n\}}{opt} \left\{ v_k(s_k, u_k) + \sum_{j=k+1}^{n} v_j(s_j, u_j) \right\}$$

$$= \underset{\{u_k\}}{opt} \left\{ v_k(s_k, u_k) + \underset{\{u_{k+1},\ldots,u_n\}}{opt} \sum_{j=k+1}^{n} v_j(s_j, u_j) \right\} = \underset{\{u_k\}}{opt}\{v_k(s_k, u_k) + f_{k+1}(s_{k+1})\}$$

(1)

Then there is a recursion equation:

$$f_k(s_k) = \underset{\{u_k\}}{opt}\{v_k(s_k, u_k) + f_{k+1}(s_{k+1})\} \tag{2}$$

Also need to have boundary conditions, generally take $f_{n+1}(s_{n+1}) = 0$. Thus, the basic equation

$$\begin{cases} f_k(s_k) = \underset{\{u_k\}}{opt}\{v_k(s_k, u_k) + f_{k+1}(s_{k+1})\}(k = n, n-1, \ldots, 2, 1) \\ f_{n+1}(s_{n+1}) = 0 \end{cases} \tag{3}$$

Similarly, for the process index function is the form of the stage index function product, the basic equation is:

$$\begin{cases} f_k(s_k) = \underset{\{u_k\}}{opt}\{v_k(s_k, u_k) * f_{k+1}(s_{k+1})\}(k = n, n-1, \ldots, 2, 1) \\ f_{n+1}(s_{n+1}) = 1 \end{cases} \tag{4}$$

The dynamic programming problem can be solved by using the basic equations and recursing backwards and forwards.

2.2 Task Allocation and Resource Scheduling Method in Cloud Environment

Services in cloud environment must be efficiently acquired, extended, and capable of dynamically discovering and combining services. While ensuring that the user's needs are guaranteed, try to minimize the completion time of all tasks and ensure that the virtual machine load participating in the calculation is balanced [12].

In this paper, the matching of tasks and servers is regarded as a multi-stage decision problem, and the success of the last task with the appropriate server can be seen as a strategy. When there is a certain range of strategies to choose from, this range is called the set of allowed policies, denoted by P. The strategy that achieves the best effect in P is the optimal strategy. Therefore, the task scheduling problem is to find an optimal strategy in P to minimize the task completion time. The task allocation and resource scheduling model proposed in this paper is shown in Fig. 2.

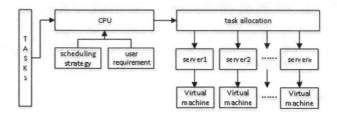

Fig. 2. Task allocation and resource scheduling model

m indicates the number of cores of the master computer cpu, n indicates the number of servers, $n > m$, and x_i indicates the number of configuration tasks that need to be delivered for each server, $i \in \{1, 2 \ldots, n\}$. The server interacts with the virtual node and delivers the configuration file. For the execution of the task, the completion time of the task is mainly based on its processing time.

$$t_i = tLong_i / mips \tag{5}$$

Where $tLong_i$ represents the length of task i and $mips$ represents the computing power of the server.

The problem can be translated into dividing the n servers into groups of m, with the total number of tasks performed by each group of servers being as even as possible. Suppose the sum of each set of data is sum_i, $i \in \{1, 2 \ldots, m\}$.

First, calculate the average of n numbers, there is

$$\bar{X} = \frac{x_1 + x_2 + \ldots + x_n}{n} = \frac{\sum\limits_{i=1}^{n} x_i}{n} \tag{6}$$

$$length = sqrt[(sum_1 - \bar{X})^2 + (sum_2 - \bar{X})^2 + \ldots + (sum_m - \bar{X})^2] \tag{7}$$

When the *length* value takes the minimum value, the group is the most average, the most efficient, and the task execution is the fastest.

Step 1, the n numbers are sorted in descending order, starting from the largest number x_{max}. If $x_{max} >= \bar{X}$, x_{max} is directly divided into a group. If there are j numbers greater than or equal to $\bar{X}$, then divided into j groups, then consider how the remaining $n\text{-}j$ numbers less than a are divided into $m\text{-}j$ groups;

Step 2, for the remaining ungrouped numbers, first put the maximum number x'_{max} into a group g, and then find the numbers p_0 and q_0 closest to Δ_0 from the remaining numbers for g, $q_0 < \Delta_0 < p_0$,

$$\Delta_0 = \bar{x} - x'_{max} \tag{8}$$

if $(\Delta_0 - p_0)^2 > (\Delta_0 - q_0)^2$, add q_0 to group g and continue to explore $= \Delta_1 - q_0$;

if $(\Delta_0 - p_0)^2 \le (\Delta_0 - q_0)^2$, then save $(\Delta_0 - p_0)^2$ and continue to explore $\Delta_0 - q_0$; recursively perform the above process, select the smallest difference squared data to join the group, continue to explore $= \Delta_1 - q_0$;

Step 3, when the sum of several numbers in the group g is greater than or equal to $\bar{X}$, the assignment of a group is completed;

Step 4, the second step is continued until $n - j$ numbers less than a are completely divided into $m - j$ groups, jumping out of the loop, and n servers are divided into m groups.

3 Experiments and Discussion

In order to verify the performance of DPSA in task allocation and resource scheduling in cloud environment, a simulation experiment was performed on a control host of an 8-core CPU. The host console sends a number of different tasks to the virtual server. The number of tasks sent by each server is generated by random numbers. The total number meets the experimental requirements. The experimental results are shown in Table 1 (20 servers), Table 2 (50 servers), and Table 3 (100 servers).

Table 1. Experimental result (20 servers)

Master computer	8-core CPU			
Number of virtual server	20	20	20	20
Number of task	500	1000	3000	10000
Task execution time/s (using DASP)	0.029	0.052	0.172	0.456
Task execution time/s (not using DPSA)	0.084	0.162	0.478	1.599

Table 2. Experimental result (50 servers)

Master computer	8-core CPU			
Number of virtual server	50	50	50	50
Number of task	500	1000	3000	10000
Task execution time/s (using DPSA)	0.026	0.051	0.129	0.509
Task execution time/s (not using DPSA)	0.089	0.161	0.481	1.549

Table 3. Experimental result (100 servers)

Master computer	8-core CPU			
Number of virtual server	100	100	100	100
Number of task	500	1000	3000	10000
Task execution time/s (using DPSA)	0.031	0.052	0.145	0.434
Task execution time/s (not using DPSA)	0.086	0.179	0.484	1.580

For more intuitively reflect the role of the proposed method in task allocation and resource scheduling in cloud environment, compare the task execution time when the method is used or not in the above three groups of experiments. The comparison results are shown in Fig. 3.

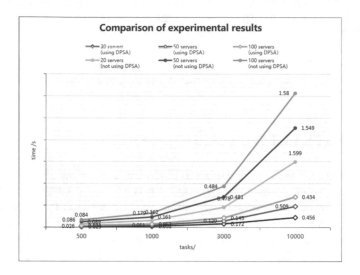

Fig. 3. Comparison of experimental results

It can be seen from the experimental results that using the task allocation and resource scheduling method based on DPSA can significantly improve the file sending speed and shorten the task execution time, especially in the case of a large number of tasks to be delivered. Task allocation and resource scheduling are related to host performance, delivered tasks, and selected task scheduling policies, and have little to do with the number of servers receiving tasks. At the same time, it also shows that DPSA have advantages in solving multi-stage decisions.

4 Conclusion

In this paper, for the task allocation and resource scheduling problem in cloud environment, using the idea of DPSA, the one-to-one mapping of task and server is regarded as a multi-stage decision process, and the minimum global scheduling time is found to obtain the task allocation optimization scheme. Compared with the traditional task scheduling algorithm, the resource load is relatively balanced, the task execution efficiency is improved, and the task completion time is shortened. However, the algorithm of this paper mainly considers the execution speed of tasks. The analysis of different types of user requests and specific resource characteristics will be the focus of future research.

Acknowlegements. The work of this paper is funded by the project of National Key Research and Development Program of China (No. 2016YFB0800802, No. 2017YFB0801804), Frontier Science and Technology Innovation of China (No. 2016QY05X1002-2), National Regional Innovation Center Science and Technology Special Project of China (No. 2017QYCX14), Key Research and Development Program of Shandong Province (No. 2017CXGC0706), and University Co-construction Project in Weihai City.

References

1. Zou, Z.: Research on application advantages of virtualization technology in cloud environment. Technol. Market **23**(12), 122 (2016)
2. Xie, R., Rus, D., Stein, C.: Scheduling multi-task agents. In: Mobile Agents, International Conference, Ma Atlanta, GA, USA, December. DBLP (2001)
3. Zheng, C., Peng, Y., Xu, Y., Liao, Y.: Improved task scheduling algorithm based on NSBBO in cloud manufacturing environment [J/OL]. Comput. Eng. pp. 1–8. 22 Mar 2019. http://kns.cnki.net/kcms/detail/31.1289.tp.20190128.1129.002.html
4. Shen, J., Luo, C., Hou, Z., Liu, Z.: Service composition and optimization method based on improved ant colony optimization algorithm. Comput. Eng. **44**(12), 68–73 (2018)
5. Shen, L., Liu, L., Lu, R., Chen, Y., Tian, P.: Cloud task scheduling based on improved immune evolutionary algorithm. Comput. Eng. **38**(09), 208–210 (2012)
6. Wang, X., Liu, X.: Cloud computing resource scheduling based on dual fitness dynamic genetic algorithm. Comput. Eng. Des. **39**(05), 1372–1376 + 1421 (2018)
7. Kong, X., Lin, C., Jiang, Y.: Efficient dynamic task scheduling in virtualized data centers with fuzzy prediction. J. Network Comput. Appl. **34**(4), 1068–1077 (2011)
8. Liu, Y., Cui, Q., Zhang, W.: A cloud scheduling task scheduling strategy based on genetic algorithm. Inf. Tech. (08), 177–180 (2017)
9. Wang, F.: Solving general assignment problem based on multi-stage dynamic programming. Inf. Technol. Inf. (04), 49–51 (2017)
10. Liao, H., Shao. X.: Principle and application of dynamic programming algorithm. Chinese Sci. Technol. Inf. (21), 42–42 (2005)
11. Zhu, D.: Optimization Model and Experiment. Tongji University Press, Shanghai (2003)
12. Shi, S., Liu, Y.: Research on cloud computing task scheduling based on dynamic programming. J. Chongqing Univ. Posts Telecommun. Nat. Sci. Edn. **24**(6), 687–692 (2012)

Impact of Human Factors in Cloud Data Breach

Monjur Ahmed[1]([⊠]), Himagirinatha Reddy Kambam[1], Yahong Liu[1], and Mohammad Nasir Uddin[2]

[1] Waikato Institute of Technology (Wintec), Hamilton 3210, New Zealand
Monjur.Ahmed@wintec.ac.nz
[2] PrideSys IT Ltd., Dhaka, Bangladesh

Abstract. In this paper, we present a study on the impact of human factors in Cloud data breach. Data breaches in Cloud platforms result in major concerns and thus the underlying reasons for such data breaches demand investigation. An incident of data breach may occur due to several reasons. The root cause for a data breach may be related to technological factors as well as human factors. While technological factors are mostly predictable, human factors may not be. Besides, human factors are dynamic that cannot be fully quantified. This leaves a room for the attackers to compromise systems through social engineering. The presented study seeks to find the extent to which human factors are contributors for data breaches. Analyses on 20 real life incidents of Cloud data breaches are carried out, and the reasons behind those breaches are explored to understand the possible implications of human factors in Cloud breaches.

Keywords: Cloud breach · Cloud computing · Cyber security · Data breach · Human factors · Security · Social engineering

1 Introduction

Security is a major concern in computing. All kinds of computing approaches are prone to security breaches and attacks. One such example is Cloud Computing [1]. Though recent emergence of Cloud Computing brings many advantages, security threats and concerns also exist [2]. There are examples of massive data breaches in the Cloud. Several factors or reasons may act behind a data breach. Thus, it is important to realise the factors that are reasons for data breach. A Cloud threat taxonomy proposed in [3] notes human factor as a core contributing one towards threats in Cloud Computing. They define human factors as those emerge as a result of actions taken by human being. Given that, other few examples of human factors as a threat metric are mentioned in [4–6] and [7].

In this paper, we present a study on the impact of human factors in Cloud data breach. The rest of the paper is structured as follows: Methodology section outlines the approach used to carry out the study. In Literature Review, we describe the concepts of several related terminologies and related work. In the Case Studies section, a summary of the incidents is described. Findings and discussion on the case studies are presented in Findings and Discussion section.

© Springer Nature Switzerland AG 2020
F. Xhafa et al. (Eds.): IISA 2019, AISC 1084, pp. 568–577, 2020.
https://doi.org/10.1007/978-3-030-34387-3_70

2 Methodology

We follow an ad-hoc mixed methodology [42, 44] approach to explore and analyse several real-life scenarios of data breaches. We study is an initial pilot study of a relatively smaller number of samples - the study is carried out by randomly collecting 20 real life incidents of data breaches from different online sources. The incidents of data breaches are collected from online news sources through web search using Internet search engines. The keywords used for the search are 'data breaches' and 'Cloud data breach'. The considered cases are then analysed to find out the percentage of total data breaches that were caused by human factors. The implications of such human factor related breaches are also analysed and then compared to the overall picture of Cloud and data breaches, to understand the implications of human factors. The findings help to understand two aspects: the emerging trend of human factors as a reason for Cloud data breaches, and whether the implications of Human Factors outweigh the implications of those in other factors. Such understanding eventually helps to focus on human factors with proper level of attention. The sample case studies are selected using random sampling method as discussed in [40]. The total population considered are the breaches in broader operational context of IT and computing. The random sampling approach we choose is also loosely described in [39, 41], and [38]. The quantitative approach of the analysis of causal relationship among variables [43] are used to analyse our findings on the case studies considered.

3 Literature Review

In computer security, human factor refers to those characteristics of human beings that can be exploited to gain unfair access to computer systems. In [3], human factor is defined as the human-centric actions that pose security threats to computing infrastructure. Human factors from which a Cloud threat may emerge are trust, compliance, regulations, competence, Service Level Agreement (SLA) misinterpretation, and social context [3]. The authors term human factor also as 'soft threat'. The human factors are described by [8] as the factors that are "concerned with applying what is known about human behaviour, abilities, limitations, and other characteristics to the design of systems, tasks/activities, environments, and equipment/technologies".

The importance of human factors in computing security is mentioned in [9–11], and [12]. The approach to attack using human factors is termed as social engineering [13]. [14] describes social engineering and human conducted attacks and the implications of human-centric attacks. [13] mentions social engineering as 'dark art' and discussed the impact of the attacks based on human factors. [15] mentions social engineering to be of profound negative impact and is likely to increase over the course of time. Similar discussions are found in [16, 17], and [18].

4   Case Studies

In this section, we briefly mention the case studies – which are real life incidents of Cloud and/or data breaches – considered for the study. Brief descriptions of the case studies are outlined in Table 1.

Table 1. Case studies.

Case Study	Description of the breach
Microsoft	A breach in 2010 due to a configuration issue in its Business Productivity Online Suite resulting in unauthorised access to employee contact info [19]
Dropbox	In 2012, hackers accessed DropBox's 68 million user accounts including their email addresses and passwords [19–21]. The reason for the breach was using same password by an employee for both DropBox and LinkedIn [20]
National Electoral Institute of Mexico	93 million of voter of the National Electoral Institute of Mexico were compromised in April 2016. It was due to poorly configured database. The institute stored its data on an insecure and illegally hosted Cloud server situated outside of Mexico [19]
LinkedIn	In 2012, LinkedIn breach resulted in 6 million user passwords being stolen. In 2016, LinkedIn suffered another breach where 167 million of users' emails and passwords were stolen [19]. The reason for the breaches is claimed to be using weak passwords and password reuse [22]
Home Depot	DIY retailer Home Depot's point-of-sale terminals at the self-checkout were exploited for months in 2014. This affected 56 million credit cards and was the biggest breach of its kind at that time [19, 23]. "Hackers used a vendor's stolen log-on credentials to penetrate Home Depot's computer network and install custom-built malware that stole customer payment-card data and e-mail addresses, …" [24]
iCloud	The breach in 2014 on Apple's cloud storage iCloud meant leaked pictures of its celebrity users [19]. Apple denied it to be architecture-wide breach, but rather is a targeted attack on celebrity accounts that used weak passwords [25]
Yahoo	In 2013, a breach on Yahoo's network resulted in disclosing information (e.g. name, email, date of birth) of more than a billion of user accounts [19, 23]. Yahoo claimed that it was carried out using forged cookies [27]
Phony phone-call	An attempt by fraudsters to capitalise the iCloud breach and launching scam calls to gain advantage by fooling people [26]
Verizon	Data breach in 2017 (exposed 14 million customer accounts) was due to the management of Verizon data by NICE Systems, a third-party vendor [28]
eBay	The attack in 2014 exposed 145 million users' names, addresses, dates of birth and encrypted password. According to eBay, the

(*continued*)

Table 1. (*continued*)

Case Study	Description of the breach
	hackers got into their system using credentials of their employees to access to eBay network [23]
Uber	In 2016, personal information of 600,000 drivers and 57 million users were exposed. The breach was due to placing Uber's Cloud server (AWS) account username and password on online code repository GitHub [23]
Deloitte	Global consultancy firm Deloitte had its clients' personal information hacked and exposed in 2017. The system was compromised through an unsecured administrator password [29, 30]
Sage	UK based accounting firm Sage was affected in 2016. An internal login was used for unauthorised access to employees' data of 300 UK firms [31]
Facebook	Facebook 'exposed 6 million users' phone numbers and email addresses to unauthorized viewers', due to technical glitch (as claimed by Facebook) [32]
Twitter	Twitter's 250,000 user accounts were hacked in 2013. The company suggested that they were aware on the incident upon detecting 'unusual access patterns across the network and had identified unauthorised attempts to access user data that had led to accounts being compromised' [33]
AFF	"Hackers collected 20 years of data … Most of the passwords were protected only by the weak SHA-1 hashing algorithm …" [23]
Equifax	An application vulnerability exposed about 147.9 million consumers' data [23]. The reason was a flaw that could be patched weeks before the attack [34]
TJX	A breach on a portion of its network exposed credit card, debit card, check and merchandise transactions [35]. One claim state that a group of hackers took advantage of a weak data encryption system, though different claims exist [23]
Scottrade	A third-party data breach inadvertently exposed 20,000 of its customers' non-public information. One of Scottrade's vendors, Genpact's employee uploaded Scottrade's database to an Amazon-hosted SQL server. It is revealed that the worker didn't take adequate safeguards to lock down the server [36]
Fashion Nexus	650,000 fashion shoppers' details including email and home addresses was exposed by a white hat hacker who breached company's web server" [37]

5 Findings and Discussion

The summary of findings from the case studies are presented in Table 2.

Table 2. Summary of Breaches.

Case Study	Source	Reason(s) for Breach	Human Factor involved?
Microsoft	[19]	Configuration issue	Not known
Dropbox	[19–21]	Employee using the same password for	Yes
National Electoral Institute of Mexico	[19]	Poorly configured database. Data stored on an insecure and illegally hosted Cloud server situated outside of Mexico	Probably yes – subject to argument
LinkedIn	[19, 22]	Using weak passwords and password reuse	Yes
Home Depot	[19, 23, 24]	Vendor's stolen log-on credentials	Probably – not known
iCloud	[19, 25]	Vendor denied breach, claimed it was account specific targeted attack	Yes (based on vendor's claim)
Yahoo	[19, 23, 27]	Using forged cookies	No
Phony phone-call on iCloud	[26]	Scam calls to victimise people	Yes
Verizon	[28]	Using third party vendor	Probably – not clear
eBay	[23]	hackers used credentials of corporate employees	Not clear
Uber	[23]	placing Uber's Cloud server (AWS) account username and password on online code repository GitHub	Yes
Deloitte	[29, 30]	System compromised through an unsecured administrator password. Weak password strategy used by administrator, no two-factor authentication	Yes
Sage	[31]	Internal login was used for unauthorised access	Not clear
Facebook	[32]	Technical glitch, as claimed by the vendor	No
Twitter	[33]	Unusual access patterns across the network, and unauthorised attempts to access user data	Not clear
AFF	[23]	Using weak hashing algorithm	Yes
Equifax	[23, 34]	Application vulnerability that the organization failed to address in good time	Partially Yes
TJX	[23, 35]	Using weak hashing algorithm	Probably
Scottrade	[36]	Inadequate safeguard by third-party vendor	Yes
Fashion Nexus	[37]	A white hat hacker breached company's server	Most likely

From the summarised information in Table-2, nine out of 20 incidents can be directly identified as human-factor related breach. This is 45% of the total incidents considered. On the other hand, only 2 incidents (i.e. Yahoo and Facebook) can be assuredly identified as not related to human factors which comprises around 10% of the total incidents considered. Even if the rest of the incidents are not taken into consideration, 45% of the total incidents for human factor as the reason for breach illustrate the significance of this factor. The rest incidents which are around 45% of all the incidents require further analysis to ascertain whether human factors were involved.

The reason for Microsoft breach is mentioned as configuration issue. It is not known whether the configuration is automatically populated or set by a human being. If the latter is the case, human factors (e.g. incompetence, fatigue, human error, lack of attention) cannot be excluded as a possible reason for the breach. The same applies to poorly configured database for the incident with National Electoral Institute of Mexico. The other reason of the breach (illegally hosted Cloud server) can be assumed as a human decision. For the cases of Home Depot, eBay and Sage, how the vendor's log-on was stolen is not known. This can well be incompetence or poor management of log-on credentials. Verizon incident also qualifies to have human factor as a probable cause for the breach. Using third party vendor is associated with contractual and other aspects that include human factors too. The reason for Twitter breach indicates that the breach was not an attack on Twitter's network; it was rather an effort to compromise users' account through brute-force approach. Using poor and guessable password could be a reason, and it would be alarming if this was the case. Thus, human factor is a strong candidate to be a reason for the breach in twitter incident. In a nutshell, even if the above 7 incidents are not related to human factors, the incidents directly identifiable as human factor related breaches, illustrate the apparent significance of human factors in cyber security breach.

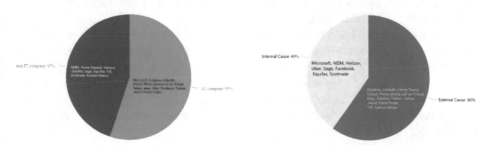

Fig. 1. Attacks to IT and non-IT company.   **Fig. 2.** Attacks originated from internal and external source.

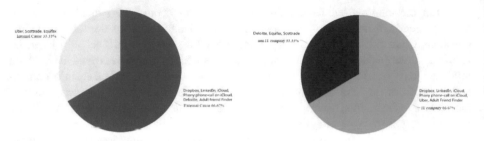

Fig. 3. Human factor-centric attack from internal and external sources/causes

Fig. 4. Human factor-centric incident on IT and non-IT Companies

The reason for Adult Friend Finder (AFF) breach is weak encryption algorithm, which is a human decision on which security mechanism to be used. This is similar to the TJX companies, Inc. scenario if the speculation of using weak encryption is true. "Equifax had nine working weeks in which to apply the patch" and "That its data breach was entirely avoidable…." [34] points to the human-centric action (or lack of action thereof) resulting in the breach taken place. The example of Scottrade is clearly due to human-factor, where the personnel from third-party vendor took inadequate safeguard to deal with server. Human factor is involved in the Deloitte case where proper strategic thought was not given to strengthen the security policy. The case of Fashion Nexus can be considered as a breached because of human factor – it can be assumed that the ethical hacker had access to the server and thus did not have to 'hack' the server to cause the breach.

Figure 1 suggests that 55% of organisations from the considered case studies are IT company, and the rest (45%) are non-IT company. Figure 2 indicated that more attacks (60% of the total case studies) are sourced from external sources compared to those took place due to internal reasons.

For human factor related breaches, External sources or causes are more responsible than internal ones. Considering the illustration in Fig. 3, we found external causes are behind 66.67% of the incidents and the rest (33.33%) are due to internal causes. Exploring how human factor related attacks are prevalent in IT and non-IT companies, we found IT companies are more prone to be a victim of human factor-centric attacks. The human factor related attacks on non-IT organisations are almost half of those in IT organisations. This is illustrated in Fig. 4.

6 Conclusions and Further Research

As discussed, several attacks are due to human errors by the professionals who manage the network, not by the end-users who may not be competent in using IT/IS. The reasons for silly mistakes by the professional demand investigation and research. The possibility for such mistakes may be linked with work pressure and overall wellbeing of the employees. Addressing human factors to any extent required would cost far less than the loss occurs as a result of data breaches.

A limitation of the presented pilot study is that, the number of samples are relatively low. The statistical significance of the finding thus may deviate with larger samples. As future research, we aim to consider more case studies. This will enable us to study how human factors are emerging as a threat factor for Cloud Computing. The future research plan also includes formulating a predictive model to determine the metrics that would minimise breaches yielding from human factors.

References

1. Jaeger, P., Lin, J., Grimes, J.: Cloud computing and information policy: computing in a policy cloud? J. Inf. Technol. Politics **5**(3), 269–283 (2008)
2. Zissis, D., Lekkas, D.: Addressing cloud computing security issues. Future Gener. Comput. Syst. **28**, 583–592 (2012)
3. Ahmed, M., Litchfield, A.T.: Taxonomy for identification of security issues in cloud computing environments. J. Comput. Inf. Syst. **50**, 79–88 (2016)
4. Gruschka, N., Jensen, M.: Attack surfaces: a taxonomy for attacks on cloud services. In: 3rd International Conference on Cloud Computing, pp. 276–279. IEEE (2010)
5. Grobauer, B., Walloschek, T., Stocker, E.: Understanding cloud computing vulnerabilities. In: IEEE Cloud Computing, pp. 14–20, May/June 2012
6. Gupta, S., Kumar, P.: Taxonomy of cloud security. Int. J. Comput. Sci. Eng. Appl. **3**(5), 47–67 (2013)
7. Srinivasan, M.K., Sarukesi, K., Rodrigues, P., Manoj, S., Revathy, P.: State–of–the–art cloud computing security taxonomies–a classification of security challenges in the present cloud computing environment. In: ICACCI 2012, pp. 470–476. ACM, India (2012)
8. National Research Council: Health Care Comes Home: The Human factors. Committee on the Role of Human factors in Home Health Care, Board on Human-Systems Integration, Division of Behavioural and Social Sciences and Education. The National Academies Press, Washington DC (2011)
9. Haniff, D.J., Baber, C.: Wearable computers for the fire service and police force: technological and human factors. In: ISWC 1999 Proceedings of the 3rd IEEE International Symposium on Wearable Computers, pp. 185–186. ACM (1999)
10. Hawkey, K., Gagne, A. Botta, D., Beznosov, K., Werlinger, R., Mukdner, K.: Human, organizational and technological factors of IT security. In: CHI 2008 Proceedings, Florence, Italy, pp. 3639–3644, 5–10 April 2008
11. Kueppers, S., Schilingno, M.: Getting our act together: human and technological factors in establishing an online knowledge base. In: SIGUCCS 1999, pp. 135–139. ACM, Denver (1999)
12. Mohamadi, M., Ranjbaran, T.: Effective factors on the success or failure of the online payment systems, focusing on human factors. In: 7th International Conference on e-Commerce in Developing Countries with Focus of e-Security, pp. 1–12. IEEE, Iran, 17–18 April 2013
13. Thornburgh, T.: Social engineering: the "Dark Art". In: InfoSecCD Conference 2004, Kennesaw, GA, USA, 8 October 2004
14. Krombholz, K., Hobel, H., Huber, M., Weippl, E.: Social engineering attacks on the knowledge worker. In: Proceedings of the 6th International Conference on Security of Information and Networks, SIN 2013, pp. 28–35. ACM, New York (2013)
15. Twitchell, D.P.: Social engineering in information assurance curricula. In: InfoSecCD Conference 2006, Kennesaw, Georgia, USA, 22–23 September 2006

16. Jagatic, T.N., Johnson, N.A., Jakobsson, M., Menczer, F.: Social phishing. In: Communications of the ACM, vol. 50, no. 10, October 2007
17. Bakhshi, T., Papadaki, M., Furnell, S.M.: A practical assessment of social engineering vulnerabilities. In: Proceedings of the Second International Symposium on Human Aspects of Information Security & Assurance (HAISA 2008), pp. 12–23 (2008)
18. Odaro, U.S., Sanders, B.G.: Social engineering: phishing for a solution. In: Proceedings of the IT Security for the Next Generation, Erfurt, Germany (2011)
19. Bradford, C.: 7 Most Infamous Cloud Security Breaches. https://www.storagecraft.com/blog//-infamous-cloud-security-breaches/. Accessed 23 May 2018
20. Gibbs, S.: Dropbox hack leads to leaking of 68 m user passwords on the internet. https://www.theguardian.com/technology/2016/aug/31/dropbox-hack-passwords-68m-data-breach. Accessed 24 May 2018
21. BBC.: Dropbox hack 'affected 68 million users'. http://www.bbc.com/news/technology-37232635. Accessed 24 May 2018
22. Schuman, E.: LinkedIn's disturbing breach notice. https://www.computerworld.com/article/3077478/security/linkedin-s-disturbing-breach-notice.html. Accessed 24 May 2018
23. Armerding, T.: The 17 biggest data breaches of the 21st century. https://www.csoonline.com/article/2130877/data-breach/the-biggest-data-breaches-of-the-21st-century.html. Accessed 24 May 2018
24. Winter, M.: Home depot hackers used vendor log-on to steal data, e-mails. https://www.usaoday.com/story/money/business/2014/11/06/home-depot-hackers-stolen-data/18613167/. Accessed 26 May 2018
25. Goldman, J.: Apple Admits Celebrity Accounts Were Hacked, But Denies iCloud Breach. https://www.esecurityplanet.com/network-security/apple-admits-celebrity-accounts-were-hacked-but-denies-icloud-breach.html. Accessed 26 May 2018
26. Fleishman, G.: Ignore that call from "Apple" about an iCloud breach. https://www.macworld.com/article/3185485/security/ignore-that-call-from-apple-about-an-icloud-breach.html. Accessed 26 May 2018
27. Condliffe. J.: A History of Yahoo Hacks. https://www.technologyreview.com/s/603157/a-history-of-yahoo-hacks/. Accessed 26 May 2018
28. O'Sullivan, D.: Cloud Leak: How A Verizon Partner Exposed Millions of Customer Accounts. https://www.upguard.com/breaches/verizon-cloud-leak. Accessed 28 May 2018
29. Burgess, M.: That Yahoo data breach actually hit three billion accounts. http://www.wired.co.uk/article/hacks-data-breaches-2017. 28 May 2018
30. Hopkins, N.: Deloitte hit by cyber-attack revealing clients' secret emails. https://www.theguardian.com/business/2017/sep/25/deloitte-hit-by-cyber-attack-revealing-clients-secret-emails. Accessed 13 June 2018
31. KCOM.: Cloud: The Data Breach Scapegoat. https://business.kcom.com/media/blog/2017/november/cloud-the-data-breach-scapegoat/. Accessed 28 May 2018
32. Shih, G.: Facebook admits year-long data breach exposed 6 million users. https://uk.reuters.com/article/net-us-facebook-security/facebook-admits-year-long-data-breach-exposed-6-million-users-idUSBRE95K18Y20130621. Accessed 28 May 2018
33. Jones, C.: Twitter says 250,000 accounts have been hacked in security breach. https://www.theguardian.com/technology/2013/feb/02/twitter-hacked-accounts-reset-security.   Accessed 28 May 2018
34. Sharwood, S.: Missed patch caused Equifax data breach. https://www.theregister.co.uk/2017/09/14/missed_patch_caused_equifax_data_breach/. Accessed 10 June 2018
35. Roberts, P.: Massive TJX Security Breach Reveals Credit Card Data. https://www.csoonline.com/article/2121609/malware-cybercrime/massive-tjx-security-breach-reveals-credit-card-data.html. 12 June 2018

36. Bisson, D.: Scottrade Confirms Third-Party Data Breach Exposed 20,000 Customers' Private Data. https://www.tripwire.com/state-of-security/latest-security-news/scottrade-confirms-third-party-data-breach-exposed-20000-customers-private-data/. Accessed 12 June 2018
37. Clark, T.: Data hacked at web provider Fashion Nexus. https://www.drapersonline.com/news/data-hacked-at-web-provider-fashion-nexus/7031553.article. Accessed 24 Oct 2018
38. Strauss, A., Corbin, J.: Basics of Qualitative Research: Grounded Theory: Qualitative Research in Nursing. Addison- Grounded Theory, Procedures and Techniques. Sage, California (1990)
39. Morse, J.M.: Strategies for sampling. In: Qualitative Nursing According, Sage, Newbury Park, California, pp. 127–145 (1991)
40. Patton, M.Q.: Qualitative Evaluation and Research Methods, 2nd edn. Sage, Newbury Park (1990)
41. Sandelowski, M.: Sample size in qualitative research. Res. Nurs. Health **18**, 179–183 (1995)
42. Johnson, R.B., Onwuegbuzie, A.J.: Mixed methods research: a research paradigm whose time has come. Educ. Res. **33**(7), 14–26 (2004)
43. Denzin, N.K., Lincoln, Y.S. (eds.): Collecting and Interpreting Qualitative Materials. Sage Publication, Thousand Oaks (1998)
44. Johnson, R.B., Onwuegbuzie, A.J., Turner, L.A.: Toward a definition of mixed methods research. J. Mixed Meth. Res. **1**(112) (2007)

Current Issues and Challenges of Security in IoT Based Applications

Mritunjay Kumar Rai[1], Rajeev Kanday[2,3], and Reji Thomas[2(✉)]

[1] School of Electronics and Electrical Engineering, Lovely Professional
University, Phagwara 144411, Punjab, India
[2] Division of Research and Development, Lovely Professional University,
Phagwara 144411, Punjab, India
rthomas.eyyalil@gmail.com
[3] School of Computer Applications, Lovely Professional University,
Phagwara 144411, Punjab, India

Abstract. The abstract should summarize the contents of the paper in short terms, i.e. 150–250 words. Automation in sensor-based network has made Internet of Things (IoTs) a reality. From wearable gadgets, smart homes, smart city, industrial internet, connected car, connected health, smart retail down to smart farming, IoT plays a major role at present. This drastic development now met with security issues such as privacy, infrastructure, leakage of information, loss of security services etc and is a major challenges. In this communication, we have surveyed literature to understand various security issues of IoT and the immediate challenges to be addressed to overcome security issues. Different factors like mobility, scalability, resource limitation etc. are considered for analyzing IoT security and current researchers can identify some open research problems and meet the challenges to promote IoT further.

Keywords: IoTs · Challenges · Problems · Issues · Security

1 Introduction

The sophistication of communication and internet technology has made human activities more concentrated on the fictional space of the virtual world [1, 2]. Internet of Things (IoTs) is a progressing technology where the network devices sense and collect data from the world around us locally or remotely, and then transmit those data across the internet to the destination, where the data are processed and utilized [3–5]. There is another important term 'Industrial Internet', often interchangeably used with IoT, deals primarily with commercial applications of IoT technology in the world of manufacturing [5–7]. IoTs are not limited to industrial applications, however they, as of now, is also associated with computing devices, mechanical and digital machines, objects, people, even animals. The network associated with IoT tags the 'things' with a unique identifier, and that reduces human-to-human or human-to-computer interaction [8–12]. The various applications of the Internet of Things are schematically shown in Fig. 1.

The equipment that is part of the IoT system primarily has one or more sensors attached. The embedded computing system in the equipment provides a unique identifier

© Springer Nature Switzerland AG 2020
F. Xhafa et al. (Eds.): IISA 2019, AISC 1084, pp. 578–589, 2020.
https://doi.org/10.1007/978-3-030-34387-3_71

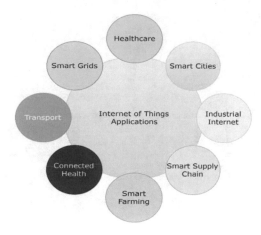

Fig. 1. Applications of Internet of Things

and operates with the existing internet infrastructure either in local or remote networks [3, 13]. The sensor monitors a specific parameter such as location, vibration, motion and temperature, moisture, pressure, etc. In IoTs, these sensors are at constant communication, collect the data as per the assigned job and send to systems (in sensor networks, they are called as sinks or base station). This BS interpret and analyze the sensor fed information and provide new information to the company or people. These processes very soon are going to be a part of technological culture [14, 15]. The main objective of IoTs is to add 'smartness' in communication technology by configuring a smart environment with automated or self-conscious independent devices. Realization of that objective provide a smart living, smart items, smart health, smart cities and many more depending on the imagination [16–20]. Current market examples include smart thermostat systems and washer/dryers that utilize Wi-Fi for remote monitoring [21]. 'Ambient intelligence' [7, 22–24] is also has become a part of IoT where different objects are unified even though they are not directly related to the users.

Figure 2 shows a schematic diagram of IoT architecture. The major modules of the architecture are people, internet, embedded hardware and environment. The embedded sensor system and its actuators sense the data from the environment with the help of suitable hardware designed for the specific task. The hardware components of the sensor system forward or store the data in clouds. People operate their devices to recreate that data for getting the required information. As you can from the figure, all modules are connected with each other to have the efficient working of IoTs. The interconnections of these devices/modules use basic network infrastructure of either wireless sensor networks, radio networks or peer-to-peer networks. All the networks are communicating with the assignment of IP addresses [25, 26]. Therefore, the basic paradigm and problems of wireless networks make challenging issues for IoTs [18]. According to a recent report [27–29], there are 25 billion connected 'things' in form of applications and another 200 billion connections in form of services and both are going to generate a revenue of approximately 700 billion Euros by the year 2020. However, the majority of these devices and applications are not secured by the design and

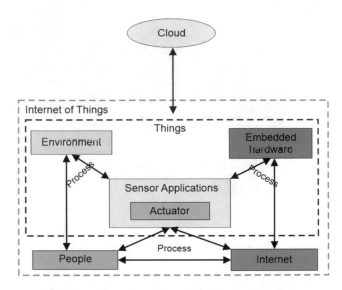

Fig. 2. Schematic architecture of Internet-of-Things

therefore vulnerable to security and privacy attacks. Therefore, IoT which depends on such devices and services, have a number of security and privacy issues. [30, 31].

IoTs often face questions about the privacy of personal data. Real-time information about our physical location or updates about the medical condition are accessed by our health care providers are streamed over wireless networks without any restrictions. Supplying power to IoT devices and their network connections is another concern. IoTs use portable devices powered with primary batteries that must be replaced when exhausted. Although many mobile devices are optimized with lower power consumption, costs to keep them running still remains high. Also, the Internet of Things opened the door for various start-ups and improved the business of corporate sector. This brought revolution in the market and also competitiveness among various solution providers. The popularity of the IoT system further demands adaptability to the different needs of the people or the configuration for many different circumstances. If people become too much dependent on this technology, any technical glitches in the system can cause serious physical and/or financial damage and hence all shortcomings should be properly studied [32–34].

2   Security Requirements in IoT

People, process and things and their inter and intra communication make a successful implementation of IoT application. IoT makes our life smarter with more ease of sustainability in term of technicalities; it also needs to be secure. IoT deals with our daily life data, health data, personal data and therefore various requirements for IoTs to be secure are significant [35].

Confidentiality: Sensor networks, wearable devices, VANETs, MANETs – all are becoming the integral parts of IoTs. All such networks contain some data which need to be protected well either in rest or in process/transaction. Therefore, confidentiality is an important factor to be included in IoT applications so that unauthorized entity should not get the understanding of the data even though forcedly accessed [36, 37].

Integrity: The data in communication throughout the IoTs and its modules need to be synchronized with mechanisms so that any modification or alteration or update-delete in ongoing data transmission can be detected and proper solution can be implemented [38, 39].

Availability: IoT is making our life simpler with more expectations of data availability with any devices at any point of time irrespective of locations. Considering this, maliciousness in the data may account for disruption of IoT services which are obviously not desired for the real-time environment such as health monitoring, location tracking, banking etc. Therefore, availability requirements must strong enough for IoT to withstand any kind of denial-of-service attacks [40–42].

Authentication: In 2020, the count for devices to be connected in IoT is predicted as 25 billion and may be more with the present status. Therefore, devices should prove their identity before becoming part of the network and transmitting or accessing any data from IoTs. Therefore, proper authentication schemes need to be incorporated in IoTs [43–46].

Non-repudiation: Extending the roles and responsibilities of the authentic devise (users), it needs to be ensured that the sender and receiver should not be able to deny the responsibility of sending or receiving the data respectively. Compromised devices may try to execute such processes with spoofed identities which need to be prevented. Enhances digital signature, hashing can be used for the purpose [47].

Access Control: Once authentication is done, authentic users must be executing their own provisions of data access and thus proper access control mechanism are required to be implemented in IoTs. These methods can be classified depending upon data access methods, users' roles or even the privacy level of the data to be accessed [48].

3 Security Challenges in IoT

Security always remains a major challenge in IoT. The application data of IoT can be diversified and always demand security and also remain confidential against stealing and tampering. Therefore, this data needs to be properly scrutinized with proper security measures [49, 50]. Several security challenges exist in IoTs which needs to be addressed and are mentioned in the following subsections.

3.1 Mobility

In various applications such as healthcare various sensors are embedded into the human body for monitoring the health information of an individual. Another example where mobility is considered as a challenge is an intelligent transportation system where network topology changes very frequently and deploying the security solution is considered as one of the stimulating tasks.

3.2 Resource Limitation

The Internet of Things comprises of tiny sensors and actuators which are used to sense the meaningful data these devices are having the limited computational capacity and thus cannot perform some complex tasks. Therefore, security will remain an open challenge with such low powered and resource-constrained devices as complex security algorithms and processes cannot be installed in such devices.

3.3 Standardization

Lack of standardization is another major security concern in IoT. Standardization is one of the major aspects that ensure interoperability such as EFC applications. Thus, using the different security solutions and not following a common guideline leaves a door open as a major concern in IoT based system.

3.4 Heterogeneity

The applicability of IoT is very vast ranging from the healthcare sector to Intelligent transportation system. Each application generates a different set of data which is heterogenous and has different security requirement. Thus, ensuring a common security solution for a vast range of applications is considered as one of the most challenging tasks.

3.5 Scalability

Due to proliferation in Internet Technology the use of smart devices increases exponentially. Scalability is one of the key concerns when it comes to security as the number of devices is continuously expanding. Thus, there should be an adaptable security solution that deals with this expansion.

3.6 Range of Attacks' Sources

The numbers of attacks (internal and external) are increasing day by day thus security solutions and policies need to be updated with new attacks introduced. The information exchanged among entities must be secured and protected from all possible attacks in order to avoid indemnities and accidents. Table 1 shows the overall summarization of various attacks and their threat level in IoT.

Table 1. Summarization of attacks in IoTs

Type	Threat level	Type	Threat level
Data Security	High	Network security	Extremely High
Insurance concerns	Low to Medium	Design considerations	Medium to High
Lack of common standards	Low to Medium	Context-aware risks	High
Technical concerns	Medium to High	Traceability/profiling	Extremely High
Security attacks	Extremely High	Decision-making problem	Medium
Application security	High		

4   Security Solutions for IoT

IoT is an emerging technology comes with lots of benefits and challenges. The device manufacturers are encouraging IoT to make daily life easy and better through numerous devices such as smart air-conditioner, smart ovens, smart cash wash, smart health monitoring, smart TVs, smart refrigerator, smartwatch, smart running machines, smart running shoes and smart band etc. IT security professionals are considering these devices unnecessary and risky for the user due to data security and privacy issues. Some of the solutions which address the security issues are as follows:

4.1   Secure the IoT Network

Security to IoT devices to the back-end systems can be achieved through employing security features such as antivirus, anti-malware, firewalls and intrusion detection and prevention system [51].

4.2   Authenticate the IoT Devices

Authentication to the IoT user can be provided by adopting multiple user management features for single IoT device and employing authentication schemes such as biometrics, three-factor security and digital certificate [52–54].

4.3   IoT Data Encryption

The privacy protection of users and prevent IoT data breaches, standard cryptographic techniques and fully-encrypted key lifecycle management mechanisms can boost up the security [55–58].

4.4   IoT PKI Security Methods

To ensure the security of IoT devices, public key infrastructure security methods such as X.509, cryptographic scheme, digital certificate and considering life-cycle skills such as public/private key generation, distribution of key, management of key and revocation can be utilized by IoT [59].

4.5   IoT Security Analytics

To detect IoT-specific attacks and malicious activity, security analytics solutions can be utilized [60].

4.6   IoT Application Programming Interface (API) Security Schemes

IoT API ensures the integrity of information association between IoT devices and back-end systems. It also ensures that only authorized devices, developers and apps are connecting with API [61].

4.7   Develop Secured IoT Apps

The security of IoT application must be emphasized by developers and also implements all above mentioned IoT technologies before developing IoT applications [62–66] (Fig. 3).

Fig. 3. Taxonomy on IoT security solution

5   Privacy and IoTs

With all those sensors collecting data on everything you do, the IoT is a potentially vast privacy headache. Let us consider the smart home which tells us about an exhaustive list of doings such as: the time to wake up (when the smart coffee machine is activated), smart toothbrush tells about your teeth using smart oven of the fridge and even what our children think about (using smart toys). The smart home alone is not smart enough. This application of IoT is installed with various applications that control the above-said functionalities. Each of those devices said above are having sensors and they collect data and transmit data accordingly. These data are significant for privacy as they relate us, our family members, our individuality and moreover our behavior. Not all smart home companies build their business model around harvesting and selling your data, but some are doing or may do. It's surprisingly easy to find out a lot about a person from a few different sensor readings [67–70].

Privacy depends upon both the parties: consumers' and service providers. Consumers need to understand what they are exchanging with the service provider and the service providers or vendors must also acknowledge the privacy concerns imposed by the consumers or users. A proper service level agreement and denial of responsibility need to be amended with existing infrastructure policies. One recent survey found that four out of five companies would be unable to identify all the IoT devices on their network [71–76]. Some of the major contributions of privacy solutions for IoT based applications have been shown below.

6 Open Research Problems

IoT applications are always a concern for various challenges as we have discussed in the previous sections. Related to those concerning issues, some of the open research problems in IoTs are identified which may help to further researchers,

- Ensuring the availability of data in IoTs
- Mitigating DoS attacks in the conjugation of spoofing and imitating attacks
- Data level privacy issues and challenges
- Data Acquisition and storing of data with optimization
- Ensuring less delay and accuracy of data
- Ensuring CIA security services with easy configurable and adaptable methods
- Privacy or Ease of Access: An on-going debate

7 Conclusion

The 'Internet of Things' at present is a technology of outmost importance and is making in roads into every 'smart' fields. Inter and intra communication between people, process and things made it a successful entity. However, without wired infrastructure, IoT devices and applications become soft targets and hence a serious security issue. Considering the importance of IoT applications, it is really necessary to resolve security issues of IoTs and its communication networks. Some of the open research problems to protect IoT services from attacks were identified and IoT researchers may be benefited.

References

1. http://saphanatutorial.com/introduction-to-internet-of-things-part-2/
2. https://en.wikipedia.org/wiki/Internet_of_things
3. Fuqaha, A.A., Guizani, M., Mohammadi, M., Aledhari, M., Ayyash, M.: Internet of Things: a survey on enabling technologies, protocols, and applications. IEEE Commun. Surv. Tutorials 17(4), 2347–2376 (2015)
4. Pfister, C.: Getting Started With the Internet of Things: Connecting Sensors and Microcontrollers to the Cloud. O'Reilly Publishers, Sebastopol (2011)
5. https://www.lifewire.com/introduction-to-the-internet-of-things-817766

6. Ghapar, A.A., Yussof, S., Bakar, A.A.: Internet of Things (IoT) architecture for flood data management. Int. J. Future Gen. Commun. Network. **11**(1), 55–62 (2018)
7. Yuan-sheng, D., Zhi-chao, Y.: Industrial IoT mining algorithm of cloud computing research. Int. J. u- e- Serv. Sci. Technol. **9**(1), 197–208 (2016)
8. Miorandi, D., Sicari, S., De Pellegrini, F., Chlamtac, I.: Internet of Things: vision, applications and research challenges. Ad Hoc Netw. **10**(7), 1497–1516 (2012)
9. Da Xu, L., He, W., Li, S.: Internet of Things in industries: a survey. IEEE Trans. Ind. Inf. **10** (4), 2233–2243 (2014)
10. Chen, S., Xu, H., Liu, D., Hu, B., Wang, H.: A vision of IoT: applications, challenges, and opportunities with china perspective. IEEE Internet Things J. **1**(4), 349–359 (2014)
11. Shrivastava, S., Shrivastava, L., Bhadauria, S.S.: Performance analysis of wireless mobile ad hoc network with varying transmission power. Int. J. Sens. Appl. Control Syst. **3**(1), 1–6 (2015)
12. Pandey, K.K., Patel, S.V.: Development of an effective routing protocol for cluster based wireless sensor network for soil moisture deficit monitoring. Int. J. Control Autom. **8**(1), 243–250 (2015)
13. Upendar Rao, R., Veeraiah, D., Mandhala, V.N., Kim, T.H.: Neighbor position verification with improved quality of service in mobile ad-hoc networks. Int. J. Control Autom. **8**(1), 83–92 (2015)
14. https://www.zdnet.com/article/25-billion-connected-devices-by-2020-to-build-the-internet-of-things/
15. Cho, M., Choi, H.R., Kwak, C.H.: A study on the navigation aids management based on IoT. Int. J. Control Autom. **8**(7), 193–204 (2015)
16. Mukherjee, S., Biswas, G.P.: Networking for IoT and applications using existing communication technology. Egypt. Inf. J. **19**(2), 107–127 (2018)
17. Kamal, Z., Mohammed, A., Sayed, E., Ahmed, A.: Internet of Things applications, challenges and related future technologies. World Sci. News **67**(February), 126–148 (2017)
18. Nalbandian, S.: A survey on Internet of Things: applications and challenges. In: 2nd International Congress on Technology, Communication and Knowledge, ICTCK 2015, pp. 165–169 (2016)
19. Shahid, N., Aneja, S.: Internet of Things: vision, application areas and research challenges. In: Proceedings of the International Conference on IoT in Social, Mobile, Analytics and Cloud, I-SMAC 2017, pp. 583–587 (2017)
20. An, Q.: Localization of sound sensor wireless network with a mobile beacon. Int. J. Sens. Appl. Control Syst. **6**(1), 1–14 (2018)
21. Kumar, M., Annoo, K., Kumar Mandal, R.: The Internet of Things applications for challenges and related future technologies & development. Int. Res. J. Eng. Technol. (IRJET) **5**(1), 619–625 (2018)
22. Özgür, L., Akram, V.K., Challenger, M., Dağdeviren, O.: An IoT based smart thermostat. In: 2018 5th International Conference on Electrical and Electronics Engineering, ICEEE 2018, pp. 252–256 (2018)
23. Pandey, S., Mahapatra, R.P.: An adaptive gravitational search routing algorithm for channel assignment based on WSN. Int. J. Sens. Appl. Control Syst. **5**(1), 1–14 (2017)
24. Kim, J.-H., Park, S.C.: IoT relationship between Korea and the Philippines. Int. J. u- e- Serv. Sci. Technol. **9**(7), 395–406 (2016)
25. Ricciardi, J., Amazonas, R., Palmieri, F., Bermudez-Edo, M.: Ambient intelligence in the Internet of Things. Mob. Inf. Syst. **2017**, 1–3 (2017)
26. Priyadarshini, R., Narayanappa, C.K., Chander, V.: Optics based biosensor for medical diagnosis. Int. J. Sens. Appl. Control Syst. **5**(2), 1–14 (2018)

27. Sastry, C., Ma, C., Loiacono, M., Tas, N., Zahorcak, V.: Peer-to-peer wireless sensor network data acquisition system with pipelined time division scheduling. In: 2006 IEEE Sarnoff Symposium (2006)
28. Kim, J.T.: Requirement of security for IoT application based on gateway. Int. J. Secur. Appl. **9**(10), 201–208 (2015)
29. Kaur, S., Hans, A., Singh, N.: An overview to Internet of Things (IOT). Int. J. Future Gener. Commun. Network. **9**(9), 239–246 (2016)
30. Mostefa, B., Abdelkader, G.: A survey of wireless sensor network security in the context of Internet of Things. In: Proceedings of the 2017 4th International Conference on Information and Communication Technologies for Disaster Management, ICT-DM 2017, vol. 2018–January, pp. 1–8 (2018)
31. Wang, L., Wang, X.V., Wang, L., Wang, X.V.: Latest advancement in CPS and IoT applications. In: Cloud-Based Cyber-Physical Systems in Manufacturing, pp. 33–61 (2017)
32. Schaumont, P.: Security in the Internet of Things: a challenge of scale. In: Proceedings of the 2017 Design, Automation and Test in Europe, DATE 2017, pp. 674–679 (2017)
33. Hang, L., Kim, D.-H : Design and implementation of IoT interworking of anchor service provider and mobius platform using RESTful API. Int. J. Control Autom. **10**(10), 101–112 (2017)
34. Eom, J.H.: Security threats recognition and countermeasures on smart battlefield environment based on IoT. Int. J. Secur. Appl. **9**(7), 347–356 (2015)
35. Kim, Y.-H., Kim, M.-S., Park, M.-H., Kang, S.-K., Eun, C.-S.: Study on the low energy consumption method for light-wight devices in IoT service environment. Int. J. Grid Distrib. Comput. **11**(8), 99–108 (2018)
36. Schaumueller-Bichl, I., Kolberger, A.: IoT as a challenge in information security. Elektrotechnik Und Informationstechnik **133**(7), 319–323 (2016)
37. Seo, J.-H., Choi, J.-T.: Customized mobile marketing platform design utilizing IoT based Beacon sensor devices. Int. J. Grid Distrib. Comput. **11**(8), 57–68 (2018)
38. Abdulrahman, Y.A., Kamalrudin, M., Sidek, S., Hassan, M.A.: Internet of Things: issues and challenges. J. Theor. Appl. Inf. Technol. **94**(1), 52–60 (2016)
39. Kim, J., Lee, Y.: A study on the virtual machine-based anti-theft system running on IoT devices. Int. J. Grid Distrib. Comput. **11**(8), 1–12 (2018)
40. Macaulay, T.: Confidentiality and integrity and privacy requirements in the IoT. In: RIoT Control, pp. 125–139 (2016)
41. Lee, S.: Communication technology and application of Internet of Things (IoT) in smart home environment. Int. J. Control Autom. **10**(3), 397–404 (2017)
42. Lee, K., Kim, D., Choi, H.R., Park, B.K., Cho, M.J., Kang, D.Y.: A study on IoT-based fleet maintenance management. Int. J. Control Autom. **10**(4), 287–296 (2017)
43. Abomhara, M., Koien, G.M.: Security and privacy in the Internet of Things: current status and open issues. In: 2014 International Conference on Privacy and Security in Mobile Systems, PRISMS 2014 - Co-located with Global Wireless Summit (2014)
44. Patil, A., Bansod, G., Pisharoty, N.: Hybrid lightweight and robust encryption design for security in IoT. Int. J. Secur. Appl. **9**(12), 85–98 (2015)
45. Jung, S.H., An, J.C., Park, J.Y., Shin, Y.T., Kim, J.B.: An empirical study of the military IoT security priorities. Int. J. Secur. Appl. **10**(8), 13–22 (2016)
46. Park, K., Kim, I., Park, J.: An efficient multi-class message scheduling scheme for healthcare IoT systems. Int. J. Grid Distrib. Comput. **11**(5), 67–78 (2018)
47. Kumar, N., Madhuri, J., Channegowda, M.: Review on security and privacy concerns in Internet of Things. In: IEEE International Conference on IoT and its Applications, ICIOT 2017 (2017)

48. Oriwoh, E., Al-Khateeb, H., Conrad, M.: Responsibility and non-repudiation in resource-constrained Internet of Things scenarios. In: International Conference on Computing and Technology Innovation (CTI 2015), 27–28 May 2015
49. Kouicem, D.E., Bouabdallah, A., Lakhlef, H.: Internet of Things security: a top-down survey. Comput. Netw. **141**, 199–221 (2018)
50. Alaba, F.A., Othman, M., Abaker, I., Hashem, T., Alotaibi, F.: Internet of Things security: a survey. J. Netw. Comput. Appl. (2017)
51. Chollet, S., Pion, L., Barbot, N., Michel, C.: Secure IoT for a pervasive platform. In: 2018 IEEE International Conference on Pervasive Computing and Communications Workshops, PerCom Workshops 2018, pp. 113–118 (2018)
52. Shah, T., Venkatesan, S.: Authentication of IoT device and IoT server using secure vaults. In: Proceedings - 17th IEEE International Conference on Trust, Security and Privacy in Computing and Communications and 12th IEEE International Conference on Big Data Science and Engineering, Trustcom/BigDataSE 2018, pp. 819–824 (2018)
53. Kim, S.H., Lee, I.Y.: IoT device security based on proxy re-encryption. J. Ambient Intell. Hum. Comput. **9**(4), 1267–1273 (2018)
54. Hwang, S.: Monitoring and controlling system for an IoT based smart home. Int. J. Control Autom. **10**(2), 339–348 (2017)
55. Xuan, S., Park, D.-H., Kim, D.: A service platform for real-time video streaming based on RESTful API in IoT environments. Int. J. Control Autom. **10**(3), 181–192 (2017)
56. Jiang, J., Yang, D., Gao, Z.: Study on application of IOT in the cotton warehousing environment. Int. J. Grid Distrib. Comput. **8**(4), 91–104 (2015)
57. Yang, H.-K., Cha, H.-J., Song, Y.-J.: A study of secure distributed management of sensing data in IoT environment. Int. J. Adv. Sci. Technol. **124**, 21–32 (2019)
58. Byun, S.: Viability-based replication management scheme for reliable IoT data services. Int. J. Adv. Sci. Technol. **124**, 89–102 (2019)
59. Schukat, M., Cortijo, P.: Public key infrastructures and digital certificates for the Internet of Things. In: 2015 26th Irish Signals and Systems Conference, ISSC 2015 (2015)
60. Matuszak, G., Bell, G., Le, D.: Security and the IoT. In: KPMG (2015)
61. Dayaker, P., Reddy, Y.M., Kumar, M.B.: A survey on applications and security issues of Internet of Things (IoT). Int. J. Mech. Eng. Technol. **8**(86), 641–648 (2017)
62. Alaa, M., Zaidan, A.A., Zaidan, B.B., Talal, M., Kiah, M.L.M.: A review of smart home applications based on Internet of Things. J. Netw. Comput. Appl. **97**, 48–65 (2017)
63. Deol, R.K.: Intruder detection system using face recognition for home security IoT applications: a Python Raspberry Pi 3 case study. Int. J. Secur. Technol. Smart Dev. **5**(2), 21–32 (2018)
64. Son, Y., Lee, Y.: A study on the interpreter for the light-weighted virtual machine on IoT environments. Int. J. Web Sci. Eng. Smart Dev. **3**(2), 19–24 (2017)
65. Kim, S.J., Min, J.H., Kim, H.N.: The development of an IoT-based educational simulator for dental radiography. IEEE Access **7**, 12476–12483 (2019)
66. Gehlot, A., Singh, R., Mishra, R.G., Kumar, A., Choudhury, S.: IoT and Zigbee based street light monitoring system with LabVIEW. Int. J. Sens. Appl. Control Syst. **4**(2), 1–8 (2017)
67. Sharma, A., Salim, M.: On the feasibility of polar code as channel code candidate for the 5G-IoT scenarios. Int. J. Future Gener. Commun. Netw. **11**(3), 11–20 (2018)
68. Jiantao, C., Xiaojun, Z.: Study of IoT terminal interface platform based on embedded technology and Zigbee protocol. Int. J. Future Gener. Commun. Netw. **9**(6), 55–64 (2016)
69. Ghapar, A.A., Yussof, S., Bakar, A.A.: Internet of Things (IoT) architecture for flood data management. Int. J. Future Gener. Commun. Netw. **11**(1), 55–62 (2018)

70. Choi, J.H., Kang, U.G., Lee, B.M.: IoT service model for measuring sleep disorder using CoAP. Int. J. Mob. Dev. Eng. **1**(1), 9–14 (2017)
71. Yang, J.H., Ryu, Y.: Design and development of a command-line tool for portable executable file analysis and malware detection in IoT devices. Int. J. Secur. Appl. **9**(8), 127–136 (2015)
72. Li, S.: Introduction: securing the Internet of Things. In: Securing the Internet of Things, pp. 1–25 (2017)
73. Bhargavi, M., Nagabhushana Rao, D.M.: Security issues and challenges in IOT: a comprehensive study. Int. J. Eng. Technol. **7**(2.32), 298 (2018)
74. Jemshit, T.: A simple energy efficient routing algorithm for the IoT environment. Int. J. Cloud-Comput. Super-Comput. **4**(2), 7–12 (2018)
75. Shaikh, M.H.: Indication of European union clients' privacy conservation in an IoT SCM system. Int. J. Private Cloud Comput. Environ. Manage. **5**(2), 7–12 (2018)
76. Vardhan, K.A., Krishna, M.M.: DGD: an intrusion detection system for providing security to web applications. J. Stat. Comput. Algorithm **1**(1), 27–32 (2017)

An Improved Localization Algorithm Based on DV-Hop

Liquan Zhao[1(✉)], Kexin Zhang[1], and Yanfei Jia[2]

[1] Key Laboratory of Modern Power System Simulation and Control & Renewable Energy Technology, Ministry of Education (Northeast Electric Power University), Jilin 132012, China
zhao_liquan@163.com
[2] College of Electrical and Information Engineering, Beihua University, Jilin 132012, China

Abstract. In wireless sensor network's localization, the distances vector per hop algorithm is typical localization, to improve the location error, an improved localization algorithm is proposed based on the distances vector per hop algorithm. Firstly, we set the different communication distances for different nodes. Each communication distance corresponds to one hop. Secondly, if the hop count is less than one, the average hop distance of the nearest anchor node is used as the hop distance of unknown node. On the contrary, the weighted average hop distance sum of the nearest four anchor nodes is used as the hop distance of unknown node. In the simulations, we compare the proposed algorithm with distances vector per hop algorithm and weighted-based distances vector per hop algorithm. The simulation results of matlab showed that the proposed method has smaller error for unknown node's localization than the other algorithm.

Keywords: Wireless sensor networks · Localization error · Distances vector per hop algorithm · Hop count

1 Introduction

In wireless sensor networks (WSN), the nodes collect information from surrounding environment [1]. It is a system which it has the function real-time acquisition, signal detection, data monitoring, data storage [2]. The nodes collect information from surrounding environment, to monitor or discovering trends in data change. Such as when a node collect the information is not in normal range, It feeds back abnormal information to the system [3]. If we want to deal with this problem, we must know the node's coordinate. So the node need to known itself position [4, 5] and the location is accurately or not, is important. This is the premise and foundation of WSN application [6–8]. In recent years, it has a great achieve in localization algorithm [9–12].

Range-based algorithm and range-free algorithm [13] are main methods to solve the node's localization in WSN. Range-based algorithm has a higher accuracy, due to it rely on the physical hardware, such as GPS [14] etc. Each node can get the more accuracy information, such as the distance and angle between the node and other nodes.

F. Xhafa et al. (Eds.): IISA 2019, AISC 1084, pp. 590–595, 2020.
https://doi.org/10.1007/978-3-030-34387-3_72

But the physical hardware is expensive, the node's energy is limited and hardware consume is serious. If we use this method to locate unknown node's coordinate, the cost more expensive and the network's lifetime is shorter [15]. So this method isn't the best way to position. Another is range-free algorithm has a lower accuracy, it rely on the connection between each nodes, so it don't need addition hardware. So the cost and the energy consumption is lower [16]. By contrast to range-based algorithm, the algorithm based on range-free has a bigger potential. In range-free algorithm, Distances Vector Per Hop (DV-Hop) algorithm is the typical algorithm to locate the unknown node's coordinate [17, 18]. Improved DV-Hop algorithm shows a differential scheme of error correction [19]. The weighted DV-Hop algorithm uses POS algorithm to correct DV-Hop algorithm [20–22]. The result is effective. To increase the accuracy of unknown nodes, at this paper, we proposed a algorithm, the new algorithm belong to range-free algorithm based on DV-Hop. Preliminary improvement the localization accuracy, make a good foundation to research in the future.

2 DV-Hop Algorithm

DV-Hop algorithm aims to known unknown node where it is and get itself coordinate, the technique rely on the connection of nodes. During the node placement phase, the nodes are distributed in workspace, the nodes consist of two sorts, the one can obtain itself location named anchor nodes, the other is unknown nodes, but they can estimate their coordinate by the relationship to anchor nodes. DV-Hop algorithm's principle has three steps as follow:

Step one: Anchor nodes broadcasts information to the whole networks, such as identifier, coordinate etc, other nodes receive those information and obtain the minimum hop counts from the same nodes. When the broadcast finish, anchor nodes calculate the distance of average per hop as follow:

$$H = \sum_{i \neq j} \sqrt{(x_i - x_j)^2 + (y_i - y_i)^2} / \sum_{i \neq j} hop_{j,i} \qquad (1)$$

where (x_i, y_i) and (x_j, y_j) are anchor nodes i and j's coordinates, and $hop_{j,i}$ is hop counts to other anchor nodes.

Step two: Each anchor nodes obtain the value of H, then flooding this information to the whole networks. Unknown nodes only gets the value at the first time and retains this information, so each unknown nodes can obtains the information and it is the only one, this value is unknown node's the distance of average per hop. That's mean that unknown node obtain the value for the nearest anchor node.

Step three: Utilize unknown node j the distance of average per hop and the hop counts to anchor nodes. We can get the estimate distance as follow:

$$d_{j,i} = hop_{j,i} \times H_j \qquad (2)$$

where $hop_{j,i}$ is unknown node j's hop counts to anchor node i, H_j represents the distance of average per hop which unknown node j obtains. After that, unknown nodes obtain the estimate coordinate by the Least Squares method.

3 Proposed Algorithm

At the part, we proposed a new algorithm to improve DV-Hop algorithm. The main work is to increase the hop counts accurately and fully considering the anchor node's distribution around unknown nodes which obtain the distance of average hop more accurately. For every nodes where in i nodes communication range, they can receive the information from node i whatever the real distance is far or near. So that setting different communication intervals are necessary. The nodes distributed in the workspace, the location is random.

Hop counts obtain principle: Setting different transmit power of nodes, the communication range is 0 to 0.3R, 0 to 0.6R and 0 to 0.9R, corresponding the hop count is 0.2, 0.5 and 0.8.

Unknown node obtains the four anchor nodes' the distance of average per hop nearest it. When the hop count is less than 1, then estimated the distance to anchor nodes is:

$$d_{j,i} = hop_{j,i} \times H_j \qquad (3)$$

When the hop count is more than 1, then estimated the distance is:

$$d_{j,i} = H_j + AH \times (hop_{i,j} - 1) \qquad (4)$$

Where AH is the average hop distance of selected anchor nodes at a weighted concept to calculate with the hop counts, it is expressed as:

$$AH = \sum_i H_i * (hop_{i,j} - 1) \qquad (5)$$

The theory of improved algorithm is as follows:

Step one: Setting different communication range, the hop count is 0.2, 0.5, 0.8. Anchor nodes broadcast information in turn to the workspace, each node only obtain the minimum hop from the same node. We can get the hop counts between each node after above processing. The distance of average per hop will be calculated for each anchor node.

Step two: Anchor nodes broadcast the value which distance of average per hop to the workspace, unknown node only retain the four nearest anchor nodes' values. Then the estimate distance can be calculated by (3) or (4).

Step three: Using Least Square method, we can get unknown node's coordinate.

4 Simulation

We simulate DV-Hop algorithm, weighted algorithm of DV-Hop [13] and proposed algorithm at a square area, and side length is 100 m, the nodes' number are 100 in total. The cover radius of node is 50 m and they are distributed in this area randomly. The anchor nodes can obtain the position information accurately. In this simulation, anchor node's amount are 10 which we assume. We also introduce a concept about the accurate of localization. The formula is expressed as follows:

$$acc = \frac{Error}{R} \tag{6}$$

Where *Error* is average error of all unknown nodes, *R* is the maximum radius of each node's communication, *acc* is the overall performance of system.

Figure 1 show the location errors for the different algorithm. From this figure, we can see the proposed method has smaller error than the others. Figure 2 shows the localization accuracy for different simulation times. From the figure, we can see that the proposed algorithm has smaller relative error than other algorithm.

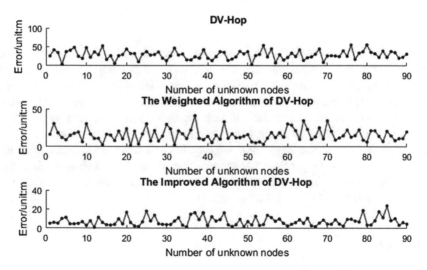

Fig. 1. Comparison of three algorithms for different number of unknown nodes

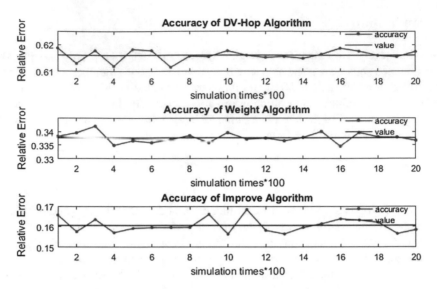

Fig. 2. Comparison of accuracy for different simulation times

5 Conclusion

At the article, we improved DV-Hop algorithm from the hop counts between anchor nodes and unknown nodes, and the distance average per hop of unknown node's obtained from anchor nodes surrounding it, the new algorithm is proposed. The proposed algorithm introduces the communication distances for different nodes and the relationship between the hop count and hop distance to the distances vector per hop algorithm. For the distance average of per hop which unknown node gets, anchor node's distributing is considered surrounding the unknown node, according to the range of hop counts, to get the distance average of per hop. The new algorithm is simulated at an ideal environment by MATLAB. The results show that, this method can increase position accuracy of unknown nodes.

References

1. Liao, W.H., Shih, K.P., Lee, Y.C.: A localization protocol with adaptive power control in wireless sensor networks. Comput. Commun. **31**(10), 2496–2504 (2008)
2. Zhao, X., Zhou, Y.: Development and application of on-line monitoring and fault diagnosis system for centrifugal PUMP based on lab VIEW. J. Northeast Electr. Pow. Univ. **37**(2), 66–72 (2017)
3. Zhu, H., Li, Y.: Distribution transformer monitoring system operating parameters and design. J. Northeast Electr. Pow. Univ. **38**(5), 74–79 (2018)
4. Mehaseb, M.A., Gadallah, Y., El-Hennawy, H.: WSN application traffic characterization for integration within the Internet of Things. In: Proceedings of IEEE International Conference on Mobile Ad-hoc and Sensor Networks, Dalian, pp. 318–323 (2013)

5. Kouche, A.E., Al-Awami, L., Hassanein, H., Obaia, K.: WSN application in the harsh industrial environment of the oil sands. In: Proceedings of International Wireless Communication and Mobile Computing Conference, Istanbul, Turkey, pp. 613–618 (2011)
6. Niclescu, D., America, N.L.: Communication paradigms for sensor network. IEEE Commun. Mag. **43**(3), 116–122 (2005)
7. Guo, X., Sun, M.: Design of communication manager for hydropower station. J. Northeast Electr. Pow. Univ. **37**(4), 94–97 (2017)
8. Jin, E., Jin, Y., Chen, Y., Zhao, Y.: Study on the new integrated protection of intelligent substations. J. Northeast Dianli Univ. **36**(6), 25–29 (2016)
9. Chatterjee, A.: A fletcher-reeves conjugate neural-network-based localization algorithm for wireless sensor networks. IEEE Trans. Veh. Technol. **59**(2), 823–830 (2010)
10. Li, M., Liu, Y.: Range-free localization in an isotropic sensor networks with holes. IEEE/ACM Trans. Network. **18**(1), 320–332 (2010)
11. Kwon, O., Song, H., Park, S.: The effects of stitching orders in patch-and-stitch WSN localization algorithms. IEEE Trans. Parallel Distrib. Syst. **20**(9), 1380–1391 (2009)
12. Katenka, N., Levina, E., Michailidis, G.: Robust target localization from binary decisions in wireless sensor networks. Technometrics 448–461 (2008)
13. Han, G., Xu, H., Duong, T.Q., et al.: Localization algorithms of wireless sensor networks: a survey. Telecommun. Syst. 2419–2436 (2013)
14. Grewal, M.S., Weill, L.R., Andrews, A.P.: Global Positioning Systems, Inertial Navigation And Integration. Wiley, Hoboken (2007)
15. Zengyou, S., Chi, Z.: Adaptive clustering algorithm in WSN based on energy and distance. J. Northeast Dianli Univ. **36**(1), 82–86 (2016)
16. Teng, Z., Xu, M., Li, Z.: Nodes deployment in wireless sensor networks based on improved reliability virtual force algorithm. J. Northeast Dianli Univ. **36**(2), 86–89 (2016)
17. Niculescu, D., Nath, B.: Ad-hoc positioning system (APS). In: Proceedings of the 2001 IEEE Global Telecommunications Conference [S1], pp. 2926–2931. IEEE Communication Society (2001)
18. Niculescu, D., Nath, B.: DV based positioning in ad hoc networks. J. Telecommun. Syst. **22**(4), 267–280 (2003)
19. Yi, X., Liu, Y., Deng, L., He, Y.: An improved DV-hop positioning algorithm with modified distance error for wireless sensor network. In: The 2nd International Symposium on Knowledge Acquisition and Modeling, vol. 2, pp. 216–218 (2009)
20. Hu, Y., Li, X.: An improvement of DV-Hop localization algorithm for wireless sensor networks. Telecommun. Syst. 13–18 (2013)
21. Singh, S.P., Sharma, S.C.: A PSO based improved localization algorithm for wireless sensor network. Wireless Pers. Commun. 1–17 (2017)
22. Tianxiao, C., Xiaolong, Z., Wenhao, L.: Gear fault diagnosis based on hilbert envelope spectrum and SVM. J. Northeast Elect. Power Univ. **37**(6), 56–61 (2017)

Simulation Study of the Improved Positive Feedback Active Frequency Shift Island Detection Method

Meng Wang, Mengda Li[⊠], and Xinghao Wang

School of Electrical Engineering, Shanghai Institute of Electrical Engineering,
No. 300, Shuihua Road, Pudong New Area, Shanghai 201306, China
254025450@qq.com

Abstract. Island detection is one of the most important problems in distributed generation systems. There are many methods of island detection, especially active island detection. In this paper, the active frequency detection method based on positive feedback (PFAFD), this paper proposes a judging load characteristic, thus modified disturbance in the direction of the new algorithm, avoids the disturbance in add a fixed initial values of load cases the contrary, delay the testing time and even lead to island failed test problems, and based on this algorithm to join the secondary disturbance quantity to speed up the detection speed and reduce the action time. Finally, MATLAB/SIMULINK software was used to build a simulation model of distributed grid-connected system, and the island detection algorithm proposed in this paper was simulated to verify the experimental results.

Keywords: Distributed generation and grid connection · Island detection · Active frequency offset

1 Introduction

With the rapid development of industrial economy, human society has accelerated the exploitation and utilization of energy. The depletion of energy resources and the emergence of environmental problems have forced countries to increase investment in new energy technologies to varying degrees. The development of distributed power supply technology has become an indispensable link in the development and utilization of new energy technology. In recent years, renewable energy has developed rapidly due to its characteristics of clean, environmental protection and renewable energy [1]. More and more Distributed Generation (DG) systems are connected to the power grid, making the problem of island effect increasingly prominent and attracting more and more attention. Island effect refers to that after the power supply is suddenly stopped, the photovoltaic power generation device fails to detect the power loss state of the power grid in time, and continues to supply power to the lost voltage grid, and forms a self-powered island system with surrounding loads [1]. Under the condition of island operation, the power quality of users will be affected, which will lead to the damage of electrical equipment. The original uncharged line is still charged, posing a threat to the

© Springer Nature Switzerland AG 2020
F. Xhafa et al. (Eds.): IISA 2019, AISC 1084, pp. 596–603, 2020.
https://doi.org/10.1007/978-3-030-34387-3_73

life and safety of line maintenance personnel [2, 3]. Therefore, how to quickly and reliably detect the island running state has become a hot topic.

Traditional active frequency shift detection methods are increasingly used in power generation sectors worldwide [4]. In this paper, a new Active Frequency shift isolated island detection scheme was proposed based on previous studies. The traditional PFAFD (Positive Feedback Active Frequency Drift) method was optimized. On the basis of ensuring power quality, the detection blind area was reduced and the detection speed of PFAFD was further improved.

2 Improved Active Frequency Shift Detection Based on Positive Feedback

2.1 Improved PFAFD Detection Method and Its Principle

Before the connection port is disconnected from the grid and the new steady state occurs, the frequency of the common coupling point will change significantly under the condition that the inverter is not disconnected in time. The steady-state frequency shall be as follows [5]:

$$\arctan\left[R\left(\omega c - \frac{1}{\omega l}\right)\right] = \frac{\omega t_z}{2} = \frac{\pi}{2} cf_k \tag{1}$$

Form (1), R, L, C for the resistance, inductance, and capacitance of the load, t_z for the duration of forced half-wave current zero crossing, T is a voltage period, $cf_k = 0.5T$ is disturbance of the power grid after loss of voltage, K indicates the moment when the action occurred.

$$Q_f = \frac{R}{\omega_0 L} \tag{2}$$

$$C_{res} = \frac{1}{\omega_0^2 L} \tag{3}$$

$$C_N = \frac{C}{C_{res}} \tag{4}$$

Of which, the quality factor set for island testing is Q_f, C is Load capacitance, C_{res} is Resonant capacitance, C_N is to mark the load capacitance, ω_0 for grid angular frequency [6].

The angular frequency of the common point between the grid and the connected end is as follows:

$$\omega = \omega_0 + \Delta\omega \tag{5}$$

According to formula (4):

$$C = C_N \bullet C_{res} = (1 + \Delta C) \bullet C_{res} \qquad (6)$$

substituting formula (6), (5), (4), (3), (2) into formula (1), and obtaining:

$$\arctan\left[Q_f \omega_0 \left(\frac{\Delta C\left(1 + \frac{2\Delta\omega}{\omega_0}\right) + 2\frac{\Delta\omega}{\omega_0} + \left(\frac{\Delta\omega}{\omega_0}\right)^2(1 + \Delta C)}{\omega_0 + \Delta\omega}\right)\right] = \frac{\pi}{2}cf_k \qquad (7)$$

Because the value $\frac{\Delta\omega}{\omega_0}$ is small, it's almost zero, to simplify, supposes $\left(\frac{\Delta\omega}{\omega_0}\right)^2 \approx 0$ and $\left(1 + \frac{2\Delta\omega}{\omega_0}\right) \approx 1$, the resulting formula (7) can be equivalent to:

$$\arctan\left[Q_f\left(\Delta C + 2\frac{\Delta\omega}{\omega_0}\right)\right] = \frac{\pi}{2}cf_k \qquad (8)$$

From the formula (8), when Q_f and ΔC at a certain time, frequency of common coupling point $\Delta\omega$ generation Post-disturbance momentum of the main off-grid the impact cf_k. If the net is broken and $\Delta\omega$ is always lower than the threshold, the islanding detection fails [7, 8].

In order to further pursue the detection speed and ensure the accuracy at the same time, we propose an active frequency offset method based on positive feedback. Experience shows that when $k > 0.05$, the detection effect can be guaranteed [9]:

$$cf_{k+1} = cf_k + k \bullet sign(f - f_g) + sign(f - f_g) \bullet [k \bullet (f - f_g)]^2 \qquad (9)$$

Of which, cf_k is a disturbance signal for outputting a current for the K-th period of the inverter, the initial value of cf_k is $0.06\Delta f$, Δf is the difference rated operating frequency for the power grid f and current frequency f_g.

Compared with the traditional Active Frequency Drift (AFD) method, the improved PFAFD method has a smaller non-detection region at the same frequency, and can accelerate the frequency offset and shorten the detection time [10]. At the same time, the load characteristic is avoided under the condition of adding fixed initial value disturbance, and the island detection time is delayed and even leads to the failure of islanding detection [11, 12].

2.2   Parameter Optimization for PFAFD Detection

In IEEE Std. 929-2000, Q_f is not more than 2.5, which is a necessary condition for all loads to realize blind-free detection in theory. The non-blind area of the detection mode also requires the size range of the load capacitance, that is, the boundary extremum of the capacitance must meet the following conditions [13]:

$$c_{norm}^+ - c_{norm}^- < 0 \qquad (10)$$

When the angle is small, $\tan \alpha = \alpha$, that is, nearly linear, then the formula (11) can be simplified as follows:

$$C_{norm} = \frac{\frac{\pi c_{f_0}}{2} + \frac{\pi k \Delta f}{2}}{Q_f} - \frac{2\Delta\omega}{\omega_0} + 1 \qquad (11)$$

Furthermore, the range of feedback gain K is deduced when there is no detection blind area:

$$k > \frac{8Q_f}{\omega_0} \qquad (12)$$

As is known from the formula (12), in the case of $Q_f = 2.5$, $f_0 = 50\,\text{Hz}$, if the PFAFD method is satisfied without blind area detection, the following conditions should be satisfied: $k > 0.0531$, island detection carried out smoothly [14].

3 Simulation Experiment of Improved PFAFD Detection Method

3.1 Simulation Model

On the basis of the above-mentioned theoretical analysis, the simulation model of 2 kW single-phase photovoltaic power generation system is established by using Matlab/Simulink. The simulation part is composed of the main circuit part and the control part of the single-phase photovoltaic power generation system. The control part includes the grid-connected control part and the island detection part [15]. The function of PFAFD module in island detection is realized by S function. Figures 1 and 2 are the main circuit and control parts of the entire single-phase photovoltaic system.

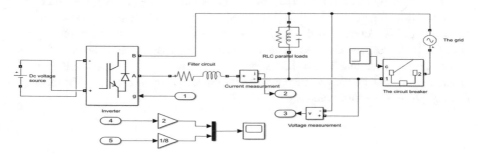

Fig. 1. Main circuit module

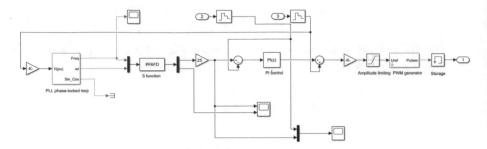

Fig. 2. Island detection and control module

3.2 Simulation Results and Analysis

In order to ensure the authenticity of the experiment, take the DC voltage as 400 V, taking the peak voltage of the power grid as 310 V, the valid value is 220 V, frequency is power frequency 50 Hz, the reference value of inverter output current is 12.5 A, the threshold of island detection frequency is set to 50 ± 0.5 HZ, the filter circuit selects the RL combination, filter inductance is 6 mH, filter resistance is 0.01 Ω, load quality factor known above $Q_f = 2.5$, the value of RLC load parameters can be obtained by calculation: R = 26.67 Ω,   L = 33.95 mH,   C = 2.98 μF;   resonant   frequency f_0 = 50 Hz. The grid is disconnected at 0.12 s.

For the convenience of observation, the actual voltage peak value is reduced to 8 times to output, and the actual inverter output current is expanded by 2 times to output. Table 1 is the simulation results of islanding detection. According to the different load properties, the traditional direction correction and secondary disturbance-free algorithm and the new algorithm proposed in this paper are simulated and analyzed, respectively. The simulation results are shown in Figs. 3 and 4. Figure 3(a) is a traditional positive feedback frequency offset method, and Fig. 3(b) is an improved positive feedback frequency offset method proposed in this paper. As we can see from Fig. 3 of the emotional load, when the power grid is disconnected at t = 0.12 s, through 8 grid cycl, t = 0.28 s, the offset of the post-frequency exceeds the pre-set threshold of frequency action for over-frequency protection, and the output of the inverter will be turned off at the same time. Figure 3(b) is an improved positive feedback frequency offset method. Since the load is inductive load, when the power grid is disconnected from the common end, the common point changes in the direction of increasing the frequency. After six power grid cycles, that is t = 0.24 s, when the power grid is separated from the common end, the frequency increases in the direction of increasing. The back frequency exceeds the threshold.

Figure 4 is an analysis of the simulation results when the load is a capacitive load, (a) a conventional positive feedback frequency offset method, and (b) the improved positive feedback active frequency offset method proposed herein. As we can see from Fig. 4(a), after the great power grid disconnects at t = 0.12 s, as the load is capacitive, the frequency of the common coupling point changes in the direction of decreasing after the power-off of the power grid. After 7 grid cycles, that is, t = 0.26 s, the offset of the post-frequency exceeds the pre-set threshold of frequency action, and the over-frequency protection is carried out, and the output of the inverter is turned off at the same time.

Figure 4(b) is the simulation result of the improved positive feedback frequency offset method. Because the improved algorithm adds disturbance signal according to the direction of common coupling point frequency offset selectively, it avoids the inundation of frequency change. After six grid cycles, that is, after t = 0.24 s, under-frequency protection occurs, the inverter is turned off. Compared with the traditional algorithm, the improved detection method is superior 0.02 s to the traditional detection method in detection time under the same conditions.

The simulation results show that the improved algorithm can avoid the slow detection caused by the capacitive load. At the same time, after adding the secondary disturbance, under the premise of ensuring the power quality, the operation time after the islanding detection is completed can be reduced. More accurate, in time to achieve the purpose of island detection.

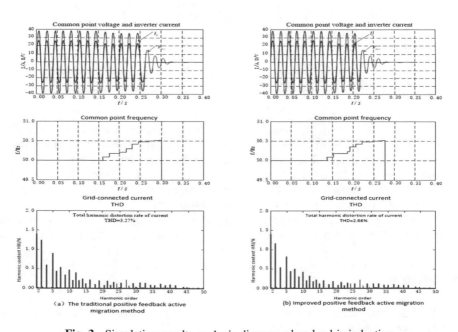

Fig. 3. Simulation results analysis diagram when load is inductive

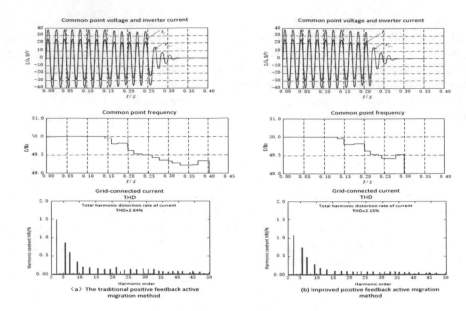

Fig. 4. Simulation results analysis diagram when load is capacitive

Table 1. Simulation results of island detection

Map title	The load type	Testing time/s	Grid-connected current THD/%
3.2.1(a)	Inductive	0.144	3.27
3.2.1(b)	Inductive	0.126	2.66
3.2.2(a)	Capacitive	0.163	2.64
3.2.2(b)	Capacitive	0.149	2.15

Acknowledgments. First of all, I would like to express my heartfelt thanks to my mentor. Mr. Li mengda gave me pertinent advice and a lot of guidance for my paper. I really appreciate his help in finishing this paper. I am also deeply grateful to all my other mentors and teachers. Translation studies have helped me directly and indirectly. I would particularly like to thank my friends for their considerable time and effort in commenting on the draft. Finally, I would like to thank my family for their continued support and encouragement.

This paper is sponsored by the applied undergraduate pilot program of Shanghai municipal universities – electrical engineering and automation (project no. g2-17-7201-008-005).

References

1. Yang, S., Han, N., Luo, N.: A review of the research on grid-connected effects of distributed generation systems. Autom. Instrum. Table **10**, 15–18 (2012)
2. Cheng, Q., Wang, Y., Cheng, Y.-m., et al.: Overview of island detection methods in distributed power generation grid-connected systems. Pow. Syst. Prot. Control **39**(6), 147–154 (2011)

3. Lu, Y., Shi, Q.: Island partition of distributed generation distribution network based on improved particle swarm optimization algorithm. New Electr. Pow. Technol. **35**(07), 17–23 (2016)
4. Yuan, J., Ding, H., Yang, Q.: Research and research of improved SMS island detection method. New Technol. Electr. Electr. Energy **36**(07), 51–56 (2017)
5. Cai, F.-H., Zheng, B.-W., Wang, W.: Island detection technology for photovolt-connected grid-connected power generation system combined with synchronous phase locking. Acta Electrotechnical Sinica **27**(10), 202–206 (2012)
6. Ding, H., Ho, Y., Lacquer, et al.: Island detection based on parabolic SMS algorithm. J. Electrotech. Technol. **28**(10), 233–240 (2013)
7. Shariatinasab, R., Akbari, M.: New islanding detection technique for DG using discrete wavelet transform. In: 2010 IEEE International Conference on Power Energy, 29 November–1 December, Kuala Lumpur, pp. 294–299 (2010)
8. Yuan, L., Zheng, J., Zhang, X.: Analysis and improvement of island detection method for grid-connected photovoltaic system. Pow. Syst. Autom. **31**(21), 72–75 (2007)
9. Yasser, M., Saadany, E.F.: Reliability evaluation for distribution system with renewable distributed generation during islanded mode of operation. IEEE Trans. Pow. Syst. **24**(2), 372–581 (2009)
10. Cao, H., Tian, Y.: Islanding control for grid-connected inverters. Pow. Syst. Prot. Control **38** (9), 72–75 (2010)
11. Wang, X., Freitas, W., Wilsun, X., et al.: Impact of DG interface controls on the Sandia frequency shift anti-islanding method. IEEE Trans. Energy Convers. **22**(3), 792–794 (2007)
12. Wang, N., Gao, P., Jia, Q., Sun, L., et al.: The active and reactive power coordination control strategy of the photovoltaic grid-connected system participating in voltage regulation research. Electr. Energy New Technol. **36**(08), 23–29 (2017)
13. IEEE Std.929-2000, IEEE Recommended Practice for U-tility Interface of PhotoVoltaic (PV) Systems [s]. Institute of Electrical and Electronic Engineers, New York Editor (2000)
14. Ren, B., Sun, X., et al.: New current disturbance method for island detection of single-phase distributed power generation system. J. Electr. Technol. **24**(7), 157–163 (2009)
15. Wang, W., Chen, A.: A modified method for detecting isolated islands with positive reactive current-frequency feedback. Proc. Pow. Supply **14**(5), 54–59 (2016)

Research on Framework Load Correlations Based on Automatic Data Extraction Algorithm

Meiwen Hu[(⊠)], Binjie Wang, and Shouguang Sun

Beijing Jiaotong University, Beijing, China
18401608297@163.com

Abstract. Actual framework loads are critical to the safety of urban rail vehicles, whose operating modes have an important impact on the fatigue of framework. When the vehicle's running states and line conditions are different, the operating modes also have obviously different characteristics. In this paper, based on the framework load-time history data of large-scale line measurement, an automatic deep learning method in data mining is used to design the program to achieve an accurate and automatic segmentation of operating conditions, thus obtaining the load-time history under different operating conditions. The calculation of the load correlations under different operating conditions is carried out, leading to corresponding correlation degree of different loads. It lays a solid foundation for building an operating mode of complex framework loads, ensuring the operational safety of rail vehicles and conducting effective reliability assessment.

Keywords: Data mining · Automatic extraction · Operating mode of loads

1 Introduction

1.1 Application Background

The current situation of dense urban population promotes the rapid development of urban rail vehicles. Subway, light rail trains, trams, straddle monorails and other forms of urban rail transit make it more convenient and faster for residents to travel. According to the "Beijing Urban Rail Transit Construction Plan", by 2020, Beijing's rail transit network will have included 30 lines with a total length of about 1,050 km and nearly 400 stations [1], which show that operating conditions and line conditions are complex and diverse. Hence, analyzing operating modes of loads under different operating conditions is critical to its safety and reliability evaluation. However, the total amount of measured data of lines in a year can reach about 20 TB. The current method of manually finding the segmentation point is inefficient and inaccurate. Therefore, it cannot meet the demand of accurate and efficient segmentation of load-time history data.

© Springer Nature Switzerland AG 2020
F. Xhafa et al. (Eds.): IISA 2019, AISC 1084, pp. 604–613, 2020.
https://doi.org/10.1007/978-3-030-34387-3_74

1.2   Operating Conditions

The operating conditions of urban rail vehicles are mainly judged by time domain characteristics of speed and angular velocity. The speed signal is used to identify the vehicle acceleration state, constant speed state, and deceleration state, while the angular velocity signal is used to identify whether the vehicle is running in a straight line or in a curvedline. This paper uses the measured data of a Beijing subway line, and segments its time series into four operating conditions: acceleration, deceleration, constant speed and turning.

When the vehicle is in the acceleration state, within error range, its angular velocity is zero and its speed linearly increases, as shown in Figs. 1 and 2.

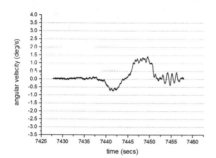

Fig. 1. Angular velocity curve   **Fig. 2.** Speed curve

When the vehicle is in the constant speed state, within error range, its angular velocity is zero and its speed remains constant, as shown in Figs. 3 and 4.

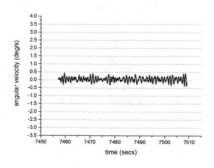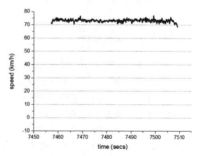

Fig. 3. Angular velocity curve   **Fig. 4.** Speed curve

When the vehicle is in the deceleration state, within error range, its angular velocity is zero and its speed linearly decreases, as shown in Figs. 5 and 6.

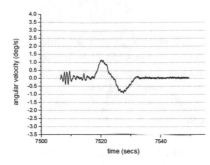

Fig. 5. Angular velocity curve

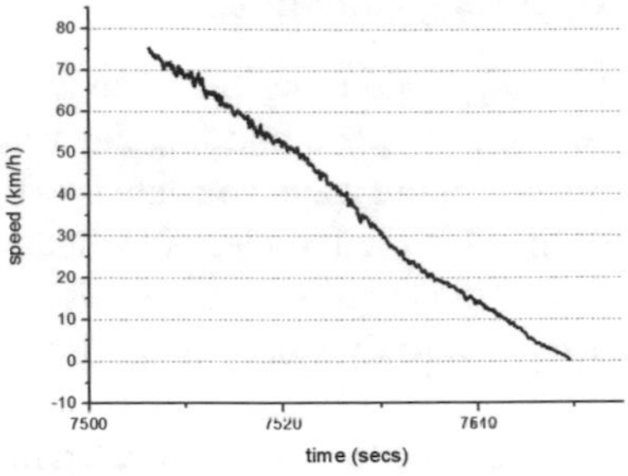

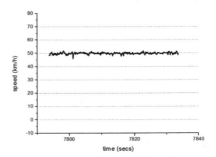

Fig. 6. Speed curve

When the vehicle is in the turning state, within error range, its angular velocity undergoes three stages: rising stage, constant stage, decreasing stage. The speed during the three stages remains constant, as shown in Figs. 7 and 8.

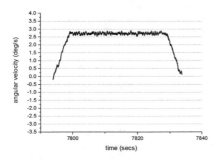

Fig. 7. Angular velocity curve

Fig. 8. Speed curve

1.3   Loads of Framework

According to the systemicity of vehicle movement, bearing characteristics and complex structural load identification, the Beijing subway framework is subject to 11 types of loads during vehicle operation. When coupling mechanism of the loads is analyzed, lateral load, floating load, roll load and torsional load shown in Fig. 9 are the main loads to analyze the bearing condition of subway framework. Therefore, this paper only calculates the correlations of these four loads.

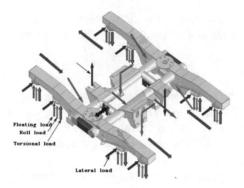

Fig. 9. Main loads of the subway framework

2 Program Design of Automatic Extraction of Operating Conditions

2.1 The Basis of the Program Theory

For the analysis of continuous data, the Gaussian distribution has good symmetry and stability, which has been confirmed in many practical applications. In this paper, the output of the operating conditions can be considered a stable behavior plus slight fluctuations or errors, which are usually described by Gaussian distribution [2, 3]. Therefore, Gaussian distribution is chosen as the theoretical basis of the program.

2.2 Parameter Selection

The analysis of the operating conditions shows that the changes in speed and angular velocity are linear within error range. Therefore, they can be described by the slope of the straight line. Since the operating conditions are segmented by the length of the time series, the length of the time series is used to describe the segmentation points.

This paper uses the speed and angular velocity data of Beijing subway as the basis for segmentation. The Gaussian distribution of the subway speed and angular velocity can be obtained by manually segmented speed and angular velocity data. On the same subway line, the difference between two stations is within error tolerance, so the unsegmented time series values comply with the same parameter Gaussian distribution. If the fluctuation of the data causes the segmentation point to be in the best matching state after crossing the sequence peak, the program's segmentation is wrong. In addition, because the time series are segmented using a single scan, it is cut from front to back. Therefore, from the wrong segmentation point, there will be problems in the subsequent segmentation points [4–8].

In this paper, a piecewise linear approximation, which approximates the time series with a more compact representation, is used to segment the time series. This can minimize the error relative to the original sequence. The root mean square error is now commonly used to measure the success of a segment. In order to avoid the above error segmentation, the mean square error (MSE) is added to the parameter [5, 8–13].

In summary, this paper selects the slope, mean, variance, time series length and MSE of the speed and angular velocity as the program decision parameters.

2.3   Programming

The data automatically segmented by the program belong to common evolution sequence. Previously, some people have studied how to segment the co-evolutionary sequence and proposed a fully automatic mining algorithm, the idea of which is to use a greedy approach [12, 14–19]. The automatic extraction program designed in this paper is based on the same idea.

In order to easily express the automatic segmented operating conditions, state IDs are assigned to each condition: acceleration state 1, constant speed state 2, deceleration state 3, and turning state 4. When the trial segmentation is performed in only one state, the probability of finding the correct state is low. Therefore, time series trial segmentation is performed in two state IDs to constrain first state by second state.

First, length expectation values of two state IDs are used as the segmentation length to trial segment time series of speed and angular velocity. For the two segments of time series of the trial segmentation, the average value of speed and angular velocity is calculated by the mean function in numpy, while the variance is calculated by the std function, and the slope of the change is calculated by the polyfit function. Since the above calculated parameters collectively determine the segmentation point of time series, the probability of the segmentation is the product of the probability of each parameter. Here the probability of each parameter is obtained by the Gaussian distribution. However, the influence of each parameter on different operating conditions is different, so the probability which is more consistent with the segmentation state may be smaller. Therefore, the program takes the - log function to convert the multiplication of probabilities into additions, while different probability coefficients are designed based on the importance of the parameters.

$$r1 = -(p0 + p1 + p2 + 1.5 * p3 + p5 + p7) \tag{1}$$

$$r2 = -(p0 + p1 + 1.5 * p2 + p3 + p5 + p7) \tag{2}$$

$$r3 = -(p0 + p1 + p2 + p3 + p5 + 1.5 * p7) \tag{3}$$

$$pid = \min(r1, r2, r3) \tag{4}$$

where $p0$ is the probability of the speed time series length, $p1$ is the speed mean probability, $p2$ is the speed variance probability, $p3$ is the speed slope probability, $p5$ is the angular velocity time series length probability, and $p7$ is the angular velocity variance probability.

Considering that the first state is the main segmentation state, the probability coefficient is set as 2 in the calculation.

$$cp2 = 2 * pid1 + pid2 \tag{5}$$

where Cp2 is the probability of the combined state, pid1 is the first state probability, and pid2 is the second state probability. pid1 and pid2 are obtained by Eq. (4).

The first state with the higher combined probability is used as the segmentation state. Within the interval of the time series length, segmentation is performed for each length, while pid, the parameter comprehensive probability, is calculated to find the segmentation length, whose calculated value of probability is the smallest. In this case, when the length is the best matching of the sequence under the state ID, the position of the segmentation point and its corresponding state ID are output. The segmentation point is used as a starting point for the next segmentation, until the time series length to be segmented is less than the average length of the combined state ID. Then the segmentation is completed.

2.4   Segmentation Results

According to the time domain characteristics of speed and angular velocity, the data obtained during one circle of running is segmented, and the result is shown in Fig. 10. In the illustration, the point on the speed time series is the segmentation point.

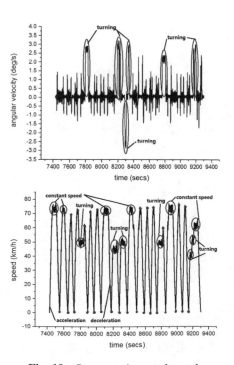

Fig. 10. Segmentation result graph

3   Correlation Calculation

After segmenting the operating conditions, the correlations between the framework loads, which are the basis of operating modes of loads, are calculated under the same operating conditions. Figure 11 is a schematic illustration of two time series S1 and S2, showing how to calculate the correlation between two time series. In the figure, the thin lines are the data under a certain operating condition, while the bold lines are the extended part data. Since the sampling frequency and time are the same, L, the total signal number of the two time series, is the same.

The proportion of the expansion in S1 is calculated as Eq. (6).

$$\eta1 = \frac{L1 + L2}{L} \times 100\% \tag{6}$$

where L1 and L2 are the number of data points in the extended part and the effective operating part of S1.

The proportion of the expansion in S2 is calculated as Eq. (7).

$$\eta2 = \frac{L3 + L4}{L} \times 100\% \tag{7}$$

where L3 and L4 are the number of data points in the extended part and the effective operating part of S2.

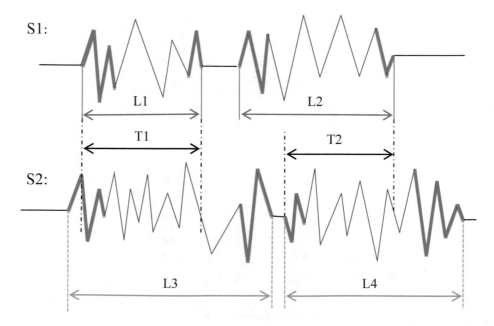

Fig. 11. Correlation calculation diagram

The correlation between the two time series is represented by the correlation percentage γ, which is calculated as Eq. (8).

$$\gamma = \frac{T1 + T2}{L \times MIN(\eta 1, \eta 2)} \times 100\% \tag{8}$$

where T1 and T2 are the number of data points in the common part of S1 and S2.

The correlation degree between two time series is judged by the percentage of correlation. It is known from experience that if γ > 0.6, two loads are considered to be strongly correlated.

4   Calculation Results

The load correlation results under different operating conditions are calculated by Eq. (8) as shown in Tables 1, 2, 3, and 4.

Table 1. Load correlation in the acceleration state

Load type	Roll load	Torsional load	Lateral load	Floating load
Roll load	1			
Torsional load	−0.464	1		
Lateral load	0.440	0.718	1	
Floating load	0.189	0.209	0.134	1

It can be seen from the data in Table 1 that the lateral load and the torsional load are strongly correlated in the acceleration state.

Table 2. Load correlation in the deceleration state

Load type	Roll load	Torsional load	Lateral load	Floating load
Roll load	1			
Torsional load	−0.811	1		
Lateral load	−0.747	0.639	1	
Floating load	0.642	−0.410	−0.363	1

It can be seen from the data in Table 2 that in the deceleration state, the roll load is strongly correlated with the other three load systems, while the lateral load and the torsional load are also strongly correlated.

Table 3. Load correlation in the constant speed state

Load type	Roll load	Torsional load	Lateral load	Floating load
Roll load	1			
Torsional load	0.727	1		
Lateral load	0.357	0.327	1	
Floating load	0.245	0.242	0.096	1

It can be seen from the data in Table 3 that the roll load is strongly correlated with the torsional load in the constant speed state.

Table 4. Load correlation in the turning state

Load type	Roll load	Torsional load	Lateral load	Floating load
Roll load	1			
Torsional load	−0.552	1		
Lateral load	−0.366	0.664	1	
Floating load	−0.194	0.398	0.133	1

It can be seen from the data in Table 4 that the lateral load and the torsional load are strongly correlated in the turning state.

5   Conclusion

This paper applied the program of automatic extraction of operating conditions to extract the operating conditions of Beijing subway. Under the extracted operating conditions, the correlations between the loads of framework are calculated, which lead to the corresponding correlation degree of the framework loads. In the acceleration state, the lateral load and the torsional load are strongly correlated. In the deceleration state, the roll load is strongly correlated with the other three load systems, while the lateral load and the torsional load are also strongly correlated. In the constant speed state, the roll load is strongly correlated with the torsional load. In the turning state, the lateral loads are strongly correlated with torsional loads.

The relevant experiments in this paper are still going on, and the program will be continuously improved according to the experiment results.

Acknowledgments. First of all, I would like to thank Professor Binjie Wang and Shouguang Sun for their guidance during the design process. In addition, I am very grateful to Mr. Wang Peng from Fudan University for his expansive explanation of relevant knowledge.

References

1. Beijing Municipal People's Government. Beijing Urban Rail Transit Construction Plan 2011–2020
2. Wang, P., Wang, H., Wang, W.: Finding semantics in time series. In: SIGMOD Conference, pp. 385–396 (2011)
3. Mueen, A., Keogh, E.J.: Online discovery and maintenance of time series motifs. In KDD, pp. 1089–1098 (2010)
4. Sakurai, Y., Papadimitriou, S., Faloutsos, C.: Braid: stream mining through group lag correlations. In: SIGMOD, pp. 599–610 (2005)

5. Shaghayegh, G., Ding, Y.F., Michael, Y., Kaveh, K., Liudmila, U.: Matrix Profile VIII: domain agnostic online semantic segmentation at superhuman performance levels. Department of Computer Science and Engineering, pp. 1–10, October 2017
6. Keogh, E.J., Chu, S., Hart, D., Pazzani, M.J.: An online algorithm for segmenting time series. In: ICDM, pp. 289–296 (2001)
7. Matsubara, Y., Sakurai, Y., Faloutsos, C., Iwata, T., Yoshikawa, M.: Fast mining and forecasting of complex time-stamped events. In: KDD, pp. 271–279 (2012)
8. Rakthanmanon, T., Campana, B.J.L., Mueen, A., Batista, G.E.A.P.A., Westover, M.B., Zhu, Q., Zakaria, J., Keogh, E.J.: Searching and mining trillions of time series subsequences under dynamic time warping. In: KDD, pp. 262–270 (2012)
9. Tatti, N., Vreeken, J.: The long and the short of it: summarising event sequences with serial episodes. In: KDD, pp. 462–470 (2012)
10. Toyoda, M., Sakurai, Y., Ishikawa, Y.: Pattern discovery in data streams under the time warping distance. VLDB J. **22**(3), 295–318 (2013)
11. Matteson, D.S., James, N.A.: A nonparametric approach for multiple change point analysis of multivariate data. J. Am. Stat. Assoc. **109**(505), 334–345 (2014)
12. Matsubara, Y., Sakurai, Y., Faloutsos, C.: AutoPlait: automatic mining of co-evolving time sequences. In: Proceedings of the 2014 ACM SIGMOD, pp. 193–204, June 2014
13. Lan, R., Sun, H.: Automated human motion segmentation via motion regularities. Vis. Comput. **31**(1), 35–53 (2015)
14. Ron, D., Singer, Y., Tishby, N.: The power of amnesia: learning probabilistic automata with variable memory length. Mach. Learn. **25**, 117–149 (1996)
15. Matsubara, Y., Sakurai, Y., Prakash, B.A., Li, L., Faloutsos, C.: Rise and fall patterns of information diffusion: model and implications. In: KDD, pp. 6–14 (2012)
16. Papadimitriou, S., Sun, J., Faloutsos, C.: Streaming pattern discovery in multiple time-series. In: Proceedings of VLDB, pp. 697–708, August-September 2005
17. Li, L., Liang, C.-J.M., Liu, J., Nath, S., Terzis, A., Faloutsos, C.: Thermocast: a cyber-physical forecasting model for data centers. In: KDD, pp. 3–8 (2011)
18. Aminikhanghahi, S., Cook, D.J.: A survey of methods for time series change point detection. Knowl. Inf. Syst. **51**, 1–29 (2016)
19. Hao, Y., Chen, Y., Zakaria, J., Hu, B., Rakthanmanon, T., Keogh, E.: Towards never-ending learning from time series streams. In: Proceedings of the 19th ACM SIGKDD International Conference on Knowledge Discovery and Data Mining, pp. 874–882, August 2013

SIP-Based Interactive Voice Response System Using FreeSwitch EPBX

Jiaxin Yue[1(✉)], Zhong Wang[2], and Yixin Ran[2]

[1] Sichuan Film and Television University, Chengdu, China
yuejiaxin2017@163.com
[2] College of Electrical Engineering and Information Technology, Sichuan
University, Chengdu, China

Abstract. In recent years, call centers have developed in full swing. The new call center based on softswitch technology not only has been effectively controlled, but also has become more powerful. As the most indispensable part of call center, IVR interactive voice response system plays a very important role. On the platform of open source soft switch (FreeSWITCH), a kind of IVR system is designed and implemented, which is convenient for initial use and secondary development. It can intelligently meet the needs of users.

Keywords: FreeSWITCH · Softswitching · Call center · IVR

1 Introduction

Call center, also known as customer service center, is a comprehensive information service system which makes full use of computer telecommunication integration technology and communication network [1, 2]. The call center includes an incoming call center and an external call center. This is inseparable from the IVR (interactive voice response system). Interactive voice response means wireless voice service increment service, similar to fixed telephone voice service [3–5]. The mobile phone user dials the assigned number, obtains the information which needs, or participates in the interactive service. In the call center, IVR is divided into front and post. Front IVR is speech into IVR processing, in the case of customer problems can not be transferred to the manual seat. Post-IVR refers to the balance between IVR and artificial seats, which can not meet the needs of customers to switch to IVR, mainly for the sake of delaying time or adding value. The market scale of domestic IVR service is about 2.5 billion RMB in 2014, which shows that IVR is one of the heavyweight services in wireless value-added service [6].

With the rise of mobile Internet and the popularity of VoIP technology, IVR voice communication system based on VoIP softswitch is designed with low cost, simple implementation and easy expansion [7, 8]. Based on the new open source VoIP switch —FreeSWITCH, we can set up a simple IVR system by adding configuration files, and can also design IVR voice flow through the supported embedded Lua script [9, 10].

© Springer Nature Switzerland AG 2020
F. Xhafa et al. (Eds.): IISA 2019, AISC 1084, pp. 614–621, 2020.
https://doi.org/10.1007/978-3-030-34387-3_75

2   Introduction of FreeSWITCH

2.1   The Basic Introduction of FreeSWITCH

As a new VoIP open source soft switch, FreeSWITCH is very powerful. Its official definition is "The world's First Cross-Platform Scalable FREE Multi-Protocol Soft Switch." Specific explanations as shown in Table 1.

Table 1. Basic characteristics of FreeSWITCH

Characteristics	Explanations
Cross platform	FreeSWITCH is native to Windows Max OS, Linux, BSD and many other 32/64 bit platforms
Scalability	FreeSWITCH can be used as a simple softphone client or a carrier using a class softswitch device, it is almost omnipotent
Free	FreeSWITCH uses MPL 1.1 protocol authorization, anyone can free access to the source code. Anyone can modify, publish, and sell their own apps
Multiprotocol	FreeSWITCH supports a variety of communication protocols such as SIPH323Skype-Google talk, and can easily communicate with various open source PBXs such as sip Xecs call Weaver BayonneneYate and Asterisk. It can also interoperate with commercial switching systems (Huawei, ZTE switch)
Softswitch platform	FreeSWITCH follows RFC and supports a number of advanced SIP features such as presenceBLAs TCPTLS and sRTP. It can also be used as a SBC transparent SIP proxy) to support other media such as T.38. FreeSWITCH to support broadband and narrowband speech coding
Communicating mechanism	FreeSWITCH adopts B2BUA mode to help both sides to communicate in real time. Any agent only communicates directly with FreeSWITCH, which controls how calls are handled

As shown in Fig. 1, FreeSWITCH has a core module called Core, which invokes modules of different functions and protocols. At the same time, any module can call the core code. Each module is independent and can be extended by adding or loading modules. Each interface is described in Table 2 below.

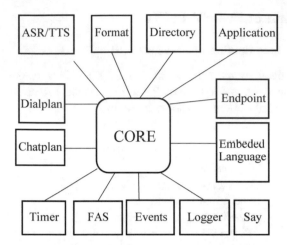

Fig. 1. FreeSWITCH general architecture

Table 2. Introduction to FreeSWITCH interface

Joggles	Function
Dialplan	Implement call status, get call data and route
Chatplan	Primarily routing text messages
Endpoint	Interfaces implemented for different protocols, such as SIP TDM, etc.
ASR/TTS	Speech recognition and synthesis
Directory	LDAP type database query
Events	Modules can trigger core events or register their own personality events
Formats	File format, such as wav.
Loggers	Console or file log
Say	A specific language module that organizes discourse from sound files
Timer	A reliable timer used for interval timing
Applications	A program that can be executed in one call, such as Voicemail.
FSAPI	A function of type CGI for XML RPC functions with input and output prototype dial-plan function variables

The core module is mainly responsible for the operation and management of the whole server. The peripheral module is responsible for extending the peripheral functions. All the modules in the diagram work together to build a rich and comprehensive communication system depending on the core API or internal events when communication is needed.

2.2   Introduction to FreeSWITCH Communication Mechanism

FreeSWITCH uses the B2BUA model, which acts as both UAS and UACs, shown in Fig. 2.

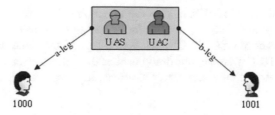

Fig. 2. B2BUA principle

The detailed flow of calls through B2BUA is shown in Fig. 3.

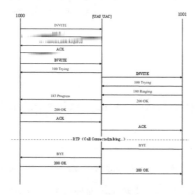

Fig. 3. Call flow based on FreeSWITCH

① User 1000 sends a message INVITE to the server requesting initiating a call with the user 1001. UAS receives a request and responds100 Trying indicates receipt.At the same time, the server verifies that the user 1000 is authorized to initiate a call through it.

② User 1000 returns ACK message to identify. Resend INVITE message to request setup call. At this point, the server sends a 100 Trying message back to the user 1000 formal response.

③ Next, the call enters the routing phase. The server finds the location of user 1001 by route, and then initiates a session to establish a connection via invite message.

④ The server successfully connected to user 1001 and sent back messages 183 to 1000 to let user 1000 know that 1001 was ringing and waiting patiently.

⑤ User 1001 answers the telephone and returns 200 OK messages to the server. The server returns the ACK message to notify the user 1001, at the same time sends the 200 OK message to the user 1000, and the user 1000 sends back the ACK message to the server after receiving the ACK message. At this point, the a-leg and b-leg of the B2Bua call have been bridged by the server and can be spoken between user 1000 and user 1001.

⑥ After the user 1001 hangs up the telephone, the user agent sends the BYE message to the server, notifies the server to disconnect the call connection, the server returns 200 OK, releases b-leg; At the same time, the user 1000 sent a BYE message, informing user 1001 to disconnect, user 1000 echo 200 OK message, after the server receives the release a-leg. the call is over.

3 Design of IVR System

3.1 Requirement Analysis

The IVR function of call center is used to promote and introduce the specialty. Introduce the professional basic knowledge to the visitors through the way of voice, self-service to meet the basic needs of the users. Main menu IVR and subordinate menu nested IVR can be used to complete a variety of functions.

3.2 System Design

The specific system set up according to the requirement analysis is shown in Fig. 4

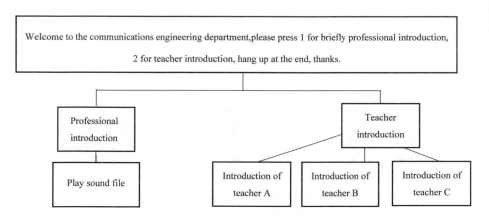

Fig. 4. Call center functional requirements

(1) Dial in the set number and play the Voice Welcome menu.
(2) Brief introduction of Communication Engineering: introduce the history, development and research direction of communication engineering by voice.
(3) Teacher introduction: first, take a brief overview of the faculty of the whole profession, then take each teacher as an entry, according to the user choice to introduce the specific teacher.

The system loads conf/autoload_configs module automatically when boot, which makes ivr.conf.xml file load into memory automatically. When a telephone calls in, FreeSWITCH routes it to the dial up plan conf/dialplan module. The default.xml file determines whether the number dialed by the user meets the criteria through the destination_number field under extension. If fit, the welcome_demo_ivr.xml file is run through the IVR system. The user can hear the macro phrase defined in the demo_ivr. xml file and play the specified sound file through playfile APP. The user can do the next operation according to the sound file prompt and FreeSWITCH will complete the corresponding function according to the user's key.

4   System Implementation

4.1   Implementation of Main Menu

In the conf/dialplan/default.xml file, the extension name of the telephone routing definition for entering the call center is ivr _ d _ demo, which enables the user to route to the extension automatically after dialing 5000. The conf/ivr_menus/welcome_demo_ivr. xml defines the whole main menu and the specific operation of the user's keystroke. Enter the call center through user 1000 dialing 5000, and the console punches out a series of system information. In addition, we can hear the clear sound "Welcome call Communications Engineering department" after calling 5000, It can be considered that the telephone routing function is normal and consistent with the design.

4.2   Realization of Professional Introduction Function

This function is implemented in conf/dialplan/default.xml, executing.

The FreeSWITCH console typed information as shown in Fig. 5 below. After the user pressed the "1" key, the system transferred the call to the call plan processing and successfully routed to 8888.xml.

```
2018-05-07 17:20:41.057813 [INFO] switch_channel.c:515 RECV DTMF 1:960

2018-05-07 17:20:43.058035 [NOTICE] switch_ivr.c:2172 Transfer sofia/internal
/10000192.168.201.128 to XML[88880default]

2018-05-07 17:20:43.058035 [INFO] mod_dialplan_xml.c:637 Processing 1000 <100
0>->8888 in context default
```

Fig. 5. Professional profile test

Through analysis and actual situation, it can be proved that the function is running normally and in accordance with the design.

4.3   Implementation of Teachers Introduction

After the user presses "2", the system routes the phone to the ivr_teacher_introduction menu through menu-sub app. Defined by the command line < entry action = "menu-sub" digits = "2" param = "ivr_teacher_introduction" / > in welcome_demo_ivr.

After routing to the ivr_teacher_introduction menu, because < entry action = "menu-exec-app" digital = "1" params = "transfer 1111 XML default" / > is defined in the conf/ivr_menus/teacher_introduction_ivr.xml, the keystroke "1" will then route to the extension matching 1111 in the dial plan. Route to 2222. Xml. after "2" (self-defined, can be changed. According to the design, 1111.xml will play the introduction of teacher A 2222.xml for teacher B.

After the user presses the "2" key, go to the teacher's introduction menu and type the information from the FreeSWITCH console as shown in Fig. 6 below.

```
2018-05-10 11:49:41.867386 [INFO] switch_channel.c:515 RECV DTMF 2:640
2018-05-10 11:49:43.967909 [NOTICE] switch_ivr.c:2172 IVR action on menu 'wel
come_demo_ivr' matched '2' param 'ivr_teacher_introduction'
2018-05-10 11:49:43.967911 [INFO] switch_ivr_play_say.c:250 Handle play-file:
[ivr/8000/my_voice/ivr_teacher_welcome.wav]
2018-05-10 11:49:43.967909 [INFO] switch_ivr.c:469 Executing IVR menu ivr_tea
cher_introduction
```

Fig. 6. Teachers introduce functional testing

After pressing "2" again, the console information shown as Fig. 7.

```
2018-05-10 11:51:25.015932 [INFO] switch_channel.c:515 RECV DTMF 2:640
2018-05-10 11:51:25.015932 [NOTICE] switch_ivr.c:2172 Transfer sofia/internal
/10000192.168.201.128 to XML[2222@default]
2018-05-10 11:51:25.015932 [INFO] mod_dialplan_xml.c:637 Processing 1000 <100
0>->2222 in context default
```

Fig. 7. Introduction of teacher B

Combined with the actual experience, the function also works normally, consistent with the design.

5   Conclusion

Based on FreeSWITCH soft switch, a kind of IVR voice communication system is designed, which is convenient for direct use and secondary development. The system uses SIP protocol for signaling interaction and realizes the function of IVR system by combining the basic function characteristics of FreeSWITCH. And can be easily combined with other modules to expand. Based on this IVR system, small call centers can be easily built for enterprises.

Acknowledgements. This work was supported by the key projects of education department in Sichuan Province under Grant No. 18ZA0308, 18ZA0307. It was also supported by science and technology project of Sichuan Province under Grant No. 2015FZ061.

References

1. Saadatizadeh, Z., Chavoshipour Heris, P., Sabahi, M., et al.: A new non-isolated free ripple input current bidirectional DC-DC converter with capability of zero voltage switching. Int. J. Circ. Theory Appl. **46**(3), 519–542 (2018)
2. Xing, L.I., Zhao, X.M., Jin, M., et al.. Design and implementation of information and communication dispatching system based on FreeSwitch platform. Electr. Power Inf. Commun. Technol. (2017)
3. Moralis-Pegios, M., Terzenidis, N., Mourgias-Alexandris, G., et al.: A 1024-port optical uni- and multicast packet switch fabric with sub-μs latency values using the Hipoλaos architecture. J. Lightwave Technol. PP(99), 1 (2019)
4. Csercsik, D., Kovács, L.: Dynamic modeling of the angiogenic switch and its inhibition by bevacizumab. Complexity **2019**(2), 1–18 (2019)
5. Liu, F., Ding, A., Zheng, J., et al.: A label-free aptasensor for ochratoxin a detection based on the structure switch of aptamer. Sensors **18**(6), 1769 (2018)
6. Wang, M., Cai, W., Zhu, D., et al.: Field-free switching of perpendicular magnetic tunnel junction by the interplay of spin orbit and spin transfer torques. Nature Electr. **1**(11), 582–588 (2018)
7. Lv, W., Tian, S.C., Ke, Q.Z., et al.: Feasible preparation and improved properties of ag-graphite composite coating for switch contact by cyanide-free electrodeposition. Mater. Corros. (2018)
8. Gupta, J., Dwivedi, V.K., Karwal, V., et al.: Free space optical communications with distributed switch-and-stay combining. IET Commun. **12**(6), 727–735 (2018)
9. Sun, C., Shum, P.P., Fu, S., et al.: Crossing-free on-chip 2×2 polarization-transparent switch with signals regrouping function. Opt. Lett. **43**(16), 4009 (2018)
10. Luo, X.R., Zhao, Z., Huang, L., et al.: A snapback-free fast-switching SOI LIGBT With an embedded self-biased n-MOS. IEEE Trans. Electr. Devices **65**(8), 3572–3576 (2018)
11. Hwang, B., Lee, J.S.: Lead-free, air-stable hybrid organic-inorganic perovskite resistive switching memory with ultrafast switching and multilevel data storage. Nanoscale **10**(18), 8578–8584 (2018). 10:10.1039.C8NR00863A

Calligraphic Stylization Learning with Active Generation Networks

Lingli Zhan$^{(\boxtimes)}$, Zihao Zhang, and Yuanqing Wang$^{(\boxtimes)}$

School of Electronic Science and Engineering, Nanjing University,
Xianlin Avenue, Nanjing, China
zhan_lingli@163.com

Abstract. In this paper, we present a calligraphic stylization learning framework which achieve reasonable results. For this purpose, we propose a novel active generation networks inspired by active shape modeling. Combined with embedding layer, we exploit information across multiple fonts and rich models of calligraphic stylization can be built once in a time. More usefully, we perform a simulation of famous calligraphist's cursive handwriting that shows our approach outperforms state-of-the-art calligraphic stylization learning.

Keywords: Deep generative networks · Style transfer · Calligraphy

1 Introduction

Chinese calligraphy is an artistic treasure like oil painting in the West. This traditional art is increasingly desolate with the convenience brought by the keyboard. Letting AI try to understand calligraphy is a challenging and meaningful work.

Traditional methods to solve the problem of font generation are to split Chinese characters into basic elements such as stroke and radicals, and then combine basic strokes to minimize the distance between them and the target Chinese characters [1–6]. The advantage of this method is that it can accurately reconstruct the structure of words. But the disadvantage is that it can't handle the situation where strokes are joined together [7, 8].

Deep learning has achieved the state-of-the-art performance in the style transfer of various works of art. Some recent work has deal with this problem based on LSTM [9] model. In [10], end-to-end GAN network is introduced to build font generation model. Although end-to-end study solves the stroke-based models dependence on preceding parsing, there are also some shortcomings. First, GAN network needs to train G network and D network at the same time, so training is difficult and easy to not converge. Secondly, as a global algorithm, GAN may lose accuracy locally compared with the stroke-based models.

Motivated by ASM(Active shape modeling), we propose a active generation networks(AGN) that will speed up the convergence. Furthermore, several regularization terms are considered in the standard GAN object function which will improve the convergence. In our approach we exploit information across multiple fonts and rich models of calligraphic stylization can be built once in a time. More usefully, we

© Springer Nature Switzerland AG 2020
F. Xhafa et al. (Eds.): IISA 2019, AISC 1084, pp. 622–628, 2020.
https://doi.org/10.1007/978-3-030-34387-3_76

perform a simulation on the cursive calligraphy with our method for the first time. This work is helpful to the restoration of calligraphic works of art.

2 Active Generation Networks (AGN)

Active shape models (ASMs) are statistical models of the shape of objects, developed by Tim Cootes and Chris Taylor in 1995 [11]. When matching a new object, the model will give a reasonable constraint to the search space. Drawing lessons from this idea in generation network, we can avoid the network falling into local optimum and accelerate the convergence speed of the model.

The framework of our proposed method includes two main parts: (1) Active generation model. (2) Optimization.

2.1 Active Generation Model

The overall structure of our networks is shown in Fig. 1. As illustrated in Fig. 1(a), Active generation networks are composed of structure encoder(), style-embedding() and active generator(). Through structure encoder, input characters are mapped to a vector in structure space. Meanwhile, the style information of different fonts is encoded by the style-embedding net. Thus, content information and style information are integrated into the net. Each layer of the generated network is obtained by the deconvolution operation on the upper layer, moreover, the content encoding layer of the corresponding space is introduced to carry out conditional constraints. These allow the generation process to be sampled from conditional distribution instead of sampling from marginal distribution. The three subnets will be described in detail below.

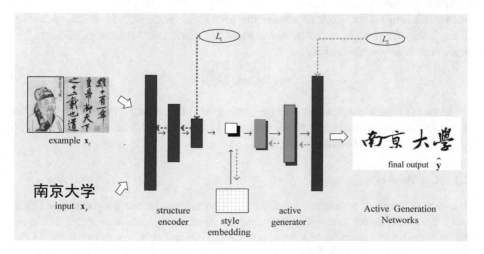

Fig. 1. Network architectures.

Structure Encoder

The purpose of this part is to enable the network to infer the character structure information by feeding small samples rather than the assemble of all Chinese characters. To achieve this goal, we need to build such a training dataset: (1) the diversity of samples is complete to cover the basic stroke types and the relative location of the Chinese characters, (2) dataset labeled with the same character should have many different fonts, this is equivalent to introducing noise information, enabling the network to have generalization and robustness. The expressed by mathematical symbols as:

$$\mathcal{F} : f = \phi(x) \tag{1}$$

$\mathbf{x}$ is a Chinese character, ϕ is a nonlinear mapping function implemented by CNN, $\mathbf{f}$ is the encoding vector in content space.

Active Generator

Character coding information is introduced into the generative model as a prior constraint of structure. This can expressed by mathematical symbols:

$$\mathcal{G} : y^i = \left[g\left(y^{i-1}, f^i\right)\right], (y^1 = g[f^0, \lambda]) \tag{2}$$

y^i is the output layer of the $\mathcal{G}$, [] is a concatenation operation, $\mathbf{f}^i$ is the corresponding feature map from the $\mathcal{F}$, g is mainly composed of the fractionally strided convolution which need the $\mathcal{G}$ to learn.

Style-Embedding

We hope to employ font labels to improve the ability of network to model various styles simultaneously. A common strategy is one-hot encoding [12]. However, considering the input fonts is not completely independent in style space, one-hot embedding is not the best solution. A word embedding technique in NLP system [12] is introduced into our network to encode the font style. With the aid of this word2vec trick, fonts will be represented so that models can learn their inherent correlation. The Style-embedding is defined as:

$$\varepsilon : \lambda = T(h_S) \tag{3}$$

Each type of input fonts is encoded into a M-dimensional vector h_S by one-hot encoding. A fully connected network with one layer is trained, and it will output an N-dimensional style-embedding vector λ.

2.2   Optimization

In order to train the parameters of different subnetworks in our proposed model and make the model easier to converge, we propose an active optimization method. We use different objective functions to learn parameters for $\mathcal{F}$ and $\varepsilon \backslash \mathcal{G}$. As shown in Fig. 1 (b), the optimization process is divided into two parts. The first part is to propagate the loss L_1 along the direction of the red arrow in the graph to optimize the parameters of the $\mathcal{F}$ network.

There are many options for L_1, such as softmax loss, triplet loss [13] or center loss [14]. In this paper, we choose softmax as the loss function.

The second part is to optimize the network parameters of ε and G by propagating the loss L_2 along the direction indicated by the green arrow in the graph. The specific definition of the cost function L_2 is as follows:

$$L_2 = \alpha_1 L_{\text{GAN\_G\_B}} + \alpha_2 L_{\text{GAN\_G\_C}} \tag{4}$$

Each sub item in the formula will be elaborated in the following paragraphs.

Adversarial training helps to illustrate the power of using a large function family in combination with aggressive regularization [16]. We refer to the auxiliary classifier GANs [15] method to achieve simultaneous processing of multiple fonts, learn from each other, and prevent over-fitting of the network. The discriminator gives both a probability distribution over sources and a class labels, $P(GAN\_G\_B|\mathbf{x})]$, $P(GAN\_G\_C|\mathbf{x})] = D(\mathbf{x})$. The objective function for G as follows:

$$L_{\text{GAN\_G\_B}} = \mathbb{E}[\log p(\text{GAN\_G\_B} = real|\mathbf{y})] \tag{5}$$

$$L_{\text{GAN\_G\_C}} = \mathbb{E}[\log p(\text{GAN\_G\_C} = i|\mathbf{y})] \tag{6}$$

3 Experiments

3.1 Training Details

Cursive scripts is one kind of Chinese calligraphic writing with both aesthetics and practicality. In this paper, cursive script is chosen as a target font with both aesthetics and practicality. Source font is set as Song typeface in our training model.

The complicated and varied structure of the cursive script poses a greater challenge to the transferring ability of the network. With our proposed algorithm, the data set consists of three parts. For the training of structure encoding network, the font style of the same Chinese character in the dataset A are very different, so that the network can learn the true structure information of the font. For the training of G, the dataset B is composed of different author' cursive scripts. This will help the network to learn the style of cursive script. Finally, in the fine tune step, the data set C is made up of the target font.

In order to verify the effectiveness of the algorithm, we chose the manuscript of Huang Tingjian, a famous calligrapher in ancient China, as the target font. The set A consists of 500 Chinese characters with 30 fonts, B consists of 500 characters written in 22 kinds of cursive script. And C consists of other 1000 characters written in Huang Tingjian.

3.2   Evaluation

We evaluate the performance of the proposed AGN and the GAN [17] on datasets. We present the output of two algorithms on verification set with 1000 samples and 20 iterations of 4 types of fonts. As shown in Fig. 2, the output of our AGN method (up at Fig. 2) has basically converged to the domain space of the target font. However, the output of the GAN algorithm is quite ambiguous.

Fig. 2.   Comparison of GAN and AGN results. Output of AGN are shown at up, Output of GAN are shown at down.

After the model learns the basic rules of writing, we fine tune the network to verify its ability to learn a certain specific handwritten style –Huang Tingjian's running script. As shown in Fig. 3, the odd columns are source characters, and even columns are generated characters. There is a distinct cursive style at the beginning and end of strokes which are marked in yellow circles in the Fig. 3(d). Huang Tingjian's running script is concise and powerful, with peculiar structure. Almost every word has some exaggerated long strokes (indicated by the green arrow in the Fig. 3(a, b, c)). The running script twists and turns, and sends it out as much as possible, forming a brand-new method of forming characters that is tightly closed in the central part (as shown in the red circle markers of Fig. 3(a, b, c)) and divergent on four edges.

Fig. 3. Fine tune result with Huang Tingjian's style as target style. the odd columns are source characters, and even columns are generated characters.

Fig. 4. More output on test set. The grey bottom part is ground truths, and the white part is generated characters by our framework.

In Fig. 4, more output on test set are given. The grey bottom part is ground truths, and the white part is generated characters by our framework. It should be noted that the ground truths are only for style reference, not pixel-level target values. This is the difference between handwritten and printed characters. According to our analysis in Fig. 3, the generated characters in Fig. 4 confirm that our framework has learned Huang Tingjian's style.

4   Conclusion

We presented a calligraphic stylization learning framework inspired by a novel active generation networks. More usefully, we perform a simulation of famous calligra-phist's cursive handwriting. In the future, we plan to test our framework on other data sets.

Acknowledgments. Key research and development program of Jiangsu province (Grant No. BE2016173) and Scientific Research Foundation of the Graduate School of Nan Jing University.

References

1. Shi, D.M., et al.: Active radical modeling for handwritten Chinese characters. In: Sixth International Conference on Document Analysis and Recognition, Proceedings, pp. 236–240. IEEE Computer Society, Los Alamitos (2001)
2. Xu, S.H., et al.: Automatic generation of chinese calligraphic writings with style imitation. IEEE Intell. Syst. **24**(2), 44–53 (2009)
3. Lake, B.M., Salakhutdinov, R., Tenenbaum, J.B.: Human-level concept learning through probabilistic program induction. Science **350**(6266), 1332–1338 (2015)
4. Lian, Z., Bo, Z., Xiao, J.: Automatic generation of large-scale handwriting fonts via style learning. In: Siggraph Asia Technical Briefs (2016)
5. Baluja, S.: Learning typographic style: from discrimination to synthesis. Mach. Vis. Appl. **28**(5–6), 551–568 (2017)
6. Shi, D., et al.: Radical recognition of handwritten Chinese characters using GA-based kernel active shape modelling. IEE Proc. Vis. Image Signal Process. **152**(5), 634–638 (2005)
7. Shi, C., et al.: Automatic Generation of Chinese Character Based on Human Vision and Prior Knowledge of Calligraphy. Springer, Heidelberg (2012)
8. Zong, A., Zhu, Y.: StrokeBank: automating personalized chinese handwriting generation. In: Proceedings of the Twenty-Eighth AAAI Conference on Artificial Intelligence, pp. 3024–3029. AAAI Press, Quebec City (2014)
9. Berio, D., et al.: Calligraphic Stylisation Learning with a Physiologically Plausible Model of Movement and Recurrent Neural Networks. arXiv:1710.01214 (2017)
10. Lyu, P., et al.: Auto-Encoder Guided GAN for Chinese Calligraphy Synthesis. arXiv:1706.08789 (2017)
11. Cootes, T.F., et al.: Active shape models-their training and application. Comput. Vis. Image Underst. **61**(1), 38–59 (1995)
12. Young, T., et al.: Recent trends in deep learning based natural language processing. IEEE Comput. Intell. Mag. **13**(3), 55–75 (2018)
13. Schroff, F., et al.: FaceNet: a unified embedding for face recognition and clustering. In: 2015 IEEE Conference on Computer Vision and Pattern Recognition, pp. 815–823. IEEE, New York (2015)
14. Wen, Y., et al.: A Discriminative Feature Learning Approach for Deep Face Recognition. In: European Conference on Computer Vision, p. 499–515 (2016)
15. Odena, A., Olah, C., Shlens, J.: Conditional Image Synthesis With Auxiliary Classifier GANs. arXiv:1610.09585 (2016)
16. Goodfellow, I., Bengio, Y., Courvile, A.: Deep Learning. p. 286
17. Goodfellow, I., Pouget-Abadie, J., Mirza, M., Xu, B., Warde-Farley, D., Ozair, S., Courville, A., Bengio, Y.: Generative adversarial networks. In: Proceedings of the International Conference on Neural Information Processing Systems, pp. 2672–2680 (2014)

A Low Power Consumption Locating Method of Wireless Sensor Network for Particle Impact Drilling Monitoring System

Lei Li[1], Jianlin Yao[1,2(✉)], and Weicheng Li[1,2]

[1] CCDC Drilling and Production Technology Research Institute, Guanghan,
Sichuan 618300, China
67933788@qq.com
[2] CCDC Petroleum Drilling & Production Technology Co., LTD, Guanghan,
Sichuan 618300, China

Abstract. In order to reduce the computing power consumption of wireless sensor network (WSN) locating technology under complex environment, such as oil drilling, space exploration, factory monitoring, etc., the article proposed a weighted probability centroid localization method for particle impact drilling monitoring system. It comprehensively considered the probability of different nodes under wireless channel transmission model, selected probability of the three biggest point as triangle, calculated the weighted centroid as the location information of unknown node. Through experiment simulation, it obtained locating error curves of the proposed method and traditional locating method in different circumstances, the comparative results demonstrated that the two methods own the same locating precision. The research shows that the proposed locating method computing amount is reduced 80% compared with the traditional locating method, the sensors power consumption have been greatly reduced. So it can effectively improve the working life of monitoring system.

Keywords: WSN · Complex environment · Weighted probability centroid · Low power consumption · Locating method

1 Introduction

Wireless Sensor Network (WSN) is a cheap wireless network, that be composed of micro sensors in the form of self-organization and multi-hop [1, 2]. It has a lot of advantages, such as rapid deployment, self-organization, strong concealment and high fault tolerance. So WSN can be widely used in the fields of Military, environmental monitoring, oil drilling, space exploration, factory monitoring, etc. [3, 4]. Now, the environment monitoring system based on WSN has been developed, that could be used to real-time monitoring environment in the wild oilfield. In the particle impact drilling system, we can use WSN to monitor the equipment's status, that making worker handle system in a safety environment. To make the system application effect best, the monitoring system based on WSN must be used in a long time. Because the sensors power consumption is the main determinants of WSN lifetime. So it is the important research for WSN system that reducing sensors power consumption. According to studies, the

© Springer Nature Switzerland AG 2020
F. Xhafa et al. (Eds.): IISA 2019, AISC 1084, pp. 629–637, 2020.
https://doi.org/10.1007/978-3-030-34387-3_77

power consumption of the sensors is determined directly by the efficiency of information processing. Under the same routing architecture, the information processing efficiency of sensors is mainly determined by the calculation amount of locating method. So it is very practical for WSN system to study low power consumption locating method, and can improve the lifetime of monitoring system in the wild.

Under complex environment, it is widely used for node locating by sensors distance measurement in WSN systems. In numerous distance measuring method, the received signal strength indicator (RSSI) model can measure distance without adding additional hardware devices [5–7]. So it is convenient, low cost and general, many scholars has carried out the related research. Because of existing signal attenuation and reflection, the measuring distance between sensors has error in actual measuring. So researchers put forward a lot of typical locating methods, such as the iterative center of mass, maximum likelihood estimate, etc. [8–11], that can improve locating precision of WSN. Aimed at the puzzle of large calculation amount of typical methods, the article proposes weighted probability centroid localization method based on analyzing the typical locating methods. That method can ensure needed locating precision, meanwhile reduce calculation amount of sensors, then may optimize the sensors power consumption effectively.

2 Analysis of Traditional Locating Method

The iterative center of mass is a common locating method based on least square estimation. Least square estimation assumes that the location precision of each anchor node are equivalent, and chose the node making error sum of squares of anchor nodes least as estimate location of unknown node, as shown in Eq. (1). Through optimizing the target area constantly and calculating multiple iterative estimate nodes, then the method chose the average value of those nodes as unknown node location [3, 4].

$$[\hat{x}, \hat{y}] = (A^T \times A)^{-1} \times A^T \times b \tag{1}$$

In Eq. (1), the coefficient of A and b are known matrix related to anchor nodes location. Although the all locating method based on least square estimation can obtain unknown node location simply and quickly, but the locating precision is too low. So it needs a lot of calculating to obtain needed locating precision.

Maximum likelihood estimation locating method chose maximum probability point as estimate location of unknown node in the joint probability distribution function, that as shown in Eq. (2) [12].

$$
\begin{aligned}
L(\theta) &= M \sum_{i=1}^{n} \left\{ \log(\alpha^{-1}((x-x_i)^2 + (y-y_i)^2)^{\beta/2}) - (\sum_{k=1}^{M} \frac{P_{ik}}{M} \alpha^{-1}((x-x_i)^2 + (y-y_i)^2)^{\beta/2}) \right\} \\
&= M \sum_{i=1}^{n} \left\{ \frac{\beta}{2} \log((x-x_i)^2 + (y-y_i)^2)) - \log \alpha - (\sum_{k=1}^{M} \frac{P_{ik}}{M} \alpha^{-1}((x-x_i)^2 + (y-y_i)^2)^{\beta/2}) \right\}
\end{aligned}
\tag{2}
$$

According to the calculating method of maximum likelihood estimation, we have to take the derivative of Eq. (2), and then we will obtain a superfine differential equation. So it is impossible to get maximum point of the superfine differential equation by calculating directly. The usual solving method is numerical computation, which finds the maximum probability point by point-by-point comparison method. So the locating method based on maximum likelihood estimation need more calculating than the iterative center of mass.

To sum up, the all traditional locating methods of WSN are almost based on iterative center of mass and maximum likelihood estimation, that need large amount of calculation to obtain the needs locating precision level of wireless sensor network. So the traditional methods increase the sensor data processing amount that would lead to high power consumption of the sensors in a long time.

This article proposes a locating method based on weighted probability centroid. According to the different estimate nodes probability in the wireless channel trans-mission model, the method structures a triangle by choosing three maximum proba-bility points, and calculates the weighted centroid of triangle as unknown node location. So it combines the merit of iterative center of mass and the maximum like-lihood estimation. firstly, it confirms the most possible area of unknown node based on the node probability distribution, secondly, it confirms the weights of different prob-ability nodes, finally, it calculates the weight centroid of most possible area. The proposed method ensures the same locating precision level as traditional locating method, but significantly reduces the amount of calculation.

3   Weighted Probability Centroid Locating Method

3.1   The Probability Distribution Function of Unknown Node

RSSI take the characteristics of signal in the transmission attenuation to estimate the distance between the nodes [14], its mathematical model as shown in Eq. (3) [13–15].

$$P_i(d_i) = P_T - P(d_0) - 10n \lg(\frac{d_i}{d_0}) + X_{\sigma_i} \tag{3}$$

In Eq. (3), d_i denotes actual distance between unknown node and the i-th anchor node, d_0 denotes known reference distance, n denotes channel attenuation index, $2 \sim 4$, X_{σ_i} denotes the anchor nodes measurement error, it is Gaussian random vari-ables and mean value is zero, P_T denotes signal strength of anchor node, $P(d_0)$ denotes the signal strength of the fixed distance (d_0) from anchor node, $P_i(d_i)$ denotes the signal strength of the fixed distance (d_i) from i-th anchor node.

By equivalence transformation, the Eq. (4) can be obtained.

$$P_i(d_i') = P_T - P(d_0) - 10n \lg(\frac{d_i'}{d_0}). \tag{4}$$

In Eq. (4), $P_i(d'_i)$ denotes the signal strength of unknown node, d'_i denotes measure distance between unknown node and i-th anchor node.

So we can obtain the probability distribution function of d_i, as shown in Eq. (5).

$$P(d_i) = P\{D_i \leq d_i\} = P\{d'_i 10^{\frac{X_{\sigma_i}}{10n}} \leq d_i\}$$
$$= P\{X \leq 10n \lg \frac{d_i}{d'_i}\} = \Phi(\frac{10n \lg \frac{d_i}{d'_i}}{\sigma_i}) \tag{5}$$

Because the measurements of all anchor nodes are mutual independence, so the probability of each node being unknown node is shown in Eq. (6).

$$P = P(d_1) \times P(d_2) \times P(d_3) \cdots P(d_m) = \prod_{i=1}^{m} P(d_i) \tag{6}$$

In Eq. (6), m denotes the number of anchor nodes be identified by unknown node.

3.2   Weighted Probability Centroid Locating Algorithm

Compared Eq. (3) with Eq. (4), we can know that d'_i is bigger than d_i usually in actual measurement. So unknown node real location must be in an intersection area, which is overlapping area of circles, that is, anchor nodes are circle centers and d'_i is radius. The overlapping area is as shown in Fig. 1 [15].

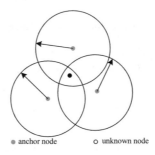

● anchor node   o unknown node

Fig. 1. Schematic diagram of overlapping area

According to usual methods, the area of Fig. 1 is too hard to solve directly. So the article divides all circles into different combinations firstly, each combination has two circles, solves intersection points of two circles at once secondly, and utilizes the triangle consisted of three maximum probability points of all intersection points as overlapping area finally. The number of combinations is shown in Eq. (7).

$$N = C_m^2 = \frac{m(m-1)}{2} \tag{7}$$

It is possible that some combinations of Eq. (7) are not intersecting, meanwhile, the combinations don't exist intersection points. For reducing unnecessary calculating, we just solve the intersection points of combinations satisfied Eq. (8).

$$d_{ij} \leq d'_i + d'_j \tag{8}$$

In Eq. (8), d_{ij} denotes distance between i-th circle center and j-th circle center, d'_i, d'_j denotes radius.

The all intersection points can be obtained by solving the intersection of two circles constantly. Because for circle the number of intersection is not more than two, so the all intersections number is not more than m × (m − 1). Take the all real number intersections into Eq. (6), and then we can obtain the probability of them. By the probability, choose three largest probability P_1, P_2, P_3, that node location are (x_1, y_1), (x_2, y_2), (x_3, y_3). So the location of unknown node is as shown in Eq. (9).

$$x = \frac{\sum_{i=1}^{3} P_i \times x_i}{\sum_{i=1}^{3} P_i} , y = \frac{\sum_{i=1}^{3} P_i \times y_i}{\sum_{i=1}^{3} P_i} \tag{9}$$

The proposed locating method has the advantages of maximum likelihood estimation and least square estimation. It reduces calculating obviously by probability distribution function optimizing possible area of unknown node, and ensures statistics locating precision by probability choosing weight. According to actual environment, the number of anchor nodes that unknown node can be indentified is not more than ten. So compared to traditional locating methods, the proposed method can reduces almost 80% calculating, meanwhile provides the same level locating precision.

4   Simulation

Table 1. Simulation conditions

Simulation parameter type	Parameter value
Size of sensor field	10 m × 10 m
Number of anchor node	N = 7
Anchor node location	Shown as in Fig. 2
Path decay exponent	N = 2
Standard deviation of anchor nodes	$\sigma_i = 5$
Number of RSS measurement	7

By MATLAB, it can build simulation experiment platform, simulation parameter is shown as in Table1, and the coordinate of unknown node is $\theta(x, y)$. Through the analysis of simulation experiment, it demonstrates that the method is feasible in above environment. Experimentation process is as the following. Firstly it builds a distribution area with $10\,\text{m} \times 10\,\text{m}$ size, and moreover chooses 20 unknown nodes randomly, Secondly it finds the estimate location $\hat{\theta}(\hat{x}, \hat{y})$ of unknown nodes by locating algorithm routine. And finally it gets the average locating error of the algorithm through the routine working 100 times repeatedly.

Through the comparison between the proposed method and least square estimation method, the locating result is shown as in Fig. 2. The least square estimation result is iterative 500 times in simulation. By means of comparative analysis, it can be seen clearly that the two methods have same locating precision, but the proposed method use lower calculating.

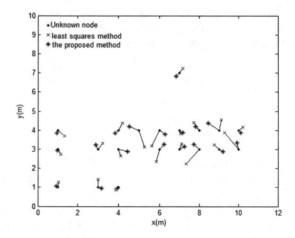

Fig. 2. Comparison of locating results for two algorithms

After the availability of the proposed method be testified, we add other simulation to find the influence rules of measurement parameters to locating precision. As shown in Fig. 3, the influence of anchor nodes measurement time has been simulated.

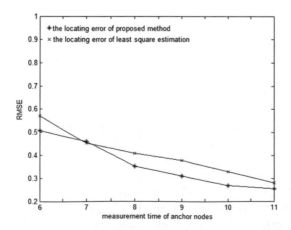

Fig. 3. Influence of anchor node measurement number of times for locating algorithm locating precision

It can be seen that the locating precision will be improved by adding anchor node measurement number of times from Fig. 3. But measurement number of times is equal power consumption of sensors, in actual, 8 times is suitable.

In Fig. 4, the influence of different measurement error to locating precision is shown, the different measurement error could be simulated by different standard deviation.

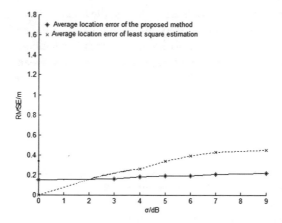

Fig. 4. Influence of standard deviation for locating precision

It can be seen that the precision will be lower with measurement error adding from Fig. 4, but the locating precision is also satisfied to locating need under larger measurement error. Finally we take the simulation of the anchor nodes number and locating precision in Fig. 5.

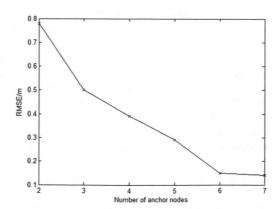

Fig. 5. Influence of number of anchor nodes

It can be seen that the locating precision will be improved by adding anchor nodes from Fig. 5. But adding anchor nodes number would improve power consumption of sensors. In actual, five to seven anchor nodes are suitable that be indentified by every unknown node.

5 Conclusion

Aimed to low power consumption need of WSN, the article proposes a low calculating locating method, which would reduce power consumption efficiently by minimization of locating calculating. Compared to traditional locating methods of WSN, the proposed method can provide the same level locating precision; meanwhile reduce almost 80% calculating in simulation. So the proposed method is more suitable for wild environment monitoring based on WSN.

References

1. Liu, X.: Research on Applied Technology of Digital Oil Field Data Transmission Based on Internet of Things. Xi'an ShiYou University (2013)
2. Liu, L.: Research on Wireless Data Collection System of Drilling Parameters Based on RFID. Northeast Petroleum University (2013)
3. Sun, L., Li, J., Chen, Y., et al.: Wireless Sensor Network. Tsinghua University Press, Beijing (2005)
4. Tian, X.-H., Zhou, Y.-J.: Theory and Technology of Wireless Location. National Defense University Press, Beijing (2011)
5. Gracioli, G., Frohlich, A.A., Pires, R.P., et al.: Evaluation of an RSSI-based location algorithm for wireless sensor networks. IEEE Latin Am. Trans. 9(1), 830–835 (2011)
6. Narzullaev, A., Park, Y., Kookeyeol, Y., et al.: A fast and accurate calibration algorithm for real-time locating systems based on the received signal strength indication. Int. J. Electr. Commun. 65(4), 305–311 (2010)
7. Wang, X., Yuan, S., Laur, R.: Dynamic localization based on spatial reasoning with RSSI in wireless sensor networks for transport logistics. Sens. Actuators A: Phys. 171(2), 421–428 (2011)
8. Nanda, M., Kumar, A., Kumar, S.: Localization of 3D WSN using Mamdani Sugano fuzzy weighted centriod approaches. In: 2012 IEEE Students' Conference on Electrical, Electronics and Computer Science, Bhopal, [s. n], pp. 1–5 (2012)
9. Tie, Q., Yu, Z., Feng, X.: A localization strategy based on n-times trilateral centriod with weight. Int. J. Commun Syst 25(1), 351–357 (2012)
10. Velimirovic, A., Djordjevic, G.L., Velimirovic, M.M., et al.: Fuzzy ring-overlapping range free localization method for wireless sensor networks. Comput. Commun. 35(13), 1590–1600 (2012)
11. Lu, K., Xiang, X., Zhang, D., et al.: Localization algorithm based on maximum a posteriori in wireless sensor networks. Int. J. Distrib. Sens. Netw. 8, 260302 (2012)
12. Miyauchi, K., Okamoto, E., Iwanami, Y.: Performance improvement of location estimation using deviation on received signal in Wireless Sensor Networks. In: 2010 Second International Conference on Ubiquitous and Future Networks, pp. 66–70. IEEE Conference Publication, Korea (2010)

13. Wang, J., Wang, F.-B., Duan, W.-J.: Application of weighted least square estimates on wireless sensor network node location. Appl. Res. Comput. **10**(1), 1–3 (2006)
14. Zhang, R.-B., Guo, J.-G., Chu, F.-H., et al.: Environmental adaptive indoor radio path loss model for wireless sensor networks localization. Int. J. Electr. Commun. **65**(12), 1023–1031 (2011)
15. Chen, S., Li, L., Zhu, B., et al.: Computing method of RSSI probability centroid for location in WSN. J. Zhejiang Univ. (Eng. Sci.) **48**(01), 100–104 (2014)

High Level Design of Power Wireless Private Network Construction

Ping Ma[1], Weiping Shao[1], Lei Zhang[1], Fengming Zhang[1], Rui Liu[2],
Jun Zou[3(✉)], and Jiyuan Sun[3]

[1] State Grid Zhejiang Electric Power Company, Hangzhou, China
[2] Information Technology and Communication Company, NARI Group
Corporation, Nanjing, China
[3] School of Electronic and Optical Engineering, Nanjing University of Science
and Technology, Nanjing, China
jun_zou@njust.edu.cn

Abstract. With the development of wireless communication technologies, LTE
networks have been widely used in various industries. For the consideration of
cost and security, private network is the best choice for the State Grid. In this
paper, we first analyze the necessity of power wireless private network and then
discuss the high level design of PWPN construction. We propose a three-stage
method to guarantee the coverage of the PWPN in the deep coverage scenario.
The comparisons of LTE 230 M and LTE 1.8G are also discussed in details.

Keywords: Power wireless private network · Coverage · Core network

1 Introduction

Energy issues have become a hot topic in China with the progress of society and the
development of science and technology. In particular, the traditional energy sources are
almost exhausted with the deterioration of the environment in recent years, which has
successfully drawn the world's attention to the exploitation and use of the emerging
green energy resources. As an important part of the emerging green energy resources,
distributed energy resources such as wind energy and solar energy are likely to become
one of the world's mainstream energy resources in the future 1.

The output power of these green energy resources is characterized by strong randomness and intermittency. Moreover, a series of problems will be caused by the
integration of large-scale new energy resources into the power grid, such as variation in
power grid's voltage, transmission power exceeding the limit, increase of system short
circuit.

To this end, real-time control and monitoring technology is needed to ensure that it
can be used effectively. In addition, more detailed collections of power grid operation
data are supposed to be done so as to support the operation of the "source-network-
load" interactive system, and to further improve the service quality of the power grid.
Information technology has gained an unprecedented development owning to the
progress of science and technology. It has been widely used in the field of energy. The
concept of energy Internet of things (IoT) came into being based on their combination.

© Springer Nature Switzerland AG 2020
F. Xhafa et al. (Eds.): IISA 2019, AISC 1084, pp. 638–645, 2020.
https://doi.org/10.1007/978-3-030-34387-3_78

The State Grid in China has carried out a number of distribution automation pilot constructions since 2010. The optical fiber communication technology is basically used as the main communication method of distribution automation access network. However, it is difficult for wired communication to achieve full coverage of distribution communication access network due to the limitation of the region and cost, that's why other means of communication are taken to overcome the disadvantage.

Wireless communication technology is known for its characteristics of flexible deployment and comprehensive coverage. It is usually taken as a powerful supplement to wired communication. Nowadays, it has become a hot research topic of energy IoT access network. With the evolution of TD-LTE (Time Division Long Term Evolution), 4G wireless communication technology has become an effective supplement to optical fiber communication depending on its advantages of high data rate and large coverage. It plays an important role in the construction of distribution automation now. At present, there are many transmissions are done by the public network LTE or General Packet Radio Service (GPRS), such as power usage information acquisition, distribution automation, and so on.

2 PWPN Construction Necessity

Currently, public network rental is a common way to deal with various electric power services. However, it has some defects. Firstly, the power grid intranet service applications cannot be extended to the external wireless terminal. For example, the project site cannot hold a remote video conference with the power grid intranet; the telephone and mobile terminal cannot achieve audio and video interoperability.

Secondly, the power applications like pictures and videos with a high requirement for bandwidth will lead to expensive traffic costs.

Thirdly, limited by the State Grid information security, video surveillance cannot be connected to the intranet. There are many inconveniences in the power applications, such as mobile visualization security supervision service, its daily construction task will be imported into the intranet "daily dynamic system" after being manually copied by the encryption disk. The operation is more cumbersome.

It is necessary to separate the power system from the existing wireless public network communication in consideration of its safety and reliability. Therefore, people have paid more attention to the establishment of the power wireless private network (PWPN) 2–4. Various researches and pilot construction work of PWPN are in implementation all over the country, especially in Jiangsu and Zhejiang province. In this paper, we work on the PWPN construction, and analyze the PWPN requirements in the energy IoT as well as the potential solutions. Figure 1 shows the architecture of power wireless private network. We can see that the PWPN is similar to the public wireless cellular network with the difference that all the terminals are power devices.

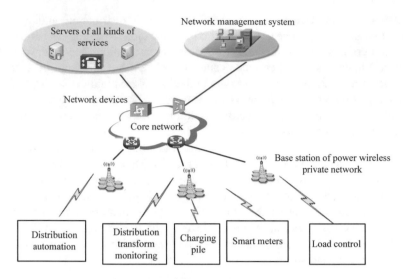

Fig. 1. Illustration of Power Wireless Private network

3 Service Category

There are a lot of different services in the PWPN. The service can be divided into basic serviceS and expanded services according to its characteristics. The basic services include user acquisition and load control, distribution automation, distributed power supply, electric vehicle charging and discharging piles (station). These kinds of services are characterized by low data rate, fixed types, and wide distributions. It has been widely used in the public network.

The extended service includes mobile inspection, distribution network repair, distribution equipment condition monitoring, power quality monitoring, robot inspection, infrastructure field video surveillance, smart home, and smart service hall. Currently, the services are not so heavy, with a large variety of types, involving data, images, voice, large granularity, strong mobility, and it is at the stage of pilot construction.

The basic service can be divided into low-latency service and latency tolerance service according to its requirements for transmission latency. For example, the data-related service like user acquisition and load control do not have a requirement for transmission latency, while the transmission of control signals such as distribution automation has a strict requirement for latency (the transmission latency is less than 0.1 s).

According to the requirements of PWPN, we summarized the system design requirements of power wireless private network construction, including frequency channel selection, safety, low latency, wide coverage methods.

4   Architecture of PWPN

4.1   Frequency Selection

At present, the license spectrum of power wireless private network in China mainly consists of 230 MHz and 1.8 GHz 4. Jiangsu Electric Power Company has set up a private network pilot for 230 MHz. The networking is carried out by using the LTE 230 M system developed by Potevio based on the public network LTE system. 1.8 GHz system basically follows the public network LTE technology. The LTE 1.8G system is gained after some improvements of the particularity of power services. These two systems with different carrier frequency have their own characteristics. It is very important to choose the appropriate frequency to build the network.

Advantages of LTE 1.8G: Firstly, LTE 1.8G system is fully based on 3GPP LTE technology, and it can evolve with the development of LTE standard technology, to provide wideband services. In addition to satisfying the existing distribution automation and power usage information acquisition services, LTE 1.8G system can also serve high data rate services, such as video, voice, and so on. Secondly, the world's mainstream operators have fully deployed the standard system based on the core network, base station, and the terminal.

Disadvantages of LTE 1.8G: Firstly, it is required to apply for the frequency points separately. Secondly, compared with LTE 230 M, its construction cost is higher.

Advantages of LTE 230 M: Firstly, 230 MHz is a private spectrum for different applications. Its bandwidth and latency can meet the basic requirements of power services. It has the advantages of wide coverage and low costs. Secondly, its overall construction cost is much cheaper than that of LTE1.8G due to the large coverage.

Disadvantages of LTE 230 M: Firstly, LTE 230 M works in 230 MHz (a private frequency channel for the power) with total 1 MHz (a non-continuous frequency) bandwidth, so it is really hard for it to meet the requirements of PWPN for multiple services with an insufficient capacity. Secondly, there is no unified standard system has been established, and the singleness of product suppliers, including the absence of a mature industrial chain, are also a problem for LTE 230 M. In addition, the system only partially makes use of 3GPP LTE standard technology, resulting in a big difference between it and LTE standard. Moreover, because of the low frequency, the multiple antenna techniques used to disturb and manage the key technologies like 4G and 5G cannot be applied in LTE 230 M. It cannot evolve to the next generation of LTE system according to the evolution route of LTE. Thirdly, LTE 230 M has the disadvantage of narrow frequency spacing, so it's specially designed for the static low-rate data service. In the mobile environment, it is greatly affected by the frequency offset.

Therefore, LTE 1.8 GHz is more suitable for the private network construction in urban areas and large granular service regions. LTE 230 MHz has certain advantages in the coverage in suburbs and rural areas, which is about 1.6 times of 1.8 GHz. It is suitable for the private network construction in the regions with less service terminals.

4.2   Security Strategy

The power industry has a high requirement for security, so does the construction of energy IoT. In order to be suitable for a mass of terminals, large-scale networks, and the hybrid networking mode of different service application scenarios, the overall design of security protection of terminal communication access network should be carried out based on an ubiquitous access network with large bandwidth, high reliability, safe and flexible access, efficient operation and maintenance. Therefore, we put forward the general principle of information security design: the protection of the service system of production control area should follow the principle of "security division, private network, horizontal isolation, vertical certification". For the management information region service system, the scheme carries out the active defense strategy based on "separate division and region, secure access, dynamic perception, comprehensive protection".

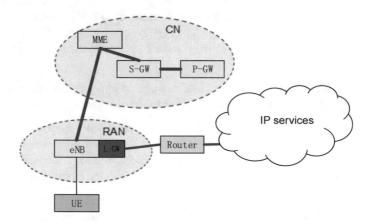

Fig. 2. Illustration of the simplified core network structure.

4.3   Simplified Core Network

The control signaling in the power grid has a high requirement for latency, so the PWPN is supposed to have the ability of low-latency communication 5. At present, the core network of the public network LTE system is quite complex 6, which will lead to a great system latency. It aims to support the mobility management of the equipment. In the case of power network, most of the equipment is fixed. In order to further reduce the transmission latency and meet the potential challenges of future ultra-low delay transmission, it is necessary to simplify the core network according to the characteristics of power wireless private network services. We designed an alternative proposal as shown in Fig. 2. In Fig. 2, the blue lines represent the data transmission of user plane, and the red lines refer to the signal transmission of control plane. The scheme follows the following main design principles:

(A) The S-GW (service gateway) of EPC (evolved packet core), and the P-GW (packet data network gateway) have the function of control plane and user plane. In the alternative, we will only remain the control plane function of S-GW and P-GW. The main function of the core network is to allocate the IP address of the UE (the function of P-GW). This design retains the original control plane process. In addition, due to the elimination of the function of the user plane, the function of the core network and the processing of the load has been effectively reduced.

(B) The data transmission of user plane will be implemented after the corresponding UE IP address is assigned to eNB by the core network through the control plane. In order to realize the process, the eNB node has done the following changes: 1. The eNB will ignore the user plane access between eNB and S-GW informed by MME (mobility management entity). 2. Deploy the L-GW (local gateway) within the eNB node. The eNB will send the user plane data directly through L-GW to the server, namely the smart grid master station.

It is important to note that this scheme does not support mobility. To this end, in the design, more attention should be paid to the problem of system error caused by abnormal switching instructions. This setting can be referred to: when the eNB receives a switching instruction sent by other eNBs, a rejection instruction can be sent forcefully.

4.4 Coverage Improvement

The coverage is very important for a communication system. In a power system, its equipment may be located in confined spaces like the depths of the building and the basement. In this context, it is necessary to take into account both the traditional wide coverage and the deep coverage.

Power usage information acquisition is a service with the largest number and a great difference in communication environment in the basic service. For this reason, we proposed a three-stage communication method to solve the communication problem when the meter is located in the deep coverage areas. The typical structure of three-stage communication method is shown in Fig. 3.

For devices with good signal quality, the transmission can be done directly through the remote channel. For the devices with poor signal quality, they will be connected to the collector or concentrator through PLC (power line communication) 7. These collectors and concentrators are put in places with good signal quality so that the data of the concentrator can be transmitted to the master station of the system through remote channel.

In addition, for wireless terminals located in the ring cabinet or in the case of building occlusion, we can also use the remote radio technology to put the antenna in a place with good signal quality to solve the problem of the coverage of the areas with poor wireless signal quality as shown in Fig. 4.

Furthermore, the NB-IoT (narrowband Internet of Things) system newly developed by 3GPP organization can provide an additional link margin of up to 20 dB based on the existing public network in response to the deep coverage scenarios without the need to add any relay devices 8. Therefore, in the construction of power wireless private

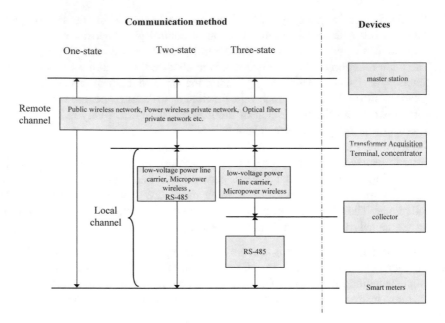

Fig. 3. Illustration of the three-stage communication method

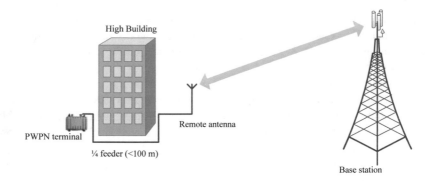

Fig. 4. illustration of remote radio for strong coverage

network, it is also necessary to take into account the compatibility of NB-IoT system and the upgrading of smooth evolution.

5   Conclusion

Wireless communication is a good choice to solve the last mile communication problem in the energy Internet of things. For the consideration of the security and cost, private network is a better choice than the public network renting for the power grid. In this paper, we firstly give the architecture of the power wireless private network. Then we discuss the requirements and the potential solutions from the service categories, security, coverage and delay aspect.

Acknowledgments. This work was supported by the National Natural Science Foundation of China (No. 61701234) and the technology project of State Grid Zhejiang Electric Power Company.

References

1. Bedi, G., Venayagamoorthy, G.K., Singh, R., Brooks, R.R., Wang, K.: Review of Internet of Things (IoT) in electric power and energy systems. IEEE Internet Things J. **5**(2), 847–870 (2018)
2. Cao, J., Liu, J., Zhang, Y., Li, X., Zeng, L.: Developing a power wireless private network based on TD LTE technology for intelligent distribution networks. In: International Conference on Automatic Control and Artificial Intelligence (ACAI 2012), Xiamen, pp. 890–893 (2012)
3. Jia, Y., Liu, R., Miao, W., Wei, L., Chen, X.: Research on network optimization in LTE power wireless private network. in: 2017 IEEE Conference on Energy Internet and Energy System Integration (EI2), Beijing, pp. 1–5 (2017)
4. Wei, L., Miao, W., Jiang, C., Guo, B., Li, W., Han, J., Liu, R., Zou, J.: Power wireless private network in energy IoT: Challenges, opportunities and solutions. In: 2017 International Smart Cities Conference (ISC2), Wuxi, pp. 1–4 (2017)
5. Jia, Y., Shi-long, C., Xue, G.: LTE wireless private network planning in electric power system. In: 2016 2nd IEEE International Conference on Computer and Communications (ICCC), Chengdu, pp. 3020–3024 (2016)
6. Thainesh, J.S., Wang, N., Tafazolli, R.: Reduction of core network signalling overhead in cluster based LTE small cell networks. In: 2015 IEEE 20th International Workshop on Computer Aided Modelling and Design of Communication Links and Networks (CAMAD), Guildford, pp. 226–230 (2015)
7. Jathar, M.R.: RS-485 based multi axis motor controller. In: 2011 3rd International Conference on Electronics Computer Technology, Kanyakumari, pp. 104–106 (2011)
8. Zou, J., Yu, H., Miao, W., Jiang, C.: Packet-based preamble design for random access in massive IoT communication systems. IEEE Access **5**, 11759–11767 (2017)

MIMO Two-Way Relaying Networks with Carrier Offsets

Guangyun Tan[1,2(✉)], Yongyi He[1], Huahu Xu[3], and Zhuo Wu[4]

[1] School of Mechatronic Engineering and Automation,
Shanghai University, Shanghai, China
fw339tgy@126.com
[2] Shanghai Qiansi Network Technology Limited Liability Company,
Shanghai, China
[3] School of Computer Engineering and Science, Shanghai University,
Shanghai, China
[4] School of Communication and Information Engineering,
Shanghai University, Shanghai, China

Abstract. In the paper, we investigate double differential detection (DD) for a two-way relaying network using multiple antennas with unknown carrier frequency offsets. The vast majority of existing research on two-way relaying hypotheses that all devices have perfect channel knowledge and no carrier frequency offsets (CFO). However, for fast changing channels, it is hard to receive certain channel state information (CSI), and the system capability is significantly reduced by reason of increasing the computational complexity of channel estimation and commonly existing CFOs. Accordingly, a signal detection method based on double differential is proposed to remove the interference caused by CFOs for a MIMO bi-directional relaying system. For superior system performance, all transmitters adopt orthogonal space-time block codes (OSTBC). Obviously, the simulation results turn out that the suggested DD scheme can not only remove the carrier offsets effectively without CSI knowledge, but also carry out full diversity order with acceptable linear computational complexity.

Keywords: Two-way relaying · Differential modulation · Orthogonal space-time block codes

1 Introduction

Two-way relaying, in which the information communication process occurs between two source nodes is completed with the assistance of the relay nodes located between them, has attracted much attention in recent years [1, 2]. It has been verified that the use of physical layer network coding (PNC) to decompose the information exchange process into the broadcast (BC) stage and the multiple-access (MA) stage to promoted the spectral efficiency [1]. Research on two-way relaying transmission has been carried out, and then put forward two relaying schemes of decode-and-forward (DF) and amplify-and-forward (AF) [3–5]. Because wireless channel naturally realized network coding by collecting wireless signals in the air, AF based two-way cooperation

© Springer Nature Switzerland AG 2020
F. Xhafa et al. (Eds.): IISA 2019, AISC 1084, pp. 646–656, 2020.
https://doi.org/10.1007/978-3-030-34387-3_79

performs exceptionally well in wireless networks Therefore, the AF based two-way cooperative transmission is the key research contents of this paper.

There has been some work studying two-way cooperation using AF [5, 7, 8]. Nevertheless, most of the existing work based on the assumption that all transmission links all have the access to the perfect channel state information (CSI). Although CSI may be acquired through pilot signals in some case, in fact, accurate CSI may not be available when channel coefficients change rapidly. In addition, the computational complexity of the rely nodes is also increased due to the channel estimation process. In the case of no noise amplification, destination is hard to acquire the source-to-relay channel excellently by booting symbol forwarding. Hence, it is worthwhile to study the differential modulation for two-way cooperation without the knowledge of CSI. Differential Modulation for two-way cooperative communications have been studied in some researches, in which perfect synchronization was assumed.

By considering two-way relaying systems with multiple antennas, one distinct feature is that each source UE transmits signal to the relay through different multiple paths and the signal is received at the relay with distinct carrier frequency offsets (CFOs) from the two sources, which may drastically degrade the system performance. Double differential modulation is one effective technique to mitigate CFO [12–16]. All the above processes, nevertheless, are performed on communication links with point-to-point, and cannot be directly used in two-way cooperative communications without the knowledge of carrier offsets, because the rely node receives the a mixed-signal composed of two source node signals, and not all nodes have the access to CSI.

As far as we know, when not all nodes have the access to CSI, there are little research on two-way relaying communications with missing carrier frequency offsets. Imperfect synchronization scenario caused by CFO was investigated in [17], using double differential modulation. However, it only considers that a single antenna is installed on both the relay and source node. Since spatial diversity, which can place multiple receive and/or transmit antennas, is very valid in reducing multipath fading and other interference, so it is worth exploring the application of multiple antennas for the two-way cooperative communications with unknown carrier frequency offsets in the case where not all devices have the access to CSI.

So, we go into the research of double differential detection (DD) for the two-way relaying communications using AF without the knowledge of offsets in this paper. To improve the property of the two-way relaying system, orthogonal space-time block codes (OSTBC) is applied at each transmitter. Simulation results turn out that the suggested DD scheme can effectively eliminate the carrier offsets very effectively, and achieve full diversity order with acceptable linear computational complexity.

2   The Model of System

We propose a network consisting of a relay node (R) and two source nodes (S_1 and S_2). Assume that a half-duplex system is used, and all nodes are configured with multiple antennas, where both source node are configured with n_t antennas and the relay is configured with n_r antennas. With the help of R, the exchange of information between S_1 and S_2 is accomplished in two stages. First, in the MA phase, the differentially encoded signals are transmitted from two source nodes to the relay node. Then in the BC stage, using side information, the relay node sends over the amplified superposition signals back to two source nodes.

3 Double Differential Modulation for MIMO Two-Way Relaying with CFO

3.1 OSTBC Based Double Differential Modulation

Let $s_i(k) = [s_i(1), s_i(2), \cdots, s_i(n_s)] \in Q$, $i \in \{1, 2\}, n_s \leq n_t$ denotes the n_s symbols sent by source node S_i at discrete-time k, where Q stands for a unity power M-PSK constellation set. After orthogonal space-time block encoding, the transmitted signal can be represented as a matrix $\mathbf{S}_i(k)(k \geq 2)$ of size $n_t \times n_t$. In addition, $\mathbf{S}_i(k)\mathbf{S}_i^H(k) = a\|s_i(k)\|^2\mathbf{I}_{n_t} = an_s\sigma_s^2\mathbf{I}_{n_t}$, where σ_s^2 is the average power of $s_i(k)$ and a is a proportionality factor and. In order to refrain from the power fluctuation and increase the power due to the usage of multiple antennas, we suppose $a = 1$ and normalize the signal constellation so that $\sigma_s^2 = 1/n_s$ and $\mathbf{S}_i(k)\mathbf{S}_i^H(k) = \mathbf{I}_{n_t}$. The first order differential matrix $\mathbf{R}_i(k)$ is shown in formula (1):

$$\mathbf{R}_i(k) = \mathbf{R}_i(k-1) * \mathbf{S}_i(k), k \geq 2, i \in \{1, 2\} \tag{1}$$

where $\mathbf{R}_i(1) \in \mathbb{C}^{n_t \times n_t}$ is an initialized matrix. And the second order differential matrix can then be acquired by the following formula:

$$\mathbf{D}_i(k) = \mathbf{D}_i(k-1) * \mathbf{R}_i(k), k \geq 2, i \in \{1, 2\} \tag{2}$$

with $\mathbf{D}_i(1) \in \mathbb{C}^{n_t \times n_t}$ is the initialization matrix. For $\rho \geq 2$, the second order differential matrix $\mathbf{D}_i(\rho)$ can be acquired by the current OSTBC matrix $\mathbf{S}_i(\rho)$ as:

$$\mathbf{D}_i(\rho) = \mathbf{D}_i(\rho - 2)\mathbf{R}_i^2(\rho - 1)\mathbf{S}_i(\rho) = \mathbf{D}_i(0)\mathbf{R}_i(1) \cdots \mathbf{R}_i(\rho - 2)\mathbf{R}_i^2(\rho - 1)\mathbf{S}_i(\rho) \tag{3}$$

It should be noted that (3) could be obtained only when $\mathbf{S}_i(\rho)$ is square with full rank [20]. Since choosing the initialization matrix is important for the double differential modulation and demodulation, $\mathbf{D}_i(0)$ and $\mathbf{R}_i(1)$ should also be full rank.

For convenience, make an assumption that the same number of antennas are assigned to both relay node and two source nodes. There are multiple channels between source nodes S_1 and S_2 and the relay R. $\mathbf{D}_1(k)$ and $\mathbf{D}_2(k)$ denote the transmitted $n_t \times n_t$ OSTBC in time k from S_1 and S_2. Due to the carrier offsets caused by doppler shift or instability relevant to the carrier oscillators belonging to S_i and R's, $\omega_{i,n} \in [-\pi, \pi)$ is assumed to be the random carrier offset (in radians) between S_i and the n-th receive antenna of relay R. Since the distances between Si and R would normally not be very small, it is further assumed that $\omega_{i,n} = \omega_i \in [-\pi, \pi), n = 1, \ldots, n_r$. For simplicity, the fading coefficient and the carrier offsets are assumed remain unchanged over frames of length L, and vary independently between frames, which is commonly assumed in the study of differential two-way relaying transmission [9, 10].

In the MA phase, the differentially encoded information is transmitted from two terminals transmit to the relay at the same time. Therefore, the signal received by the relay is shown in formula (4):

$$Y_r(k) = \sqrt{P_1}e^{jn_t\omega_1 k}HD_1(k)\Omega_1 + \sqrt{P_2}e^{jn_t\omega_2 k}GD_2(k)\Omega_2 + N \tag{4}$$

where P_i represents the transmit power of source S_i, $\Omega_1 \in \mathbb{C}^{n_t \times n_t}$ is a diagonal matrix with $\Omega_i = \text{diag}\left[1, e^{j\omega_i}, \cdots, e^{j\omega_i(n_t-1)}\right]$, and N is a $n_r \times n_t$ matrix, of which the element $n_{n,m}(k)$ denotes a zero mean complex Gaussian random variable with variance N_0.

$$H = \begin{bmatrix} h_{1,1} & h_{1,2} & \cdots & h_{1,n_t} \\ h_{2,1} & h_{2,2} & \cdots & h_{2,n_t} \\ \vdots & \vdots & \ddots & \vdots \\ h_{n_r,1} & h_{n_r,2} & \cdots & h_{n_r,n_t} \end{bmatrix} \quad \text{and} \quad G = \begin{bmatrix} g_{1,1} & g_{1,2} & \cdots & g_{1,n_t} \\ g_{2,1} & g_{2,2} & \cdots & g_{2,n_t} \\ \vdots & \vdots & \ddots & \vdots \\ g_{n_r,1} & g_{n_r,2} & \cdots & g_{n_r,n_t} \end{bmatrix} \quad \text{represent}$$

the flat Rayleigh fading channel matrices between source node S_1 and the relay R, and source node S_2 and the relay R, which are normalized as $\frac{1}{n_t}\sum_{m=1}^{n_t}\left|h_{m,n}\right|^2 = \frac{1}{n_t}\sum_{m=1}^{n_t}\left|g_{m,n}\right|^2 = 1, n = 1, 2, \cdots, n_r$ so that the received signal power is not increased through using multiple antennas.

In the BC phase, the relay R magnifies the received signal Y_r (as shown in formula (4)) by a facient β and then return back to both source S_1 and S_2 with transmit power P_r. The corresponding signal received by S_1 on m-th antenna at time k, represented $y_m^1(k)$, can then be written as:

$$y_m^1(k) = \sqrt{P_r}e^{-jn_t\omega_1 k}\beta h_m Y_r(k)\Omega_1^{-1} + q_m^1(k), k \geq 0 \tag{5}$$

where $y_m^1(k) \in \mathbb{C}^{1 \times n_t}$, $h_m = \left[h_{m,1}, \cdots, h_{m,n_t}\right]$ is a row vector containing the channel coefficients from all the transmit antennas to the m-th antenna of node S_1, and $q_m^1(k) \in \mathbb{C}^{1 \times n_t}$ contains zero mean complex Gaussian random variables with variance N_0. The amplification factor β can be calculated as:

$$\beta = \sqrt{\frac{P_r}{P_1\sigma_1^2 + P_2\sigma_2^2 + N_0}} \tag{6}$$

Therefore, formula (5) can be rewritten as:

$$\begin{aligned} y_m^1(k) &= \beta\sqrt{P_1 P_r}h_m HD_1(k) + \beta\sqrt{P_2 P_r}e^{jn_t(\omega_2-\omega_1)y}h_m GD_2(y)\Omega_2\Omega_1^{-1} + \hat{q}_m(y) \\ &= \mu D_1(k) + \beta\sqrt{P_2 P_r}e^{jn_t(\omega_2-\omega_1)k}h_m GD_2(k)\Omega_2\Omega_1^{-1} + \hat{q}_m(k) \end{aligned} \tag{7}$$

where $\mu = \beta\sqrt{P_1 P_r}h_m H, \hat{q}_m(k) = \sqrt{P_r}e^{-jn_t\omega_1 k}h_m N + q_m^1(k)$.

From (7), it can be observed that $y_m^1(k)$ is a complex superimposed signal containing both the node S1's own transmitted signal multiplied by an unknown factor μ and inter-symbol interference generated by carrier offset, therefore, traditional double differential detection is not straightforward to be used for point-to-point communication link. When μ is unknown, if self-information $D_1(k)$ can not be subtracted from $y_m^1(k)$, it is hard to decode the prospective message $S_2(k)$, because there is no CSI at S_1. Thereby, a differential detection and interference suppression method for the bi-directional relaying cooperative communication system is proposed in the following subsection.

In the same way, the received signal by S_2 can be denoted as

$$y_m^2(k) = \sqrt{P_r}e^{-jn_t\omega_2 k}\beta h_m Y_r(k)\Omega_2^{-1} + q_m^2(k), k \geq 0 \tag{8}$$

Considering that S_1 and S_2 receive mathematically symmetric signals, as described in formulas (5) and (8), we will only explore the signal received by S_1 for simplicity.

3.2   Double Differential Detection with CFO

The proposed signal detection of double differential in MIMO bi-directional cooperative communications is segmented into two procedures. The first procedure is to subtract the self-information of $\mu D_1(k)$ from $y_m^1(k)$, which is the foremost step in the whole detection process. Since the double differential modulation has the feature of being insensitive to unknown frequency offset, it is not necessary to acquire and track frequency offset. Therefore, in the second procedure of double differential detection, as long as the self-information $\mu D_1(k)$ is removed from the received signal at nodes S_1, the double differential detector for OSTBC can be applied.

Since S_1 does not know the channel state information between itself and the relay, therefore, μ cannot be directly obtained. Here, we propose a blind estimation scheme based on the orthogonality characteristics of OSTBC, where $D_1(k)$ is orthogonal space-time encoded, $D_1(k)$ and $D_2(k)$ are independent with the same distribution. The following can be obtained.

$$D_1(k)D_1^H(k) = I_{n_t} \tag{9}$$

$$\mathbb{E}\{D_1(k)D_2^H(k)\} = 0 \tag{10}$$

Multiplying (7) with $D_1^H(k)$, we can obtain formula (11):

$$y_m^1(k)D_1^H(k) = \mu D_1(k)D_1^H(k) + \sqrt{P_2 P_r}e^{jn_t(\omega_2-\omega_1)k}h_m G D_2(k)\Omega_2\Omega_1^{-1}D_1^H(k) + \bar{q}_m(k)$$
$$= \mu + \sqrt{P_2 P_r}e^{jn_t(\omega_2-\omega_1)k}h_m G\Omega_2\Omega_1^{-1}D_2(k)D_1^H(k) + \bar{q}_m(k) \tag{11}$$

The expectation of (11) can be written as:

$$\mathbb{E}\{y_m^1 D_1^H\} \approx L\mu + \sqrt{P_2 P_r}\sum_{k=1}^{L}e^{jn_t(\omega_2-\omega_1)k}h_m G\Omega_2\Omega_1^{-1}D_2(k)D_1^H(k) + \mathbb{E}\{\bar{q}_m\} \approx L \tag{12}$$

$$\mu \approx \mathbb{E}\{y_m^1 D_1^H\} \approx \frac{\sum_{k=1}^{L} y_m^1(k) D_1^H(k)}{L} \tag{13}$$

Hence, μ can be estimated using (13). Since the transmitted information of node S_1 already known, $\mu D_1(k)$ can be subtracted from the received signal, and formula (7) can be represented as:

$$\widehat{y}_m^1(k) = y_m^1(k) - \mu D_1(k) = \sqrt{P_2 P_r} e^{jn_t(\omega_2-\omega_1)k} h_m G D_2(k) \Omega_2 \Omega_1^{-1} + \widehat{q}_m(k)$$
$$= \sqrt{P_2 P_r} e^{jn_t\omega k} \widehat{h}_m D_2(k) \Omega + \widehat{q}_m(k) \tag{14}$$

where $\widehat{h}_m = h_m G, \omega = \omega_2 - \omega_1, \Omega = \Omega_2 \Omega_1^{-1}$.

The second step of the proposed double differential detection is described in detail as follows. For convenience, let $P_1 = P_2 = P_r = 1$. Thereby, the distribution of $\widehat{y}_m^1(k)$ can be represented as:

$$f\left(\widehat{y}_m^1(k)|\omega, \widehat{h}_m, D_2(k)\right) = \frac{1}{\pi^{n_t}\det(\Lambda)} \times \exp\left(-\left[\widehat{y}_m^1(k) - e^{jn_t wk}\widehat{h}_m D_2(k)\Omega\right]^* * \Lambda^{-1}\right.$$
$$\left. * \left[\widehat{y}_m^1(k) - e^{jn_t\omega k}\widehat{h}_m D_2(k)\Omega\right]^T\right) \tag{15}$$

where Λ is the covariance matrix of $\widehat{q}_m(k)$ with $\Lambda = \mathbb{E}\{\widehat{q}^T(k)\widehat{q}^*(k)\}$. Assume $\widehat{q}_m(k)$ is AWGN with $\Lambda = N_0 I_{n_t}$. Let $\widehat{y}_m^1 = \left[\widehat{y}_m^1(k-2), \widehat{y}_m^1(k-1), \widehat{y}_m^1(k)\right] \in \mathbb{C}^{1\times 3n_t}$, and $\widehat{y}_m^1$ contains the vector samples received in 3 consecutive time intervals. When $\widehat{h}_m$, ω, $D_2(k-2)$, $F_2(k-1)$ are known,

$$f\left(\widehat{y}_m^1|\omega, \widehat{h}_m, D_2(k-1), F_2(k-1), S_2(k)\right) = \frac{1}{\pi^{3n_t} N_0^{3n_t}} \times \exp\left(-\frac{1}{N_0}\sum_{l=k-2}^{k}\left\|\widehat{y}_m^1(l)\right.\right.$$
$$\left.\left. - e^{jn_t\omega l}\widehat{h}_m D_2(l)\Omega\right\|^2\right) \tag{16}$$

In order to obtain the maximum likelihood estimation of $S_2(k)$, then the joint probability distribution function (16) is first maximized with respect to unknown factors $\omega, \widehat{h}_m, D_2(k-1), F_2(k-1)$, then maximized over $S_2(k)$. Since (16) is in an exponential form, maximizing it over $S_2(k)$ can be simplified by minimizing (17) with respect to $\omega, \widehat{h}_m, D_2(k-1), F_2(k-1)$ and $S_2(k)$.

$$\Gamma_m(k) = \sum_{l=k-2}^{k}\widehat{y}_m^1(l) - e^{jn_t\omega l}\widehat{h}_m D_2(l)\Omega^2, k \geq 2 \tag{17}$$

However, minimizing (17) is computationally complicated and time consuming, therefore, the following suboptimal detection method is introduced, which is independent of carrier offsets and channel state information and, with only linear computational complexity.

It has been demonstrated that when $X(k)$ is a square OSTBC matrix, so $\widehat{X}(k) \triangleq \Omega^H X(k)\Omega$ is also a square OSTBC matrix. To streamline the decision-making process, we deliberate over a degenerated metric from as:

$$\mathcal{D}_m(k) = \sum_{l=k-2}^{k-1} \left\| \widehat{\boldsymbol{y}}_m^1(l) - e^{jn_t\omega l}\widehat{\boldsymbol{h}}_m D_2(l)\Omega \right\|^2, k \geq 2 \tag{18}$$

Let $\boldsymbol{g}_m(k) = e^{jn_t\omega(k-2)}\widehat{\boldsymbol{h}}_m D_2(k-2)\Omega$, then minimizing (18) over $\boldsymbol{g}_m(k)$, we can obtain

$$\widehat{\boldsymbol{g}}_m = \frac{1}{2}\left(\varepsilon_\omega^* \widehat{\boldsymbol{y}}_m^1(k-1)\widehat{\boldsymbol{F}}_2^H(k-1) + \widehat{\boldsymbol{y}}_m^1(k-2)\right) \tag{19}$$

where $\varepsilon_\omega = e^{jn_t\omega}$, $\widehat{\boldsymbol{F}}_2^H(k-1) = \Omega^H \boldsymbol{F}_2(k-1)\Omega$. Replace (19) into (18), with the unitary property of OSTBC, (18) can be simplified as

$$\mathcal{D}_m(k) = \left\| \widehat{\boldsymbol{y}}_m^1(l) - \varepsilon_\omega \widehat{\boldsymbol{y}}_m^1(k-2)\widehat{\boldsymbol{F}}_2(k-1) \right\|^2, k \geq 2 \tag{20}$$

From (20), it is shown that if $\widehat{\boldsymbol{F}}_2(k-1)$ is unknown, then ε_ω cannot be estimated through (20). However, it should be noted that when $\boldsymbol{F}_2(k-1) = \boldsymbol{I}_{n_t}$, $\widehat{\boldsymbol{F}}_2(k-1) = \boldsymbol{I}_{n_t}$. Hence, (20) can be simplified as:

$$\mathcal{D}_m(k) = \left\| \widehat{\boldsymbol{y}}_m^1(l) - \varepsilon_\omega \widehat{\boldsymbol{y}}_m^1(k-2) \right\|^2, k \geq 2 \tag{21}$$

Expand (21), we can obtain (22):

$$\mathcal{D}_m(k) = \left\| \widehat{\boldsymbol{y}}_m^1(k-1) \right\|^2 + \left\| \widehat{\boldsymbol{y}}_m^1(k-2) \right\|^2 - \varepsilon_\omega^* \widehat{\boldsymbol{y}}_m^1(k-1)\widehat{\boldsymbol{y}}_m^{1\,H}(k-2) \\ - \varepsilon_\omega \widehat{\boldsymbol{y}}_m^1(k-2)\widehat{\boldsymbol{y}}_m^{1\,H}(k-1) \tag{22}$$

Let $\varepsilon_\omega = e^{j\theta}$, then differentiate (22) over θ:

$$e^{j\theta} = \frac{\widehat{\boldsymbol{y}}_m^1(k-1)\widehat{\boldsymbol{y}}_m^{1\,H}(k-2)}{\widehat{\boldsymbol{y}}_m^1(k-2)\widehat{\boldsymbol{y}}_m^{1\,H}(k-1)} = \frac{\exp\left(j\arg\{\widehat{\boldsymbol{y}}_m^1(k-1)\widehat{\boldsymbol{y}}_m^{1\,H}(k-2)\}\right)}{\exp\left(-j\arg\{\widehat{\boldsymbol{y}}_m^1(k-2)\widehat{\boldsymbol{y}}_m^{1\,H}(k-1)\}\right)} \\ = \exp\left(j2\arg\{\widehat{\boldsymbol{y}}_m^1(k-1)\widehat{\boldsymbol{y}}_m^{1\,H}(k-2)\}\right) \tag{23}$$

The estimation of ε_ω can then be obtained with $\widehat{\varepsilon}_\omega(k-1) = \exp(j\arg\{\widehat{\boldsymbol{y}}_m^1(k-1)\widehat{\boldsymbol{y}}_m^{1\,H}(k-2)\})$.

Due to the property of OSTBC, we cannot assume that $\boldsymbol{F}_2(k-1)$ is an identity matrix, however, we can assume $\boldsymbol{F}_2(1) = \boldsymbol{I}_{n_t}$. Therefore, when $k = 2$, $\widehat{\varepsilon}_\omega(1)$ can be estimated, while when $k > 2$, $\widehat{\varepsilon}_\omega(1)$ can replace ε_ω in latter analysis.

According to (1) (2) and (14), the received signal $\widehat{\boldsymbol{y}}_m^1(k)$ on the m-th antenna of node after self-interference subtraction can be expressed in terms of ε_ω, $\boldsymbol{g}_m(k)$, $\boldsymbol{F}_2(k-1)$ as:

$$\widehat{\boldsymbol{y}}_m^1(k) = \varepsilon_\omega^2 \boldsymbol{g}_m(k) \boldsymbol{\Omega}^H \boldsymbol{F}_2^2(k-1) \boldsymbol{S}_2(k) \boldsymbol{\Omega} + \widehat{\boldsymbol{q}}_m(k) \tag{24}$$

Thereby, by minimizing the following metric, the estimation of $\boldsymbol{S}_2(k)$ can be obtained:

$$\widehat{\boldsymbol{S}}_2(k) = \arg^{min}_{\boldsymbol{S}_2(k) \in \mathcal{Q}} \widehat{\boldsymbol{y}}_m^1(k) - \widehat{\varepsilon}_{\omega,1}^2 \widehat{\boldsymbol{g}}_m(k) \widehat{\boldsymbol{\Omega}}^H \widehat{\boldsymbol{F}}_2^2(k-1) \widehat{\boldsymbol{\Omega}} \widehat{\boldsymbol{S}}_2(k)^2 \tag{25}$$

where $\mathcal{Q}$ is the constellation set of the M-PSK modulation. $\widehat{\boldsymbol{\Omega}}$ can be obtained through $\widehat{\varepsilon}_{\omega,1}^2$ with $\widehat{\boldsymbol{S}}_2(k) = \widehat{\boldsymbol{\Omega}}^H \boldsymbol{S}_2(k) \widehat{\boldsymbol{\Omega}}$.

It should be also noted that for the same frame of data, in decision metric (25), the first estimation of ε_ω, $\widehat{\varepsilon}_\omega(1)$ is used throughout the frame, instead of $\widehat{\varepsilon}_\omega(k)$. According to (1) (2) and (25), when $k > 2$, $\widehat{\boldsymbol{F}}_2(k-1)$ can be reconstructed by using the estimated $\widehat{\boldsymbol{S}}_2(k)$, then the value of ε_ω can be estimated again. However, when $k > 2$, the reconstructed $\widehat{\boldsymbol{F}}_2(k-1)$ includes the effect of residue noise. As a consequence, using $\widehat{\boldsymbol{F}}_2(k-1)$ estimate ε_ω again will result in worse estimation accuracy. Therefore, we always use the first estimation of $\widehat{\varepsilon}_\omega(1)$ in the decision metric.

Let $\widehat{\boldsymbol{h}}_m = \widehat{\varepsilon}_{\omega,1}^2 \widehat{\boldsymbol{g}}_m(k) \widehat{\boldsymbol{\Omega}}^H \widehat{\boldsymbol{F}}_2^2(k-1) \widehat{\boldsymbol{\Omega}}$, then (25) can be further written as:

$$\widehat{\boldsymbol{S}}_2(k) = \arg^{min}_{\boldsymbol{S}_2(k) \in \mathcal{Q}} \left\| \widehat{\boldsymbol{y}}_m^1(k) - \overline{\boldsymbol{h}}_m \widetilde{\boldsymbol{S}}_2(k) \right\|^2 \tag{26}$$

Given that $\widetilde{\boldsymbol{S}}_2(k)$ is a square OSTBC matrix, thereby $\widetilde{\boldsymbol{S}}_2(k)$ can be represented as $\widetilde{\boldsymbol{S}}_2(k) = \sum_{n=1}^{n_s} (\bar{s}_n \boldsymbol{A}_n + j\breve{s}_n \boldsymbol{B}_n)$, where $\bar{s}_n$ and $\breve{s}_n$ are the real part and imaginary part of a complex signal s_n respectively, and $\boldsymbol{A}_n$ and $\boldsymbol{B}_n$ are matrices of size $n_t \times n_t$. Expanding (26) and utilizing the orthogonality property of $\widetilde{\boldsymbol{S}}_2(k) \widetilde{\boldsymbol{S}}_2^H(k) = \sum_{n=1}^{n_s} |s_n|^2 \boldsymbol{I}_{n_t}$, it is not difficult to see that the decoding of $\boldsymbol{S}_2(k)$ is with linear computational complexity. In addition, compare with the coherent decoder, the decision process on account of three consecutive received block data, so it has the lower decoding complexity. The former uses FFT, specific training matrices, and maxima searching procedure to estimate carrier offsets [23], and training channel coefficients estimator before decoding the OSTBC data [24]. The decoding of (25) can be generalized for $k \geq 2$ to the n_r receive antennas case as:

$$\widehat{\boldsymbol{S}}_2(k) = \arg^{min}_{\boldsymbol{S}_2(k) \in \mathcal{Q}} \sum_{m=1}^{n_t} \left\| \widehat{\boldsymbol{y}}_m^1(k) - \widehat{\varepsilon}_{\omega,1}^2 \widehat{\boldsymbol{g}}_m(k) \widehat{\boldsymbol{\Omega}}^H \widehat{\boldsymbol{F}}_2^2(k-1) \widehat{\boldsymbol{\Omega}} \widehat{\boldsymbol{S}}_2(k) \right\|^2 \tag{27}$$

4 Simulation Results

This section exhibits the simulation results for the suggested double differential detection scheme for a MIMO two-way relaying system with carrier frequency offsets. Suppose that both the source nodes and the relay are allocated 2 antennas. The OSTBC applied are Alamouti Codes. Both source nodes and the relay node have the same noise variance N_0. And the variance of complex channel coefficient of all links is 1. It is assumed that the MIMO channel coefficients between the sources and the relay is independent complex Gaussian random variables with the covariance matrix of identity matrix. The random carrier offsets $\omega_1, \omega_2 \in [-\pi, \pi)$. The initialization matrices, D_0, F_1, are chosen to be identity matrix with $D_0 = F_1 = I_2$. All simulations were implemented using BPSK modulation and the frame length was set to 100.

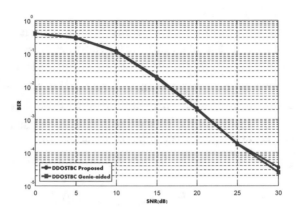

Fig. 1. BER performances of double differential detection with μ under estimation and perfectly known.

Figure 1 shows the BER performance of estimating μ as depict in (10) and (11). For comparison of results, we also contained the result of Genie-aided, in which assumes completely known μ at source node so then traditional differential decoding without the knowledge of carrier offsets can be applied to double differential. The result show that the estimation method with carrier offsets has almost no performance loss, which clearly proves the robustness of the proposed scheme.

It can be observed from Fig. 2 that the effect of interference generated from carrier offsets has been greatly removed with the double differential detection scheme, showing no error floor in the BER performance with both single and multiple receive antennas. By comparing the results of 1 receive antenna and 2 receive antennas, it can be observed that full diversity order of $n_t n_r$ could be gained by using the suggested double differential OSTBC scheme. Compared to [12] where on STBC block could only send one M-PSK modulated symbol, the OSTBC here can transmit n_s symbols within one STBC block. Therefore, the transmission rate and spectrum efficiency are improved. We also include the result of double differential for bi-directional two-way relaying communications with only single antenna equipped at all nodes within the system. It is shown that diversity order gain has been gained by equipping multiple antennas at both the sources and the relay node.

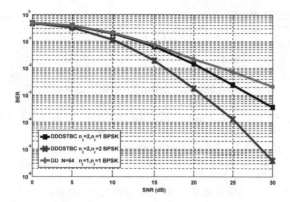

Fig. 2. Simulated BER performance of the proposed DDOSTBC with BPSK modulation under different numbers of receive antennas.

5 Conclusions

This paper proposed a double differential modulation scheme to eliminate the unknown frequency offset in a two-way relaying communication system with multiple antennas without knowing the GIS at both the sources and the relay. The simulation results reveal that, in the absence of channel knowledge, the proposed algorithms can effectively eliminate the influence of frequency offsets and achieve full diversity order with lower computational complexity.

Acknowledgment. This work was supported in part by Shanghai Technology and Science Administration Project under Grant No. 16511104800, and National Key R&D Program 2016YFC0800405.

References

1. Zhang, S., Liew, S.C., Lam, P.P.: Physical layer network coding. In: Proceedings of MobiCom 2006, pp 358–365, New York, NY, USA (2006)
2. Popovski, P., Yomo, H.: Physical network coding in two–way wireless relay channels. In: Proceedings of IEEE ICC 2007, pp. 707–712 (2007)
3. Xu, C., Song, L., Han, Z., Zhao, Q., Wang, X., Jiao, B.: Efficient resource allocation for device-to-device underlaying networks using combinatorial auction. IEEE J. Sel. Areas Commun. **31**(9), 348–358 (2013). Special issue "on Peer-to-Peer Networks"
4. Yuen, C., Chin, W.H., Guan, Y.L., Chen, W., Tee, T.: Bi-directional multi-antenna relay communications with wireless network coding. In: IEEE Proceedings of Vehicular Technology Conference, pp. 1385–1389, May 2008
5. Popovski, P., Yomo, H.: Wireless network coding by amplify-and-forward for bi-directional traffic flows. IEEE Commun. Lett. **11**(1), 16–18 (2007)
6. Zhang, S., Liew, S.C., Lam, P.P.: Hot topic: physical-layer network coding. In: Presented at the ACM MobiCom, Los Angeles, CA (2006)

7. Wang, H., Xia, X., Yin, Q.: A linear analog network coding for asynchronous two-way relay networks. IEEE Trans. Wirel. Commun. **9**(12), 3630–3637 (2010)
8. Jing, Y.: A relay selection scheme for two-way amplify-and-forward relay networks. In: IEEE WCSP 2009 (2009)
9. Cui, T., Gao, F., Tellambura, C.: Differential modulation for two-way wireless communications: a perspective of differential network coding at the physical layer. IEEE Trans. Commun. **57**(10), 2977–2987 (2009)
10. Song, L., Li, Y.: Differential modulation for bidirectional relaying with analog network coding. IEEE Trans. Signal Process. **58**(7), 3933–3938 (2010)
11. Fang, Z., Liang, F., Li, L., Jin, L.: Performance analysis and power allocation for two-way amplify-and-forward relaying with generalized differential modulation. IEEE Trans. Veh. Technol. **63**(2), 937–942 (2014)
12. Liu, Z., Giannakis, B., Hughes, B.L.: Double differential space-time block coding for time-selective fading channels. IEEE Trans. Commun. **49**(9), 1529–1539 (2001)
13. Liu, J., Petre, S., Simon, M.: Single differential modulation and detection for MPSK in the presence of unknown frequency offset. In: IEEE conference on Signals, Systems and Computers, pp. 1440–1444 (2006)
14. van Alphen, D.K., Lindsey, W.C.: Higher order differential phase-shift keyed modulation. IEEE Trans. Commun. **42**, 440–448 (1994)
15. Divsalar, D., Simon, M.K.: Multiple symbol differential detection of MPSK. IEEE Trans. Commun. **38**(3), 300–308 (1990)
16. Simon, M.: Multiple-symbol double-differential detection based on least-square and generalized-likelihood ratio criteria. IEEE Trans. Commun. **52**(1), 46–49 (2004)
17. Gao, Z., Sun, L., Wang, Y., Liao, X.: Double differential transmission for amplify-and-forward two-way relay systems. IEEE Commun. Lett. **PP**(99), 1. http://doi.org/10.1109/LCOMM.2014.2350478

Research and Design of High Speed Data Transmission System Based on PCIE

Zhang Kun, Li Keli, Bi Fanghong, Liang Ying, and Yang Jun[✉]

School of Information Science and Engineering, Yunnan University,
Kunming 650091, Yunnan, China
junyang@ynu.edu.cn

Abstract. In view of the traditional hardware to realize high speed data transmission system based on PCIE bus in the data cache and transmission speed is not high, using Xilinx FPGA chip MIG IP core and PCIE IP core, design the DDR3 based on time division multiplexing high-speed data caching module and PCIE interface module based on the structure of the DMA, cache and data transmission is realized. Experimental results show that the designed PCIE high-speed data transmission system has a data transmission capacity of 3.6 GB/s.

Keywords: PCIE · FPGA · DDR3 · DMA

1 Introduction

High-speed data Transmission is widely used in many fields, such as radar signal processing, biological spectrum analysis and Transmission of UAV intrusion detection data [1]. Therefore, it is very important to design a data Transmission system with high bus transmission rate [2]. Compared with traditional bus. This paper studies a high speed data Transmission system based on PCIE2.0 bus. By means of FPGA technology, the time-division multiplexing cache controller and PCIE bus transmission controller with DMA structure are designed to realize efficient cache and high-speed reliable transmission of system data.

2 Overall System Design

This design adopts the modular design method to realize the high-speed data Transmission system based on PCIE [3]. The designed PCIE high-speed data Transmission system mainly includes high-speed I/O interface module, FPGA configuration module, clock management module, DDR3 large-capacity cache module, FPGA master control module and PCIE interface module [4]. The overall design block diagram of the system is shown in Fig. 1. The key modules are FPGA main control module, DDR3 large-capacity cache module, PCIE bus interface module and high-speed I/O interface module.

© Springer Nature Switzerland AG 2020
F. Xhafa et al. (Eds.): IISA 2019, AISC 1084, pp. 657–663, 2020.
https://doi.org/10.1007/978-3-030-34387-3_80

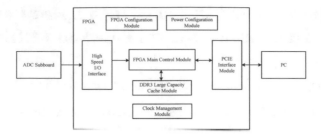

Fig. 1. System overall design block diagram

(1) FPGA main control module: composed of FPGA chip and related configuration circuit, responsible for the overall scheduling and control of each module of the whole system. Considering the large number of modules in this system, its data Transmission and up-down transmission rate are very high, so there are quite high requirements on the number of I/O ports, port speed and on-chip logical storage resources of FPGA chips. This system chooses Xilinx company's FPGA chip XC7K410TFFG900. The chip has 350 I/O ports, 36 Kb dual-port block RAM storage units, meeting the system design requirements [5].

(2) DDR3 large-capacity cache module: the main design purpose of large-capacity cache module is to cache the data sampled by ADC timely and ensure the stability of the system under high-speed reading and writing. In this system, DDR3 is selected as high-capacity cache device.

(3) PCIE bus interface module: PCIE bus interface is responsible for communication and data transmission between FPGA control motherboard and PC upper computer. This system uses PCIE 2.0x8 channel as the data transmission bus of the system, which can transfer the data of the FPGA control motherboard to the PC in real time, and also transfer the data of the PC control motherboard to the FPGA control motherboard in a fast manner, with the theoretical data transmission up to 4 GB/s.

(4) high-speed I/O interface module: high-speed I/O interface is used for communication and data transmission between ADC Transmission subboard and FPGA control motherboard. In this design, FMC (FPGA Mezzanine Card) interface with 400-pin pin was used as the I/O interface for high-speed communication and data transmission between the system subboard and the control motherboard, guaranteeing the output of digital signal from ADC Transmission subboard to FPGA control motherboard.

Among the above 4 modules, DDR3 large-capacity cache module and PCIE bus interface module are the most important ones that determine the performance of the whole system. The design of DDR3 large capacity cache module and PCIE bus interface module will be introduced in detail below.

3   System Key Module Design

3.1   DDR3 Large-Capacity Cache Module Design

DDR3 Large-Capacity Cache Hardware Module Structure
The DDR3 hardware module is divided into four parts: user interface module, time division multiplexing bridge circuit module, DDR3 controller provided by Xilinx company and DDR3 chip. The user interface module, as a user-oriented interface, contains two independent read-write interfaces. This module is the source of the read and write operations. Time division multiplexing (TDM) bridge module can convert double read-write ports to single read-write ports. This module contains four FIFO for DDR3 to read and write two-channel temporary cache of data. DDR3 controller is generated by IP core MIG and can complete the operation of writing data and reading data to DDR3. It is the main controller of DDR3, but it only has a set of independent read-write channels. DDR3 chip belongs to external hardware device and realizes data storage function. The hardware design structure of the whole module is shown in Fig. 2.

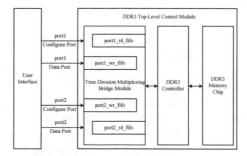

Fig. 2.   Hardware design structure diagram of DDR3 module

Time Division Multiplexing Bridge Circuit Design
This design adopts the idea of time division multiplexing. Through the data switcher, two sets of read-write interfaces are connected with DDR3 controller interfaces at different times. Each group of read-write interfaces is responsible for two tasks: reading and writing.

In order to make the four tasks independent of each other in the process of execution without conflicts, a TDM state machine is designed in this design to assign the execution of each task. The state machine determines the jump to the task state issued by the user through the conditional judgment structure in the idle state; After entering the task state, start the task execution timing; When the task is full, the loop is completed by jumping to an idle state with an additional end state, regardless of whether the task is completed or not. The time-division multiplexing state machine state transition of this design is shown in Fig. 3.

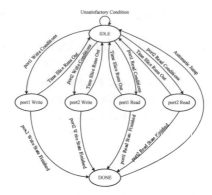

Fig. 3. Time division multiplexing state machine state transition diagram

The state machine shown in Fig. 3 can realize four kinds of task command detection and divide the execution order of multiple task instructions issued by the user. Task execution sequence is divided by a 2 bit wide unidirectional counter flag. When flag value is 0, the corresponding port1 write data task; When flag value is 1, it corresponds to the port1 read data task; When flag value is 2, the corresponding port2 write data task; When the flag value is 3, it corresponds to the port2 read data task. It is specified that flag jumps to the idle state after a task has been timed for eight clock cycles, regardless of whether the task is completed or not.

Four FIFO are used in the user interface circuit as the data cache, and the data to be written to DDR3 is cached in the write data FIFO. Write data FIFO input bit width is 256 bit, output bit width is 512 bit, in order to facilitate DDR3 controller docking. The data to be read is cached in the read data FIFO. The input bit width of read data is 512 bit, and the output bit width is 256 bit, so as to facilitate user interface docking. When a read to a FIFO is empty or not ready due to a full write, the entire system task is suspended until the FIFO is back to normal.

3.2   PCIE Bus Interface Module

Overall Design of PCIE DMA Write Logic
In this design, PCIE DMA control structure is divided into six modules, the relationship of each module is shown in Fig. 4.

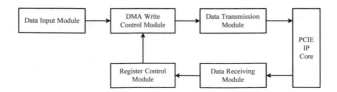

Fig. 4. PCIE DMA control structure

Among them, the data input module is mainly used to store front-end data, while the data receiving module is used to receive configuration information of high-speed data transmission system and relevant parameters of DMA transmission mechanism from users, and unpack according to PCIE byte specification protocol, and put forward useful information to register control module for storage. The register control module is mainly used to store the base address information, transmission packet length and size information parsed by the data receiving module, and to regroup the information parsed by the data receiving module and control the data writing address. DMA write control module mainly completes DMA write data transfer process packet length control core corresponding DMA write offset address control. According to the packet header information and address information sent by DMA write control module, the data transmission module groups each packet according to the PCIE protocol specification, and then sends the packet to the PCIE core, which is transmitted to the computer for storage.

Register Control Module Design

The register control module repackages the parsed information of the data receiving module and controls the data writing address. The offset addresses and values corresponding to the data information types in this design are shown in Table 1. The offset address is determined by the FPGA control part of PCIE and the PCIE driver part through negotiation.

Table 1. Register control allocation table.

Offset Address	Information types	Numerical value
0x08	Base_Addr	It depends on the hardware of the system
0x0C	TLP_Size	0xXXXX_0020
0x10	TLP_Count	0xXXXX_2800
0x14	TLP_User_Cmd	Set according to user requirements
0x04	DMA_Start	0x0000_0001/0x0000_0000

The signal block diagram of the register control module is shown in Fig. 5. In the register control module, if the memory write signal pcie_reg_wr is detected to be valid, and the data received is 1, the corresponding address offset address is 0x01, it means that the DMA write data start signal dma_en is valid and sent to the DMA data transfer request module DMA_REQ. In the DMA_REQ module, if dma_en is detected to be valid, a DMA write enable signal is generated and sent to the DMA_WRITE module of the DMA write data control module. If the DMA_WRITE module detects that the dma_wr_enable signal is valid, the tpl_send_req transmission data request signal is generated in this module, which causes the state machine to jump to SEND_TPL to send the data state, and generates dma_write_req DMA write request signal, which is sent to TX module. In the TX module, if the state machine is in the initial state and a DMA write request signal is received, a DMA write request feedback signal dma_-write_ack is generated and sent to the DMA_WRITE module again.

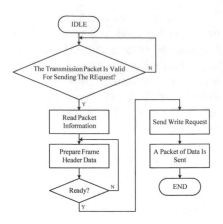

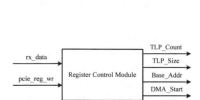

Fig. 5. Register control module signal block diagram

Fig. 6. DMA write control module control flow

DMA Write Control Module Design

DMA write control module completes the control of packet length and the corresponding DMA write offset address during DMA write data transmission. The flow chart of DMA write control module is shown in Fig. 6. Wait for the data transmission packet request signal in the idle state. If the transmission packet sends a valid signal, send the data transmission packet request signal and read the packet information. In this state, according to the TLP_Size information of the maximum data length that can be carried by the transmission packet fed back by the system, the data length carried by each packet is determined.

4   Experimental Results and Analysis

According to the above research and design, the logic program of each module is written on the Xilinx FPGA development platform. PCIE bus is PCIE 2.0 x8, its theoretical data transmission rate is 4 GB/s, and the FPGA chip model is XC7K410TFFG900. The simulation signal waveform in DMA write control module is shown in Fig. 7.

As can be seen from Fig. 7, DMA write Offset Address signal in the DMA write control module waveform is the sent data signal. Each packet of valid signals takes up 30 clock cycles, each clock cycle is 4 ns, and each packet of data size is 128Bytes, so the transmission bandwidth can be estimated to be 3.6 GB/s.

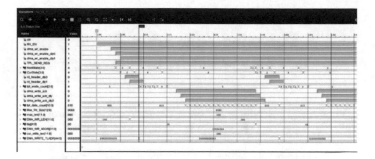

Fig. 7. DMA write control module simulation waveform

5 Conclusion

In this paper, Xilinx FPGA chip is used to realize data transmission on PCIE high-speed bus, DDR3 large-capacity data cache module is designed by time-division multiplexing, and the logic program of DMA writing control for each module is completed. The test results indicate that the designed PCIE high-speed data Transmission system has a data transmission capacity of 3.6 GB/s. If a higher version of the PCIE protocol is adopted, the data Transmission capability will increase further. The designed PCIE high-speed data Transmission system can be widely used in radar signal processing, biological spectrum analysis and UAV intrusion detection data Transmission and other fields.

Acknowledgments. The author, Zhang Kun, thanks the postgraduate project of Yunnan university school of information for supporting the publication of this paper "research on key technologies of intelligent home robot voice control system based on FFT dual-core processor".

References

1. Du, T., Jia, Q.: Design of Universal PCIE interface module based on Vs. In: IOP Conference Series: Materials Science and Engineering, vol. 449, no. 1, (2018)
2. Shim, C., Shinde, R., Choi, M.: Compatibility enhancement and performance measurement for socket interface with PCIe interconnections (2019)
3. Kavianipour, H., Muschter, S., Bohm, C.: High performance FPGA-based DMA interface for PCIE. IEEE Trans. Nucl. Sci. **61**(2), 745–749 (2014)
4. Gong, J., Wang, T., Chen, J., et al.: An efficient and flexible host-FPGA PCIE communication library. In: 2014 24th International Conference on Field Programmable Logic and Applications (FPL). IEEE (2014)
5. Preußer, T.B., Spallek, R.G.: Ready PCIE data streaming solutions for FPGAs. In: 2014 24th International Conference on Field Programmable Logic and Applications (FPL). IEEE (2014)

Communication Base-Station Antenna Detection Algorithm Based on YOLOv3-Darknet Network

Ying Xu[✉], Ruiming Li, Jihua Zhou, Yu Zheng, Qirui Ke,
Yihang Zhi, Huixin Guan, Xi Wu, and Yikui Zhai

Department of Intelligent Manufacturing, Wuyi University,
Jiangmen 529020, China
yingxu117@163.com

Abstract. Aiming at improving the efficiency and the accuracy of the base station antenna inspection, a new method is proposed to use the deep learning algorithm YOLO v3 algorithm combined with geometric mathematics to detect the downtilt angle of the antenna in our paper. The method uses Darknet-53 network to extract antenna features and multi-scale border prediction. The feature extraction is based on Darknet-53 network, and the network is inspired by the residual network. The method proposed can detect target in real time and has obvious advantages in speed and accuracy compared to Fast-RCNN. The method proposed also has weak point in the detection of antenna.

Keywords: Base station antenna detection · Feature extraction · Multi-scale prediction · YOLO v3 · Convolutional network

1 Introduction

Object detection is become more and more popular technique in the world. R-CNN [1], Fast-RCNN [2], Faster-RCNN [3], Mask-RCNN [4] and SSD [5] play an important role in object detection. The obtainment of downtilt angle of mobile communication base station require the technique of object detection. In recent years, with the rapid development of mobile communication technology, people have higher requirements for the quality of mobile communication. The traditional base station maintenance inspection is mainly for professional staff to obtain various parameters of the required downtilt angle by installing sensors on the base station antenna arm, but the method has low inspection rate and is not practical [8]. A new object detection named YOLO [6] has been first proposed by Redmon et al. at the CVPR in 2016. Aiming at improving the performance of YOLO, YOLOv2 [9] and YOLOv3 [7] have been proposed in recent time. Compared with the method which are mentioned earlier, YOLO can detect object in real time. The output of YOLO has higher accuracy. The average accuracy of Fast-RCNN is 74.64% for the target antenna test set, while the average accuracy of YOLO v3 is 90%. Subsequent paragraphs, however, are indented.

© Springer Nature Switzerland AG 2020
F. Xhafa et al. (Eds.): IISA 2019, AISC 1084, pp. 664–668, 2020.
https://doi.org/10.1007/978-3-030-34387-3_81

2   Base Station Antenna Detection Method of Mobile Communication Based on YOLOv3

The algorithm in this paper detects the video of the antenna of the mobile communication base station collected by the drone. The detection antenna can be divided into two modules, which are the target antenna bounding box prediction module and the target antenna feature extraction module.

2.1   Prediction of Antenna Boundary Frame

Using the logistic regression to predict the bounding box, divide the whole image into N * N grids, and scan all the grids and then locate target. When the target antenna is at the center of the grid, the target antenna is predicted by this grid. The corresponding grid can produce several bounding box. Each bounding box contains four coordinate prediction values: (tx, ty, tw, th). There is another value called confidence [6]. The upper left corner offset of each target cell is denoted as (cx, cy), and the bounding box's frame height and weight is denoted as (px, py). Then the network's prediction formula is as follow:

$$b_x = \sigma(t_x) + c_x \tag{1}$$

$$b_y = \sigma(t_y) + c_y \tag{2}$$

$$b_w = p_w e^{t_w} \tag{3}$$

$$b_h = p_h e^{t_h} \tag{4}$$

$$confidence = Pr(object) * IOU_{prd}^{truth} \tag{5}$$

2.2   Multi-scale Prediction of the Antenna Bounding Box

YOLO v3 predicts the bounding box by using three different scales in the accuracy of the target. The image is detected by using three different detection layers for the antenna, and different detection layers are realized by controlling the step size. In each layer of prediction, three target frames and anchors [3] are obtained, and finally 9 cluster center points are obtained. Downsampling the first detection layer, using a step size of 32. In order to connect with the previous same feature map, the layer is upsampled, and high resolution can be obtained. Secondly use the detection layer with a step size of 16 has the same feature processing as the first layer; in the third layer, the step size is set to 8, and the feature prediction is performed.

2.3   Target Classification

The prediction formula and loss function for the target category are as follows:

$$Pr(\text{Class}_i|\text{Object}) * Pr(\text{Object}) * IOU_{\text{pred}}^{\text{truth}} = Pr(Class_i) * IOU_{pred}^{truth} \qquad (6)$$

$$loss = \sum_{i=0}^{s^2} coordErr + iouErr + clsErr \qquad (7)$$

$Pr(Class_i|object)$ is the possibility of labeling the object category, if the value is 1, it means that the correct rate of the object which displayed by the current label is 100% 5, and when the value of Pr(Object) = 1 is close to 1, the overlap of current predicted bounding box and the real bounding box are larger and better than the previous prediction. The weight of CoordErr (coordinate error) is λ_{coor}. It is set to 0.5. IouErr (Iou error) is the corrected value of the object's predicted center point. The logistic regression layer is used to identify the object target.

3   Experimental Results and Analysis

3.1   Drone Target Antenna Detection Dataset

This experimental is based on NVIDIA GeForce GTX 1060 with Max-Q Design graphics card in Windows 10 environment. It is composed of Visual Studio2013, CUDA9.1, cuDNN7.0, Anaconda3-5, Opencv3.4. The experimental data set is collected by the drone. After the information of the mobile base station antenna is captured by the flight, the video is transmitted back to the server, and then the video is transformed into a series of frame images, and then the image is manually selected. We select images which include the side of antenna. Configuration of drone antenna target detection data set is as follow (Table 1):

Table 1. Drone antenna target detection data set

Category	Training set	Testing set	Validation set	Total
Number of antenna	1563	3065	1573	6201

3.2   Experimental Settings

The training model network is Darknet-53. In training step, Our network threshold is set to 0.4, 0.5, 0.6 and 0.75 respectively. The number of iterations is set to 10000, 9000, 8000, 5000 respectively. The initial learning rate of this experiment is 0.001. In the testing step, we use the model which is obtained by the training step. We compare different model according to the accuracy. The performances of network which is affected by different thresholds are shown in the Fig. 1. The detection accuracy is shown in Fig. 2. The comparison between Faster RCNN and YOLO v3 is shown in

Table 2. When the threshold is set to 0.5, the detection effect of the model is optimal. Compared with the antenna as the unique detection target, the accuracy measured by the Faster-RCNN optimal algorithm model and the YOLO V3 optimal model is shown in Table 2. As can be seen from Table 2, the YOLO v3 model antenna is more efficient than the Faster RCNN and it can be detected in real time.

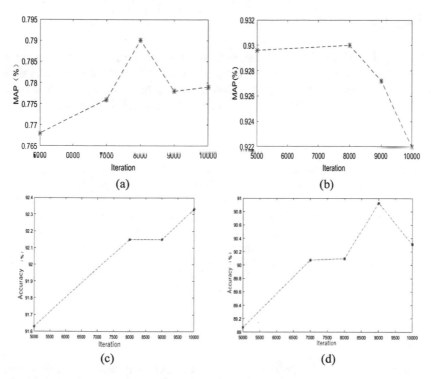

Fig. 1. The proposed network model test data (a) threshold is 0.4; (b) threshold is 0.5; (c) threshold is 0.6; and (d) threshold is 0.75.

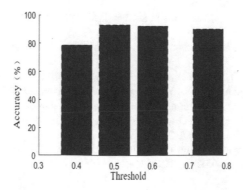

Fig. 2. Relationship between threshold and model accuracy

Table 2. Faster RCNN vs. YOLO V3 algorithm

Algorithm model	Test accuracy (%)	Performance
Faster RCNN	89.85	Not real-time detection
YOLO v3	92.72	Real-time detection

4   Conclusion

In this paper, YOLOv3-Darknet network structure is applied to the detection of antennas. real-time target detection is achieved and good results are obtained through model learning and parameter optimization. At the same time, the model is not only suitable for the detection of antenna, but also suitable for real-time detection of antenna video. The practicability and detection rate of YOLO are better than the Faster-RCNN, and it has higher detection accuracy. During the detection process, an object that is close to the antenna may appear in the image, and it may be mistakenly recognized as an antenna. In the future work, the network algorithm will be optimized, and the parameters of the network will be selected better to solve the problem.

Acknowledgments. This work is supported by National Natural Science Foundation of China (No. 61771347), Characteristic Innovation Project of Guangdong Province (No. 2017KT SCX181), Youth Innovation Talent Project of Guangdong Province (No. 2016KQNCX171), National Natural Science Foundation of China (No. 61771347), Basic Research and Applied Basic Research Key Project in general colleges and Universities of Guangdong Province (No. 2018KZDXM073).

References

1. Girshick, R., Donahue, J., Darrelland T, et al.: Rich feature hierarchies for object detection and semantic segmentation. In: 2014 IEEE Conference on Computer Vision and Pattern Recognition. IEEE (2014)
2. Girshick, R.: Fast R-CNN. Computer Science (2015)
3. Ren, S., He, K., Girshick, R., et al.: Faster R-CNN: towards real-time object detection with region proposal networks (2015)
4. Kaiming, H., Georgia, G., Piotr, D., et al.: Mask R-CNN. IEEE Trans. Pattern Anal. Mach. Intell. (2018)
5. Redmon, J., Divvala, S., Girshick, R., et al.: You only look once: unified, real-time object detection (2015)
6. Liu, W., Anguelov, D., Erhan, D., et al.: SSD: single shot multibox detector (2015)
7. Redmon, J., Farhadi, A.: YOLOv3: an incremental improvement (2018)
8. Zhen, C., Jianzhong, J.: Detection, inspection and maintenance for safe use of the existing communication base station steel towers. Building Structure (2016)
9. Redmon, J., Farhadi, A.: YOLO9000: Better, Faster, Stronger. In: IEEE 2017 IEEE Conference on Computer Vision and Pattern Recognition (CVPR), Honolulu, HI, 21 July 2017–26 July 2017, pp. 6517–6525 (2017)

Applications

Comprehensive Evaluation on Influencing Factors of Inventory Financing Under the Background of Financial Deleveraging

Zhanhai Wang[✉] and Zhipeng Han

School of Management, Northwestern Polytechnical University,
Xi'an 710072, China
jackwzh@126.com

Abstract. Supply chain finance is an important way to solve the financing difficulties of enterprises under the background of deleveraging, which is of great theoretical and practical significance to the overall performance of the enterprise supply chain. This paper introduces the history and current situation of inventory pledge financing in supply chain, analyzes the main modes of inventory financing in supply chain, and introduces the influence of periodic factors on inventory financing. The index system of influencing factors of inventory financing in supply chain is constructed, and the weight of each index is reasonably determined based on the practice of each participant in supply chain finance by using the analytic hierarchy process (AHP). Finally, the fuzzy comprehensive evaluation is applied to analyze the influencing factors. In this paper, AHP and fuzzy comprehensive evaluation are combined to provide a scientific and feasible method for the study of the factors affecting inventory pledge in the supply chain financial environment, and it has important guiding significance for the practice of enterprises.

Keywords: Supply chain finance · Supply chain performance · Inventory financing · Analytic hierarchy process · Fuzzy comprehensive evaluation

1 Introduction

With the standardization and modernization of the supervision services provided by the third party logistics, inventory financing has become one of the main financing methods for enterprises in the supply chain finance. Under the background of deleveraging, the supply chain financial financing system, which is dominated by banks and other financing institutions and composed of the third-party logistics supervisory service providers and enterprises, has become the most important financing way for SMEs to activate the stock assets in the supply chain effectively. Supply chain finance is the research direction in the field of micro finance, which is different from the creditor's rights financing such as bank lending and other forms of financing such as equity financing and venture capital investment. The most important difference between SCF and traditional bank financing is that in addition to relying on its own financing attributes, it also relies heavily on the status of enterprises in the supply chain

© Springer Nature Switzerland AG 2020
F. Xhafa et al. (Eds.): IISA 2019, AISC 1084, pp. 671–679, 2020.
https://doi.org/10.1007/978-3-030-34387-3_82

and accumulated credibility, which makes supply chain finance financing have systematic characteristics.

The main influencing factors of banks' financing credit rating for SMEs are the financial situation of individual enterprises and assets situation [1]. Therefore, small-scale assets and poor risk-resistant ability are the main obstacles faced by SMEs in the process of financing. Eliminating information asymmetry in the process of supply chain financing is the core to improve the efficiency of supply chain financing. Based on the perspective of traditional credit theory, the mode of supply chain finance financing can obtain the hard and soft information of enterprise financing through the closed loop of real transaction process and the embedding relationship of supply chain network members. Thus, banks can effectively control the process and results to reduce the information asymmetry in the credit process and reduce the credit risk of financial institutions [2]. In the mode of supply chain financial inventory financing, the real transactions between SMEs and members of the supply chain can more fully reflect the dynamic indicators such as the profitability of enterprises.

Relying on the supply chain finance background, on the one hand, it can improve the credit level of the loan enterprises and enhance the loan ability; on the other hand, it can increase the double default cost of the bank and the supply chain system, and promote the bank and the enterprise credit behavior to form a virtuous circle, and ultimately increase the overall financing ability of the supply chain system [3, 4]. In this paper, the inventory financing which is widely applied in supply chain finance is studied, and the influencing factors of inventory financing are systematically analyzed and discussed by using AHP and fuzzy comprehensive evaluation.

2   Development Status of Inventory Financing

The traditional investment mode of banking industry in China is that banks chase large-scale enterprises with assets, while SMEs' financing needs cannot be satisfied because of the small-scale assets. The difficulty of financing has hindered the development of enterprises. The initial inventory pledge business is a dual system model composed of banks and borrowing enterprises. With the development of specialized division of labor, especially the third-party logistics industry, it gradually formed a modern supply chain financial inventory financing model with bank-led, third-party logistics enterprises providing regulatory services and borrowing enterprises pledging cargo rights financing [5]. Shenzhen Development Bank officially launched its brand products of supply chain finance in 2006, serving SMEs in financing. Four state-owned commercial banks and four major banks have launched their own supply chain financial products. Supply chain finance plays an active role in promoting supply chain coordination and integration of industry and finance [6].

The State Council officially promulgated the Guiding Opinions on Actively Promoting Supply Chain Innovation and Application in October 2017. The introduction of the supply chain national policy accelerated the rapid development of supply chain finance theory and practice. Although domestic scholars have started late research on supply chain finance, they have made rapid progress. Relevant scholars of supply chain initially studied the conceptual background of supply chain finance, and analyzed centrally three major application modes of supply chain finance: accounts receivable financing, financing warehouse financing and confirming warehouse financing, and built a preliminary index system around the risk factors of supply chain finance and conducted a quantitative study [7–10].

With the deepening of financial deleveraging, it is becoming increasingly difficult for enterprises to finance through debt or equity, especially for SMEs. Since 2018, there has been an increase in the number of bonds being cancelled or postponed. According to WIND statistics, 531 bonds have been cancelled or postponed this year, with a total planned issuance of nearly 330 billion Yuan. After the financial crisis, people have realized that the cash flow existing in all links of the supply chain can enhance the competitiveness of the supply chain. The large-scale inventory financing in the supply chain financing is an important means and way to reduce the financing cost and cash flow pressure of enterprises [11, 12].

3 Analysis of Factors Affecting Inventory Financing

Under the background of supply chain finance, there are both micro-level and macro-level influence factors. Scholars who studied inventory pledge before mostly focused on the micro-evaluation, focusing on the main content of the enterprise and inventory itself. In this paper, the economic cycle, product cycle, third-party logistics supervision and other factors are included in the scope of research, trying to more comprehensively analyze the factors affecting inventory pledge.

Financing of supply chain financing is closely linked to the asset side, as the main body of the loan, the enterprises link the capital end with the high-quality assets, the main body of borrowing enterprise is the main grasp of information credit and risk control [13]. The real right foundation of financing relationship between banks and borrowing enterprises is the pledge itself. The risk of inventory pledge is the most important factor to consider the risk control of pledge financing. The promotion and standardization of logistics supervision services have made the application of inventory financing more and more extensive. Inventory financing requires banks to select the logistics service providers whose prices and services meet the conditions [14]. Transaction structure is a series of arrangements determined by the parties of pledge financing in the form of contract terms. In order to coordinate and realize the interests of both parties, risk and cost should be considered. The specific influencing factors are shown in Table 1.

4 Using Analytic Hierarchy Process to Determine Index Weight

4.1 Constructing Hierarchical Structure of Evaluation Index

The environment of inventory financing of supply chain finance is complicated, and it involves many factors. In order to reflect the actual situation objectively, we follow the principles of scientific rationality, comprehensiveness, purposiveness and practicality in the process of constructing the system. Based on the practice of each participant in supply chain finance and the discussion of industry experts, the index system of influencing factors of inventory financing as shown in Table 1 is reasonably determined. The target level is the influencing factor A of inventory pledge, and the criterion level is the 6 main influencing factors.

Table 1. Index system of influencing factors of inventory financing

Target layer	First level index (Standard layer)	Secondary index (Sub-criteria layer)
Comprehensive index A of influencing factors of inventory financing	The influencing factors B1 of the main body of borrowing enterprises type	Total operating revenue C11
		Net operating profit rate C12
		Asset liability ratio C13
		Operating cash flow C14
	The influencing factors B2 of inventory pledge type	Price volatility C21
		Storage characteristics C22
		Difficulty of realization C23
		Time value of pledged goods C24
	The influencing factors B3 of logistics supervision type (3PL)	Basic business level C31
		Operation and management ability C32
		Synergistic ability C33
		Enterprise qualification C34
	The influencing factors B4 of transaction structure type	Buyback Contract C41
		Guarantee measure C42
		Underwriting C43
		Insurance Clause C44
	The influencing factors B5 of cycle type	Inventory cycle factors C51
		Product cycle factors C52
		Economic cycle factors C53
	The influencing factors B6 of operating type	The transaction structure of inventory financing C61
		Supervision difficulties of 3PL and self managed warehouse C62
		Operation process C63
		Supervision technology application and illegality cost C64

4.2   Establishing Judgement Matrix

After completing the establishment of the evaluation index system, it is necessary to compare the various factors in the system, construct a judgment matrix, and determine the ranking of its importance. Judgment matrix takes the elements of the higher level as the criterion of comparative evaluation, combining with relevant data, practical experience of the evaluation subject and expert's opinions, makes a one-to-one comparison of the elements of this level scientifically and rationally, and finally gets the judgment matrix. The element in the judgement matrix adopts 1–9 scale. Experts analyze and compare the influencing factors in the first-class indicators get the A–B judgment matrix as shown in Table 2.

Table 2. A–B judgement matrix

A	B1	B2	B3	B4	B5	B6
B1	1	1/2	2	3	3	5
B2	2	1	3	4	4	6
B3	1/2	1/3	1	2	3	4
B4	1/3	1/4	1/2	1	2	3
B5	1/3	1/4	1/3	1/2	1	2
B6	1/5	1/6	1/4	1/3	1/2	1

4.3   Index Weight Determination and Consistency Check at All Levels

Determining index weight: First, the judgement matrix is quadrature according to the row elements, and then the 1/n power is obtained.

$$\overline{W}_i = \sqrt[n]{\prod_{j=1}^{n} a_{ij}} \tag{1}$$

Finally, normalization is done to get weight coefficients.

$$W_i = \frac{\overline{W}_i}{\sum\limits_{i=1}^{n} \overline{W}_i} \tag{2}$$

The weight vectors of each index are obtained. The weight of each factor in the first level index is as follows.

$W = (0.2436, 0.3758, 0.1632, 0.1028, 0.0714, 0.0432)^T$

Consistency check: In general, consistency index C.I and consistency ratio C.R are used as the criteria for consistency check.

The largest eigenvalue of the judgement matrix A–B is:

$$\lambda_{\max} \approx \frac{1}{n}\sum_{i=1}^{n}\frac{(AW)_i}{W_i} = \frac{1}{n}\sum_{i=1}^{n}\frac{\sum_{j=1}^{n}a_{ij}W_j}{W_i} = 6.1442 \tag{3}$$

The index C.I of consistency test is:

$$C.I = \frac{\lambda\max - n}{n-1} = 0.0288 < 0.1 \tag{4}$$

The ratio C.R of consistency is:

$$C.R = C.I/R.I = 0.0233 < 0.1 \tag{5}$$

According to the above method, taking each first level index as criteria respectively, the weights of each secondary index relative to the primary index can be calculated. The weights have passed the consistency test. The weights are shown in Table 3.

5   Making Comprehensive Evaluation of Influencing Factors by Fuzzy Comprehensive Evaluation Method

5.1   Determining the Evaluation Factors and Scale

The above has reasonably determined evaluation index weight through the AHP, and then we will use fuzzy comprehensive evaluation to make further comprehensive evaluation for the various factors. In order to carry out the evaluation, we need to identify the evaluation factors and evaluation criteria. The evaluation factors are the elements in the inventory financing influencing factors index system. The influence degree of each factor on inventory pledge is divided into five grades. Thus the evaluation scale set V = {extremely important, very important, important, slightly important, and unimportant} = {0.9, 0.7, 0.5, 0.3, 0.1}.

5.2   Giving the Judgement Matrix

Experts are invited to evaluate the inventory financing. According to the information of inventory financing in supply chain and the evaluation standard and criteria, the experts evaluate 23 secondary indicators that affect the inventory financing in sub-criteria level. The statistical normalization of the evaluation results is summarized as shown in Table 3.

Table 3. Evaluation table of influencing factors of inventory financing in supply chain

Target layer	Standard layer		Sub-criteria layer		Evaluation results (R_k)				
	Content	Weight (W)	Content	Weight (W_k)	Extremely important	Very important	Important	Slightly important	Unimportant
Comprehensive index system A for influencing factors of inventory financing	B1	0.2436	C11	0.057	0	0	0.1	0.3	0.6
			C12	0.2976	0.1	0.4	0.4	0.1	0
			C13	0.1222	0	0	0.2	0.5	0.3
			C14	0.5232	0.7	0.2	0.1	0	0
	B2	0.3758	C21	0.1415	0	0.1	0.4	0.3	0.2
			C22	0.2355	0.1	0.4	0.4	0.1	0
			C23	0.5705	0.7	0.2	0.1	0	0
			C24	0.0525	0	0	0.1	0.3	0.6
	B3	0.1632	C31	0.2622	0.1	0.3	0.5	0.1	0
			C32	0.0553	0	0	0.1	0.4	0.5
			C33	0.1175	0	0.1	0.2	0.3	0.4
			C34	0.565	0.5	0.4	0.1	0	0
	B4	0.1029	C41	0.0882	0	0	0.1	0.3	0.6
			C42	0.4828	0.6	0.3	0.1	0	0
			C43	0.157	0	0.1	0.3	0.4	0.2
			C44	0.272	0.2	0.4	0.3	0.1	0
	B5	0.0714	C51	0.6483	0.5	0.3	0.2	0	0
			C52	0.2297	0	0.2	0.2	0.5	0.1
			C53	0.122	0	0	0.3	0.3	0.4
	B6	0.0432	C61	0.5232	0.5	0.3	0.2	0	0
			C62	0.2976	0.1	0.2	0.4	0.3	0
			C63	0.057	0	0	0.1	0.2	0.7
			C64	0.1222	0	0	0.5	0.4	0.1

5.3 Comprehensive Evaluation

Computing the comprehensive evaluation vectors B_k of each secondary index, in which weight is obtained by using the analytic hierarchy process, and the evaluation matrix R is shown in Table 3.

$$B_k = W_k \cdot R_k \qquad (6)$$

B_1–B_6 can be calculated as follows:

$B_1 = (0.3960, 0.2237, 0.2015, 0.1080, 0.0709),$
$B_2 = (0.4229, 0.2225, 0.2131, 0.0818, 0.0598),$
$B_3 = (0.3087, 0.3164, 0.2166, 0.0836, 0.0747),$
$B_4 = (0.3441, 0.2694, 0.1858, 0.1165, 0.0843),$
$B_5 = (0.3242, 0.2404, 0.2122, 0.1515, 0.0718),$
$B_6 = (0.2914, 0.2165, 0.2905, 0.1496, 0.0521).$

From this, the final evaluation result of fuzzy comprehensive evaluation is obtained, and the information for decision-making is provided according to the evaluation result.

$$N_k = B_k \cdot V^T \tag{7}$$

$$N_1 = 0.6532, N_2 = 0.6734, N_3 = 0.6402, N_4 = 0.6346, N_5 = 0.6187, N_6 = 0.6091$$

Based on the comprehensive analysis of AHP and fuzzy, it can be concluded that many influencing factors have an important impact on inventory financing. The influencing factors of inventory pledge type have the greatest impact, followed by the influencing factors of the main body of borrowing enterprises type.

6   Conclusion

Aiming at the present situation of inventory financing in supply chain and combining with practice, the index system of influencing factors of inventory financing in supply chain is constructed in this paper, and the weight of each influencing factor is reasonably determined by AHP. Finally, the influencing factors are comprehensively analyzed by fuzzy comprehensive evaluation, and the influence of each factor on inventory financing is obtained. In this paper, cycle factors and other factors are introduced into the evaluation index system, which makes the evaluation results more comprehensive and accurate. It is of great theoretical and practical significance to provide a set of scientific and feasible methods for the study of influencing factors of inventory financing in the supply chain financial environment.

References

1. Guo, N.: Government, market, which one is more effective? - A study of effectiveness of solving mechanism. J. Financ. Res. **3**, 194–206 (2013)
2. Song, H., Lu, Q., Yu, K.: A comparative study on the effects of supply chain finance and bank lending on the financing performance of SMEs. J. Manag. **6**, 897–907 (2017)
3. Li, X.L., Xin, Y.H.: Credit market evolution analysis of SMEs based on supply chain finance. Oper. Manag. **10**, 101–105 (2017)
4. Chandima Ratnayake, R.M.: Small and medium enterprises project finance: identifying optimum settings of controllable factor. Int. J. Appl. Decis. Sci. **7**(2), 136–150 (2014)
5. Sheng, Q., Wu, Y.: Fuzzy comprehensive evaluation of supply chain inventory financing risk based on AHP. Sci. Technol. Manag. Res. **11**, 52–57 (2012)
6. Song, Y.F., Huang, Q.Y.: Research progress of domestic supply chain finance-an analysis of CSSCI from 2005 to 2017. Chin. Bus. Market **1**, 47–54 (2018)
7. Hu, Y.F., Huang, S.Q.: Supply chain finance: background, innovation and concept. Res. Financ. Econ. Issues **8**, 76–82 (2009)
8. Yan, J.H., Xu, X.T.: Analysis of SMEs financing model based on supply chain finance. Shanghai Financ. **2**, 14–16 (2007)
9. Xiong, X., Ma, J., Zhao, W.J., Wang, X.Y., Zhang, J.: Credit risk assessment under supply chain finance model. Nankai Manag. Rev. **4**, 92–98 (2009)
10. Qiu, Y.P.: Internet of things boosted supply chain finance development. Financ. Technol. Era **3**, 45–46 (2011)
11. Randall, W.S., Theodore Farris, M.: Supply chain financing: using cash-to-cash variables to strengthen the supply chain. Int. J. Phys. Distrib. Logist. Manag. **39**(8), 669–689 (2009)

12. Kerle, P.: The necessity for supply chain finance. Credit Control **31**(1), 39–44 (2003)
13. Song, H.: Innovative trend of supply chain finance based on industrial ecology. China Circ. Econ. **12**, 85–91 (2016)
14. Wang, Z.H., Liang, G.Q.: Logistics service provider selection based on AHP-TOPSIS method. Stat. Decis. **9**, 62–64 (2017)

Study on Short-Term Load Forecasting Considering Meteorological Similar Days

Zhaojun Lu[1], Zhijie Zheng[2], Hongwei Wang[1], Yajing Gao[3],
Yuwei Lei[3], and Mingrui Zhao[3(✉)]

[1] State Grid Shandong Electric Power Company, Jinan, Shandong, China
[2] State Grid Shandong Electric Power Company, Economic & Technological
Research Institute, Jinan, Shandong, China
[3] CEC Electric Power Development Research Institute, Beijing, China
15810833709@163.com

Abstract. A short-term load forecasting method based on meteorological similar days and error correction is proposed in this paper. First, SPSS software is used to carry on the regression analysis of meteorological factors, select most significant meteorological factors in each season, and determine the weight of each factor as the basis for selecting the weather similar days. Then the historical forecast error data sample set is set up. For a certain forecast date, the error data samples from the similar days are extracted to establish a set, and the probability density distribution model is established. Finally, the error fluctuation of the forecast point is analyzed to get the compensated value of forecast error. The sampling error closest to the error compensation value is selected as the fitted error values and added to the predicted value to improve the forecast accuracy.

Keywords: Meteorological similar days · Probability density distribution · Error correction · Volatility analysis

1 Introduction

The short-term load forecasting has become an important part of system scheduling, operation and planning. The results of load forecasting have become the key for power enterprises to make production and marketing plans and improve economic benefits. The load of power system has its own inherent periodic rule, and is also affected by many factors, such as climate conditions, economic development level, energy supply mode, etc. Based on the analysis of load characteristics, the factors affecting the load should be considered, and then the appropriate method should be selected to improve the forecast accuracy [1]. Meteorological factors are one of the important reasons for the short-term load change of power. There are two main ways to deal with meteorological factors: first, starting from the load characteristics, the idea of load decomposition or hierarchical modeling is adopted [2]. The second approach is to start from the forecast algorithm and model, and take the influence of daily meteorological factors into consideration by using the effectiveness of the algorithm [3]. In the research of forecast technology, the idea of error correction is widely used. In literature [4], the error is stratified according to the ultra-short-term wind power forecast error probability

© Springer Nature Switzerland AG 2020
F. Xhafa et al. (Eds.): IISA 2019, AISC 1084, pp. 680–688, 2020.
https://doi.org/10.1007/978-3-030-34387-3_83

density characteristics, and then classified according to the error volatility and error amplitude characteristics in different layers.

The idea of error correction is introduced in this paper. Firstly, the meteorological factors to find the main meteorological factors affecting the load change, and the weight are assigned to them so as to select the meteorological similarity days. Then, the forecast errors of similar days were extracted and the fitting model of probability density distribution was established. For a certain forecast point, the error fitting value was obtained through the volatility analysis of the error, and the final forecast load value was obtained after correcting the predicted value.

2 Regression Analysis of Meteorological Factors Based on SPSS

2.1 Establishment of Multiple Regression Analysis Model

When the relationship between independent variables $(X_1, X_2, \ldots, X_k)$ and dependent variables (y) is considered, the establishment of multiple linear regression model based on principle of least square method is as follows [5]:

$$y = y' + \mu = b_0 + b_1 X_1 + b_2 X_2 + \ldots + b_i X_i + \ldots + b_k X_k \tag{1}$$

In Eq. (1), b_0 is constant term, representing the intercept of the equation; b_i is the partial regression coefficient, representing the change in y for every change of 1 unit of independent variable (X_i) when other variables remain unchanged.

The multivariate linear regression requires not only the test of regression coefficient, the estimation of confidence interval of regression coefficient, the discussion of prediction and hypothesis testing, etc., but the consideration of the relationship between each independent variable. Before the multivariate regression analysis by using SPSS, data should be organized first, and then complete in the function menu of SPSS multivariate linear regression analysis.

In order to overcome the collinearity problem, simplify the model and increase the prediction accuracy, the Stepwise Regression Method is selected in this paper. Description of output results: (1) R shows the regression effect. The closer it gets to 1, the better; (2) F value test the regression effect in variance analysis. The higher F value is, better regression effect will be; (3) F test value Sig. < 0.05 indicates that the regression coefficient of at least one independent variable is not zero, and the regression model established has statistical significance. (4) the significance of t-test for the regression coefficient determines whether the corresponding variables can enter the regression equation as explanatory variables; (5) the t-test value Sig. < 0.05 indicates that coefficient of corresponding variable has the statistical significance.

2.2 Analysis of the Influence of Meteorological Factors

The relationship between the various meteorological factors and the daily maximum load, daily minimum load, daily average load of each quarter from 2013–2014 in a certain region was analyzed by Multiple Stepwise Regression. Main meteorological factors affecting the daily maximum load, daily minimum load and daily average load in different quarters in the region 2 are as follows: (1) Main meteorological factors affecting the daily maximum load and daily minimum load in spring are the minimum temperature and the average temperature, while the main factors affecting the daily average load in spring are the average temperature; (2) Main meteorological factors affecting daily maximum load in summer in average temperature and relative humidity, while the main factors affecting daily minimum load in summer are average temperature and rainfall; (3) Main meteorological factors affecting the maximum load in autumn are average temperature and relative humidity, while the main factors affecting the daily minimum load in summer are average temperature and rainfall; (4) Main meteorological factors affecting the maximum load in winter are the minimum temperature and relative humidity, while the main factor affecting the daily average load is the maximum temperature.

2.3   Selection Method for Meteorological Similarity Days

The results of multiple regression analysis show that the main meteorological factors influencing load changes in different quarters are different. The partial correlation coefficient is used to conduct a quantitative analysis of the load factors. In this paper, five meteorological indexes, namely daily maximum temperature, daily minimum temperature, daily average temperature, daily relative humidity and daily rainfall, are selected to form the daily feature vector for the selection of similar days. partial correlation coefficient is selected as an important basis for calculating the weight of each meteorological factor.

$$\alpha_i = \gamma_i / \sum_{i=1}^{n} \gamma_i \tag{2}$$

Of which, γ_i is the absolute value of the partial correlation coefficient between each index and the daily average load in this area.

Determine the weight coefficient of meteorological factors by season in Table 1:

Table 1. Weighting factor of meteorological factors

	Daily maximum temperature	Daily minimum temperature	Daily average temperature	Daily relative humidity	Daily rainfall
Spring	0.283	0.313	0.307	0.027	0.071
Summer	0.223	0.231	0.230	0.152	0.164
Autumn	0.213	0.218	0.217	0.172	0.179
Winter	0.209	0.213	0.212	0.180	0.185

The weighted similarity formula is adopted to calculate weighted similarity sim (*i*) between the daily eigenvectors of day i and the daily eigenvector of the predicted day:

$$sim(i) = \frac{1}{\sum_{j=1}^{n} \alpha_j |p_j(i) - p_j(0)| + \varepsilon} \tag{3}$$

Where, $p_j(i)$ is uniformed value of influence factor j on day i. ε is a small number.

The weighted similarity between all the historical days and the forecast days in the sample was calculated and sorted in descending order. The previous historical days were selected as the similar days of the load forecast.

3 Generation of Forecast Error Sampling and Compensation Value

3.1 Distribution of Forecast Error

The regression analysis of meteorological factors can indicate the significant correlation between meteorological factors and load changes in this area. The statistical study shows that meteorological conditions also affect the distribution of prediction errors. The error distribution generated by the same load prediction method also varies due to different climatic characteristics.

According to season type, this paper analyzed the error distribution characteristics generated in the prediction in this region, established the sample set of historical error, and fitting error by season. The distribution of relative errors approximately symmetric with respect to the *Y* axis, and the probability density function is generally decreasing with increasing $|x|$, and tend to be the *X* axis. In the winter and summer, the error distribution is concentrated, that is, the variance $\sigma_{2winter}$, $\sigma_{2summer}$ is the smallest with the "high and thin" graph, and the expectation value μ summer is the closest to zero, and μ winter is obviously greater than zero. In contrast, the distribution in spring and autumn is scattered, and the error distribution curve in spring is the "chunkiest" (Fig. 1).

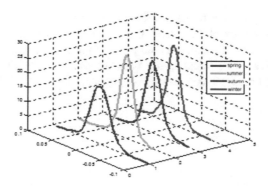

Fig. 1. Corresponding error distribution in different seasons

3.2   Sampling the Forecast Error

Based on System Sampling Method, the fitting relative error probability density curves in sample set of similar daily errors under different seasons are sampled. The principle of the system sampling method is as follows [6]: When extracting a sample of size n from a population of size N, the population is divided into uniform parts, and then an individual is extracted from each part according to predetermine rules to obtain the required sample. Its characteristics are: (1) The overall capacity N is large; (2) The gap of the overall sections shall be equal (generally $K = N/n$); (3) Random sampling is adopted in each sample segment to ensure the equal probability of each sample being selected.

When sampling, the error data in the sample set of similar daily errors are divided according time periods, and the 24 h of a day are divided into 6 time periods to ensure uniform sampling in each time period. Since the probability of large error is low, in order to avoid the phenomenon of error overcompensation caused by excessive sampling results, the space with the peak value as the center and the probability of 85% is selected as the actual sampling interval in each fitting curve.

3.3   Prediction Error Compensation Based on Volatility Analysis

By analyzing the fluctuation direction and amplitude regularity of the historical day prediction error in the sample, the variation trend of the error can be predicted and the corresponding compensation value can be generated with the help of the prediction method [7].

In volatility analysis, n long-term relative errors are selected as the measurement criteria for recent relative errors analysis. Define the critical value (k_1) of the long-term variance level (σ_1) and the absolute value of the slope of the fitting line:

$$\sigma_l^2 = \frac{1}{n} \sum_{i=1}^{n} \left(\delta_i - \frac{1}{n} \sum_{i=1}^{n} \delta_i \right)^2 \tag{4}$$

$$k_l = \frac{|k_1 - k_2|}{4} \tag{5}$$

Where, k_1 and k_2 are the upper and lower critical values of unilateral confidence interval determined by the fitting model and confidence level respectively.

Get the relative error of the first 3 data pints of predicted point as the sample value of the recent relative error; calculate the variance (σ_s) and the absolute value (k_s) of slope of the fitting line. Specific analysis methods are shown in Table 2.

Table 2. Volatility analysis method of relative error

Condition	Fluctuating cases	Methods
$\sigma_s < \sigma_1$ & $k_s < k_1$	The amplitude is small and stable	Moving average method
$\sigma_s < \sigma_1$ & $k_s > k_1$	The amplitude is small and non-stable	Autoregressive moving average method
$\sigma_s > \sigma_1$ & $k_s < k_1$	The amplitude is large and stable	Weighted moving average method
$\sigma_s > \sigma_1$ & $k_s > k_1$	The amplitude is large and non-stable	Linear method

The error compensation principle is established in order to avoid introducing new error in volatility analysis. On the basis of the relative error estimate, the relative error compensation method is determined according to the difference between the predicted point error, (δ_i) and the previous moment error, (δ_{i-1}), and the threshold (δ_0), which is the difference between large error and small error threshold, as shown in Table 3.

Table 3. Relative error compensation mode

Conditions	Compensation mode
$\delta_i < \delta_0$ and $\delta_i < \delta_0$	No compensation
$\delta_i < \delta_0$ and $\delta_i > \delta_0$	Equal amplitude compensation
$\delta_i > \delta_0$ and $\delta_i < \delta_0$	Equal amplitude compensation
$\delta_i > \delta_0$ and $\delta_i > \delta_0$	Half amplitude compensation

The sample results of the probabilistic model were sorted based on the compensation error value of the volatility analysis, and the sample results closest to the compensation value of the volatility analysis were selected as the correction of the relative error to realize the error correction of the prediction, so as to improve the accuracy of the short-term prediction.

4 Short-Term Load Forecasting Model Based on Error Correction

In this paper, the idea of error correction is introduced into the study of short-term load forecasting. The error fitting value s generated through the simulation of the forecast error and superimposed with the predicted load. Finally, the predicted load after error correction is obtained.

The main meteorological factors affecting the load change in this region are analyzed by regression analysis based on SPSS. The seasonal characteristics of meteorological conditions are taken into account in the analysis, the load prediction is carried out on this basis. The specific ideas are shown as follows:

In the load forecasting process of establishing historical error sample set, this paper chooses wavelet neural network prediction algorithm with simple structure, good self-learning ability and fast convergence speed. The training steps of wavelet neural network algorithm are as follows: Step 1: network initialization. At the beginning of the training, the wavelet neural network should be initialized first, that is, the parameters of the wavelet function should be initialized; Step 2: sample classification, which includes training samples for network training and test samples for test accuracy; Step 3: output prediction. The training samples are input into the network, and the error (e) between the predicted output and the expected output of the network is compared; Step 4: update the weights. After obtaining the error (e) of the previous step, the parameters of the wavelet function and the weight of the middle layer are further updated to make the predicted result more in line with expectation; Step 5: determine whether the predicted output is close enough to the expected value, and if not, return to step 3.

5   The Example Analysis

In order to verify the applicability of the load forecast principle with simulation relative error, the example uses the load data and meteorological data of a certain region for prediction, and MATLAB programming was used for simulation. In this paper, each day of the four seasons in this certain region is selected as the forecast day, which is recorded as A1, A2, A3, A4. For a certain forecast day, determine its season, take real-time meteorological data as the basis to select similar days, and then establish the error sample set of similar days prediction. Uniform sampling was conducted based on the system sampling method, and the sampling results were sorted according to the volatility analysis. Figure 2 shows the compensation error results obtained from the volatility analysis of A1–A4 and the final fitting error after sorting.

The relative fitting error is used to modify the prediction result of wavelet neural network and the load prediction value with fitting error is obtained. Figure 3 shows the final prediction results of the spring prediction day (A1). It shows that the prediction results of each season can improve the prediction accuracy to some extent after combining with the fitting error. It can be seen form Figs. 2 and 3 that the sampling value of relative error fitting distribution can effectively eliminate the extreme values appearing in the volatility analysis and avoiding phenomenon of the error divergence and over-compensation. The volatility analysis has the preliminary estimation ability for the compensation value of relative error. Based on the sorting of sampling error, the load prediction value can be accurately matched with the fitting relative error value, so as to compensate the error and improve the prediction accuracy.

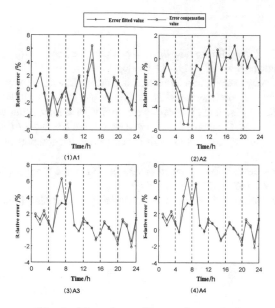

Fig. 2. Flow chart of forecasting method

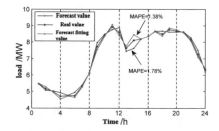

Fig. 3. Comparison of predicted A1 forecast results before and after correction

6 Conclusion

In this paper, the correlation between meteorological factors and load data in different seasons is analyzed based on SPSS. Based on this, the selection of similar days is carried out, and the probability density function of relative errors is established by using the historical forecast errors of similar days. The sample set of historical forecast error was sampled by the system sampling method, and the sampling results were sorted according to the volatility analysis, and finally the fitting value of the load forecast relative error was obtained, which was superimposed with the predicted value to correct the predicted value and improve the prediction accuracy. The practicability and effectiveness of the proposed method are verified by a practical example.

References

1. Kang, C., Xia, Q., Zhang, B.: Review of power system load forecasting and its development. Autom. Electr. Power Syst. **28**(17), 1–11 (2004)
2. Tai, N., Hou, Z.: New short-term load forecasting principle with the wavelet transform fuzzy neural network for the power systems. Proc. CSEE **24**(1), 24–29 (2004)
3. Liu, W., Yang, K., Da, L., et al.: Day-ahead electricity price forecasting with error calibration by hidden markov model. Autom. Electr. Power Syst. **33**(10), 34–37 (2009)
4. Li, X., Chen, Z.: Correctly using SPSS software for principal components analysis. Stat. Res. **27**(8), 105–108 (2010)
5. Gundersen, H.J.G., Jensen, E.B.: The efficiency of systematic sampling in stereology and its forecast. J. Microsc. **147**(3), 229–263 (1987)
6. Fan, J., Wang, Y.: Multi-scale jump and volatility analysis for high-frequency financial data. J. Am. Stat. Assoc. **102**(480), 1349–1362 (2007)
7. Chen, Y., Yang, B., Dong, J.: Time-series forecast using a local linear wavelet neural network. Neurocomputing **69**(4–6), 449–465 (2006)

The Application of Computer Science in Chemical Engineering: Intelligent Solvent Selection in Liquid-Liquid Extraction Process

Jingyi Tang[1], Jiying Li[2], Xinlu Li[1], and Shun Yao[1(✉)]

[1] School of Chemical Engineering, Sichuan University, Chengdu 610065, China
Cusack@scu.edu.cn
[2] Department of Biological Medicine, Changjiang Polytechnic,
Wuhan 430074, China

Abstract. Chemical process simulation is a new intelligent technology developed in recent decades, which has become a powerful computing tool for chemical engineering design and original engineering transformation and optimization. As a representative, ChemCAD is one of the commonly used simulation system. In this paper, optimal solvent in liquid liquid extraction process was firstly screened by computing and then process simulation software ChemCAD5.2 was used to simulate the working process. In order to verify the effectiveness of the solvent extraction method selected by the computer optimization method in the liquid-liquid extraction process, the furfural-water liquid-liquid extraction system was selected as the sample system. Amyl propionate, *n*-octyl acetate, butyl acetate and propyl acetate were used as the research subjects. Compared with the performance of ethyl acetate used in the literature, the results showed that the extraction effect of these four solvents was superior to that of ethyl acetate.

Keywords: Computer science · Chemical engineering · Solvent selection ·
Liquid-liquid extraction · Process simulation

1 Introduction

Currently, more and more computer and intelligent technologies are being applied in the following fields of chemical engineering, including (1) modeling, numerical analysis and simulation; (2) process and product synthesis/design; (3) process dynamics, monitoring and control; (4) cyberinfrastructure, informatics and intelligent systems; (5) mathematical programming (optimization); (6) abnormal events management and process safety; (7) factory operations, integration, supply chain and planning/scheduling so on. Chemical process simulation (also known as process simulation) technology, whose basis is mechanism model of the process, describing the chemical process with the use of mathematical methods, through the application of computer-aided computing means to complete process material balance, heat balance, equipment size design and energy analysis, environmental and economic evaluation, etc. [1]. This system is a combination of chemical thermodynamics, chemical engineering, system engineering, calculation method and computer application technology.

© Springer Nature Switzerland AG 2020
F. Xhafa et al. (Eds.): IISA 2019, AISC 1084, pp. 689–695, 2020.
https://doi.org/10.1007/978-3-030-34387-3_84

It is a new technology developed in recent decades and has become a powerful tool for chemical engineering design and original engineering transformation and optimization [2].

Chemical process simulation software has developed gradually with the application of computer in chemical industry in the late 1950s. Initially people just designed and developed it for some special processes (e.g., ammonia synthesis, hydrocarbon cracking to ethylene, etc.), then with improvements in its function, the enhancement of accuracy together with reliability, it has gradually developed to apply to various processes of general-purpose simulation system. To the late 1960s, chemical process simulation system has been widely used and has become a main conventional means of chemical process development, design and existing improvement of production operation [3, 4]. At present, Aspen Plus of AspenTech company in United States, which is the main chemical process simulation software abroad, is a powerful chemical design, dynamic simulation and various calculation software, it has the most strict and the latest calculation method, not only for calculation of the unit and the whole process, to provide accurate unit operation models, but also for the assessment of the optimization of the existing device and the optimization design of device built and rebuilt [5]. The chemical process simulation software PRO/II developed by SimSci - Esscor companies in the United States, mainly used in complicated chemical simulation, is able to complete the new process design, evaluation in the different devices, optimization and improvement of existing device, eliminate the bottleneck processing equipment, optimize production, increase income on the basis of environmental assessment [2]. HYSYS was originally a product of Canadian company Hyprotech, which has become a part of AspenTech in July 2002. Compared with other simulation software, HYSYS has the most advanced integrated engineering environment, powerful dynamic simulation engineering; it can provide powerful physical property calculation package and data regression package, with physical property prediction system and built-in artificial intelligence, etc. [6]. Another famous tool, ChemCAD chemical simulation system, was developed by Chemstations Company of United States for unit or full-process calculations and simulations [7]. The simulation software is mainly used in computer simulation of process in chemical, petroleum industry, oil refining, oil and gas processing and other fields. The built-in powerful standard physical property database provides a large number of calculation methods for thermal equilibrium and phase equilibrium, including 39 K value calculation methods and 13 enthalpy calculation methods [8]. More than 50 common operation units are provided for users to choose [9], which can meet the needs of different design aspects.

Several representative chemical process simulation softwares have been introduced above. They have different characteristics, functions and application fields. In our country, ASPEN, PRO/II and ChemCAD simulation software have been used widely for its excellent fit of the thermodynamic system, high and fast convergent algorithm. This research adopts the ChemCAD5.2 simulation software as a simulation software for separating furfural - water system.

2   Simulation Process

In this study, ChemCAD (Version 5.2) was mainly adopted for process simulation. It is a powerful tool for material balance and energy balance accounting, which provides theoretical guidance for process development, engineering design and optimization operation, and can design more effective and new processes to maximize benefits. The workflow can be represented by Fig. 1.

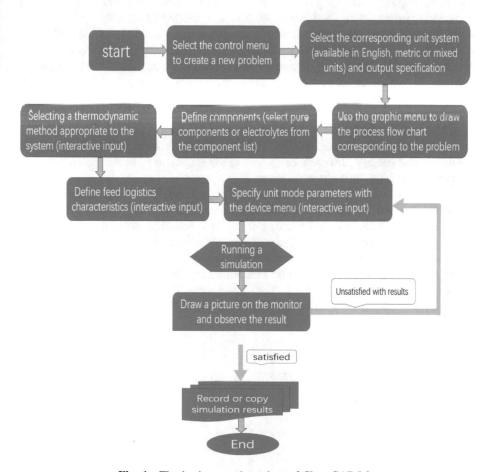

Fig. 1. The basic procedure chart of ChemCAD5.2

The selection of extractant is the key of extraction operation. It directly affects whether extraction can be carried out, and it is also the main factor to the yield, quality and economy of the extraction process. In the process of extraction from aqueous phase, the extraction agent is usually selected under the condition that the aqueous phase medium has been determined. The conventional extraction agent selection method is the combination of experience and experiment. However, the experience is mainly from the accumulation of engineers' work, which is always limited by individual level and preference, with uncertainty and blindness. At present, there are three main methods to choose extraction agent: one is experimental method. According to the physical and chemical properties of the extracted components, the extraction agent was selected by experience, and the extraction agent was determined by measuring the activity coefficient of infinite dilution. The second is the property constraint method. It is to delimit the general scope of solvent according to some principles, and to determine a kind of extraction agent which may be suitable for the preliminary screening of extraction agent, including empirical screening method and activity coefficient method. The third is the computer optimization method, which refers to design or selection of the optimal solvent according to various selection indicators through computer optimization algorithm; and it includes computer optimization screening and computer-aided molecular design method generally. Group contribution method and molecular design were used in this research to pre-select alternative extraction solvent, which means to pre-select the functional categories to find out good candidate functional groups and combine with the basic principles to determine the potential molecular composite solvent molecular structure, finally calculate the liquid-liquid equilibrium data of furfural aqueous solution with an extractant of the same quality as furfural aqueous solution. Thus, the selectivity coefficient and solute distribution coefficient of the extractant were calculated to find the best extractant for simulation verification.

In order to verify the effectiveness of this method in the selection of solvent extraction method in the liquid-liquid extraction process, the furfural-water liquid-liquid extraction system was selected, and the four extractive solvents of amyl propionate, *n*-octyl acetate, butyl acetate and propyl acetate were used as the research objects, and were compared with ethyl acetate used in the literature [10]. The process of establishing a liquid-liquid extraction column using the process simulation software ChemCAD5.2 is shown in Fig. 2. The liquid-liquid extraction process of furfural extraction from aqueous solution was simulated and verified.

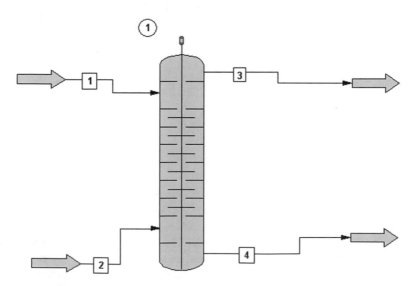

Fig. 2. Flowsheet of liquid-liquid extraction of furfural-water

3 Results and Discussion

Four potential extractive solvents of amyl propionate, *n*-octyl acetate, butyl acetate and propyl acetate were firstly screened through our self-written program and then compared with previously used ethyl acetate. The program was based on Matlab platform and also has been designed as App with Python language for convenient use, which originated from the strategy of UNIFAC Group Contribution method combined with molecular design. The solvent selectivity was divided into two parts: the combinatory solvent selectivity and the residual solvent selectivity. The former represents the contribution of the shape and size of the extractive solvent molecule and the extracted component molecule to the selectivity, while the latter represents the contribution of the interaction force between the extractive solvent molecule and the extracted component molecule to the selectivity. According to calculate solvent selectivity and distribution coefficient, a series of appropriate potential solvents can be obtained. According to the computing results, amyl propionate and *n*-octyl acetate should be more ideal. In order to verify this finding, the extraction process was simulated according to the following procedures:

The theoretical plate number was set as 8, raw material and extraction solvent feed flow rate was 10 kg/h, a phase of furfural water content was 7% (wt%); material flow from the top of the tower, extractive solvent from the bottom; "UNIFAC LLE" was selected as model equation, the tower pressure was 1 ATM, the tower pressure drop was 0.5 ATM, the default value of tray efficiency was "1". After these parameters setting, the process simulation began and the whole data including flow temperature, enthalpy, mass flow rate together with composition of each flow unit are shown in Table 1 in detail.

Table 1. Stimulated results of extraction process of five investigated systems

Furfural-water-amyl propionate	Flow number	1	2	3	4
	Temperature (°C)	25	25	25	25
	Enthalpy (kcal/h)	−35574.00	−806.43	−1494.80	−34886.00
	Mass flow rate (kg/h)	10	10	10.7883	9.2117
	Component mass fraction				
	Furfural	7	0	6.4885	0
	Water	93	0	0.8342	99.9826
	Amyl propionate	0	100	92.6773	0.0174
Furfural-water-*n*-octyl acetate	Flow number	1	2	3	4
	Temperature (°C)	25	25	25	25
	Enthalpy (kcal/h)	−35574.00	−8744.20	−9415.20	−34903.00
	Mass flow rate (kg/h)	10	10	10.7851	9.2149
	Component mass fraction				
	Furfural	7	0	6.4904	0
	Water	93	0	0.7918	99.9967
	n-octyl acetate	0	100	92.7177	0.0033

(continued)

Table 1. (*continued*)

Furfural-water-n-butyl acetate	Flows name	1	2	3	4
	Temperature (°C)	25	25	25	25
	Enthalpy (kcal/h)	−35574.00	−10879.00	−11851.00	−34602.00
	Mass flow rate (kg/h)	10	10	10.8498	9.1502
	Component mass fraction				
	Furfural	7	0	6.4517	0
	Water	93	0	1.5751	99.7694
	n-butyl acetate	0	100	91.9731	0.2306
Furfural-water-n-propyl acetate	Flows name	1	2	3	4
	Temperature (°C)	25	25	25	25
	Enthalpy (kcal/h)	−35574.00	−11794.00	−12916.00	−34452.00
	Mass flow rate (kg/h)	10	10	10.8587	9.1413
	Component mass fraction				
	Furfural	7	0	6.4464	0
	Water	93	0	2.0712	99.2758
	n-propyl acetate	0	100	91.4824	0.7242
Furfural-water-ethyl acetate	Flows name	1	2	3	4
	Temperature (°C)	25	25	25	25
	Enthalpy (kcal/h)	−35574.00	−13017.00	−14290.00	−34301.00
	Mass flow rate (kg/h)	10	10	10.7944	9.2056
	Component mass fraction				
	Furfural	7	0	6.4848	0
	Water	93	0	2.9904	97.5189
	Ethyl acetate	0	100	90.5247	2.4811

Above stimulated computing data from Table 1 show that the extraction performance of ethyl acetate as the extractive solvent for furfural-water liquid-liquid extraction proposed in the literature is really worse than that of the other four kinds of extractants selected by our program under the condition of the same ratio of feed and solvent. Furthermore, among amyl propionate, n-octyl acetate, butyl acetate and propyl acetate, the former two were more efficient than the latter two, which was same as the prediction of our program. Obviously, the method was feasible and it has been proved to be a rapid and effective way for chemical engineers to look for their most useful solvent.

4　Conclusion

In this study, our self-developed intelligent method and process simulation software ChemCAD5.2 were used for the selection of ideal extractant, so as to establish the process of liquid-liquid extraction tower. Group contribution method and molecular design were adopted to select several alternative extraction solvents. In the furfural-water liquid-liquid extraction system, amyl propionate, n-octyl acetate, butyl acetate and propyl acetate, which were screened by ChemCAD5.2, were used as the research objects and compared with ethyl acetate used in the literature. The liquid-liquid extraction process for furfural extraction from aqueous solution was simulated by ChemCAD5.2. The results show that under the same ratio of feed and solvent, the extraction solvent designed in this paper has a better extraction effect than the furfural-water liquid-liquid extraction solvent ethyl acetate proposed in the literature. For furfural containing wastewater solution, complete separation of furfural aqueous solution can be achieved after liquid-liquid extraction with amyl propionate, n-octyl acetate, butyl acetate and propyl acetate, providing more feasible candidate solvents for the recovery of furfural from various objects in chemical industry. In the future, it is expected that intelligent systems and computer science will promote chemical engineering to a new level.

Acknowledgements. Preparation of this paper was supported by 2017 "Stars of Chemical Engineering" outstanding young talent training program of Sichuan University.

References

1. Yang, G.H.: Chemical process simulation technology and application. Shandong Chem. **37**, 35–38 (2008)
2. Chen, H., Li, M., Wang, T., Zhang, T.: Application progress of chemical process simulation software. J. Sichuan Univ. Sci. Technol. **23**(5), 580–585 (2010). (natural science edition)
3. Xu, J.R.: Chemical simulation system. Tianjin Chem. **17**(4), 42–43 (2002)
4. Casauant, T.E., Raymond, P.C.T.: Using chemical process simulation to design industrial ecosystem. J. Cleaner Prod. **12**(8–10), 901–908 (2004)
5. Sun, H.X., Zhao, T.Y., Cai, G.L.: Application and development of chemical simulation software. Comput. Appl. Chem. **24**(9), 1285–1288 (2007)
6. Guo, G.Z.: Petrochemical dynamic simulation software HYSYS. Petrochemical Des. **14**(3), 29–33 (1997)
7. Li, H.M.: Analysis on the key of common chemical process simulation. Gansu Petrol. Chem. **24**(3), 45–48 (2010)
8. Luo, H.: The application of chemical process simulation software ChemCAD. Zhejiang Chem. **36**(10), 40–41 (2005)
9. Wang, S., Wu, H.X.: Typical application examples of ChemCAD (part one)-basic application and dynamic control. Chemical Industry Press (5) (2006)
10. Zhang, Y.M., Huang, H.J.: Study on the extraction agent selection for furfural aqueous solution. Guangxi Chem. (1), 25–29 (1990)

The Complex System of Geographical Conditions and Its Hierarchical Regionalization Analysis

Xie Mingxia[1,2,3](✉), Guo Jianzhong[4,5], and Zhang Li[1,3]

[1] Changjiang Institute of Survey, Planning, Design and Research,
Wuhan 430010, China
xmx0424@whu.edu.cn
[2] Key Laboratory of Urban Land Resources Monitoring and Simulation,
Ministry of Land and Resources, Shenzhen 518034, China
[3] Changjiang Spatial Information Technology Engineering Co. Ltd,
Wuhan 430010, China
[4] Henan Institute of Spatiotemporal Big Data Industry Technology,
Kaifeng 475000, China
[5] The College of Environment and Planning of Henan University,
Kaifeng 475000, China

Abstract. This investigation analyzes the basic characteristics of the geographical conditions and the complex system. Based on this, the concept of complex system of geographical conditions (CSGC) is put forward. The hierarchical regionalization model of CSGC is constructed combined with DPSIR model and the method of TOPSIS. Based on the data of Henan Province's geographic and national conditions census, the statistical yearbook and urban planning data of Henan Province, prefecture-level cities and cities, the coordination degree evaluation and analysis experiments of CGCS in Henan Province are carried out. The results show that: (1) There is little difference in the development status of social and resource subsystems among each city in Henan Province, and there is great difference in the development status of economic and ecological subsystems. (2) On the whole, the coordination of CGCS in Henan Province shows a zonal pattern of development, which gradually develops from both ends to the middle, and is positively correlated in the overall geographic space. In coordinating regional development of geographical conditions, scientifically recognizing geographical conditions and promoting regional coordination and sustainable development of geographical conditions, feasible strategies for balanced regional development of geographical conditions can be formulated according to the similarities and differences of the development structure of geographical conditions of various cities.

Keywords: Geographical conditions · Complex system · Coordination degree · Pattern analysis · Hierarchical regionalization analysis · Henan province

© Springer Nature Switzerland AG 2020
F. Xhafa et al. (Eds.): IISA 2019, AISC 1084, pp. 696–705, 2020.
https://doi.org/10.1007/978-3-030-34387-3_85

1   Introduction

Geographical national conditions, as a complex spatial complex of man and nature, is a complex system that integrates many elements such as resources, ecology, environment, economy and society. Geographical and national conditions complex system can break the traditional single factor independent research state, systematically comb the theoretical system of resources, environment, ecology, economic and social subsystems, and integrate them organically. Many countries and organizations have carried out research projects or projects related to geographical conditions in order to better serve the resources, environment, energy and social fields of their countries or regions. The useful experience of foreign geographic national condition monitoring has a good reference significance for our country to carry out relevant work, but also reflects the necessity of carrying out geographic national condition survey and monitoring.

A system is an integral whole constituted by several parts, and the concept of a system emphasizes that the "whole" is made up of correlative, interactive and mutually restraining parts. A complex system generally refers to the research object of a complex science, which is usually an emerging cross-disciplinary branch of knowledge that aims to reveal the dynamic behaviors of the complex system which are difficult to be interpreted by existing scientific methods. The main research task of complex science is the complex system theory, which emphasizes the analysis of the system by the approach of combining holism and reductionism. With research being continuously undertaken on binary, ternary, and quaternary systems. As a new ontological and epistemological point of view, the complex system theory has, so far, been gradually and widely accepted. However, depending on the perspectives taken, the characteristics of a complex system may vary significantly and, restrained by the division of disciplines, no consensus has been reached in academia with regard to the range of a complex system. The basic characteristics of a complex system include integrity, complexity, multi-component structure, multi-dimensionality, dynamics, nonlinearity and coordinativity.

In summary, the basic characteristics of geographical conditions and complex systems are the same, such as multi-dimensional structural characteristics, non-linear characteristics, sequential characteristics, coordinate symbiosis characteristics and so on.

(1) Multi-dimensional structural characteristics

The multi-dimensional structural characteristics are apparent in two ways. Firstly, the information of the geographical conditions is multi-dimensional. Secondly, the comprehensive analytical research of the geographical conditions is a multi-dimensional decision-making process. The information of geographical conditions at the same position can have an information structure of multiple themes and attributes.

(2) Non-linear characteristics

Linear thinking dominates the traditional way of thinking, and it usually seeks the optimization of single elements in terms of setting goals. This way of thinking usually results in restraining and tightening of resources, the intensification of environmental pollution, and the degeneration of ecosystems and other global issues.

(3) Sequential characteristics

The sequential characteristics of geographical conditions are apparent in two ways. Firstly, the world is in constant motion and change, and the geographical conditions reflect these changes. Secondly, there are widespread flows of materials, energy and information, both among and within various elements of the geographical conditions, and the degrees of coordination among the various elements, and the overall functions of the geographical conditions constantly change.

(4) Coordinate symbiosis characteristics

There are two aspects to describe the coordinate symbiosis characteristics of geographical conditions. Firstly, a coordinate symbiosis exists among the constituent indexes of the various elements. Secondly, there is also a coordinate symbiosis between the elements.

In terms of research methods and corresponding research objectives, coordination degree is one of the important methods to study complex systems, and its purpose is global coordination and optimization design, so that the subsystems of complex systems can coordinate and cooperate with each other to accomplish complex systems together. The overall task is to achieve the overall goal, which is consistent with the specific goal of statistical analysis of geographical conditions in the direction. The interrelation and intrinsic relationship between geographical conditions and complex systems are shown in Fig. 1. Therefore, we say that geographical conditions are essentially a complex system that integrates resources, environment, ecology, economy and social factors. The object of geographical conditions statistical analysis is virtually the complex system of geographical conditions, and that its specific goal is to quantitatively reflect and express the relationship of "resources-environment-ecology-economy-society" complex system, which has not been paid enough attention to in the past research on how to measure the relationship. The complex system theory has provided new theoretical guidance for geographical conditions statistical analysis and, as an important method for the measurement analysis of the complex system, the

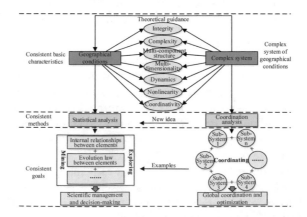

Fig. 1. Correlations and internal relations between the complex system and geographical conditions

degree of coordination provides a new way for measuring the internal and the relationship between the elements of geographical conditions, as well as providing the opportunity for this research. The geographical conditions statistical analysis may go through a theoretical transformation under the scientific perspective of the complex system, based on which theoretical basis of such analysis will be significantly enriched and perfected, and the statistical analysis of the geographic elements will be guided by theories which will have more rational contents, set more definite goals and make more persuasive achievements.

2   Materials and Methods

2.1   Study Area and Data Sources

Our study area is the Henan province which is located in the central part of China, bounded by $31°23'–36°22'$ in north latitude, $110°21'–116°39'$ in east longitude, Anhui and Shandong in the east, Hebei and Shanxi in the north, Shaanxi in the West and Hubei in the south. Henan Province has a total area of 167,000 km². Henan Province has the tendency of looking northward to the South and connecting the East with the west. The terrain is high in the West and low in the east. It is composed of plains and basins, mountains, hills and water surfaces. It is located in four major river systems, namely, the Haihe River, the Yellow River, the Huaihe River and the Yangtze River. Most of Henan Province is located in the warm temperate zone, and the south is trans-subtropical, which belongs to the continental monsoon climate transiting from the north subtropical zone to the warm temperate zone. Henan Province is located in the combination of the coastal open areas and the central and Western regions, which is the middle zone of China's economic development from east to west. As of May 2018, Henan Province has 17 provincial municipalities and 1 provincial municipality directly under its jurisdiction. Land use data are derived from the geographic census data in 2015. The data of resources, environment, ecology, society and economy come from the statistical yearbook and urban planning of Henan province and city in 2013.

2.2   Methods

Definition 1: Multivariate complex systems. A complex system consisting of three or more $(n \geq 3)$ subsystems.

Definition 2: *Geographical conditions hierarchical regionalization.* It is based on Geographic Conditions census data and deeply integrates thematic statistical data such as resources, environment, ecology, economy and society. Based on the index system of Geographic National Conditions grading zoning, it analyses and evaluates the coordinated development of CSGC in China, thus objectively, accurately and thoroughly analyzing and evaluating regional resources, ecological environment, economic and social development. The exhibition and other elements reflect the national and regional conditions macroscopically and integrally from the perspective of geographical and national conditions.

Definition 3: *Coordination Degree of Complex Systems.* It is an important feature of reflecting the relationship between the internal indicators and subsystems of complex systems. Coordination degree is a quantitative indicator of the degree of coordination, which measures the degree of harmony and consistency between the elements of the system or the internal elements of the system in the process of development. It reflects the trend of the system from disorder to order. It is a comprehensive description of the overall coordination state of complex systems. Quantitative indicators. Assuming that the state space of a complex system is x and any state y, it can be seen from the definition of coordination that the coordination of a complex system is actually a scalar function describing the state of the system.

The following is an example of ternary complex system $(n = 3)$ to design and explain the specific calculation process and method of coordination degree of multivariate complex system.

(1) Constructing a regular triangle with radius 1 as a circle and subsystems S_1, S_2 and S_3 as vertices.

(2) Taking point O as the starting point and according to the comprehensive development level of each subsystem, the corresponding lengths are measured on line OS_1, OS_2 and OS_3 respectively. That is, the lengths of OS_1, OS_2 and OS_3 are taken as the development level of corresponding subsystems, and the triangle $S_1'S_2'S_3'$ is constructed with points S_1', S_2' and S_3' as the vertices.

(3) The coordination degree formula for ternary complex systems is

$$CD = \sqrt{S_{S_1'S_2'S_3'} \big/ S_{S_1S_2S_3}} \tag{1}$$

Among them, $S_{S_1S_2S_3}$ and $S_{S_1'S_2'S_3'}$ represent the area of triangle $S_1S_2S_3$ and $S_1'S_2'S_3'$; $CD \in [0, 1]$, when $S_{S_1'S_2'S_3'} = S_{S_1S_2S_3}$, the subsystems are in the optimal state of development, and the coordination degree of ternary complex systems is 1, that is, the subsystems are in a completely consistent state.

At present, there is no specific method to select the geographical conditions index. The index in this paper is selected manually according to the characteristics of CSGC, and the index system for CSGC based on the DPSIR model is adopted in Xie et al. [13]. The model of geographic classification and zoning is shown in Fig. 6. Firstly, on the basis of the index system, the comprehensive development level of each subsystem is calculated by TOPSIS. Then, the coordination degree of geographic national conditions is calculated according to the coordination degree calculation method of the designed multi-complex system. On this basis, the coordination degree is classified into different levels. Making thematic maps of hierarchical zoning of geographic and national complex systems, and analyzing the results of hierarchical zoning of geographic and national complex systems (Fig. 2).

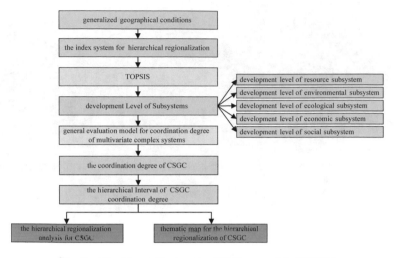

Fig. 2. The hierarchical regionalization model of CSGC

3 Results and Discussion

3.1 The Coordination Degree of CSGC

According to the index system and evaluation model of coordination analysis of geographical conditions, the development level of subsystems and coordination degree of CSGC in Henan Province are calculated. Among them, Jiyuan, Sanmenxia, Luoyang, Zhengzhou and Nanyang ranked the top five cities in the coordination degree of geographical conditions, while Shangqiu, Kaifeng, Zhoukou, Xinxiang and Puyang ranked the bottom five cities. Around the development level of subsystems in each city, the spatial distribution map of subsystems is constructed (Fig. 3). As can be seen from Fig. 3, the overall development status of social subsystems in Henan Province is the worst, followed by economic and ecological subsystems, and the coordination among the elements of resource and environment subsystems is ideal. This phenomenon shows that the development and utilization of resources and environment in various cities of Henan Province has not brought about a healthy and reasonable economic growth situation, and the unsatisfactory development of the economic subsystem itself has directly hindered the stable, orderly, fair and just development of society, and further affected the process of ecological civilization construction. The development patterns of resource, environment and ecological subsystems in Henan Province are similar. The development status of economic subsystems varies greatly among different cities, while the development of social subsystems is relatively balanced.

With the ArcGIS platform to construct the spatial distribution map of the coordination degree for CGCS in Henan province (Fig. 4). As can be seen from Fig. 4, the coordination of CGCS showed belt spatial development pattern as a whole, which appeared in the form of gradual coordinated development from the northeast to the southwest, as indicated by the arrows. According to the belt spatial development pattern of CGCS in Henan Province to build the coordination development level line of the geographical conditions, as shown in the blue line.

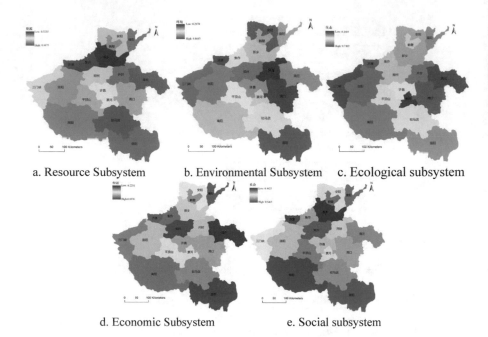

a. Resource Subsystem   b. Environmental Subsystem   c. Ecological subsystem

d. Economic Subsystem   e. Social subsystem

Fig. 3. Spatial distribution map of evaluating development status of geographical and national conditions complex system subsystem in 18 cities of Henan Province

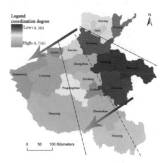

Fig. 4. The spatial distribution map of the coordination degree for CGCS in Henan province

3.2   The Pattern Analysis of Classification Index (D, P, S, I, R)

Based on the classification index evaluation, the pattern analysis of coordinated development for geographical conditions is carried on. The index evaluation of DPSIR in 18 cities of Henan Province is analyzed by radar charts (Fig. 5). According to similarities in the classification index development patterns (Fig. 6), the 18 cities in Henan province can be divided into six groups. Group 1: Anyang, Luohe, Xuchang and Kaifeng; Group 2: Nanyang, Shangqiu, Xinyang and Puyang; Group 3: Luoyang,

Sanmenxia and Zhengzhou; Group 4: Jiaozuo and Xinxiang; Group 5: Zhoukou and Zhumadian; Group 6: Hebi, Pingdingshan and Jiyuan. From the spatial distribution map (Fig. 6), it can be seen that the urban distribution with similar development pattern is not random, and its distribution has strong spatial dependence, that is, similar cities are often adjacent to each other. The regional relevance of coordinated development is expounded.

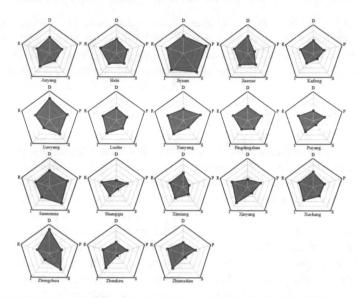

Fig. 5. Radar charts of the classification index evaluation in 18 cities of Henan province

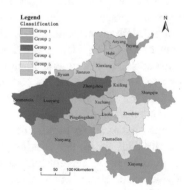

Fig. 6. Spatial correlation analysis of regions with similar classification index development patterns in Henan province

4 Conclusions

In coordinating regional development of geographical conditions, scientifically recognizing geographical conditions and promoting regional coordination and sustainable development of geographical conditions, feasible strategies for regional coordinated development of geographical conditions can be formulated according to the similarities and differences of the development structure of regional geographical conditions.

(1) The research content of CSGC hierarchical regionalization includes not only the basic research content of geographical conditions in geographic conditions census, but also other fields closely related to geographical conditions, such as resources, ecology, environment, economy and society. That is to say, the regionalization can not only embody the characteristics of geographical conditions, but also express them well in related fields Contact.

(2) By analyzing the relationship between geographical conditions and complex systems, the concept of CSGC is proposed and quantitatively described. On this basis, with the support of complex system theory, a coordinated analysis method of geographical conditions is proposed. Complex system provides theoretical guidance for statistical analysis of geographical conditions, and coordination degree analysis provides a new way to measure the relationship among the elements of geographical conditions. Realizing the coordinated development of resources, environment, ecology, economy and social subsystems in CSGC is an important subject in the study of geographical conditions, and also an effective way to solve various problems faced by statistical analysis of geographical conditions.

(3) According to the pattern analysis and spatial correlation analysis of geographical conditions, we found that the coordinated development of Henan province's geographical conditions is not only relevant in space, but also in development pattern. Based on this, we can adopt similar adjustment policies combined with the coordinated development characteristics for each pattern group for the similar regions in the coordinated development pattern in order to promote coordinated and orderly development of CSCG in Henan province, thus further service in national strategies for sustainable development. Each city in the Henan province should adapt procedures to local conditions to coordinate the development of CSGC. This must be based on the current situation and pressure of regional development, highlight problems and focus on governance, and should not pursue comprehensive improvement measures or blindly draw lessons from them.

(4) The coordination degree of CSGC in Henan Province is divided into four levels: coordination, moderate coordination, uncoordinated and extremely uncoordinated. Among them, Jiyuan, Sanmenxia, Luoyang and Zhengzhou are the coordination areas of geographical development; Nanyang, Xuchang, Xinyang, Hebi and Anyang are the moderate coordination areas; Pingdingshan, Jiaozuo, Zhumadian, Luohe and Puyang are the uncoordinated areas; Xinyang is the new one. Township, Zhoukou, Kaifeng and Shangqiu. The development status of social and resource subsystems differs slightly among cities, while that of economic and

ecological subsystems differs greatly. On the whole, the coordination of geo-graphical and national conditions in Henan Province shows a zonal pattern of development, and in morphology it shows a gradual and coordinated development from northeast to southwest.

Acknowledgments. This paper is supported by the project supported by the Open Fund of Key Laboratory of Urban Land Resources Monitoring and Simulation, Ministry of Land and Resources (KF-2018-03-050), National Key R&D Program of China (2017YFC1502600), and China Postdoctoral Science Foundation (2018M642800).

References

1. Armon, R.H.: Desertification and desertification indicators classification (DPSIR). In: Armon, R., Hänninen, O. (eds.) Environmental Indicators, pp. 277–290. Springer, Netherlands (2015)
2. Boone, R.B., Galvin, K.A.: Simulation as an approach to social-ecological integration, with an emphasis on agent-based modeling. In: Manfredo, M., Vaske, J., Rechkemmer, A., Duke, E. (eds.) Understanding Society and Natural Resources, pp. 179–202. Springer, Netherlands (2014)
3. Chen, J.Y.: The thinking of geographic national condition survey. Geospatial Inf. **12**(2), 1–3 (2014)
4. Cui, H.R., Wang, D.: A study of China's energy-economy-environment system based on VAR model. J. Beijing Inst. Technol. **12**, 23–28 (2010). (Social Science Edition)
5. Espinosa, A., Porter, T.: Sustainability, complexity and learning: insights from complex systems approaches. Learn. Organ. **18**(1), 54–72 (2011)
6. Guo, Y.T., Wang, H.W., Nijkamp, P., et al.: Space–time indicators in interdependent urban environmental systems: a study on the Huai River Basin in China. Habitat Int. **45**, 135–146 (2015)
7. Li, Y., Liu, Y., Yan, X.: Studying DPSIR model based evaluation index system for watershed ecological security. J. Peking Univ. (Nat. Sci.) **48**(6), 971–981 (2012)
8. Liu, Y.Q., Xu, J.P., Luo, H.W.: An integrated approach to modelling the economy-society-ecology system in urbanization process. Sustainability **6**, 1946–1972 (2014)
9. Miguel, M.S., Johnson, J.H., Kertesz, J., et al.: Challenges in complex systems science. Eur. Phys. J.-Spec. Top. **214**(1), 245–271 (2012)
10. Vattam, S., Goel, A.K., Rugaber, S., et al.: Understanding complex natural systems by articulating structure-behavior-function models. Educ. Technol. Soc. **14**(1), 66–81 (2011)
11. Wang, H., Liu, F., Yuan, Y., et al.: Spatial-temporal analysis of the economic and environmental coordination development degree in Liaoning Province. Geogr. J. **1–8**, 2013 (2013)
12. Wang, J.Y., Xie, M.X.: Geographical national condition and complex system. Acta Geodaetica et Cartographica Sinica **45**(1), 1–8 (2016)
13. Xie, M.X., Wang, J.Y., Chen, K.: The coordinating evaluation and spatial correlation analysis of CSGC: a case study of Henan province China. PLoS ONE **12**(6), e0174543 (2017)
14. Yu, Q.Y., Wu, W.B., Tang, H.J., et al.: Complex system theory and agent-based modeling: progresses in land change science. Acta Geographica Sinica **66**(11), 1518–1530 (2011)

Solution to the Conformable Fractional System with Constant Variation Method

Liangsong Li, Yongfang Qi$^{(\boxtimes)}$, Chongci Tang, and Kecheng Xiao

Sponge City Research Institute, Pingxiang University, Pingxiang 337000,
Jiangxi, People's Republic of China
qiyongf2007@163.com

Abstract. The main purpose of the paper is to solve conformable fractional system with Constant Variation Method. The main contents are as follows: in the first section, some useful definitions and lemmas are introduced; In the second section, the Constant Variation Method is proposed and main results are achieved; Finally, examples are presented.

Keywords: Conformable fractional · Solution · Constant Variation Method

1 Introduction

In the last decade, fractional derivative was very popular for the researchers, that is because fractional systems play an important role in solving practical problems. The most popular fractional derivatives are Riemann-Liouville definition or Caputo definition. It is known to all that both Riemann-Liouville derivative and Caputo derivative do not satisfy the simple rules, such as Multiplication Law and Division Law. Hence, conformable fractional derivative was proposed by Khalil [1]. Chain rule, Gronwall Inequality, exponential functions and Laplace transform have been introduced by Abdeljawad [2].

There are many achievements since the conformable fractional derivative is defined. the Numerical simulations were introduced [3], the stability of conformable fractional systems were researched through Lyapunov Function [4], conformable fractional Cauchy problem was studied through the stochastic method [5] and dynamic cobweb models with conformable fractional derivatives were studied [6].

In addition, in order to solve the conformable fractional equations, more and more methods have been proposed, such as the invariant subspace method [7], the new extended direct algebraic method [8] and the fractional transformation method [9].

As you know, Constant Variation Method also plays an important role in solving differential systems. It is a powerful tool not only in solving low-order equation but also in solving higher-order equation. But until now, this method has not been used in solving the conformable fractional systems. The main purpose of the paper is to solve conformable fractional system with Constant Variation Method. The main contents are as follows: in the first section, some useful definitions and lemmas will be introduced; In the second section, the Constant Variation Method will be proposed and main results will be achieved; Finally, examples will be presented.

© Springer Nature Switzerland AG 2020
F. Xhafa et al. (Eds.): IISA 2019, AISC 1084, pp. 706–711, 2020.
https://doi.org/10.1007/978-3-030-34387-3_86

2 Definitions and Lemmas

In this section, the definitions will be presented, and the important rules about conformable fractional derivative will be introduced. All of them will be used in this manuscript.

Definition 2.1 [1]: The fractional definition of conformable derivative of $u(\tau)$ is defined by

$$T_\alpha u(\tau) = \lim_{\theta \to 0} \frac{u(\tau + \theta \tau^{1-\alpha}) - u(\tau)}{\theta},$$

Where $0 < \alpha \le 1$, $0 < \tau < +\infty$, $u(\tau) \in R$. If $u^{(n)}(\tau)$ exists and $u^{(n)}(\tau)$ is continuous, then the high order of conformable fractional definition is written by

$$\mathbf{T}_\alpha^{(n)} u(\tau) = \overbrace{T_\alpha \ldots T_\alpha}^{n \ times} u(\tau).$$

Definition 2.2 [1]: The fractional definition of conformable integrals of y is defined by $u(\tau)$ is defined by

$$I_\alpha u(\tau) = \int u(\tau) \tau^{\alpha-1} d\tau, \text{ where } 0 < \alpha \le 1.$$

Lemma 2.1 [2]: Let $u'(\tau)$ is continuous when $0 < \tau < +\infty$, then the following equation holds $T_\alpha I_\alpha u(\tau) = u(\tau)$ Where $0 < \alpha \le 1$.

Lemma 2.2 [2]: When $0 < \alpha \le 1$, if $T_\alpha g(\tau)$, $T_\alpha h(\tau)$ exist, then the following rules hold:

(1) $T_\alpha(ag + bh) = aT_\alpha g + bT_\alpha h$ (2) $T_\alpha(hg) = hT_\alpha g + gT_\alpha h$
(3) $T_\alpha\left(\frac{g}{h}\right) = \frac{hT_\alpha g - gT_\alpha h}{h^2}$ (4) $T_\alpha h(\tau) = \frac{dh(\tau)}{d\tau} \tau^{1-\alpha}$

With the help of the fourth rule, the conformable fractional derivatives of some certain functions can be obtained:

(1) $T_\alpha(c) = 0$ (2) $T_\alpha(e^{\frac{1}{\alpha}\tau^\alpha}) = e^{\frac{1}{\alpha}\tau^\alpha}$
(3) $T_\alpha(\sin\frac{1}{\alpha}\tau^\alpha) = \cos(\frac{1}{\alpha}\tau^\alpha)$ (4) $T_\alpha(\cos\frac{1}{\alpha}\tau^\alpha) = -\sin(\frac{1}{\alpha}\tau^\alpha)$
(5) $T_\alpha(\ln\frac{1}{\alpha}\tau^\alpha) = \frac{\alpha}{\tau^\alpha}$ (6) $T_\alpha(\frac{1}{\alpha}\tau^\alpha) = 1$

3 Main Results

In this section, the Constant Variation Method will be proposed, main results will be achieved and examples will be presented.

Lemma 3.1. If $u(\tau), v(\tau) : [0, \infty) \to R$ are differentiable functions, let $w(\tau) = u(v(\tau))$, then

$$T_\alpha w(\tau) = \frac{dw}{dv} T_\alpha v(\tau), \text{ where } 0 < \alpha \leq 1.$$

Proof. Using the conditions that $u(\tau), v(\tau) : [0, \infty) \to R$ are differentiable functions, we have

$$
\begin{aligned}
&T_\alpha w(\tau) \\
&= \lim_{\theta \to 0} \frac{w(\tau + \theta \tau^{1-\alpha}) - w(\tau)}{\theta} \\
&= \lim_{\theta \to 0} \frac{u(v(\tau + \theta \tau^{1-\alpha})) - u(v(\tau))}{\theta} \\
&= \lim_{\theta \to 0} \frac{u(v(\tau + \theta \tau^{1-\alpha})) - u(v(\tau))}{v(\tau + \theta \tau^{1-\alpha}) - v(\tau)} \frac{v(\tau + \theta \tau^{1-\alpha}) - v(\tau)}{\theta} \\
&= \frac{dw}{dv} T_\alpha v(\tau),
\end{aligned}
$$

the proof is completed.

Lemma 3.2. The conformable fractional system $T_\alpha u(\tau) = v(\tau) u(\tau)$ has the solution

$$u(\tau) = c e^{I_\alpha v(\tau)}, \text{ where } 0 < \alpha \leq 1.$$

Proof. With the help of Lemmas 2.1, 2.2 and 3.1, it is obvious that

$$
\begin{aligned}
T_\alpha(c e^{I_\alpha v(\tau)}) &= c T_\alpha(e^{I_\alpha v(\tau)}) \\
&= c e^{I_\alpha v(\tau)} [T_\alpha I_\alpha v(\tau)] \\
&= c e^{I_\alpha v(\tau)} v(\tau),
\end{aligned}
$$

hence, $u(\tau) = c e^{I_\alpha v(\tau)}$ is a solution of the system $T_\alpha u(\tau) = v(\tau) u(\tau)$, where c is constant.

Theorem 3.1. For $0 < \alpha \leq 1$, if $p(\tau)$ and $q(\tau)$ are continuous functions for $\tau \in R$, then the solution of system $\mathbf{T}_\alpha u(\tau) = p(\tau) u(\tau) + q(\tau)$ is

$$u(t) = e^{I_\alpha p(\tau)} [I_\alpha(q(\tau) e^{-I_\alpha p(\tau)}) + \tilde{c}],$$

where $\tilde{c}$ is constant.

Proof. Assume that $u(\tau) = c(\tau)e^{l_\alpha p(\tau)}$ be solution of the system

$$T_\alpha u(\tau) = p(\tau)u(\tau) + q(\tau),$$

based on the Lemma 2.2, it is clear that

$$
\begin{aligned}
&T_\alpha(c(\tau)e^{l_\alpha p(\tau)}) \\
&= c(\tau)T_\alpha(e^{l_\alpha p(\tau)}) + c'(\tau)e^{l_\alpha p(\tau)} \\
&= c(\tau)e^{l_\alpha p(\tau)}v(\tau) + c'(\tau)e^{l_\alpha p(\tau)} \\
&= p(\tau)u(\tau) + c'(\tau)e^{l_\alpha p(\tau)},
\end{aligned}
$$

Inserting the above equation into $T_\alpha u(\tau) = p(\tau)u(\tau) + q(\tau)$, we have

$$p(\tau)u(\tau) + c'(\tau)e^{l_\alpha p(\tau)} = p(\tau)u(\tau) + q(\tau),$$

or
$$c'(\tau)e^{l_\alpha p(\tau)} = q(\tau),$$
therefore, $c(\tau) = I_\alpha(q(\tau)e^{-l_\alpha p(\tau)}) + \tilde{c}$ which implies that the general solution of system $T_\alpha u(\tau) = p(\tau)u(\tau) + q(\tau)$ is

$$u(\tau) = e^{l_\alpha p(\tau)}[I_\alpha(q(\tau)e^{-l_\alpha p(\tau)}) + \tilde{c}].$$

Theorem 3.2. For $0 < \alpha \leq 1$, if $p(\tau)$ and $q(\tau)$ are continuous functions for $\tau \in R$, then the solution of system $T_\alpha u(\tau) = p(\tau)u(\tau) + q(\tau)u^n(\tau)$ is

$$u^{1-n}(t) = e^{I_\alpha((1-n)p(\tau))}[I_\alpha((1-n)q(\tau)e^{-I_\alpha((1-n)p(\tau))}) + \tilde{c}],$$

where $\tilde{c}$ is constant.

Proof. The system $T_\alpha u(\tau) = p(\tau)u(v) + q(\tau)u^n(\tau)$ can be transformed to

$$u^{-n}(\tau)T_\alpha u(\tau) = p(\tau)u^{1-n}(\tau) + q(\tau), \tag{1}$$

let $w(\tau) = u^{1-n}(\tau)$, Lemma 3.1 is applied, then the following equation holds

$$T_\alpha w(\tau) = (1-n)u^{-n}(\tau)T_\alpha u(\tau), \tag{2}$$

considering the Eqs. (1) and (2), we have

$$T_\alpha w(\tau) = (1-n)p(\tau)w(\tau) + (1-n)q(\tau),$$

with the help of Theorem 3.1, it is clear that

$$w(\tau) = e^{I_\alpha((1-n)p(\tau))}[I_\alpha((1-n)q(\tau)e^{-I_\alpha((1-n)p(\tau))}) + \tilde{c}],$$

which implies that

$$u^{1-n}(\tau) = e^{I_\alpha((1-n)p(\tau))}[I_\alpha((1-n)q(\tau)e^{-I_\alpha((1-n)p(\tau))}) + \tilde{c}],$$

the proof is completed.

4 Examples

Example 4.1. Let us consider the following equation $\mathbf{T}_\alpha u(\tau) = u(\tau) + \frac{1}{\alpha}\tau^\alpha$, here $p(\tau) = 1$, $q(\tau) = \tau$, Definition 2.2 and Theorem 3.1 is applied, it is obvious that the system has the following general solution

$$
\begin{aligned}
u(\tau) &= e^{I_\alpha p(\tau)}[I_\alpha(q(\tau)e^{-I_\alpha p(\tau)}) + \tilde{c}] \\
&= e^{\int \tau^{\alpha-1}d\tau}[\int \tau^{\alpha-1}\frac{1}{\alpha}\tau^\alpha e^{-\int \tau^{\alpha-1}d\tau}d\tau + \tilde{c}] \\
&= e^{\frac{\tau^\alpha}{\alpha}}[\int \tau^{\alpha-1}\frac{1}{\alpha}\tau^\alpha e^{-\frac{\tau^\alpha}{\alpha}}d\tau + \tilde{c}] \\
&= e^{\frac{\tau^\alpha}{\alpha}}[\int (-1)\frac{1}{\alpha}\tau^\alpha de^{-\frac{\tau^\alpha}{\alpha}} + \tilde{c}] \\
&= -e^{\frac{\tau^\alpha}{\alpha}}[\int \frac{1}{\alpha}\tau^\alpha de^{-\frac{\tau^\alpha}{\alpha}} + \tilde{c}] \\
&= -e^{\frac{\tau^\alpha}{\alpha}}[\frac{\tau^\alpha}{\alpha}e^{-\frac{\tau^\alpha}{\alpha}} - \int e^{-\frac{\tau^\alpha}{\alpha}}d(\frac{1}{\alpha}\tau^\alpha) + \tilde{c}] \\
&= -e^{\frac{\tau^\alpha}{\alpha}}[\frac{\tau^\alpha}{\alpha}e^{-\frac{\tau^\alpha}{\alpha}} + e^{-\frac{\tau^\alpha}{\alpha}} + \tilde{c}] \\
&= -\frac{1}{\alpha}\tau^\alpha - 1 - \tilde{c}e^{\frac{\tau^\alpha}{\alpha}}.
\end{aligned}
$$

Example 4.2. Let us consider the equation $\mathbf{T}_\alpha u(t) = 6\frac{\alpha}{t^\alpha}u(t) - \frac{1}{\alpha}t^\alpha u^2(t)$, here $n = 2$, Definition 2.2 and Theorem 3.2 is applied, it is obvious that the system has the following general solution $\frac{(\frac{1}{\alpha}t^\alpha)^6}{u(t)} - \frac{(\frac{1}{\alpha}t^\alpha)^8}{8} = c$.

Acknowledgments. The Natural Science Foundation (11661065) of China, the Scientific Research Foundation of Jiangxi Provincial Education Department (GJJ181101) and Youth Fundation of Pingxiang University.

References

1. Khalil, R., Al Horani, M., Yousef, A., Sababheh, M.: A new definition of fractional derivative. J. Comput. Appl. Math. **264**, 65–70 (2014)
2. Abdeljawad, T.: On conformable fractional calculus. J. Comput. Appl. Math. **279**, 57–66 (2015)
3. Morales Delgado, V.F., GmezAguilar, J.F., Atangana, A.: Fractional conformable derivatives of Liouville-Caputo type with low-fractionality. Physica A Stat. Mech. Appl. **503**, 424–438 (2018)

4. Souahi, A., Makhlouf, A.B., Hammami, M.A.: Stability analysis of conformable fractional-order nonlinear systems. Indagationes Mathematicae **28**(6), 1265–1274 (2017)
5. Cenesiz, Y., Kurt, A., Nane, E.: Stochastic solutions Of conformable fractional Cauchy problems. Stat. Probab. Lett. **124**, 126–131 (2017)
6. Bohner, M., Hatipoglu, V.F.: Dynamic cobweb models with conformable fractional derivatives. Nonlinear Anal. Hybrid Syst. **32**, 157–167 (2019)
7. Hashemi, M.S.: Invariant subspaces admitted by fractional differential equations with conformable derivatives. Chaos Solitons Fractals **107**, 161–169 (2018)
8. Rezazadeh, H., Tariq, H., Eslami, M., Mirzazadeh, M., Zhou, Q.: New exact solutions of nonlinear conformable time-fractional Phi-4 equation. Chin. J. Phys. **56**(6), 2805–2816 (2018)
9. Feng, Q.: A new approach for seeking coefficient function solutions of conformable fractional partial differential equations based on the Jacobi elliptic equation. Chin. J. Phys. **56**(6), 2817–2828 (2018)

Homology Bifurcation Problem for a Class of Three - Dimensional Systems

Ruiping Huang[1(✉)], Lei Deng[2], Xinshe Qi[1], and Yuyi Li[3]

[1] Public Training Courses Office in College of Information and Communication of NUDT, Xi'an, China
249431969@qq.com
[2] Common Military Courses Office in Army Academy of Border and Coastal Defense, Xi'an, China
[3] Foreign (Taiwan) Army Office in College of Information and Communication of NUDT, Wuhan, China

Abstract. By combining the central manifold theorem and planar bifurcation theory, this paper studies the bifurcation limit cycles of a class of three dimensional systems with homology rings, gives the conditions for the existence of stable limit cycles, and generalizes the results of existing planar systems.

Keywords: Central manifold theorem · Homoclinic orbits · Branch · Limit cycle

1 Preliminary Knowledge

In recent decades, a large number of researchers have devoted themselves to the study of limit cycles for planar quadratic differential systems [1, 2], but not many for high-dimensional systems research. In paper [3], the existence of limit cycles of planar quadratic systems for a class of quadratic differential systems (III) is studied. In this paper, we study the existence of limit cycles for a class of planar cubic systems which can be transformed into plane cubic systems by using the central manifold theorem.

First, we give some Lemma, and consider the equation

$$\begin{cases} \dot{x} = Ax + F(x,y) \\ \dot{y} = Bx + G(x,y) \end{cases} \tag{1}$$

A, B is a constant matrix of order $n \times n$ and $m \times m$ respectively.

Lemma 1 (Center Manifold Theorem) [4]: Let all eigenvalues of A have zero real parts, and eigenvalues of B have negative real parts. $F(x,y)$ and $G(x,y)$ are continuous differentiable functions of second order and satisfy $\begin{cases} F(0,0) = 0, DF(0,0) = 0 \\ G(0,0) = 0, DG(0,0) = 0 \end{cases}$.

© Springer Nature Switzerland AG 2020
F. Xhafa et al. (Eds.): IISA 2019, AISC 1084, pp. 712–717, 2020.
https://doi.org/10.1007/978-3-030-34387-3_87

$DF(0,0), DG(0,0)$ is the Jacobi matrix of $F(x,y), G(x,y)$ at $(0,0)$. Then the system has a local C^2 center manifold $y = h(x) (||x|| < \delta)$.

Considering the plane autonomous system $\begin{cases} \dot{x} = f(x,y) \\ \dot{y} = g(x,y) \end{cases}$ (2)

and its perturbed system $\begin{cases} \dot{x} = f(x,y) + pf_0(x,y,p,q) \\ \dot{y} = g(x,y) + pg_0(x,y,p,q) \end{cases}$ (3)

in which f, g, f_0, $g_0 \in C^1$, x, $y \in R^1$, $p \in R$, $q \in R^k$, $k \geq 0$.

Lemma 2 Hypothesis 1. The system (2) has a homoclinic orbit Γ at the saddle point $O(0,0)$, and P_0 is any point on Γ, Over P_0 for system (2) and the transverse line l is collinear with the direction of the outer normal at point Γ. 2. The saddle point of the perturbation system (3) near point $O(0,0)$ is $\overline{O}$, and the intersection points of the over-stable manifold $W_{\overline{o}}^s$ and the unstable manifold $W_{\overline{o}}^{us}$ are respectively P_s and P_u, then, the directed distance $d(P_s, P_u)$ from P_s to P_u (positive in the same direction as) under small perturbation is

$$d(P_s, P_u) = \frac{1}{\sqrt{f^2(P_0) + g^2(P_0)}} pM + O(p)$$

in which $M = \int_{-\infty}^{+\infty} e^{-\int_0^t (f_x + g_y)dx} (fg_0 - gf_0)|_{p=0} dt$ is the Melnikov function [5].

Lemma 3 (Hartman Theorem) [5]: Considering a system $\dot{x} = f(x)$ in which, $x \in W$, $W \in R^n$ is open and $f : W \to R^n$ is continuously differentiable. The system is balanced at the origin O, and $A = Df(0)$. if the real part of all eigenvalues $\lambda_k (k = 1, 2, \cdots, n)$ is non-zero, that is $Re\lambda_k \neq 0$. There is a single continuous transformation of both sides $x = u(\varepsilon)$. It Is defined in the neighborhood of $\varepsilon = 0$, $u(0) = 0$, and maps the solution of a linear equation $\dot{\varepsilon} = A\varepsilon$ to the solution of $\dot{x} = f(x)$.

2 Main Results

Considering system

$$\begin{cases} \dot{x} = -y + mxy + y^2 + \delta xz \\ \dot{y} = x(1 + ax + by) \\ \dot{z} = -z + x^2 + y^2 \end{cases}$$ (4)

According to Lemma 1, there is a local C^2 class center manifold $z = h(x, y)$ in system (4). Let

$$z = h(x, y)$$
$$= a_{10}x + a_{01}y + a_{20}x^2 + a_{11}xy + a_{02}y^2 + a_{30}x^3 + a_{21}x^2y + a_{12}xy^2 + a_{03}y^3 + \ldots$$

By substituting the partial differential equation [4] satisfied by it and determining the right end coefficient by comparison coefficient method, the local central manifold of the system (4) can be obtained as $z = h(x, y) = x^2 + y^2 + O(\rho^3)$, $(\rho = \sqrt{x^2 + y^2})$.

Substitute into system (4) to get

$$\begin{cases} \dot{x} = -y + mxy + y^2 + \delta x^3 + \delta xy^2 + \delta O(\rho^4) \\ \dot{y} = x(1 + ax + by) \end{cases} \tag{5}$$

Let $\quad p = sign(\delta)\sqrt{m^2 + \delta^2}, \quad q = \delta/p, \quad a = pr, \quad b = ps, \quad$ then $\quad \delta = pq,$ $m = sign(m)|p|\sqrt{1 - q^2}$, System (5) can be reduced to

$$\begin{cases} \dot{x} = -y + y^2 + p[sign(m)sign(p)\sqrt{1 - q^2}xy + qx^3 + qxy^2 + qO(\rho^4)] \\ \dot{y} = x + p(rx + sy)x \end{cases} \tag{6}$$

Let $x = u$, $y = v + 1$, System (6) becomes to:

$$\begin{cases} \dot{u} = v + v^2 + p[sign(m)sign(p)\sqrt{1 - q^2}u(v + 1) + qu^3 + qu(v + 1)^2 + qO(\rho^4)] \\ \dot{v} = u + p[ru + s(1 + v)]u \end{cases} \tag{7}$$

The cubic approximation system of system (7) is

$$\begin{cases} \dot{u} = v + v^2 + p[sign(m)sign(p)\sqrt{1 - q^2}u(v + 1) + qu^3 + qu(v + 1)^2] \\ \dot{v} = u + p[ru + s(1 + v)]u \end{cases} \tag{8}$$

Let

$$\begin{cases} P_1(u, v) = v + v^2 + p[sign(m)sign(p)\sqrt{1 - q^2}u(v + 1) + qu^3 + qu(v + 1)^2 + qO(\rho^4)] \\ Q_1(u, v) = u + p[ru + s(1 + v)]u \end{cases}$$

$$\begin{cases} P_2(u, v) = v + v^2 + p[sign(m)sign(p)\sqrt{1 - q^2}u(v + 1) + qu^3 + qu(v + 1)^2] \\ Q_2(u, v) = u + p[ru + s(1 + v)]u \end{cases}$$

Then the result can be obtained.

$$D_1 = \det\left(\frac{\partial(P_1, Q_1)}{\partial(u, v)}\right)_{(u,v)=0}$$

$$= \det\left(\begin{matrix} p[\sqrt{1-q^2}(v+1)+3qu^2+q(v+1)^2+qO(\rho^3) & 1+2v+p[\sqrt{1-q^2}u+qu(2v+1)+qO(\rho^3)] \\ 1+2pru+ps(1+v) & ps \end{matrix}\right)_{(u,v)=0}$$

$$\neq 0,$$

$$D_2 = \det\left(\frac{\partial(P_2, Q_2)}{\partial(u, v)}\right)_{(u,v)=0} = \left(\begin{matrix} p[\sqrt{1-q^2}(v+1)+3qu^2+q(v+1)^2 & 1+2v+p[\sqrt{1-q^2}u+qu(2v+1)] \\ 1+2pru+ps(1+v) & ps \end{matrix}\right)_{(u,v)=0} \neq 0$$

$$T_1 = div(P_1, Q_1)_{(u,v)=0} = p[\sqrt{1-q^2}(1+v)+3qu^2+q(v+1)^2+qO(\rho^3)+s]_{(u,v)=0} \neq 0$$

$$T_2 = div(P_2, Q_2)_{(u,v)=0} = p[\sqrt{1-q^2}(1+v)+3qu^2+q(v+1)^2+s]_{(u,v)=0} \neq 0$$

According to Lemma 3, homomorphic T_1 and T_2 exist respectively in the neighborhood of singularities G. Let systems (7) and (8) are homomorphic in systems

$$\begin{cases} \dot{u} = v+v^2 \\ \dot{v} = u \end{cases} \tag{9}$$

Therefore, the existence of homomorphism T_3 makes the flows of systems (7) and (8) in the domain G equivalent. So we can directly analyze system (8).

$$\begin{cases} \dot{x} = -y+mxy+y^2+\delta x^3+\delta xy^2 \\ \dot{y} = x(1+zx+by) \end{cases} \tag{10}$$

Let $p = sign(\delta)\sqrt{m^2+\delta^2}$, $q = \delta/p$, $a = pr$, $b = ps$, then $m = sign(m)|p|\sqrt{1-q^2}$, $\delta = pq$, System (10) can be reduced to

$$\begin{cases} \dot{x} = -y+y^2+p[sign(m)sign(p)\sqrt{1-q^2}xy+qx^3+qxy^2] \\ \dot{y} = x+p(rx+sy)x \end{cases} \tag{11}$$

Let $x = u$, $y = v+1$, System (11) changed to:

$$\begin{cases} \dot{u} = v+v^2+p[sign(m)sign(p)\sqrt{1-q^2}u(v+1)+qu^3+qu(v+1)^2] \\ \dot{v} = u+p[ru+s(1+v)]u \end{cases} \tag{12}$$

The undisturbed system of system (12) is system (9)——Hamilton system.

Through the qualitative analysis of system (9) and system (12), we can obtained that system (9) has a saddle point $A(0,0)$ and a center $B(0,-1)$; System (12) has saddle point $A(0,0)$ and unstable focus $B(0,-1)$; Systems (10) and (11) have saddle points $A_0(0,1)$ and unstable focal points $B_0(0,0)$.

The first integral of system (9) is

$$H(u, v) = u^2 - v^2 - \frac{2}{3}v^3 = h \tag{13}$$

As can be seen from Eq. (13), when $h = 0$, (13) is the homology orbit of the saddle-crossing point $A(0,0)$ of system (9), denoted as Γ, that is

$\Gamma = \{u = u(t), v = v(t), t \in (-\infty, +\infty)\} = \{(u, v) : H(u, v) = 0\}$, when $-\frac{1}{3} < h < 0$, (13) is the closed orbit of system (9) in a cluster surrounds the center $B(0, -1)$ of Γ. According to calculation, the intersection point of Γ and V is $P(0, -\frac{2}{3})$, we may record this moment as the zero moment, that is $u(0) = 0, v(0) = -\frac{3}{2}$, Thus we can obtain:

Theorem 1: Assume that $\delta > 0, m > 0$, when $r < \frac{35}{8}(\frac{2}{5}\sqrt{1 - q^2} - \frac{10}{11}q - \frac{81}{32}s)$, the system (10) and (11) (system (12)) have at least one stable limit cycle around in the neighborhood of Γ.

Further we can obtain:

Theorem 2: Assume that $\delta > 0, m > 0$, when $r < \frac{35}{8}(\frac{2}{5}\sqrt{1 - q^2} - \frac{10}{11}q - \frac{81}{32}s)$, the system (4) has at least one stable limit cycle around $B_0(0, 0, 0)$ in the neighborhood of Γ on its local central manifold $z = x^2 + y^2 + o(\rho^3)\rho = \sqrt{x^2 + y^2}$.

3   Theorem Proving

Proof of Theorem 1: from formula (3):

$$\begin{cases} f = v + v^2 \\ g = u \end{cases} \text{ and } \begin{cases} f_0 = \sqrt{1 - q^2}u(v + 1) + qu^3 + qu(v + 1)^2 \\ g_0 = [ru + s(1 + v)]u \end{cases}.$$

According to Eq. (13), the expressions of Γ^+ (the part on the right side of the V) and Γ^- (the part on the left side of the V) are $\Gamma^+ : u = -v\sqrt{1 + \frac{2}{3}v}$ and $\Gamma^- : u = v\sqrt{1 + \frac{2}{3}v}$. Γ is counterclockwise, then according to Lemma 2 and the symmetry of the system, we can obtain that

$$M(t) = 2\int_{-\infty}^{0} \left\{ (v + v^2)[ru + s(1 + v)]u - u\left[\sqrt{1 - q^2}u(v + 1) + qu^3 + qu(v + 1)^2\right] \right\}dt$$

$$= 2\int_{0}^{-\frac{3}{2}} \left\{ (v + v^2)[rv\sqrt{1 + \frac{2}{3}v} + s(1 + v)] - \sqrt{1 - q^2}v\sqrt{1 + \frac{2}{3}v}(v + 1) \right.$$

$$\left. + qv^3(1 + \frac{2}{3}v)\sqrt{1 + \frac{2}{3}v} + qv\sqrt{1 + \frac{2}{3}v}(v + 1)^2 \right\}dv$$

$$= \frac{8}{35}r + \frac{81}{32}s - \frac{2}{5}\sqrt{1 - q^2} + \frac{10}{11}q$$

In order to simplify the calculation, it is advisable to take $P_0 = P$, then $f(P_0) = \frac{3}{4}$, $g(P_0) = 0$. By calculation, $d(P_s, P_u) = \frac{4}{3}p(\frac{8}{35}r + \frac{81}{32}s - \frac{2}{5}\sqrt{1 - q^2} + \frac{10}{11}q) + O(p)$, because $p > 0$, therefore $d(P_s, P_u)$ and $M(t)$ have the same sign. When $r < \frac{35}{8}(\frac{2}{5}\sqrt{1 - q^2} - \frac{10}{11}q - \frac{81}{32}s)$, $d(P_s, P_u) < 0$, the stable manifold W_A^s of system (9) about saddle point $A(0, 0)$ is outside the unstable manifold W_A^u after the homoclinic orbit Γ of system (9) breaks.

Now let's prove the existence of limit cycles by using the ring theorem.

First of all, from the above analysis, we know that $B(0,1)$ meets the condition of inner boundary of ring domain D required by the ring domain theorem, so $B(0,1)$ is taken as the inner boundary of D.

Secondly, when $r < \frac{35}{8}\left(\frac{2}{5}\sqrt{1-q^2} - \frac{10}{11}q - \frac{81}{32}s\right)$, W_A^s is in the external of W_A^u, when the rail line of the system (12) passes through $\overline{P_sP_u}$, $u = v + v^2 > 0$. So it can be known that these rail lines all pass through $\overline{P_sP_u}$ from left to right, and (12) satisfies the existence and uniqueness condition of the solution, so it can be selected $\overline{P_sP_u} \cup P_uA \cup AP_s$ as the external boundary of D.

So far, according to Lemma 2 and ring domain theorem, system (12) has at least one stable limit cycle in ring domain D. Theorem 1 is proved by the invertibility of the previous transformation.

Proof of Theorem 2: from the previous analysis, Lemma 1 and Theorem 1, the conclusion of Theorem 2 can be obtained directly.

Note: In this paper, the three-dimensional system is reduced to the two-dimensional system by the central manifold theorem. And quadratic differential system (III) three times the number of times to class equation. In system (10) when $\delta = 0$, it is quadratic differential system (III) equation, thus it can be seen that the system in [3] study is a special case of this article.

References

1. Ye, Y.: Limit Cycle Theory. Shanghai Science and Technology Press, Shanghai (1984)
2. Ye, Y.: Qualitative Theory of Polynomial Differential System. Shanghai Science and Technology Press (1995)
3. Jin, Y., Zhuang, W.: Branch problems of quadratic differential system (III) equation. J. Chang De Normal Univ. (Nat. Sci. Ed.) (2002)
4. Ma, Z., Zhou, Y.: Qualitative and Stability Methods for Ordinary Differential Equations. Science Press, Beijing (2001)
5. Zhang, J., Feng, B.: Geometric Theory and Bifurcation of Ordinary Differential Equations. Peking University Press, Beijing (2000)

Experimental Research on Disyllabic Tone of Liangzhou Dialect in Gansu Province

Ying Li and Yonghong Li[✉]

Key Laboratory of China's Ethnic Languages and Information Technology of
Ministry of Education, Northwest Minzu University, Lanzhou 730030,
Gansu, China
lyhweiwei@126.com

Abstract. This paper mainly uses the combination of acoustic experiment and auditory perception to analyze the disyllabic tone of Liangzhou dialect, focusing on revealing the phonetic facts and rules of its transposition. There are 8 basic transposition forms in the two-syllable phrase of Liangzhou dialect, including 3 kinds of Yinping, 2 kinds of Yangping, and 3 kinds of Qu.

Keywords: Liangzhou dialect · Disyllabic tone · Tone sandhi

1 Introduction

Liangzhou is the political, economic and cultural center of Wuwei City, Gansu Province. From the dialect partition, Liangzhou dialect belongs to the Hexi section of Lanyin Mandarin in the Northwest of China. The current research focus is on discussing the rhythm system, revealing its phonetic changes and rules from the perspective of diachronic and synchronic. With the rapid development of experimental phonetics, Yang (2013) used the experimental method to study Liangzhou Mandarin in his postgraduate thesis, so as to better learn and promote the Stand Chinese. Therefore, it is of great significance to further study the Liangzhou dialect by means of acoustic experiments.

When the static tone, that is, the citation tone, enters the context of the language combination, the original tone and mode are often changed, resulting in the tone sandhi. Disyllabic tone is the smallest unit of tone sandhi, and it is also the starting point of investigating the mode of tone sandhi in a dialect. In this paper, based on the experimental study of the citation tone, continue to explore the disyllabic tone of Liangzhou dialect.

2 Experimental Description

2.1 Pronunciation Partner

This study selected a 62-year-old male speaker who has been living in the Liangzhou district of Wuwei city. He has a certain education and is familiar with Liangzhou dialect. Above all, he pronounces clearly.

F. Xhafa et al. (Eds.): IISA 2019, AISC 1084, pp. 718–723, 2020.
https://doi.org/10.1007/978-3-030-34387-3_88

2.2 Experimental Equipment

The recording location is selected in a professional studio. The recording software is Adobe Audition 3.0, the sampling rate is 44100 s, with 16-bit resolution. The analysis software includes Pratt5.0 and Matlab, which Praat5.0 is used to segment and mark speech samples and extract the experimental data.

2.3 Pronunciation Table

There are three citation tones in Liangzhou dialect, which are Yinping (44), Yangping (24) and Qu (51). The three tones match each other and we can get 9 combinations. Seven common phrases are selected for each combination, for a total of 63 two-syllable words. The following is a part of the typical pronunciation words of the selection, the front is the Chinese pinyin, followed by the English (Table 1).

Table 1. Vocabulary of Liangzhou dialect

Former	Latter		
	Yinping (44)	Yangping (24)	Qu (51)
Yinping (44)	hua sheng (peanut) qing wa (frog)	hui chen (dust) san shi (thirty)	wan dou (pea) feng mi (honey)
Yangping (24)	mu zhu (sow) mao jin (towel)	zuo shou (the left hand) qian tou (at the head)	huang dou (soybean) yi fu (uncle)
Qu (51)	fa shao (fever) jie hun (marry)	dou fu (bean curd) du ji (jealousy)	wei dao (taste) zai jian (good-bye)

2.4 Data Extraction and Analysis

Firstly, using the speech analysis software Praat, manually extract the fundamental frequency parameters of all samples. Secondly, using the Matlab to run the normalized script program, and take the tone category as a unit, and normalize all experimental data according to the trend of the fundamental frequency curve. Then through formula operation, according to the way of five-degree value, the T value is transformed into T value. Finally, the T values of each combination are averaged, and the corresponding T-value graph is made. Combined with the sense of hearing, the disyllabic tone of Liangzhou dialect is combined and classified to determine the final tone sandhi modes. When normalizing the fundamental frequency, we use the T value calculation method proposed by Shi Feng. The formula is:

$$T = [(\lg f0 - \lg min)/(\lg max - \lg min)] * 5 \tag{1}$$

Among them, F0 represents the fundamental frequency value. Min refers to the lower limit of the fundamental frequency. Max is the upper limit of the modulation range. T represents the final normalized result. When the T value is converted to the fifth degree, we use the "precinct" strategy proposed by Liu (2008). Based on the fundamental frequency perception, there is a floating range of ±0.1 per degree boundary. The specific corresponding relations are: In the range of 0–1.1, the corresponding fifth value is 1; in the range of 0.9–2.1, the corresponding fifth value is 2; in the range of 1.9–3.1, the corresponding fifth value is 3; in the range of 2.9–4.1, the corresponding fifth value is 4; and in the range of 3.9–5, the corresponding fifth value is 5.

3 Experimental Result and Discussion

3.1 Experimental Result

This paper mainly discusses the tone sandhi forms of Liangzhou dialect from two aspects of tone type and tone value. There are two types of tone sandhi: transposition and non-transposition. Transposition means the type and value of tone changed, such as the level tone changed to a falling tone. Non-transposition means the type of tone do not changed but only the value changed. For the convenience of writing, we use T1, T2, and T3 to represent the Yinping, Yangping, and Qu, respectively. For example, we use "T1+X" and "X+T1" to represent the combinations of T1, and so on. The details are as follows:

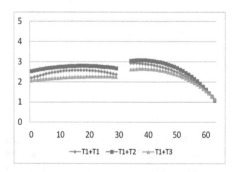   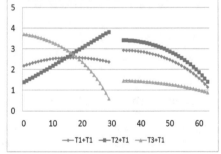

Fig. 1. T value curve of "T1+X"   **Fig. 2.** T value curve of "X+T1"

It can be seen from the Fig. 1 that when T1 is used as the former word, the three curves are relatively flat and balanced, and the starting point and the end point are distributed in the third level. Compared with the result of citation tone, the tone trend of T1 is unchanged, but the tone value from 44 to 33. It indicates that T1 is the non-transposition when it is located in the former word. In the combination of "T1+X", when T1, T2 and T3 are located in the latter word, the three tone curves have the same trend. The combination of T1+T1/T2/T3 was changed from 44+44/24/51 to 33+31, respectively. Although the values of the three combinations are the same, they are different in hearing, so they cannot be merged.

As can be seen from the Fig. 2, when T1 is used as the latter word, the distribution of the T value curve is not concentrated. In the combination of T1+T1 and T2+T1, the trend of latter word T1 is consistent, but the tone value is different, from a flat tone 44 to the falling tone 31 and 42, respectively. In the T3+T1 combination, the latter word T1 is obviously affected by the former word T3, and the type and value of tone are changed, from 44 to 21. It can be seen that T1 belongs to the transposition when it is located in the latter word.

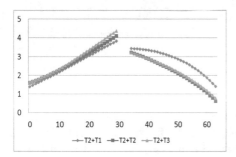

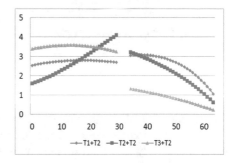

Fig. 3. T value curve of "T2+X"   **Fig. 4.** T value curve of "X+T2"

It can be seen from the Fig. 3 that when T2 is in the former word, the slopes of the three curves are almost identical to the level at which they are located, starting at the second level, finally at the fourth and fifth levels. Compared with the result of citation tone, the tone type of T2 has not changed due to the influence of the combination, and still maintains the rising tone. In the "T2+X" combination, the curve of the latter word T1 is consistent with the curves of T2 and T3, but there is a difference in the slope. The combination of T2+T1 is changed from the original value of 24+44 to 24+42. The combination of T2+T2/T3 is changed from the original value of 24+24/51 to 25+31. Combined with the sense of hearing, the two combinations can be combined.

As can be seen from the Fig. 4, when T2 is used as the latter word, it is easy to be affected by the combination, and the type and value of tone are changed. Compared with the result of citation tone, the three curves of T2 are changed from rising to falling, and the slope of the curve is different. In the combinations of T1+T2 and T2+T2, the starting point of the T2 curve is at the third level, and the end point is at the first level, and the slope is different, from the rising tone 24 to the falling tone 31. In the combination of T3+T2, the latter word T2 is affected by the former word T3, and is changed from 24 to 21, indicating that T2 belongs to the transposition when it is in the latter word.

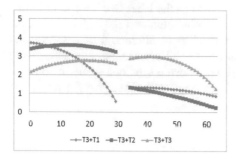

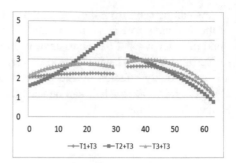

Fig. 5. T value curve of "T3+X"   **Fig. 6.** T value curve of "X+T3"

It can be seen from the Fig. 5 that when T3 is in the former word, the T value curve is relatively scattered. Compared with the result of citation tone, in the combination of T3+T1, the former word T3 has the same type, but the tone value is decreased by one degree, from 51 to 41, indicating that T3 is the non-transposition. In the combination of T3+T2/T3, the former word T3 is a flat tone, but is in the fourth level and the third level, respectively. The T3 from the falling tone 51 to the flat tone 44 and 33, indicating that T3 is the transposition. In the "T3+X" combination, the starting points of the latter word T1 and T2 are the same, but the slope of the T2 curve is larger than the T1 curve. The combination of T3+T1 is changed from the original value 51+44 to 41+21, T3+T2 is changed from 51+24 to 44+21, and T3+T3 is changed from 51+51 to 33+31.

It can be seen from the Fig. 6 that when T3 is used as the latter word, the T value curves are concentrated. Although there is a slight difference in the slope, it does not affect the final value judgment. The latter word T3 is better adapted and connected to the former word, the value will changed from 51 to 31, which is the non-transposition.

3.2   Discussion of Result

It can be seen from the above table that the combinations of T1+T1/T2/T3 and T3+T3 are 33+31, but the four combinations have obvious differences in the sense of hearing, so they cannot be combined. The combinations of T2+T2 and T2+T3 are 25+31. It is difficult to distinguish between the two groups in auditory sense and can be classified into one category. The specific tone sandhi rules of Liangzhou dialect are: When the T1 is used as the former word, the type of tone is unchanged, and the tone value is changed to 33. When T2 is used as the former word, the type of tone is unchanged, and the values are 24 and 25 respectively. When T3 is used as the former word, except for the

Table 2. Basic forms of disyllabic tone in Liangzhou Dialect

Former	Latter		
	T1 (44)	T2 (24)	T3 (51)
T1 (44)	33+31	33+31	33+31
T2 (24)	24+42	25+31	25+31
T3 (51)	41+21	44+21	33+31

T3+T1 combination, the tone value of T3 becomes 41, and in the combination of T3+T2/T3, T3 is changed from a falling tone to the flat tone. When T1, T2 and T3 are used as the latter words, each type of tone becomes a falling tone, and the value is mostly at 31. Each type of tone will be influenced by the former word, making appropriate adjustments and changes (Table 2).

4 Conclusion

This paper mainly uses the combination of acoustic experiment and auditory perception to explore the disyllabic tone of Liangzhou dialect. There are 8 basic forms of tone sandhi. When it is the former word, Yinping and Yangping are relatively stable, and Qu is easily affected by the combination. On the contrary, when it is the latter word, the Qu is relatively stable, and the Yinping and Yangping are easily affected by the former word and then change the tone type and value.

Acknowledgments. This work was financially supported by Fok Ying Tung Education Foundation fund (Grant No. 151110) and NSFC grant fund (No. 11564035).

References

Kong, J.: Basic Course of Experimental Phonetics. Peking University Press, Beijing (2013)
Wu, Z., Maocan, L.: Summary of Experimental Phonetics. Higher Education Press, Beijing (1989)
Zhu, X.: Shanghai Tone Experiments. Shanghai Education Press, Shanghai (2004)
Shi, F.: Tone analysis of Tianjin dialect double characters. Lang. Res. (01), 77–90 (1986)
Yang, Z.: Experimental Study on the Pronunciation of Liangzhou Mandarin. Yunnan Normal University (2013)

Dynamic Variance Analysis
of One-Dimensional Signal in Gaussian Noise
Based on Recursive Formula

Lang Yu[1], Wenqing Wu[1], Gang He[2(✉)], and Wenxin Yu[2]

[1] School of Science, Southwest University of Science and Technology,
Mianyang, China
[2] School of Computer Science and Technology, Southwest University of Science
and Technology, Mianyang, China
ganghe@swust.edu.cn

Abstract. Gaussian noise is a statistical noise having a probability density function (PDF) equal to that of the normal distribution (which is also known as the Gaussian distribution). In telecommunications and computer networks, communication channels may be affected by broadband Gaussian noise from many natural sources, such as thermal vibrations of atoms in a conductor (called thermal noise or Johnson-Nyquist noise), shot noise, from Earth and other warm objects, as well as celestial bodies such as the sun. Therefore, Gaussian noise is particularly common in signal communication. It is determined by two parameters, mean and variance. Therefore, when filtering a random signal containing a Gaussian noise, it is necessary to perform variance analysis and mean calculation. As the most commonly used basic statistic, variance describes the distance between the sample data and the sample mean. It is widely used in finance, aerospace, communications and other fields. When calculating the variance of the Gaussian noise with real-time high-speed variation, the traditional definition algorithm adopts the full-sample algorithm, and the algorithm has the disadvantages of inefficiency when dealing with such real-time changing data. Aiming at this problem, this paper proposes a new real-time variance dynamic recursive algorithm based on the sliding window idea, which makes the workload of computer reading data significantly reduced. Finally, through simulation experiments, it is applied to the mean and variance estimation of Gaussian noise with real-time high-speed variation. The experimental results show that the algorithm significantly improves the efficiency of calculating the variance and mean of Gaussian noise. This verifies the correctness and practicability of this algorithm.

Keywords: Time-varying variance · Recursive formula · Algorithm complexity · Sliding window · Gaussian noise

1 Introduction

Conventional signal processing, especially in the field of communication signal processing, Gaussian distribution is a commonly used noise interference model. On the one hand, Gaussian distribution can be derived from the central limit theorem, which

© Springer Nature Switzerland AG 2020
F. Xhafa et al. (Eds.): IISA 2019, AISC 1084, pp. 724–732, 2020.
https://doi.org/10.1007/978-3-030-34387-3_89

can be used to describe a large amount of background noise in communication and signal processing systems. On the other hand, signal processing algorithms based on Gaussian distribution model design usually have good linear characteristics and are easy to implement in engineering. The Gaussian noise interference signal is determined by two parameters, mean and variance. Therefore, when filtering a random signal containing a Gaussian noise, it is necessary to perform variance analysis and mean calculation.

Sample variance is a basic statistical item that describes the fluctuation of sample data, data analysis, and is also an important basic statistic for studying the distribution characteristics of random variables. By calculating the variance of the sample data, the degree of fluctuation of the data sequence from the center can be reflected, thereby performing mass analysis on the sample data. The sample variance is particularly widely used in practical problems and plays a very important role in data analysis. Such as satellite timing system and real-time data analysis [1–4]. This led to a group of scholars to study such statistics.

In the actual calculation, although the definition of the sample variance can be used for calculation, it is necessary to traverse each sample multiple times to calculate the mean of the sample, and then traverse the entire sequence to calculate the second central moment of the sample. Of course, when the sample data size is small, such calculation is not problematic, but when the sample data volume is large, the calculation efficiency is very low, which not only occupies a large amount of computer storage resources, but also has a long calculation time. This traditional sample variance algorithm does not have real-time performance when faced with real-time changing data sequences. Based on this, in the literature [5], a real-time calculation method for variance of non-zero mean measurement and control data is proposed. This algorithm reduces data storage and improves computational efficiency, and is the same as the global calculation result. Literature [6] studied the calculation rules of the common discrete distributed k-order origin moment, and proposed a unified recursive formula for the common discrete distributed k-order origin moment. Literature [7] proposed a one-step recursive algorithm for variance, which is well applied in the sequence of data length changes. In literature [8], starting from binomial, poison and geometric distribution of k-order origin moment, the recursive formula of discrete distribution of k-order origin moment is studied. In literature [9], inspired by binomial, poison and geometric distribution of k-order central moment, a recursive formula of discrete distribution of k-order central moment is proposed. By constructing a class of special polynomials and series in literature [10], the recursive formula of k-order central moment of binomial, poison and geometric distribution is derived. In literature [11], the recursive calculation formulas of sample mean, sample variance and covariance matrix after sample size change are derived.

Through reading the research results of the above literature, it is found that they are all studying the recursive formula of sample mean and sample variance under the change of sequence length. At present, there is no recursive algorithm with fixed sequence length in time-varying variance calculation. Based on the above research, this paper proposes a dynamic time-varying variance calculation algorithm. Combined with the idea of sliding window, this paper studies the calculation of sample variance of this type of data series and deduces its recursive formula. And apply it to high-speed real-time varying one-dimensional Gaussian noise. The time-varying variance recursive calculation through the sliding window improves the efficiency of analyzing the one-dimensional Gaussian noise and reduces the storage resources of the computer.

2 The Dynamic Recursion Formula of Variance

2.1 One-Step Recursive Calculation of Time-Varying Variance

According to the knowledge of probability and statistics, the formula for calculating the variance of the sample is

$$V = E\left((X - \mu)^2\right). \tag{1}$$

Deformation of the above formula to obtain the calculation formula of the origin moment

$$V = E\left((X - \mu)^2\right) = E(X^2) - 2E(X)\mu + \mu^2 = E(X^2) - \mu^2. \tag{2}$$

Where $\mu = \frac{1}{n}\sum_{i=1}^{n} x_i$ and $E(X^2)$ represent the second central moment of the sample.

Consider a sequence $A = (a_1, a_2, \cdots, a_M)$ with a fixed length of M, the sample mean of the sequence at time k

$$\mu_{k,1} = \frac{1}{M}\sum_{i=1}^{M} a_i. \tag{3}$$

Similarly, the sample mean of the sequence at time $k + 1$

$$\mu_{k+1,1} = \frac{1}{M}\left(\sum_{i=1, j\neq i}^{M} a_i + \tilde{a}_j\right). \tag{4}$$

Where $\tilde{a}_j$ represents the j-th data of the sequence $A = (a_1, a_2, \cdots, a_M)$ at time $k + 1$, and is used to replace the a_j of the sequence at time k.

From the above Eqs. (3) and (4), the recursive formula of the sample mean is obtained after the sequence $A = (a_1, a_2, \cdots, a_M)$ is changed twice before and after.

$$\mu_{k+1,1} = \frac{1}{M}\left(\sum_{i=1, j\neq i}^{M} a_i + \tilde{a}_j\right) = \frac{1}{M}\left(\sum_{i=1}^{M} a_i + \tilde{a}_j - a_j\right) = \mu_{k,1} + \frac{\tilde{a}_j - a_j}{M}, (1 \leq j \leq M). \tag{5}$$

The formula for calculating second origin moments of the sample sequence $A = (a_1, a_2, \cdots, a_M)$ at time k is

$$\mu_{k,2} = \frac{1}{M} \sum_{i=1}^{M} a_i^2. \tag{6}$$

The formula for calculating second origin moments at time $k+1$ is

$$\mu_{k+1,2} = \frac{1}{M} \left(\sum_{i=1, j \neq i}^{M} a_i^2 + \tilde{a}_j^2 \right). \tag{7}$$

By using the recursive idea of the sample mean above, the second origin moments recursion formula of the sample sequence $A = (a_1, a_2, \cdots, a_M)$ can be obtained as

$$\mu_{k+1,2} = \frac{1}{M} \left(\sum_{i=1, j \neq i}^{M} a_i^2 + \tilde{a}_j^2 \right) = \frac{1}{M} \left(\sum_{i=1}^{M} a_i^2 + \tilde{a}_j^2 - a_j^2 \right)$$

$$= \mu_{k,2} + \frac{\tilde{a}_j^2 - a_j^2}{M}, (1 \leq j \leq M). \tag{8}$$

It can be known from the above formula (2) that the sample variance of the sample sequence $A = (a_1, a_2, \cdots, a_M)$ at time k is

$$V_k = \mu_{k,2} - \left(\mu_{k,1} \right)^2. \tag{9}$$

The sample variance at time $k+1$ is

$$V_{k+1} = \mu_{k+1,2} - \left(\mu_{k+1,1} \right)^2. \tag{10}$$

The difference between Eqs. (9) and (10) above can be obtained

$$V_{k+1} - V_k = \mu_{k+1,2} - \mu_{k,2} + \left(\mu_{k,1} \right)^2 - \left(\mu_{k+1,1} \right)^2$$

$$= \mu_{k,2} + \frac{\tilde{a}_j^2 - a_j^2}{M} - \mu_{k,2} + \left(\mu_{k,1} \right)^2 - \left(\mu_{k,1} + \frac{\tilde{a}_j - a_j}{M} \right)^2 \tag{11}$$

$$= \frac{\tilde{a}_j - a_j}{M} \left(\tilde{a}_j + a_j - 2\mu_{k,1} - \frac{\tilde{a}_j - a_j}{M} \right), (1 \leq j \leq M).$$

Through the formula (11), a one-step recursive formula for time-varying variance can be obtained

$$V_{k+1} = V_k + \frac{\tilde{a}_j - a_j}{M} \left(\tilde{a}_j + a_j - 2\mu_{k,1} - \frac{\tilde{a}_j - a_j}{M} \right), (1 \leq j \leq M). \tag{12}$$

2.2   Multi-step Recursive Calculation of Time-Varying Variance

From the one-step recursion formula, we can extend the recursive calculation of sample variance to multi-step recursion. Suppose the mean value of sequence $A = (a_1, a_2, \cdots, a_M)$ at time k can be calculated by formula (3). Replace $m - p + 1$ points data in sequence $A = (a_1, a_2, \cdots, a_M)$ with sequence $\tilde{A} = (\tilde{a}_1, \tilde{a}_2, \cdots, \tilde{a}_{m-p+1})$. Let the index of the replaced data in sequence $A = (a_1, a_2, \cdots, a_M)$ be $R = [p, p+1, \cdots, m]$, $(1 \leq m \leq M, 1 < p < m \leq M)$.

Let sequence $A = (a_1, a_2, \cdots, a_M)$ replace a part of the data with sequence $\tilde{A} = (\tilde{a}_1, \tilde{a}_2, \cdots, \tilde{a}_{m-p+1})$, and get the sample mean at time $k + m$

$$\mu_{k+m,1} = \frac{1}{M} \left(\sum_{\substack{i=1, j \notin R}}^{M} a_i + \sum_{q=1}^{m-p+1} \tilde{a}_q \right). \tag{13}$$

From the one-step recursive formula of the previous sample mean, the multi-step recursive formula for the mean value of the sequence sample can also be obtained as

$$\mu_{k+m,1} = \frac{1}{M} \left(\sum_{\substack{i=1, j \notin R}}^{M} a_i + \sum_{q=1}^{m-p+1} \tilde{a}_q \right) = \frac{1}{M} \left(\sum_{i=1}^{M} a_i + \sum_{q=1}^{m-p+1} \tilde{a}_q - \sum_{l=p}^{m} a_l \right)$$
$$= \mu_{k,1} + \frac{\sum_{q=1}^{m-p+1} \tilde{a}_q - \sum_{l=p}^{m} a_l}{M}. \tag{14}$$

With the second central moment recursive process of the previous sequence $A = (a_1, a_2, \cdots, a_M)$, the second central moment multi-step recursion formula of the sequence can be obtained by the same reason

$$\mu_{k+m,2} = \frac{1}{M} \left(\sum_{\substack{i=1, j \notin R}}^{M} a_i^2 + \sum_{q=1}^{m-p+1} \tilde{a}_q^2 \right) = \frac{1}{M} \left(\sum_{i=1}^{M} a_i^2 + \sum_{q=1}^{m-p+1} \tilde{a}_q^2 - \sum_{l=p}^{m} a_l^2 \right)$$
$$= \mu_{k,2} + \frac{\sum_{q=1}^{m-p+1} \tilde{a}_q^2 - \sum_{l=p}^{m} a_l^2}{M}. \tag{15}$$

Let the sample variance of sequence $A = (a_1, a_2, \cdots, a_M)$ at time k be calculated by the above formula (9), then the formula for calculating the variance of the sample at time $k + m$ is

$$V_{k+m} = \mu_{k+m,2} - (\mu_{k+m,1})^2. \tag{16}$$

From the formula of the sample variance origin moment, after the difference between the above Eqs. (9) and (16), the multi-step variance recursion formula of the sequence $A = (a_1, a_2, \cdots, a_M)$ can be obtained as

$$V_{k+m} - V_k = \mu_{k+m,2} - \mu_{k,2} + \left(\mu_{k,1}\right)^2 - \left(\mu_{k+m,1}\right)^2$$

$$= \mu_{k,2} + \frac{\sum\limits_{q=1}^{m-p+1} \tilde{a}_q^2 - \sum\limits_{l=p}^{m} a_l^2}{M} - \mu_{k,2} + \left(\mu_{k,1}\right)^2 - \left(\mu_{k,1} + \frac{\sum\limits_{q=1}^{m-p+1} \tilde{a}_q - \sum\limits_{l=p}^{m} a_l}{M}\right)^2$$

$$= \frac{\sum\limits_{q=1}^{m-p+1} \tilde{a}_q^2 - \sum\limits_{l=p}^{m} a_l^2}{M} - 2\mu_{k,1} \frac{\sum\limits_{q=1}^{m-p+1} \tilde{a}_q - \sum\limits_{l=p}^{m} a_p}{M} - \left(\frac{\sum\limits_{q=1}^{m-p+1} \tilde{a}_q - \sum\limits_{l=p}^{m} a_l}{M}\right)^2$$

$$= \frac{\sum\limits_{l=1,q=1}^{m-p+1} \left(\tilde{a}_q^2 - a_l^2\right)}{M} \quad 2\mu_{k,1} \frac{\sum\limits_{l=1,q=1}^{m-p+1} \left(\tilde{a}_q \quad a_l\right)}{M} - \frac{\left(\sum\limits_{l=1,q=1}^{m-p+1} \left(\tilde{a}_q \quad a_l\right)\right)^2}{M^2}.$$

$$(17)$$

According to Eq. (17) above, the final expression of multi-step variance recursive calculation can be written as

$$V_{k+m} = V_k + \frac{\sum\limits_{l=1,q=1}^{m-p+1} \left(\tilde{a}_q^2 - a_l^2\right)}{M} - 2\mu_{k,1} \frac{\sum\limits_{l=1,q=1}^{m-p+1} \left(\tilde{a}_q - a_l\right)}{M} - \frac{\left(\sum\limits_{l=1,q=1}^{m-p+1} \left(\tilde{a}_q - a_l\right)\right)^2}{M^2}.$$

$$(18)$$

3   Variance Analysis of Gaussian Noise

In recent years, with the advancement of society and the rapid development of digital information, various random signals have been widely used in daily life. However, due to the imperfect performance of various signal processing devices, these random signals are susceptible to various types of noise during acquisition, transmission, and storage. Direct processing of noisy signals will affect subsequent processing such as feature recognition and classification. The purpose of signal denoising is to retain the original signal information as much as possible while removing the noise. Gaussian noise is particularly common in actual signals. Gaussian noise is a noise with a probability distribution function of a normal distribution (also known as a Gaussian distribution). In other words, the value of a gaussian noise follows a gaussian distribution or its energy at each frequency component has a gaussian distribution. The normal sinusoidal signal is shown in Fig. 1 below, and the Gaussian noise is added to the sinusoidal signal in Fig. 2.

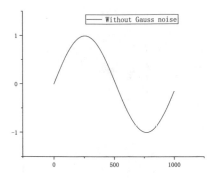

Fig. 1. Sinusoidal signal without Gauss noise.   **Fig. 2.** Sinusoidal signal in Gaussian noise.

Gaussian noise is determined by two parameters, mean and variance. Therefore, when filtering random signals of Gaussian noise, it is necessary to perform variance analysis and mean calculation. In high-speed information processing and analysis, the real-time performance of computing results is often required. If this kind of signal data is subject to Gaussian noise, when carrying out variance analysis and mean value calculation on it, if the calculation is carried out according to the definition of variance and mean, it often fails to meet the real-time requirements. Therefore, this paper proposes to use the above recursive formula to calculate the difference and mean, which can not only improve the calculation efficiency and reduce the workload of the computer to read the data in the disk, but also meet the real-time requirements of the calculation results and improve the noise reduction efficiency.

In order to verify the practicability and correctness of the proposed algorithm, a 1×1000000 Gaussian noise data sequence was randomly generated using professional statistical software, and the sequence was used for numerical simulation experiments. The idea of arithmetic progression is used to simulate Gaussian noise with real-time high-speed changes. That is, the number of data to be replaced dynamically each time is an arithmetic sequence with 10000 as the first term and 10000 as the tolerance, 100 groups in total. The calculation time consumed each time at different scales is recorded and compared with the definition algorithm of variance. Figure 3 is the comparison between the variance definition algorithm and the recursive algorithm in the time required to calculate the variance of the Gaussian noise under different input scales.

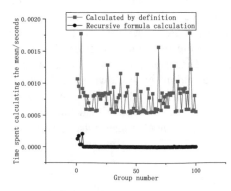

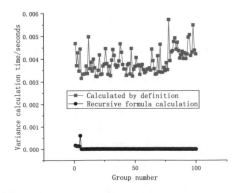

Fig. 3. Mean calculation time consumption under different change scales.

Fig. 4. Difference calculation time consumption under different change scales.

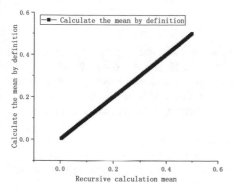

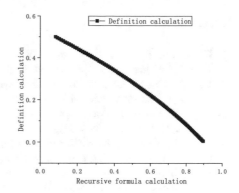

Fig. 5. Mean calculation error under different scales.

Fig. 6. Difference calculation error under different change scales.

Figures 5 and 6 is an error bar graph of the calculation results of the mean and variance of 100 Gaussian noise in the defined calculation mode and the recursive calculation mode. It can be clearly seen that with the continuous expansion of the input scale, the mean and variance of the Gaussian noise are consistent with the calculation method of the full sample (that, the definition calculation method) using the real-time variance recursion formula proposed in this paper. Moreover, as can be seen from Figs. 3 and 4 above, the recursive algorithm proposed in this paper significantly improves the calculation efficiency of variance and mean value of real-time high-speed Gaussian noise, shortens the calculation time and saves the computational storage resources. This verifies the correctness and practicability of the proposed algorithm.

4 Conclusion

The variance analysis of Gaussian noise based on traditional definition of variance and mean would require too much time in a high speed, massive data and real-time situation. A dynamic recursive algorithm for the variance analysis of Gaussian noise is proposed in this paper and it can greatly improve the computational efficiency and decrease the calculation time. The proposed recursive algorithm demonstrates good efficiency in the variance analysis of the Sinusoidal signal in Gaussian noise.

Acknowledgments. This paper has been supported by the Doctoral Research Foundation of Southwest University of Science and Technology (No. 16zx7108, No. 15zx7151. No.15zx7118), the Educational Reform Research Project of Southwest University of Science and Technology (No. 15xnzd05) and the Undergraduate Innovation Fund Project Accurate Funding Special Project by Southwest University of Science and Technology (No. JZ19-057).

References

1. Chen, A.S., Chang, H.C., Cheng, L.Y.: Time-varying variance scaling: application of the fractionally integrated ARMA model. N. Am. J. Econ. Financ. **47**, 1–12 (2019)
2. Chu, A.M.Y., Li, R.W.M., So, M.K.P.: Bayesian spatial–temporal modeling of air pollution data with dynamic variance and leptokurtosis. Spat. Stat. **26**, 1–20 (2018)
3. Yang, Y., Wang, S.: Two simple tests of the trend hypothesis under time-varying variance. Econ. Lett. **156**, 123–128 (2017)
4. Wang, J., Wang, L., He, X.: The study of high accuracy time keeping based on FPGA when navigation satellite losing connection. Chin. J. Electron Devices **39**(01), 140–143 (2016)
5. Liu, J., Sun, X.M.: Real-time calculation method of non-zero mean measurement and control data variance. Math. Res. Study **21**, 148–150 (2016)
6. Chen, H.R., Gao, H.: A unified recurrence formula for k-order origin moments of common discrete distributions. J. Hunan Bus. Coll. **02**, 110–111 (2002)
7. Deng, H.B., Liu, J.F., Wang, Y.N.: Recursive algorithm of mean variance and its application. Comput. Mod. **04**, 9–11 (1996)
8. Chen, Q.B.: Recursive formula of k-order origin moments of discrete distribution. Stat. Inf. Forum **01**, 26–28 (2000)
9. Chen, Q.B., Ma, Y.H.: Recursive formula of k-order central moment of discrete distribution. Stat. Inf. Forum **02**, 36–38+72 (1999)
10. Chen, C.G.: The counting method of k-order central moment of three kinds of discrete random variable. J. Jianghan Univ. (Nat. Sci.) **06**, 66–68 (2000)
11. Fan, S.F.: Recurrence formula calculating the statistics for changed sample size. J. Inn. Mong. Coll. Agric. Anim. Husb. **01**, 153–162 (1990)

Database Watermarking Algorithm Based on Difference Expansion

HaoYang Gao and De Li[(⊠)]

Department of Computer Science, Yanbian University, Yanji, China
leader1223@ybu.edu.cn

Abstract. Database watermarking technology is an important means to protect database information security. Firstly, the key components of the database are divided into several sub-groups by using the key hash function. Then, the size of the Pearson correlation coefficient is used in the sub-group to be selected the attribute column to be embedded in the watermark, and then the threshold is set by the differential expansion algorithm to determine the final embedded watermark. The pair of attributes, embedding in the virtual watermark repeatedly in the position that satisfies the condition. By setting the threshold, the embedding amount of the watermark is reduced, and the robustness of the watermark is improved. The loss of the original data of the database is guaranteed, and the watermark and the watermark extraction rate are better when attacked. When the attack ratio is low, the data recovery effect is better.

Keywords: Database watermark · Correlation coefficient · Hash function · Difference expansion algorithm

1 Introduction

Database watermarking technology is an effective method to protect database information security. Last few years, it has also achieved fruitful research results. However, compared with multimedia data, due to the structure of its specification and sensitivity to minor changes, there are problems such as small embeddable redundancy and narrow application range [1]. Therefore, there are still many areas for improvement in database the watermarking technology, and it is still a relatively new technical means, which deserves further research and exploration. R. Agrawal proposed the first relational database watermarking algorithm in 2002 [2]. R. Sion proposes a watermarking algorithm based on secret sorting [3]. Yong Zhang, Dongning Zhao, Deyi Li proposes a watermarking algorithm that can be used to implement database copyright protection [4], which can embed meaningful information in the database. Xia mu Niu, liang Zhao, Wenjun Huang proposes Reversible database watermarking algorithm based on time difference and difference expansion [5]. The Jawad and Khan used genetic algorithm with difference expansion watermarking to propose a robust and reversible database watering approach called GADEW [6]. The method used DEW technique to embed the watermark bits into the database of a reversible manner. Ftikhar S proposed a robust reversible watermarking technique [7], which first uses data from mutual information algorithms. G. Gupta implements the watermark embedding operation by applying the differential extension technique proposed by Alattar scholars to the relational database [8]. Zhang proposed a reversible watermarking scheme [9], which uses the histogram expansion method to implement reversible watermarking. Scholar Zhou realizes the protection of database copyright by embedding bitmap image in relational database [10], This method uses error correction mechanism to correct the wrong watermark bit on watermark extraction and improve watermark robustness.

2   Related Theory and Technology

The details of the difference expansion watermarking technique, and Pearson correlation coefficient used in the proposed method are given in this section.

2.1   Relevance Determines the Attribute Column to Be Embedded

Correlation between a_j and a_k any two columns of attribute vectors $cor_{j,k}$ in the database, Expressed by the Pearson correlation coefficient, the formula is as shown in (1):

$$cor(j,k) = \frac{a_j a_k - \frac{1}{v}\sum_{i=1}^{v} a_k(i)}{\sqrt{\|a_j\|_2^2 - (\frac{1}{v}\sum_{i=1}^{v} a_j(i))^2}\sqrt{\|a_k\|_2^2 - (\frac{1}{v}\sum_{i=1}^{v} a_k(i))^2}} \tag{1}$$

Among them $\|a_j\|_2^2 = |a_j(1)|^2 + |a_j(2)|^2 + \ldots + |a_j(v)|^2$ Evaluate the attribute column correlation of the database by correlation calculation. The v Indicates the number of elements in the data column a_j, a_k Represents two columns of data, The $a_j(i), a_k(i)$ representing values corresponding to the two columns of data.

Determining the attribute column of the database to be embedded by correlation can improve safety performance. The concealment of the algorithm is also improved.

2.2   Differential Extension Technique

Difference Expansion (DE) was proposed by Jun Tian et al. [10] in 2002 and is mainly used in the field of image processing.

Relational database, Where pk is the primary key, The A_j $(1 \leq j \leq n)$ is jth row numeric attribute, The R_i $(1 \leq i \leq m)$ is of Tuples in the database, The $r_i A_j$ represents value of r_i attribute in A_j tuple.

Step 1: Assume that the two attribute values of the database to be embedded in the watermark A_1, A_2.

$$avg = \left\lfloor \frac{|A_1 + A_2|}{2} \right\rfloor \tag{2}$$

$$d = A_1 - A_2 \tag{3}$$

The operator $\lfloor . \rfloor$ indicates a rounding down operation.

Step 2: According to formula (3), calculate the attribute values $\tilde{A}_1$ and $\tilde{A}_2$ after embedding the watermark bit b and obtaining the embedded watermarking, and the difference between the modified attribute values is given as:

$$\tilde{d} = 2d + b \tag{4}$$

$$\tilde{A}_1 = avg + \left\lfloor \frac{\tilde{d}+1}{2} \right\rfloor \tag{5}$$

$$\tilde{A}_2 = avg - \left\lfloor \frac{\tilde{d}}{2} \right\rfloor \tag{6}$$

Step 3: Calculating the average value of the modified attribute value according to the following formula (7) and (8):

$$avg' = \left\lfloor \frac{\tilde{A}_1 + \tilde{A}_2}{2} \right\rfloor \tag{7}$$

$$\tilde{S} = \tilde{A}_1 - \tilde{A}_2 \tag{8}$$

Step 4: Extract the watermark bit b using the attribute difference according to the following formula (9):

$$b = \tilde{S} - \left\lfloor \frac{\tilde{S}}{2} \right\rfloor \tag{9}$$

Step 5: Restore the original values A_1, A_2.

The Pearson correlation coefficient is used to determine the attribute column to be extended, and then the size of the change after embedding the watermark by the difference expansion establishes the objective function as Eq. (10):

$$\min obj = \left(\sum_{k=1}^{row_w} \left| TA_k^x - CA_k^x \right| + \left| TA_k^y - CA_k^y \right| \right) \cdot w_a + (row_w).w_c \big/ row_{total} \tag{10}$$

Where $x \in (1, \cdots, M), y \in (1, \cdots, M)$ and $x \neq y$.

Assume that the grouped DS after the group contains the row_{total} tuples, Each row of tuples contains M attributes, and the objective function row_w refers to the number of rows that can be embedded the watermark in the non-distortion range.

The TA_k^x refers to the data value of the k row x column in the original database, and CA_k^x refers to the k row x column data value of the watermark added after applying DEW technology. The objective function value can be divided into two parts. The first part is the absolute value of the sum of all data values after embedding the watermark; the second part is the watermark capacity value. w_a, w_c respectively indicate the weight value of the two parts in the objective function, and $w_a + w_c = 1$, selects the minimum target value on the premise of large capacity. In turn, a pair of attribute columns that are finally embedded in the watermark can be selected. The w_a, w_c are determined by humans.

3 Algorithm Process Description

3.1 Watermark Embedding Algorithm

Figure 1 Watermark embedding flow chart. The specific steps are as follows:

Step 1: Use the MD5 method to find the hash value of the database primary key value and key, and hash the value on the watermark length to group the database tuple.
Step 2: Select the column with the smallest coefficient according to the Pearson correlation coefficient in the subgroup as the extended attribute.
Step 3: Select the lowest digit of the attributes.
Step 4: embedding the watermark by the difference expansion algorithm.
Step 5: Set the threshold [0, 9] to calculate the capacity of the embedded watermark.
Step 6: Determine a pair of attribute columns of the final embedded watermark by the objective function.

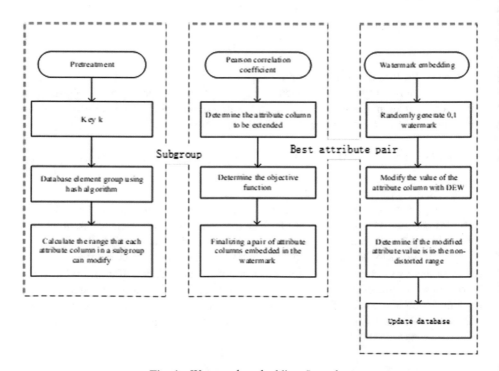

Fig. 1. Watermark embedding flow chart

3.2   Watermark Extraction and Data Recovery Process

Step 1: The same as the embedding process, first find the hash value, then group the tuples.

Step 2: extract the least significant bit of a pair of attribute values determined at the time of embedding, and use the DEW algorithm to find the last bit of the target attribute column, and extract the pair simultaneously while satisfying [0, 9], and the others remain unchanged.

Step 3: Extract the watermark embedded in a set.

Step 4: the watermark extracted from each group uses the majority election method to determine the extraction bit of the watermark.

Step 5: Use the differential extension algorithm to recover the embedded watermark and the original data values.

4   Experimental Results and Analysis

The watermark of this experiment is randomly generated 29 binary bits of 0 and 1. The hash function uses MD5 and the key is 15. The algorithm repeatedly embeds the watermark into different tuples in the same subgroup, and the number of tuples in each group is about 68. According to the requirements of this algorithm, the average number of embedded in each group is 40. Since the primary key values are randomly generated, the number of tuples in each group and the number of embedded watermarks are different. Figure 2 shows the number of tuples in each group and the numbers of watermarks embedded in the group.

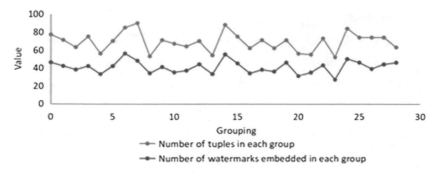

Fig. 2. The number of tuples in each group and the number of watermarks embedded in each group

4.1   Attack Experiment and Algorithm Robustness

4.1.1   Tuple Deletion Attack Experiment

We can see from Fig. 3 that when the tuple deletion attack tuple is less than 50%, the watermark can be completely extracted, and the data recovery rate is also about 70%.

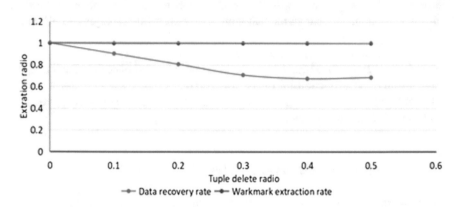

Fig. 3. Tuple deletion attack data recovery rate and watermark extraction rate

4.1.2   Tuple Change Attack Experiment

We can see from Fig. 4 that when the tuple change attack is less than 50% watermark can be fully extracted, the data recovery rate of at least 80%.

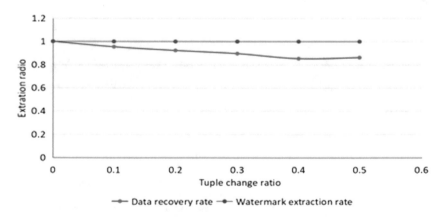

Fig. 4. Tuple change attack the watermark extraction rate and data recovery rate

4.1.3   Tuple Addition Attack Experiment

We can see from Fig. 5 that when the tuple addition attack, watermark extraction rate and the data recovery rate are both 100%.

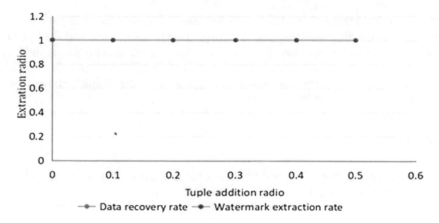

Fig. 5. Tuple addition attack watermark extraction rate and data recovery rate

4.1.4   Tuple Sorting Attack Experiment

We can see from the Fig. 6 that the tuple order attack watermark extraction rate and the recovery rate of the last bit of the data are both 100%.

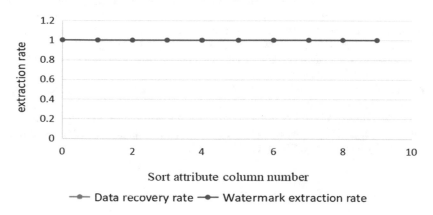

Fig. 6. Tuple sort attack watermark extraction rate and data recovery rate

5   Conclusion

In this paper, the watermark difference expansion algorithm is used to map the virtual embedded watermark 0 or 1 in database, and the tuple is grouped by the hash algorithm. The embedding capacity of 2000 tuples is about 1200. when the tuple deletion attack is less than 50%, the watermark can be completely extracted, and the data recovery rate is also about 70%. When the tuple change attack is less than 50% watermark can be fully extracted, the data recovery rate of at least 80%., Adding tuple attack and sorting attack has no effect using this method, the watermark extraction rate and the data recovery rate are 100%, and the packet is embedded multiple times. The robustness of database watermarking is strong. This article can recover the last bit of data, and the error becomes lower, achieving a low difference from the original data.

Acknowledgments. This research project was supported by the National Natural Science Foundation of China (Grant No. 61262090).

References

1. Tsai, P., Hu, Y., Yeh, H.: Reversible image hiding scheme using predictive coding and histogram shifting. Sig. Process. **89**(6), 1129–1143 (2009)
2. Voyatzis, G., Piatas, I.: The use of watermarks in the protection of digital multimedia products. Proc. IEEE **87**(7), 1197–1207 (1999)
3. Zhang, M.: Current status and prospects of database security research. Proc. Chin. Acad. Sci. **26**(3), 303–309 (2011)
4. Feng, D.G., Zhang, M., Li, W.: Big data security and privacy protection. Chin. J. Comput. **37**(1), 246–258 (2014)
5. Mehta, B.B., Aswar, H.D.: Watermarking for security in database: a review. In: IT in Business, Industry and Government, Indore, India, pp. 1–6. IEEE (2014)
6. Jawad, K., Khan, A.: Genetic algorithm and difference expansion based reversible watermarking for relational databases. J. Syst. Softw. **86**(11), 2742–2753 (2013)
7. Iftikhar, S., Kamran, M., Anwar, Z.: A robust and reversible watermarking technique for relational data. IEEE Trans. Knowl. Data Eng. **27**(4), 1132–1145 (2015)
8. Jun, T.: Reversible data embedding using a difference expansion. IEEE Trans. Circ. Syst. Video Technol. **13**, 890–896 (2003)
9. Zhang, Y., Yang, B., Niu, X.M.: Reversible watermarking for relational database authentication. J. Comput. **17**(2), 59–66 (2006)
10. Zhou, X., Huang, M., Peng, Z.: An additive-attack-proof watermarking mechanism for databases copyrights protection using image. In: Proceedings of the ACM Symposium on Applied Computing (SAC 2007), pp. 254–258 (2007)

Optical Fiber Pressure Sensor Based on Corrugated Diaphragm Structure

Lin Zhao[✉], Jiqiang Wang, Long Jiang, and Lianqing Li

Laser Institute, Qilu University of Technology (Shandong Academy of Sciences),
Jinan 250103, China
linzhao1225@126.com

Abstract. Aiming at the current pressure monitoring requirement in the energy and chemical industry, combined with the principle of fiber double grating sensing, a fiber optic pressure sensor is designed and developed to meet practical engineering applications. Using corrugated diaphragm pressure sensor structure, the sensor monitoring sensitivity is improved, reducing the monitoring error caused by pressure multi-stage conduction and material expansion and contraction of the traditional diaphragm pressure sensor, and improving the overall stability of the sensor. At the same time, the sensor adopts double grating temperature compensation model, which solves the cross-sensitivity of temperature to pressure measurement and realizes temperature self-compensation of pressure measurement. The experimental results show that the sensor monitoring sensitivity is 0.078 kPa/pm, the monitoring error is <0.5 kPa, and it has good linearity and long-term reliability. It is especially suitable for pressure monitoring of flammable and explosive environments such as petroleum, coal and chemical industry, and has good application prospects.

Keywords: Fiber Bragg grating · Pressure sensor · Corrugated diaphragm · Linearity · Reliability

1 Introduction

With the rapid development of petroleum, chemical, coal and metallurgy industries in China, the flammable and toxic gas leakage accidents occur frequently in the production, storage and transportation of enterprises. The traditional monitoring and early warning system uses the sensing terminal-electronic sensor, which has short service life and easy to drift. Especially the sensor itself is charged. It is not an intrinsically safe sensing device. It introduces new security risks while monitoring hidden dangers [1]. Therefore, it is more and more important to use reliable and safe sensors to monitor the pressure of inflammable and explosive environment such as urban underground natural gas and oil pipelines.

At present, there are mainly two types of general optical fiber pressure sensors based on F-P cavity and fiber Bragg grating. In the fiber-optic sensor with F-P cavity as the sensitive component, the capillary structure glass is difficult to weld, the drift is large, the repeatability is poor, and the service life is short [2, 3]. However, there are

© Springer Nature Switzerland AG 2020
F. Xhafa et al. (Eds.): IISA 2019, AISC 1084, pp. 741–747, 2020.
https://doi.org/10.1007/978-3-030-34387-3_91

some shortcomings in the micro-electro-mechanical chip pressure sensor, such as the difficulty of optical fiber coupling, encapsulation and so on [4–6].

Pressure sensor in the form of fiber grating structure, is mainly through fiber grating as a sensitive component for pressure detection [7]. Among them, the Bourdon tube pressure sensor has poor resolution, and the Bourdon tube itself has high processing precision, the fiber grating package is difficult, and the drift is large [8, 9]. Thin-walled cylindrical structures have disadvantages such as poor resolution and low detection sensitivity [10]. The traditional elastic diaphragm package structure is affected by factors such as metal expansion and rubber packaging. The long-term drift problem of the sensor is outstanding and the stability is poor.

Aiming at the above problems, this paper proposes a fiber grating pressure sensor based on corrugated diaphragm, which adopts double grating temperature compensation model and simplified pressure conduction structure, effectively reduces the interference of metal and packaging factors on the sensor, and has high sensitivity and long-term reliability [11]. At the same time, since the optical fiber sensor itself is not charged and intrinsically insulated, it is more suitable for pressure monitoring in the case of dangerous situations such as inflammable and explosive.

2   Sensor Basic Structure

Combined with the pressure monitoring needs in the flammable and explosive environment of petroleum, chemical, coal, metallurgical and other industries, the structure of the fiber optic pressure sensor is designed as shown in the figure (Fig. 1):

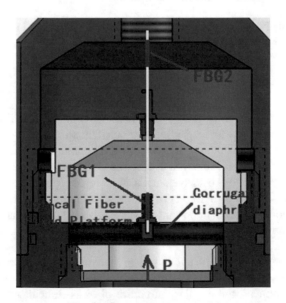

Fig. 1. Sensor structure model

When the gas or liquid is filtered through the filter mesh inside the joint, it reaches the bottom of the corrugated diaphragm through the base. The pressure change will cause the corrugated diaphragm to deform in the axial direction, the corrugated diaphragm deformation pushes the fiber-fixed platform to pull the grating FBG1 to deform, which causes the FBG1 reflection center wavelength to change. At the same time, the grating FBG2 is suspended without force, and monitors the change of ambient temperature in real time. According to the central wavelength variation of the two gratings FBG and the temperature compensation algorithm, the current internal pressure value of the sensor can be obtained.

3 Sensor Principle

It is known from the sensor structure and the sensitivity of the grating to temperature and strain. When the temperature and strain change simultaneously, the gratings FBG1 and FBG2 have the following specific effects:

$$\frac{\Delta \lambda_{B1}}{\lambda_{B1}} = (\alpha + \zeta)c\Delta T + (1 - P_e)\varepsilon_1 \tag{1}$$

$$\frac{\Delta \lambda_{B2}}{\lambda_{B2}} = (\alpha + \zeta)\Delta T \tag{2}$$

Among them, λ_{B1} is the central wavelength of FBG1, λ_{B2} is the central wavelength of FBG2, and c is the influence coefficient of sensor structure on the wavelength change of fiber when temperature changes. From formula (1) and formula (2):

$$\varepsilon_M = \varepsilon_1 = \frac{\left(\frac{\Delta \lambda_{B1}}{\lambda_{B1}} - c\frac{\Delta \lambda_{B2}}{\lambda_{B2}}\right)}{1 - P_\varepsilon} \tag{3}$$

When the pressure rises or falls, the liquid pressure on the elastic diaphragm changes. The flexural strain ε_M can be simplified as the deformation of the elastic diaphragm inside the sensor multiplied by a coefficient.

$$\varepsilon_M = k \times X \tag{4}$$

In the formula, k is the coefficient and X is the deformation of strain. According to Hooke's law

$$F = K \times X \tag{5}$$

K is the stiffness coefficient of an object and X is the elastic deformation value. The stiffness coefficient of an object is related to the material, thickness and diameter of the elastic diaphragm. When the relevant size of the elastic diaphragm is determined, the stiffness coefficient K is also determined. According to the pressure conversion formula

$$F = P \times S_{\text{Diaphragm}} \tag{6}$$

From Formula (3), (4), (5), (6):

$$\varepsilon_M = \frac{\left(\frac{\Delta\lambda_{B1}}{\lambda_{B1}} - c\frac{\Delta\lambda_{B2}}{\lambda_{B2}}\right)}{1 - P_\varepsilon} = \frac{kPS_{\text{Diaphragm}}}{K} \tag{7}$$

The temperature coefficient C is calibrated after the sensor is encapsulated, and P, λ_B, λ_{B1}, λ_{B2} is calibrated by high precision pressure transmitter, pressure gauge and optical fiber demodulator. Calculate $kS_{\text{Diaphragm}}/K$, and $kS_{\text{Diaphragm}}/K$ is a constant, so we can directly monitor the pressure (pressure) change by demodulating the wavelength change.

4   Sensor Design

The wavelength variation of the fiber center is related to the thickness of the diaphragm, the diameter, and the pressure of the liquid. To ensure the design range, the diaphragm should not be too large or too thin in practical applications, too large, not easy to package, too thin, not easy to machine, but also easy to cause grating breakage.

Due to the low sensitivity of the flat diaphragm, the corrugated diaphragm is used in the design process. The corrugated diaphragm is 0.2 mm high, the wave spacing is 1.05 mm, the diaphragm thickness is 0.1 mm, and the diameter is 30 mm. The simulation test results are shown in Fig. 2, and sensor prototype as shown in Fig. 3.

Fig. 2. Corrugated diaphragm strain simulation   **Fig. 3.** Fiber optic pressure sensor

5   Experimental Test

5.1   Calibration Experiment

The sensor calibration adopts standard piston liquid pressure gauge developed by Xi'an Institute of Special Instruments. The pressure range is 0–1 MP, and the measuring accuracy is 0.02% FS. The experimental device is shown in the figure. The center wavelength is demodulated using the FBG demodulator produced by Shandong Micro-Sensor Electronics Co., Ltd. with a minimum resolution of 1 pm (Figs. 4 and 5).

Fig. 4. Sensor calibration experiment device

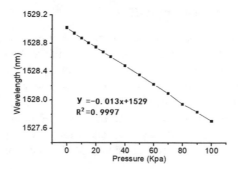

Fig. 5. Sensor linearity curve

During the experiment, adjust the pressure value applied to the sensor, record center wavelength of the sensor pressure grating, and establish the corresponding relationship between pressure and wavelength. The experimental test sensor calibration curve is shown in the Figure. The experimental results show that the new type of optical fiber pressure sensor has a monitoring sensitivity of 0.078 kPa/pm and good linearity.

5.2   Stability Test

After the sensor calibration is completed, input the sensor coefficient into the software, adjust the pressure of the piston pressure gauge, make a difference between the fiber pressure sensor display value and the piston standard pressure value, and test the error between the measured value of the pressure sensor and the standard value. The test result is shown in the figure (Fig. 6).

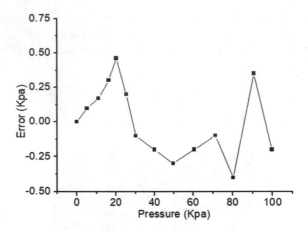

Fig. 6. Sensor monitoring error curve

From the test data, it can be seen that the maximum measurement error of the optical fiber pressure sensor in the full range is <0.5 kPa.

5.3   Long-Term Environmental Testing

The sensor is placed in a water tank with a depth of about 1.6 m. The water tank is placed outdoors, the water temperature changes with the ambient temperature. Adjust sensor parameters, monitor sensor pressure in real time, and test sensor adaptability to the environment and long-term reliability. The experimental results show that the sensor monitors the pressure change by 1 kPa and the water level drops by about 10 cm within one month. Considering the influence of natural conditions such as evaporation of water during one month, the sensor monitors the pressure change within the allowable range of error, so the sensor has good environmental adaptability and long-term stability (Fig. 7).

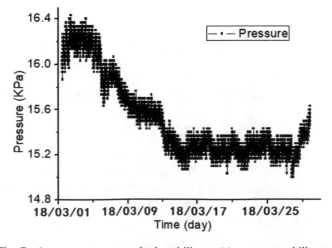

Fig. 7. Sensor environmental adaptability and long-term stability test

6 Conclusion

This paper designs and develops a fiber optic pressure sensor suitable for practical engineering applications in the fields of petroleum, chemical, pipeline transportation and other fields, and the overall characteristics of the sensor were deduced and tested in detail. The results show that the fiber grating pressure sensor based on corrugated diaphragm has the advantages of high monitoring sensitivity, good linearity and repeatability, high environmental adaptability and long-term reliability. At the same time, the fiber sensor is intrinsic insulated, especially suitable for flammable and explosive environments, has good application prospects.

Acknowledgments. This work was financially supported by Shandong key research and development plan (2017GSF20102) and Shandong Natural Science Foundation (ZR2016QZ006).

References

1. Jun, H.: Development and Application of Fiber Bragg Grating. Master's Degree Thesis of Wuhan University of Technology (2013)
2. Chen, L., Zhu, J., Li, Z., Wang, M.: Optical fiber fabry-perot pressure sensor using corrugated diaphragm. Acta Optica Sinica **36**(3), 0306002 (2016)
3. Zhang, W., Jiang, J., Wang, S.: Fiber-optic fabry-perot high-pressure sensor for marine applications. Acta Optica Sinica **37**(2), 0206001 (2017)
4. Ge, Y., Wang, M., Yan, H.: Optical MEMS pressure sensor based on a mesa-diaphragm structure. Opt. Express **16**(26), 21746–21752 (2008)
5. Wang, X., Li, B., Russo, O.L., et al.: Diaphragm design guidelines and an optical pressure sensor based on MEMS technique. Microelectron. J. **37**(1), 50–56 (2006)
6. Wang, Y., Wang, M., Ni, X., Xia, W.: An optical fiber MEMS pressure sensor using microwave photonics filtering technique. In: 25th International Conference on Optical Fiber Sensors, p. 1032368 (2017)
7. Liao, Y.: Optical Fiber of Light, pp. 197–202. Tsinghua University, Beijing (2000)
8. Fu, T.: Research on FBG Pressure Sensor With Temperature Compensation. Master's Degree Thesis of Chengdu University of Electronic Science and Technology (2014)
9. Li, X., Zhang, B., Yao, J., Hu, J.: Theoretical analyses on parameter option of a pressure fiber sensor. Chin. J. Sens. Actuators **1**, 133–135 (2004)
10. Tan, B.: A high sensitivity fiber grating pressure sensor. J. Optoelectron. Laser **23**(11), 2012–2105 (2012)
11. Gao, Y., Liu, C., Mu, H.: Pressure response of fiber Bragg grating based on plate diaphragm and equal-strength cantilever. Opt. Instrum. **36**(4), 333–336 (2014)

Operating Modal Analysis of the Turntable on a Working Road-Header Using DTCWT

Yuanyuan Qu[1(✉)], Xiaodong Ji[1], and Fuyan Lv[2]

[1] China University of Mining and Technology (Beijing), Beijing 100083, China
qyy2014@oumtb.edu.cn
[2] Shandong University of Science and Technology, Qingdao 266590, China

Abstract. In order to identify the main vibration modes of the turntable on a working road-header, the Duall-tree Complex Wavelet Transform (DTCWT) is carried out firstly over the acquired vibration data, to achieve frequency division and noise reduction. Operating modal analysis is then carried out using Auto-Regressive and Moving Average model (ARMA). This data processing composed of DTCWT and ARMA is proposed to monitor the state of the turntable in the aspect of vibration modes changing. The processing results based on vibration data from different test points and time periods show that this proposal is feasible for the assembled turntable. Also, according to the identified first 12 intrinsic frequencies, it is suggested to avoid drive signal at frequencies around 311 Hz, 425 Hz and 540 Hz to decrease mechanical noise and abrasion of the turntable.

Keywords: Road-header · Turntable · Vibration test · DTCWT · Operation modal analysis

1 Introduction

The cantilever type road-header is one of the main equipment in roadway shaping in coal mining industry in China. The turntable is one of the key components on the cantilever road-header. It is usually working under complex stress and vibration, thus it is easy to damage with invisible sudden crack [1]. Study on the dynamic characteristics of turntable is meaningful and helpful to optimal design or maintain long-term lifecycle of the road-header.

Modal analysis is an effective approach used in dynamic analysis of the large scale mechanical equipment. For the road-header working underground, due to the tough working environment and complicated operating conditions, real data acquisition for both inputs and vibration response are difficult to achieve. Literatures [2] and [3] are widely referred to when deal with analysis of road-header based on vibration test, however the data were measured under a simulated mining environment instead of real mining site underground.

To acquire data close to the real working situation as much as possible, our research group had carried out a series of measurements over the working road-header using a set of vibration and operating data acquisition devices especially for equipment working in mining underground. The shocks onto the working road-header are usually complex and hard to measure, therefore only vibration signal were obtained. This paper presents a set of operating modal analysis of the turntable using only the vibration response of the structure on assembled position. Since the main vibration modes of the turntable may locate widely along the frequency scale [4], to save data processing cost, it is reasonable to narrow down the frequency scale of the vibration data to a certain level. Vibration modes at lower frequency band are concerned firstly in this paper, therefore the Duall-tree Complex Wavelet Transform (DTCWT) is applied to the raw vibration signal for frequency division and noise reduction. Operating modal analysis then is carried out using Auto-Regressive and Moving Average model (ARMA).

2 Vibration Signal Acquisition

The test object is a cantilever road-header for hard rocks. The machine can shape a roadway with maximum cutting height of 5.8 m and width of 8.8 m. More details about data acquisition refer to the literature [5].

Due to the limitation of test conditions, only a representative set of test points over the surface of the road-header were taken. Some of which locating around the turntable are numbered as: No. 5 and No. 6 near the junction between the cutting arm and the turntable; No. 7 and No. 8 near the junction of the turntable and hydraulic cylinder for cutting arm lifting; No. 9 and No. 10 on the up-top of the turntable; No. 11 and No. 12 near the junction of the turntable and the main body of the road-header [6]. The vibration data from each test point last for a certain period of time as the road-header operates different actions, such as idling, swing and cutting, drilling, etc.

It is difficult to summarize the characteristics of the original vibration data by primary signal processing. To study the rare vibration data from the working machine as much as possible, an idea is to monitor the variation of the main vibration modes of the turntable for a very long period. To save processing cost, only the prominent natural frequencies of the static turntable within 700 Hz are concerned in this paper. Thus, it is necessary to remove components at higher frequency band from the raw vibration data and reduce noise appropriately before model identification.

3 Signal Decomposition and Reconstruction Using DTCWT

As a novel type of wavelet transform, the DTCWT is widely used due to its nice characteristic of low frequency aliasing, and proximate translation invariance [7]. DTCWT applies 2 wavelets at the same time over the real and imaginary parts, respectively, and forms a complex wavelet, which, could be presented as:

$$\varphi(t) = \varphi_h(t) + i\varphi_g(t) \tag{1}$$

According to the wavelet transform theory, the wavelet coefficients and scale coefficients of the real part of the DTCWT can be obtained by inner product operation, respectively:

$$dI_j^{\mathrm{Re}}(n) = 2^{j/2} \int_{-\infty}^{+\infty} x(\mathrm{t})\varphi_h(2^j\mathrm{t} - n)dt \,; cI_J^{\mathrm{Re}}(n) = 2^{J/2} \int_{-\infty}^{+\infty} x(\mathrm{t})\varphi_h(2^J\mathrm{t} - n)dt, \tag{2}$$
$$j = 1, 2, 3\ldots J$$

j is the scale factor and J is the decomposition level. Similarly, the wavelet coefficients $dI_j^{\mathrm{Im}}(n)$ and scale coefficients $cI_J^{\mathrm{Im}}(n)$ of the imaginary part can be obtained.

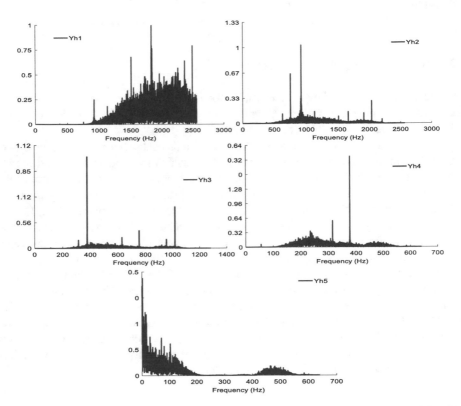

Fig. 1. Frequency spectrum of the single branch reconstructed signal (Yh1–Yh5) at sub-bands..

The decomposition using DTCWT is carried out along the real and imaginary parts side by side:

$$d_j^{\varphi}(n) = dI_j^{\mathrm{Re}}(n) + \mathrm{i} \cdot dI_j^{\mathrm{Im}}(n); c_j^{\varphi}(n) = cI_j^{\mathrm{Re}}(n) + \mathrm{i} \cdot cI_j^{\mathrm{Im}}(n), j = 1, 2, 3\ldots J \tag{3}$$

To reconstruction, the following formulation is used:

$$
\begin{aligned}
d_j(t) &= 2^{(j-1)/2}\left[\sum_n dI_j^{\mathrm{Re}}(n) \cdot \varphi_h(2^j\mathrm{t} - n) + \sum_n dI_j^{\mathrm{Im}}(n) \cdot \varphi_g(2^j\mathrm{t} - n)\right] \\
c_J(t) &= 2^{(J-1)/2}\left[\sum_n cI_J^{\mathrm{Re}}(n) \cdot \varphi_h(2^J\mathrm{t} - n) + \sum_n cI_J^{\mathrm{Im}}(n) \cdot \varphi_g(2^J\mathrm{t} - n)\right]
\end{aligned}
\tag{4}
$$

Single branch reconstruction is realized by keeping only one of the branches but set the coefficients of the rest to be zero during the reconstruction. Similarly, for the vibration data after decomposition using DTCWT, if the branches at lower frequency levels were taken while the ones at higher frequency levels were abandoned during reconstruction, the reconstructed data would contain only partial of the vibration modes.

In this paper, the signal acquisition frequency was 10 kHz, and the modal components within 700 Hz are concerned mainly, therefore, DTCWT with 4 levels is adopted. Therefore 8 decomposed branches are obtained in total. Figure 2 shows the frequency spectrum of some branches (Yh1–Yh5). It is shown that branch Yh4 together with branch Yh5 occupy the most components that locate at the frequency band below 700 Hz. Therefore, only branch Yh4 and Yh5 are kept in data reconstruction for the subsequent operating modal identification.

4 Operating Model Analysis

Time domain identification methods are often adopted in operating modal analysis considering its many advantages as follows. Before operating modal analysis, it is generally necessary to extract free vibration response from the vibration signal [8]. Common pretreatment methods include random decrement method (RDT) and natural excitation technique (NExT) [9]. Once the free vibration response is available, time domain methods like ITD (Ibrahim Time Domain) or STD (Spare Time Domain), ARMA model (Auto-Regressive and Moving Average model) [8], SSI (Stochastic Subspace Identification), LSCE (least squares complex exponent), ERA (Eigen-system Realization Algorithm), as well as modern signal processing methods like EMD (Empirical Mode Decomposition), Hibert transform, and Laplace wavelet filter are adoptable to identify modal parameters according to different applicative conditions [10, 11].

Considering the dimension and amount of the measured vibration, according to the layout of those test points, ARMA model time domain method is a suitable choice in this paper.

4.1 ARMA Model

ARMA model is also known as time sequence analysis method [8]. It is applicable when the system input is unavailable, or the system itself is ambiguous, or the input and response are available but with serious noise. Its concept can be explained briefly as follows:

Suppose a linear system has n degrees of freedom, the relationship between inputs and response usually can be described by a set of differential equations of high order. Represent these differential equations into discrete form, a set of equations about time sequences at discrete time moments are obtained, namely, the ARMA model as following:

$$\sum_{k=0}^{2n} a_k x_{t-k} = \sum_{k=0}^{2n} b_k f_{t-k} \tag{5}$$

x_t is the sequence of response, and x_{t-k} is the historical value of x_t. The left hand side of the Eq. (5) is known as the auto regressive difference polynomial, i.e. AR model, while the right hand side is called the moving average differential polynomial, i.e. the MA model.

f_t denotes the white noise excitation. 2n is the order in ARMA model. a_k, b_k are autoregressive coefficients and moving average coefficients, respectively. They are unknown parameters waiting for identification. The transfer function of the identified ARMA model gives the characteristic roots (poles) of the system, from which the modal parameters of the structure can be calculated, using the relationship between the system poles and modal frequency ω_k and damping ratio ξ_k:

$$\begin{cases} z_k = e^{s_k \Delta t} = e^{(-\xi_k \omega_k + j\omega_k \sqrt{1-\xi_k^2})\Delta t} \\ R_k = \ln z_k = s_k \Delta t \\ \omega_k = |R_k|/\Delta t \\ \xi_k = \sqrt{1 / 1 + (\mathrm{Im}(R_k)/\mathrm{Re}(R_k))^2} \end{cases} \tag{6}$$

4.2   Modal Identification of the Turntable

In this paper the operating modal identification is carried out following the procedure as below:

(1) Vibration acceleration acquisition (Sect. 2)
(2) DTCWT decomposition and reconstruction using partial branches (Sect. 3)
(3) Converse acceleration to displacement (by using quadratic integration in frequency domain)
(4) Obtain free vibration response using NExT (Sect. 4.2)
(5) Operating modal identification using ARMA (Sect. 4.1).

The reconstructed signal by DTCWT is still the acceleration data of the turntable. The corresponding displacement of the vibration is usually obtained by quadratic integration in frequency domain. Figure 2(a) shows the obtained vibration displacements at the test point No. 5 and No. 7.

Recall that the NExT method is used to exact free vibration response from the origin vibration of the structure. Its basic principle is that the cross-correlation function of the response signals from two points of the structure against ambient excitation is very similar to the impulse response function, i.e. free response of the system.

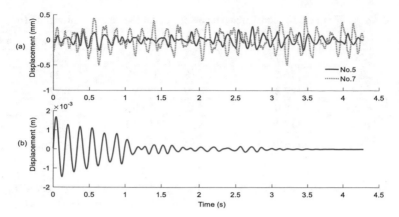

Fig. 2. (a) Vibration displacements at the test points No. 5 and No. 7, respectively; (b) Free vibration response curve obtained by NExT method.

Therefore, the cross-correlation of the two responses instead of the free vibration response is theoretically adoptable in modal identification [8].

The test points No. 5 and No. 7 locate on the turntable and own the same vibration direction during measurement. Therefore, the vibration data from those two test points are taken to exact free vibration response. The obtained free vibration response sequence is presented in Fig. 2(b) and is obviously damping as time goes.

As introduced previously, the ARMA time sequence model is applied based on the obtained free vibration response to achieve local modal identification of the turntable. In the process of calculation, the modal order closely affects the fitting accuracy. Generally, AIC criterion function is used to figure out the order. The first 12 modal frequencies are concerned and the calculated results are presented in Table 1 and Fig. 3.

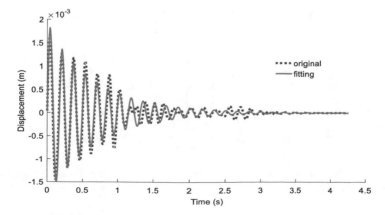

Fig. 3. Comparison of the free vibration response and the fitted curve according to identified modal parameters.

Table 1. The first 12 modal frequencies by modal identification using ARMA.

Order	Frequency (Hz)	Damping ratio (%)	Order	Frequency (Hz)	Damping ratio (%)
1	6.07	3.78	7	251.53	6.22
2	10.95	16.51	8	311.39	2.24
3	36.37	47.48	9	375.97	0.17
4	87.97	26.02	10	425.58	4.44
5	140.49	13.19	11	537.40	22.55
6	208.74	9.76	12	541.20	3.44

Figure 3 shows the free vibration response and the fitted curve according to the identified modal parameters using ARMA model. The two curves have a high degree of agreement.

From the numerical results shown in Table 1, the following statements are also given:

(1) The obtained modal frequency around 375 Hz is rather a forced vibration component than one of the nature frequencies of the turntable, because the corresponding damping ratio is at 0.17%, which is close to 0;

(2) The nature frequencies with low damping ratio need to pay more attention, which are the orders at 1st, 8th, 10th and 12th frequencies, i.e. 6 Hz, 311 Hz, 425 Hz and 540 Hz. In order to decrease mechanical noise and abrasion of the turntable, either to reshape the turntable, or to avoid frequency around those values when design the drive system for the road-header.

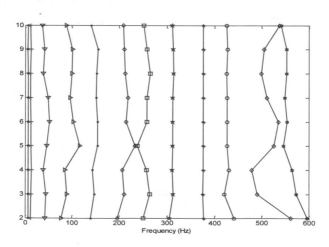

Fig. 4. Collection of the identified modal frequencies based on vibration data from different test points at different time intervals.

Additionally, the processing is carried out over origin vibration data from different test points on the turntable and at different time intervals. There are totally 10 sets of identified modal frequencies obtained. Figure 4 presents them all together to demonstrate if they are consistent to each other. This is an alternative way to evaluate the identification results when lack of actual operating vibration modal parameters of the turntable. It is shown in Fig. 4 that most of the calculated modal frequencies are consistent to each other except the ones at the 11th and 12th order.

5 Conclusion

This paper presents a tentative study on local vibration data that were acquired from a working road-header under the shaft. According to the vibration test condition, a procedure composed of DTCWT and ARMA model is proposed to identify the operating modal parameters of the constructional turntable. The DTCWT is used to remove the components at higher frequency bands from the original vibration signal since only the main vibration modes at lower frequency bands are concerned firstly in this paper. The ARMA model for operating modal analysis is simple and suitable for the acquired vibration data considering their dimensions and positions during vibration tests.

By implementing the aforementioned procedure over vibration data from different test points at different time period, groups of nature frequencies of the turntable are calculated and present high consistency to each other. Thus, it proves that the ambient excitation-based modal identification, which is more often been used for huge scale constructions, is also feasible for the road-header. The obtained modal parameters are valuable for the design of the structure of road-header, as well as damage assessment of working structures.

Acknowledgments. Great gratitude is sent to the National Natural Science Foundation of China, for the financial supports to this work through projects No. 61803374 and No. 51874308.

References

1. Zhao, L.J., Zhu, S., Liu, X.N.: Research on reliability of longitudinal road-header's rotary table by ADAMS. J. Mech. Strength **38**(1), 128–132 (2016)
2. Huang, M., Wu, M., Wei, R.Z., et al.: Vibration characteristics analysis and structural optimization simulation of AM50 road-header. J. China Coal Soc. **22**(2), 192–196 (1997)
3. Huang, M., Wu, M., Wei, R.Z., et al.: Study on vibration characteristics of transverse cutting head of the road-header. Chin. J. Mech. Eng. **38**(8), 89–93 (2002)
4. Zhang, N., Liu, Y.T., Wen, J.B.: FEM analysis of road-header EBZ260 turning platform based on ANSYS workbench. Coal Mine Mach. **34**(2), 95–97 (2013)
5. Yang, Y., Hu, Y.L., Yang, J., et al.: Vibration signal test and analysis of horizontal axial hard rock road-header. Meas. Control Technol. **34**(10), 58–60 (2015)
6. Yang, Y., Qu, Y., Liu, W.J., et al.: Vibration analysis of a working tunnel boring machine in underground mine. Coal Technol. **35**(7), 240–242 (2016)
7. Wang, F., Ji, Z., Peng, C.L.: Research on ECG signal de-noising based on dual-tree complex wavelet transform. Chin. J. Sci. Instrum. **34**(5), 1160–1166 (2013)

8. Cao, S.Q., Zhang, W.Z., Xiao, L.X., et al.: Modal Analysis of Vibration Structure, 2nd edn. Tianjin University Press, Tianjin (2014)
9. James, G.H., Carne, T.G., Lauffer, J.P.: The natural excitation technique (NEXT) for modal parameter extraction from operating structures. Int. J. Anal. Exp. Modal Anal. **10**(4), 260–277 (1995)
10. He, Q.Y.: Ambient Exciting Modal Parameter Identification Based on Modern Time-frequency Analysis. Chongqing University, Chongqing (2009)
11. Qi, K.Y., Xiang, J.W., Zi, Y.Y., et al.: Accurate structure modal identification method based on laplace wavelet correlation filtering. Chin. J. Mech. Eng. **43**(9), 167–172 (2007)

Multi-objective Optimization Based on NSGA-II Algorithm for 3DP Process

Zijia Niu, Weidong Yang$^{(\boxtimes)}$, Xiangyu Gao, and Xiyuan Tu

Hebei University of Technology, Tianjin, China
yangweidong@hebut.edu.cn

Abstract. The forming direction and layer thickness of the model in 3D inkjet printing technology are the key factors that affect the surface accuracy, processing time and processing cost of the part. In order to reduce the processing time and improve the surface precision, a hierarchical optimization method based on improved fast non-dominated sorting genetic algorithm is proposed. Two objective functions of surface precision and processing time are established, and iterative solution is realized by selection, crossover and mutation. Experiments show that this method can effectively solve the optimization problem of layering direction in 3D printing process.

Keywords: 3D inkjet printing · NSGA-II algorithm · Forming direction · Layer thickness · Multi-objective optimization

1 Introduction

The 3D inkjet printing process is a kind of rapid prototyping technology. Using the idea of layered manufacturing, the 3D model is dispersed in a layered direction into a set of 2D graphics, i.e. a set of thin layers (also known as slicing or layering)., layer-by-layer processing to form a three-dimensional model entity, this technology has been introduced into many practical applications [1]. The STL model is used as a file format for rapid prototyping services. However, STL data cannot be directly used as input data for the 3D printing process, and must be converted into a data format that can be directly processed by the printer through layered software. The factors that determine the stratification result are the stratification direction and the stratification thickness. The layered thickness refers to the longitudinal distance between two adjacent slices. The layering direction is the normal vector direction of the slice plane. The smaller the layer thickness is, the finer the constructed model and the longer the construction time are. At the time of layer thickness, the layering direction uniquely determines the effect of the model construction. Different layer thicknesses and molding directions have a great influence on the quality of three-dimensional printed parts. Finding the right forming direction and layer thickness not only reduces the time required to print the part, improves the printing efficiency, but also significantly improves the accuracy of the part [2].

Optimizing the 3D inkjet printing process to improve part surface finish quality and printing efficiency is a typical multi-objective optimization problem. In recent years, domestic and foreign scholars have conducted in-depth research on multi-objective

F. Xhafa et al. (Eds.): IISA 2019, AISC 1084, pp. 757–765, 2020.
https://doi.org/10.1007/978-3-030-34387-3_93

optimization algorithms [3–7]. Currently, the commonly used multi-objective optimization algorithms are NSGA-II, SPEA2, PAES and DMOEA. The non-dominated set sorting genetic algorithm (NSGA-II) algorithm has become one of the benchmarks of multi-objective optimization algorithms because of its fast running speed, low computational complexity and easy implementation. It is a scientific research field and engineering practice in solving multi-objective problems widely used in the field. However, the NSGA-II algorithm has a non-uniform distribution of population convergence, poor global search ability, and easy to fall into local optimum, which make it less effective in solving some multi-objective optimization problems. Therefore, two improved methods for NSGA-II algorithm are proposed: stepped adaptive genetic algorithm (SAGA) and double sorting genetic algorithm (DSGA), and the improved genetic algorithm is combined into a new multi-target based SD-NSGA. -II genetic algorithm and verified the effectiveness of the algorithm.

2   Genetic Algorithm Improvement

2.1   Improved Adaptive Genetic Algorithm

Among genetic algorithms, crossover and mutation are the most obvious features that distinguish them from other heuristic algorithms. In the genetic algorithm, the two pairs of individuals perform the exchange of internal genes according to the cross-rules formulated by the decision makers, thus generating two new individuals. Crossover probability is essential for population diversity and population optimization. Excessive crossover probability leads to high fitness in new populations, and excellent individuals are not retained. However, if the value is too small, it will be difficult to cross-over. It will delay the evolution of the population and is also very unfavorable to the population.

In the genetic algorithm, the crossover rate and the mutation rate are fixed values, that is to say, the same crossover rate and mutation rate are used in the front, middle and the end of the algorithm, which is obviously not conducive to the algorithm seeking the optimal solution. In order to find the appropriate crossover rate and mutation rate at the appropriate stage, this paper proposes an improved stepped adaptive genetic algorithm. The crossover and mutation rate of the algorithm.

$$
P_m = \begin{cases}
\frac{P_{mmax}+P_{mmin}}{2} - \frac{\pi}{f_{max}-f_{avg}} \cdot \frac{P_{mmax}-P_{mmin}}{2}\left(f - \frac{f_{max}+f_{avg}}{2}\right) \\
\frac{f_{max}+f_{avg}}{2} - \frac{f_{max}-f_{avg}}{\pi} \leq f \leq \frac{f_{max}+f_{avg}}{2} + \frac{f_{max}-f_{avg}}{\pi} \\
P_{mmax} \qquad\qquad f < \frac{f_{max}+f_{avg}}{2} - \frac{f_{max}-f_{avg}}{\pi} \\
P_{mmin} \qquad\qquad f > \frac{f_{max}+f_{avg}}{2} + \frac{f_{max}-f_{avg}}{\pi}
\end{cases}
\tag{1}
$$

$$
P_c = \begin{cases}
\frac{P_{cmax}+P_{cmin}}{2} - \frac{\pi}{f_{max}-f_{avg}} \cdot \frac{P_{cmax}-P_{cmin}}{2}\left(f' - \frac{f_{max}+f_{avg}}{2}\right) \\
\frac{f_{max}+f_{avg}}{2} - \frac{f_{max}-f_{avg}}{\pi} \leq f' \leq \frac{f_{max}+f_{avg}}{2} + \frac{f_{max}-f_{avg}}{\pi} \\
P_{cmax} \qquad\qquad f' < \frac{f_{max}+f_{avg}}{2} - \frac{f_{max}-f_{avg}}{\pi} \\
P_{cmin} \qquad\qquad f' > \frac{f_{max}+f_{avg}}{2} + \frac{f_{max}-f_{avg}}{\pi}
\end{cases}
\tag{2}
$$

Where: the value of the crossover operator P_c is 0.2–0.8, P_{cmax} is the maximum crossover rate, taking a larger value in the range of P_c interval, P_{cmin} is the minimum crossover rate, taking a smaller value in the range of P_c interval, The variation operator P_m has a value range of 0.01–0.3, P_{mmax} is the maximum mutation rate, taking a larger value within the P_m interval, and P_{mmin} is the smallest mutation rate, taking a smaller value within the P_m interval.

2.2 Improved Double Sorting Genetic Algorithm

Since the genetic algorithm often appears premature, the solution fails, etc. This section optimizes the sorting function of NSGA-II algorithm, and proposes a double sorting improved genetic algorithm DSGA algorithm to improve the accuracy of the algorithm.

The double-sorting improved genetic algorithm is a new algorithm based on the existing improved selection and crossover method. It uses two sortings: global sorting and local sorting. The global sorting uses the GA genetic operator in the NSGA-II algorithm., and the local fast non-dominated sorting uses the PIM genetic operator [8].

Global sorting plays a leading role in the whole operation process, while local sorting is to re-cross and mutate individuals with insufficient fitness, and then re-order all individuals to form a new population. Sorting is equivalent to playing an auxiliary role. For local re-crossing and mutation we use the sigmoid function [9] $f(x) = \frac{1}{1+e^{-x}}$. The range of the sigmoid function output is [0,1], Where x is the difference between the maximum fitness value and the individual fitness value of the variation, and it can be seen that x > 0, so a 1/2 compensation term needs to be added. The expression of the local mutation operator PIM is as follows:

$$PIM(j) = \left(\frac{1}{1+e^{-x}} - \frac{1}{2} \right) \cdot k_1 \tag{3}$$

$$x = f_{max} - f(j) \qquad j = 0, 1, 2 \ldots m; \; m < N \tag{4}$$

Where f_{max} is the maximum fitness value of the population, the proportionality coefficient k_3 = 0.08, the local variation rate PIM ranges from 0–0.4, and the value of $f(j) = f_{max}$ is 0, and the maximum value is 0.4. Therefore, under the action of PIM, the initial stage of evolution can effectively increase the diversity of the population, while the evolutionary approach to convergence has little effect on the population.

2.3 Multi-objective Genetic Algorithm Based on SD-NSGA-II

The above two algorithms are only an improvement on a certain aspect of the genetic algorithm. The following two improved algorithms based on the NSGA-II algorithm are combined, which is an adaptive mutation double-sorting fast non-dominated sorting genetic algorithm with elite strategy SD-NSGA-II. The improved SD-NSGA-II algorithm flow chart is shown in the Fig. 1.

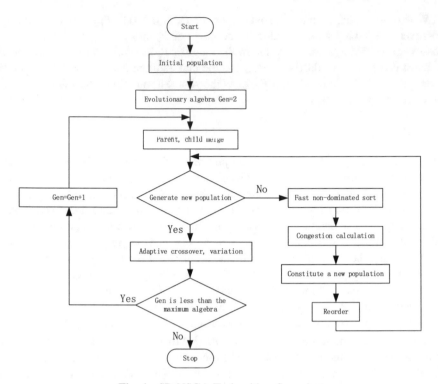

Fig. 1. SD-NSGA-II algorithm flow chart

3   Example Analysis

In order to verify the effectiveness of the improved SD-NSGA-II algorithm, the Bunny model was selected as the test object. The Bunny model is a commonly used model for multi-objective optimization selection in the 3D inkjet printing process. According to the flow of the SD-NSGA-II algorithm, the Bunny model is first converted into a readable STL data file, and then placed in the folder specified by Matlab, so that it can be read uniformly in the future. The parameters in the algorithm are set as follows: the population size is set to 400, and the maximum number of iterations is set to 400. The thickness of the layer is initially selected to be 0.200–0.300 mm, and the range is 0–180° the range is 0–360°. According to the optimization results of impact and penetration in the previous paper, the printer uses furan resin as a binder and sanding with a BSS100 sand bed. The Bunny model is shown in Fig. 2.

Fig. 2. Bunny model

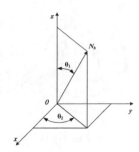

Fig. 3. Model coordinate system

3.1 Multi-Objective Optimization Model

(1) Function establishment aiming at forming quality

Assume that the bunny model has a layer thickness of h, the forming direction is N_b, and then the model is layered according to the forming direction N_b and the layer thickness h to obtain the sectional profile information of the part. The vector coordinate of the molding direction N_b is (x_b, y_b, z_b) shown in Fig. 3 and

$$\begin{cases} x_p = \sin \theta_1 \cos \theta_2 \\ y_p = \sin \theta_1 \sin \theta_2 \\ z_p = \cos \theta_1 \end{cases} \tag{5}$$

After the 3D model is converted into an STL file, there are n triangular patches, and m layers are obtained through layering. Figure 4 shows the volume error caused by the i-th step. The volume error produced by the entire model is:

$$\Delta V = \sum_j \sum_i \frac{h^2 \cos \theta_j}{2 \sin \theta_j} l_{ij} = \sum_j \Delta V_j \tag{6}$$

And

$$\Delta V_j = \sum_i \frac{h \cos \theta_j}{2} dA_{ij} = \frac{h \cos \theta_j}{2} A_j \tag{7}$$

Then

$$\Delta V = \sum_j \frac{h \cos \theta_j}{2} A_j, j = 1, 2, 3 \ldots s \tag{8}$$

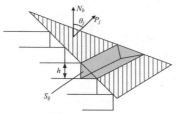
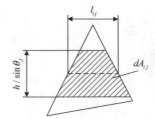

(a) Layering of the i layer   (b) Layered in the case of triangular patches

Fig. 4. Geometry of the volume error produced by the i layer and the triangular patch j

$$
\begin{cases}
A_j = \dfrac{1}{2}\left\{ \left| \begin{matrix} x_{j2}-x_{j1} & y_{j2}-y_{j1} \\ x_{j3}-x_{j1} & y_{j3}-y_{j1} \end{matrix} \right|^2 + \left| \begin{matrix} y_{j2}-y_{j1} & z_{j2}-z_{j1} \\ y_{j3}-y_{j1} & z_{j3}-z_{j1} \end{matrix} \right|^2 + \left| \begin{matrix} z_{j2}-z_{j1} & x_{j2}-x_{j1} \\ z_{j3}-z_{j1} & x_{j3}-x_{j1} \end{matrix} \right|^2 \right\}^{\frac{1}{2}} \\[12pt]
\cos\theta_j \begin{cases} 0 & \theta_j = 0, \pi \\ |N_b \cdot P_j| & \theta_j \ne 0, \pi \end{cases}
\end{cases}
$$

$$(9)$$

(2) Function establishment aiming at molding efficiency

Analysis of the three-dimensional inkjet printing process shows that the time for forming parts mainly includes powder coating time, inkjet bonding time, table down time and other auxiliary time. The ink sticking time is related to the forming direction and the layer thickness, while the powder paving time and the table falling time are only related to the number of layers. Therefore, the molding time can be reduced by reducing the number of layered layers.

And $Q = x_{ja} \cdot \sin\theta_1 \cos\theta_2 + y_{ja} \cdot \sin\theta_1 \sin\theta_2 + z_{ja} \cdot \cos\theta_1$, the optimization model targeting molding efficiency is:

$$
T = \{\max\{x_{ja}\sin\theta_1\cos\theta_2 + y_{ja}\sin\theta_1\cos\theta_2 + z_{ja}\cos\theta_1\} \\
- \min\{x_{ja}\sin\theta_1\cos\theta_2 + y_{ja}\sin\theta_1\cos\theta_2 + z_{ja}\cos\theta_1\}\}/h
$$

$$(10)$$

(3) Multi-objective optimization model establishment

Combining the above mathematical expressions with the aim of molding precision and molding efficiency, this paper establishes a multi-objective mathematical model with the molding direction and layer thickness as the variables and the molding precision and molding efficiency as the target. The mathematical expression is

$$Find : X = [\theta_1, \theta_2, h]$$

$$min : \begin{cases} \Delta V(X) = \sum_j \frac{h \cos \theta_j}{2} A_j \\ T = \{\max\{x_{ja} \sin \theta_1 \cos \theta_2 + y_{ja} \sin \theta_1 \cos \theta_2 + z_{ja} \cos \theta_1\} \\ \quad -\min\{x_{ja} \sin \theta_1 \cos \theta_2 + y_{ja} \sin \theta_1 \cos \theta_2 + z_{ja} \cos \theta_1\}\}/h \end{cases} \quad (11)$$

3.2 Optimization Results and Discussion

For the verification of the calculation results of NSGA-II algorithm and SD-NSGA-II algorithm, the solution with the forming time of 250 min is selected from the optimal solution set obtained by the two algorithms, and the forming direction of the two solutions and the layer thickness are extracted for printing. That is, the NSGA-II algorithm uses a variable of 99. The variable is 87. The layer thickness is 0.25 mm for printing, and the printing volume error is 1206.10 mm^3. The SD-NSGA-II algorithm uses a variable of 104. The variable is 93. The layer thickness is 0.22 mm for printing, and the printing volume error is 1063.06 mm^3.

In this paper, the ZGScan intelligent handheld laser 3D scanner of Wuhan Zhongguan Automation Technology Co., Ltd. is used to scan and measure the printed parts. In order to compare the accuracy of the NSGA-II algorithm and the SD-NSGA-II algorithm more conveniently, the accuracy of the model is obtained by finding 10 sampling points in the scan map, wherein the error range of the sampling point is ± 0.3000 mm, when the range is exceeded. To be out of tolerance, the sampling point is judged to be a failure state, otherwise it is a pass state.

The error of the 10 sampling points corresponding to the forming entity based on the NSGA-II algorithm is shown in Table 1:

Table 1. NSGA-II algorithm prints ten sample point errors for a formed entity

Name	Deviation (mm)	Status	Deviation X (mm)	Deviation Y (mm)	Deviation Z (mm)
A019	0.5822	fail	0.2360	−0.3871	−0.3653
A020	0.7288	fail	−0.4573	−0.5546	0.1205
A021	0.0530	pass	−0.0070	−0.0525	0.0033
A022	0.6131	fail	−0.0650	−0.2973	0.5323
A023	−0.2051	pass	0.0035	0.1776	0.1024
A024	−0.1467	pass	−0.0128	0.1461	−0.0004
A025	−0.1043	pass	0.0488	0.0872	−0.0299
A026	−0.0916	pass	−0.0784	0.0464	−0.0098
A027	0.3052	fail	0.0137	−0.0889	0.2917
A028	0.2418	pass	0.0716	−0.2193	0.0726

The error of the 10 sampling points corresponding to the model based on the SD-NSGA-II algorithm is shown in Table 2:

Table 2. SD-NSGA-II algorithm prints ten sample point errors of a model

Name	Deviation (*mm*)	Status	Deviation X (*mm*)	Deviation Y (*mm*)	Deviation Z (*mm*)
A019	0.3131	fail	0.0782	0.2941	−0.0736
A020	0.0200	pass	0.0015	0.0192	−0.0052
A021	0.0226	pass	0.0089	0.0206	0.0029
A022	−0.1869	pass	0.0745	−0.1713	0.0055
A023	−0.2384	pass	0.0506	−0.2296	0.0395
A024	−0.1467	pass	−0.0128	0.1461	−0.0004
A025	0.0832	pass	0.0330	0.0658	0.0387
A026	−0.0804	pass	0.0182	−0.0763	−0.0174
A027	0.1320	pass	0.0944	−0.0017	0.0923
A028	0.0658	pass	0.0041	0.0460	0.0469

The SD-NSGA-II can be known through experimental calculation. The absolute deviation of the algorithm is 0.1288 mm, and the absolute deviation obtained by the NSGA-II algorithm is 0.3065 mm. The accuracy of the SD-NSGA-II algorithm is 57% higher than that of the NSGA-II algorithm. It can be seen that the results obtained by the SD-NSGA-II algorithm are higher than those of the NSGA-II algorithm, and the parameter optimization of the multi-objective model based on SD-NSGA-II is also verified.

4   Conclusion

Through the comparison of simulation tests and the evaluation of performance indicators, it is proved that the new SD-NSGA-II algorithm performs better in the convergence and distribution of the obtained solutions. Then through the example verification, the experiment shows that compared with the NSGA-II algorithm, the SD-NSGA-II algorithm can solve a better solution for the specific model, thus completing the optimization of the molding direction and layer thickness of the model, and finally effective. Improve the molding accuracy and molding efficiency of the model.

Acknowledgments. This research was supported by Natural Science Foundation of Hebei Province, China under Grant No. E2016202297.

References

1. Tilford, T., Bruan, J., Janhsen, J., Burgard, M.: SPH analysis of inkjet droplet impact dynamics (2018)
2. Huang, R., Dai, N., Li, D., Cheng, X.: Parallel non-dominated sorting genetic algorithm-II for optimal part deposition orientation in additive manufacturing based on functional features. Proc. Inst. Mech. Eng., Part C: J. Mech. Eng. Sci. **232**(19), 3384–3395 (2018)

3. Zhao, Z., Liu, B., Zhang, C., Liu, H.: An improved adaptive NSGA-II with multi-population algorithm. Appl. Intell. **49**(2), 569–580 (2019)
4. Han, Z., Wang, S., Dong, X., Ma, X.: Improved NSGA-II algorithm for multi-objective scheduling problem in hybrid flow shop. In: Zhu, Q., Na, J., Wu, X. (eds.) Innovative Techniques and Applications of Modelling, Identification and Control. LNEE, vol. 467, pp. 273–289. Springer, Singapore (2018). https://doi.org/10.1007/978-981-10-7212-3_17
5. Chatterjee, S., Sarkar, S., Dey, N., Ashour, A.S.: Hybrid non-dominated sorting genetic algorithm: II-neural network approach. In: Advancements in Applied Metaheuristic Computing: IGI Global, pp. 264–286 (2018)
6. Yi, J.-H., Deb, S., Dong, J., Alavi, A.H.: An improved NSGA-III algorithm with adaptive mutation operator for big data optimization problems. Fut. Generat. Comput. Syst. **88**, 571–585 (2018)
7. Safi, H.H., Mohammed, T.A., Al-Qubbanchi, Z.F.: Minimize the cost function in multiple objective optimization by using NSGA-II. In: International Conference on Artificial Intelligence on Textile and Apparel, pp. 145–152 (2018)
8. Fattahi, E., Bidar, M., Kanan, H.R.: Focus group: an optimization algorithm inspired by human behavior. Int. J. Computat. Intell. Appl. **17**(01), 1850002 (2018)
9. Daliakopoulos, I.N., Coulibaly, P., Tsanis, I.K.: Groundwater level forecasting using artificial neural networks. J. Hydrol. **309**(1–4), 229–240 (2005)

Label Propagation Algorithm Based on Improved RA Similarity

Haitao Cui, Lingjuan Li$^{(\boxtimes)}$, and Yimu Ji

School of Computer Science, Nanjing University of Posts and
Telecommunications, Nanjing 210023, China
`lilj@njupt.edu.cn`

Abstract. Community detection is an effective method to analyze the structure of social network. In order to improve the accuracy and efficiency of community detection, a label propagation algorithm based on improved RA similarity named IRA-LPA is designed. The algorithm introduces the simplified SimRank algorithm into the RA similarity algorithm and combines the improved RA similarity algorithm with the label propagation algorithm (LPA). The improved RA algorithm is used to calculate node similarity and generates seed communities, and LPA is used to complete the final community detection based on the seed communities. So that IRA-LPA algorithm can solve the problem of randomness and resource consumption when LPA initially allocates labels to nodes. The community detection results on Dolphin network and Football network show that IRA-LPA can effectively divide the community structure with higher accuracy and lower time complexity.

Keywords: Community detection · SimRank · RA similarity · LPA

1 Introduction

Social network is a networked generalization of social relations, it's an important content of complex networks. In recent years, researchers have proposed many algorithms for community detection, such as GN algorithm based on the edge betweenness [1], FN algorithm based on modularity optimization [2], label propagation algorithm [3]. However, with the development of society, the complexity of social networks is increasing, and many algorithms are faced with various problems. For example, the GN algorithm performs a split operation, which greatly increases the time complexity of the algorithm. FN algorithm uses the idea of greedy algorithm to make the algorithm time complexity significantly lower than the GN algorithm, but its accuracy is lower than the GN algorithm. When dealing with a network which contains many nodes, the time complexity of the label propagation algorithm is linear, but the randomness of the algorithm in setting label and label propagation is too high, and the results are not accurate enough.

Aiming at the above problems, this paper designs a label propagation algorithm based on improved RA similarity named IRA-LPA. The algorithm combines RA similarity with label propagation algorithm. Firstly, to solve the problem that there are many related nodes and their similarities are zero or cannot be calculated in the results

© Springer Nature Switzerland AG 2020
F. Xhafa et al. (Eds.): IISA 2019, AISC 1084, pp. 766–772, 2020.
https://doi.org/10.1007/978-3-030-34387-3_94

of RA similarity algorithm, the idea of SimRank algorithm is introduced to improve the RA similarity algorithm, and the improved RA algorithm is used to obtain the initial communities; then the LPA is used to complete the final community detection based on the initial communities.

2 Design of IRA-LPA Based on Improved RA Similarity

2.1 Improvement of RA Similarity Algorithm

(1) RA similarity algorithm

The similarity between nodes is often used to determine if nodes belong to a same community. There are many methods for calculating similarity, such as Jaccard, Salton, RA [4], etc., but the stability and accuracy of each algorithm are quite different. The researchers found that the RA similarity algorithm has better effect and higher stability. Therefore, the RA similarity algorithm is chosen to calculate the similarity.

The RA similarity algorithm measures the similarity between nodes by simulating the process of transmitting information through nodes in the cyberspace. Suppose there is a pair of target nodes (i, j), Node i sends information to node j through their common neighbor nodes as relay nodes. At the same time, the number of channels of each relay node is specified to be constant. Therefore, the amount of information that each relay node sends to node j depends on its degree. The amount of information that node j receives from node i can be defined as the similarity between node i and node j, as shown in Eq. 1.

$$S_{ij} = \sum_{z \in \Gamma(i) \cap \Gamma(j)} \frac{1}{k(z)} \tag{1}$$

In Eq. 1, $\Gamma(i)$ represents the neighbor nodes of node i (excluding i itself), $\Gamma(i) \cap \Gamma(j)$ represents the set of common neighbor nodes of nodes i, j, and $k(z)$ represents the degree of node z.

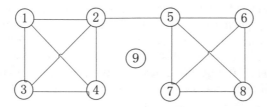

Fig. 1. The sample network

In the results of similarity which calculated by the Eq. 1, there are many cases where the similarity does not exist or is zero. Taking Fig. 1 as an example, the RA similarity algorithm has the following problems:

(1) The algorithm only considers the similarity between nodes from the local structure and neglects the overall structure of the network.
(2) The similarity cannot be calculated under the following conditions:
   (a) The pair of nodes i, j can be connected through other nodes, but they do not have any common neighbor nodes, such as (1, 6);
   (b) The pair of nodes i, j is directly connected, so there are no common neighbor nodes between them, such as (2, 5);
   (c) The pair of nodes i, j is independent of each other, such as (1, 9);
   (d) The similarity of node i to itself.

(2) Improvement of RA similarity algorithm

To solve those problems, this paper makes the following improvements to RA algorithm:

(1) Take the shortest path between nodes into account to solve the problem that the original RA similarity algorithm only focuses on local structure and ignores the overall structure of the network.
(2) Introduce the SimRank algorithm [5] as shown in Eq. 2 and simplifying it as Eq. 3 to calculate the similarity S_{ij} of i, j that can be connected through other nodes but do not have any common neighbor nodes.

$$S_{ij} = \frac{C}{k(i) * k(j)} \sum_{t \in \Gamma(i)} \sum_{t \in \Gamma(j)} S_{tr} \tag{2}$$

$$S_{ij} = \frac{C}{k(i) + k(j)} \sum_{t,r \in d(i,j) \, and \, t \neq r} S_{tr} \tag{3}$$

In Eqs. 2 and 3, C is the damping coefficient, also known as the attenuation coefficient. It means that the more nodes are separated between two nodes, the more the similarity decreases. Obviously, $C \in (0,1)$, this paper takes the empirical value $C = 0.8$. $k(i)$ represents the degree of node i. $\Gamma(i)$ represents the neighbor nodes of node i (excluding i itself), and $d(i, j)$ represents the set of nodes which distribute in the shortest path of i, j (including i, j). Equation 3 uses the similarity of the nodes which belong to the shortest path of target nodes i and j instead of using the similarity of all neighbor nodes of i and j to calculate the similarity of i, j.

Taking (1, 6) in Fig. 1 as an example, $d(1,6) = (1, 2, 5, 6)$, so the similarity of node (1, 2), (2, 5,), (5, 6) can be used to calculate the similarity of (1, 6) according to Eq. 3.

(3) Extend the definition of $\Gamma(i)$ in Eq. 1 to include i itself to calculate the similarity S_{ij} that i and j directly connected but they do not have any neighbor nodes.
(4) Set the similarity of two independent nodes to zero because they are not related.

(5) Set the similarity of i with itself to zero because this similarity is not important and setting it to 1 may affect the results.

In conclusion, combined with the simplified SimRank, the improved RA similarity algorithm (IRA) is as Eq. 4.

$$
S_{ij} = \begin{cases}
\dfrac{n}{d_{ij}} \sum\limits_{z \in \Gamma(i) \cap \Gamma(j)} \dfrac{1}{k(z)} & \text{If } i \text{ and } j \text{ are connected, and they have common neighbor nodes.} \\
\dfrac{C}{k(i) + k(j)} \sum\limits_{t,r \in d(i,j) \text{ and } t \neq r} S_{tr} & \text{If } i \text{ and } j \text{ are connected, and they do not have common neighbor nodes.} \\
0 & \text{If } i \text{ and } j \text{ are not connected.} \\
0 & \text{If } i = j.
\end{cases}
$$

$$(4)$$

In Eq. 4, n is the number of nodes in the network. The results are properly amplified by the amplification coefficient n to prevent the too small results and the floating-point underflow problem. d_{ij} represents the length of the shortest path between i and j. $d(i, j)$ indicates the nodes which distribute in the shortest path of i, j (including i, j). $k(i)$ represents the degree of i, and C represents the attenuation coefficient, $C = 0.8$. The bigger S_{ij}, the greater the probability that the two nodes belong to the same community; Conversely, the smaller the S_{ij}, the greater the probability that the two nodes belong to two different communities.

2.2 Idea and Process of IRA-LPA

(1) The basic idea of IRA-LPA

Label Propagation Algorithm (LPA) is a community detection algorithm for non-overlapping communities. Its basic idea is to add a label for each node randomly, and a new label of each node will be obtained through iteration each time. The iterative method of obtaining the label is as follows: the target node uses the label which belongs to the majority of the neighbor nodes as the label of itself. The label obtained after multiple iterations represents the community which the node belongs to. After multiple labels 'propagation', the nodes with the same label are allocated to the same community. The advantage of LPA over other community detection algorithms is that its time complexity is closer to linear. When dividing a large network, it is faster, the result is more accurate, and the effect is better. However, LPA also has shortcomings. Because of the random strategy, many resources are consumed in the initial label allocation, and the stability of LPA is low when dividing large community networks.

In this paper, the IRA-LPA algorithm uses the IRA algorithm to calculate the nodes similarity and obtain the initial communities, then it treats initial communities as different seed communities and uses LPA algorithm to allocate labels to them. It solves the problem that many resources are consumed in the process of label allocation and the randomness of initial label allocation.

(2) The process of the IRA-LPA

Input: Undirected and weightless network G with n nodes.

Output: The community structure after the division.

Step 1: Calculate the similarity between nodes using Eq.4.

Step 2: Divide each node and its node with the highest similarity into the same community. In this step, n small communities with only two nodes can be obtained.

Step 3: For these small communities, do the merge-remove operation. The operation is to determine whether there are common nodes within the two small communities, and if there are, merge the small communities and remove the duplicate nodes. Do this operation until the generated communities do not contain the same nodes. In this step, the initial seed communities can be obtained.

Step 4: Allocate different labels to the seed communities, and the nodes which belong to same seed communities have the same label.

Step 5: Set the number of updates $t=1$.

Step 6: Generate a random sequence X, and iteratively update the label of $Xuxa$ in turn. The iteration's rule is the label with the highest occurrences of the x neighbor nodes is used as the label of x. If there are many candidate labels for x, it will randomly select one label l_t from the candidate labels $L = \{l_1, l_2, ..., l_n\}$ for x.

Step 7: If the labels of all nodes in the network no longer change, the algorithm ends, otherwise, $t=t+1$, and return to Step 6.

Step 8: The nodes with the same label are divided into the same community.

2.3 Analysis of IRA-LPA

In terms of applicability, IRA-LPA solves the randomness problem of LPA initial label allocation and the problem of resource consumption. Compared with LPA, IRA-LPA improves the stability and accuracy.

In terms of time complexity, for IRA-LPA, to divide an undirected and weightless graph G with n nodes, the time complexity of calculating the similarity is about $O(n^2)$, and the time complexity of the merge-remove operation is about $O(n^2)$, the time complexity of LPA is approximately $O(n)$, so the time complexity of IRA-LPA is approximately $O(n^2)$. This is the same as the FN algorithm (the time complexity is about $O(n^2)$), which is better than the GN algorithm (the time complexity is about $O(n^3)$).

3 Experiments and Results Analysis

In order to verify the performance of IRA-LPA, this paper carries out corresponding experiments on the classic social network data sets which are Dolphins network dataset [6] and Football network dataset [7] and uses the modularity function Q [8] and NMI (Normalized Mutual Information) [9] as two indicators to measure the quality of the community detection. Modularity function Q indicates the degree of intimacy within the community. The closer Q is to 1, the closer the internal links of community are, the clearer the division between communities is. NMI measures the degree of similarity between the community structure detected by the algorithm and the real community structure. If the two structures are identical, the value of NMI is 1. In order to ensure the accuracy of the algorithm, this paper performs 100 tests on the above datasets respectively and calculates the NMI and Q of the community detection results; compares the results with the GN algorithm, FN algorithm, LPA, LPA-E, IPLPA, and analyzes the results.

The community detection results obtained by IRA-LPA are shown in Table 1.

Table 1. The results of Dolphin network & Football network

Network	Number of nodes	Number of edges	Number of communities
Dolphins	62	159	4
Football	115	616	12

Table 1 shows that IRA-LPA divides the networks into 4 communities and 12 communities which are in line with the real situation of the network basically. NMI and Q are listed in Table 2.

Table 2. Algorithm performance comparison on Dolphin network & Football network

Network	Dolphins		Football	
Algorithm	NMI	Q	NMI	Q
LPA	0.376	0.425	0.753	0.459
GN	1	0.517	0.904	0.603
FN	0.693	0.515	0.513	0.549
IPLPA	0.711	0.451	0.890	0.483
LPA-E	0.382	0.446	0.767	0.477
IRA-LPA	0.731	0.517	0.973	0.589

Table 2 shows that the performance of IRA-LPA is better than other algorithms, especially the NMI of Football network, which is much better than other algorithms.

4 Conclusions

This paper designs a new label propagation algorithm based on improved RA similarity (IRA-LPA). The algorithm simplifies the SimRank algorithm and reduces its computational difficulty; uses the simplified SimRank to make up for the shortcomings of the RA algorithm in practical applications; combines the improved RA similarity algorithm and label propagation algorithm to take their respective advantages. It uses the improved RA similarity algorithm to generate the seed communities as the input of the label propagation algorithm, so that it solves the problem of randomness and high consumption of resources when LPA allocates labels to nodes initially. The results of experiments show that IRA-LPA proposed in this paper has higher community detection accuracy and lower time complexity.

Acknowledgments. This work is supported by the National Key R&D Program of China (2017YFB1401302, 2017YFB0202200).

References

1. Hui, D.: Application of GN algorithm in product module partition. Int. J. Dig. Content Technol. Appl. **6**(14), 236–245 (2012)
2. Newman, M.: Fast algorithm for detecting community structure in networks. Phys. Rev. E **69**(6), 066133 (2004)
3. Raghavan, U.N., Albert, R., Kumara, S.: Near linear time algorithm to detect community structures in largescale networks. Phys. Rev. E **76**(3), 036106 (2007)
4. Zhou, T., Lü, L., Zhang, Y.C.: Predicting missing links via local information. Eur. Phys. J. B-Condens. Matter Complex Syst. **71**(4), 623–630 (2009)
5. Cai, Y., Li, P., Liu, H., et al.: S-SimRank: Combining content and link information to cluster papers effectively and efficiently. J. Front. Comput. Sci. Technol. **3**(4), 378–391 (2009)
6. Lusseau, D.: The emergent properties of a dolphin social network. Proc. Royal Soc. London, Ser. B: Biol. Sci. **270**(Sup2), 186–188 (2003)
7. Girvan, M., Newman, M.E.J.: Community structure in social and biological networks. Proc. Nat. Acad. Sci. USA **99**(12), 7821–7826 (2002)
8. Peng, W., Hengshan, W.: Evaluation and improvement method on division of community structure in network. Appl. Res. Comput. **31**(7), 722–729 (2013)
9. Fortunato, S.: Community detection in graphs. Phys. Rep. **486**(3), 75–174 (2010)

Mechanical Modeling of Local Variable Load and Deform of Valve Slice

Yijie Chen[✉], Yafeng Zhang, Mengyan Xu, Fu Du, and Baoqiang Li

China North Vehicle Research Institute, Beijing 100072, China
chenyijie1206@163.com

Abstract. The damping valve structure of shock absorber usually takes throttle slice as the main part, when the pressure difference at both ends of the valve and the pretightening force of the valve slice can be overcome, annular gap will be formed between the valve body and slice to form throttle channel, so as to achieve the purpose of attenuating external vibration. In this case, the valve slice mechanical model is established, and the deflection formula under the nonuniform load is deduced, which can calculated the deformation of valve slice at any position, and the analysis of the influence factors is carried out, which lays a foundation for study the performance characteristics of the damping device.

Keywords: Damping valve · Annular valve slice · Nonuniform load · Flexural deformation

1 Introduction

There are many methods to produce damping force in the shock absorbers of vehicle suspension, and one of the methods which oil is throttled through deformation aperture of valve slice is widely used in engineering practice. It is easier for processing and assembling, and its cost is lower. Since the amount of deformation of the throttle plate is related to the size of the oil flow area, it will directly affect the external characteristics of the vibration damping device and the suspension damping ratio, and ultimately change the overall driving performance of the vehicle, so accurate calculation is required. In the prior art, the deformation of the annular valve plate is mainly solved by the uniform solution load solution formula and the finite element numerical calculation method of the mechanical design manual, which limits its application in engineering design [1, 2].

When the throttle plate is subjected to special loading mode, there is no corresponding deformation solution formula for people's reference. The design process can only be calculated by approximate and simplified methods, but practice shows that it is easy to cause the actual vibration damping device. The large error between the damping force and the theoretical value does not meet the product requirements well.

Based on this situation, this paper intends to use the relevant theory of thin plate mechanics to conduct mechanical derivation and simulation analysis of the deformation analytical formula of the valve plate subjected to local variable load in the typical assembly mode of the damping valve.

© Springer Nature Switzerland AG 2020
F. Xhafa et al. (Eds.): IISA 2019, AISC 1084, pp. 773–780, 2020.
https://doi.org/10.1007/978-3-030-34387-3_95

2 Physical Model of Throttle Valve

Figure 1 is a schematic view showing the installation and deformation of the annular throttle plate. The slice is mounted outside of the piston end, which belongs to the structure of unilateral valve. During the piston moves, if the pressure of above cavity is larger than the below cavity, there will form annular aperture between the slice and the end of the piston to throttle the oil and produce damping force in order to reduce the vehicle vibration when the slice is bended. After opening the valve, due to the influence of the deformation gap, the throttle plate will be subjected to a uniform load.

The force diagram of the throttle valve piece is extracted, the inner ring of the valve piece is fully constrained by the piston, and the outer ring is freely deformed. The specific structural parameters are defined as shown in Fig. 2. In the figure, q is the local uniform load, q_1 is the load on the outer diameter of the valve, h is the valve thickness, r_a is the valve inner diameter, r_b is the valve outer diameter, r_c is the piston inner diameter.

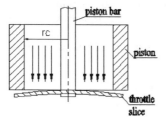

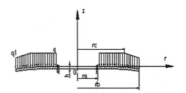

Fig. 1. Throttle valve plate installation deformation.

Fig. 2. Throttle valve force distribution

3 Valve Slice Flexural Deformation Derivation

From the thin plate mechanics, the differential equation of the deformation potential energy of the throttle plate under polar coordinates: [3].

$$U = \frac{D}{2} \iint \left[r \left(\frac{d^2w}{dr^2} \right)^2 + \frac{1}{r} \left(\frac{dw}{dr} \right)^2 + 2u \frac{dw}{dr} \frac{d^2w}{dr^2} \right] dr d\theta \qquad (1)$$

And the differential equation of the external force on the valve can be expressed as: [4].

$$W = \iint Fwr dr d\theta \qquad (2)$$

The total potential energy of the valve plate when the axis is symmetrical is as follows:

$$\Pi = \pi D \int \left[r \left(\frac{d^2 w}{dr^2} \right)^2 + \frac{1}{r} \left(\frac{dw}{dr} \right)^2 + 2u \frac{dw}{dr} \frac{d^2 w}{dr^2} \right] dr - 2\pi \int Fwrdr \qquad (3)$$

Using the principle of minimum energy, the total potential energy is minimized when the valve plate is in a stable equilibrium state:

$$\delta \Pi = 0 \qquad (4)$$

From Eqs. (2) and (3), the total potential energy is the radius of the functional, and the number of deflections of the valve plate is represented by the series: [5].

$$w(r) - \sum_m C_m w_m(r) \qquad m = 1, 2, 3 \cdots \qquad (5)$$

In which, C_m is m pending coefficients, $w_m(r)$ is shape function, each of its principles should in principle meet the geometric boundary conditions of a given valve.

Bringing the formula (5) into (4) for variation and deriving the following specific forms:

$$\frac{\partial U}{\partial C_m} = 2\pi \int Fw_m(r)rdr \qquad (6)$$

Through the above boundary conditions and various trade-offs, the following flexible surface function expression is used to solve:

$$w = \left(1 - \frac{r^2}{r_a^2} \right)^2 \left[C_1 + C_2 \left(1 - \frac{r^2}{r_a^2} \right) + C_3 \left(1 - \frac{r^2}{r_a^2} \right)^2 \right] \qquad (7)$$

When $r \leq r_c$, the load on the annular sheet is:

$$F = q_2 \qquad (8)$$

When $r_c \leq r \leq r_b$, The load expression for deriving the annular sheet is:

$$F = \frac{q_2 - q_3}{r_c - r_b} (r - r_c) + q_2 \qquad (9)$$

Bring the above two formulas into (2)–(7) for sequential derivation, and sort out the first-order, second-order derivatives and deformation potential expressions of w versus r;

$$\frac{dw}{dr} = -\frac{8C_3 r \left(r_a^2 - r^2\right)}{r_a^8 A_1}(A_2 + A_3) \tag{10}$$

$$A_1 = r_a^2(1+u)\left(r_a^2 - 3r_b^2\right) + 2r_b^4(2+u)$$

$$A_2 = 2r_b^4(u+2)\left(r_a^4 + r^4\right) - r_b^6(11+5u)\left(r_a^2 + r^2\right)$$

$$A_3 = r_a^2(u+1)\left(r_a^2 r^4 - 3r_b^2 r_a^2 r^2 - 3r_b^2 r^4\right) + 2r_b^4 r_a^2 r^2(4u+5) + 3r_b^8(3+u)$$

$$\frac{d^2 w}{dr^2} = -\frac{8C_3}{r_a^8 A_1}\{B_1 + B_2\} \tag{11}$$

$$B_1 = 2r_b^4(u+2)\left(r_a^6 - 7r^6\right) + r_b^6(5u+11)\left(5r^4 - r_a^4\right) + 3r_b^8(u+3)\left(r_a^2 - 3r^2\right)$$

$$B_2 = r^2 r_a^2(1+u)\left[3r_b^2\left(7r^4 - 3r_a^4\right) + r_a^2\left(18r_b^4 - 7r^4\right) + 5r^2\left(r_a^4 - 6r_b^4\right)\right]$$

Bring into Eqs. (2) and (3) and organize them:

$$U = -\frac{128\pi D C_3^2 \left(r_a^2 - r_b^2\right)^5}{35 r_a^{16} B_3^2}(B_4 + B_5 + B_6 + B_7) \tag{12}$$

$$B_3 = r_a^4(1+u) + 2r_b^4(2+u) - 3r_a^2 r_b^2(1+u)$$

$$B_4 = 3r_a^{12}\left(1 + u^2 + 2u\right) - 24 r_a^{10} r_b^2\left(u^2 + 1 + 2u\right)$$

$$B_5 = 8r_a^8 r_b^4\left(13 + 8u^2 + 21u\right) - 2r_a^6 r_b^6\left(31u^2 + 117 + 148u\right)$$

$$B_6 = r_a^4 r_b^8\left(313 + 25u^2 + 158u\right) - 2u r_a^2 r_b^{10}(9u - 39)$$

$$B_7 = 6r_b^{12}\left(15 + 2u^2 + 15u\right)$$

At the same time, the work equation of the external force on the valve piece is derived:

$$W = C_3(q_2 X + q_3 Y) \tag{13}$$

In which,

$$X = \frac{\pi}{3 r_a^8 B}(X_1 + X_2 + X_3 + X_4 + X_5 + X_6)$$

$$X_1 = \frac{2A(r_b^{11} - r_c^{11})}{11(r_c - r_b)} + \frac{A(r_b^{10} - r_c^{10})}{5}\left(1 - \frac{r_c}{r_c - r_b}\right) + \frac{2(r_b^9 - r_c^9)}{9}\frac{(C - 2r_a^2 A)}{r_c - r_b}$$

$$X_2 = \frac{(r_b^8 - r_c^8)}{4}(C - 2r_a^2 A)\left(1 - \frac{r_c}{r_c - r_b}\right) + \frac{2(r_b^7 - r_c^7)}{7(r_c - r_b)}(r_a^4 A - 2r_a^2 C + D_1)$$

$$X_3 = \frac{(r_b^6 - r_c^6)}{3}(r_a^4 A - 2r_a^2 C + D_1)\left(1 - \frac{r_c}{r_c - r_b}\right) + \frac{2(r_b^5 - r_c^5)}{5(r_c - r_b)}(r_a^4 C - 2r_a^2 D_1)$$

$$X_4 = \frac{(r_b^4 - r_c^4)}{2}(r_a^4 C - 2r_a^2 D_1)\left(1 - \frac{r_c}{r_c - r_b}\right) + \frac{2r_a^4 D_1(r_b^3 - r_c^3)}{3(r_c - r_b)}$$

$$X_5 = D_1 r_a^4(r_b^2 - r_c^2)\left(1 - \frac{r_c}{r_c - r_b}\right) + \frac{1}{5}A(r_c^{10} - r_a^{10}) + \frac{1}{4}(-2r_a^2 A + C)(r_c^8 - r_a^8)$$

$$X_6 = \frac{1}{3}(r_a^4 A - 2r_a^2 C + D_1)(r_c^6 - r_a^6) + \frac{1}{2}(r_a^4 C - 2r_a^2 D_1)(r_c^4 - r_a^4) + r_a^4 D_1(r_c^2 - r_a^2)$$

$$Y = 2\pi(Y_1 + Y_2 + Y_3 + Y_4)$$

$$Y_1 = \frac{-A(r_b^{11} - r_c^{11})}{33r_a^8 B(r_c - r_b)} + \frac{A(r_b^{10} - r_c^{10})}{30r_a^8 B}\frac{r_c}{r_c - r_b} - \frac{(r_b^9 - r_c^9)}{27r_a^8 B}\frac{(C - 2r_a^2 A)}{r_c - r_b}$$

$$Y_2 = \frac{(r_b^8 - r_c^8)}{24r_a^8 B}(C - 2r_a^2 A)\frac{r_c}{r_c - r_b} - \frac{(r_b^7 - r_c^7)}{21r_a^8 B(r_c - r_b)}(r_a^4 A - 2r_a^2 C + D_1)$$

$$Y_3 = \frac{(r_b^6 - r_c^6)}{18r_a^8 B}(r_a^4 A - 2r_a^2 C + D_1)\frac{r_c}{r_c - r_b} - \frac{(r_b^5 - r_c^5)}{15r_a^8 B(r_c - r_b)}(r_a^4 C - 2r_a^2 D_1)$$

$$Y_4 = \frac{(r_b^4 - r_c^4)}{12r_a^8 B}(r_a^4 C - 2r_a^2 D_1)\frac{r_c}{r_c - r_b} - \frac{D_1(r_b^3 - r_c^3)}{9r_a^4 B(r_c - r_b)} + \frac{D_1(r_b^2 - r_c^2)}{6r_a^4 B}\frac{r_c}{r_c - r_b}$$

$$A = \left(12r_b^4 + 3r_a^4 - 9ur_a^2 r_b^2 + 3ur_a^4 - 9r_a^2 r_b^2 + 6ur_b^4\right)$$

$$B = \left(r_a^4 + ur_a^4 - 3ur_a^2 r_b^2 + 4r_b^4 - 3r_a^2 r_b^2 + 2ur_b^4\right)$$

$$C = \left(2r_a^6 - 18r_a^4 r_b^2 + 48r_a^2 r_b^4 - 20ur_b^6 - 44r_b^6 - 18ur_a^4 r_b^2 + 36ur_a^2 r_b^4 + 2ur_a^6\right)$$

$$D_1 = \left(r_a^8 - 40ur_a^2 r_b^6 + ur_a^8 - 9r_a^6 r_b^2 + 30ur_a^4 r_b^4 + 54r_b^8 - 9ur_a^6 r_b^2 + 18ur_b^8 + 48r_a^4 r_b^4 - 88r_a^2 r_b^6\right)$$

Solve the undetermined coefficients and bring in the formula (7) to extract the common factors:

$$w = \frac{G_{L3}(r)}{Eh^3}(q_2 X + q_3 Y) \tag{14}$$

In which, G_{L3} is the special load deformation coefficient. E is the elastic modulus of the throttle slice.

Equation (14) is the analytical formula for the deflection of the annular throttle plate when it is subjected to local transformation. By collation, the coefficients G_{L3}, X and Y both include the throttle plate inner diameter, the outer diameter dimension, and the material Poisson's ratio, and G is also a function of the deflection radius, and X and Y are functions of the local average load radius.

Verify the correctness of the above mechanical derivation by finite element modeling. Firstly, the related parameters are assumed as follows: the inside radius $r_a = 27.5 \times 10^{-3}$ m, the outside radius $r_b = 40 \times 10^{-3}$ m, the thickness of slice $h = 0.5 \times 10^{-3}$ m, the load $q = 0.2$ Mpa, $q_1 = 0.1$ Mpa, the effecting radius $r_c = 30 \times 10^{-3}$ m, 35×10^{-3} m, the elasticity modulus $E = 2.06 \times 10^{11}$ Pa, the Poisson's ratio $u = 0.3$. The results are shown in Figs. 3 and 4.

Again, the local uniform distribution of load is changed as $q = 0.1$ Mpa, 0.4 Mpa, the edge loads are $q_1 = 0.05$ Mpa 0.2 Mpa, the effecting radius is $r_c = 33 \times 10^{-3}$ m, and the other parameters are the same as above. The results are shown in Figs. 5 and 6.

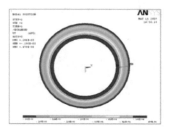

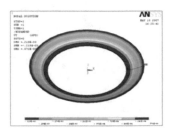

Fig. 3. The effecting radius is $30 \times 10^{-3} m$.   **Fig. 4.** The effecting radius is $35 \times 10^{-3} m$.

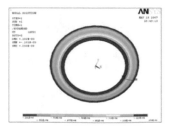

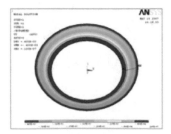

Fig. 5. The edge load is 0.05 Mpa.   **Fig. 6.** The edge load is 0.2 Mpa.

Analytical calculation and finite element simulation of the deformation error of the valve plate, as shown in the following tables.

Table 1. Comparison of different outer diameter deformations of valve plates

r_c/m	f_A/m	f/m	$\Delta f/\%$
30×10^{-3}	0.186×10^{-3}	0.1831×10^{-3}	1.56
35×10^{-3}	0.219×10^{-3}	0.2158×10^{-3}	1.46

Table 2. Comparison of different load deformations of valve plates

q_1/Mpa	f_A/m	f/m	$\Delta f/\%$
0.05	0.101×10^{-3}	0.0999×10^{-3}	1
0.2	0.405×10^{-3}	0.3995×10^{-3}	1.3

In the tables, f_A is the finite element calculation of the outer edge deformation of the valve, f is the analytical calculation of the outer edge deformation of the valve, and Δf is the calculation error (Tables 1 and 2).

The analytical results calculated by the formula are shown in Figs. 7, 8 and 9, the results show that as the load and its radius of action increase, the deflection of the throttling at any radius also shows an increasing trend.

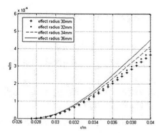

Fig. 7. The results in different load effecting radius.

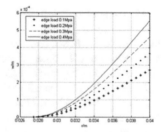

Fig. 8. The results in different edge loads.

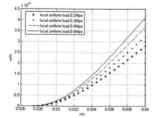

Fig. 9. The results in different local uniform loads.

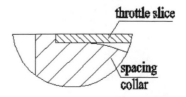

Fig. 10. Valve slice and spacing collar.

In addition, the formula (23) also can be used to design accurate spacing collar contour (shown in Fig. 10) to limit the maximum deformation of the slice at any radius in order to avoid stress concentration inducing the slice fatigue break when excessive deformation happens.

4 Conclusions

(1) Based on the relevant theory of the special loading method of thin plate mechanics, the analytic formula of the deflection deformation of the throttle valve piece under special load is derived, and the expression form which is convenient for theoretical analysis and engineering application are summarized.
(2) Compared with the finite element numerical solution, the correctness of the deformation analytical formula under the special load of the annular valve plate is verified, which provides theoretical support for product engineering design.
(3) The deformation analytical formula under different loading modes can accurately solve the deflection deformation at any radius of the throttle plate, which provides a theoretical basis for the accurate design of the limit retaining ring, and can also accurately calculate the gap flow area.

References

1. Da-xian, C.: Mechanical design handbook the first book. Bei Jing Chemic Industry Publishing House, p. 1-168-6 (2002)
2. Yuan-bing, H.: Design and calculation of elastic valve block for shock absorber valve train. Des. Test Study **5**, 13–15 (2002)
3. Duflou, J.R., Nguyen, T.H.M., Kruth, J.P.: Geometric reasoning for tool selection in sheet metal bending operations. In: International Conference on Integrated Design and Manufacturing in Mechanical Engineering (IDMME 2004), p. 15 (2004)
4. Hirota, K., Mori, Y.: Precision small angle bending of sheet metals using shear deformation. JSME Int. J. Ser. A (Solid Mech. Mat. Eng.) **48**(4), 352–357 (2005)
5. Alinger, M.J., Van Tyne, C.J.: Evolution of die surfaces during repeated stretch-bend sheet steel deformation. J. Mat. Process. Technol. **141**(3), 411–419 (2003)

Analysis of Hotspot Knowledge Graph of Chinese and Foreign Border Studies Since the 21st Century

Fang Du[1] and Lingmin Sun[2(✉)]

[1] School of International Studies, Collaborative Innovation Center for Security and Development of Western Frontier China, School of Public Administration, Sichuan University, Chengdu, China
[2] College of History and Culture, Chengdu, China
308969898@qq.com

Abstract. As an important geo-strategic fulcrum and marginal zone of political power, border areas are not only the geographical advantage zone of geo-cooperation between countries, but also the sensitive zone of geo-conflict between countries. Since the 21st century, domestic and foreign border research has presented the strategic characteristics and evolutionary trend of advancing with the times. This paper uses Citespace software to compare and analyze the hotspot maps and evolution trends of border research at home and abroad. The analysis shows that the development trajectories of border areas at home and abroad are roughly the same, and they have experienced three periods of "slow-fast-stable". Comparing the research hotspots in the border areas at home and abroad, its homogenization performance that the research hotspots at home and abroad is focused on the boundary connotation and political attributes, location advantages and border trade, cross-border cooperation and conflicts. At the same time, the heterogeneity of domestic border areas is mainly manifested in: domestic border research is more focused on the game of "national identity and national identity" and the game of "inter-ethnicism and regionalism". Finally, it is concluded that every country must face and make full use of the geographical advantage of the border and weakening the geographical disadvantage is the long-term responsibility and obligation of every border country.

Keywords: 21st century · Border area · Research hotspot · Citespace software

1 Introduction

The study of the border began with the discussion of national organisms in Rattzal (1844). He believes that the country is a growing space organism, the capital is the mind, the heart and the lungs, and the border is the terminal part. As a dividing line between sovereign states, it can be either tangible or intangible. Natural, it can also be artificial. The slogan: "When the king does not take the brave to the side, he is safe." It can be seen that the border as the pivot point of the national geostrategy is not only the battlefield of geopolitical and economic games, but also the sensitive zone of cross-border cooperation and conflict. China has a total of 22,000 km of land borders,

F. Xhafa et al. (Eds.): IISA 2019, AISC 1084, pp. 781–787, 2020.
https://doi.org/10.1007/978-3-030-34387-3_96

adjacent to 14 land countries and 135 border counties. The total area of the border area is 2.12 million square kilometers, accounting for 22% of the country's total area. The total population of the border area is 20.5 million, accounting for about the whole country. 1.6% of the total population. Therefore, it is of great theoretical and practical significance to fully understand the research trends of border areas at home and abroad. This paper uses Citespace software to draw the hot spot trend map of border research at home and abroad.

2 Data Sources and Analysis Methods

2.1 Data Sources

First of all, the retrieval period is set to 2000–2017. The foreign data mainly come from Web of Science database, the subject word is "Border or Borderland". After data processing, 4720 effective scientific papers are obtained. The main source of domestic data is CNKI database, the key word is "border", and 1259 effective scientific papers are obtained, as shown in Fig. 1.

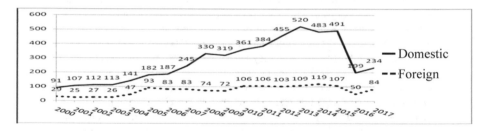

Fig. 1. Time distribution of literature in border areas at home and abroad

Comparative analysis of the above domestic and foreign border research development trajectory, domestic and foreign development context is roughly the same, has experienced three stages: (1) slow growth period. The main time period was around 2000–2004, and the number of published articles in this stage was generally not large. The number of published articles in foreign countries was about 105, and nearly 30 were published in China. (2) rapid growth period. From 2005 to 2010, the number of published articles in this stage increased rapidly. In 2008, the number of published articles in foreign countries exceeded 300 for the first time, reaching 330. The number of domestic publications exceeded 100 for the first time in 2010, reaching 106. Subsequently, the average number of articles published per year was basically maintained at 106. (3) stable growth period. After 2010, the average number of foreign publications is about 400, and the average number of domestic publications is about 100.

2.2   Analytical Methods

The scientific analysis tool used in this paper is Citespace, which is a visual knowledge mapping software developed under the JAVA language environment. The main principle is to measure hotspots and development trends in the field. Firstly, introduce Citespace domestic and foreign border research literature for statistical collation; secondly, the literature is collated and the hotspot time zone map is drawn, the hotspots and evolution of border research are visually observed.

3   Country Distribution of Border Studies

Since the advent of democracies, research on borders has never stopped. In particular, since the 21st century, with the deepening of globalization, there are more and more interactions and connections between countries, and the research on border is becoming more and more important. The research fields and topics of concern are becoming more and more prominent with their own characteristics. Table 1 lists the country distribution of border research.

Table 1. List of foreign border research countries

Rank	Country	Frequency of occurrence	Centrality
1	USA	1520	0.48
2	England	566	0.19
3	Germany	357	0.17
4	Canada	310	0.10
5	France	190	0.14
6	Australia	189	0.09
7	Netherlands	182	0.07
8	Spain	158	0.03
9	Italy	138	0.05
10	China	118	0.06

It is known from the above table: First, the North American region, led by the United States and Canada, has the strongest research strength, with 1,520 articles in the United States and 310 articles in Canada, accounting for 43.7% of the total. Among them, the US border research is extremely strong and possesses absolute Status, the centrality distribution of the two is 0.48 and 0.1, the higher the centrality, indicating that the research theme is more concentrated. Secondly, the number of representatives from European countries such as the United Kingdom, Germany, France, the Netherlands, Spain and Italy has reached 1,791, accounting for 48.0%, which indicates that the overall research strength of European countries is comparable to that of North America; Oceania is mainly Australia, 189 The center degree is 0.09. Finally, as a representative of the Asian region, China is ranked 10th in total, with 118 appearances and a center of 0.06. This shows that the Chinese border research in the international perspective has begun to develop rapidly.

In recent years, China has adjusted its border foreign policy, and the border area has changed from the traditional security buffer zone to the front position of opening to the outside world. Therefore, the research on the border has developed rapidly. Among the 14 border countries bordering China, due to their different degrees of development, such as polity, economy, culture, nationality, religion, security and so on, the research content and research paradigm of the border show extensive and intersecting integration. Table 2 lists the top 10 border researchers in China. Through the analysis, it can be concluded that the overall influence of domestic scholars on border research is relatively weak, and the citation frequency is generally not high.

Table 2. Analysis of domestic researchers

Sequence number	Author	Cited frequency	Number of articles
1	Li Xiaolin	41	93
2	Hu Zhiding	12	11
3	Long Yao	10	9
4	Chen Dingbo	10	24
5	He Yue	10	5
6	Luo Huasong	8	7
7	Wang Jian	8	10
8	Jiang Xiaolong	7	9
9	Li Hong	7	11
10	Li Cansong	7	6

4 The Hot Topic Analysis of Border Research

Import Wos data and CNKI data in Citespace, and draw time zone maps of research hotspots in border areas at home and abroad (Figs. 2 and 3). In the time zone diagram, the color in the inner circle of the node represents the change from blue (early) to warm (nearly). The more frequently the keywords appear, the larger the nodes and the higher the centrality, which indicates that it is a research hotspot in the field. According to Fig. 2, research hotspots in foreign border areas focus on border concepts and theories, border trade, border policy, border politics, border models, geopolitics, border immigration, border markets, and US border research. According to Fig. 3, research hotspots in domestic border areas focus on border area concepts, location advantages and geo-economics, identity and geopolitics, cross-border conflicts and cooperation, and border cooperation models.

Compared with the domestic and foreign border time zone maps (Figs. 2 and 3), the mainstream research fields of the border are hot at home and abroad, such as border definition and political attributes, location advantages and border zone map and trade in the domestic border areas. Border conflict and cooperation. At the same time, the domestic border areas under the international vision show their own uniqueness, such

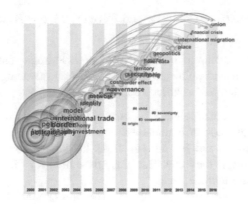

Fig. 2. Time zone map of foreign border areas

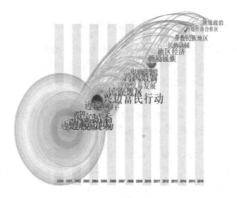

Fig. 3. Time zone map of domestic border areas

as the game of national identity and national identity, the game of nationalism and regionalism, and so on. The similarities and differences are analyzed as shown in the following figure (Table 3):

Table 3. Analysis of similarities and differences of research hotspots in border areas

Analysis in border areas	Heterogeneity	Homogeneity
Foreign	Border immigration, border mode	Border definition and political attributes Location advantage and border trade Cross-border conflicts and cooperation
Domestic	National identity and national identity Internationalism and regionalism	

4.1   The Homogeneity of Domestic and Foreign Border Research Hotspots

The homogeneity of research hotspots in border areas at home and abroad is embodied in: border connotation and political attributes, location advantages and border trade, cross-border cooperation and conflicts. details as follows:

(1) Border connotation and political attribute. The main focus was on the study of the definition and division of boundaries.
(2) Location advantage and Border Trade. As the border zone of national political power, the border is the initial and final space field for the exchange of interests between the state and the state.
(3) Cross-border cooperation and conflict. The border area is not only a regional advantage zone for cross-border cooperation, but also a sensitive zone for cross-border conflicts, as a space field for cross-border cooperation and conflict.

4.2   Analysis of Heterogeneity of Hotspot Research at Home and Abroad

As one of the countries with the largest number of countries in the world, China faces a complicated border situation and borders with 14 countries. Most of these countries are at the level of development, and a few countries have a civil war and political instability. Therefore, on the basis of maintaining its homogeneity, its research hotspots also show heterogeneity, mainly in the game of "the game of national identity and national identity" and "inter-ethnicism and regionalism".

(1) The game between "national identity" and "national identity". Multi-ethnic countries often face the dilemma of "national identity" and "national identity", and China is no exception.
(2) The game between "inter-ethnicism" and "regionalism". There has always been a game between "chauvinism" and "regionalism" in the governance tendency of border areas. The orientation of inter-ethnic governance is to equate border areas with ethnic areas, focusing on solving conflicts and contradictions between ethnic groups, mainly around ethnic issues.

5   Conclusions

Based on the research literature of the border areas, this paper sorts out the hotspot time maps of research papers in the border areas since the 21st century. Compared with foreign border research topics, domestic and foreign research has its homogeneity and heterogeneity. Among them, (1) homogeneity is manifested in: border definition and political attributes, location advantages and border trade, cross-border conflicts and cooperation; (2) Heterogeneity is mainly manifested in the game of "national identity and national identity" and the game of "inter-ethnicism and regionalism". Generally speaking, China's overall level and strength of border research has been continuously improved in Asian countries. when relying on Citespace for research hot spot analysis, we should pay attention to the thinking of a large number of related subject literature as the basis.

Acknowledgments. This research was financially supported by the "Sichuan University", Chengdu, China;

① Grant No. 2018hhf-22, "Research on Modern Construction of Non-Traditional Security Governance System and Governance Capacity in Southwest Frontier".

② Grant No. skbsh201814, "Study on the Competence and Competence of Cadres of Grassroots Nationalities and its Influence on Work Performance".

③ Grant No. 2018ggxy-03, "Research on Competency Evaluation and Promotion Mechanism of Grass-roots Cadres in Ethnic Minority Areas from the Perspective of Rural Revitalization".

④ Grant No. 2018skzx-pt54, "Study on the Evaluation System and Optimizing Path of the Ruling Ability of Grass-roots Leading Cadres in Southwest Border Areas".

⑤ Grant Self-study in 2019-School of International Studies 03. "A Study on the Competency of the First Secretary for Poverty Alleviation in Ethnic Minority Areas and Its Impact on the Effectiveness of Poverty Alleviation".

This research was still supported by the "National Social Science Fund", China.

⑥ Grant No. 17AZD018. "The Principle of Chinese Frontier Science".

References

1. Mao, H.: Geopolitical and geoeconomic patterns and countermeasures in China. Adv. Geogr. **03**, 289–302 (2014)
2. Luo, Z.: Reflections on some major issues in the study of borderland in Western China. J. Sichuan Univ. (Philos. Soc. Sci.) **01**, 5–12 (2015)
3. Li, J., Li, H.: China's transnational ethnic groups and governance of frontier public affairs. Public Manag. News **01**, 1–10 (2015)
4. Song, T., Liu, W., Li, W.: Research progress and enlightenment of foreign countries from the perspective of geographical perspectives. Adv. Geogr. **03**, 276–285 (2016)
5. Yang, X.: Influencing factors and regional patterns of China's border trade. Reform **06**, 110–117 (2013)
6. Zhang, W.: "Paradise is very far away, but China is very close"—the law and characteristics of geopolitical interaction between China and neighboring countries and regions. World Econ. Polit. **01**, 5–39 (2013)
7. Ping, Z.: Analysis of national identity issues in multi-ethnic countries. Polit. Sci. Res. **01**, 26–40 (2013)
8. Ping, Z.: China's borderland governance: inter-ethnicism or regionalism? Thought Front **03**, 25–30 (2008)
9. Aker, J.C., Klein, M.W., O'Connell, S.A., et al.: Borders, ethnicity and trade. J. Dev. Econ. **107**, 1–6 (2014)
10. Lan, H.F., Sin, L.Y.M., Chan, K.K.C.: Chinese cross-order shopping: an empirical study. J. Hosp. Tour. Res. **29**(1), 110–133 (2005)

Optimality Conditions for a Class of Major Constraints Nonsmooth Multi-objective Programming Problems

Xuanwei Zhou[✉]

School of Basic Courses, Zhejiang Shuren University, Hangzhou 310015, China
wdzxw@163.com

Abstract. The optimality conditions of major constraints nonsmooth programming with multi-objectives are studied. The objective function of this class of multi-objective programming is the sum of a convex vector function and a differentiable vector function. The constraint condition is major cone constraint in Euclid space. By using of the structure representation of major cone constraint of the given problem, the Fritz John condition of Pareto weakly effective solution is obtained by Gordan theorem. Meanwhile, the Kuhn-Tucker condition for Pareto weakly effective solution under Slater constraint qualification is given. The results are very useful to design its numerical methods.

Keywords: Nonsmooth multiobjective programming · Gordan theorem · Fritz John condition · Slater constraint qualification · Kuhn-Tucker condition

1 Introduction

Multi-objective programming is an interdisciplinary of operations research and decision sciences. It is an emerging discipline and has been rapidly developed in nearly 50 years. Because most realistic optimization problems involve multiple objectives, people have paid great attention to the study of multi-objective programming since the 1970s (see Ref. [1–9]). Nonsmooth problems are a common phenomenon in nature. Since the 1960s, more and more problems involve nonsmooth optimization due to the continuous development of computing technology and the needs of research and actual calculation of civil engineering, mechanical engineering, management science, economics and national defense construction(see Ref. [10–15]). While an actual optimization model is established, decision makers give these constraint conditions in the different needs, thus it inevitably occurs that the constraint conditions of the optimization model are incompatible. Therefore, a new class of nonsmooth multiobjective programming is considered here. It is called major constraints nonsmooth multiobjective programming. Relatives to the general multiobjective programming model, the major constraint conditions are compatible.

In Ref. [16], with the help of major cone, a major constrained programming problem was proposed, which solved such a case of incompatibility of constraint conditions. In Ref. [17], the structural representation of major constraint set was given. For the major constraint single objective programming problem, the definition of the

© Springer Nature Switzerland AG 2020
F. Xhafa et al. (Eds.): IISA 2019, AISC 1084, pp. 788–795, 2020.
https://doi.org/10.1007/978-3-030-34387-3_97

major constraint optimal solution was given, and the optimality conditions (OC) of major optimal solution were obtained. This paper studies the OC of major constraints nonsmooth programming with multi-objectives. The objective function of this class of multi-objective programming is the sum of two vector functions, one of them is convex and another one is differentiable; the constraint condition is major cone constraint in Euclid space. By using of the structure representation of major cone constraint of the given problem, the Fritz John condition of Pareto weakly effective solution is obtained by Gordan theorem. Kuhn-Tucker condition for Pareto weakly effective solution under Slater constraint qualification is given also. The results are very useful to design its numerical methods.

2 Definitions and Lemmas

Let $w = f + \varphi$, g be two vector functions, they are defined on R^n, where $w = (w_1, \cdots, w_m) = (f_1(x) + \varphi_1(x), \cdots, f_m(x) + \varphi_m(x))$, $g = (g_1(x), \cdots, g_p(x))$. Let $f_i(x)$ (i from 1 to m), $g_j(x)$ (j from 1 to p) be differentiable, and let $\varphi_i(x)$ (i from 1 to m) be convex.

The general multiobjective programming is $\min f(x)(x \in X)$. The constraint set $X = \{x|g(x) \leq 0\}$ satisfies $g_j(x) \leq 0$ for all $j \in \{1, \cdots, p\}$.

While an actual optimization model is set up, the various constraint conditions are often independently given by decision makers respectively, and sometimes the inequality constraints will appear incompatible, the constraint set X of this multi-objective programming problem then is an empty set. Thus the multi-objective programming problem loses the significance of further study in this case, so we consider the problem as follow. This is called as major constraints nonsmooth programming with multi-objectives.

$$\begin{cases} \min w(x) \\ s.t. (g_1(x), \cdots, g_p(x)) \underset{clH}{\leq} 0, x \in C, \end{cases} \qquad \text{(MCMP)}$$

where $C \subset R^n$ is convex with $riC \neq \emptyset$.

The major constraints condition of problem (MCMP) $(g_1(x), \cdots, g_p(x)) \underset{clH}{\leq} 0$ means that there exist at least $[\frac{p+1}{2}]$ numbers $i_1, \cdots, i_{[\frac{p+1}{2}]} \in \{1, \cdots, p\}$ such that x satisfies the inequality system $g_{i_1}(x) \leq 0, \cdots, g_{i_{[\frac{p+1}{2}]}}(x) \leq 0$.

The major constraints set X_M is defined by

$$X_M = \{x|g(x) \underset{clH}{\leq} 0\}.$$

Definition 2.1. If $x^* \in X_M \cap C$ and there exists no $x \in X_M \cap C$ meeting

$$w_i(x) < w_i(x^*) \ \forall i \in \{1, \cdots, m\},$$

then x^* is called as major constraints Pareto weakly effective solution of (MCMP). The set of all major constraints Pareto weakly effective solution is denoted by S_M^w.

In order to give the representation of major constraints set, there are the following two lemmas by Ref. [17].

Let $v = (v_1, \cdots, v_p)^T \in R^p$, and let $|v|_+$, $|v|_-$ and $|v|_0$ be the number of components of the greater than zero, the less than zero and the equal to zero respectively.

Lemma 2.1. Let $H \subset R^p$ be the major cone. Then its closed cone

$$clH = \{v \in R^p \mid |v|_+ + |v|_0 \geq [\frac{p+1}{2}]\}.$$

Let $i_1, \cdots, i_{[\frac{p+1}{2}]} \in \{1, \cdots, p\}$, $K_{i_1, \cdots, i_{[\frac{p+1}{2}]}} = \{v \in R^p \mid v_{i_1} \geq 0, \cdots, v_{i_{[\frac{p+1}{2}]}} \geq 0\}$.

Lemma 2.2. Let $H \subset R^p$ be the major cone. Then its closed cone

$$clH = \bigcup_{i_1, \cdots, i_{[\frac{p+1}{2}]} \in \{1, \cdots, p\}} K_{i_1, \cdots, i_{[\frac{p+1}{2}]}}.$$

Now denote

$$X_{i_1, \cdots, i_{[\frac{p+1}{2}]}} = \{x \in R^n \mid -g(x) \in K_{i_1, \cdots, i_{[\frac{p+1}{2}]}}\}.$$

Then we have the structure representation of the major constraints set X_M. This is

$$X_M = \cup X_{i_1, \cdots, i_{[\frac{p+1}{2}]}},$$

for all $i_1, \cdots, i_{[\frac{p+1}{2}]} \in \{1, \cdots, p\}$.

3   Optimality Conditions (OC)

Now we give two necessary OC of major constraints Pareto weakly effective solution of (MCMP).

Theorem 3.1. (Fritz John condition) Let $x^* \in S_M^w$ be a major constraints Pareto weakly effective solution of (MCMP) and let $f_i(x)$ (i from 1 to m), $g_j(x)$ (j from 1 to p) be differentiable at x^* and $\varphi_i(x)$ (i from 1 to m) be convex on R^n and continuous at x^*. Then there exist $(\tilde{w}_1, \cdots, \tilde{w}_m)^T \in R^m$, $(\tilde{\lambda}_{i_1}, \cdots, \tilde{\lambda}_{i_{[\frac{p+1}{2}]}}) \in R^{[\frac{p+1}{2}]}$ satisfying

$$0 \in \sum_{i=1}^{m} L_i(x^*) + R(x^*) + \sum_{k=1}^{\lceil \frac{p+1}{2} \rceil} \tilde{\lambda}_{i_k} \nabla g_{i_k}(x^*) + N_C(x^*), \tag{1}$$

$$\tilde{\lambda}_{i_k} g_{i_k}(x^*) = 0 (k \in \{1, \cdots, \lceil \frac{p+1}{2} \rceil\}), \tag{2}$$

$$(\tilde{w}_1, \cdots, \tilde{w}_m)^T \geq 0, (\tilde{\lambda}_{i_1}, \cdots, \tilde{\lambda}_{i_{\lceil \frac{p+1}{2} \rceil}})^T \geq 0, (\tilde{w}_1, \cdots, \tilde{w}_m, \tilde{\lambda}_{i_1}, \cdots, \tilde{\lambda}_{i_{\lceil \frac{p+1}{2} \rceil}})^T \neq 0, \tag{3}$$

where $L_i = \tilde{w}_i \nabla f_i, R = \partial(\sum_{i=1}^{m} \tilde{w}_i \varphi_i)$.

Proof. Since $x^* \in S_M^w$ is a major constraints Pareto weakly effective solution of (VMCP), By Definition 2.1, there is no $x \in X_M \cap C$ meeting

$$w_i(x) < w_i(x^*) (\forall i \in \{1, \cdots, m\}).$$

Hence there are $i_1, \cdots, i_{\lceil \frac{p+1}{2} \rceil} \in \{1, \cdots, p\}$ satisfying $x^* \in X_{i_1, \cdots, i_{\lceil \frac{p+1}{2} \rceil}} \cap C$ by the structure representation of the major constraints set X_M. So there is no $x \in X_{i_1, \cdots, i_{\lceil \frac{p+1}{2} \rceil}} \cap C$ meeting $w_i(x) < w_i(x^*) (\forall i \in \{1, \cdots, m\})$.

We are first to prove that the following inequalities system

$$\begin{cases} M_i(x) + \varphi_i(x) - \varphi_i(x^*) < 0, i = 1, \cdots, m \\ N_j(x) < 0, j \in J \end{cases} \tag{4}$$

has no solution x in riC, where $M_i(x) = \nabla f_i(x^*)^T (x - x^*)$, $N_i(x) = \nabla g_j(x^*)^T (x - x^*)$, $J = \{j | g_j(x^*) = 0, j \in \{i_1, \cdots i_{\lceil \frac{p+1}{2} \rceil}\}\}$. Assume that there is $\hat{x} \in riC$ meeting system (4). Since $x^* \in C$ and $\hat{x} \in riC$, it deduces $\theta(v) = x^* + v(\hat{x} - x^*) \in affC$. We can prove that $\theta(v) \in riC$ if v is sufficiently small, that is, there exists $\delta_1 > 0$ meeting $\theta(v) \in riC$ for $v \in (0, \delta_1)$. Since $\hat{x}$ solves the system (4),

$$N_j(\hat{x}) < 0, j \in J.$$

It means that

$$\lim_{v \to 0^+} \frac{g_j(\theta(v)) - g_j(x^*)}{v} < 0, j \in J.$$

Hence, there is $\delta_2 > 0$ meeting

$$g_j(\theta(v)) - g_j(x^*) < 0, \forall j \in J, \forall v \in (0, \delta_2). \tag{5}$$

If $j \in \{i_1, \cdots i_{\lceil \frac{p+1}{2} \rceil}\} \backslash J$, then $g_j(x^*) < 0$ by the assumption. It implies that there is $\delta_3 > 0$ meeting

$$g_j(\theta(v)) < 0, \forall v \in (0, \delta_3). \tag{6}$$

Since $\hat{x}$ also satisfies the inequalities system (4), for $i \in \{1, \cdots, m\}$

$$M_i(\hat{x}) + \varphi_i(\hat{x}) - \varphi_i(x^*) < 0.$$

Hence there is $\varepsilon > 0$ meeting

$$M_i(\hat{x}) + \varphi_i(\hat{x}) - \varphi_i(x^*) < -\varepsilon. \tag{7}$$

By the assumption, $\varphi_i(x)$ is convex, we have that each function $\frac{\varphi_i(\theta(v)) - \varphi_i(x^*)}{v}$ $(i \in \{1, \cdots, m\})$ is a monotone increasing about $v > 0$. Therefore for $v \in (0, 1)$,

$$\frac{\varphi_i(\theta(v)) - \varphi_i(x^*)}{v} \leq \varphi_i(\theta(1)) - \varphi_i(x^*) = \varphi_i(\hat{x}) - \varphi_i(x^*).$$

Since $\lim_{v \to 0^+} \frac{f_i(\theta(v)) - f_i(x^*)}{v} = \nabla M_i(\hat{x})$, there is $\delta_4 > 0$ meeting

$$\frac{f_i(\theta(v)) - f_i(x^*)}{v} < M_i(\hat{x}) + \varepsilon, \forall v \in (0, \delta_4).$$

So from (7)

$$\frac{f_i(\theta(v)) - f_i(x^*)}{v} + \frac{\varphi_i(\theta(v)) - \varphi_i(x^*)}{v}$$

$$< M_i(\hat{x}) + \varepsilon + \varphi_i(\hat{x}) - \varphi_i(x^*) < 0. \tag{8}$$

Taking $\delta = \min\{\delta_1, \delta_2, \delta_3, \delta_4\}$, we then deduce from (8) that

$$f_i(\theta(v)) + \varphi_i(\theta(v)) - f_i(x^*) - \varphi_i(x^*) < 0.$$

It is a contradiction to that x^* is a major constraints Pareto weakly effective solution of (VMCP), since $\theta(v) \in X \cap riC$ by (5) (6). So, (4) has no solution in riC. From Theorem 21.2 in Ref. [18], there are $(\tilde{w}_1, \cdots, \tilde{w}_m)^T \in R^m$, $(\tilde{\lambda}_{i_1}, \cdots, \tilde{\lambda}_{i_{\lceil \frac{p+1}{2} \rceil}}) \in R^{\lceil \frac{p+1}{2} \rceil}$ with $(\tilde{w}_1, \cdots, \tilde{w}_m, \tilde{\lambda}_{i_1}, \cdots, \tilde{\lambda}_{i_{\lceil \frac{p+1}{2} \rceil}})^T \neq 0$ satisfying

$$\sum_{i=1}^{m} \tilde{w}_i M_i(x) + \sum_{i=1}^{m} \tilde{w}_i(\varphi_i(x) - \varphi_i(x^*)) + \sum_{j \in J} \tilde{\lambda}_j N_j(x) \geq 0, \forall x \in riC,$$

By Theorem 27.4 in Ref. [18], there is $\xi \in \partial(\sum\limits_{i=1}^{m} \tilde{w}_i \varphi_i)(x^*) = R(x^*)$ meeting

$$\sum_{i=1}^{m} \tilde{w}_i M_i(x) + \xi^T(x - x^*) + \sum_{j \in J} \tilde{\lambda}_j N_j(x) \geq 0, \forall x \in riC.$$

Taking $\tilde{\lambda}_j = 0 (j \notin J)$, it means

$$\sum_{i=1}^{m} \tilde{w}_i M_i(x) + \xi^T(x - x^*) + \sum_{k=1}^{\lceil \frac{p+1}{2} \rceil} \tilde{\lambda}_{i_k} N_{i_k}(x) \geq 0, \forall x \in riC,$$

$$\tilde{\lambda}_{i_k} g_{i_k}(x^*) = 0 (k \in 1, \cdots, \lceil \frac{p+1}{2} \rceil). \tag{9}$$

By the definition of $N_{riC}(x^*)$, (9) implies

$$0 \in \sum_{i=1}^{m} L_i(x^*) + R(x^*) + \sum_{k=1}^{\lceil \frac{p+1}{2} \rceil} \tilde{\lambda}_{i_k} \nabla g_{i_k}(x^*) + N_{riC}(x^*).$$

Since C is convex and riC is nonempty, it means $N_{riC}(x^*) = N_C(x^*)$. Therefore, (1) (2) and (3) hold true.

Next we give the Kuhn-Tucker condition for Pareto weakly effective solution under Slater constraint qualification (SCQ).

SCQ (See Ref. [19]):

Each $g_j(j \in J)$ is pseudoconvex at x^*, each $g_j(j \in \{i_1, \cdots, i_{\lceil \frac{p+1}{2} \rceil}\} \backslash J)$ is continuous at x^* and there is an $\bar{x} \in riC$ meeting $g(\bar{x}) < 0$.

Theorem 3.2. Let $x^* \in S_M^w$ and let $f_i(x)$ (i from 1 to m), $g_j(x)$ (j from 1 to p) be differentiable at x^* and $\varphi_i(x)$ (i from 1 to m) be convex on R^n and continuous at x^*, and $g_j(j \in J)$ satisfy SCQ at x^*. Then there are $(\tilde{w}_1, \cdots, \tilde{w}_m)^T \in R^m$, $(\tilde{\lambda}_{i_1}, \cdots, \tilde{\lambda}_{i_{\lceil \frac{p+1}{2} \rceil}}) \in R^{\lceil \frac{p+1}{2} \rceil}$ satisfying

$$0 \in \sum_{i=1}^{m} L_i(x^*) + R(x^*) + \sum_{k=1}^{\lceil \frac{p+1}{2} \rceil} \tilde{\lambda}_{i_k} \nabla g_{i_k}(x^*) + N_C(x^*), \tag{10}$$

$$\tilde{\lambda}_{i_k} g_{i_k}(x^*) = 0 (k \in \{1, \cdots, \lceil \frac{p+1}{2} \rceil\}), \tag{11}$$

$$(\tilde{w}_1, \cdots, \tilde{w}_m)^T \geq 0, (\tilde{\lambda}_{i_1}, \cdots, \tilde{\lambda}_{i_{\lceil \frac{p+1}{2} \rceil}})^T \geq 0, (\tilde{w}_1, \cdots, \tilde{w}_m)^T \neq 0. \tag{12}$$

where $L_i = \tilde{w}_i \nabla f_i, R = \partial(\sum\limits_{i=1}^{m} \tilde{w}_i \varphi_i)$.

Proof. By Theorem 3.1, there are $(\tilde{w}_1, \cdots, \tilde{w}_m)^T \in R^m$, $(\tilde{\lambda}_{i_1}, \cdots, \tilde{\lambda}_{i_{\lfloor \frac{p+1}{2} \rfloor}}) \in R^{\lfloor \frac{p+1}{2} \rfloor}$ with $(\tilde{w}_1, \cdots, \tilde{w}_m, \tilde{\lambda}_{i_1}, \cdots, \tilde{\lambda}_{i_{\lfloor \frac{p+1}{2} \rfloor}})^T \neq 0$ meeting (10) (11). We just have to prove that $(\tilde{w}_1, \cdots, \tilde{w}_m)^T \neq 0$. By contradiction, suppose that $(\tilde{w}_1, \cdots, \tilde{w}_m)^T = 0$, then we have from (10)

$$0 \in \sum_{k=1}^{\lfloor \frac{p+1}{2} \rfloor} \tilde{\lambda}_{i_k} \nabla g_{i_k}(x^*) + N_C(x^*). \tag{13}$$

From (10) (11) (13), we have

$$\sum_{j \in J} \tilde{\lambda}_j N_j(x) \geq 0, \forall x \in riC. \tag{14}$$

Since each $g_j(j \in J)$ satisfies SCQ at x^*, it means $g_j(j \in J)$ is pseudoconvex at x^* and there is an $x \in riC$ meeting $g_j(\bar{x}) < 0$ $(\forall j \in J)$. By the definition of pseudoconvex (see page 51 of Ref. [19]), we have

$$N_j(\bar{x}) < 0, \forall j \in J.$$

Since $(\tilde{\lambda}_{i_1}, \cdots, \tilde{\lambda}_{i_{\lfloor \frac{p+1}{2} \rfloor}})^T \geq 0$ and $(\tilde{\lambda}_{i_1}, \cdots, \tilde{\lambda}_{i_{\lfloor \frac{p+1}{2} \rfloor}})^T \neq 0$, the above inequalities imply $\sum_{j \in J} \tilde{\lambda}_j N_j(\bar{x}) < 0$. Which is a contradiction to (14). Therefore, (10) (11) and (12) hold true.

4   Conclusion

Here the constraint conditions of the nonsmooth multi-objective programming model are incompatible, but its major constraint conditions are compatible. So this paper studies a new class of nonsmooth multi-objective programming. This new type of nonsmooth multi-objective programming is called as be major constraints nonsmooth multi-objective programming.

The establishment of OC is very important to an optimization problem. This paper gives the concept of major constraints Pareto weakly effective solution. Then using of the structure representation of major cone constraint, the Fritz John condition and Kuhn-Tucker condition of Pareto weakly effective solution are obtained.

Furthermore, we will study some sufficient OC for this new type of nonsmooth programming with multi-objective. We will also discuss the stability and sensitivity analysis of the major constraints Pareto weakly effective solution.

References

1. Bao, T.Q., Gupta, P., Mordukhovich, B.S.: Necessary conditions in multiobjective optimization with equilibrium constraints. J. Optim. Theory Appl. **135**, 179–203 (2007)
2. Bao, T.Q., Mordukhovich, B.S.: Variational principles for set-valued mappings with applications to multiobjective optimization. Control Cybernet. **36**, 531–562 (2007)
3. Bao, T.Q., Mordukhovich, B.S.: Necessary conditions for super minimizers in constrained multiobjective optimization. J. Glob. Optim. **43**, 533–552 (2009)
4. Bao, T.Q., Mordukhovich, B.S.: Relative Pareto minimizers for multiobjective problems: existence and optimality conditions. Math. Program. **122**, 301–347 (2010)
5. Luc, D.T.: On duality theorems in multiobjective programming. J. Optim. Theory Appl. **48**, 557–582 (1984)
6. Ben-Tal, A., Zowe, J.A.: Unified theory of first and second order conditions for extremum problems in topological vector spaces. Math. Program. Study **19**, 39–76 (1982)
7. Michel, V.: Theorems of the alternative for multivalued mappings and applications to mixed convex\concave systems of inequalities. Set-Valued Anal. **18**, 601–616 (2010)
8. Yang, X.M.: Generalized Preinvexity and Second Order Duality in Multiobjective Programming. Springer, Singapore (2018)
9. Luc, D.T.: Multiobjective Linear Programming: An Introduction. Springer, Cham (2016)
10. Mond, B.: A class of nondifferentiable mathematical programming problems. J. Math. Anal. Appl. **46**, 169–174 (1974)
11. Mond, B., Schechter, M.A.: On a constraint qualification in a nonlinear programming problems. New Res. Logic Q. **23**, 611–613 (1976)
12. Xu, Z.K.: Constraint qualifications in a class of nondifferentiable mathematical programming problems. J. Math. Anal. Appl. **302**, 282–290 (2005)
13. Luo, H.Z., Wu, H.X.: K-T necessary conditions for a class of nonsmooth multiobjective programming. OR Trans. **7**, 62–68 (2003)
14. Wu, H.X., Luo, H.Z.: Necessary optimality conditions for a class of nonsmooth vector optimization problems. Acta Math. Appl. Sinica **25**, 87–94 (2009)
15. Yang, X.M., Yang, X.Q., Teo, K.L.: Higher-order generalized convexity and duality in nondifferentiable multiobjective mathematical programming. J. Math. Anal. Appl. **297**, 48–55 (2004)
16. Hu, Y.D.: Classes of major order of vector space. Chin. Ann. Math. **11**, 269–280 (1990)
17. Hu, Y.D., Zhou, X.W.: Major constraint programming and its optimality conditions. OR Trans. **6**, 69–75 (2002)
18. Rockfellar, R.T.: Convex Analysis. Princeton University Press, New Jersey (1972)
19. Bazaraa, M.S., Sherali, H.D., Shetty, C.M.: Nonlinear Programming: Theory and Algorithms. Wiley, Hoboken (2006)

Study on Activated Carbon Fiber Reactivation Process

Liyuan Yu, Bo Zhu$^{(\boxtimes)}$, and Kunqiao

Shandong University, Jinan 250014, China
zhubo@sdu.edu.cn

Abstract. The large specific surface area of activated carbon fiber (ACF) is the main reason for its excellent adsorption performance. In the production process of ACF the activation process is the key step which affect the ACF specific surface area, and the ACF specific surface area produced by different activation processes varies greatly. In this paper, a self-generated activation process was used for the secondary activation of the commercial viscose based activated carbon fiber. According to different activation factors, the specific surface area was used as the evaluation standard for the activation performance. The orthogonal test method was used to optimize the activation process, and the best process for the preparation of ACF with high specific surface area was finally obtained, which was of certain significance for the improvement of the adsorption performance of ACF.

Keywords: Active Carbon Fiber · Activated · Surface area · Orthogonal test

1   Introduction

Active Carbon Fiber (ACF) is a new type of activated carbon product developed after powdered, granular and fibrous activated carbon [1], which is produced by high-temperature activation of carbon fiber (such as phenolic fiber, PAN fiber, viscose fiber, asphalt fiber, etc.), with a large number of nanometer pore size on the surface, large specific surface area and unique physicochemical characteristics. Activated carbon fibers are amorphous carbon structures of amorphous state, which mainly exist in the form of disordered graphite microcrystals and amorphous carbon [2], and carbon is their basic constituent element.

Activated carbon fiber belongs to porous carbon material. According to the classification of pores by IUPAC (International Union of Pure and Applied Chemistry) [3], there are micropores, mesopores and macropores in activated carbon material, and 90% of the surface area is micropores. Compared with traditional activated carbon (GAC), ACF has the following characteristics:

- There is a great difference between the pore structure of ACF and GAC, pore distribution of ACF, which contains a large number of micropores with diameter less than 2.0 nm, are monodisperse state, and the orifice opening directly on the fiber surface [4]. The adsorbate has a short diffusion path to reach the adsorption

F. Xhafa et al. (Eds.): IISA 2019, AISC 1084, pp. 796–802, 2020.
https://doi.org/10.1007/978-3-030-34387-3_98

site, and the fiber diameter is fine. Therefore, the contact area with the adsorbed substance is large, which increases the adsorption probability and enables uniform contact.

- The specific surface area of ACF can up to 2500 m^2, about 10 to 100 times of GAC, so adsorption capacity is large, about 1.5 100 times that of GAC, and the adsorption capacity was 400 times more than GAC. The adsorption and desorption of ACF are fast, and the adsorption of ACF on gas can reach equilibrium within several minutes.

- The internal structure of GAC is divided into micropores, transition pores and macropores, while the distribution range of ACF pore diameter is narrow. The vast majority of the pore diameter in ACF is below 10 nm, and 90% of the pore diameter is below 2 nm. ACF only has micropores and a few transition pores, with no macropores in, so its adsorption selectivity is good.

- ACF not only has an obvious adsorption capacity for high concentration adsorbents, but also has an excellent adsorption capacity for low concentration adsorbents. For instance, ACF can also adsorb toluene gas when its content is below 10 ppm, while GAC can only adsorb toluene when the content is higher than 100 ppm.

The unique adsorption structure of ACF is derived from a large amount of unsaturated carbon on its surface. ACF is a typical microporous carbon, which is considered as "a combination of ultrafine particles, irregular surface structure and extremely small space" [5]. ACF contains many irregular structures and microstructures of surface functional groups, with a large surface area, which enable the micropores interact with the molecules on the pore wall to form a strong molecular field, and provide a high-pressure system of molecular physical and chemical changes in adsorption state.

Specific surface area is an important indicator to measure the performance of ACF and is a key factor affecting its adsorption performance as well as the pore structure and adsorption property of activated carbon fiber [6]. In the production process of ACF, the activation stage, which include the diffusion rate of activator and the activation reaction rate [7], has a huge impact on the relative surface area, which can directly determine the specific surface area of ACF.

ACF activation methods mainly include physical activation method and chemical activation method [8]. In this paper, a new type of secondary activation process was developed for the commercial viscose-based activated carbon fiber, and four factors (type of activator, concentration of activator, soaking time and soaking temperature) in the activation process were systematically screened by orthogonal test method to optimize the optimal activation process for the production of high specific surface area.

2 Experimental Section

- Experimental materials and equipment
  Viscose-based activated carbon fiber; NaOH; Na_2HPO_4; H_2O_2; Deionized water; GSL-1700X-S60 Tube furnace; N_2 gas; Vacuum drying oven; Glassware for chemical experiment.

- Experimental Process

  In this experiment, the reactivation modification process for VACF has three main steps: pretreatment, ultrasonic impregnation and high temperature treatment.

(1) Pretreatment

  Before impregnation, the impurity on the surface of VACF need to be removed first, this article adopts the method of heating stripping to remove the impurity of VACF, the specific operating conditions are as follows: first soak VACF in 50 °C water bath for 0.5 h, then put samples in drying oven of 120 °C for 2 h, and make them natural cooling to room temperature.

(2) Ultrasonic impregnation

  Dip a certain mass fraction of samples after pretreatment into different kinds of activator solution, under the condition of ultrasonic in different temperature, and the ultrasonic power is 100 W. Wash samples to neutral with deionized water after impregnation, then put samples in drying oven of 120 °C for 2 h.

(3) High temperature treatment.

  High temperature treatment is processed in the tube furnace (as shown in Fig. 1), with 3 °C/min heating rate heated to 750 °C, heat preservation for 0.5 h, then get the sample cooled to room temperature, all the process is under nitrogen protection.

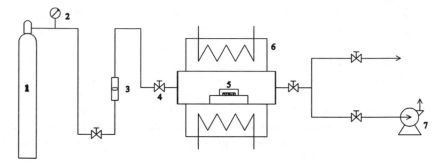

1-High purity nitrogen gas bottle; 2-Nitrogen pressure reducing valve;

3-Flowmeter; 4-Gas regulating valve; 5- Corundum crucible;

6-Vacuum tube furnace; 7-Vacuum pump

Fig. 1. High temperature treatment device for reactivated carbon fibers

3   Orthogonal Test

In the secondary activation process, the main factors that affect the ultrasonic impregnation are the type of activator, the concentration of activator, the impregnation temperature and the impregnation time. With HNO_3, Na_2HPO_4 and H_2O_2 as activator, the concentration (mass fraction) gradient is set to 3 wt%, 10 wt%, 15 wt%, the activation time of 1 h, 3 h, 5 h respectively, dipping temperature is 20 °C, 50 °C and

80 °C respectively. The specific surface area was used as the evaluation criterion for the activation process. Orthogonal experimental design shown as Table 1.

Table 1. Orthogonal experimental design

Sample number	Horizontal array	Factors			
		Concentration of activator (wt%)	Dipping temperature (°C)	Dipping time (h)	Type of activator
1	$A_1B_1C_1D_1$	3	20	1	Na_2HPO_4
2	$A_1B_2C_2D_2$	3	50	3	H_2O_2
3	$A_1B_3C_3D_3$	3	80	5	HNO_3
4	$A_2B_1C_2D_3$	10	20	3	HNO_3
5	$A_2B_2C_3D_1$	10	50	5	Na_2HPO_4
6	$A_2B_3C_1D_2$	10	80	1	H_2O_2
7	$A_3B_1C_3D_2$	15	20	5	H_2O_2
8	$A_3B_2C_1D_3$	15	50	1	HNO_3
9	$A_3B_3C_2D_1$	15	80	3	Na_2HPO_4

4   Results and Discussion

Nitrogen adsorption tests were performed on samples prepared by nine different secondary activation processes. The pore structure parameters of each sample are shown in Table 2. According to the test results that except sample 8 the specific surface area of all the samples is higher than that of ACF0. The specific surface area, total pore volume and micropore volume of sample 7 are all the largest. Compared with ACF0, the specific surface area of sample 8 decreased, for the reason that the high concentration of activator HNO_3 led to the destruction of ACF pore structure, and a large number of micropores were expanded into macropores, which lead to the collapse in heating process, resulting in the reduction of pore volume and the increase of pore diameter.

Figure 2 shows the pore size distribution of each sample. It can be seen from the figure that the pore size distribution of ACF0 and 9 reactivated ACF samples was mainly within 1.2 nm, which is a typical characteristic of microporous material. Consist with the test results of BET, the average pore size of sample 5 was the smallest, with its pore size mainly concentrate at 0.5 nm and 0.6 nm. Except sample 8, the pore size distribution within 0.6 nm of each reactivated ACF was improved to varying degrees, in addition, sample 5 was also partially distributed at 0.7 nm, 1.1 nm and 1.4 nm.

The adsorption performance of ACF is closely related to its specific surface area, so this experiment takes specific surface area as the parameter standard for evaluating the effect of secondary activation process. The orthogonal experimental results are shown in Table 3.

According to the result of R, maceration temperature has the largest influence on ACF specific surface area among the four secondary activation factors, followed by

Table 2. Pore structure parameters of ACF0 and 9 reactivated ACF samples

Sample	BET surface area (m²/g)	Micropore surface area (m²/g)	Total pore volume (cm³/g)	Micropore volume (cm³/g)	Mesoporous volume (cm³/g)	Average aperture (nm)	Average Micropore aperture (nm)
ACF0	1190.8	1183.8	0.4371	0.4124	0.0247	1.8323	0.8858
1	1332.3	1125.0	0.4652	0.4095	0.0557	1.9979	0.9169
2	1305.5	1197.6	0.4920	0.4015	0.0905	1.8158	0.9118
3	1378,8	1190.8	0.5821	0.4650	0.1171	1.6887	0.8479
4	1382.4	1171.0	0.5894	0.4590	0.1304	1.7056	0.8585
5	1372.4	1184.3	0.5755	0.4580	0.1175	1.6773	0.8388
6	1426.2	1180.1	0.5877	0.4586	0.1291	1.7536	0.8292
7	1445.8	1187.8	0.6181	0.4683	0.1498	1.7102	0.8675
8	932.5	536.7	0.4011	0.2876	0.1135	2.0715	1.3073
9	1390.7	1197.2	0.5802	0.4306	0.1496	1.6914	0.8551

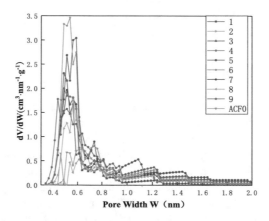

Fig. 2. Pore size distribution of ACF0 and 9 reactivated ACF samples

maceration time and type of activator, while maceration concentration has a relatively small influence. According to K1, K2 and K3, the best activation conditions are as follows: H_2O_2 of 10wt%, ultrasonic immersion under 80 °C for 5 h, and we can obtain the highest specific surface area of ACF after reactivated. We prepared ACF sample (marked as sample 10) under this condition, and conducted nitrogen adsorption test on this sample. The relevant test results were recorded as shown in Table 4.

It is easy to learn from the test results that the specific surface area ratio of the samples prepared by the reactivated process obtained from the orthogonal test is 26.43% higher than that of ACF0. In addition, the total pore volume, micropore volume and specific surface area are the largest among the 10 samples, and the ideal activation results are obtained.

Table 3. Specific surface area of orthogonal experimental designed samples

Sample number	Factors				Specific surface area
	Concentration of activator (wt%)	Dipping temperature (°C)	Dipping time (h)	Type of activator	
1	3	20	1	Na_2HPO_4	1332.3
2	3	50	3	H_2O_2	1305.5
3	3	80	5	HNO_3	1378.8
4	10	20	3	HNO_3	1382.4
5	10	50	5	Na_2HPO_4	1372.4
6	10	80	1	H_2O_2	1426.2
7	15	20	5	H_2O_2	1445.8
8	15	50	1	HNO_3	932.5
9	15	80	3	Na_2HPO_4	1390.7
I	4016.0	1160.5	2601.0	4005.4	
II	4181.0	3610.4	4078.6	4177.5	
III	3769.0	4195.7	4197.0	3693.7	
K1	1338.9	1386.8	1230.3	1365.1	
K2	1393.7	1203.5	1359.5	1392.5	
K3	1256.3	1398.6	1399.0	1231.2	
R	137.4	195.1	168.7	161.3	

Table 4. Pore structure parameters of sample 10

Sample	BET surface area (m^2/g)	Micropore surface area (m^2/g)	Total pore volume (cm^3/g)	Micropore volume (cm^3/g)	Mesoporous volume (cm^3/g)	Average aperture (nm)	Average Micropore aperture (nm)
10	1505.5	1183.8	0.6220	0.4917	0.1303	1.8058	0.8110

5 Summary and Conclusions

In this paper, ACF was reactivated on the basis of a self-created secondary activation process, and the reactivated condition for preparation of ACF with high specific surface area was obtained by orthogonal test: H_2O_2 of 10wt%, ultrasonic immersion under 80 °C for 5 h. The activation effect was verified by nitrogen adsorption test. The specific surface area of the sample 10 was 26.43% higher than that of ACF0, and the total pore volume and micropore volume were both the largest among all samples.

Project Number. 2016YFC0301402, Ocean project of ministry of science and technology, AUV Intelligent composite shell design and processing; 2017CXGC0407, Key research and development plan of Shandong province, Key preparation technology of high performance T1000 carbon fiber.

References

1. Guo, K., Yuan, C.: Chemical progress **5**, 36 (1994)
2. Zeng, H., Fu, R.: Study of asphalt based activated carbon fibers Preparation, structure and properties. Synth. Fibre Ind. **5**, 13–21 (1987)
3. Fuhe: Carbon Fiber and Application Technology, vol. 9, pp. 160–161. Chemical Industry Press, Beijing (2004)
4. Deng, B.: Study of Modified Activated Carbon Fibers on Adsorption Properties and Mechanism for Toluene. Kunming University of Science and Technology, Kunming (2007)
5. Zheng, J.: Active carbon fiber. New Carbon Mater. **15**(2), 80 (2000)
6. MartíN-Gullón, I., Andrews, R., Jagtoyen, M., Derbyshire, F.: PAN-based activated carbon fiber composites for sulfur dioxide conversion: influence of fiber activation method. Fuel **80**(7), 969–977 (2001)
7. Yu, C., Ren, J., Fu, M., Wang, R., Peng, Y., Tan, X.: Modification of activated carbon fiber and its microporous structure. J. Environ. Sci. **28**(4), 714–719 (2008)
8. Guo, K., Xie, Z., Ye, Z., Hou, L.: Adsorption Technology and Application in Environmental Engineering of Activated Carbon. Chemical Industry Press, Beijing (2016)

Research on Semantic Composition of Smart Government Services Based on Abstract Services

Youming Hu, Xue Qin[✉], and Benliang Xie

School of Big Data and Information Engineering,
Guizhou University, Guiyang, China
xqin@gzu.edu.cn

Abstract. Smart government widely integrates heterogeneous business systems to provide intelligent applications, and improving interdepartmental government coordination capability is a key technical issue that needs to be solved urgently. Service composition technology based on SOA architecture is a good solution. However, due to the inherent heterogeneity and huge scale of smart government, the state space of service composition is very large, which is a great challenge to the efficiency and accuracy of problem solving. Based on analysis of the existing service composition research methods, combined with the domain characteristics of smart government, this paper studies and constructs an abstract government Web service template to reduce the scale of state space for service composition problem. Furthermore, a semantic-based service composability measurement is studied to select service solutions to improve the accuracy.

Keywords: Smart government · Service composition · AND/OR graph · Abstract services

1 Introduction

With the development of big data, cloud computing, Internet of Things, artificial intelligence, 5G and other technologies, E-government has been pushed forward to the direction of smart government, which integrate multiple services functions of different application systems to meet the complex needs of government coordination, and realize intelligent management and smart government services. Thus, the ability to coordinate government affairs across departments is a significant symbol of measuring the development level of smart government. The development of government affairs informatization in the past twenty years has determined that there must be a large number of heterogeneous business systems, which have different operating environments, different technical standards and functional characteristics, making the interconnection and collaboration between systems very difficult.

Web service technology based on SOA architecture can seamlessly interconnect heterogeneous systems, which has been widely studied and applied in the integration e-government systems. Smart government system integration and business collaboration is the development of traditional government data exchange platform. It evolves from

© Springer Nature Switzerland AG 2020
F. Xhafa et al. (Eds.): IISA 2019, AISC 1084, pp. 803–814, 2020.
https://doi.org/10.1007/978-3-030-34387-3_99

simple data exchange to business capability reuse, exposes and shares business functions in a service way, realizes interconnection and business collaboration of application systems, and solves the problem of duplicate construction. From a technical point of view, the system integration and business collaboration problems in the field of smart governments have been transformed into the semantic composition of Web services under the SOA architecture. However, smart government covers a wide range of business systems with intelligent application requirements, which will inevitably lead to an extremely large number of exposed shared service functions. In other words, the size of the state space that the service portfolio needs to search is also very large, which will be a huge challenge to the efficiency and accuracy of problem solving.

Based on the analysis of existing service portfolio research methods, combining with the characteristics and practical experience of smart government, this paper will reduce the size of state space for service composition problems by combing and identifying abstract government Web service templates, and will study a semantic-based service composition method and composability indicators to improve the accuracy of service composition.

2 Theories and Previous Literature

2.1 Previous Research on Service Composition

Since the emergence of the concept of service-oriented computing, especially after the composition of service and Semantic Web, the automatic composition of services immediately became a hot topic in this field [1]. In essence, automatic service composition must be associated with semantic issues. That is, all services must give explicit semantic descriptions of functional and non-functional attributes (i.e. Semantic Web Services), to provide sufficient semantic support to discover candidate service sets, plan the execution process and execute, etc. [2].

At present, most of the methods used in the service automatic composition research come from the relevant research results of artificial intelligence [3]. The common method is to convert the service composition problem into the problem of artificial intelligence, using some mature and effective methods to solve it and adjusting it according to the characteristics of the service to get a better solution. Some common methods include: graph-based spatial search [4], planning problems [5], logic-based proofs [6], model detection [7], and rule systems [8], etc. These methods have their own characteristics when dealing with service compositions. Planning problems are closer to service portfolio issues, but it involves less data in the process. Logic-based methods have advantages in expression capabilities, but are more complex and inefficient, especially when the number of services is large. Graph-based spatial search is more balanced in expression and efficiency, but there also exists the search efficiency problem.

2.2   Application of AND/OR Graph in Service Composition

Liang and Su [9] studies the use of AND/OR graph search to discover automatic composite services. This research constructed a service dependency graph (SDG) as the search space at first. Nodes in the graph are divided into service (operation) nodes and data nodes, which correspond to service and IO data respectively. The arcs in the graph are divided into two categories: the input data node points to the service and the service points to the output data node. Since the input data is necessary for the service call, service nodes are AND nodes. On the other hand, data nodes, generated by the output of multiple services, are OR nodes. Then, the combined request (target) is expressed as a list of known input data and a list of desired output data and optional services (operations). Next, use the AND/OR graph search algorithms to search the SDG for the subgraph g, which contains the input data node, the output data node, and the optional service. Finally, a candidate solution for the combined service is obtained. There are two types of search for AND/OR graph: top-down search (such as AO* algorithm) and bottom-up search (such as REV* algorithm). Gu, Li, and Xu [10] improves this method through enhancing the expression ability and performance of the graph.

In the research of service composition, it is necessary to search a diagram from the list of input data that users request, including specified operations and expected output, so bottom-up is a better choice in this paper. This combining services method based on AND/OR graph is more versatile in description of the request and construction of the search space, as it will not only characterize the IO function, but also characterize the PE function of services. Based on the characteristics of smart government, this study will maintain the advantages of the combined method, and strive to improve its service function description and execution efficiency.

3   Abstract Services

In the research and practice of e-government, there has been a consensus and progress in compiling a unified data resource catalogue and business table catalogue to classify government information and business [11]. Combining the practical experience of e-government, through combing and identifying the administrative business and service items, this section describes and defines the administrative business and service in the way of web service, and finally forms the definition of abstract service.

3.1   Abstract Service Identification Methods and Examples

Sorting out administrative business and service items is an overall expansion of the existing administrative examination and licensing system. The sorting content includes the data flowing in, processing, and outflowing of each link of each administrative transaction, which can be classified into basic data, subject data, and business data. The data can be incorporated into the corresponding business classification system according to their logical relationship. From the perspective of SOA, these operations are functions that are exposed from the relevant assistance systems in the form of services. The data is the input and output data streams of these services. These services will be described, classified and registered in the government cloud center in accordance with the government service semantic model for complex collaborative applications such as query, call and composition.

The following is an example of the subsistence allowances and approvals in social assistance to illustrate the process of business and service grooming, and to form service definitions and descriptions. The subsistence allowance belongs to the business scope of the civil affairs department, and the completion of this business requires data sharing and business collaboration across dozens of departments (including community street service centers, public security, housing management, industry and commerce, taxation, banks, social security, civil affairs, etc.). For example, checking family's economic situation is an important process for the business which requires the assistance of other departments, such as the Social Security Administration verifies employment and labor remuneration, the Housing Authority verifies the real estate, the Transportation Bureau verifies the ownership of vehicles, etc. That is, the business departments involved need to disclose the information query service (or information checking service). For vehicle information query service, its basic grammatical attribute description is shown in Table 1, and the input and output messages and semantic description are annotated according to the corresponding domain ontology, as shown in Table 2.

Table 1. Basic information table of vehicle information query service

Name	CheckVehicle
Note	Query for vehicle information owned by individuals
Usability	Information
Supplier	Traffic Administration (Government)
User	Government departments (such as banks, civil affairs, community service centers, etc.)
Business	Vehicle Information Query

Table 2. IOPE semantics of vehicle information query service

	Paras	Local	Expressions
I	pId	×	IDCNumber(pId)
O	pId, owner, brand, productModel, licensePlate, purchaseDate	×	IDCNumber(pId), IDCNumber(owner), NameOfArticle(brand), IDCCommodity(productModel), Number(licensePlate), Date(purchaseDate)
P	pId	×	isValid(pId)
E	pId, owner, brand, productModel, licensePlate, purchaseDate	{x}	car(x), hasCar(owner, x), = (owner, pId), hasBrand(x, brand), hasModel(x, productModel), hasLicense(x, licensePlate), hasPurchaseDate(x, purchaseDate)

According to the service IOPE semantic description definition, the input of the service in Table 2 is the personal identification number (pId), and the output is the vehicle information (brand, model, license plate number, purchase date, owner, etc.), The semantic types of which are annotated according to the general e-government element. The PE semantics are described by predicates and general arithmetic operators in the domain ontology.

3.2 The Definition of Abstract Service in Smart Government

In the integration of smart government systems and resource integration, all the business functions that participate in collaboration and sharing exist in the form of Web services and are marked with service categories. Obviously, the functions of services belonging to the same business category are very similar, and the semantic types of their input and output parameters are roughly the same. Therefore, we can make statistics and analysis on services of each business category, extract the main input and output, and construct an abstract service template for each business category.

Abstract service refers to extracting the common part of the semantic type of input and output parameters according to the business category of registered service. These common semantic types are used to construct Abstract Service Templates (AST). Assuming that B is a business category, the corresponding AST is defined by formula 1–3:

$$AST_b = (In_b, Out_b) \tag{1}$$

$$In_b = \{d_i | P(d_i) \geq \tau_1, d_i \in \bigcup_{s_j \in S(b)} PIs(s_j)\} \tag{2}$$

$$Out_b = \{d_i | P(d_i) \geq \tau_2, d_i \in \bigcup_{s_j \in S(b)} POs(s_j)\} \tag{3}$$

Where S(b) represents a service set of all service types b in the service registry. PIs(sj) and POs(sj) represent a list of input and output parameter semantic types (from the service registry) of the service sj, respectively. P(di) represents the probability that the semantic type di appears, which is the quotient of the number of services containing di in S(b) and the total number of services in S(b). Considering that input data is necessary for service calls, and output data can be derived from the output of many services, the requirements for selecting the input and output semantic type lists for AST are different. Select more input data in the allowed range and select less output data, i.e. $\tau_1 \leq \tau_2$. On the one hand, the abstract service template can describe as many specific services as possible. Another advantage is that when the specific service is selected later, it has a larger range of choices and reduces the possibility of missed selection.

4 Smart Government Service Composition Method

It is a better choice to construct the problem space of service composition of intelligent government by using AND/OR graph, and to obtain the composite scheme by searching diagrams. However, when the number of services is large, the scale of the state space of the problem will be huge, which brings great challenges to search.

Another problem is that the AND/OR graph describes the semantic type (IO semantics) of the service input and output, but does not describe the premise or effect of the service (PE semantics), what is likely to result in the failure of composite services at runtime due to the absence of PE. This section will discuss the method based on AND/OR graph to realize the semantic composition of smart government service, which will solve these two problems centrally.

4.1 Smart Government Service Composition Based on AND/OR Graph

Firstly, the abstract service template in e-government system will be identified to construct the problem space of graph, the advantage of which is that the original service relationship is basically maintained, while the scale of the problem is greatly reduced meantime. On the other hand, it makes the problem space stable and does not need to rebuild a problem space every time there is a new registration service.

The second problem needs to be solved through the composability check of services. Once the composite service template is generated, the choice of each specific component service must depend on the composability calculation between these specific services throughout the combined process [12]. That is to say, two adjacent specific services must meet the requirements of composability, and at the same time make the selection of the entire composition scheme satisfy the overall maximum composability. From the definition of composability and calculation rules, PE semantics is an important link, which ensures the matching of PE semantics with PE services. On the other hand, the overall maximum composability also makes the error possibility reduced greatly in the operation of the combined service.

4.2 Dependency Diagram of Abstract Services

Abstract service dependency graph (ASDG) is a special AND/OR graph that describes the dependency between a service and its input and output, and indirectly expresses the calling relationship between services through output and input. The nodes in the graph are divided into two categories: nodes that represent services and nodes that represent input and output data. Due to the different characteristics of service input and output, the graph restricts each node to only one k connector: For a service node, all its input data nodes are connected to the service node through an "and" k connector, indicating that the input data must be required for the service at the same time, so the service node can also be called And node. For a data node, connect all services that have this data in the output to the node through an "or" k connector, indicating that the data node can be generated by any of these services, so the data node can also be or node. To describe the IO semantics of a service, the data nodes in the graph are the semantic types of input and output parameters.

Figure 1 depicts a service scenario for traffic violation handling. The circle is the data node and the rounded rectangle is the service node. The arc of the service node represents the input, the connection on the arc represents the relationship between these arcs, and the arc exit represents the output of the service. It can be seen from the figure that the data node is connected to the service, indicating the input and output relationship between the data node and the service node; Services are not directly connected, but indirectly represent the dependencies between services by correlating data nodes.

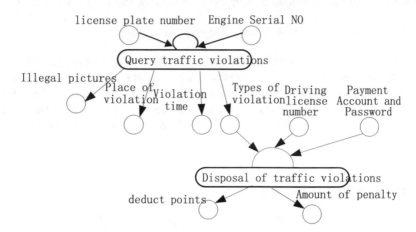

Fig. 1. An ASDG sample diagram

4.3   Dependency Diagram of Abstract Services

When a service is represented as an ASDG diagram, the service composition request can be represented as a set of data nodes (semantic types) as input, a set of data nodes as output, and an optional designated service. At this time, the problem of generating composite service template is transformed into the search problem of AND/OR graph. In order to facilitate the implementation of the search process, we will make some modifications to the ASDG diagram, denoted ASDG'. Firstly, an "or" node s is added and connected to the input data node in the composite request from the s emission arc. Since these input nodes are considered known, the OR node s is also marked as Solved, which will serve as the starting node for bottom-up searches. Then, add another node g to connect the output node specified in the composite request and the optional service emission arc to the node g. Since these output data and the specified service are required to be met in the request, node g is the AND node, which will be the destination of the search, marked as the target node.

Due to the existence of the OR node, the search process may generate multiple candidate solutions. Therefore, some evaluation methods are needed to compare these different solutions to select the optimal one. In general, the solution with fewer nodes is relatively better. Because for composite services, more AND nodes (services) mean more service invocation costs, and more data nodes mean more input costs. Therefore, there must be a suitable evaluation function to estimate the cost of the extended node in the search process. On the one hand, it is extended by selecting the smallest cost node to provide search efficiency; on the other hand, in the case of a solution, it is guaranteed that the lowest cost solution can be obtained.

(1) Evaluation Function

Since this algorithm is a bottom-up search, that is, extending from a known node. Therefore, the evaluation of the expansion cost of a node n is a function of the cost of the node itself and its parent node, denoted by f(n). This can be defined as formula 4:

$$f(n) = \varphi(h(n), f(p_1), f(p_2), \ldots, f(p_k)) \tag{4}$$

Among them, h(n) is the cost of node n itself. {p1, p2, ..., pk} are the k parent nodes of node n, respectively. h(n) is written as two different constants according to the type of node n. If n is a service node, h(n) takes the value d. If n is a data node, h(n) takes the value d/Z, and Z is a large number. This is based on the fact that the cost of service invocation is far greater than that of data nodes. In addition, the H(n) value of all data nodes and service nodes in the composite request is set to 0, because these nodes will appear in all the diagrams and will not affect the cost calculation of the diagrams. The function φ differs depending on the type of node n. When n is an OR node, the calculation is defined as Formula 5:

$$f(n) = h(n) + \min\{f(p_1), f(p_2), \ldots, f(p_k)\} \tag{5}$$

When n is an AND node, the calculation is defined as Formula 6:

$$f(n) = h(n) + \sum\{f(p_1), f(p_2), \ldots, f(p_k)\} \tag{6}$$

(2) Bottom-up search

The search process (the algorithm shown in Fig. 2) will proceed from the starting node s of the graph ASDG', gradually expanding and marking the solvability of the successor node until the target node g is marked as solvable or failed. In the process, an "or" node is marked as Solved if and only if it has a parent node that is Solved; an "and" node is marked as Solved if and only if all its parent nodes are Solved.

In the algorithm described in Fig. 2, first set the initial value of f(n) for all nodes: f (s) = 0. The other initial values are set to INF (a relatively large value). The second step is to initialize the node table OPEN and find the solution of Figure G, At the beginning of the algorithm, both contain only the starting node s. The third step is to take the node to be extended from the "OPEN" table, and then search all its child nodes, calculate and update its f(n) value for each child node. Once the child node is found to be Solved, it is added to the OPEN table and SG. Repeat step 3 until the OPEN table is empty or the target node is labeled Solved. Finally, it is judged and the optimal solution SG is obtained.

When the solution SG is obtained, it should contain the input and output data nodes and service nodes specified in the composition request, as well as the data and service nodes obtained by the search algorithm. The nodes obtained by these searches connect the input and output data nodes and the service nodes. Next, a composite service template is generated according to the solved graph SG, and the dependencies between the services in the template are determined by the data nodes between them. If the

```
// s is the starting node in ASDG' and g is the target node
// The OPEN table is the set of nodes to be extended currently
// SG is the current optimal solution
AOSearch(){
1  f(s)=0; f(n)=INF;// The starting node is set to 0, and the other is set to INF
2  Mark(s,Solved); SG=OPEN={s}; // Initialization
3  while( OPEN≠Φ){
4    n = Remove(OPEN); // Take out node n, so that f(n) is the smallest in the OPEN table
5    if(g==n){ Mark(g, Solved); break;}
6    CL = Expand(n); // Extend n to get its child node set CL
7    foreach(k∈CL){ // Search every child node of n
8      if(k is the AND node){
9        if(All parent nodes of K are Solved){
10         f(k)=h(k)+SUM_{p is the parent of k}{f(p)};
11         if((k∉OPEN) AND (k∉SG)){
12           Mark(k, Solved);OPEN=OPEN ∪ {k};SG=SG ∪ {k}; } }
13       }else{ // k is the OR node
14         cost = h(k) + f(n);
15         if(f(k)>cost) {
16           f(k)=cost; //Update the cost value of node k
17           setParent(k, n); // Under the current cost value, set the parent node of k to n }
18         if((k∉OPEN) AND (k∉SG)){
19           Mark(k, Solved);OPEN=OPEN∪ {k};SG=SG ∪ {k}; }
20       }//end if
21     }//end foreach
22   }//end while
23   if(Mark(g, Solved)){ SG is the smallest diagram and f(s) is the smallest cost value; }
24   else{ No diagram; } }//end AOSearch
```

Fig. 2. AND/OR graph bottom-up search algorithm

output of service SERV1 is the input of service serv2, then SERV1 is in the precursor position of serv2 in the composite template.

4.4 Selection of Composition Scheme

Since the search space ASDG of the service composition problem is constructed based on the abstract service, it is necessary to replace each component service with a specific service in order to generate a specific composite service scheme and orchestrate it as a service execution process. Due to the limitations of the AND/OR graph, ASDG describes the IO semantics of the service well, but lacks the characterization of PE semantics, which may cause the final composite service to fail during execution. Therefore, when selecting a specific registration service, we will introduce the level of composability between services to screen candidate services [12]. This is because the PE semantics and matching degree of the service are important links in the calculation of horizontal composability. In this way, the probability of composite service execution errors caused by PE mismatch can be reduced.

Since the abstract service template is based on the statistics of the registration service for the same service category, there are generally a plurality of specific services to choose from, which results in a lot of replacement service candidate solutions. Assuming that there are n abstract services in the composite service template, and each abstract service has k specific services to replace, then there will be n × k combined service plans to choose from after the replacement. Assuming that the n service sequence in the composition scheme cpt is {s1, s2, ..., sn}, then the overall composability TDegree of cpt is defined, as in Formula 7:

$$\text{TotalDegree}(cp_t) = \sum_{i=1}^{n-1} HDegree(s_i, s_{i+1}) \tag{7}$$

Obviously, the overall composability of each composition scheme will be different, and the probability of success in execution will be different. Therefore, when choosing a composition scheme, it is preferred to select the scheme with the largest overall combinability, which will make the composition scheme less likely to fail during execution to a certain extent.

The following continues to illustrate the selection of the maximum overall composition using the Dijkstra algorithm with traffic violations as an example. In order to better illustrate the selection process, a short message notification service is added after the violation processing. Considering that the violation inquiry and processing service can be provided by the relevant government management department, it can also be provided by a profitable or other agent. The same SMS notification service can also be provided by multiple parties. It may be agreed that each abstract component service template has three candidate replacement services, as shown in Fig. 3. Considering that the optimal path search in directed graph is closely related to the number of arcs, in order to reduce the time cost of the algorithm, the threshold τ = 0.8 is set for the horizontal composability between services, and all arcs below τ are removed.

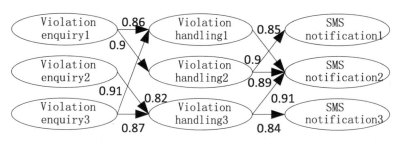

Fig. 3. Composition service composition scheme for traffic violation treatment

In order to conveniently apply the Dijkstra algorithm to solve, the weights in Fig. 3 have been modified. First use 1 to subtract the weight of each arc. Secondly, two pseudo nodes are added, which respectively represent the start node START and the end node END, as shown in Fig. 4. At this point, solving the problem is converted to the shortest path problem. The solution result is as shown in Fig. 5. The final selected composition scheme consists of three registration services: violation inquiry service, violation treatment service and SMS notification service. The Composition degree at this time is 1.8.

The method of selecting the composition scheme based on the maximum com-posability studied here is carried out from the perspective of reducing the possibility of execution failure of the combined service. Another common method is to choose from the perspective of QoS [13]. This method focuses on considering the user's needs, so the selected composition process is closer to it. These two methods have their own advantages, which are complementary and can be used together.

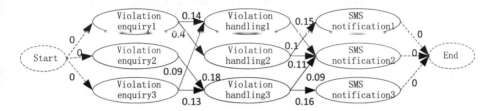

Fig. 4. Conversion of traffic violation treatment composition scheme

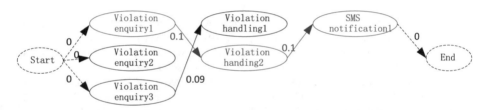

Fig. 5. Selection results of combined service scheme for traffic violation treatment

5 Conclusion

This paper studies the issue of smart government service portfolio. Since the service category is specified at the time of service registration, it creates favorable conditions for generating a combined template using the AND/OR graph search method. The construction of abstract services greatly reduces the search space of AND/OR graph, which makes it possible to apply this method to solve the problem of smart government service composition. For the method based on AND/OR graph, there is a shortage of service PE function description, but it can be solved by introducing composability to calculate and select the composition scheme. Another important use of composability is to choose a reference standard that replaces service selection, which reduces the likelihood of a composite service execution error after replacement. Regardless of service composition or replacement, another issue of concern is the mapping of parameters between execution services. Establishing an accurate parameter mapping relationship is an important condition for the allocation and automatic calling of the combined process. Further research is to introduce QoS to improve the combined effect, and test and verify the performance of the method through the actual smart government operating environment.

Acknowledgments. This work is supported by the Provincial Joint Fund of Guizhou (Grant LH20147631), the Talent Introduction Project of Guizhou University (Grant No. 2014-33) and the Talent Introduction Project of Guizhou University (Grant No. 2015-29).

References

1. Medjahed, B., Bouguettaya, A.: Service Composition for Semantic Web. Springer, Berlin (2011)
2. Dong, H., Hussain, F.K., Chang, E.: Semantic web service matchmakers: state of the art and challenges. Concurr. Comput. Pract. Exp. **25**(7), 961–988 (2013)
3. Lemos, A.L., Daniel, F., Benatallah, B.: Web service composition: a survey of techniques and tools. ACM Comput. Surv. (CSUR) **48**(3), 33 (2016)
4. Gu, Z., Li, J., Xu, B.: Automatic service composition based on enhanced service dependency graph. In: Proceedings of IEEE International Conference on Web Services, Washington, D.C., USA, pp. 246–253. IEEE (2008)
5. Ben Lamine, R., Ben Djemaa, R., Amous, I.: A framework for the composition and formal verification of adaptable semantic web services. Association for Computing Machinery (2018)
6. Shi, Z.-Z., Chang, L.: Reasoning about semantic web services with an approach based on dynamic description logics. Chin. J. Comput. **31**(9), 1599–1611 (2008)
7. Tang, X.F., Jiang, C.J., Ding, Z.J., Wang, C.: A Petri net-based semantic web service automatic composition method. J. Softw. **18**(12), 2991–3000 (2007)
8. Meditskos, G., Bassiliades, N.: A semantic web service discovery and composition prototype framework using production rules. In: OWL-S: Experiences and Future Developments workshop in conjunction with the 4th ESWC, Innsbruck, Austria, June 2007
9. Liang, Q.A., Su, S.Y.: AND/OR graph and search algorithm for discovering composite web services. J. Web Serv. Res. **2**(4), 48–67 (2005)
10. Gu, Z.F., Li, J.Z., Xu, B.: Automatic service composition based on enhanced service dependency graph. In: 6th IEEE International Conference on Web Services (ICWS 2008), Beijing, pp. 246–253 (2008)
11. Pasini, A., Pesado, P.: Quality model for e-government processes at the university level: a literature review. In: Proceedings of the 9th International Conference on Theory and Practice of Electronic Governance, pp. 436–439, March 2016
12. Qin, X., Liu, K., Tang, S.: Service composition plan selection based on maximum composability degree. J. Comput. Inf. Syst. **9**, 5477–5484 (2013)
13. Wagner, F., Ishikawa, F., Honiden, S.: QoS-aware automatic service composition by applying functional clustering. In: IEEE International Conference on Web Services (ICWS) (2011)

Security Issues for Smart Grid EC System

Aidong Xu[1], Yunan Zhang[1], Yixin Jiang[1], Jie Chen[2], Yushan Li[2],
Lian Dong[2], Yujun Yi[2], Fei Pan[2], and Hong Wen[2(✉)]

[1] EPRI, China Southern Power Grid Co., Ltd., Guangzhou, China
[2] National Key Lab of Communication, University of Electronic Science,
and Technology of China, Chengdu, China
sunlike@uestc.edu.cn

Abstract. As a typical industry auto control system, security issue is a necessary task in order to guarantee safe and secure grid operation. The edge devices the local data processing and storage that provide a real time control, a low latency and also can play a role of the security gate, in which the terminal authentication, data privacy and data security can be provide. Due to the terminals of the SG are resource and computing constrained devices, the encryption security measurements are hard to perform on these devices. Physical-layer (PHY) security methods use the characteristics of the channel or the device fingerprints, which can run on the edge devices. Therefore, the terminals almost do nothing, which means these kinds of PHY security methods are light weight to the terminals. This paper introduced several PHY light weight authentication methods for the smart grid based on the edge computing and illustration their efficiency.

Keywords: Security · Smart grid · Edge computing · Authentication · Terminals

1 Introduction

The smart grid consists of sensing, communication, control, and actuation systems which represent an evolution transformation in the way electricity is delivered from suppliers and enable pervasive control and monitoring of the power grid [1]. Due to the cyberattack on Ukraine's power grid, more than 100,000 people in and around the Ukrainian city of Ivano-Frankivsk were left without power for six hours on December 2015 [2]. The most famous example is the Stuxnet malware used to destroy equipment in the Iranian power transmission facility and cause an outage lasting an hour [3]. As a typical industry auto control system, security issue is a necessary task in order to guarantee safe and secure grid operation [4].

In the smart grid (SG) system, a huge amount of computation and storage need to afford for the variety of information, huge user service requests and massive data. Cloud computing are recognized as one to the potential solution for such huge information and data processing, as shown in Fig. 1, however, which can not satisfy the real time control and privacy requirement due to the long transmission and network congestion. Therefore, the edge computing (EC) becomes an alternative solution be performing the local data processing and storage, as in Fig. 1, the edge devices perform a

© Springer Nature Switzerland AG 2020
F. Xhafa et al. (Eds.): IISA 2019, AISC 1084, pp. 815–821, 2020.
https://doi.org/10.1007/978-3-030-34387-3_100

real time control in a small region and get the cloud support by connecting with the cloud layer. As a result, a low latency can be guaranteed. The edge devices also can play a role of the security gate [5], in which the terminal authentication, data privacy and data security can be provide.

As a new computing architecture, edge computing has gradually penetrated into many fields and played an important role. Because of the proximity layout, the network architecture of the edge computing can provide better security than the cloud computing architecture, but as a new type of node in the network, the security protection of the edge computing equipment itself also has become a very challenging task.

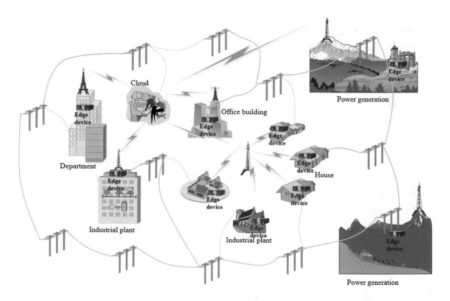

Fig. 1. Smart grid EC system

Due to the terminals of the SG are resource and computing constrained devices, the encryption security measurements are hard to perform on these devices. For meet the high security protection requirement, instead, strong security algorithms should run in the edge devices. Physical-layer (PHY) security methods [6–12] use the characteristics of the channel or the device fingerprints, which can not be faked and clone. By performing the characteristics identification on the edge devices, the terminals almost do nothing, which means these kinds of PHY security methods are light weight to the terminals [13–15]. This paper introduced several PHY light weight authentication methods for the smart grid based on the edge computing and illustration their efficiency.

2 PHY Authentication Method Based on CSI

Authentication is the first line of defense for the SG system security, which is to verify whether the user has the right to access the system or not. Without authentication, malicious identity can easy to access the system and launch the attack, as shown in Fig. 2. Generally speaking, authentication means that the authenticated identity needs to provide a "token" agreed with the legitimate sender and receiver in advance to prove his legitimacy. In SG systems, existing authentication mechanisms are based on cryptography. However, the authentication system based on cryptographic algorithm is facing the challenge of being cracked by attackers with powerful computing power. And in some terminals or sensor nodes with insufficient computing power, it is not conducive to completely rely on the implementation of cryptographic algorithm authentication scheme.

Fig. 2. Authentication is necessary

As the access gate of data, the edge computing has good real-time performance as it is naturally distributed near the terminal and can well meet the real-time response requirements. At the same time, it can provide asymmetric resources. Compared with the terminal, edge computing can carry more complex computing and storage functions. When a terminal or node transmits data to the edge side, the edge side authenticates the data packet legitimately by extracting the physical characteristics from the data packet to authenticate on the edge side, which makes the terminal or the micronode do not bear any computational load. Meanwhile, the computing resources on the edge side make it possible to use the machine learning to do authentication. Now the two main catalogues of the PHY authentication are PHY authentication methods based on the channel state information (CSI) and PHY authentication methods based on the radio frequency fingerprints (RFF) of the devices.

PHY authentication methods based on CSI take advantages of the channel uniqueness, which is a new way to realize security authentication. By extracting channel information directly from pilot sequence, it avoids the compatibility requirement of upper protocol and has a flexible access authentication method and can provide lving, suitable for mass heterogeneous terminal access. The basic principle of this method is descripted as following. The legal terminals and attackers send the data packet to the edge device in which the CSI is extracted in time slot k and $k + 1$, respectively. Then In the null hypothesis, $\mathcal{H}_0$, the claimant is the original send. Otherwise, in the alternative hypothesis, $\mathcal{H}_1$, the claimant terminal is someone else.

$$\mathcal{H}_0 : \underline{\hat{\mathcal{H}}}_{k+1}^{UR} \rightarrow \underline{\hat{\mathcal{H}}}_k^{LR}$$
$$\mathcal{H}_1 : \underline{\hat{\mathcal{H}}}_{k+1}^{UR} \mapsto \underline{\hat{\mathcal{H}}}_k^{LR}$$

(1)

which means as following:

$$\mathcal{H}_0 : diff\left(\underline{\hat{H}}_{k+1}^{UR}, \underline{\hat{H}}_k^{LR}\right) < \eta$$
$$\mathcal{H}_1 : diff\left(\underline{\hat{H}}_{k+1}^{UR}, \underline{\hat{H}}_k^{LR}\right) > \eta$$

(2)

where $\eta \in [0, 1]$ is the determination threshold. The normalized likelihood ratio test (LRT) statistic as:

$$\Lambda_0 = \frac{K_{co}\left\|\underline{\hat{H}}_{k+1}^{UR} - \underline{\hat{H}}_k^{LR} e^{j\varphi}\right\|^2}{\left\|\underline{\hat{H}}_k^{LR}\right\|^2} \begin{array}{c} >\mathcal{H}_1 \\ <\mathcal{H}_0 \end{array} \eta$$

(3)

where K_{co} is the normalization factor. This method is called as hypothesis test-based method. The drawback of this method is its low identification rate (Fig. 3).

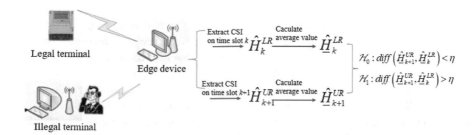

Fig. 3. The basic principle of PHY authentication method based on CSI

Machine learning, as one of the research hotspots in academia and industry, aims to improve computer algorithm and performance through continuous data analysis and learning. Integrating machine learning with channel characteristics is a new approach to improve physical layer authentication rate. In these methods [18], the edge device collects and preprocesses the channel information of the legitimate access nodes and the illegal access nodes, which are known beforehand, to form a two-class data set. The first class of data set is the training set that will be trained by machine learning classification algorithm to build the model and get the classifier. Then the second data is identified as test set that will be used to test the classifier. If the classifier past the test, it can be used to identify the access nodes.

By using the toolbox of MATLAB, four typical classification algorithms, simple tree, support vector machine (SVM), ensemble and K-nearest neighbors (KNN) are selected. The total repeated number of experiences are 100, 200, 500, respectively. The performances results are shown in Fig. 4 and are compared with hypothesis test-based method (HTBM).

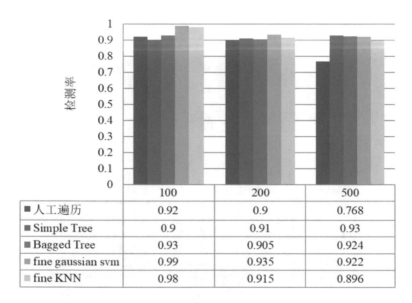

	100	200	500
■ 人工遍历	0.92	0.9	0.768
■ Simple Tree	0.9	0.91	0.93
■ Bagged Tree	0.93	0.905	0.924
■ fine gaussian svm	0.99	0.935	0.922
■ fine KNN	0.98	0.915	0.896

Fig. 4. The performances results

3   PHY Authentication Methods Based on RFF

PHY authentication method based on CSI can identify weather the data packet is legal or illegal, but it cannot identify the identity of the terminals. Due to the manufacturing process, the manufacturing accuracy of the component cannot be infinitesimal, Differences exist in the hardware of any two wireless devices, and will reflect in signal waveforms [17], which becomes the mechanism of RF fingerprinting. The common decision methods in RFF include threshold decision method and classification algorithm based on machine learning or deep learning, in which the classification algorithm based on machine learning or deep learning has better performance with higher computing complexity. However, due to the computing task can be performed on the edge device, the terminals do almost nothing on the EC smart grid system. RFF based on machine learning realizes lightweight access authentication of terminals through off-line training and on-line judgment. However, it needs a large amount of computing resources to support off-line learning. The EC devices provide such computing resources; therefore, such identification method is a typical asymmetric secure access method.

The machine learning algorithm for the radio frequency fingerprint identification mainly involves several types of algorithms such as classification and clustering. KNN algorithm is a commonly used supervised learning algorithm. It finds the nearest sample from the training sample set based on distance measure, and then predicts according to the class information of the sample. By taking this algorithm, 99 different kinds of terminals are set as identification terminals and the experiences are repeated 100 times, the results are illustration on Fig. 5.

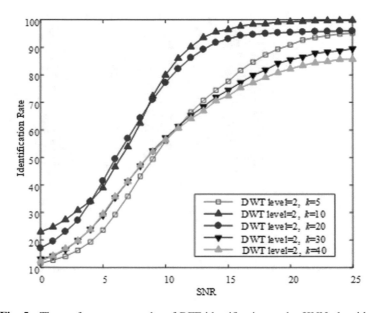

Fig. 5. The performances results of RFF identification under KNN algorithm

4 Conclusions

Authentication is the first line of defense for the SG system security, which is the most important security issue in the SG system. Without authentication, malicious identity can easy to access the system and launch the attack. In SG systems, existing authentication mechanisms are based on cryptography. Due to the terminals of the SG are resource and computing constrained devices, such security measurements are hard to perform on them. This paper surveys the Physical-layer (PHY) security methods to against the faking and cloning attack by performing the characteristics identification on the edge devices. The traditional method and novel machine-learning approaching authentication methods are introduced and simulations are provided.

Acknowledgments. This work was supported by National major R&D program (2018YFB0904900 and 2018YFB0904905).

References

1. Li, F., Qiao, W., Sun, H., et al.: Smart transmission grid: vision and framework. IEEE Trans. Smart Grid **1**(2), 168–177 (2010)
2. Tuptuk, N., Hailes, S.: The cyberattack on Ukraine's power grid is a warning of what's to come. https://phys.org/news/2016-01-cyberattack-ukraine-power-grid.html. Last accessed 13 Jan 2016
3. Kushner, D.: How Kaspersky Lab tracked down the malware that stymied Iran's nuclear-fuel enrichment program. https://spectrum.ieee.org/telecom/security/the-real-story-of-stuxnet, Oza N C. Online Ensemble Learning. The AAAI Conference on Artificial Intelligence, 2000. Last accessed 26 Feb 2013
4. Pan, F., Pang, Z.B., Luvisotto, M., et al.: Physical-layer security for industrial wireless control systems: basics and future directions. IEEE Ind. Electron. Mag. **12**(4), 18–27 (2018)
5. Xie, Y.P., Wen, H., Wu, B., Jiang, Y.X., Meng, J.X.: A modified hierarchical attribute-based encryption access control method for mobile cloud computing. IEEE Trans. Cloud Comput. **7**(2), 383–391 (2019)
6. Wen, H., Li, S., Zhu, X., Zhou, L.: A framework of the PHY-layer approach to defense against security threats in cognitive radio networks. IEEE Netw. **27**(3), 34–39 (2013)
7. Hu, L., Wen, H., Wu, B., Pan, F., Liao, R.F., Song, H.H., Tang, J., Wang, X.: Cooperative jamming for physical layer security enhancement in Internet of Things. IEEE Internet Things J. **5**(1), 219–228 (2018)
8. Song, H.H., Wen, H., Hu, L., Chen, Y., Liao, R.F.: Optimal power allocation for secrecy rate maximization in broadcast wiretap channels. IEEE Wireless Commun. Lett. **7**(4), 514–517 (2018)
9. Tang, J., Monireh, D., Zeng, K., Wen, H.: Impact of mobility on physical layer security over wireless fading channels. IEEE Trans. Wireless Commun. **17**(12), 7849–7864 (2018)
10. Hu, L., Wen, H., Wu, B., Tang, J., Pan, F.: Adaptive secure transmission for physical layer security in cooperative wireless networks. IEEE Commun. Lett. **21**(3), 524–527 (2017)
11. Wen, H., Ho, P.H., Wu, B.: Achieving secure communications over wiretap channels via security codes from resilient functions. IEEE Wireless Commun. Lett. **3**(3), 273–276 (2014)
12. Wen, H., Tang, J., Wu, J.S., et al.: A cross-layer secure communication model based on discrete fractional fourier transform (DFRFT). IEEE Trans. Emerging Topics Comput. **3**(1), 119–126 (2015)
13. Wen, H., et al.: Lightweight and effective detection scheme for node clone attack in WSNs. IET Wireless Sensor Syst. **1**(3), 137–143 (2011)
14. Wen, H., Gong, G., Lv, S.C., Ho, P.H.: Framework for MIMO cross-layer secure communication based on STBC. Telecommun. Syst. **52**(4), 2177–2185 (2013)
15. Wen, H., Wang, Y., Zhu, X., Li, J., Zhou, L.: Physical layer assist authentication technique for smart meter system. IET Commun. **7**(3), 189–197 (2013)
16. Chen, Y., Wen, H., Song, H.H., et al.: Lightweight one-time password authentication scheme based on radio frequency fingerprinting. IET Commun. **12**(12), 1477–1484 (2018)
17. Xie, F.Y., Wen, H., Li, Y.S., et al.: Optimized coherent integration-based radio frequency fingerprinting in Internet of Things. IEEE Internet Things J. **5**(5), 3967–3977 (2018)
18. Chen, S., Wen, H., Wu, J., et al.: Physical-layer channel authentication for 5 g via machine learning algorithm. Wireless Commun. Mob. Comput. **2018**, 1–10 (2018)

A Novel Smart Grid Model for Efficient Resources

Aidong Xu[1], Wenjin Hou[2], Yunan Zhang[1], Yixin Jiang[1],
Wenxin Lei[2], and Hong Wen[2(✉)]

[1] EPRI, China Southern Power Grid Co., Ltd., Guangzhou, China
[2] National Key Lab of Communication,
University of Electronic Science, and Technology of China, Chengdu, China
sunlike@uestc.edu.cn

Abstract. It is a key issue to efficiently manage resources in the smart grid (SG) network that is a dynamic distributed grid, in which the production, storage and users of electricity will work together under specific control. Therefore, in such network an important challenge is how to achieve unified control of distributed equipment on coordinating generators and users distributed in different geographical locations. This paper proposes an innovation model that is the edge computing nodes can be used to collect, compute, and store data while the edge computing nodes are being connected via peer-to-peer. By this way, the peered edge devices can communicate with each other after data processing. The experimental results of the proposed model show that there is a significant improvement to energy resources management due to the abilities of real time control. As results, the economic cost is decreased while the utilization of renewable energy is increased.

Keywords: Smart grid · Edge computing · Peer-to-peer

1 Introduction

Smart grids are power grid networks that aim to provide high level of adaptability, scalability, economy, self-healing, robustness, security and protection in the high dynamic environments [1, 2], which can support the efficient and real time response for energy delivering and energy utility requirements by considering both electricity and information [3, 4]. By employing the edge computing model which enables computation to be performed at the edge of the network, the real time control, security and privacy of the smart grids can be improved. Edge computing shifts cloud computing applications, data, and services from centralized nodes to edge of the network [5, 6], which can perform computing of loading, data storage, caching and processing, as well as distribute request by well-designed edge network that meet the reliability and security requirements [7, 8], and privacy protection. Edge computing could be an ideal platform for the distributed smart grid system considering following characteristics. Firstly, the edge computing could be an efficient solution for large data quantity by processing the data at the edge of the network. The edge computing paradigm can be flexibly expanded from a single energy community to several communities, or even

© Springer Nature Switzerland AG 2020
F. Xhafa et al. (Eds.): IISA 2019, AISC 1084, pp. 822–830, 2020.
https://doi.org/10.1007/978-3-030-34387-3_101

huge scale communities. Secondly, edge computing claims that computing should happen as close as possible to the data source, which is also an appropriate paradigm since it could save the data transmission time and permit low latency. Finally, in edge computing, data could be collected and processed based on geographic location without being transported to the privacy grid cloud that aware the geographic location.

Although the edge computing architecture can provide the such advantages to the smart grid, how to control unify the distributed equipment become a new challenge. Compare with the cloud-based architecture, the edge computing network lacks of a central management characteristic. An elegant solution to these challenges is the combination of P2P networks with edge computing applied into the electrical energy distribution networks. The P2P technology is suitable for the distributed system with the advantages of reliability, robustness, controllability, extensibility and availability [9–12].

In this paper, a novel smart grid network model is introduced, which combines the P2P network with the edge computing smart grid model for coordination of electric power resource. A flexible and economic way is provided with the facility of reliable and efficient communication systems for edge devices by taking advantages of perform computing at the adjacent peer grid data. By this way unified control of distributed equipment and be reliably and efficiently achieved.

2   The Proposed Novel Smart Grid Model

After edge computing are introduced, it has attracted a heavy attention. Many of research work has been done by combing the smart grids with the edge computing. In such model, the edge computing is a networking architecture which dispenses computing, storages, transmission and system management closing to the terminals. Figure 1 illustrates a typical edge computing model with three-layer architecture for the smart grid. The first layer consists of different communities and the power utilities, such as smart meters, wind power, solar energy, battery and other elements of power generators. It is necessary to communicate between devices in the smart grid, such as smart appliances, smart meters, and generator devices. The second layer is edge computing nodes that interact with intelligent electronic devices locally in its coverage area, collect the data coming from terminals and provide data analysis, storages, transmission and administrations. The intelligent electronic devices is also connected to the cloud and each edge computing nodes by these edge computing nodes. It processes the collected data and analyze the data to reduce the amount of transmitted and stored data. The third layer is a traditional cloud server, which is located and is responsible for further analysis and storage of huge amount data that selected and sent by edge computing nodes [5]. There are four types of communication: (1) intelligent device to intelligent device, (2) intelligent device to edge nodes, (3) edge nodes to edge nodes, (4) edge nodes to cloud server.

Unlike cloud computing, the edge computing model is a distributed system that edge nodes act as a bridge between cloud and the smart terminal devices, and computing of loading, data storage, caching and processing take place at edge of the networks [6–8]. The edge computing is well positioned for large data quantity analytics and can be flexibly expanded from a single energy community to several communities, or even huge scale communities. Comparing cloud computing systems it can provide optimized SG system by performing data processing at the proximity of grid data sources as well as save the data transmission time and permit low delay. In edge computing, data could be collected and processed based on geographic location without being transported to the privacy grid cloud that aware the geographic location. It brings many benefits to enhance the overall data processing performance for smart grids in terms of reliability, low latency and bandwidth, processing time, reduced cost, and higher storage capacity. Moreover, edge computing will facilitate the interactions between power facilities and make it possible that power equipment collaborate very effectively with each other. Thus, edge computing could be an ideal platform for the distributed smart grid system. However, some problems still need to be addressed. Firstly, the logical communication topologies among power devices become complex when the number of power devices connected to smart grid increases. Secondly, the state of the distributed energy generators, distributed energy storage systems and users are always changing in the smart grid system, communications between devices and users are changing and result in the logical dynamic communication topologies all the time. There is a need for extensible and controllable service in edge computing layer. Thirdly, the production, storage and usage of electricity are supposed to work together under specific control in smart grids. It is a critical issue to achieve unified control of distributed equipment coordinating power energy resources among distributed generators, distributed energy storage systems and users in different geographical locations. Finally, electric power resource is difficult to control due to the dynamicity of power facilities accessing and quitting unpredictably in result of weather, temperature, operating personnel in the smart grid system.

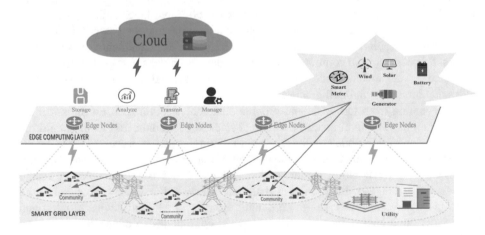

Fig. 1. The edge computing smart grid model

3   The Proposed Model and Its Using Case Scenario

As shown in Fig. 2, a P2P network based edge computing smart grid model is proposed. In the novel model, the P2P network [9–12] is put on the top of the edge computing layer. Comparing to the edge computing smart grid model, edge computing nodes can be used to collect, compute, and store data which produced by distributed generators, distributed energy storage systems, smart meters and controllable smart appliances while the edge computing nodes are connected by P2P networks, by which they can communicate with each other after data processing. From Fig. 3, we can know that communities employ some local distributed generators and produce energy from renewable sources the energy may be use for self-consumption or for sale to the power utility or other communities under the control of edge computing nodes. The P2P network establishes a special mobile communication network based on relevance of nodes on edge computing layer providing reliable communication on electrical energy transactions between community and community or community and the power utility. It is online platform for communication and share information between edge computing nodes in essence. Without accessing to the cloud, edge computing nodes establish communication links directly for each other to improve flexibility and convenience of the edge computing layer. Due to the support of edge computing, the edge node is no longer restricted to share information, cache resource and computing resource also can be shared without accessing to the cloud storage or cloud computing, which can not only reduce the link transmission delay, but also improve the resource utilization [13–15] for device.

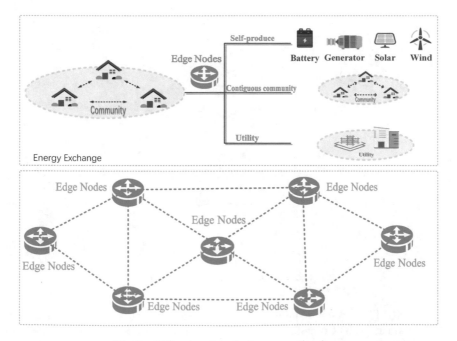

Fig. 2. P2P networks of edge computing layer

Aa an example scenario of our new model that illustrated in Fig. 3, different edge computing modules manage the resource distribution of the different smart communities. Smart communities bring many conveniences to the development of energy coordination and sharing, which allows each community can manage the production and consumption of energy. Because each community that equipped with distributed generators and distributed energy storage equipment. can be self-consumption or trade with other communities via P2P networks [21]. Besides, the proposed P2P network based edge computing smart grid model also provides reliable communication on electrical energy transactions between community and community or community and the power utility. Next, we depict the processing of transactions in Fig. 4, which aims to coordinate energy. Figure 4 shows the following business scenarios: Firstly, all remaining energy associated with community production and consumption is sold or stored in distributed energy storage systems. Secondly, energy trading with other communities or power utilities/get all the energy needed for smart appliances in the community from energy storage batteries. Finally, the role of the balance node here is that if the energy consumed by the community is greater than the energy produced, then the energy stored in the battery will be used, and then concentrators publish commands to purchase energy from distributed energy storage systems of other communities or power utilities to meet the energy needs of the community. Conversely, if the energy produced by the community is greater than the self-consume energy, the remaining energy will be stored in the battery firstly. When the battery reaches its maximum capacity, the excess energy will be sold to distributed energy storage systems in other communities or power utilities.

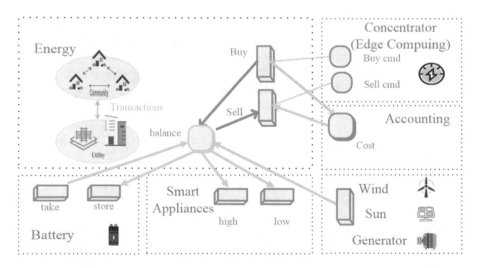

Fig. 3. The energy balance model

Similarly, different communities aggregate together to be also a smart micro-grid system in smart grids. Communities are composed of multiple users, distributed generators (including clean energy generators such as wind generators, solar generators and traditional generators), distributed energy storage systems, controllable smart appliances and so on. Each community is equipped with intelligent power concentrators developed based on edge computing, whose functions mainly involve the intelligent storage and analysis of the entire community's power information, intelligent communication with neighboring community nodes, and user-friendly intelligent management of electricity information. Therefore, In smart communities, under the P2P network based edge computing smart grid and energy balance model, we aim to minimize the overall economic costs based on each day, the time period of daily operation is denoted as the set of time intervals $\Pi := \{1, 2, \cdots, T\}$, and we use $\tau = 1, 2 \cdots T$ to represent a specific time slot.

A. Smart Communities

The collection of all communities can be denoted by P, and one of the communities is represented by an integer i. The set of other communities that can be traded with community i is denoted as N_i^{relate}, where i represents a specified community and $i \in P$, and $n \triangleq P$ represent the number of communities. If a community i is about to trade with another community and the electricity needs of community i are met by community $j(j \in N_i^{relate})$, then the transaction is considered a valid event. Conversely, if the power demands of a community i is not met by the community j, the transaction is invalid. In any community i, the data for electricity consumption demand of all users is $\{r_i^\tau, i \in P\}$ including the data for the electricity demand of all smart home appliances.

When the energy generated by community i does not meet the requirement for energy consumption or is more economical to trade with other communities, the community sends a request for a power transaction to the neighboring community via the P2P network, and the published electricity transaction price is ς_i. If a transaction occurs between the community and the neighborhood, $\{x_{ij}^\tau, j \in N_i^{relate}\}$ is used to indicate the amount of electricity traded between community i and community j in τ time slot. In addition, for $\{x_{ij}^\tau = -x_{ji}^\tau, j \in N_i^{relate}, i \in N_j^{relate}\}$, a positive value indicates that community i purchases energy from community j, and a negative value indicates that community i sells energy to community j. If no power transaction occurs, the value is 0. Since the generated energy and consumed energy are dynamically changed, the actual physical link capacity is limited, so the following is the constraints.

$$x_{ij,\min}^\tau \leq x_{ij}^\tau \leq x_{ij,\max}^\tau, i \in P, j \in N_i^{relate}, \tau \in \Pi \tag{1}$$

B. Utility Company

In addition to other communities, smart communities can trade with the utility company. When power transactions between communities can not satisfy the demand of consumers or there is surplus energy after power trading between communities, community i can make power transactions with local power utility, and $\{y_i^\tau, i \in P, \tau \in \Pi\}$ indicates the energy of electricity trading. If $y_i^\tau \geq 0$, it indicates that community i

purchase energy from the power utility. If $y_i^\tau \leq 0$, it indicates that community i sells energy to the power utility.

$$y_{i,\min}^\tau \leq y_i^\tau \leq y_{i,\max}^\tau, i \in P, \tau \in \Pi \qquad (2)$$

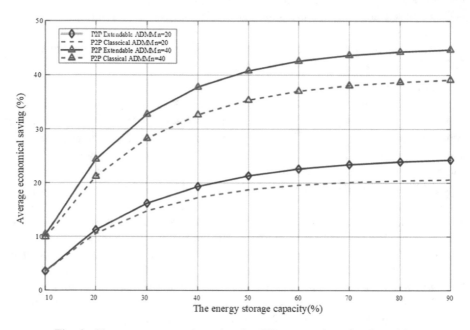

Fig. 4. The energy economic saving for different number of communities

Figure 8 illustrates that the average economical saving is higher when the energy storage capacity is larger. Here, we assume that the number of communities being $n = 20, 40$, and the capacity of energy generated is same, the economical saving increase when the number of communities rises. In other words, when the transactions of communities are more, the community benefits more from resources sharing and coordination under the P2P network based edge computing smart grid model. It means that the sharing and coordination among communities enable them to pool and distribute energy, which increases the utility of renewable energy and thus increases the economical profits.

4   Conclusions

In this paper, a P2P network based edge computing smart grid model is presented. By putting a P2P overlay to the edge nodes, the novel model can be applied to the distributed smart grids. Compared with conventional edge computing smart grid model, the proposed model enhanced the robustness, sustainability, reliability and extensibility of the smart grid system to manage and distribute power resources in a flexible and economic way with the facility of reliable and efficient communication systems for edge devices. The experiments show that communities can make transactions with others communities and power utilities to reduce economic cost and achieve effective electric power balancing.

Acknowledgments. This work was supported by National major R&D program (2018YFB0904900 and 2018YFB0904905).

References

1. Brown, R.E.: Impact of smart grid on distribution system design. In: IEEE Power and Energy Society General Meeting - Conversion and Delivery of Electrical Energy in the 21st Century, pp. 1–4 (2008)
2. Rahimi, F., Ipakchi, A.: Demand response as a market resource under the smart grid paradigm. IEEE Trans. Smart Grid 1(1), 82–88 (2010)
3. Okay, F.Y., Ozdemir, S.: A fog computing based smart grid model. In: 2016 International Symposium on Networks, Computers and Communications (ISNCC), Yasmine Hammamet, pp. 1–6 (2016)
4. Zahoor, S., Javaid, N., Khan, A., Ruqia, B., Muhammad, F.J., Zahid, M.: A cloud-fog-based smart grid model for efficient resource utilization. In: 14th International Wireless Communications & Mobile Computing Conference (IWCMC), Limassol, pp. 1154–1160 (2018)
5. Xie, Y.P., Wen, H., Wu, B., Jiang, Y.X., Meng, J.X.: A modified hierarchical attribute-based encryption access control method for mobile cloud computing. IEEE Trans. Cloud Comput. 7(2), 383–391 (2019)
6. Xie, Y.P., Jiang, Y.X., Wu, J.S., Wen, H., et al.: Three-layers secure access control for cloud-based smart grids. In: IEEE VTC 2015-Fall, September 2015
7. Xie, Y.P., Jiang, Y.X., Liao, R.F., Wen, H., et al.: A hierarchical key management system applied in cloud-based smart grid. In: IEEE ICCC 2015, November 2015
8. Xie, Y.P., Jiang, Y.X., Liao, R.F., Wen, H., et al.: User privacy protection for cloud computing based smart grid. In: IEEE ICCC 2015, November 2015
9. Han, Q.Y., Wen, H., Ma, T., Wu, B.: Self-nominating trust model based on hierarchical fuzzy systems for peer-to-peer networks. In: 2014 IEEE/CIC International Conference on Communications in China, Shanghai, pp. 199–204 (2014)
10. Han, Q.Y., Wen, H., Ren, M.Y., et al.: A topological potential weighted community-based recommendation trust model for P2P networks. Peer-to-Peer Netw. Appl. 8(6), 1048–1058 (2015)
11. Han, Q.Y., Wen, H., Wu, J.S., Ren, M.Y.: Rumor spreading and security monitoring in complex networks. In: The 4th International Conference on Computational Social Networks (CSoNet), pp. 48–59 (2015)
12. Han, Q.Y., Wen, H., Feng, G., et al.: Self-nominating trust model based on hierarchical fuzzy systems for peer-to-peer networks. Peer-to-Peer Netw. Appl. 4(7), 1020–1030 (2016)
13. Zhan, M., Wu, J., Zhang, Z.Z., Wen, H., Wu, J.J.: Low-complexity error correction for ISO/IEC/IEEE 21451-5 sensor and actuator networks. IEEE Sens. J. 15, 2622–2630 (2015)
14. Xiao, J., Wen, H., Wu, B., et al.: Joint design on DCN placement and survivable cloud service provision over all-optical mesh networks. IEEE Trans. Commun. 62(1), 235–245 (2014)

15. Fu, S., Wu, B., Wen, H., et al.: Transmission scheduling and game theoretical power allocation for interference coordination in CoMP. IEEE Trans. Wireless Commun. **13**(1), 112–123 (2014)
16. Zhan, M., Wu, J., Wen, H.: Reduced memory decoding schemes for turbo decoding based on storing the index of the state metric. IET Commun. **8**(12), 2095–2105 (2014)
17. Wen, H., Li, S., Zhu, X., Zhou, L.: A framework of the PHY-layer approach to defense against security threats in cognitive radio networks. IEEE Netw. **27**(3), 34–39 (2013)
18. Hu, L., Wen, H., Wu, B., Pan, F., Liao, R.F., Song, H.H., Tang, J., Wang, X.: Cooperative jamming for physical layer security enhancement in internet of things. IEEE Internet Things J. **5**(1), 219–228 (2018)
19. Xie, F.Y., Wen, H., Li, Y.S., et al.: Optimized coherent integration-based radio frequency fingerprinting in internet of things. IEEE Internet Things J. **5**(5), 3967–3977 (2018)

The Idea of Deep Neural Network Algorithms Application in Sharing Economic Dynamic Value CO-creation Mechanism

Meiwen Guo and Liang Wu[⊠]

School of Management, Xinhua College of Sun Yet-sen Universtiy,
Guangzhou 510520, Guangdong, China
andywuliang@126.com

Abstract. In recent years, deep neural network algorithm has shown great potential in the fields of economy, industry, finance, academia and other fields. Its deep information processing idea, which highly simulates human brain, has been applied in various fields. In this study, the deep neural network algorithm is used to reveal the operating mechanism of dynamic value co-innovation in the context of shared economy. It provides solutions to the problems of the dynamic fitting relationship between dynamic value creation process and factors in the context of shared economy, the influence of multi-participants in dynamic environment caused by role changes and the construction of dynamic adaptive adjustment mechanism of weight parameters, in order to provide ideas for the construction of a new framework model for the mechanism of sharing dynamic economic value.

Keywords: Deep Neural Network · Shared economy · Dynamic value · Co-creation

1 Introduction

Artificial neural network originated in the 1940s. The first neuron model was proposed by McCulloch and Petts in 1943. It is called threshold logic. It can realize some functions of logic operation. Deep neural network is also the focus of artificial intelligence research. Depth neural network combines low-level features to form high-level features, so as to find out the data distribution characteristics. It is widely used. It can effectively utilize the rapidly growing data in various fields and tap its value.

Shared economy has become an important force in social service industry. Globally, the shared economy has penetrated into many fields. In accommodation, there are Airbnb, Piggy Short Rent; in transportation, there are DiDi company; in skills, such as Zaihang, and in life services, such as Eleme, etc., and a large number of excellent sharing economic enterprises have emerged. With the further development of the shared economy, enterprises and consumers should regard the shared economy as a sustainable and profitable activity. New ways of resource allocation and economic concepts will inevitably impact the original business model, and the value creation mechanism under the traditional economic background will also change in the context

© Springer Nature Switzerland AG 2020
F. Xhafa et al. (Eds.): IISA 2019, AISC 1084, pp. 831–840, 2020.
https://doi.org/10.1007/978-3-030-34387-3_102

of the shared economy. However, the change of value realization mode, the change of participants' contribution weight to the process of value realization and the change of fitting relationship between participants need timely intervention. Otherwise, it will affect the maximization of utility and sustainable development of shared economic value creation to a certain extent.

Then, how does the dynamic fitting relationship between the process of value co-creation and the factors of shared economy? How does the role change of multi-participants affect participants in a dynamic environment? How to realize the dynamic adaptive adjustment mechanism of weight parameters of multi-participant co-creation value? In order to grasp the dynamic operating mechanism of sharing economic, the solution of such problems has important practical application value. The idea of deep neural network algorithm can solve the above problems better. Using the idea of deep neural network algorithm, we can simulate the real relationship function between shared economic characteristics and objectives, with larger capacity of main body relationship fitness function and stronger function simulation ability. The dynamic value co-creation mechanism of shared economy is deeply studied from two aspects: the relationship fitting dynamically between process and factors and multi-participants role differentiate and its weight adjustment.

Based on the perspective of dynamic value co-creation process, this study uses the deep neural network algorithm to analyze and refine the dynamic value co-innovation framework model under the background of shared economy, which provides universal suggestions and guidance for enterprise practice.

2   Literature Review

The concept of Deep Neural Network (DDL) and Deep Learning (DL) was put forward in 2006, which opened a new era for neural network. Deep neural network combines low-level features to form high-level features, so as to find out the form of data distribution characteristics. The term was named by the Geoff Hinton Research Group at the University of Toronto in 2006 [1].

Based on the theory of human ecology, the concept of shared economy has been proposed and proved to affect people's lives in all aspects [2, 3]. Contemporary research on shared economy can be roughly divided into two categories. The first is related research from the perspective of customer experience of products or services in general shared environment [4, 5]. The second category is related research from the perspective of sharing and utilizing products or services in the Internet environment [6]. For example, the research on sharing value realization of representative network travel platforms [7]. Professional sharing platform under the Internet is a shared economic model that can benefit the general public and maximize the creation of social value.

The idea of value co-creation mainly appears in the field of service economics. It can be divided into two types. The first one is the research of value co-creation between producers and customers from the perspective of static value co-creation [8]. The second kind is value co-creation research from the perspective of dynamic value co-creation. Scholars adopt different research methods to prove and test it. Such as scale analysis and verification [9], case study [10], based on DART model analysis [11]. At this stage, the process of value co-creation has been proved to be composed of three important stages: invitation, product and service, and payment [12]. Some scholars have proposed that the process of dynamic value co-creation is realized by multi-participants [7]. The contemporary shared environment is complex and changeable. The above research on value co-creation based on static environment is out of date. The research in dynamic environment also needs to be combined with the dynamic relationship between value co-creation process and factors.

The research of role theory can be divided into two types: one is the role study of enterprises and customers from the perspective of static service process [13]. Second, based on the dynamic role will produce different value perspectives. Starting from the shared economic background, it is considered that the role of participants changes according to the dynamic environment [14]. The classification of roles has also been proposed by scholars that participants will have two kinds of behavior in value co-creation: participation behavior and citizenship behavior, also known as in-role behavior and out-of-role behavior [7, 9]. The role theory lays a theoretical foundation for identifying and distinguishing the roles of participants in the shared economy. However, there is no further analysis and demonstration on the impact of the role changes of multi-participants in the dynamic environment.

The idea of deep neural network algorithm in artificial intelligence goes through neuron model, single-layer neural network, two-layer neural network and multi-layer (deep) neural network. In the process of evolution from single-layer to multi-layer (deep) neural network algorithm, deep neural network algorithm has better processing ability in relation fitting and function simulation for multiple objects [1, 15]. However, the application of deep neural network algorithm is mostly in the fields of information, medicine, machinery, film and television, and seldom in the field of economic management.

Participants in the shared economy are the main body in the process of value creation. Today's complex and changeable sharing environment and simple and easy-to-use sharing platform make participants no longer the former bilateral model of enterprises and customers [16]. Most of the studies have not made further analysis and demonstration on the influence of role change of multi-participants in dynamic environment. However, the different and changeable roles of multi-participants in dynamic environment make the regulation of value-creating interest policy uncertain and blind. This research is based on dynamic and changeable shared environment, and proceeds from role theory to recognize participants' roles in depth. It analyses the influence of multi-participants' role changes in dynamic environment, and constructs a dynamic adaptive adjustment mechanism of multi-participants' weight parameters by using auxiliary algorithm to help multi-participants improve value co-creation.

Shared economy is a new mode of value realization, since the emergence of value co-creation theory of service logic and experience value [8]. Academic circles have emerged a number of researches and literatures on the theory of new value creation model. Such as, from the perspective of process and factor evolution, some papers explores the driving factors of value co-creation for consumers at the bottom of the pyramid [17], describes the value co-creation process between customers and suppliers based on system dynamics, and studies the value co-creation process mechanism of shared travel platform [9], Customer loyalty in the shared economy [18]. The study of evolution mechanism based on process and factors is the main feature of the study of shared economy. The advancement of information technology provides objective conditions for the realization of sharing economic value. In artificial intelligence algorithm technology, the idea of deep neural network algorithm goes through neuron model, single-layer neural network arithmetic, two-layer neural network arithmetic and multi-layer (deep) neural network arithmetic. In the process of development and evolution, the idea of single-layer to multi-layer (deep) neural network algorithm is constantly deepened and improved. The deep neural network algorithm has better processing ability in relation fitting and function simulation for multiple objects [1, 15, 19].

However, most studies lack dynamic analysis of multi-participants in the shared economy. With the deepening trend of socialization, specialization and individualization of the shared economy, it is indispensable to study the role of multi-participants in the evolution of the shared economy process and factors. In addition, the deep neural network algorithm in artificial intelligence has been widely used. Most of the research results are applied in the fields of information, medicine, machinery, film and television, and seldom in the field of management and economy. Based on this, this study attempts to bridge the shortcomings of existing research, focusing on the dynamic environment, using on the idea of deep neural network algorithm to derive the dynamic fitting relationship between the process of value creation and factors of shared economy, and reveal its characteristics. Based on the role theory, this paper analyses the influence of the role change of multi-participants in dynamic environment, reveals the dynamic adaptive mechanism of multi-participants' weight parameters, and constructs the dynamic value creation mechanism model of shared economy. In short, this study combines the latest research results of shared economy development, enriches the theory of shared economy, and lays a theoretical foundation for further research.

Combining with the current global research situation, the research on shared economy still needs to be further deepened in the following two aspects, which is also the starting point of this study.

a. Shared economy should take into account the dynamic fitting relationship between process and factors in value creation. Based on dynamic environment and complementary interdisciplinary advantages, the dynamic fitting relationship between process and factors in value creation can be identified by using advanced algorithm technology, and the law of action can be grasped.

Nowadays, the growing maturity of cloud-moving technology has dramatically reduced the cost of information matching resources, providing favorable conditions for sharing economy to enter human daily life [9]. Based on the complex and changeable dynamic environment of shared economy and the latest research progress of the theory of dynamic value creation of shared economy, this study uses the deep neural network algorithm to solve the dynamic fitting relationship between process and factors in the process of dynamic value creation of shared economy, and grasps its development law.

b. The process of shared economic behavior should take into account the dynamic process of multi-participant value creation. This paper identifies the roles of multi-participants, analyses the influence of multi-participants' role changes in the dynamic environment, and constructs a dynamic adaptive adjustment mechanism model of the weights of the main participants' value co-creation.

3 Research Construction

3.1 Research Thinking

In this study, the deep neural network algorithm is used to analyze the dynamic fitting relationship and characteristics between the process and factors of value co-creation of shared platform. Based on the analysis of the role of multi-participants and the impact of the role changes in the process of dynamic value co-creation, a dynamic adaptive fine-tuning mechanism model of multi-participants' weight parameters is constructed. On the basis of solving the above two problems, a new framework model of dynamic value co-creation mechanism in the shared economy is constructed. Suggestions are put forward to clarify the direction of value promotion and realize the sustainable and scientific development of the shared economy.

Firstly, it analyses the process and reasons of dynamic value co-creation under the shared economy, and the characteristics of the factors of shared value co-creation. The relevant literature resources and typical survey data of sharing economic platform are studied. The dynamic fitting relationship between value co-creation process and elements and its characteristics are analyzed and discussed by using deep neural network algorithm.

Secondly, according to the analysis conclusion of the fitting relationship between dynamic value creation process and factors, the roles of multi-participants in the shared economic platform are studied. The roles are identified and the impact of the role changes of multi-participants in the dynamic environment is discussed in depth. Based on this, the Markov chain Monte Carlo calculation is used. A dynamic adaptive fine-tuning mechanism model of multi-participant weighting parameters is constructed by using the method and the idea of constrained Boltzmann machine model of depth neural network algorithm.

Finally, based on the characteristics of dynamic fitting relationship between process and elements, and multi-participant role identification and dynamic adaptive adjustment mechanism of weight parameters, a new framework model of dynamic value creation mechanism of shared economy is constructed, which is proposed for the promotion of dynamic value creation in the shared economy environment.

3.2 Around the Era Background of Shared Economy and the Latest Research Literature at Home and Abroad, the Specific Research Content Consists of Four Parts

Firstly, Stage Fitting Relation and Characteristic Analysis in the Process of Dynamic Value Co-creation of Shared Platform.

The dynamic value creation process of shared platform is: sharing invitation, product and service sharing and payment. Among them, sharing invitation and payment are carried out on the online platform, and products and services are shared offline [9, 12]. The dynamic value co-creation process is located in the initial input layer of the neural network graph. As shown in Fig. 1, the dynamic value creation process matches product and service resources to customers by sharing platform, and links providers with customers through direct contact. Sharing platform service providers, products and service providers serve end customers together. As shown in Fig. 2, the attributes of invitation, product and service and payment are represented by al, A_2 and a_3, respectively. From left to right, the attributes are represented by neural network decomposition graph. The connection lines of input layer A–Z represent the connection between "neurons". Each connection line corresponds to a different weight parameter, i.e. trained. The weight vector W. At the end of the connection, the value becomes a * w, and the value of the target Z of the unknown attribute is obtained. Because of the complexity of three-stage attribute characteristics, the target of prediction is a vector, and the output layer is two "output units". The weights of $W_1 = g (a_1 * W_{11} A_2 * W_{12} A_3 * w_{13})$ and $W 2 = g (a_1 * W_{21} A_2 * W_{22} A_3 * w_{23})$ are obtained by training, which show the fitting relationship of three input elements on two dimensions, so as to make in-depth analysis and push forward layer by layer. The results of three-stage neural network analysis are taken as the basis for the analysis of the characteristics of each stage in the process of dynamic value co-creation.

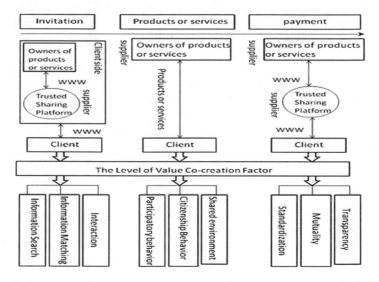

Fig. 1. Dynamic value co-creation process and factor diagram of shared platform

Secondly, the fitting relationship and characteristic analysis of the process and factors of value co-creation of shared platform.

According to the three stages of the process, there are three levels of value co-creation factors: information search, information exchange and information matching in the invitation stage, participation behavior, citizenship behavior and sharing environment in the product and service stage, and standardization, reciprocity and transparency in the payment stage. According to the construction idea of the neural network structure in the first part, the second factor layer is constructed, and the process layer is fitted with the factor layer. Because the environment is dynamic, the fitting process is derived from the single layer to the multi-layer neural network algorithm. As shown in Fig. 3, a (1) is the initial input layer and a (3) and a (5) are the post-derivation layers of the

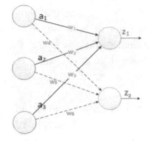

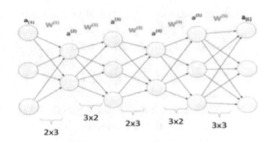

Fig. 2. Three-stage ANN anatomy

Fig. 3. Value co-creation process and dynamic fitting hierarchy

initial input layer, and a (2), a (4), a (6) are hidden layers. Among them, a (1) and a (3) prediction targets are two-dimensional vectors, and a (5) the prediction target is a three-dimensional vector. The number of parameters of the neural network is $2(2 * 3) + 2(3 * 2) + 3 * 3 = 33$, as shown in Fig. 4, which is the corresponding deep neural network model. Through the number of parameters generated in the process of value co-creation and dynamic fitting of factors, the matrix of group characteristics of parameters can be further deduced. Combining the different types of representation vector structures and group characteristics of the two, the structure of the fitting relationship between them can be obtained finally, and the characteristics of the relationship can be understood through the correlation structure.

Thirdly, role recognition of multi-participant dynamic value creation, influence of dynamic role change and adaptive adjustment mechanism of weight parameters in shared environment.

(1) Multi-participant Dynamic Value Co-creation Role Recognition and Dynamic Role Change's Impact on Participants
   The roles of participants are closely interdependent in the process of value creation. Participants' behavior in value co-creation can be divided into participatory behavior and citizen behavior, i.e. intra-role behavior necessary for value co-creation and extra-role behavior necessary for non-value co-creation that can bring additional value to enterprises [9]. In this part, the behavior of multi-participants is analyzed in depth, the role of multi-participants is identified, the main participants are identified, and the impact of role changes in dynamic value creation of each participant is analyzed.

(2) Making use of Markov chain Monte Carlo algorithm and constrained Boltzmann machine model-assisted depth neural network algorithm to deduce and determine the weight parameters of the main participants, and building a dynamic adaptive fine-tuning mechanism model of multi-participant weight parameters.
   In the dynamic value co-creation depth neural network model, there is still uncertainty in accurately deriving the weights of participants. In order to achieve accurate estimation, the feature expression of the final layer is deduced according to the content characteristics of the factors in the model, and then the posterior

distribution of the target object is deduced. With the hidden layer, the weight is estimated by using Markov chain Monte Carlo algorithm. In order to determine the weight parameters of participants, we obtain the data of factors in the shared environment to conduct in-depth learning of high-level features, that is, the deviation and the weight matrix of adjacent layers contained in the in-depth learning model.

Constructing multi-participant constrained Boltzmann machine model in the shared platform can smoothly complete the pre-training process of each level step by step, and fine-tune the weight parameters reflected by the content characteristics of the factors in the model, so as to make the weight changes of each factor coincide with the value creation. The dynamic self-adaptive regulation mechanism is formed by the change of process.

$$Q(h_t = 1|v) = \frac{1}{1 + \exp\left(-b_t - \sum v_s W_{s,t}\right)}$$

In the formula, v_s is the state value of the s-th neuron in the visible layer v, h_t is the state value of the t-th neuron in the hidden layer h, b_t is the deviation from h_t, W_s and t are the weight of the link between v_s and h_t, and matrix W is symmetrical. When the hidden layer V is known, the conditional distribution of the s-th neuron's value 1 in the visible layer is a constrained Boltzmann machine, and the approximate solution is obtained by comparing the bifurcation sampling. In this way, a multi-participant adaptive weight parameter of fine-tuning mechanism model is constructed.

Fourth, construct a new framework model of dynamic value creation mechanism under the background of shared economy.

The dynamic value co-creation mechanism under the background of shared economy should be studied in many dimensions [7, 18]. On the basis of solving the first two problems, this stage combines the dynamic fitting relationship and characteristics of process and factors, the change of roles of multi-participants and the analysis results of the heavy parameters of human rights.

Emphasizing the dynamic environment, the dynamic adaptive mechanism model of each participant's weight parameters is constructed through the nine factors associated with the three stages of invitation, product and service, payment and the positive or negative factors brought by the dynamic role change of multi-participants. Based on the solution of the two major problems, the mechanism of dynamic value co-creation of shared economy is revealed, and the model of dynamic value co-creation mechanism of shared economy is constructed by using the idea of deep neural network algorithm. Finally, suggestions and guidance are put forward for the dynamic value creation and promotion under the shared economic environment.

4   Conclusion

The rapid rise of shared economy maximizes the value of social resources. However, the monopoly of platform resources, the trust crisis of shared environment, the instability of product and service providers, and the imbalance of regional shared resources have brought uncertainty and blindness to the dynamic value creation of shared economy. Therefore, the case study of the sharing platform at this stage is conducive to revealing its commonalities. Then, we can combine the general case with the characteristics of each sharing platform to form a scheme of dynamic value creation and promotion. In this study, the idea of deep neural network is proposed to interpret the dialectical relationship among the objects of the sharing platform. In addition, in the dynamic environment, the influence of role changes and the construction of dynamic adaptive adjustment mechanism of weight parameters of multi-participants also give suggestions to solve the problem, and put forward construction strategies for the new framework model of dynamic value-sharing mechanism.

Acknowledgements. This paper was supported by "13th Five-Year" plan research project of philosophy and social sciences of Guangdong province; "Research of the co-creation mechanism of shared economic dynamic value based on deep neural network" (Project number: GD17YGL03); Humanity and social science youth foundation of Ministry of Education of China "Business agglomeration attractiveness to consumers based on the "Internet plus": an empirical study on the evolution process and mechanism" (Project number: 16YJCZH119).

References

1. Hinton, G.E., Salakhutdinov, R.R.: Reducing the dimensionality of data with neural networks. Science **313**(5786), 504–507 (2006)
2. Hollingshead, A.B.: Human ecology: a theory of community structure. Am. Sociol. Rev. **15** (05), 684–685 (1950)
3. Felson, M., Spaeth, J.L.: Community structure and collaborative consumption. Am. Behav. Sci. **21**(04), 614–624 (1978)
4. Lamberton, C.P., Rose, R.L.: When is ours better than mine? A framework for understanding and altering participation in commercial sharing systems. J. Mark. **76**(4), 109–125 (2012)
5. Andruss, P.: What's mine is yours. Entrepreneur **43**(1), 78–85 (2015)
6. Botsman, R.: Sharing's not just for start-ups. Harvard Bus. Rev. **92**(9), 23–25 (2014)
7. Yang, X., Tu, K.: Research on dynamic value co-creation under the background of shared economy: a case study of travel platform. **28**(12), 258–268(2016)
8. Prahalad, C.K., Ramaswamy, V.: Co-creating unique value with customers. Strategy Leadersh. **32**(3), 4–9 (2004)
9. Yi, Y., Gong, T.: Customer value co-creation behavior: scale development and validation. J. Bus. Res. **66**(9), 1279–1284 (2013)
10. Frederic, P., Philipp, K., Roger, S.M.: Experience co-creation in financial services: an empirical exploration. J. Serv. Manage. **26**(2), 295–320 (2016)
11. Zhang, J., Hong, C., Zhao, H.: Research on customer participation value co-creation model based on DART model in network virtual environment—take Japanese enterprise Muji as an example. Sci. Technol. Prog. Countermeasure **18**, 88–92 (2015)

12. Marco, T., Tiziana, R., Claudia, C.: Being social for social: a co-creation perspective. J. Serv. Theory Pract. **25**(2), 198–219 (2015)
13. Lin, J.: Service Marketing and Management. Peking University Press, Beijing (2014)
14. Normann, R., Ramirez, R.: From value Chain to value constellation: designing interactive strategy. Harvard Bus. Rev. **71**(4), 65–77 (1993)
15. Zhang, L., Zhang, Y.: Infinite depth neural network method for large data analysis. Comput. Res. Dev. **01**, 68–79 (2016)
16. Agrawal, A.K., Kaushik, A.K., Rahman, Z.: Co-creation of social value through integration of stakeholders. Procedia-Soc. Behav. Sci. **189**(25), 442–448 (2015)
17. Kumkum, B., Rajat, A., Sharma, V.: Literature review and proposed conceptual framework. Int. J. Market Res. **57**(4), 571–603 (2015)
18. Yang, Shuai: Types, elements and impacts of shared economy: a perspective of literature research. Ind. Econ. Rev. **3**(25), 35–45 (2016)
19. Chang, F.: Research on positive linear functions in deep neural networks. Comput. Eng. Des. (03), 759–762, 801 (2015)

Author Index

© Springer Nature Switzerland AG 2020
F. Xhafa et al. (Eds.): IISA 2019, AISC 1084, pp. 841–845, 2020.
https://doi.org/10.1007/978-3-030-34387-3

Printed in the United States
By Bookmasters